BANG-BANG BOYS, JEDBURGHS, and the HOUSE OF HORRORS

A History of OSS Training and Operations in World War II

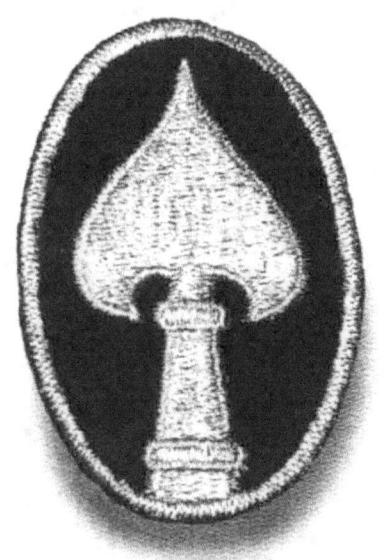

Gold and black spearhead insigna proposed by Donovan for the OSS

John Whiteclay Chambers II

To Those Who Served

Bang-Bang Boys, Jedburghs, and the House of Horrors
A History of OSS Training and Operations in World War II

John Whiteclay Chambers II

Printed in the United States of America
ISBN 978-1951682224

AN ORCHARD INNOVATIONS BOOK
February 2020

Originally published in 2008 as *OSS Training in the National Parks and Service Abroad in World War II*, by the U.S. National Park Service, Washington, D.C.

TABLE OF CONTENTS

Author's Preface v

Acknowledgments. ix

Abbreviations xi

Introduction 1

CHAPTER 1. "Wild Bill" Donovan and the Origins of the OSS 5

CHAPTER 2. A Wartime Organization for Unconventional Warfare. 22

CHAPTER 3. Catoctin and Prince William Parks Join the War Effort 59

CHAPTER 4. Converting Catoctin Mountain Park into Military Camps 73

CHAPTER 5. Transforming Prince William Forest Park into Military Camps . . . 95

CHAPTER 6. Instructing for Dangerous Missions 123

CHAPTER 7. Daily Life in Camp, Park and Town 164

CHAPTER 8. OSS in Action: Mediterranean and European Theaters. 205

CHAPTER 9. OSS in Action: The Pacific and the Far East 277

CHAPTER 10. Postwar Period: End of the OSS and Return to the Park Service. . . 353

CHAPTER 11. Summary and Conclusion 411

Endnotes. 449

Bibliography 607

The Author 623

Illustrations. 625

AUTHOR'S PREFACE

How could I not be interested when the National Park Service invited me to write about the OSS?

The fascinating men and women in the Office of Strategic Services (OSS) in World War II included daring spies and rowdy daredevils who parachuted behind enemy lines to undermine their ruthless regimes. Movie star and working seaman Sterling Hayden evaded German patrol boats to supply anti-Nazi guerrillas in Yugoslavia; Maryland socialite Virginia Hall ran espionage networks and resistance groups among the *maquis* in France. Twenty-five-year old William Colby from St. Paul, Minnesota, who had dropped out of Columbia Law School to enlist, destroyed a major railroad line and bottled up thousands of German troops in Norway. Reginald ("Reg") Spear, a 20-year-old Californian, was sent on several dangerous missions in the Far East: spying on Japanese fortified islands, determining conditions of American prisoners of war in the Philippines and tracking down a missing agent in China. When Tokyo announced Japan's surrender, John Singlaub, leading a small OSS team, parachuted into a POW camp holding captured Allied servicemen, rescuing them from possible annihilation by fanatical Japanese officers.

But it was not only the dangerous, daring work in the field by agents, saboteurs, guerrilla leaders and their radio operators that comprised the activity of America's first centralized intelligence agency. Foreign area specialists from universities combed published and unpublished records for valuable economic, political, and strategic data, among them William L. Langer, H. Stuart Hughes, and Gordon A. Craig. German-born actress and singer Marlene Dietrich recorded songs to accompany the OSS's propaganda beamed at undermining the morale of German soldiers. A tall, lanky young woman named Julia McWilliams worked as a file and registry clerk in the OSS regional headquarters in China. Later, she married a fellow OSSer and accompanied her husband, Paul Child, to Paris where she learned French cooking, returned to the United States and became famous as author and TV chef, Julia Child.

OSS had cryptological and communications specialists who developed and employed the codes and ciphers, the short-wave transmitting equipment, and the clandestine communications networks that enabled the secret organization to function effectively. OSSers built a portable, short-wave, wireless telegraphy (W/T) transmitter/receiver system that fit into a suitcase, and by 1944, they had developed an ultra-high frequency voice communication system which could link an agent on the ground with a hand-held transceiver with an aircraft circling six miles above. A young electrical engineer named Jack Kilby, who served with the OSS Communications Branch in India and China, continued the OSS philosophy of thinking outside the box, and in 1958 while working for Texas Instruments, he invented the integrated circuit, which formed the basis for the microchip industry.

During the war, OSS developed scores of gadgets, secret devices, weapons, and munitions. They ranged from flexible swim fins and self-contained underwater breathing devices to buttons, shoes, and pipes with secret compartments, and a variety of lethal inventions including single-shot, cigarettes and fountain pens as well as flashless pistols and machine guns. Among the special munitions, one innovation was a batter nicknamed "Aunt Jemima" that came packed in Chinese flour sacks, to deceive the Japanese. It could be harmlessly baked in an oven, but with a fuse attached, it became a powerful explosive that OSS saboteurs could blow up a radio tower, railroad line, or even a bridge with it.

Modeled in part after the British Special Operations Executive and the Secret Intelligence Service, OSS shared the kind of reputation for derring-do and innovative gadgets made famous by the James Bond films based on the novels of Ian Fleming. A member of the British Special Operations Executive during the war, Commander Fleming had been one of the consultants that London had sent temporarily to the United States to advise the Americans on the creation of such secret wartime services. The character of Commander James Bond (007) may be fictitious, but the organization was not and neither was the gadget specialist of the books and films, "Q," who was based on the head of SOE's technical branch. "Q's" counterpart in the United States was chemist Stanley P. Lovell, head of OSS Research and Development, whose invention of so much lethal weaponry and diabolical gadgets earned him the sobriquet, "Professor Moriarty."

The OSS was a combination of socialites, business and professional people, scholars and scientists, movie stars and athletes, and mostly just regular although highly intelligent Americans, as well as even a few safecrackers and forgers, who were sprung from prison to apply their skills to the secret war effort. It included a host of then well known names such as Henry Ringling North of Circus fame, Russian Prince Serge Obolensky, who had married into the Astor fortune, two of the three Alsop brothers, and numerous Vanderbilts, Mellons, Du Ponts, and Morgans, but also baseball star and linguistic genius Moe Berg and champion fullback and professional wrestler "Jumping Joe" Savoldi. OSSers ranged from right to left in their political views, and they worked with a political spectrum of resistance groups in Nazi-occupied Europe.

To conduct a new kind of warfare, an unconventional war in the shadows, the OSS recruited very few professional soldiers. It was a partly military and partly civilian operation, and even most of those in uniform had been shortly earlier been civilians. But these "glorious amateurs" prided themselves on being able to think innovatively and act independently. Their leader, who encouraged this, was, William J. ("Wild Bill") Donovan, a legend himself. On his uniform, he wore the star-spangled white and blue ribbon of the Medal of Honor for heroism in the Meuse-Argonne offensive in World War I. This self-made Irish American, who rose from the shadow of the dockside grain elevators in Buffalo to a top floor view on Wall

Street, proved a courageous and inspiring leader in both world wars.

So when the National Park Service asked me to find out what Donovan's organization had done in two of its National Parks in the Second World War, I was more than happy to accept that challenge.

The Park Service knew that the OSS had taken over what is now Catoctin Mountain Park, north of Frederick. Maryland and Prince William Forest Park near Quantico, Virginia, but it wanted to know what the super-secret organization had done in those parks between 1942 and 1945. It sought to fill in that missing part of the parks' history and inform the public as part of its interpretive mission. What precisely had Donovan's organization done inside the parks in those years? How that had related to the war effort? And how it had affected the parks both during and after the war?

My quest for answers led to many different sources of information. Most voluminous were the records of the OSS itself, held for nearly half a century by the CIA and only relatively recently declassified and deposited—more than 5,000 cubic feet of them—in the National Archives II at College Park, Maryland. The National Park Service records at the National Archives and at the parks themselves were also useful. Additionally, my search led me to the papers of several key figures, from Donovan himself at the U.S. Army Military History Institute in Carlisle, Pennsylvania, and the National Archives II, to Presidents Franklin D. Roosevelt at the Roosevelt Presidential Library at Hyde Park, New York, and Harry S Truman at the Truman Presidential Library at Independence, Missouri, the papers of Wallace Reuel and Arthur Goldberg at the Library of Congress, Allen Dulles at Princeton University, and M. Preston Goodfellow and General Albert Wedemeyer at the Hoover Institution at Stanford, California. There were valuable oral histories, not just from OSS veterans at the Rutgers Oral History Archives of World War II, the Veterans Project of the Library of Congress, but also from wartime National Park Service Director Newton Drury at Berkeley and Prince William Forest Park superintendent Ira Lykes at Harper's Ferry. From the National Personnel Records Center in St. Louis I obtained the personnel files of by Lykes and Catoctin park superintendent, Garland B. ("Mike") Williams. Crucial insights were provided by surviving OSS veterans. I personally interviewed nearly forty of them plus more than a dozen sons and widows, and I toured the two parks with several of the veterans.

The OSS had used the parks as primarily as training camps. In their isolated woodlands, secured from public scrutiny, recruits for Donovan's organization, especially operatives who were going to be sent overseas into the war zones, were put through courses of physical toughening, psychological preparation, and skills that would help them to survive and accomplish their missions, many of which would be behind enemy lines. My research quickly revealed that not only was there no study of the OSS in the National Parks during the war and its legacy afterwards, but there was no scholarly study of either OSS training or its relationship to the missions of the OSS overseas and its contribution to victory.

Consequently, this study does not simply explore the training camps of "Wild Bill" Donovan's organization in the two National Parks. It also assesses the effectiveness of the training OSS agents received and how they utilized it in their dangerous and often heroic exploits overseas. It explores the legacies of the OSS, upon the parks themselves, upon the veterans, and upon the organizational heirs of the OSS. For the OSS in its training, its organization, and its missions, was a forerunner of both the Central Intelligence Agency and the today's Special Operations Forces. Understanding such origins and legacies as well as the seemingly paradoxical relationship, at least during World War Two, of the ordinarily peaceful sanctuaries of the National Parks and the murky, violent world of spies, saboteurs, and guerrillas is a major aim of this book.

John Whiteclay Chambers II
New Brunswick, New Jersey
September 2008

ACKNOWLEDGMENTS

This study was commissioned by the National Park Service, and I am indebted to a number of its members who have provided invaluable advice and assistance in its preparation. Gary Scott, Chief Historian, National Capital Region, of the NPS, gave me his expert advice all along the way. This study would not have been possible without his help and also without the strong support and advice of Bob Hickman, Superintendent of Prince William Forest Park, and Mel Poole, Superintendent of Catoctin Mountain Park.

Brian Carlstrom, Resource Manager at Prince William Forest Park in 2004 when I began my research, had initiated this project. I am grateful to him and also to other members of the NPS staffs at the two parks, who provided assistance on many occasions and who read all or parts of the manuscript and offered valuable suggestions. At Prince William Forest Park, I received help from, in addition to Superintendent Hickman, Assistant Superintendent George Liffert, Resource Manager David Hayes, Chief of Interpretation Laura Cohen, and Museum Technician Judy Volonoski. At Catoctin Mountain Park, in addition to Superintendent Poole, I had assistance from Resource Manager James Voigt and his successor, Sean Denniston, as well as Supervisory Park Ranger Sally Griffin, Chief Ranger Holly Rife, and Biologist Becky Loncosky.

I wish also to acknowledge the support of Charles Roman, National Park Service, North Atlantic Coast Cooperative Ecosystem Studies Unit, located at the University of Rhode Island Bay Campus, Narragansett, Rhode Island; of Carole Daye, National Park Service, Boston, Massachusetts; and of the Office of Research and Sponsored Programs at Rutgers University, New Brunswick, New Jersey.

Much of my research was done at the National Archives II in College Park, Maryland. There I was fortunate to have the guidance of the venerable John E. Taylor of the Modern Military Records Branch, who among his may other achievements convinced the CIA to open the OSS files, and of Lawrence H. McDonald, archivist and author who oversaw the OSS collection from 1985 until his retirement in 2008, and who always welcomed me and guided me in my many visits to the central research room. Archivist Joseph D. Schwartz helped with the records of the National Park Service.

Bringing to life the story of training and service in the OSS would have been much more difficult without the help of OSS veterans themselves. I am grateful to all the veterans who shared their stories with me. Their names are listed in the Bibliography, as are the names of relatives of deceased OSS veterans who provided me with useful material. I would be remiss, if I did not give special thanks to those veterans, who time after time provided additional information or other material: John W. Brunner, Caesar J. Civitella, Frank A. Gleason, Joseph E. Lazarsky, Elizabeth P. McIntosh, Albert Materazzi, Charles M. Parkin, Arthur F. Reinhardt, and Reginald G. Spear.

Several non-veterans who have been engaged in work about the OSS also generously offered assistance. Charles T. Pinck, President of the OSS Society, encouraged me to join the society's online discussion group. Jonathan Clemente and Lynn Philip Hodgson shared material. Toni L. Hiley, curator of the CIA Museum, provided insights. Troy J. Sacquety, U.S. Army's Special Operations Command's History Office, helped generously. No study of Donovan, the OSS, and the origins of the CIA, would be possible without the meticulous early work of CIA historian Thomas F. Troy.

My wife, Amy P. Chambers, read my entire manuscript and gave me her always insightful advice. The present work is the better for it.

I gratefully acknowledge the assistance I have received from so many people, but any errors or inaccuracies that remain in the work are, of course, my own responsibility.

J.W.C.

ABBREVIATIONS

CB	Communications Branch
CIA	Central Intelligence Agency
CO	Commanding Officer
COI	Coordinator of Information
Commo	Communications
CW	Continuous [Electromagnetic] Wave of Constant Frequency
DCI	Director of Central Intelligence
FBI	Federal Bureau of Investigation
FNB	Federal Bureau of Narcotics
G-2	Army Intelligence
HQ	Headquarters
JCS	Joint Chiefs of Staff
MO	Morale Operations Branch
MIS	Military Intelligence Service
MU	Maritime Unit
NPS	National Park Service
OG	Operational Group
ONI	Office of Naval Intelligence
OSS	Office of Strategic Services
R&A	Research and Analysis Branch
R&D	Research and Development Branch
RG	Record Group of Documents in National Archives II
SI	Secret Intelligence Branch
SIS	Secret Intelligence Service, MI-6, British
SO	Special Operations Branch
SOE	Special Operations Executive, British
S&T	Schools and Training Branch
WJD	William J. Donovan
W/T	Wireless Telegraphy
X-2	Counter-Intelligence Branch

INTRODUCTION

Because of the secrecy that enveloped the U.S. Office of Strategic Services in World War II, it surprises most people, including nearby residents, to learn that spies were trained in some of the National Parks—not just spies but guerrilla leaders, saboteurs, clandestine radio operators and others who would be infiltrated behind enemy lines. What are today known as Catoctin Mountain Park in Maryland and Prince William Forest Park in Virginia played a vital role in the training of the operatives of the U.S. Office of Strategic Services. Known by its initials, the OSS was a specially created wartime military agency that fought a largely invisible and covert war against the Axis powers between 1942 and 1945. America's first national centralized intelligence agency with thousands of clandestine operatives, spies, and intelligence analysts, the OSS is acknowledged as the predecessor of the Central Intelligence Agency.[1] With its Special Operations troops and Operational Group commandos, the OSS is also widely considered a forerunner of today's Special Forces.[2]

From 1941 to 1945, the men and women of the OSS were part of a "shadow war," a war largely behind the scenes and often behind enemy lines around the world. Highly secret during the war and in many instances for years thereafter, that effort was designed to help undermine the conquests of Nazi Germany, fascist Italy, and militaristic Japan. The "shadow warriors" sought to supply and guide local resistance movements, demoralize the enemy through "black propaganda," and to gather intelligence and commit sabotage in enemy occupied territory to contribute to the victory of the Allies' armed forces as they overcame the totalitarian, Axis aggressors.

The OSS was a civilian wartime organization with many military personnel assigned to it. It reported to the Joint Chiefs of Staff (JCS), although its director, William J. ("Wild Bill") Donovan, had direct, personal access to President Franklin D. Roosevelt. The actual number of men and women recruited and trained by the super secret organization may never be known. Traditionally it was generally believed that the OSS had 13,000 members.[3] But this figure seems to have been based on the number of American personnel employed by the OSS at its peak in late 1944.[4] In August 2008, just as the present study was completed, the CIA released 750,000 newly declassified OSS documents, including personnel files that seemed to indicate that there may have been 24,000 people employed by the OSS at one time or another during the war.[5] But the status of these 24,000, whether permanent or temporary, member or consultant, American or foreigner, remained to be determined.

OSSers worked in half a dozen different branches of the organization and in a variety of missions, some of which were highly dangerous. Half of the OSSers served overseas, and perhaps as much as one-third of the OSS recruits were trained as agents.[6] OSS Special Operations agents in small teams or commandos in larger Operational Groups infiltrated behind enemy lines in nighttime parachute drops, small plane landings or submarine or swift boat embarkations in Axis occupied countries in North Africa, Europe, and the Far East to help organize, arm, and help lead local guerrilla resistance fighters. Some of these OSS agents and commandos also engaged in sabotage—blowing up railroads, bridges, and tunnels as well as power plants, communications centers, and weapons and munitions depots. Accompanying them were radio operators, members of the OSS Communications Branch, who kept the field operatives and headquarters informed through encoded messages via short-wave radios and temporarily strung antennas. In the field, they had to be on the move to avoid detection by enemy direction-finding and surveillance units. OSS Secret Intelligence operatives engaged in "cloak and dagger" espionage, ferreting out and relaying information about industrial and military production, transportation, war plans, deployment of enemy units, and the status of supplies and morale.

Donovan's organization also employed thousands of men and women engaged in equally important but less hazardous work in regional stations overseas and especially at the OSS's headquarters in Washington, D.C. Among these were scholars in the Research and Analysis Branch (R&A) pouring through evidence of political and economic developments in enemy and enemy occupied countries; scientists in the Research and Development laboratories, developing new weapons and means of communications; members of the Counter-Intelligence Branch, who sought to apprehend enemy agents sent or left behind Allied lines; and the men and women of the Morale Operations Branch MO), who specialized in foreign-language radio broadcasting and leafleting sought actively to undermine enemy military and civilian morale.

Among the many highly talented, ambitious and often well-connected members of the OSS, a number later achieved national prominence. Among them were subsequent CIA directors Allen Dulles, Richard Helms, William Colby and William J. Casey; Supreme Court Justice Arthur J. Goldberg; Ambassador David Bruce and two dozen other U.S. ambassadors; presidential advisers such as political scientist Roger Hilsman, economist Walt Rostow, and historian Arthur M. Schlesinger, Jr.; ornithologist and wildlife specialist S. Dillon Ripley II, and Treasury Secretary C. Douglas Dillon. The glamorous aura, independence and social standing of the OSS brought in representatives of some of the wealthiest families in America—DuPonts, Mellons, Morgans, and Vanderbilts—a fact that led some critics to suggest that the agency's initials, OSS, actually meant "Oh, So, Social!"[7] Despite the socialites, however, in sheer numbers, the vast majority of OSS personnel came from middle-class backgrounds. Most of them had never heard of OSS, and were selected mainly on merit. They did the bulk of the work in the new organization and did it well. Little wonder that so many of the OSS members proved so successful in their postwar careers as well.

Thousands of men and women served in the OSS, but it has been mainly those of the operational branches, especially those engaged clandestinely behind enemy lines, that have been the most heralded. Their work remained largely secret during the war, but almost immediately afterwards, through published accounts and fictionalized Hollywood films starring Alan Ladd, Gary Cooper, and James Cagney, they achieved a celebrated place in the public imagination.[8] Americans thrilled to the daring exploits and cocky, devil-may-care attitudes of the OSS spies and commandos as they fought with courage, skill and initiative against the Nazis and the Japanese. Working in the shadow of enormous, depersonalized armies, navies, and air forces, the solo OSS agents, fighting behind enemy lines, surviving through their courage and wit, seemed to exemplify a continuing role for the daring and able individual even in an age of industrialized mass warfare.

The story of the OSS and its charismatic director, Wild Bill Donovan, has been told many times.[9] But in the past two decades, there has been an outpouring of books about other aspects of the OSS. This resulted from the extensive declassification of OSS organizational records at the National Archives.[10] It also stemmed from the willingness of many OSS Veterans, sworn to secrecy during the war, to tell about their exploits as part of what was being hailed in the 1990s, fifty years after the war, as the "Greatest Generation."[11] The publishing avalanche has gone beyond autobiographies of OSS operatives who subsequently achieved notable careers.[12] As part of the new social history "from the bottom up," many former operatives, without prior claim to public attention, have been publishing their memoirs.[13] Historians and biographers, some of them OSS veterans, are also mining the OSS records, examining virtually every aspect of the organization—its leaders, its exotic weapons, its female members, its relationships with the Allies, and its secret operations in countries around the world.[14] The Central Intelligence Agency (CIA) and the U.S. Special Forces Command (USSOCOM) have celebrated their connections with the OSS.[15]

One largely neglected subject in this outpouring, however, has been the training of the OSS agents. Historical accounts or memoirs of the OSS generally ignore this or dispense with it in only a few paragraphs or a couple of pages. Consequently, there has been no adequate study of the recruitment and preparation of the men and women of the OSS for what were demanding and often dangerous activities.[16]

Seeking to help rectify this omission, the present study provides a detailed account of OSS training, particularly of the Special Operations, Operational Group, and Communications Branches but also including when relevant, the training of personnel of Secret Intelligence, Morale Operations, and the Maritime Unit. The geographical focus is on three of the main OSS training facilities, the ones located in two forested areas of the National Park Service. The three camps were OSS Training Areas A and C in what is today named Prince William Forest Park, some 15,000 forested acres on the watershed of the Quantico Creek in northern Virginia, and OSS Training Area B in Catoctin Mountain Park, some 5,800 acres of wooded mountain terrain in northern Maryland. These National Park Service areas were the sites of OSS training camps during most of World War II.

In the following account of the OSS training in the two National Parks and subsequent service overseas, this historical study addresses a number of questions. Why did Donovan's organization choose these parks as training sites? What were the OSS's training aims and methods? Did they change over time? Who ran the camps, and who was trained there? What were typical experiences at the camps and in OSS overseas missions How effective was the training? What relationship did the National Park Service have with the OSS and the military during the war? How did military leasing and use affect the parks, their operation, and their relationship with their surrounding areas during and after the war? What roles and responsibilities did NPS's park managers have at the two parks during the war, and how did they manage their relationship with the military? After the end of the war, how did the military and the National Park Service handle the return of the parks to civilian control and use? Finally, what has been the legacy of the wartime training in the parks—to the parks, to the National Park Service, to the veterans who trained there, to the OSS, and to the institutional heirs of Donovan's organization, the CIA and the Special Forces?

The present study attempts to answer these and other questions as it addresses the significant roles that the National Park Service's Catoctin Mountain Park, Maryland, and Prince William Forest Park, Virginia, played in connection with the Office of Strategic Services during the Second World War.

CHAPTER 1

"WILD BILL" DONOVAN AND THE ORIGINS OF THE OSS

The origins of the Office of Strategic Services (OSS) lay in the dark early days of World War II in Europe. Employing a new, highly mobile form of warfare called *Blitzkrieg* ("lightning war"), Nazi Germany had quickly and brutally conquered Poland in 1939, then Denmark, Norway, Belgium, the Netherlands, and finally France by the middle of 1940. That summer German dictator Adolf Hitler launched a massive air offensive against Great Britain, and many believed that England would be the next to fall. Not until mid-September 1940 was it clear that the *Luftwaffe* had failed, and Britain would remain an island bastion against the Nazis' expanding empire.[1]

Faced with the German onslaught against the western democracies, President Franklin D. Roosevelt committed the United States to their aid, and then, when Britain was left standing alone against Germany, to all-out assistance—short of war—to the British under their new prime minister, Winston Churchill. Like many others at the time, both Roosevelt and Churchill believed that Hitler's shockingly swift military victories were due not simply to prowess of the German Army and its Blitzkrieg tactics, but also by the effective use of demoralizing propaganda and internal subversion by Nazi sympathizers called "fifth columnists," who engaged in espionage and sabotage for the German military intelligence services.[2] In the summer of 1940, one of the special envoys President Roosevelt sent to London to encourage the beleaguered British, to assess their ability to withstand the German onslaught, and to find out what London had learned and was doing about new methods of warfare, especially unconventional warfare, was a prominent New York lawyer and former war hero, William J. ("Wild Bill") Donovan.[3]

As a young man, Donovan had acquired the nickname "Wild Bill," but in 1940, at age 57, however, he seemed anything but wild. He was a rather stocky, silver-gray haired, highly successful senior partner in a Wall Street law firm. Despite his social position and his somewhat reserved, soft-spoken, fatherly manner, however, there was, in fact, an adventurous, daring, driving, and inspiring side to the man. His intellectual ability, steady determination and fertile imagination had led him from a tough, working-class neighborhood in Buffalo to Niagara University and then via scholarship to Columbia College and Columbia Law School. Ultimately, he became part of the nation's economic, political, and foreign policy elite. Donovan was a fine athlete. He participated in boxing, rowing, and track, and a football and was a quarterback while an undergraduate at Columbia. With his law degree, Donovan returned to his home town in upstate New York and became a practicing attorney there in 1908. The handsome, blue-eyed, Irish Catholic with a quick wit and winning smile, who neither smoked nor drank, remained a bachelor until 1914 when, at the age of thirty-one, he married Ruth Rumsey, daughter of one of the leading and wealthiest families in Buffalo. She was a Protestant; he a Roman Catholic, but they both were Republicans.[4]

Donovan had some National Guard and other military training and service but no combat experience until World War I, when he became one of the most decorated heroes of the war. As a 35-year-old major and then lieutenant colonel in New York City's legendary Irish-American, National Guard Regiment, the "Fighting 69th," Donovan bravely directed his outnumbered battalion, despite being wounded, in attacking and overcoming a superior German force during the Second Battle of the Marne in France in July 1918. For that, he received the Purple Heart and the Distinguished Service Medal. Afterwards, in October during the Meuse-Argonne Offensive, when the units he led forward to attack the Hindenberg Line became dispirited and faltered under heavy machine-gun and artillery fire, Donovan rallied and regrouped them, waving his pistol overhead and urging his men to a more advantageous position. He continued to lead them even when he received a severe wound in his right leg and had to be carried. He refused to be evacuated and instead continued in command for five hours although in pain from a smashed knee and tibia and dizzy from loss of blood, until he and his men had halted a German counterattack. For his heroism, coolness under fire, and efficient leadership, Donovan was awarded the French *croix de guerre*, a bronze oak leaf cluster to his Distinguished Service Medal, and the Medal of Honor, the highest award in the U.S. military. He also received another Purple Heart (he had been wounded three times), and by the end of the war, had been promoted to full colonel and was a national hero.[5] With attendant publicity, he became nationally known as "Wild Bill" Donovan. He did not mind the publicity, but he disliked the nickname, preferring "Colonel Donovan" or to his close friends, simply "Bill."

Returning home from France in 1919, Donovan resumed his corporate law practice and became active in the new veterans' organization, the American Legion, and in the Republican Party. As a rising young Irish Catholic political star in the largely Anglo-Saxon Protestant Republican establishment, Donovan received appointments as U.S. Attorney in western New York and subsequently Assistant U.S. Attorney General in charge of the Anti-Trust Division under President Calvin Coolidge. Returning to private practice, he moved his firm from Buffalo to Wall Street, and soon had a list of corporate clients that was longer than before. He supported Herbert Hoover in 1928, but failed to receive an appointment as Attorney General of the United States as he had hoped.[6] He ran unsuccessfully for governor of New York in 1932. Donovan supported the Republican Party's unsuccessful candidates against Democrat Franklin Roosevelt in the Presidential elections of 1932 and 1936.

Concomitant with his connections with Republican political, legal and financial figures, Donovan ran his Wall Street law firm and also continued an active interest in the military and in world affairs. He obtained overseas clients, including some of the London banks and some British politicians, including Winston Churchill. He traveled overseas frequently and obtained contacts around the world. During fascist Italy's invasion of Ethiopia, Donovan toured the battle lines there in 1935, and in 1936. Subsequently, in the Spanish Civil War, he studied the new modern weaponry and tactics being used by both sides. He reported his findings back to President Roosevelt. In the foreign policy debate in the United States between isolationists and interventionists in the late 1930s and early 1940s, Donovan was an active interventionist. In 1940, after the fall of France, he joined the Fight for Freedom organization, one of the new foreign policy pressure groups created to counter isolationists and to build support for vigorous assistance to Britain and other allies against Hitler.[7]

Donovan gained a major supporter in the Roosevelt Administration in June 1940, when the President appointed Frank Knox, a Republican newspaper publisher and a friend and admirer of Donovan's, to be Secretary of the Navy. Knox encouraged the President to draw upon the international lawyer and war hero. Roosevelt would do so throughout the war, but although Donovan had personal access to the President, he was never a member of Roosevelt's inner circle.[8] In the early summer of 1940, the British, under attack by the German Air Force and Navy, sent representatives of their secret services to see Donovan. The key contact was an old friend of Donovan's, William Stephenson, a Canadian air ace in World War I, who had become a wealthy steel magnate, and was in 1940 appointed the New York station chief for the British Secret Intelligence Service (SIS, also known as MI6). In June 1940, Churchill dispatched Stephenson (code named "Intrepid") to New York to encourage increased American military aid. At a meeting at the St. Regis Hotel, Stephenson invited his old acquaintance, Donovan, to visit Britain and personally evaluate its military and intelligence capabilities and assess its chances for surviving the German attack. Donovan went as Roosevelt's personal envoy in July 1940.[9]

It was a fateful visit. Churchill understood the importance of Donovan's mission and sought to use it to persuade American leaders that Britain would not fall and that they should increase their aid for the island bastion against Hitler. To reinforce his position, he provided Donovan with extraordinary access to British military and intelligence secrets. Following meetings with Churchill and King George VI, Donovan was introduced to leading figures in British intelligence and special operations agencies, including Sir Stuart Graham Menzies, head of the civilian Secret Intelligence Service (SIS),[10] and Admiral John H. Godfrey, director of the Royal Navy's intelligence service. Godfrey was especially attentive to Donovan, a kindred spirit. The President's envoy was shown the latest Spitfire fighter planes, the new radar warning system, and Britain's secret coastal defense arrangements. In fact, the only major secret Donovan was not shown was "Ultra," the new and evolving, super-secret process that was beginning to decipher radio transmissions encoded by the German military's Enigma coding machine. Donovan also gathered information about the role of propaganda and subversion in countries that had been conquered by the *Wehrmacht*.[11]

Equally important, Donovan was briefed on the new Special Operations Executive (SOE), a commando agency formed in July 1940 at Churchill's personal insistence. The prime minister had been fascinated by the way the Wehrmacht had used specially trained units, paratroopers and others, to infiltrate behind enemy lines and sabotage and otherwise disrupt their enemy's lines of communication and supply, confusing and throwing off balance the defending forces. After the British Army was driven off the continent at Dunkirk, Churchill needed to demonstrate that Britain could still lash back at Germany. One way was a conventional strategic bombing campaign that the prime minister launched against Hitler's heartland. But Churchill also adopted a new form of warfare, an unconventional subversive effort in German-occupied countries designed in his words to "set Europe ablaze."[12]

Setting German-occupied Europe ablaze with sabotage and guerrilla resistance was the mission that Churchill assigned to two different types of organizations. Commandos—new elite, highly trained, and independently acting units, such as the Royal Air Force's new Special Air Service, and similar units created in the British Army and the Royal Marines—would stage raids to cause havoc around the edges of Hitler's empire. Meanwhile, through a different organization, covert special operations teams, operating under the new Special Operations Executive (SOE), would be infiltrated into occupied Europe to help organize local anti-Nazi resistance groups, supply them with weapons, clothing, food, medical supplies and funds, and direct them in attacks against the German military's lines of communication and supply. Through subversion, sabotage, and the direction of local guerrilla forces, these SOE British agent teams had the mission of keeping the Germans off balance and impeding their military efforts, using primarily groups of young local men and women, like those of the French *maquis*, who actively resisted the German occupation of their country. It was optimistically hoped at the time that these efforts alone might help the conquered peoples of Europe overthrow the Nazi occupiers. Later, a more realistic appraisal redefined the mission to harassment designed to impede the effectiveness of occupiers and the *Wehrmacht*.[13]

Back in the United States, Donovan asserted publicly both the desirability of American aid for Britain and the need for a new, centralized American strategic intelligence agency to be prepared to wage unconventional warfare against the German challenge. Donovan warned that Hitler's shocking military successes were not due solely to the German Army but to new forms of warfare exploited by the Nazis: systematic propaganda to undermine civilian and military morale and active treachery by "fifth columnists" including espionage and sabotage.[14] The United States, he argued needed a national agency to coordinate its efforts to combat and to employ these new forms of warfare. He urged a centralized intelligence agency capable of gathering, analyzing and acting on intelligence, including the use of propaganda to undermine the will of America's enemies and sabotage and guerrilla operations to impede their military capabilities.

Espionage and Guerrilla Warfare in the American Past

Although what Donovan advocated in 1940, America's first centralized intelligence and special operations agency, was unique, it was also imbedded within an American wartime history of sporadic intelligence gathering and subversion. In the colonial period, Europeans had imported to America their own tradition of using spies and other secret agents. Both were employed in the French and Indian War of 1754-63, along with an American innovation, frontier soldiers called "rangers," who took the lead as scouts as well as raiders harassing the enemy's rear. A New England frontiersman, Robert Rogers, organized and led "Rogers' Rangers" in a series of dangerous expeditions behind enemy lines, employing ambushes and other surreptitious, Indian-style warfare rather than European open-field battle in which lined-up ranks of infantrymen directly confronted each other. During the American Revolution, cavalryman Francis Marion, the "Swamp Fox," and rifleman Daniel Morgan, led similar operations. In the Civil War, Confederate cavalryman John S. Mosby organized a similar group, "Mosby's Partisan Rangers." In regard to intelligence gathering, General George Washington, commander-in-chief of the Continental Army, was his own chief intelligence officer, personally supervising a network of spies, including Nathan Hale. The Continental Congress established a Committee of Secret Correspondence to establish direct and clandestine connections with American sympathizers in a number of European countries—including Great Britain.[15]

Later, as President of the United States, Washington obtained from Congress the prerogative to conduct intelligence-gathering operations and to spend appropriated funds for unstated purposes. (Those purposes were especially for espionage, and by 1793, during the growing crisis with Revolutionary France, the secret funds had grown to 12 per cent of the federal government's budget.[16]) Various U.S. Presidents authorized covert operations, from rescuing Americans held by the Barbary states of North Africa in the 1790s and 1800s to providing additional secret missions for military officers such as John Charles Fremont, who helped Americans settlers in California rebel successfully against Mexican rule in 1846. During the nineteenth century, the United States largely forsook foreign espionage, except during wartime. In the Civil War, both North and South employed numerous spies, women as well as men. Rose O'Neal Greenhow spied for the Confederacy in Washington. Allan Pinkerton, detective and northern spy master, quickly established a network of Union agents behind Confederate lines. In 1863, he and attorney and Union Army officer George Henry Sharpe helped create a temporary Bureau of Military Information to centralize intelligence for the general in command of the Army of the Potomac. Not until the late nineteenth century, did the U.S. Army and Navy create permanent military intelligence units, the Office of Naval Intelligence (ONI) in 1882, and Military Intelligence Division (MID) in 1885. Both were small and worked mainly through American attachés overseas. Deferring to Congress, ONI and MID excluded domestic counter-intelligence as incompatible with American civil-military traditions. There was little U.S. intelligence or counter-intelligence operation in the brief war against Spain in Cuba in 1898, but the U.S. Army employed both in the long, counter-insurgency effort to suppress the Philippine Insurrection, 1899-1902.[17]

Sizable American intelligence bureaucracies emerged only with the First World War, in which all the major powers conducted extensive intelligence and counter-intelligence operations. In the period of American neutrality, 1914-1917, both Britain and Germany established extensive espionage and propaganda operations in the United States to aid their cause and hinder their enemy's. The British scored a major coup in early 1917 by decoding and revealing the infamous January 1917 telegram by German Foreign Minister Arthur Zimmermann seeking an alliance with Mexico against the United States. During American intervention in the war, 1917-1918, the United States expanded its own intelligence operations, at least temporarily. The Military Intelligence Division, for example, expanded from four to 1,300 persons. Another precedent was set in 1917, when MID established a secret code and cipher unit (the Cipher Unit, nicknamed the "Black Chamber"). That marked the beginning of U.S. military signals intelligence to augment information obtained from human sources, including spies. Fearing German subversion, espionage and sabotage within the United States, the Army and Navy, for the first time, created domestic counterintelligence bureaus. In addition, the Justice Department established a General Intelligence Division headed by an ambitious, young lawyer, J. Edgar Hoover, in its Bureau of Investigation (which had been created in 1908). The State Department sought temporarily to centralize intelligence from its various geographical area desks in a unit known only as "U-1" because it was located in the office of the Undersecretary of State. But it consisted of only a few analysts. In another major departure, the United States began joint intelligence cooperation with Great Britain.[18]

When peace returned, cutbacks became inevitable. The Justice Department's General Intelligence Division was shut down, although the Bureau of Investigation, later renamed the Federal Bureau of Investigation (FBI), was retained and soon came under the leadership of J. Edgar Hoover. MID and ONI were reduced in size. State established a central department to coordinate secret information obtained by its staff around the world. In regard to signals intelligence, MID's "Black Chamber," was turned into a joint operation, run by the War and State Departments. Its chief continued to be the idiosyncratic Herbert O. Yardley, a former telegrapher, code clerk, and brilliant, self-taught cryptologist. In 1919, Yardley, working under a professional cryptographer Ralph Van Deman, broke the main Japanese diplomatic code. This coup enabled U.S. delegates to the Washington Naval Arms Limitation Conference of 1921-22 to be aware of Tokyo's maximum concessions and confidently to push the Japanese down to that figure, a major success.[19]

With budget cuts by economy minded administrations in the 1920s, intelligence staffs were further reduced. State closed down its U-1 office in 1927, and in 1929, Herbert Hoover's new Secretary of State, Henry L. Stimson, closed down Yardley's "Black Chamber." As Stimson explained his action to his biographer eighteen years later, "Gentlemen do not read each other's mail." Despite the end of the War and State Departments' Cipher Unit, the Army continued its work on code-breaking.[20]

American Intelligence Agencies at the Outset of World War II

When World War II broke out in Europe in 1939, U.S. intelligence operations were splintered among nearly a dozen federal agencies, often suspicious bureaucratic rivals accustomed to competing for appropriations in tight peacetime budgets. The primary intelligence agencies were the Army's Military Intelligence Division, the Navy's Office of Naval Intelligence, the State Department's various geographic area divisions, and the Justice Department's Federal Bureau of Investigation, but there were a number of others.[21] Beyond this array of civilian and military units, however, there was no central coordinating body for sorting, evaluating and summarizing the masses of information that they obtained.

Inter-service rivalry hampered effective cooperation between the intelligence officers of the Army and Navy. The military also tended to focus on tactical rather than strategic information, and also unlike Donovan's proposed agency, they were reluctant to use employ spies or conduct other undercover operations, and they were relatively unconcerned with psychological, social, or economic intelligence. Yet the armed forces did achieve major successes in signals intelligence. Despite its limited, peacetime size and budget, the Army's code-breaking team in the Signal Intelligence Service finally broke the Japanese "Purple" cipher, Tokyo's main new diplomatic code, in September 1940. This top secret decoding process was labeled "Magic."[22] The Navy had established its cryptographic unit, OP-20-G, later than the Army and suffered comparative lack of experience. Under a joint agreement, the two services had worked on breaking the Japanese diplomatic code. But once the Army had made the breakthrough, their rivalry prevented them from cooperatively exploiting it.[23] Such Army-Navy rivalry would continue throughout the war, hampering U.S. intelligence assessments and also complicating co-operation with other countries' intelligence services.[24]

The outbreak of World War II and the shockingly quick conquest of most of Western Europe by the German Army had revealed the paucity and fragmentary nature of U.S. intelligence operations. Roosevelt had sent Donovan to Britain in July 1940 to provide an alternative source of information. In December 1940, a month after Roosevelt had been elected to an unprecedented third term, the President sent Donovan across the Atlantic again, this time on a much broader inspection tour. From December 1940 to March 1941, Donovan, with Churchill's support, inspected British military and diplomatic efforts in the Mediterranean, the Balkans, and the Middle East. Returning home, Donovan, at Roosevelt's request, gave a nationwide radio address praising American aid to countries fighting the Germans and Japanese, from England to Greece to China. He said "lend-lease" military assistance could in effect transform many of these local resistance movements into surrogate "expeditionary forces" that could block the growing enemy threat to America.[25]

Donovan concluded that that the United States needed a central national intelligence and covert operations agency. As he wrote to Secretary of the Navy Knox a month after his return from the Mediterranean: "Modern war operates on more fronts than battle fronts. Each combatant seeks to dominate the whole field of communications. No defense system is effective unless it recognizes and deals with this fact. I mean these things especially: the interception and inspection (commonly and erroneously called censorship) of mail and cables; the interception of radio communication; the use of propaganda to penetrate behind enemy lines; the direction of active subversive operations in enemy countries." Donovan recommended that such an agency be headed by a Coordinator of Strategic Information, reporting directly to the President.[26]

Roosevelt, who had only begun to think about a more effective intelligence establishment after the outbreak of the war, was initially more cautious about such a consolidation. Donovan's suggestion was vigorously opposed by existing intelligence agencies, particularly those in the Army and Navy and also by the State Department under Cordell Hull and the FBI under J. Edgar Hoover.[27] Beginning in 1939, fears in the United States of Nazi, fascist, or Communist espionage, sabotage, or other subversion by possible "fifth columnists" as well as some bumbled attempted arrests that were highly publicized had led the President to request greater coordination among the various intelligence and counter-intelligence agencies. The agencies resisted, and Roosevelt became increasingly irritated at being forced to arbitrate their bureaucratic squabbles. He also strongly advised the intelligence agencies to take a broad strategic perspective rather than their narrow parochial views of the growing international challenges facing the nation.[28]

Germany and Japan continued their expansion throughout 1941. Hitler's forces invaded Greece and Yugoslavia and then the Soviet Union. Roosevelt and Churchill, despite their distaste for Joseph Stalin's brutal communist regime, agreed to aid the Soviet empire, because it now bore the brunt of the fighting against Germany. In North Africa, Benito Mussolini's Italian forces were reinforced and led by German General Erwin Rommel and his *Afrika Korps*, which pushed the British Army back toward Egypt. In the Far East, in an attempt to restrain Japanese expansion, Roosevelt extended U.S. lend-lease assistance to China and began to curtail America's trade with Tokyo. However, Imperial Japan Army continued its expansion in China and increased pressure on French Indochina in an attempt to cut off British and American supplies to the Chinese Nationalist government and Army under Chiang Kai-shek (Jiang Jieshi).[29]

To help meet the challenges and as part of the increasing Anglo-American cooperation under Roosevelt and Churchill, American and British military staffs began holding joint meetings in January 1941. Eventually the U.S. Joint Chiefs of Staff (JCS), which included the chiefs of the Army, Navy, and Army Air Forces, as well as the military adviser to the President, and the Combined [American and British] Chiefs of Staff, worked out the broad concepts of Allied strategy. Priority was given to the defeat of Nazi Germany, seen as a greater threat because of its domination of the resources of Europe. The defeat of fascist Italy and imperialist Japan, while important, were viewed as secondary and, therefore, were allocated fewer resources than those against Hitler's Germany. The Allied offensive plan envisioned obtaining naval control of the sea lanes, conducting strategic air bombardment of enemy industry, encouraging resistance movements to undermine the enemy's control in occupied countries, and finally launching massive ground campaigns to defeat the enemy armies.[30] In August 1941, Churchill and Roosevelt met secretly on warships off Newfoundland, pledged greater cooperation, and committed themselves to an "Atlantic Charter," a statement of broad, idealistic postwar aims, including freedom of speech and religion, freedom from fear and want, freedom of the seas, free trade, self-determination of peoples, and the disarmament of aggressor nations.[31]

Amidst these rapidly unfolding international events and faced with the apparent inability of the traditional intelligence agencies to cooperate, Roosevelt finally directed his attention to creating coordinated intelligence. The groundwork had been laid over the past year, but in the spring or early summer of 1941, the President had asked a Cabinet committee consisting of Secretary of War Stimson, Secretary of the Navy Knox, and Attorney General Robert Jackson, who oversaw the FBI, to make a recommendation. According to Donovan his meeting with the committee led it to recommend his idea to the President.[32] But it was not really that simple. There had been a flurry of activity before the final decision was reached. Donovan had submitted his first formal proposal to the President at the end of May 1941.[33]

After Donovan discussed his proposal fully with Knox, a copy was sent to Stimson. A number of people, with whom Donovan discussed the idea, endorsed it. Among them were certainly Knox, Presidential speechwriter Robert Sherwood, and probably Felix Frankfurter, a Supreme Court Justice and Presidential confidant.[34] Added to those pressing for it were two British intelligence officials, William Stephenson and Admiral John Godfrey, who had arrived in the United States on 25 May 1941 (accompanying Godfrey was his personal assistant, Commander Ian Fleming, later to become famous as the author of the *James Bond* novels, but who in late June 1941 wrote memoranda to Donovan on how to get his organization up and running by December).[35]

Despite increasing American material support, the British felt frustrated by rivalries among the U.S. armed forces and various intelligence offices that hampered effective Anglo-American cooperation. Consequently, they joined in encouraging a centralized American intelligence agency. Godfrey met Donovan in New York and then journeyed to Washington, where the British admiral found the chiefs of the various intelligence agencies still unwilling to cooperate with each other. In desperation, Godfrey obtained an appointment with the President and on 10 June urged Roosevelt to create a centralized intelligence service. The President was typically loquacious, but Godfrey persevered, recalling, "At last I got a word in edgeways. I said it a second time, and a third time—one intelligence security boss, not three or four."[36] Given their good working relationship with Donovan, the British actively, if carefully, championed him to direct it.[37] Meanwhile, Donovan on 10 June 1941 submitted directly to the President a slightly expanded version of his draft memorandum of late May. He had added two new paragraphs: one on the desirability of common, accurate economic information, and the other emphasizing the need for accurate foreign information in order to use radio most effectively as a weapon to demoralize the enemy.[38]

Roosevelt Agrees to a Coordinator of Information (COI)

The President made his decision on 18 June 1941. At 12:30 p.m. that day, Donovan, accompanied by his friend Secretary of the Navy Frank Knox, met with Roosevelt and Ben Cohen, one of the President's most trusted legal advisers. It was a lengthy meeting and no minutes were taken, but it is clear afterward that Roosevelt had agreed to a new intelligence organization, and he had chosen Donovan to head it. At the meeting's conclusion, the President scrawled his approval in a hand-written note on the cover sheet of Donovan's 10 June memorandum. America would have its first centralized intelligence agency, the Office of the Coordinator of Information (COI).[39]

Back in New York that night, Donovan met with Stephenson and reported on what had happened. According to the Stephenson, Donovan said he had accepted appointment as Coordinator of Information "after [a] long discussion wherein all points were agreed. He said the new agency would coordinate intelligence and would also conduct offensive special operations; Donovan would report directly to the President and would have the rank of major general." In a cable to London a week later, Stephenson was elated by the results that were so much to Britain's liking.[40]

Later, there would be considerable uncertainty, especially among Donovan's foes, concerning precisely what authority and roles Roosevelt had given him. But it was clear that the President had taken an historic step in establishing a central, national intelligence agency and, at least indirectly, authorizing it to use clandestine funds as well as covert methods. It would also become clear that Donovan would interpret and employ the President's authorizations to the fullest extent, particularly once the United States entered the war. He would include both gathering and analysis of intelligence as well as acting upon that information through special covert operations.

The establishment of America's first centralized intelligence agency owed much to the interest, ability and political sensibilities of William J. Donovan, who is quite properly called the "father" of the Central Intelligence Agency.[41] Yet, the impetus for the organization and particularly its initial credibility owed much to the support of the British, who intelligence operations dating back hundreds of years and who had recently created much heralded commando and special operations teams.[42] What remains debated and somewhat controversial is the degree of British influence in Donovan's selection and the nature of his organization, the COI and subsequently the OSS. Some sources, mainly British and Canadian, assert that both Donovan's selection and his organization were influenced by the British experience and actions of William Stephenson and Admiral John Godfrey. Others, mainly Americans, have belittled such assertions and argued that Donovan's selection was solely a Washington decision, that the agency was Donovan's own idea, although the COI and subsequently the OSS were genuinely American institutions with historic roots in the nation's history of sporadic intelligence gathering and guerrilla operations.[43]

The Office of the Coordinator of Information (COI) was officially created by Roosevelt on 11 July 1941 under his executive authority. The President simultaneously appointed Donovan to head it.[44] The agency's name had been considerably altered since Donovan's original proposal. The Army and Navy had objected to any use of "strategic" or "defense" in the title. The resulting title, Coordinator of Information, was, as the director of the Bureau of the Budget complained in presenting it to the President, "vague and is not descriptive of the work Colonel Donovan will perform."[45] Pressures from the armed services had resulted in the new agency being stripped of any military character. The agency would be entirely civilian until the inclusion of military personnel after became the Office of Strategic Services (OSS) in June 1942.[46] Even the OSS remained a civilian agency throughout the war, although some three-quarters of its personnel were assigned from the military or were awarded military rank.[47]

The press release of 11 July 1941 announcing the creation of the new office and Donovan's appointment offered a rather vague and circumscribed view of the COI's authority. Prepared by White House press secretary Stephen Early, it stated that "Mr. Donovan will collect and assemble information and data bearing on national security from the various departments and agencies of the Government and will analyze and collate such materials for the use of the President and such other officials as the President may designate. Mr. Donovan's task will be to coordinate and correlate defense information, but his work is not intended to supersede...or involve any direction of or interference with, the activities of the General Staff, the regular intelligence services, the Federal Bureau of Investigation, or of other existing departments or agencies."[48] What Steve Early did not explain, however, was that the authority under the executive order itself was broad enough to allow Donovan not just to collect information but to engage in all sorts of actions to obtain and act upon information vital to national security. Donovan would ultimately make the most of the executive order and include what he and the British had envisioned: offensive action, including psychological warfare, political warfare, subversive operations, guerrilla warfare, espionage, and sabotage.[49]

Although to the public, the ostensible purpose of the COI was to gather and coordinate information and to project propaganda overseas, both of which were true in part, the secret activities of Donovan's organization, espionage, counter-espionage, and special operations, were purposely covered up, as the country was not yet at war.[50] Beside propaganda, the COI was intended as an intelligence-gathering agency with additional authority to engage in unorthodox warfare. The vague but operable paragraph in Roosevelt's executive order issued under his authority as commander-in-chief, gave the COI "authority to collect and analyze all information and date, which may bear upon national security; to correlate such information and data, and to make such information and data available to the President, and to such departments and officials of the Government as the President may determine; *and to carry out, when requested by the President, such supplementary activities as may facilitate the securing of information important for national security not now available to the Government* [emphasis added]."[51] It was to be an unprecedented action in American history—an attempt to establish an effective unified organization that would handle research, intelligence, propaganda, subversion, and commando and guerrilla operations in modern war. It had the potential, like the intelligence, propaganda, and subversive operations in Britain, to have an independent strategic role, a "Fourth Arm" added to the Army, Navy, and air force.[52]

Not unexpectedly, the new organization came under renewed attack from other American intelligence agencies, jealous of their prerogatives and suspicious of Donovan, whom they believed had achieved his coup through political connections on both sides of the Atlantic. These rivals derided the Wall Street lawyer and most of the civilian associates he recruited to run the COI and later the OSS as rank amateurs in the fields of intelligence, counter-intelligence, and military operations. To discredit him, rumors were planted with certain columnists that Donovan was seeking to establish an American *"Gestapo,"* the Nazis' secret police, and that he wanted to emulate Nazi propaganda minister Joseph Goebbels in having strict control over the mass media. Rumors were floated that Donovan had created a "super spy agency" that would give him power over the Army, the Navy, and the FBI, and that his influence on strategy would allow him to dictate to the General Staff. Even six months later, Roosevelt was still trying to explain to the press that such rumors were totally false.[53]

Viewing Donovan as a potential intelligence "czar," FBI director J. Edgar Hoover had fought vigorously against the COI, and its creation represented one of the major bureaucratic defeats of his long career. But Hoover was able to protect the bureau's territory by getting the President to prohibit Donovan's agency from conducting espionage or counter-espionage operations within the United States or in Latin America, where Hoover had obtained jurisdiction for the FBI.[54] Later, the bureau was able to infiltrate the COI/OSS with an informant, a woman in the message center, who kept Hoover informed of activities in Donovan's organization. The female "mole" was finally identified and dismissed, but the Hoover had already used her information to outflank or embarrass Donovan's organization on a number of occasions.[55] After U.S. entry into the war, the COI staged a number of night-time break-ins at foreign embassies in Washington, illegal operations conducted with the President's knowledge, to photograph codebooks and other important documents. The Vichy French embassy break-in was successful, but when Donovan's agents clandestinely entered the Spanish embassy, opened the safe, and took out the code books, and photographed their contents, Hoover's men arrived, sirens screaming, and accompanied by the press. The FBI publicly arrested Donovan's agents and confiscated their copied materials. Hoover refused to release the agents until Donovan himself appeared before him and took personal responsibility for the break-in, but the FBI director would not turn over the copies of the codebooks to the Donovan, giving them instead to another COI rival, the Army's Military Intelligence Division.[56]

Donovan Builds the COI

Without any of the usual fanfare accompanying a new agency, Donovan began to set up shop. In the space-tight capital, he obtained a few rooms and telephones, and with half a dozen assistants began to recruit an organization. After several moves, each into larger quarters, Donovan, in September 1942, consolidated into what would be his organization's headquarters throughout the war. It was a 13-acre, six-building complex at the far western end of E Street between that and Constitution Avenue, which ran parallel to E street, and bordered by 23rd Street to the east and 25th Street to the west. The buildings were those formerly occupied by the National Institutes of Health and the Navy Bureau of Medicine and Surgery. Donovan, who for security purposes was referred to in coded messages as "109" had his office on the southwest corner on the second floor of the South Building. Several large wooden huts, called temporary buildings, although some dated back to World War I, housed more offices, including most famously within the OSS, "Q" Building at 2430 E Street, N.W., the main personnel administration center where most new recruits reported. As the organization expanded during the war, OSS established additional administrative and storage facilities in a nearby former public skating rink and warehouses down the hill. Motorists driving along the then Rock Creek Park Drive generally paid no attention to the anonymous looking governmental structures scattered around a generally, rather disreputable industrial area.[57]

Donovan had an anti-bureaucratic philosophy. Because he saw members of his agency as learning their way in new forms of warfare, he was more interested in initiative, innovation, and results than abidance by the rules and being held to strict accountability. He told subordinates that he would rather have them use their imagination, try new things, and take risks, even if it meant that they would make mistakes and sometimes fail, rather than simply to stick cautiously to traditional ways of doing things. Donovan was not interested in military expertise as much as people who could think quickly and clearly and find innovative solutions to difficult situations. He asked for bold, new thinking and action, and, to a surprising extent, he got them. The organization was infused with Donovan's own spirit of energy, experimentation, and possibility. He was an inspiring leader: visionary, bright, brave, quick to make decisions, open and fair. "He was open-minded," recalled Arthur M. Schlesinger, Jr., historian and veteran of the Research and Analysis Branch. "He listened to anything. He'd try anything. He was adventuresome. He was not a conventional figure."[58] The innovators, explorers, and point people in his organization probed new frontiers in the war against the Axis powers. They felt a sense of uniqueness, of special quality, of membership in an elite group. The members of Donovan's organization viewed themselves as an advanced guard leading at the point of attack against the Axis threat to civilization. No wonder that the OSS selected for its emblem, its shoulder patch, a golden spear point.[59]

With a free hand in hiring, Donovan began by enlisted a number of his able associates and then began recruiting Americans who had traveled abroad or were otherwise well versed in world affairs. In the early 1940s, that often meant educated or affluent members of the American elites or foreign émigrés. Donovan relied upon his personal contacts with people he or his subordinates trusted, and he drew most of his top aides from prestigious colleges and universities, businesses, and law firms, including his own.[60] As the war approached and particularly after the United States entered the war following the Pearl Harbor attack in December 1941, many Americans volunteered to serve their country. In that rush to service, Donovan's COI and its successor, the OSS, drew such a disproportionate number of socially prominent men and women that some wags claimed the initials of O.S.S. stood for "Oh-So-Social." Although prominent people held a number of high level positions in the agency, the vast majority of the men and women recruited by the OSS were neither prominent nor listed in the Social Register.[61]

First priorities were to obtain experts to evaluate incoming intelligence and also propagandists who would use some of that research to undermine enemy morale abroad. As early as June 1941, Donovan had obtained the support of the Librarian of Congress, poet Archibald MacLeish, to allow the prospective organization to use the library's extensive materials in analyzing the Axis' strengths and weaknesses. In July, Donovan hired the President of Williams College, James Phinney Baxter III, an historian, to head the COI's Research and Analysis (R&A) Branch. Baxter and Donovan quickly recruited noted scholars in various disciplines from prestigious colleges and universities and put them to work in the Library of Congress Among the early recruits were Harvard historian William L. Langer; Edward Meade Earle from the Institute for Advanced Study in Princeton; economist Edward S. Mason from Harvard; Joseph Hayden, a University of Michigan political scientist and former vice governor of the Philippines; historian Sherman Kent of Yale; Wilmarth S. Lewis, millionaire Yale biographer of Horace Walpole; and James L. McConnaughy, President of Wesleyan University, and many others. Within a few months, Donovan began sending to Roosevelt summaries of detailed R&A reports on strategic economic, political, social, and military information about conditions and strategic prospects in Europe, North Africa, and the Middle East.[62] Robert E. Sherwood, noted playwright, pacifist turned interventionist and a speech writer for the President, enthusiastically endorsed the idea of undermining enemy morale and bolstering resistance via short-wave radio broadcasts and other media aimed at Nazi Germany and German-occupied countries, and Donovan quickly chose him to head COI's Foreign Information Service. Within a few months, Donovan added a Visual Presentation Branch, which would include Hollywood directors John Ford, famous for his westerns and other epics, and Merian C. Cooper, adventurer/filmmaker and creator of King Kong.[63] To facilitate COI's work in Europe and the German-occupied countries there, Donovan, with the permission of Roosevelt and Churchill, set up an office in London in October 1941, the first of many overseas regional headquarters.[64]

Donovan's organization expanded dramatically. When the COI was established in July 1941, planners at the Bureau of the Budget had estimated that it would need only a small staff and an annual budget of about $1.5 million. At the same time, Donovan warned that additional funds for the secret operations would be needed later. Still, the overall estimate was $5 million. The Budget planners certainly underestimated Donovan. In November 1941, Budget Director Harold Smith was shocked by Donovan's budget request for $14 million for fiscal year 1942. Roosevelt concurred in most of Donovan's requests, and by December 1941, COI had 600 staffers and a current budget of $10 million, the major outlays of which were for international short-wave and medium-wave broadcasting to Europe and the Far East, counterespionage and secret activities in Europe, Research and Analysis, and the creation of a War Situation Room for the President. On December 8, 1941, the day after the Pearl Harbor attack. Roosevelt authorized an immediate additional $3 million for the COI.[65]

More importantly, although the majority of expenditures for COI/OSS—payrolls, supplies, and other regular expenses—were paid with vouchered funds, subject to government audit, COI and later OSS also obtained authority to use "unvouchered" funds (U.V.F.) from the President's emergency allocations. Congress granted these to the President and a few other designated officials to spend solely on their personal responsibility. They did not have to disclose the specific purpose for which the funds were used, and these secret expenditures were not subject to detailed audit. In practice, Donovan had only to sign a note certifying that the funds had been used properly for national security purposes. This fiscal authority, augmented by the espionage authority that Donovan received from the armed forces, allowed him to conduct a wide variety of secret activities, from hiring foreign spies, to sending American agents behind enemy lines with bags full of currency, gold or silver coins, or other inducements for recruiting and purchasing supplies for indigenous guerrillas, for bribes for guards or turncoat officers, for theft, assassination attempts and a host of other clandestine purposes. As a CIA historian later put it, unvouchered funds were "the lifeblood of clandestine operations."[66]

During the war, considerable sums were paid for secret ends. Neither the names of the OSS personnel in the field who made the secret payments, nor the identity of those who received them was revealed for the record. There was no detailed accounting of that kind of disbursement. "U.V.F. was dollar dynamite," recalled Stanley P. Lovell, chief of Research and Development. "Always haunting us was the specter of some postwar Congressional committee, which might well be empowered by the Congress to ignore all wartime secrecy and which, assuming a hostile attitude, might make a Teapot Dome type of thing [scandal] out of these large sums, for which no accounting whatever existed."[67] Consequently, Donovan placed responsibility for the unvouchered funds in the hands of a triumvirate of individually wealthy and highly respected financiers: Junius S. Morgan of J.P. Morgan and Company in New York; Robert H. Ives Goddard, an immensely wealthy financier from Providence, Rhode Island; and W. Lane Rehm, financial genius of one of the largest investment trusts in the United States. Together they performed the delicate task of approving or denying requests for use of unvouchered funds and assessing reports on their expenditure in secret activities.

When the COI was established in July 1941, Donovan focused first on building an administrative staff and then on recruiting college faculty, who were area experts, for research and analysis of available information, and setting up a propaganda system. But even before the U.S. entered the war, he had begun to plan a secret operational division that would engage in espionage, counter-espionage, and, as he confided to a representative of the Bureau of the Budget, "very secret activities dealing with sabotage and other ideas which might be developed as the program progresses."[68]

In the fall of 1941, Donovan set up a small working group in COI called "Special Activities," instructing its members to study clandestine activities, not just espionage but also subversive special operations' activities by saboteurs, commandos or guerrilla units. His primary American advisers at COI and then OSS on espionage and subversion would be two old and trusted friends. One was David K.E. Bruce, a diplomat married to one of the Pittsburgh Mellons, who was allegedly the wealthiest woman in America. In early 1942, Donovan put Bruce in charge of a fledgling espionage unit known first as Special Activities, Bruce (or SA/B), and then once the COI became the OSS in June 1942, the Secret Intelligence Branch (SI).[69]

The other man was M. Preston Goodfellow, a Brooklyn newspaper publisher, who in 1942 would head the Special Operations Branch, and would play an important role in the creation of the training camps in the National Parks. As befitting his last name, Preston Goodfellow was a jolly, good-natured man, an executive with ability to charm and even to ingratiate himself to diverse people while, keeping his eye on the main chance. Born and raised in Brooklyn, he spent a career in New York newspapers. After graduation from New York University with a degree in journalism, he had worked his way up in the city's newspapers from copy boy to reporter to city editor. He had also joined the Army reserves as an officer, and in World War I, he served in the Army Signal Corps in the United States. After the war, Goodfellow rejoined the *Brooklyn Eagle* but this time on the business rather than the editorial side. A successful advertising manager there, he left to spend three years as assistant publisher of Hearst's *New York American*, then resigned in 1932 to become co-owner and publisher of the *Brooklyn Eagle*. Six years later, he formed his own business, which he ran until July 1941 when he was recalled by the Army to active duty. Major Goodfellow was assigned to duty in the office of the Assistant Chief of Staff for Intelligence (G-2) in Washington, D.C. There, the friendly 49-year-old New Yorker, who was at ease with both civilian and military personnel, became sympathetic to Donovan's ideas about unconventional warfare. Consequently beginning in September 1941, Goodfellow, by then a lieutenant colonel, was assigned by G-2 to work informally as liaison between Army Intelligence and the new Coordinator of Information.[70]

At COI, Goodfellow became in effect head of the special covert operations planning side of Donovan's organization after Donovan had a falling out with the first head of that activity, Robert Solberg. In October 1941, Donovan has sent Solberg to England for three months to study the British Special Operations Executive. Goodfellow served as acting chief while Solberg was away and succeeded him in January 1942 when Solberg returned and proposed a plan to replicate British SOE that Donovan rejected.[71] With Solberg's departure, the office became known as Special Activities/ Goodfellow (or SA/G) until OSS was established in June 1942, when it became the Special Operations Branch. For nearly a year, from the fall of 1941 through August 1942, Goodfellow had divided his time between the two intelligence agencies, Donovan's and the Army's, before being assigned fulltime as deputy director of OSS.[72] Goodfellow's main impact on Donovan's organization in 1942 was in launching Special Operations and the first American-based training program for agents of the OSS.

In October 1941, when Donovan had sent Solberg to Britain to study the organization, training, and effectiveness of the British Special Operations, the

Coordinator of Information did not believe that it was either wise or practical for the Office of COI, being a civilian agency, to seek formal authorization for commandos or guerrilla units when the United States was not officially at war. Consequently, special operations planning in the Office of the COI had not gone beyond rudimentary ideas and an informal title by November1941.[73] That would change dramatically, as would Donovan's entire organization, after the Japanese attacked Pearl Harbor, December 7, 1941.

CHAPTER 2

A WARTIME ORGANIZATION FOR UNCONVENTIONAL WARFARE

On a clear and brisk afternoon on Sunday, December 7, 1941, William J. Donovan was enjoying an exhibition football game at the New York's Polo Grounds, when unexpectedly he heard an announcement on the stadium's loud speaker system that he had an important phone call. He soon learned that the Japanese had attacked Pearl Harbor and that the President wanted him right away. He took the next train to Washington. "It's a good thing you got me started on this," Roosevelt told the Coordinator of Information that night at the White House. Then they talked about getting Donovan's agency ready for war and about America's will and ability to fight.[1] Within a few days, the United States was officially in the war against all three Axis powers—Japan, Germany, and Italy.

U.S. entry into the Second World War soon led to major changes in Donovan's fledgling organization. In June 1942, its name was changed from the Office of the Coordinator of Information (COI) to the Office of Strategic Services (OSS). It was placed under the Joint Chiefs of Staff (JCS), but Donovan, now the Director of OSS, reported directly to the President.[2] The change also involved a reorganization and a redefinition of its mission. Donovan lost his authority to conduct counter-espionage in the Western Hemisphere to the FBI, and his Foreign Information Service, COI's propaganda branch, with nearly half of Donovan's 2,300 personnel, went to the new Office of War Information.[3] The loss of his overseas propaganda branch was most disappointing to Donovan because that unit was key to his concept of psychological warfare encouraging captive peoples to rebel.[4] Despite his losses to the FBI and OWI, Donovan saw his organization's secret operations—espionage, counter-intelligence, disinformation, and guerrilla leadership—expanded. Under the auspices of the Joint Chiefs of Staff and the protection of President Roosevelt, the OSS grew in size and stature to become America's primary espionage and unconventional warfare agency during the war.

The Joint Chiefs of Staff, composed of the heads of the armed services, had been created in February 1942 to coordinate the American armed forces and to expedite cooperation with the British chiefs of staff.[5] At the urging of British Prime Minister Winston Churchill and with the concurrence of the President, psychological warfare, subversion and sabotage behind enemy lines had been added to the planning agenda of the U.S. military. Previously, the American armed forces forces had largely ignored such unconventional warfare. Since Donovan's civilian agency was the American authority in that area, the Army and the newly formed JCS decided that it might be useful to include the Donovan's group under their authority.[6] With Donovan's free-wheeling style clashing with the military's inherent belief in the sanctity of the chain of command, Lieutenant General Dwight D. Eisenhower, head of the War Plans Division, recommended that be made directly responsible to the JCS.[7] But Donovan was able to reach an agreement under which his organization would come under JSC jurisdiction but still retain its autonomy instead of being subsumed under Army Intelligence (G-2). Both organizations benefited: Donovan's group gained military support and resources, and the armed forces acquired a useful agency in intelligence gathering and analysis, psychological warfare, and special operations, in short, for unconventional warfare.[8]

OSS and the Army

Still, the relationship between military leaders and Donovan's organization remained strained throughout the war. The head of Army Intelligence, Brigadier General George V. ("George the Fifth") Strong, led an effort to stymie the OSS, until Army Chief of Staff General George C. Marshall forged a compromise that limited overt interference with the OSS.[9] Most of the professionals continued to be suspicious of Donovan's citizen-soldiers, their neglect of normal military routine, and their unorthodox methods. There was also resentment about "direct commissions," for civilians in the OSS with military ranks from lieutenant to colonel, without any previous training or service in the military, and pressures to appoint prominent civilians as generals, "political generals," Marshall called them. Because the General Staff considered him an amateur not a professional soldier, Donovan did not get the general's stars that Roosevelt had promised him until midway through the war. The Army recognized Donovan as an active duty colonel, but General Marshall resisted promoting him to flag rank until March 1943 when Donovan became a brigadier general and November 1944, when he received his second star as a major general.[10] Despite such difficulties, Donovan put a good face on the creation of the OSS, crowing to a senior British general, about his success "in having our Joint Chiefs of Staff do something which has never been done in our military history. That is to take in as part of their organization a civilian unit. There had been great neglect of the new elements in modern warfare and we have succeeded in getting them set up and all under one tent, including special intelligence, special operations, and psychological warfare."[11]

Donovan's organization grew dramatically during World War II to fulfill those missions. The COI, which had only 100 persons working for it in September 1941 had reached 2,300 in June 1942.[12] OSS reached 5,000 personnel by September 1943.[13] At its peak strength in late1944, it had almost 13,000 men and women on its rolls, alhough the number of people who served at various times with COI and OSS between 1941 and 1945 may have totaled 21,600 or even 24,000.[14] In the spring of 1945, of the approximately 13,000 OSS personnel, nearly three-quarters, some 9,000 were uniformed members of the armed forces. Of those, 8,000 were in the Army, 2,000 of them officers.[15] Of the total of 21,640 persons who served in COI/OSS at one time or another, 4,000, or nearly one-quarter, were women. Most of these college-educated women served in the United States, the majority working in Washington, but 700 women went overseas, and a few served behind enemy lines.[16]

Those women who did serve overseas were primarily at the organization's main base stations—London, Cairo, Naples, Kunming, China and Kandy near Colombo, Ceylon/Sri Lanka. Most of them continued to do much of the kind of paper, communications, and linguistic work they had done in Washington. After graduating from Smith College, Barbara Hans was living in Mount Kisco, New York with her mother, when she was hired by OSS's Communications Branch in 1942. "My mother said, 'Oh, how can you leave me?' I said, '"There's a war on. I'm going.' It was quite exciting for a young woman."[17] OSS decided that since she and her group of "Smithies" were good in math and cross-word puzzles, they would be trained in coding and decoding.[18] She accompanied eight other women to Ceylon. "I spent half my time in Kandy locked up in a cage on night duty, receiving and sending cable traffic to and from the Arakan [coast of Burma], OSS drops and Detachment 101."[19]

Organizationally, OSS during the war had several main branches: operational, supportive, and administrative in nature. Although Donovan's organization was divided into more than a dozen different branches, plus other divisions and sections, the two largest categories were intelligence and special operations. Intelligence gathering and analysis had several different components: Secret Intelligence, Counter-Intelligence, Foreign Nationalities, and Research and Analysis, but all together at the peak strength of OSS in October 1944, the intelligence branches amounted to over 3,000 persons, representing 25 per cent of the people in Donovan's organization. Special Operations had the same number and percentage at that time. Technical operations, which included a variety of support services from Communications Branch to Research and Development and the like, amounted to just over 2,000, representing 18 percent of the OSS in October 1944, and Administrative Services had the same numbers and percentage. Administration included about 1,800 persons and represented 14 percent of Donovan's organization at that time.[20]

By October 1944, 65% of OSS personnel were serving overseas. That percentage remained relatively constant, but the theater distribution changed. The initial buildup was for the Mediterranean Theater and then the European Theater, especially for the invasion of France in 1944. By the end of that year, with the Wehrmacht being driven back into Germany, the OSS was shifting many of its assets to the Far East. The percentage of the agency's personnel in the China-Burma-India Theater expanded from 14% in October 1944 to 36% (more than 4,000 persons) in June 1945. It remained at that level until the war ended in September 1945.[21]

Main Branches of OSS Secret Intelligence

Most of the glory was won by the operational branches, especially the bold and daring agents of the Secret Intelligence and Special Operations Branches and the commando-style military units of the Operational Groups. Most mysterious were the "cloak and dagger" operatives of the Secret Intelligence Branch (SI).[22] These were the men and some women who ran intelligence operations and rings of indigenous spies primarily in enemy or enemy-occupied countries. Intelligence gathering an analysis was a major function of OSS.

"The information came in from agents. We never called them spies," recalled Dorothy Hayes Stout, who typed up incoming coded messages.[23] The military was particularly interested in obtaining intelligence about the enemy, and the main supplier of military-related intelligence in the OSS was the Secret Intelligence Branch. SI played its first important role in helping ensure the success of the American invasion of Vichy French North Africa in November 1942. As early as June when planning for the operation had begun, a female SI agent, Amy E. Thorpe (code-named "Cynthia"), a 32-year-old socialite, seduced a Vichy official and copied valuable naval codes from the French Embassy in Washington.[24] In North Africa itself, OSS agents like diplomat Robert Murphy, World War I war hero and American professor at Cairo University William Eddy and Harvard Arabist and anthropologist Carleton Coon worked successfully to minimize resistance to an Allied landing by the colonial forces of Vichy France which was collaborating with the Nazis. The OSS may have had some complicity in the assassination of Vichy French commander Admiral Jean Darlan.[25] SI continued its work throughout the Mediterranean and in Western and Central Europe. On the other side of the world, in China, SI networks behind Japanese lines provided information on bombing targets for U.S. Army Air Forces and on Japanese shipping for the U.S. Navy.

The majority of SI personnel worked in Allied or neutral countries, from which they made contacts with informants or, more frequently, sent foreign speaking Americans or indigenous agents, into enemy areas by night parachute drops or submarine landings to carry out their missions of obtaining information directly or through paid or unpaid informants about military, economic, political, and morale conditions. Among SI station chiefs, none was more successful than Allen Dulles, whose headquarters was in Bern in neutral Switzerland. This scion of a family of international lawyers and diplomats, Dulles had served as an intelligence agent in Switzerland in the First World War, returning in 1942, he built a ring of more than one hundred agents in Germany, including lawyers, businessmen, labor leaders and socialists, and learned about the development sites for V-1 and V-2 weapons, the organization of opposition to Hitler within the German officer corps prior to their attempt to assassinate him in July 1944, and other valuable information. Dulles also became the contact for the most valuable single human intelligence source of the war, Fritz Kolbe, an anti-Nazi bureaucrat in the German Foreign Office with liaison to the German General Staff. British intelligence rejected him, but Dulles trusted him, and from 1943 to 1945, Kolbe smuggled to the OSS station chief in Switzerland some 1,600 valuable military and foreign policy documents providing important insights into military, economic, and political conditions in wartime Germany. Donovan included summaries of that information in the intelligence reports he gave regularly to President Roosevelt.[26] Dulles later became Director of the Central Intelligence (DCI) under President Dwight Eisenhower. The SI chief in London in the later part of the war, William J. Casey, became DCI under President Ronald Reagan. In World War II, Casey, then a young New York lawyer who joined the OSS, underwent close combat and demolitions training at OSS Training Area B in Catoctin Mountain Park, became the last chief of SI in London. In the final five months of the war in Europe, Casey, using parachuted German agents, achieved a penetration of Hitler's Third Reich by infiltrated spies, something British intelligence said could not be done.[27]

Although the Secret Intelligence Branch produced some valuable information during the war, OSS was excluded by both the U.S. and British military from the most important signals intelligence source of the war—the intercepted and decoded Axis radio communications (code named Ultra and Magic). However, beginning in 1943, Donovan was able to persuade the British intelligence services to share some of the Ultra intercepts for the purpose of counter-espionage, the identification of German spies and their elimination or conversion to "doubled" agents, who would report deceptive information to Berlin for the Allies. Headed by James Murphy in London and with station chiefs like the brilliant but ultimately controversial James Jesus Angleton in Rome (Angleton had taken his OSS basic training course at Area B at Catoctin before his more specialized training at SI and X-2 schools in Maryland and Virginia), the Counter-Intelligence Branch (X-2) was the most secretive of the OSS branches because of its access to some of the Ultra intercepts; it was also one of the most powerful, as its access enabled it to cancel SI or SO operations without explanation.[28]

Like the majority of women who enlisted in the OSS, Aline Griffith, from Pearl River, New York and a tall, statuesque Hattie Carnegie model, was assigned to the office in Washington. But in December 1943, after a month's secret intelligence training at Area RTU-11, "the Farm," the adventurous, 21-year-old woman, codenamed "Tiger," was sent overseas to Madrid to work in the capital of neutral, if pro-Axis Spain. In addition to coding and decoding, he mission was to counter-espionage, initially to learn the identity of a top German agent there. The bright and beautiful young woman proved a great success, gaining access to information while dining and dancing in an international social circle that included diplomats, attachés, businessmen, undercover agents and the like. On weekend visits to country mansions, she was not above rummaging through drawers and cracking safes and photographing documents. She became a top counter-espionage agent. After the war, she remained there and married a Spanish nobleman, becoming the Countess of Romanones.[29]

Only a relative handful of OSS women were trained for Secret Intelligence or Special Operations behind enemy lines. Most of them attended one of the OSS parachute classes taught at Fort Benning, Georgia or in North Africa or China by Colonel Lucius O. Rucker, U.S. Army paratrooper and a veteran of 119 jumps himself. The 38 women he instructed as parachutists represented only 1 percent of the 3,800 people he trained to jump out of airplanes, including Americans, British, French, Italian, Chinese and Thais. Rucker supervised more than twenty thousand jumps in his career, and he reported that only 50 trainees had refused at the last minute to jump out of the plane. None of those who refused was a woman. The women jumped, but they complained that their breasts were badly bruised by the severe snap back of the harness when the parachute opened.[30]

A number of American women in SI worked just behind the Allied lines, among them Betty Lussier, who set up an extensive double-agent network in southern France after the Americans had liberated Nice, and Wanda Di Giacomo, who had started in the Personnel Division of the OSS in Washington but ultimately served in SI in Italy. She received her overseas espionage training at a converted warehouse in Roslyn, Virginia. "After my work in Personnel, this was the pits," she recalled, contrasting the quiet, comfortable facilities in OSS headquarters with the hurly-burly of a temporary OSS training site. "Bathroom facilities were terrible. I had to share them with men. And what a collection! The place was always full of agents: Chinese, Arabs, French, even Germans, coming and going."[31] She was taught clandestine entry, safe-cracking, steaming letters open. However, neither Lussier nor De Giacomo was allowed to operate behind enemy lines. The only American OSS women who were sent behind enemy lines were apparently those who had already been in German-occupied countries before they joined the OSS.[32]

The women who served behind enemy lines were the heroines of the OSS. Often they were indigenous agents, such as Hélène Deschamps (code named "Anick"), who joined the French Resistance as a teenager and who reported to the OSS on German mines and camouflaged weapons, and helped downed fliers and persecuted Jews to escape. She was interrogated and beaten and suffered hearing loss from a bomb explosion, but she survived the war, married an American officer, and moved with him to the United States. Her advice for spies: "You have to think. If you look scared, you're dead. So smile."[33]

The most famous woman spy for the OSS was Virginia Hall, an American socialite from Baltimore, known to the Gestapo in France as the limping lady, because of her wooden leg. Fluent in French and living in Paris when the Nazis invaded in 1940, Hall served first as a volunteer ambulance driver until the French surrendered. Then she fled to London where she was recruited by the British Special Operations Executive (SOE). They sent her to Lyons, where she launched an operational resistance unit. Her unit was betrayed to Klaus Barbie, the "Butcher of Lyons," but she escaped across the Pyrenees Mountains. When SOE declined to send her back to France as too dangerous, Hall joined the OSS, which smuggled her into northern France, disguised as an elderly peasant. Becoming a major Resistance leader, Hall (now code named "Diane") directed espionage and guerrilla operations that established "safe houses" for intelligence agents and downed airmen, located drop zones for supplies for the Resistance. After the Allied invasion of France, her teams helped impede German reinforcements by blowing up bridges and supply trains, and ambushed truck convoys. They killed more than 150 German soldiers and captured nearly a thousand. After the war, Virginia Hall was awarded the French *croix de guerre*, the Order of the British Empire title, and in September 1945, the U.S. Army's Distinguished Service Cross "for extraordinary heroism in connection with military operations against the enemy." She was the first women and civilian awarded that medal, the highest after the Medal of Honor. She married a French-born, former OSS agent, Paul Goillot in 1950, joined the CIA the next year, and served in the agency until her retirement in 1966.[34]

Special Operations and Operational Groups

If OSS SI aimed at intelligence information, OSS Special Operations Branch (SO) aimed at destruction. Its members were trained to blow up bridges and railroad lines and to lead guerrilla attacks on enemy outposts and lines of communication and supply. Initially, the Army did not view Special Operations as useful, in contrast to the Secret Intelligence Branch, which proved its effectiveness in the North African invasion of November 1942. "Our endless trouble child," one top OSS executive, labeled SO in his diary in June 1943.[35] The Army was not yet convinced of the usefulness of SO's planned guerrilla and sabotage campaigns. Later, between 1943 and 1945, Special Operations Branch would prove its value in Europe and in Asia, but it took time.[36]

Like the U.S. Army paratroopers and rangers, there was a high degree of *sang froid*, and a gutsy, cocky, devil-may-care attitude among many OSS special operations agents. They were serious about their business and the risk of fighting behind enemy lines, but there was also a swagger to men who saw themselves as part of an elite unit, physically and mentally at top form, ready to jump out of airplanes into the dark behind enemy lines, risking capture and death in daring, hazardous secret missions that as far as they knew the public would never hear about, but which, they believed, might help win the war. Some of the people in Secret Intelligence derided the SO operatives, with their emphasis on explosives and automatic weapons, as the "Bang-Bang Boys."[37] Unlike SI's spies, who were usually civilians, male or female, and often worked alone, Special Operations combat operatives were uniformed, military personnel, men who worked in teams.

Those American Special Operations teams generally consisted of an SO officer and an enlisted radio man trained by the Communications Branch (CB). But beginning in May 1943, OSS augmented SO by establishing an Operational Group Branch (OG).[38] The Operational Groups differed from commandos in that they were recruited by language. They also usually operated far behind enemy lines on a sustained basis working with indigenous resistance groups. Donovan's idea was to recruit among various nationality groups in America's multi-ethnic society, individuals who knew the culture and language of their forbearers' country, and were willing to fight against its occupiers. Such ethnic soldiers in American military uniforms parachuted behind enemy lines, would be welcomed, Donovan believed, by indigenous resistance groups. By 1944 the OGs included ethnic Norwegians, Frenchmen, Italians, Greeks and other ethnic groups. Each consisted of about half a dozen officers and about 30 enlisted men.

Like the two-to three-man Special Operations teams, but in larger units, the Operational Group units were trained to work with local resistance forces and to engage in sabotage, hit-and-run raids, and other guerrilla operations to disrupt enemy lines of communication and supply in conjunction with directives from the Allied theater commander to assistant the invasion or other offensives of the main Allied forces.[39] In the Mediterranean and European theaters of operations, OSS Special Operations and Operational Groups were infiltrated by submarine, small craft, or parachute drop into North Africa, Italy, Yugoslavia, Albania, Greece, France, Belgium, the Netherlands, and Norway far in advance of the main Allied forces. In 1944, as a prelude to the Allied invasion of Normandy and subsequently the South of France, dozens of SO and OG teams were dropped behind German lines there with instructions to link up with indigenous guerrillas and impede German resistance through sabotage and hit-and-run raids. Among them were nearly one hundred multi-national SO teams, usually American, British, and French, code named Jedburghs.[40] The "Jeds" were a handpicked, colorful, capable, and adventurous lot, who received considerable publicity after the war. Most of the Americans picked to be "Jedburghs" had initially trained at Catoctin Mountain Park before undergoing Jed training in Britain. By August 1944, in addition to the "Jeds" and the other SO teams, approximately 1,100 Americans were operating in OG units throughout Europe.[41]

In all, there may have been up to 2,000 members of OSS Operational Groups.[42] This was in addition to the 1,600 operatives that Special Operations sent behind enemy lines.[43] During the war, almost all the American Special Operations teams received most of their training at the OSS Training Area B at Catoctin Mountain Park in Maryland or Training Area A at Prince William Forest Park in Virginia. The radio operators who accompanied the SO officers and the Operational Groups received their training area Area C at Prince William Forest Park. The Operational Groups themselves generally received their initial OSS training at Area F, the former Congressional Country Club in Bethesda, Maryland, but they usually then were sent on to Area B, or sometimes Area A, for further training, before being shipped overseas.

The first Special Operations unit sent into the field was Detachment 101, which was sent to northeastern India in the middle of 1942 to begin guerrilla operations against the Japanese in Burma. The group of some two dozen Americans had received their training for the mission in the spring of 1942. Their commander, Colonel Carl F. Eifler, and half of the initial contingent was trained at Camp X founded by British SOE on the edge of Lake Ontario outside of Toronto, Canada. The other half of the contingent, under Lieutenant William R. ("Ray") Peers, trained at Area B in Catoctin Mountain Park. Detachment 101, a group that never exceeded 120 Americans in the field, recruited and directed nearly 11,000 native Kachin and other Burmese tribesmen. Operating as guerrillas deep behind Japanese lines in the jungles of Burma from 1942 to 1945, they harassed and weakened the Imperial Japanese Army and Air Force.[44] Later in the war, other OSS units operated in Thailand and eventually in Japanese-occupied French Indochina.[45] Beginning in 1943, OSS special operations forces in conjunction with SI agents also operated in China, providing target information for U.S. General Claire L. Chennault's "Flying Tigers" pursuit planes and ultimately his bombers. SO and OG officers working with Chinese troops, the first Chinese OGs, as well as Chinese guerrillas sought to impede Japanese advances by blowing up supply depots, railroads and bridges After Tokyo surrendered in 1945, flying SO rescue teams into POW camps to prevent Allied prisoners from being harmed by fanatical Japanese militarists.[46]

Because OSS teams became involved in a number of seaborne projects in which operatives would be infiltrated by boat or which would involve the demolition of ships, harbor facilities, or landing obstacles, the OSS in June 1943 established a Maritime Unit (MU). Earlier, some seaborne infiltration training took place on a lake at Prince William Forest Park, but most of such training occurred at a Maritime training camp (Training Area D) established in late April 1942 at Smith Point on the eastern bank of the Potomac River in southern Charles County, Maryland. Members of SO and SI continued to receive seaborne landing instruction there, but the OSS Maritime Unit emphasized for its own personnel training not just in small craft handling and underwater explosives, but also sustained underwater combat swimming. Since the Potomac proved unsatisfactory for various reasons (lack of surf to simulate sea landings, iced over in winter, too murky and polluted for underwater swimming), MU moved its facilities to the Caribbean and California (Area D continued to be used by for advanced SO training until it was closed in April 1944). In the meantime, the Maritime Unit produced underwater explosive devices such as the Magnetic Limpet that saboteurs could use to blow a hole in a ship's steel hull. It also developed specialized boats, such as collapsible kayaks that could be launched from submarines, and it was a pioneer in the invention of underwater, self-contained breathing devices, like the Lambertsen Unit, which did not leave any tell-tale trail of bubbles on the surface. The Maritime Unit also developed the flexible swim fins used by the combat swimmers or "frogmen" of the OSS and the U.S. Navy. The Maritime Unit deployed and supplied Special Operations teams and Operational Groups by small craft across the Adriatic Sea to Yugoslavia, over the Aegean Sea to Greece, and across the Bay of Bengal to the coasts of Burma, Thailand and Malaya. In the Pacific, OSS frogmen assisted the Navy in scouting the shores and defenses of Japanese held islands prior to invasion by the Marines.[47]

Other Operational Branches

In regard to psychological warfare, having lost the Foreign Information Service to the Office of War Information in spring 1942, Donovan used the distinction between regular information, so-called "white" propaganda, and disinformation, "black propaganda," and created a Morale Operations Branch (MO) in January 1943. Its mission was to use "black" propaganda to spread confusion, dissension, and disorder among enemy troops and civilians. It was most effective if such rumors at least appeared to originate in the enemy territory; consequently, MO units operated in the war zones, albeit behind Allied lines, as well as in regional headquarters. Some of the men in MO were trained at Area A in Prince William Forest Park, Virginia; the women and some other men of MO received their training at Areas F or E in Maryland. Defeatism among enemy troops and encouragement among captive populations were encouraged through material stressing the deteriorating position of the occupiers, inciting internal dissension and suspicion of leaders. Leaflets and radio programs, mingling lies, rumors, and deceptions with truth, were the most common media for such disinformation cued to recent developments and often to regional conditions. In Italy, Barbara Lauwers, a private in the Women's Army Corps, stationed with the MO unit in Rome, helped develop "Operation Sauerkraut," an innovative psychological program that temporarily dispatched disaffected German POWs back behind enemy lines to persuade their fellow soldiers to surrender. Initiated following the July 1944, German officers' attempt to assassinate Hitler, the project's propaganda declared that Germans were in revolt against the Nazis. Hundreds of German soldiers did surrender as a result. Later, the Czech-born Lauwers created a program that induced 500 Czech soldiers conscripted into the Germany Army to give up to the Western Allies. She was awarded the Bronze Star Medal for her achievements.[48]

In China, when civilian MO operative Elizabeth McDonald (later Elizabeth McIntosh} arrived in Kunming in early 1945, she reported that the crude printing presses using hand-carved characters that MO field units were using to wage psychological warfare were being effectively replaced by new lightweight aluminum offset presses developed in OSS in Washington. She was assigned to a project providing leaflets for Chinese and Korean agents with instructions on how surreptitiously to place OSS incendiary devices shaped like a piece of coal into railroad coal bunkers so that they would be shoveled into a locomotive's firebox and explode at the proper time, thus disrupting the transportation of Japanese troops.[49] The primary MO role in China, however, was directed by socialite and media man Gordon Auchincloss, who arrived from the European Theater in August 1944. MO set up a powerful radio transmitter and beamed programs in various dialects to different regions of China encouraging guerrilla action by Chinese against Japanese occupiers and providing discouraging news to Japanese soldiers.[50]

One of the most respected and successful units of OSS was the Research and Analysis Branch (R&A).[51] Headed for most of the war by noted Harvard historian William L. Langer and supported by a large staff, it consisted of more than 900 scholars who used materials in the Library of Congress—and other facilities abroad—to collect and analyze economic, political, social, and military information about Axis or Axis-occupied nations or countries or regions for which the military needed information, such as North Africa prior to the Allied invasion.[52] Its Enemy Objectives Unit in London analyzed the German economy and war production, recommended particular targets and ultimately helped convince Allied air commanders that the key objectives of the bombing campaign should be first, German aircraft factories and second, German oil and synthetic oil production facilities.[53] R&A reports from World War II on crucial targets as well as other aspects of the industrial and transportation systems of particular countries in Europe, the Mediterranean, the Middle East, and the Far East, continued to prove useful to the American military even during the Korean War and the Cold War.[54] OSS R&A was later described by McGeorge Bundy, head of the Ford Foundation, as the first great center of area studies "not located in any university." For the government, it proved that scholarly research and analysis could produce important even vital information unavailable by espionage.[55]

Communications Branch

"The importance of communications to secret activities cannot be over-stated," declared the OSS official report at the end of the war.[56] It was essential that OSS operatives in the field be able to send and receive secure, coded, radio messages. Thus, the Communications Branch (CB or "Commo") included cryptography as well as communications operations. Indeed without secure and effective communications, the entire process of planning, coordinating, and implementing OSS operations, whether involving espionage, counter-espionage, covert operations, or psychological warfare, would have been jeopardized. As OSS Communications Branch veteran Arthur ("Art") Reinhardt expressed it: "Commo underpinned everything OSS did."[57]

To deal efficiently and securely with a global network serving initially primarily the SI, SO, and R&A branches, Donovan created a unified OSS Communications Branch on 22 September 1942.[58] Lawrence ("Larry") Wise Lowman, vice President in charge of operations at the Columbia Broadcasting System (CBS), with fifteen years experience in handling the technical aspects of that radio network, was named Chief of the Communications Branch.[59] The "Commo" Branch had an engineering section that built and operated the main message center and the regional relay base stations, purchasing and development sections that supplied specialized equipment, and a training section that developed a supply of operators for field and relay base work as well as code clerks for cryptography.[60] OSS communications agents in the field required both linguistic skills, particularly in foreign languages, as well as, as Donovan expressed it, "a higher degree of self reliance than those assigned to a normal military operation, due to the autonomous nature of each mission."[61]

The Communications Branch developed a system capable of rapid and secret communication among OSS locations around the globe. OSS headquarters was able to maintain wireless and cable communication with fixed and mobile relay base stations located in secure areas behind Allied lines in almost every major theater of operations. Those base stations communicated with agents in the field, whose transmissions were restricted in range by the limitations of their small, portable equipment. The relay stations, therefore, were necessary to interact between those field agents and OSS headquarters, forwarding coded instructions from headquarters and receiving and decoding information and requests from the agents. The primary coding system used in the field was the "one-time pads," a double-transposition system, based on random key text printed on the back and a memorized Vigenere Square.[62]

Given its highly specialized needs, the Communications Branch recruited and trained technical personnel in the skills required for its communications and its cryptography. The "Commo" branch radio operators, all of whom were ultimately military servicemen, underwent training at the Communications Branch's Training Area C at Prince William Forest Park, Virginia. (CB cryptographers, men and women, were trained in Washington, D.C., but men going into the war zone were sent to Area C for close combat and weapons instruction.) Living in tents and cabins there in the woods, the radio trainees became familiar with OSS communications equipment, its operation and maintenance, codes and ciphers, direction finding, and they learned or improved their telegraphy using International Morse code. Between 1942 and 1945, perhaps as many as 1,500 Communications Branch personnel (plus numerous other OSS personnel taking shorter radio and code courses) trained for generally three-month periods at Area C in Prince William Forest Park.[63]

The OSS's global communications network included base stations in Washington and around the world that operated twenty-four hours a day, seven days a week. The "Commo" men and women at those stations had to code and decode quickly and accurately messages and ensure their duplication and distribution. Indicative of the enormous volume of radio and cable traffic handled by OSS, the Communications Branch, which in addition to Washington, D.C., maintained message centers in some 25 major locations in 15 countries, handled 60,500 messages comprising 5,868,000 code groups in the single month of April 1945.[64] Despite the enormous number of OSS messages during the war, there was no known compromise of OSS's secure coding system or its cable traffic.[65]

Although the Communications Branch purchased most of its equipment from the civilian market or the Army Signal Corps, it had its own Research and Development Division. It developed wiretap devices, "Eureka" electronic beacons to identify drop sites, direction finding set, early "squirt" technology, and highly effective portable radios. Most widely used was what would be known as one of the most outstanding long-range, clandestine field radio sets of era, the SSTR-1 (for Strategic Services Transmitter/Receiver Number One). It was a transmitter, receiver, spare parts kit and a power supply compact enough to fit into a small and innocuous looking suitcase. OSS purchased 8,000 of these wireless telegraphy (WT) sets. Even more innovative was the so-called "Joan-Eleanor" communications system that OSS got into the field in the Netherlands and Germany by late 1944. The new equipment employed for the first time the Very High Frequency (VHF) band operating at 200 to 300 MHz, a frequency much more difficult for the enemy to monitor (most radio equipment worked under 100 MHz). With a small hand-held radio (an SSTC 502/ nicknamed "Joan"), an agent on the ground could talk with a radio operator in a plane circling in the dark six miles above and equipped with a substantial transmitter/receiver (an SSTR-6, nicknamed "Eleanor"), expediting the transmittal of information because the voice communication replaced telegraphy and the VHF obviated the need for coding and decoding.[66] The Joint Chiefs of Staff credited the OSS' Joan-Eleanor communication system one of the "most successful wireless intelligence gathering operations" of the war.[67]

OSS Research and Development Branch

For their war of insurgency, sabotage, and espionage, OSS operatives needed a number of highly specialized weapons, explosives and other deadly devices. The OSS had its own workshops and laboratories in its Research and Development Branch headed by Stanley P. Lovell, a chemist and former business executive from Boston.[68] Among the specialized weaponry were silenced, flashless pistols and submachine guns, dart guns, even hand-cranked crossbows, as well as various styles of knives and clubs. The OSS also adapted the "Liberator" pistol or $2 "Woolworth" gun, a cheaply made, one-shot .45 caliber pistol, that was distributed to indigenous guerrillas, especially in China and the Philippines (the single shot was to be used to kill an enemy soldier and obtain his rifle or pistol). For demolition work, OSS adopted some existing explosives such as Torpex, and a plastic explosive, Composition C, paradoxically a moldable, gelatin-like substance that was also much more stable than TNT and which exploded so effectively that it could blow a hole through one-inch steel. It also developed "Aunt Jemima," an explosive powder disguised as flour that could be baked as biscuits without exploding (although poisonous if eaten), but when ignited by a timed or contact fuse was powerful enough to blow up a bridge. For espionage, OSS R&D developed a variety of devices from 16mm cameras hidden in matchboxes or even jacket buttons to maps concealed in playing cards, plus a variety of invisible inks as well as faked identification cards, passes, and counterfeit currency.[69]

Beginnings of OSS Training

During the first six months after U.S. entry into the war, Donovan and his staff also worked to establish training schools that would prepare personnel to carry out the new clandestine missions, particularly those in intelligence and special operations. Part of the justification for this elite organization was that special skills, self-confidence, initiative and ingenuity were needed in the unconventional warfare that Donovan envisioned. Consequently there was a great emphasis on the kind of training that would provide agents with the skills and the spirit required for success in operations behind enemy lines.

In early January 1942, M. Preston Goodfellow hired Garland H. Williams to help organize military style training camps to produce agents who would work undercover in enemy occupied territory as spies, saboteurs, or guerrilla leaders. Reflecting the civilian orientation of Donovan's organization, Garland H. Williams was a career law enforcement official. By 1942, he had spent a dozen years with the Federal Bureau of Narcotics (FBN), rising to become director of the bureau's New York office, one of its top positions. He had been one of the first agents that FBN Commissioner Harry J. Anslinger had hired when the drug control agency was established in 1930.[70]

Besides his civilian law enforcement career, Williams was also a longtime Army reservist. He had joined the National Guard of his native Louisiana as a 2nd lieutenant in 1928, later transferred to the Army Reserves and was a major by 1936. When the Army decided to create a "Corps of Intelligence Police" (later designated the Counter-Intelligence Corps) to identify any narcotics peddlers, criminals, or enemy agents who operated in Army camps or other facilities, Williams was chosen to organize such a unit and prepare its training school.[71] Called to active duty in January 1941, Major Williams quickly accomplished this assignment, and subsequently, at his request, served as an instructor at the Infantry

School at Fort Benning, Georgia and then the Chemical Warfare School at Aberdeen, Maryland.[72] Preston Goodfellow, working in early 1942 with both Military Intelligence and COI, was impressed with Williams' performance in establishing the new military intelligence organization and school and with his long experience in police, intelligence and undercover work, which it was assumed would be applicable to Donovan's clandestine operatives. Williams joined Donovan's organization in early January 1942, and he and Goodfellow immediately set to work to establish training schools and instructors and to obtain and train a force of special operations troops.[73]

By the second week in February, they had ready for Donovan's signature a letter to Secretary of War Henry Stimson, requesting authorization for recruiting 2,000 men at the rank of sergeant, staff sergeant, technical sergeant and master sergeant. Most of these would be the trainees for the foreign-speaking units Donovan envisioned, but some would help to staff the training schools. For the main instructional and camp staff at the COI special operations training camps, Donovan requested slots for 48 commissioned officers. The War Department would help with obtaining the training camps and their staffs, but it was slow in giving Donovan the large numbers of troops he had requested.[74]

Mysterious Camp X

Meanwhile, early in 1942, Williams and a number of the other men joining the new wartime special operations or secret intelligence components of Donovan's organization underwent periods of hands-on training at a secret British training camp newly established for that purpose just across the border in Canada. While the United States had been official neutral, Donovan had been reluctant to recruit and prepare special operations commandos. This changed after the U.S. entered the war. But as early as September 1941, the British, with Donovan's encouragement, had agreed to build a secret, paramilitary training camp out side of Toronto to replicate SOE schools in Great Britain. It had a dual purpose. "A really efficient training school would impress the Americans," an SOE official reported in October 1941. "It would also provide us with valuable propaganda in obtaining their cooperation in the realm of subversive activities."[75] Camp X, as the facility was called, was erected behind an isolated shoreline of Lake Ontario, an hour from Toronto. On 275 acres of rolling, sparsely settled farmland between the lake not far from the villages of Oshawa and Whitby, the half dozen buildings and other training facilities were completed by the first week of December 1941. Local residents were informed that it was a military storage depot; its true purpose, the training of special operations agents, commandos, remained undisclosed for the next forty years.[76]

Camp X opened on December 9, 1941, the day after the United States declared war. It was ready to provide expertise to "the commandos," a new type of combatant that the British had been developing since 1940. "The Commandos," a British training document declared, "started free of all the conventions which surround a traditional [military] Corps. From the very beginning, the aim was to combine all the essentials of irregular bands [of guerrillas] with the superior training, equipment, and intelligence of regular troops. With his Bren [gun], his grenades and his Tommy gun, the Commando soldier had to be able to scale a cliff like a Pathan [in Afghanistan], to live like a Boer [in South Africa] with no transport columns or cookhouses, and to disperse and break away like an Arab before the enemy could pin him to his ground."[77]

The first group of about a dozen Americans from Donovan's organization arrived at Camp X in February 1942, for the basic four-week course. Garland H. Williams was among them. By April, Camp X would be training not just COI administrators and instructors but American SO agents who would be sent out into the field—to India, Burma, China, North Africa and Europe.[78] More from both SO and Secret Intelligence would follow over succeeding months. The basic special operations course lasted four weeks, although some Americans attended for shorter periods, some administrators even attending for only a weekend. In small groups of a dozen or fewer, the main training classes learned the principles of special operations warfare: infiltration, field craft, concealment, the use of various Allied or enemy weapons, hand-to-hand combat, guerrilla leadership and sabotage. A stiff British professional soldier ran the camp and, like him, many of his regular officers were formal and reserved. More appealing to the Americans was the chief instructor, Major R.B. ("Bill") Brooker, a businessman before the war, whose open and outgoing personality, dismissal of pomp and pretension as well as his eloquent and convincing lectures, made him "a born salesman."[79] The SOE knew of American resistance to the strict discipline and rigid class system of the British Army, but Brooker was able to win over many of the "Yanks" by speaking to them enthusiastically as equals and emphasizing camaraderie in a vital cause.

In its first year of operation, Camp X turned out dozens of American agents and instructors, SO and SI alike, and helped to make OSS a partner of SOE in covert warfare. Beginning in May 1942, Brooker would make frequent advisory trips to OSS headquarters in Washington, and in the winter of 1942-1943, he would be instrumental in reorganizing the entire OSS training program. But although many in the OSS found him colorful, engaging and experienced, others considered him showy, aggressive, and dogmatic. Brooker's critics were also put off by his casting of himself as a highly successful commando and spy, when in fact he had never had any actual operational experience behind enemy lines.[80]

The British role in shaping COI and OSS, especially in molding training, went beyond Camp X.[81] British officers advised the COI/OSS in establishing its own training and then loaned to the Americans, experienced British instructors, manuals, course outlines, and lecture books, as well as British acquired equipment, weaponry, and explosive devices for training in covert operations. They also made available SOE advanced training schools in Great Britain.[82] The postwar report of OSS's Schools and Training Branch concluded that: "Too much credit cannot be given to the aid received from British SOE at this [early] stage of the game. British SOE played a great part if not the greatest in the planning of the new SA/G [Special Activities/Goodfellow, i.e. Special Operations] schools."[83]

Establishing Secret Training Camps in the United States

Initially, the American training program consisted of a general curriculum that provided preliminary, basic, and advanced training courses to both SO and SI operatives before they were subsequently prepared for their different missions. The program provided elasticity and allowed for varying the instruction according to a person's previous experience, special qualifications or mission. Williams believed that the preliminary two-week, "toughing up" course of demanding physical exercise, obstacles, night marches, and tryouts in close combat and weapons skills would "weed out" the unqualified and help to classify accepted individuals for future instruction and assignment. This was followed by two weeks of basic SO training drawing on more intellectually demanding skills taught by the SOE at Camp X, including identification of targets of opportunity, observation, intelligence gathering, and sabotage. In addition to learning new skills, Williams explained, "the students will also be physically and mentally conditioned during these two courses for the aggressive and ruthless action which they will be called upon to perform at later dates."[84]

After completing these, the student would go on to either parachute or seaborne infiltration training and then one of the Advanced Schools in intelligence work, propaganda, sabotage or guerrilla leadership. Throughout all of the training, the emphasis was not only to impart skills and build physical condition and confidence but to develop the student's individual initiative, personal courage, and resourcefulness. Williams emphasized that all instruction should be as practical as possible. Theory should be deferred to practical application. Lecture periods should be short; rather, instructors should employ the "discussion or conference method of instruction" and use "interest-provoking equipment and materials." (Indeed, OSS did ultimately produce several hundred training films.[85]) Classroom instruction should alternate with outdoor instruction. As Williams summarized the pedagogical philosophy, "Whenever possible, the system of instruction will follow the principles of explanation, demonstration, application, and examination."[86]

Later, the advanced courses for saboteurs, for example, included assigning some students to conduct mock demolition exercises against nearby facilities: placing imitation explosives under railroad bridges or radio towers, or for spies or saboteurs, infiltrating defense plants in nearby cities from Baltimore to Pittsburgh. There were special courses for American ethnic groups or foreign nationals who would be infiltrated into enemy occupied countries.[87] Each Special Operations training camp was run along military lines, with a commanding officer in charge of both the instructional staff with its chief instructor and the camp complement with its responsibility for the administration, supply, maintenance and security of the facility. Williams stressed that the emphasis was on the individual:

"Constant thought will be given to the building of a high state of morale and a high esprit de corps. However, the military indoctrination will be so handled as to develop to the maximum extent his individual initiative, personal courage and resourcefulness. Emphasis will be constantly placed on the development of this agent as an individual and not as a fighter who is only effective when under close leadership. The guerrilla concept of warfare will be the guiding principle."[88]

Locating the First U.S. Training Camps in the National Parks

Since British SOE had established its training camps in isolated country estates in Scotland and England, Williams scouted out a number of country estates not far from Washington, D.C., but he rejected these mansions and their grounds as training sites for American Special Operations. He was looking for locations "situated in the country[side] and thoroughly isolated from the possible attention of any unauthorized persons" with plenty of land, at least several hundred acres, and located "well away from any highway or through-roads and preferably far-distant from other human habitations." The plan was to locate the training schools within a radius of about 50 miles from Washington, D.C. in order, in Williams' words, to "facilitate inspection and supervision by higher authority."[89]

Most readily available were sites in two nearby National Parks, what were then called Recreational Demonstration Areas (RDA) operated by the U.S. National Park Service.

Catoctin Mountain Park: Training Area B

Catoctin Mountain Park, north of Frederick, Maryland, a mountainous, forested region of some 9,000 acres about 75 miles north of Washington, was selected as the initial site for OSS basic paramilitary training for SO and SI personnel. It as designated Training Area B because the basic paramilitary course would be given there. With existing cabins, dining halls and kitchens, and other facilities in several rustic cabin camps built by the New Deal's Works Progress Administration (WPA) in the late 1930s, plus a Civilian Conservation Corps (CCC) work camp, the site for Area B already had a capacity of handling some three hundred trainees, plus a station cadre of staff and instructors of seventy five.[90] The War Department obtained a lease on the property from the Department of the Interior in March 1942, and in early April, Area B opened to receive SO trainees.[91] The first group represented half of the original contingent of Detachment 101 which would leave shortly for India and Burma (the other half were training at Camp X in Canada). Area B thereby became the first operative training camp for Donovan's organization in the United States. During the next two years, Area B would go through many phases, including several temporary hiatuses in training, when only the camp staff remained. Although established as a basic SO training camp, Area B courses were attended not only by SO and, in 1943-1944, OG trainees, but various numbers of men (there were no female staff or students there) from the other operational branches of COI/OSS: especially Secret Intelligence and the Operational Groups.[92]

Prince William Forest Park: Training Area A

The second National Park selected was Prince William Forest Park, then called Chopawamsic RDA, near Quantico, Virginia, 35 miles south of Washington, D.C. The park consisted of some 9,000 acres of hilly, forested land in what was then a sparsely settled rural area west of the Potomac River. Scattered around the wooded park were a Civilian Conservation Corps work camp and five summer cabin camps. The wooded, rugged terrain was judged well suited for paramilitary training.[93] The large western sector of the park, consisting of about 5,000 acres, was chosen as the site for advanced SO training and was designated Training Area A. It contained a CCC work camp and three cabin camps (OSS sub-camps A-4, A-2, A-3, and A-5 respectively). Although Area A, like Area B opened in early April 1942, advanced students were not immediately available, so instruction in Advanced Special Operations did not begin until late April or early May. Like Area B, the use of OSS Area A would go through several phases, including handling both basic and advanced courses for SO personnel and beginning in 1943 for the new OGs, and also serving as a holding area. Occasionally paramilitary instruction would be given there for members of Secret Intelligence, Morale Operations, the Maritime Unit, and Counter-Intelligence (X-2). Area A was composed of three rustic cabin camps, plus a CCC work camp, which eventually became the Area A headquarters. It had a station complement of 130 and with the capacity to handle about five hundred trainees at a time. It sometimes also served as a holding area for personnel who were waiting assignment either before or after training.[94]

Prince William Forest Park: Training Area C

Within the northeastern sector of Prince William Forest Park, Goodfellow and Williams established training for clandestine radio operation in the two cabin camps there beginning in mid-April 1942.[95] Because it was the site for communications training, and because it was the third camp to be founded, it was designated Area C. The two cabin camps (sub-camps C-1 and C-4), located comparatively close to each other were surrounded by 4,000 wooded acres in the eastern section of the park. Although not ideal for communications training, the area offered a great degree of security because it was isolated and all approaches to it could be kept under 24-hour guard.[96] At this time, before the Communications Branch was established in September 1942, the organization of the area was informal, and communications ("commo") courses were offered Morse Code, secret ciphers, and the operation of clandestine wireless telegraphy equipment. After the Communications Branch (CB) was established, it ran the communications training at Area C. For most of the war, Area C had a station complement of about 75 and a capacity of about two hundred students. (The demand for clandestine radio operators became so great by the fall of 1943, that the Communications Branch temporarily opened another, but much smaller, communications school, Area M, at Fort McDowell, Illinois.) Administration of Area C remained under the Special Operations Branch until January 1943, when the new Schools and Training Branch was created and soon assumed responsibility for administration. Nevertheless, Special Operations periodically used some of the Area C facilities as a holding area or to train foreign nationals, such as Thais and Koreans, for special operations activities in the Far East. Communications instruction at Area C, however, remained under the Communications Branch for the duration of the war.[97]

"We got three camps from the Department of Interior for a dollar a year [lease]," Goodfellow recalled in 1945, "provided we'd clear up [purchase] the farm areas which they contained. We had the War Department real estate people condemn them, then bought them up."[98]

Obtaining Instructors and Trainees

In recruiting instructors as well as prospective SO agents, Garland Williams began by working through law enforcement officers he knew and trusted, men experienced in the use of firearms and undercover work. They came initially not only from his own agency, the Federal Bureau of Narcotics but also from the Customs Service, the Border Patrol, and other federal, state, and local law enforcement agencies. As Williams wrote to one prospective instructor early in his planning, the training, as he envisioned it, would involve "... the use of various types of weapons, including pistol, Tommy gun [Thompson submachine gun], rifle, bayonet, knife, and black jack" plus boxing, jujitsu and various forms of physical toughening.[99] For other skills, however, Williams drew from the military: The Corps of Engineers for instruction in explosives and demolition work; the Military Police for pistol shooting and close combat techniques; Infantry officers for map-reading, tactical maneuvers and other field craft, and the use of rifles, hand grenades, machine guns, mortars, and later bazookas; the Signal Corps for wireless telegraphy, coding and decoding; the Medical Corps for first aid; Airborne units for parachute instructors; and the Navy for small craft handling and seaborne landing instructions.

There were some problems particularly with the use of law enforcement officers, however. Despite their applicable qualifications for weaponry and undercover work, law enforcement officers were deeply imbued with a respect for the law and a belief that lawbreakers and fugitives should and would be apprehended. But the aim of Donovan's organization working behind enemy lines was to break the law and not get caught. Donovan himself recognized this by frequently recruiting daring individualists, including, bold risk-taking, rule-bending or breaking journalists, adventurers, professionals, entrepreneurs or others. One OSS recruiter remembered looking for activists such as trade union organizers. "What seemed liked faults to rigid disciplinarians of the regular services often appealed to us as evidence of strong will power and an independent cast of mind."[100]

Beginning in early summer 1942, after the creation of OSS under the Joint Chiefs of Staff, recruiting for Special Operations drew almost entirely upon the military— not so much career military as former civilians now in the wartime armed forces. Recruiters from the OSS's newly formed Personnel Procurement Branch scoured training camps and advanced schools of the Army, Navy, Marine Corps and Army Air Corps looking for daring, intelligent, able recruits who were willing to volunteer for unspecified challenging and hazardous duty. Instructors in small arms and close combat skills often came from the Military Police, demolitions teachers from the Corps of Engineers. The Operational Groups (OGs) initiated in 1943 were secured entirely from the military and mainly from among individuals with particular ethnic backgrounds, particularly in the Infantry or Engineers. Radio operators came largely from the Signal Corps. Qualifications for OSS special operations service included mental and physical ability, linguistics, and the capability to operate under considerable stress.

OSS procurement officers first checked personnel records, then called men for personal interviews. In the interest of security, potential recruits were not given details, often not even the name of the OSS. They were asked if they were willing to volunteer for "hazardous duty behind enemy lines." (Such duty was made more hazardous by the fact that although most Special Operations teams and Operations Groups operating behind enemy lines were soldiers in uniform, Hitler refused to accord them POW status if captured and ordered that they be shot as "spies.") Only an estimated 10 per cent of the initial, potential recruits who were interviewed subsequently volunteered to join the OSS.[101] "The type of individual who was recruited for the saboteur teams can best be described as a complete individualist," an OSS report concluded at the end of the war. "When he volunteered to do this work, it was with the expectation of giving up his life, if need be, to accomplish his project. These men unquestionably represented some of the bravest men in the entire war effort of the United States."[102]

Training Sites outside the Parks: Areas D, E, F.

Donovan's organization also acquired a number of other properties for training purposes in or not far from the nation's capital but outside of the National Parks. Secret Intelligence Branch's assistant director of training Kenneth H. Baker, a former psychology professor from Ohio State University who had first been in Research and Analysis Branch, had begun training a handful of students in undercover work, intelligence gathering, security and reporting in a room at OSS headquarters in Washington. He started with one student in January, four in February and half a dozen in March 1942. Given the pressure for major expansion of the SI training program, Baker and his superior, J.R. Hayden, a former political science professor at Michigan who had also served as vice-chancellor of the U.S. territory of the Philippines, met with Garland Williams at the end of March and asked to what extent his Special Operations Schools, then ready to open, could help train SI personnel. Williams concluded that the paramilitary SO training could help SI agents with certain useful skills and would also build up a feeling of self-confidence, and he agreed to include SI trainees in the SO training program, albeit in a manner in which their identity or branch would not be disclosed (all the trainees were given fictitious names and told not to discuss their affiliation or missions). Meanwhile during the first two weeks in April, Dr. Baker was sent to the British training school at Camp X in Canada, from which he returned with the complete lecture book, which would play and important role in the future SI curriculum as it had in the SO training program. Hayden and Baker also decided that in addition to using the SO camps, they should establish an independent SI training school in a country estate on the British model.

Lothian Farm, an estate of about 100 acres in the Maryland tobacco-growing countryside near Clinton, about twenty miles south of the nation's capital, was recommended by Garland H. Williams. He had scouted it earlier but rejected it for SO training. SI Branch liked the site, and the secluded private estate with manor house and out buildings was quickly leased from its owner, a Pittsburgh industrialist. Designated RTU-11, but known informally as "the Farm," the SI school opened on 5 May 1942, for its first class of eight students.[103] Ultimately, it had a staff complement of nine and was capable of handling about 15 students.[104] The cover story to local residents was that it was the headquarters for a small military unit testing new equipment. In reality, however, espionage students were, in an informal civilian atmosphere quite different from the military-oriented SO camps, took classes over a three-to four-week period dealing with observation, concealment, cover stories, safecracking, bribery, recruiting and handling agents, communication and ciphers, and some use of weaponry and unarmed combat. Beginning in June 1942, SI members began attending the SO schools as well, and ultimately most of the SI graduates would also attend special operations training. Also in June, Marine Captain Albert Jenkins was assigned to the SI School at RTU-11 to teach the use of pistols and other small arms as well as map reading and field craft. A month later, when Hayden was replaced as director of the SI training school by Baker, Jenkins became his executive officer, the only military professional on the staff.[105] Later in 1942, Jenkins would be assigned as commanding officer of OSS Training Area C in Prince William Forest Park, a position he would hold for more than two years. RTU-11 ("the Farm") continued to be the main SI training school until its closure in July 1945.

Williams also played the key role in creating a OSS Training Area D, the fourth created by Special Operations Branch. It was first established to instruct SO teams or agents from SI, MO, or other operational branches, on seaborne infiltration on enemy shores. Some similar instruction was done on a lake in training Area A, but Williams decided that a more specialized facility was required. Consequently in late April 1942, the Special Operations Branch leased some 1,400 acres on and back from the east bank of the Potomac River some 40 miles south of the capital, probably on Smith's Point at Clinton Beach, bordered to the north by what is today Purse State Park, in Charles County, Maryland.[106] A British naval officer, experienced in clandestine landings Commander H. Woolley, who had been loaned to Donovan's organization in February, directed the establishment of the facility, assisted by U. S. Navy personnel led by Lt. Jack Taylor. Some of the initial naval personnel were sent to OSS Training Area B for testing and training.[107] Temporary prefabricated building had been obtained from the Army (possibly one of the CCC work camps that the War Department had supervised) and erected on the property. Eventually, these included seven wood frame and tarpaper barracks, plus officers' quarters, mess hall, classrooms, infirmary, latrines, garages and motor shop. The first class of eighteen SO and SI students learning small rubber boat, raft, canoes and kayak handling for clandestine landings and transfers of equipment began at Area D in August 1942. In the fall the facility was expanded and winterized.[108]

In January 1943, a Marine Section of SO was created. Its subsequent development of new specialized equipment such as magnetic mines to put on ship's hulls, undetectable underwater breathing devices (the Lambertsen Unit and collapsible eight-man kayaks for use from submarines) and of underwater swimming units ("frogmen") led OSS to designate it as an independent branch, the Maritime Unit (MU) in June 1943. There were two sub-camps there for SO as well as MU with a total station complement of 50.

Area D had the advantage of being not too far from Washington and even closer to OSS Areas A and C and the Marine Base at Quantico, the latter of which was only half a dozen miles up and across the river. The isolated wooded area provided good security. But being on a river location it did not have the surf and beach conditions comparable to those MU personnel would face in actual theaters of operation. Furthermore the murky and polluted nature of the Potomac and cold and ice in the winter hindered training of the frogmen. Consequently, in November 1943, MU moved its staff, students, and training activities first to Florida and then to Nassau in the Bahamas and in the summer of 1944 to Marine Camp Pendleton and to Catalina Island in California. Meanwhile Area D was also used by Special Operations Branch in 1942-1943 as an additional Advanced SO training site, when Area A-4 was filled, and in 1943-1944 it served as one of the holding areas for SO teams and Operational Groups that had completed their training and were awaiting shipment overseas. Area D was closed in April 1944 and the property returned to its owner.[109]

When it became clear in late 1942 that OSS was going to expand, new training areas were acquired. Area E was opened in November 1942 to provide basic Secret Intelligence Training (and later Counter-Intelligence). RTU-11 ("The Farm") then became the advanced SI school. Again reflecting the British influence, Area E consisted of two country estates and a former private school, each with considerable grounds. They were located near Towson and Glencoe in the rolling Maryland countryside about thirty miles north of Baltimore.[110] Beginning in 1944, a basic, two-week course for all OSS operational personnel—SI, SO, MO and X-2—was developed and taught at Area E, which in all had a station complement of nearly 50 persons and could handle 165 trainees.[111] The Schools and Training Branch later sought with mixed success to have this so-called "E-type" basic course taught by the various branches of OSS at their own training facilities. Like the other SI schools, but unlike the SO and CB schools in the parks, women as well as men received instruction there. Dissatisfaction by the operational branches to the E-type basic course led to the closure of Area E in July 1944.[112]

In April 1943, faced with a dramatic expansion of the OSS, particularly with the creation of Operational Groups for the planned invasion of France the following year, Donovan's organization obtained its final training area in the East, Area F. It was the least isolated and the most spectacular of the OSS training camps. For Donovan had leased the clubhouse, golf course, and the more than 400 acres of grounds of the Congressional Country Club located only 13 miles northwest of Washington in Bethesda, Maryland.[113] Since the OG military units would be training in uniform, like other soldiers in the Washington region, Donovan did not think they needed to be so isolated. Furthermore, this proved much more accessible than

Area B or A for taking visiting dignitaries to see the rough and tumble training of OSS SO and OG teams.[114] A former watering hole for the Washington power elite, the club had fallen onto hard times as a result of the depression and the war, and its board of directors was happy to lease it to the OSS for the duration. The lease required a monthly rental, which more than met the mortgage payments, and full restoration of the property when the OSS relinquished it.[115]

For two years, the formerly manicured grounds and golf links were used by Operational Groups, along with some Special Operations agents, for their initial paramilitary training. Most of them spent two weeks or more at Area F before going on to Area B, or Area A, for final training before being shipped overseas. "The water holes and sand traps made excellent areas for demolition training, especially for the OGs," veteran Joseph Kelley recalled. "And of course, the beautiful greens were also targets."[116] The mission of the OGs and SOs was to create mayhem behind enemy lines, and they trained at it on the once glorious grounds. An obstacle course was erected from the swimming pool to the first tee. Submachine gun and pistol ranges and a concrete bunker for observing the effect of new weapons and munitions were built across River Road on the north 80 acres. A mock fuselage of a C-47 used for parachute training sat on the putting green in front of the clubhouse. Inside, the ballroom was converted into a lecture hall, the elegant dining room into an Army mess hall. Near the fifteenth tee, machine guns fired live ammunition over the heads of trainees crawling forward. During the two year period, three trainees were killed in accidents at Area F. Two were felled by machine gun bullets on the range, one when he bolted up when he came upon a snake. Another was mortally wounded when in a nighttime mock attack on a bridge, a trainee playing a guard backed into the razor sharp stiletto of the attacking trainee who had taken him prisoner.[117]

The officers and enlisted men of the Operational Groups were housed in tents with wooden floors, more than seventy of them, each holding five men and, for warmth in winter, a pot-bellied stove. They stretched in parallel rows, accompanied by latrines, along entrance road between the clubhouse and the main entrance on River Road. Area F could hold more than 400 trainees in tents, plus the OSS staff of 80, whose quarters were mainly in the clubhouse.[118]

Area F was also used to train other OSS branches. Morale Operations used it as well as Area A, and because of its closeness to the OSS headquarters as well as its hotel-like facilities in the red-brick main building, the former Congressional Country Club was used to house most of the female members of OSS who received instruction in weaponry and other basic military training before being sent overseas. "With a real gun, yes," affirmed Elizabeth MacDonald, a civilian OSS employee in the Moral Operations Branch who was sent to Ceylon and then China.[119] OSS women in MO, SI, Communications Branch, or other operational OSS units, received their stateside training at OSS Headquarters complex, the clubhouse at the Congressional Country Club or the country estates at Area E or RTU-11 ("the Farm"). Apparently no women trainees were sent to the tents and cabins of the OSS Special Operations training camps in the forests at Catoctin Mountain Park or Prince William Forest Park.[120]

Early Problems and Reorganization

Despite a promising start in the spring of 1942, OSS training soon ran into difficulties. The Joint Chiefs of Staff at first took a skeptical view of Donovan's request for more than 2,000 sergeants' and officers' positions for men who would be recruited and trained for the Special Operations Branch. Since JSC waited until almost the end of 1942 before approving that many positions, many graduates of the SO training program could not be given missions and sent overseas. This was demoralizing to the graduates and also caused a problem of what to do with them in the meantime. Area A, for example, was converted in May 1942 to a holding area for them; SO cut back on recruiting, and classes at Area B were shut down temporarily.[121] Many of the instructors initially brought in from law enforcement for both SO and SI instruction were returned to their agencies in this period, and in SO instruction, more reservists and career soldiers were brought in to replace them. A debate emerged within Donovan's organization over the suitability of longtime soldiers as instructors. Critics argued that many of these regulars were too rigid and rule-bound to be effective instructors in this new, innovative, and developing form of non-conventional warfare. Supporters, including the JCS, asserted that they had the necessary military and technical skills.[122] Differences also emerged about the degree to which SO and SI could benefit from any common basic training course. Some thought the differences between the goals and operations of the two branches were so great as to negate any common training. There were also philosophical differences. SI favored preparing individuals for particular missions, but in SO, the predominant view was that the global war effort and the mass mobilization of sixteen million members of the armed forces necessitated mass production of OSS agents as well. Consequently, after some initial cooperation, SI and SO had each set up its own training program. Yet, there were commonalities—in some of the subject areas, in the need for training materials, curricula, instructors, security, food and other basic supplies—and the fact that they were both part of a single organization—the OSS. Most disturbingly, the SO emphasis on quantity over quality, as its critics in SI put it, had led to temporary overproduction of SO operatives, given that the JCS and many theater commanders were not yet convinced of the effectiveness of such agents and the consequent delay in approval of military slots or authorization of transportation overseas meant that the supply provided by the initial SO training system had not been accurately aligned with demand. Despite its earlier start, SO lagged behind SI in organization, coordinated training and overseas deployment.

Concerns within the OSS itself over the nature and effectiveness of the initial training efforts led to major reorganizations and a continuing effort to establish more centralized and effectual direction for OSS training. Instigation for a more coordinated program came especially from the Secret Intelligence branch advised by Maj. Bill Booker and others in British SOE.[123] The first reorganization came in August 1942. M. Preston Goodfellow's reputation had fallen considerably by then. Donovan still considered him an effective liaison with the military if not always a good judge of character, remained his friend and even promoted him to head of a new OSS unit called the Strategic Services Command, which both hoped would become a worldwide commando organization, independent of military theater commanders.[124] (The JCS ultimately rejected such as sweeping proposal, but in 1943 it would accept a greatly reduced version in the Operational Groups, which, however, would be outside Goodfellow's purview.[125]) Despite Donovan's friendship, Goodfellow was in effect pushed aside by the group of planners and administrators who largely ran the organization on a daily basis.[126]

Although Goodfellow remained with the OSS for the duration of the war, Garland H. Williams, his right-hand man, was pushed out entirely, accepting reassignment to the U.S. Army's Airborne Command, where he provided advice on establishing its paratrooper training program.[127] Williams had brought talent, vision, and enthusiasm for the creation of a an elite special operations force and unified preliminary training, his impatience with military routine and his tendency to cut corners and use unorthodox methods to expedite action caused problems, particularly when, after June 1942, the military had begun to dominate the special operations side of the organization. Looking back at the end of the war, the OSS Schools and Training Branch was gracious in its appraisal of his contribution. "It could be said that he was one of the early sacrifices to the order which was deemed necessary for an ultimately successful organization. Certainly it can be said that he was a year and a half ahead of his time."[128]

In the change in command, Goodfellow was succeeded as head of the Special Operations Branch by Ellery C. Huntington, Jr. Replacing Williams in charge of training was his former assistant, Philip G. Strong, one of the first American graduates of Camp X.[129] Strong, 41, was an eminent banker, the son of Benjamin Strong, governor of the Federal Reserve Bank of New York. He had also been a reserve Marine officer since 1926, and had been recruited from the Marines in 1942 by Donovan himself.[130] The 49-year-old Huntington was an old friend and squash partner of Donovan's. Born and raised in Tennessee, Huntington had become an All-American quarterback at Colgate University, earned a law degree from Harvard, served in France in World War I, and became a highly respected Wall Street lawyer. Widely traveled and fluent in half a dozen languages, he joined the COI hoping for overseas adventure. Instead he was assigned first as chief of Security and subsequently as Goodfellow's successor as head of Special Operations.[131] Associates recalled Huntington as an able, highly personable man, with "gung-ho" enthusiasm and a sense of daring that fitted him for Special Operations and also made him highly compatible with Donovan (later he would get the overseas assignments he desired in Italy and Yugoslavia).[132] Major Strong, in contrast, disagreed with much of the civilian leadership of the OSS and its training program, and returned to the Marine Corps, where served as an intelligence officer with two Battleship Squadrons.[133]

Faced with the difficulties in SO, the OSS began what would be a long process of trying to establish an effective centralized training system that would ultimately emerge under a separate Schools and Training Branch (S&T). But it took several steps to get there, and even with the establishment of the Schools and Training Branch in 1943, there was a struggle for control over training between S&T and the operational branches for the rest of the war. In August 1942, Donovan created the first combined training office, the Strategic Services Training Unit (SSTU), and appointed

Huntington to head it. Advised by Bill Brooker of the British SOE, Huntington supported the idea of a unified training program with a common syllabus and the linking of individual trainees with geographical desks in their branches so that their training could be geared toward their future missions. In September, SSTU was officially superseded by a new Training Directorate. It was headed by Dr. Kenneth H. Baker. A native Vermonter, Baker had been a professor of psychology at the Ohio State University and a reserve Army officer. Donovan recruited him for the Research and Analysis Branch in 1941 where he served for several months, but in 1942, this educator was appointed chief of training for the SI Branch. Baker headed the new Training Directorate as well as remaining chief of SI training. The Directorate also included the chiefs of the SO and Communications Branches, as well as Bill Brooker.[134]

Under the new plan, all SO and SI resources including instructors were to be pooled and a combined instructional program adopted by the new Training Directorate. The plan called for a Preliminary or Basic School, where all agents would be given physical conditioning as well as introduced to the rudiments of espionage and sabotage. Undercover and other techniques would be detailed at Advanced School. Finally, there would be different Specialist Schools (for SO, SI, MU, or Communications).[135] The plan sought for the first time to tailor the training of individual students to their ultimate assignments by seeking to have geographical desk officers coordinate with instructors to tailor training to particular missions.[136] The lack of such specific linkage had been a major complaint by the students, many of whom did not have the slightest idea of the role their training would play, because they did not know their ultimate assignment. "I don't see what this has to do with me, especially since I am not certain that I will ever be using it," was how one instructor summarized the students' complaint.[137] The Training Directorate's plan represented the first major step in the evolution of OSS training from its early phase where each branch wound up largely training its own agents but with considerable overlap to a more coordinated, centralized, standardized approach. The aim was flexible, yet standardized type training to accommodate all needs, which the budding OSS training establishment believed would be a more effective use of the organization's resources.

The Training Directorate did try to coordinate schooling and the roles of current and planned training areas. In its plan of November 1942, Area A, actually sub-camp A-4 in Prince William Forest Park, was designated as the Basic SO Training Area; Area B, in Catoctin Mountain Park, would be made available as a holding area for unassigned SO graduates, and for training SOs' new Operational Groups (OG) combat teams, when they were created beginning the following year.[138] Area C, in Prince William Forest Park, would remain the site of the Communications School for the duration of the war. When established in 1943, Area E, north of Baltimore, would serve, the Training Directorate hoped, as the Advanced School for SI and SO training. Area F, the former Congressional Country Club, it was expected, would provide training for the new Operational Group combat teams as well as some SO, SI, and MO personnel.[139]

Planning was one thing, implementing was another. Suspicious of the value of putting their students under training geared at least in part toward information and skills unrelated to their mission as saboteurs and guerrilla leaders on one hand or spies on the other, SI and SO continued to do most of their own training and in their own training areas. There was occasional crossover, mainly by SI sending its spies in training to a few weeks of SO's basic course with its introduction to physical toughening, close combat skills, and demolition work.

Creation of the Schools and Training Branch (S&T)

Because each of the branch chiefs in the Training Directorate had equal authority, the inter-branch rivalry prevented effective coordination. Consequently, Baker and Brooker convinced Donovan to replace it with a new unit, one with branch status equal to the operational branches, and whose chief would report directly to Donovan himself. In January 1943, Donovan established the Schools and Training Branch (S&T); Baker, its chief, would report directly to Donovan.[140] Baker did not last long in his new position, however, and even Brooker, his British advisor and champion of combined training, was soon dispatched to an SOE assignment far away in Algiers. There were a number of problems. The fundamental one was essential policy differences between Baker and the operational branches, which continued to resist his attempts at standardization and centralization of training for their agents. But Baker also alienated the military, particularly by his informal record keeping and irregular, out of channels methods for rapidly procuring supplies. In the end, the showdown came over military procedures and control, which Baker had resisted. Even Donovan's obtaining of a commission as a lieutenant colonel for the former psychology professor did little to help Baker's tendentious relations with the Army. Adding to these difficulties was the fact that Baker and the new Schools and Training Branch were simply overwhelmed by the massive influx of SO and SI recruits beginning in the spring of 1943, and then by the OGs, all in preparation for the Allied offensives in the Mediterranean that year and the invasion of France in 1944. In late June 1943, Baker was forced out.[141]

Even without Baker, OSS training faced significant problems in the summer of 1943. The remnants of Baker's use of informal requisition procedures continued to rankle the military. The operational branches continued to complain about unsatisfactory attempts at centralized direction. Baker's departure left the Schools and Training Branch under temporary leadership at a time when OSS faced dramatically increased demands caused by a flood of new recruits. He had sent many of the best instructors overseas to establish OSS schools in North Africa, England, and the Far East, which helped the OSS effort in those regions, but weakened instruction in the United States. This made it difficult for the remaining instructors and staff to handle and train the vastly increased numbers of recruits who descended upon the American training camps in 1943.[142] These developments all contributed to a series of internal investigations by OSS headquarters in the summer and the reorganization of the Schools and Training Branch in September 1943.[143] The branch was put under direct supervision of OSS headquarters, and as its new chief, Donovan appointed Lt. Col. Henson L. Robinson, a career Army officer from the field artillery, who had been serving as commanding officer of the OSS Headquarters and Headquarters Detachment. Robinson would head the Schools and Training Branch for the next two years, until the end of the OSS.[144] Subsequently, an advisory board was established composed of the heads of the geographic area desks but under Dr. John L. McConaughy, a former college President who had successfully headed a nationwide United China Relief effort. McConaughy was also appointed a Deputy Director of OSS for Schools and Training to add to his authority and that of Schools and Training Branch.[145]

A recurring problem in OSS training had been the friction between military and civilian viewpoints. The kind of irregular training characteristic of most OSS instruction did not lend itself to a strictly military approach. The new leadership of Schools and Training Branch proved a workable arrangement, because Robinson, the career soldier, chose as his executive officer, an Army officer with a civilian background. The new training executive was Lt. Col. Ainsworth Blogg, a former insurance executive from Seattle and a longtime reserve officer in the Military Police, who had served as commanding officer at Area B in Cactoctin Mountain Park from its founding in the spring of 1942.[146] Soon Blogg came to deal primarily with administrative matters, personnel, transportation, supply and the like, and an executive for instruction was appointed to deal with the instructors and the curriculum of the training schools. The latter was Phillip E. Allen, who had begun as an instructor in the Secret Intelligence Branch and then served as chief instructor at Area E near Baltimore from June to October 1943, before his appointment in charge of instruction for Schools and Training Branch.[147]

From this arrangement in September 1943, things began to improve in the Schools and Training Branch, in part because Blogg was able to work satisfactorily with both former civilians and the professional military. Robinson and Blogg oversaw the "militarization" of Schools and Training that Baker had resisted. This was probably inevitable because the basic directive for the OSS decreed that its wartime functions were to be in support of the military, and the military, of course, was the largest and most important customer for OSS's intelligence, sabotage, and guerrilla operations. In the field, most of the OSS personnel had to work with the military and were often part of the military. There were, in fact, a number of advantages to the militarization of the training program. In the United States, it solved problems of supply at a time of civilian shortages and a cumbersome civilian rationing system of allocation. The military simply had priority over civilian needs. Furthermore, OSS trainees and instructors were now largely drawn from the military, which had a practical monopoly on young, able-bodied men. However, the militarization of training also had a number of disadvantages according to its critics in the OSS. Complaints continued about inflexible, rulebook attitudes by some career officers. The use of instructors from the military also caused turnover and consequent instability in the training program when they were suddenly reassigned overseas. There was dissatisfaction about dual control in the training areas between the commanding officer and the chief instructor, particularly when the two men were of different rank or if the commanding officer were a military officer and the chief instructor a civilian.[148]

Militarization and effective leadership of the training program expedited the mass production of OSS agents required in the massive build up of 1943-1944, a doubling of the total size of Donovan's organization.[149] These increases in production led to major alterations in training. The changes usually incorporated, but sometimes overrode, the desires of the different OSS branches. They enabled courses to become more standardized to satisfy what Schools and Training considered the common requirements of all OSS branches. The names of the courses seldom changed, but their content was revised by the instructors or by the Schools and Training Branch as a result of new information from the field or new missions (such as the expanded role in the China-Burma-India Theater in 1944-45). The following course titles remained fairly constant, despite the changing course outlines and training manuals of the OSS:

- Intelligence Gathering and Reporting Techniques
- Order of Battle [the enemy's military units]
- Enemy Identification [of rank, unit, branch, etc.]
- Agent Material [Undercover Techniques, Recruitment & Organization of Indigenous Agents]
- Enemy Organizational Background
- Foreign Background [of the country in which the agent might serve]
- Weapons Instruction [in enemy as well as Allied weapons; plus unarmed combat]
- Demolition and Special Sabotage Devices
- Fieldcraft, Tactics, Scouting, Reconnaissance
- Mapping and Map Interpretation
- Radio Training [wireless transmitting and receiving, coding and decoding]
- Police Methods [understanding them so as to avoid detection and capture]
- Morale Operations and ["Black"] Propaganda Composition and Dissemination[150]

In addition, the SO and SI schools often capped their training courses with a field problem or "scheme." This involved sending a trainee or a team of trainees into a nearby city–Philadelphia, Pittsburgh, or Baltimore were the most utilized–with the assignment of infiltrating major industrial defense plants and obtaining inside information about them (by SI students) or arranging a plan for sabotaging them (by SO trainees).

Remembering Training

Recollections of veterans of the OSS training camps provide differing perspectives and judgments. Some, like University of Michigan graduate John Waller, who went into counter-espionage (X-2), recalled the series of different OSS training courses in 1943, as "quite entertaining and remarkable good" with the dynamics of the situation creating a sense of camaraderie, "a feeling of loyalty to the OSS [and] an esprit de corps."¹⁵¹ John K. Singlaub, an ROTC infantry lieutenant from UCLA who had volunteered for hazardous duty with the OSS, and later became part of Jedburgh team in France and a leader of a mission rescuing POWs in China, recalled his SO training at Area B. By the end of the month, he said, it "had become a grueling marathon. We crawled through rain-soaked oak forest at night to plant live demolition charges on floodlit sheds. We were introduced to clandestine radio procedure and practiced tapping out code and encrypting messages in our few spare moments. Many mornings began with a run, followed by a passage of an increasingly sophisticated and dangerous obstacle course. The explosive charges under the rope bridges and wire catwalks no longer exploded to one side as exciting stage effects. Now they blasted directly below, a moment before or after we had passed."¹⁵²

Lt. Albert Materazzi, a 26-year-old Italian American with a chemistry degree from Fordham University who was being prepared for Operational Group action in Italy, remembered night exercises around Area A at Prince William Forest Park in late April 1943. One was planting dummy demolition charges on the main railroad bridge over the Rappahannock. "There was a poor night watchman. We scared him half to death. Nobody had told him." Another field exercise was to sabotage a dam. A third problem seemed ridiculous to Materazzi. "They put us on a truck, dropped us a hundred yards apart on a straight stretch of road adjoining a woods, pointed and said 'that's were the camp is, go find it.' It was a dark night. It was late April, the trees were covered with leaves. There were no stars. We couldn't have seen them if we had tried…. [When the instructors left], we said 'this is crazy.' We hopped a ride into Triangle [the nearest town], found a bar. We shot pool, drank beer until about midnight. We took cabs within a half a mile [of the camp], went in and signed in - coming from different directions, of course—and went to sleep. We never heard any more about it. We had expected that we would be chastised for it, but we weren't.¹⁵³

Lt. Col. Francis B. ("Frank") Mills, an Oklahoma City native and an officer in the Field Artillery, had been recruited by Special Operations and trained in Area A and other camps in 1942 before being sent behind enemy lines in France in 1944 and then China in 1945. He remembered that at those camps they had been trained "primarily in self-defense combat. We were not trained in any way in the intelligence gathering activity. We were in special operations, fighting with sympathetic forces behind enemy lines. We knew that. We were given very marginal, almost no real training in how guerrillas were supposed to operate. So we were given what little training the Army or OSS had to offer."¹⁵⁴

In contrast, Raymond Brittenham of SI, who went to various training schools in the summer of 1943, found the instructors, particularly those teaching the use of weapons and explosives to be what he called "real pros." He also commended the ability and perseverance of his fellow students in the training program. "You were there with a bunch of . . . very bright, able people," he recalled. "It was a pretty pressured deal. . . . In almost all of these [training] camps, a few people would just drop out by the wayside."¹⁵⁵

OSS women going overseas also had to take a basic training type of course including the use of weapons. Elizabeth ("Betty") MacDonald (later McIntosh), who would work with Morale Operations in India and China, remembered that at Area F, the Congressional County Club, she had "learned how to handle weapons and throw grenades out on the golf course....The members were furious [after the war] because we ruined the greens. I don't remember the training being particularly rigorous. There was a lot of writing stuff and sometimes we had to trail people, so that we would not lose track of them when we were in cars....There were a lot of drinking parties which I think were deliberate, to see how we acted in social situations and how we behaved after drinking alcohol. We figured this out, because we were plied with liquor. You had to be pretty smart, on your toes and know what you were doing."[156]

Assessing Recruits to Avoid Disasters in the Field

As the Allies expanded their offensives in 1943, a growing demand for OSS personnel overseas led to a massive recruiting effort to sustain the agency's rapid expansion. Overall, the expansion was successful, but the accelerated recruitment program produced problems. A significant number of the recruits proved unfit to handle the training or slowed down the training due to their limitations. More importantly, some who made it through training, proved psychologically or emotionally unsuited for dangerous field operations overseas.[157] By the middle of 1943, OSS headquarters began to receive worrisome complaints of incompetence in the field. There were even reports of a few dramatic mental breakdowns.[158]

To deal with this problem, Colonel Robinson, the new chief of the Schools and Training Branch, requested a plan in November 1943 for assessing prospective OSS personnel as to their physical, mental and emotional capabilities for their intended assignments. This would be done in separate assessment areas, so that the unsuited could be weeded out before they became a part of OSS and learned about its secret operations. Seeking assessment methods that would reveal the guts, savoir-faire, and intelligence desired for OSS agents, psychologists on the OSS planning staff led by Dr. James A. Hamilton and Dr. Robert C. Tryon, both from the University of California, rejected the military's written tests as inadequate and concluded that a pragmatic and holistic approach was required. They drew upon the system employed by the British Officer Selection Boards, which used observation and testing of officer candidates during an intensive, three-day, military house party at a country-estate.[159] In January 1944, Hamilton and his associates opened the first OSS assessment school, designated Station S (for Secret), in a manor house on a 100-acre private country estate leased from the Willard Hotel family in Fairfax, Virginia, thirty minutes west of the nation's capital.[160]

Groups of 15 to 20 recruits would spent three and a half days there being observed by a team of psychologists and others as they underwent a series of tests and situational problems designed to evaluate mentality, personality, emotional stability, and aptitude. It proved so successful that Donovan soon ordered assessments for all OSS personnel going overseas, and an additional location, Station W, a house at 19th and I Streets in Washington was acquired to provide a condensed, one-day evaluation for the majority of the OSS personnel going abroad who would not be facing hazardous duty. Working at these stations, and at Area F, where veterans returning from western Europe were evaluated, the assessment teams processed nearly 4,000 men and women in 1944 alone.[161]

A number of techniques were employed to analyze a candidate's ability to perform highly demanding assignments under stressful conditions. Candidates for the OSS arrived anonymously, were given cover names and identical Army fatigue uniforms, and told not to reveal their real identities to the other candidates. Each then underwent routine written tests providing a general index of their intellectual capabilities and aptitudes. Over the next three days, while they lived at the house, they were constantly being tested, observed, and evaluated by the staff. "They had these characters around all the time that were watching you all the time. This was the most tedious part about it," recalled Raymond Brittenham, a Harvard Law School graduate and fledgling Chicago lawyer. "Then, you'd be sitting in the dining room and the guy would come back and drop a whole plate of trays right behind you. Then, if you jumped, he would write that down. I mean, it was kind of kooky, you know."[162]

The most famous or infamous problem was designed to test capacity for leadership and self-control under stress and how they would deal with frustration. The candidate was told to direct two assistants in setting up a complicated wooden frame construction, a kind of giant tinker-toy. It had to be done in ten minutes, and a bell would ring every minute to indicate the passage of time. Unknown to the candidate, however, the assistants were staff members whose job was to impede and frustrate him or her. The task was, in fact, made impossible. Reactions to this frustrating situation varied widely. Some candidates tried to discipline the assistants, others relinquished authority to them and followed their suggestions, a few completely lost control and beat the assistants or tried to construct it themselves. Others like John Waller, soon realized after initial frustration, that a "ringer" had been planted to sabotage them and just relaxed and chuckled about it.[163] In another test, a candidate was told to assume he or she had been caught riffling through secret files in a government office and would by interrogated in ten minutes. Then, facing a spotlight in a dark cellar room, the accused spy would be cross examined by an expert (at Station S that expert was the sole attorney on the staff, Sidney L. Harrow, a Philadelphia trial lawyer who had been recruited by the OSS after being drafted as a private).[164] There were also numerous tests of identification and instant and accurate memory recall of specific sequential or random images or spoken or written data. At the end of the process came a clinical interview that in addition to the results of the previous three days played a decisive role in the assessment of a candidate.[165]

OSS's assessment program proved most effective in providing a psychological evaluation of the candidate; it was less effective in determining the individual's suitability for a particular job. There were several reasons for the latter difficulty: one was because until the end of the war, few members of the assessment staff knew field conditions firsthand; another was that too often men were sent overseas fitted for one assignment but then were shifted to another assignment for which they were not suited.[166]

The data from the assessment program provides a rough profile of the kind of people the OSS obtained, at least in the last year and a half of the war. From January 1944 to July 1945, the OSS assessment schools evaluated 5,300 men and women, both recruits and candidates for overseas service. They scored above average on intelligence tests. Approximately half had visited foreign countries; one-eighth had visited two or more continents. Twenty percent had spent at least five years abroad, as many were foreign born or political refugees. One out of every four spoke a foreign language fluently. Many had college degrees, some held doctorates, at a time when only 40 percent of Americans had gone past elementary school.[167] The profile with its emphasis on college education and foreign travel, reinforced the perception that many in the organization came from middle or upper socioeconomic backgrounds.

The official history of the OSS concluded that the assessment program screened out the 15 to 20 percent of recruits who were obviously unfit for the stressful missions overseas. Of the first 300 candidates who passed the assessment, completed training, and then were sent to OSS assignments overseas, only 6% proved unsatisfactory and only one of those 18 individuals was found to be psycho-neurotic and yet had gotten through the assessment program.[168] During the war, 52 emotional breakdowns occurred among OSS personnel, but of those, only two had gotten through the screening process at assessment Station S.[169]

The effort to assess an individual's total personality had never before been attempted in the United States. The psychologists and psychiatrists, who developed and ran the assessment program for the OSS—appraising the personality of individuals and predict that person's performances on unpredictable jobs, particularly those involving considerable stress—employed the methodology after the war for other government agencies, including the Veterans Administration. Aspects were also applied by businesses.[170] *Assessment of Men*, a book resulting from the OSS program, became a classic, employed in the private and public sectors. It has been reprinted many times, most recently as *Selection of Personnel for Clandestine Operations: Assessment of Men*. The armed forces continue to use some of its principles and even its techniques—the construction situation, the stress situation and the like—in making assessments of candidates for junior officers, the special forces, or other positions requiring clear judgment under pressure.[171]

Continuing Differences about Standardized Training

Colonel Henson Robinson instituted other changes as well as he sought to improve Schools and Training Branch's performance. In addition to the assessment program, he sought increased coordination between the training program and the geographic desk officers of the operational branches. Most controversially, at least with the operational branches, Robinson continued the effort toward standardization and centralization. Schools and Training Branch developed a common, basic OSS agent course. Labeled the "E-type" course, because it began at Area E north of Baltimore, it was designed to give every agent some familiarity with the skills and methods of the other operational branches, a combination of spy saboteur, and psychological warfare training. It was firmly established by the fall of

1944, and by the following spring, two thousand agents of various branches had taken the basic E-type course.[172]

Nevertheless, there remained continual resistance from the operational branches to S&T's centralized training. SI, SO, and X-2 argued bitterly in 1944 and 1945 that Schools and Training Branch remained frozen into a 1942 standardized, centralized training model that had become badly outdated. For example, SI and SO had established field offices and additional training schools overseas, which recruited indigenous nationals in the occupied countries. Consequently, they already knew specific conditions and could recruit and train new agents there. With SI, since only a few of its agents who went through the S&T programs, had gone into enemy territory by spring 1944 (the majority of its spies in occupied lands already resided there), there was less need for training in the old "Cloak and Dagger approach" in the OSS schools in the United States.[173]

Most importantly, the operational branches claimed Schools and Training's program had become woefully inadequate because it did not incorporate important changes in skills, and methods recommended by agents returning with experience from the field. As a result, the branches did much of their own training both in the United States and abroad. Special Operations Branch, for example, considered S&T's basic E-type course such a waste of time that it sent very few of recruits to take it at Area E. Instead, Special Operations taught its own basic course at the SO school in Area A-4 in Prince William Forest Park. So frustrated were SO, SI, and X-2 with the resistance of Schools and Training to respond to their complaints that in June 1944, training representatives of the three operational branches held a highly unusual joint meeting on their own. They learned that each was dissatisfied with S&T's program at Area E, "its curriculum, its method of teaching, and its misplaced emphasis on unwanted subjects." Most disturbingly, the three discovered that "each [operational] Branch had been blocked in its efforts to win changes by being told that the other Branches would tolerate no revision."[174] Consequently, in an angry memo, SI urged "strong action" to remove "the stranglehold that Schools & Training has been able to exercise to prevent the various Branches from making the schools serviceable instruments."[175]

With Schools and Training Branch under such intense criticism and still plagued with problems, John McConaughy, the deputy director and head of the Schools and Training Board, wrote an apology and explanation in July 1944 to Colonel Ned Buxton, Assistant Director of the OSS:

> Many of our difficulties stem from the haste with which OSS was organized, the fact that the concept of training followed a program of operations (ideally, it should have preceded it). Schools and Training was the "tail" of the OSS "dog." For a long time, it was not given strong leadership, it did not achieve Branch status until recently, etc. Not very long ago, the "chief indoor sport" of some persons in some Branches was to pick on Schools and Training—and our record probably justified their doing so.[176]

Donovan wanted a unified OSS Basic Course to be attended by all new male OSS personnel, but changes had to be made to bring training into line with the field experience and demands of the operational branches. Consequently, Area E was closed down, in July 1944, and its instructors transferred to other training areas. S&T Branch agreed to an OSS Basic Course that adopted, with some modifications, the basic SI training course.[77] The new two-week OSS Basic Course was given at Area A-3 in Prince William Forest Park beginning 24 July 1944. It was accepted immediately by SI, X-2, and MO, but SO agreed to send its personnel only after the curriculum had been substantially modified meet the needs of Special Operations recruits.[78]

Increased Demands on Schools and Training Branch

During the big OSS buildup between the summer of 1943 and the fall of 1944, the training camps had operated at a breakneck pace as field activities of Donovan's organization expanded along with the U.S. military effort, first in Europe and then in Asia. Increased demands were imposed on Schools and Training Branch. S&T's staff at its headquarters numbered some 50 men and women and its instructional and other staff at the more than a dozen schools and training camps in the United States amounted to nearly 500 men.[179] Schools and Training Branch did not have authority over the training of every unit of the OSS, but it was increasingly used even by units outside its jurisdiction, like Research and Analysis and OSS Services Branch, for example, particularly for their personnel who were being sent overseas.[180] In August 1944, at the insistence of OSS headquarters, S&T was given authority over the numerous overseas schools that the operational branches had established mainly to train indigenous agents. These included SO and SI schools and an SO parachute school in Algeria, SO and SI schools in England, Italy, France, India, Ceylon (Sri Lanka) and China as well as the SO parachute school transferred from Algeria to China in late 1944.[181] Within the United States, the number of OSS training facilities more than doubled to sixteen in the last year of the war, as the original training areas and assessment stations in Maryland and Virginia, were augmented by the establishment of eight new training and assessment facilities on the West Coast including Santa Catalina Island, Newport Beach, Camp Pendleton and other sites to handle the increased OSS presence as the focus of the American war effort shifted to the Far East.[182]

On the West Coast, Phillip Allen, Training Executive for S&T Branch, who was made chief of training there in late 1944, established a training program that S&T called "probably the most efficient that was given by Schools and Training, since it combined the best features of the training that had been given in the East and eliminated some of the weaknesses that experience had brought to light."[183] In addition, by the time the OSS training camp on Santa Catalina Island near Los Angeles was established, a significant number of veterans with experience in the field were available as instructors who could provide realism and information on current conditions to the courses. Allen's program included a basic two-week course, a modification of the E-type course, which was followed by an advanced course in SI, SO, or MO, or a combination of them. Maritime Unit had its own training on the island. Last came field problems. Advanced SI students, for example, were sent with radio operators into southern California with the mission of obtaining information about activities in northern Mexico and relaying it to their base station. Advanced SO men were sent on survival problems, dispatched into desolate areas with only a minimum of food and forced to life on what they could hunt, to test how well they could survive while at the same time preparing an effective plan to sabotage a bridge, railroad, utility tower or other facility.[184]

As OSS veterans returned from Europe in late 1944 and early 1945, many of them were lodged, at least temporarily, in the East Coast training camps, which became holding areas where the veterans were processed, assessed, or simply held pending decision on their future use. In January 1945, there were 1,700 such veterans in stateside camps run by Schools and Training Branch.[185] With the end of the war in Europe only a few months away, most of these men were simply being held in those areas. Only 330 men were actually being trained in the United States at that point. Nearly half of them were in East Coast schools like Training Area C, the Far Eastern Background School at Georgetown University and the Japanese Language School at the University of Pennsylvania. The other half, composed to a great extent of Korean or Chinese nationals, or Japanese Americans, were being trained in the West Coast camps.[186] In terms of shear numbers, most of the OSS trainees in 1945 were the hundreds of Chinese commandos being trained at the OSS schools in China.

Like the OSS itself, the Schools and Training Branch, was without precedent in the American experience. It was conceived in haste, created from whole cloth, experimental, and subject to change without warning. Efforts at standardized and centralized training had been resisted or given lip service by key operational branches of OSS, such as Special Operations, Secret Intelligence, and Counter-Intelligence. Even though they finally accepted a common basic, two-week introductory OSS course the operational branches, with their different missions and diverse personnel, continued to retain substantial control over the bulk of their own training, preliminary and advanced, throughout the war.[187]

Yet, the logic of a centralized supply, administration, and coordination for the schools and training camps of what was after all a single organization, despite its diverse branches, led to increased authority and jurisdiction for the OSS's Schools and Training Branch. By the end of 1944, S&T gained complete authority, in principle if not always in practice, over the training schools at home and abroad of the main operational branches, SO and SI. It had at least divided control over the training of operational personnel in highly specialized technical units such as the Communications Branch and the Maritime Unit. It provided at least preliminary training for the most secret unit, the Counter-Intelligence Branch. By the last year of the war, even those units which remained outside of S&T's jurisdiction because of their different experience, like the Research and Analysis Branch, often drew upon it for certain needs, such as the Basic OSS Training Course for their personnel being sent overseas. By the end of the war, S&T's training programs in the East and West Coast schools for the OSS efforts against Japan in the Far East demonstrated an effective maturity: A common basic course, a combination of advanced specialized courses, and practical field problems, with instructors who had personal experience in the field.

The path to effective coordinated training had been as difficult as crossing a minefield. The journey had been filled with unexpected dangers and obstacles. It had also been filled with frustration. At the end of the war, Schools and Training Branch compared the task it had been given with the frustration test with the giant tinker-toy apparatus and the uncooperative "helpers" given to prospective OSS personnel. Like those candidates, S&T was "assigned the construction problem which was nearly impossible to achieve, were heckled from the sidelines, annoyed by meaningless bells, observed by critical reporters, and discouragingly limited as to time."[188]

"Many things would be done differently," the Schools and Training Branch concluded after the war, if it had to be done again. But the reference was mainly to the timing rather than the final result. The lessons drawn were that centralized control of training would be established from the very beginning. There would be from the beginning one head of training and that person would be, in effect, a staff officer of the Director of the entire organization. Similarly, the assessment program would be set up at the beginning and would apply to everyone recruited for the organization from top to bottom. Overseas training would come under the jurisdiction of the Washington training headquarters, and that training headquarters would be organized as S&T's was after 1943 with a director and deputy directors for administration and for training, and in each school a commander responsible for the smooth running of the service functions necessary in the school, such as supplies, food, and transportation, and a chief instructor, who was responsible for all of the training at the school.[189]

"If we ever have to do it again," the final report of Schools and Training Branch concluded, "it is hoped that the experiences of OSS training will be exhumed from the Archives along with the old training schedules, lecture notes, training films, and the rest, even though another war will be fought much differently. Much remains that could be modified and used as an example for training in an agency similar to the OSS."[190] Indeed, that is exactly what happened, as the successors to the OSS—the CIA and the Special Forces—drew in subsequent years upon the training lessons of the OSS in World War II.

CHAPTER 3

CATOCTIN AND PRINCE WILLIAM PARKS JOIN THE WAR EFFORT

The decision of Donovan's organization to establish its first U.S. training camps at Catoctin Mountain Park, Maryland, and Prince William Forest Park, Virginia, had been based on their' rural, isolated location yet comparative proximity to the nation's capital. Each site should occupy several hundred acres, be located away from highways and be "preferably far distant from other human habitations," Major Garland H. Williams, deputy chief of Special Operations, put it.[1] Given the need for speed, secrecy and isolation, these particular sites were obtained in March 1942 because they were readily available and met the OSS criteria at that time.[2] It was particularly helpful that the two sites were already federal property and that they combined rugged, wooded terrain with existing camping, administrative, and maintenance facilities. Those favorable circumstances were, of course, the result of their being part of the U.S. National Park Service.

Origins of Catoctin and Prince William Parks

Both parks had been developed in the 1930s by the National Park Service as part of President Franklin D. Roosevelt's "New Deal" program in response to the Great Depression. They were part of a federal project for converting land that had become sub-marginal in agricultural production and was accessible from metropolitan areas into parks and recreational facilities. Under the leadership of Secretary of the Interior Harold L. Ickes, the federal government in the mid-1930s began acquiring land for what were called Recreational Demonstration Areas (RDAs). The original intention was for the federal government to establish the RDAs and subsequently to relinquish them to the states. Most were deeded back as state parks, but a few—Prince William and Catoctin among them—would remain permanently within the national park system.[3] Early on, leadership in the RDA program was assumed by the U.S. National Park Service (NPS), a permanent government agency created in 1916, to preserve and oversee the federal parks.[4]

Enthusiastically Conrad L. ("Connie") Wirth, assistant director of the National Park Service, championed recreational development in the parks.[5] He led the

National Park Service's dramatic expansion into development-oriented, recreational programs in the parks to provide healthy, outdoor, recreational facilities for urban dwellers. NPS studies identified a special need for such facilities by social organizations for urban youths.

The National Park Service sought sites within about 50 miles from an urban center, with at least 5,000 acres, and with good water, available building materials, and an interesting environment. The plan was to construct picnic areas, family campgrounds, and an organized camping area. The latter would contain three to six group camps, each of which would contain a general dining hall, an office, and an infirmary. Each group camp would have three to four sub-units, each with a lodge and half a dozen four-bed cabins plus a cabin for the group leader. Thus each camp was designed for 70 to 100 campers and a group staff of 20 to 30 persons. In general, the camp buildings were to be rustic style constructed of lumber or unpeeled logs on a stone foundation, with these materials obtained from the site.[6] As Wirth recalled forty years later, "We proposed to build campsites primarily for group camping but also to provide year-round camping and recreation for individuals, small groups, and families. Our general objective was to provide quality outdoor recreation facilities at the lowest possible cost for the benefit of people of lower and middle incomes."[7]

In 1936 President Roosevelt signed an executive order giving the National Park Service control over the entire Recreational Demonstration Area program. Eventually it would establish 46 RDAs in 24 states. Drawing on financial and labor support from such temporary New Deal agencies as the Works Progress Administration (WPA), the Civilian Conservation Corps (CCC) and others, the National Park Service's large-scale movement into outdoor recreational development contributed in no small part to the dramatic growth of the NPS from managing primarily a limited number of large national parks of the West to responsibility for dozens of parks across the country.[8]

Acquiring Land and Building Cabin Camps at Prince William Forest Park

Chopawamsic Recreational Demonstration Area (later renamed Prince William Forest Park) with 14,000 acres in Virginia and Catoctin Recreational Demonstration Area (later Catoctin Mountain Park), with nearly 10,000 acres in Maryland were two of the seven largest of the RDAs.[9] They were also among the first selected by the National Park Service. Because of Chopawamsic's proximity to Washington only an hour away, and accessibility to top government officials and the national press corps, Wirth wanted to make it a model experiment. He believed that its success or failure would affect the entire RDA program.[10]

The sprawling wooded area west of Quantico, Virginia, seemed perfect for the RDA program. It was only 35 miles from Washington D.C. which badly needed additional camping facilities for urban youth. The land was economically unproductive, the soil having been depleted by farming and mining practices. The area was sparsely populated, and residents suffered high unemployment, more than half of the 150 resident families were on relief and only 40 families had any regular income. The forest itself was attractive, with second growth trees, abundant wildlife, three lakes, 18 miles of fresh water streams—Chopawamsic Creek and Quantico Creek—and timber and stone for constructing access roads and park facilities for the organized group camping that the youth and charitable organizations of Washington needed.[11]

Land acquisition, or more accurately obtaining options to purchase land, began early in 1935.[12] Many of the several hundred residents, whites and blacks, did not want to leave, some of the families having lived there for generations. Few had electricity, telephones, or indoor plumbing, but most of the people did not consider themselves poor, as they existed on subsistence farming, barter, or occasional work, often at the U.S. Marine Corps base at nearby Quantico, the Army Corps of Engineers Post at Fort Belvoir or the Navy Yard in Washington. Illicit bootleg whisky stills operated in the backwoods. Many of the rural residents did not see why they should be forced to leave in order for group camping facilities to be created for the urban youth from Washington. Yet, they felt helpless against the power of the federal government, as even those property owners who took their case to the courts ultimately had to sell their land. While the Resettlement Administration acquired the land, the NPS project managers helped relocate 40 of the poorest families and connect them with relief and retraining agencies. But after leaving their homes, many of the residents simply moved to just beyond the boundaries of the park, where many of their descendants continue to live to the present day.[13]

In contrast, the Washington Council of Social Agencies was overjoyed, calling the National Park Service "our fairy godfather" because it was not only providing the park but was incorporating the suggestions of the council and its social agencies for the kinds of wholesome outdoor facilities they wanted.[14] NPS made plans for Chopawamsic RDA as a model, character-building camp. The idea was to take underprivileged youth, who had been forced to live in the "artificial living conditions" in the "sprawling, overcrowded cities" and through organized camping help produce a "human crop" of "sturdy citizens."[15]

Construction began in 1935, and was performed, under overall NPS supervision, primarily by jobless local workers and some artisans hired by the Works Progress Administration (WPA) and the 18 to 25-year-old unemployed urban youths in the Civilian Conservation Corps (CCC).[16] The CCC units of about 200 young men each, working under quasi-military regimen, built or brought their own portable, wood and tarpaper work camps, then joined with WPA men to construct the access roads and the first two cabin camps, using mainly local materials and working with their own portable sawmill and a rock crushing gravel maker.[17] Despite working through 1935 and early 1936, the CCC and WPA crews were not able to complete the two cabin camps when they were officially opened in June 1936.[18]

By 1939, the CCC and other work crews had essentially completed the five cabin camps at Chopawamsic RDA. Instead of turning it over to the state of Virginia, however, it was decision to retain it as a permanent part of the National Park System and make it part of the National Capital Parks. The reason given was its value as "an ideal recreational and camping area needed to organized camping facilities for various social service agencies and other organizations" in the Washington area.[19]

Since Chopawamsic RDA (Prince William Forest Park) was virtually complete and because it would remain within the National Park Service, the NPS, in late 1939, sent a new manager to take over the facility from the project managers who had overseen its acquisition by the Resettlement Administration and its construction by the CCC and WPA. The new manager, Ira B. Lykes, would manage Chopawamsic RDA for the next dozen years, from 1939 to 1951. Lykes was originally from Trenton, New Jersey, and after graduating from high school there, he had worked in engineering and construction companies in the state capital, while taking some night courses in business administration at nearby Rider College. Subsequently, after five years researching titles and purchasing land for the State Highway Commission, he joined the National Park Service in 1933 as a foreman in charge of general park development at Voorhees State Park in High Bridge, New Jersey. The next year, he was called to the NPS regional office in Richmond, Virginia, and over the next several years, he had various assignments with the National Park Service in the southeastern states.[20] In December 1939, now that Chopawamsic Recreational Demonstration Area was ready for full operation, he was sent there as its manager to run the park.[21] At 33, Lykes was still slender and of medium height and he looked trim and fit in his NPS ranger's uniform and broad-brimmed hat.[22] Three decades later, he recalled the appointment. "I said, yes I would [accept the position]. I wanted some more field experience. So I went up there ... and established myself, living in a CCC barracks and commuting back to Richmond until the [manager's] residence was finished in 1940."[23]

Lykes took over an operating park. Some of the CCC units were still working, but most of the organized group cabin camps had been completed. The cabin camps were numbered in the order in which they were built. They were segregated by race. Racial segregation remained a fixture in southern society until the mid-20th century, and the NPS at that time deferred to "local custom" in the racially segregated camps.[24] Camps 1 and 4 in the northeastern part of the park were then known as the "Negro camps." They were maintained by organizations for underprivileged African-American youths from Washington, D.C. Camp 1 (originally known as Camp Lickman, later Goodwill) was for boys from what was then called the "Negro YMCA." Camp 4 (Pleasant) was originally designated for African-American mothers with children up to age three, but it was used only one year for that purpose and then became the camp for black girls and young women. The other three camps were maintained by organizations for white youths. In the southwestern quadrant of the park, Camp 2 (Mawavi) in the western quadrant was ultimately maintained by the white Girl Scouts of Arlington, Virginia. Camp 3 (Orenda) in the southeastern quadrant, near the park headquarters, was to be maintained by the Family Service Association for White Children, for white mothers and their children to age three, but this lasted only one year; thereafter, Camp 3, which held 120 campers plus staff that brought it to a total of 160 persons, was used by a variety of white groups.[25]

Camp 5 (Happy Land), on the west side, close to Camp 2 (Mawavi) had not been finished when Lykes took over at the end of 1939, but it was completed the next year.

Originally, Camp 5 was the smallest of the five camps, accommodating 96, but later it was expanded to hold 120 persons. It was maintained and operated by the Salvation Army.[26] The abandoned CCC camp with its barracks rather than cabins was not designed for organized camping, but it could be used for group purposes. There were separate entrances to the white and black camping facilities. The black camps were entered from the Dumfries Road (Route 234) on the north side of the park. The white camps were entered from the old Joplin Road (old Route 629; today Route 619). Although NPS planned for a main entrance that would connect both white and black camping areas, lack of funding for land acquisition and construction delayed the establishment of a single main entrance and interior connecting roads until 1951.[27]

Although by 1940, the five cabin camps were essentially complete, a system of roads and bridges needed to be built along with a main entrance. In addition, day-use facilities had not been constructed. Unfortunately, the Depression emergency funds and agencies were being shut down, and beginning in 1940, the federal funding focus shifted to a defense buildup. Lykes had to work creatively for funding to get work done that was not related to defense needs. Throughout the twelve years that Lykes ran the park, he demonstrated great energy and imagination and proved to be an effective manager in particularly difficult times.

Like other park superintendents, Lykes was dedicated to the park and a combination of preservation and development. Public support is crucial for National Park Service, which, despite being a government bureaucracy, has high public approval ratings due to its role in preserving parks, monuments, and historic sites for the American people. Lykes understood he had to ensure influential public support for the new park. Some former park employees remembered him as strong-willed, and demanding as well as creative, but in interacting with the public, Lykes emphasized friendly persuasion. He sought, he said, to "convince people."[28] Through numerous acts of attention and kindness, he built a reservoir of good will toward the park among neighbors as well as officials and other influential individuals and groups, from mayors, chambers of commerce and tourist agencies to the charitable and youth organizations in Washington, conservationists, and wildlife groups, and the commanders at the adjacent Marine base at Quantico and the nearby Army post at Fort Belvoir.[29] Indeed, in his primary mission of preserving and developing the park, Lykes, for much of his tenure, would be dealing with the military.

Although the main function of the park was for civilian use, the military had begun using the park as early as 1938. Both the Corps of Engineers post at Belvoir and the Marine Corps Base at Quantico had conducted maneuvers that year in the emerging Chopawamsic RDA (Prince William Forest Park).[30] Such field exercises continued, and after U.S. entry into the war in December 1941, they became so frequent, particularly from Quantico, that by early 1942, Lykes complained to his superiors that the Marines "have assumed the right to enter upon the area without [even] advising or consulting this office."[31]

Land and Cabin Camps at Catoctin Mountain Park

Catoctin Mountain Park, north of Frederick in central western Maryland, also originated as an NPS Recreational Demonstration Area in the mid-1930s. It was a mountain area with submarginal land on which the soil was badly eroded and the forests overharvested. Unemployment was widespread among the mainly poor, white residents, who qualified for resettlement. There was fresh water, roads, and excellent fishing, and the National Park Service believed that conservation could restore the area and that recreational facilities could be provided for metropolitan areas including Baltimore, 60 miles away, and even Washington, D.C., 75 miles distant.[32] The Catoctin Recreational Demonstration Area (RDA) was authorized in 1935, and land acquisition began under the Federal Emergency Relief Administration (FERA), which had the funds.[33] Although the nearest town, Thurmont, supported the establishment of a park to boost tourism, many of the Appalachian people who lived on and around Catoctin Mountain were unwilling to sell, and FERA began appraising the properties in preparation for going to court.[34]

Meanwhile, NPS appointed a project manager to develop and supervise the new Catoctin Recreational Demonstration Area. The man who would run the park for its first two decades, from 1935 until his retirement in 1957, was Garland B. ("Mike") Williams. Despite the similarity of their first and last names there was no family relationship between Garland B. ("Mike") Williams of the National Park Service and Garland H. Williams of the Federal Narcotics Bureau and the Office of Strategic Services.[35] Mike Williams, the son of a yardmaster for the Chesapeake and Ohio Railway, was born in Chesterfield County, Virginia in 1893. He graduated from John Marshall High School in Richmond, took correspondence course in civil engineering, and then worked on railroad construction before joining the U.S. Army in World War I. During the war, his leadership abilities led to his promotion through the ranks from private to second lieutenant. Returning to civilian life in 1919, Mike Williams worked as a civil engineer on a variety of projects and then as a real estate appraiser for the railway. Laid off by the railway in the Depression, he soon obtained a position with a New Deal emergency relief employment agency, the Civil Works Administration, becoming manager for CWA projects in the City of Petersburg, Virginia.[36]

Mike Williams' experience in dealing with difficult issues of land and labor impressed NPS assistant director Conrad Wirth, and in January 1935, he offered the 42-year-old Virginian the position of managing the development of the new Catoctin RDA.[37]

Accepting the position, Williams immediately went to work to counter the local opposition which had spread because of unfounded rumors that the government wanted to use the land as an artillery range for nearby Fort Ritchie and fears of Republican county commissioners about government intervention and the loss of property taxes. Williams emphasized that plans were for a recreational park and that increased tourism would enhance local revenue. NPS reduced the land acquisition goal from 20,000 to 10,000 acres, and Williams assured residents that condemnation would only be used as a last resort. Meanwhile, he decided to obtain five-year leases for the government to use the land while attempting to convince owners to sell. He also began to line up support groups including fishing and conservation organizations, and several social welfare youth groups in Baltimore.[38] By the fall of 1935, Williams' campaign had produced results, and the commissioners of Frederick and Washington counties publicly praised the government's program of purchasing the land for a park as "an asset to the nearby local communities."[39] With the five-year leases in hand and as the land acquisition program slowly proceeded, Mike Williams launched the construction program for the park in January 1936. Recruiting unemployed men and others with skills, Williams first put in fire break trails as a fire prevention measure, then erected a work camp, including a saw mill, blacksmith's forge, garage and administration building at what would later be known as "Round Meadow." The workers then constructed roads and bridges across steams to provide access to the planned recreational facilities.[40]

The facilities at Catoctin RDA were built over a three year period, between January 1936 and May 1939. Plans called for two picnic areas and four group cabin camps, and Williams requested information, including plans for the latter, from Chopawamsic RDA, which had already completed its first camps.[41] The rustic architecture and layout of the two camps were similar in the two parks, although the nearly 2,000 foot high forested Catoctin Mountain provided a different environment than the low-country, hilly woodlands of Chopawamsic. At Catoctin, the first two cabin camps were built of local chestnut logs and native stone. Cabin Camp 1, later known as "Misty Mount," was erected on a rather rugged, wooded southern slope of the mountain. It had a capacity of 96 campers, plus counselors and other staff.

Cabin Camp 2, "Greentop," was established farther up, near the top of the mountain on a less rugged area that was already partly cleared and partly wooded. Greentop was created as a special needs camp for disabled children as a result of an intensive effort by the Baltimore-based Maryland League for Crippled Children, most of whose young clients had paralyzed legs as a result of poliomyelitis. Because of its special needs purpose, Cabin Camp 2 included an in-ground swimming pool and particular sports and therapy facilities. The League helped to purchase special equipment required for the children. The camp included a dining hall and kitchen, a kitchen staff quarters with bathrooms attached, two recreation lodges with porches, wash houses, a craft shop, and an infirmary, and quarters for the leaders and the staff. Since it was a co-ed camp, there were separate units of cabins for boys and for girls. It had a capacity for 72 campers plus counselors and staff members. Cabin Camp 3 located atop the mountain, was to be called "Hi-Catoctin."[42] It was planned as an organized group camp for boys with a capacity for 72 campers and nine counselors in three subunits; each subunit in the standard NPS boys' camp comprised six, four-cot cabins. In addition, the standard camp had an infirmary, a dining hall, and an administrative unit housing six staff members, one nurse, and four kitchen helpers. Plans called for a swimming pool, craft shop, playing field, camp fire circle, camp office, central shower, recreation hall and nature lore building. Instead of the logs used in the other two camps, Hi-Catoctin was constructed with frame construction and rough board siding. A lodge was constructed with a beautiful view of the valley below the mountaintop. The planned fourth cabin camp, a girls' camp, was never built, due to lack of funding and changed conditions resulting from preparations for World War II.[43]

Catoctin Recreational Demonstration Area opened fully to the public in the spring of 1939, although public use had begun in 1937 when Camp 1 was completed. The NPS press release lauded "the 10,000-acre preserve, of which 90 per cent is forested, embraces a rugged terrain ranging from 523 feet to 1,890 feet in elevation. Some of the higher summits command excellent views of the surrounding countryside. Approximately 50 miles of nature trails lead the hiker through picturesque woodlands and beside sparkling streams." The area was being adapted, the press release stated, for service as "a low cost organized camping center available to accredited character and health-building groups and institutions, but there also are facilities for picnicking and other recreational activities by the general public."[44]

With the opening of the third of the cabin camps in spring 1939 and with land acquisition proceeding rapidly after a favorable decision by a U.S. District Court had cleared the way for condemnation, Williams turned to other improvements in the park.[45]

From 1939 to 1941, he was assisted by a unit of the Civilian Conservation Corps (CCC) which arrived from Quantico, Virginia.[46] Supervised by two Army officers, the 200 young CCC men brought along some two dozen prefabricated portable structures, including dormitories, a mess hall, recreation hall, administrative center, residence for the camp superintendent, and a garage for their vehicles. They erected the CCC work camp on a site just north of the Central Garage Unit and project office of Catoctin RDA, a site later known as Round Meadow.[47]

At Catoctin between 1939 and 1941, the CCC workers engaged in a massive conservation and construction program. One of their first tasks in 1939 was to build a project manager's residence on the Park Central Road for Mike Williams, his wife, and their three sons, who had been living in Frederick. In April 1941, they completed the Blue Blazes Contact Station at the entrance to the park on Route 77.[48] In the meantime, they installed a water supply system, sewage systems, installed power lines, blazed miles of foot trails, improved roadways, planted 1,500 trees and shrubs, and enhanced Hunting and Owens Creeks by clearing obstructions and building small dams for fishing pools. In addition, they fought numerous brush fires successfully, but lost a battle with one in the fall of 1941 that destroyed their garage and trucks. With the economy booming under defense spending, Congress cut back CCC, and the unit at Catoctin was terminated in November 1941. With the help of the CCC boys and the FERA workers before them, NPS manager Mike Williams had created a magnificent new park.[49]

The park was put to use as quickly as it became available. The polio-afflicted boys and girls from low-income families were brought by the Maryland League for Crippled Children to the park each summer between 1937 and 1941.[50] At the camp, the physically-handicapped children, ages 7 to 16, spent eight-week periods, improving, it was hoped, their physical condition and mental outlook through healthy foods, exercise, arts and crafts, and camaraderie in a nature setting. The Salvation Army used Cabin Camp 1 (Misty Mount) every summer from 1938 to 1941.

A number of other organizations, including Girl Scouts from Frederick and Washington counties, used Misty Mount for short-term camping. Cabin Camp 3 (Hi-Catoctin) was used each summer from 1939 to 1941 by the Federal Camp Council and by white Boy Scout and Girl Scout troops from Washington, D.C. There were no facilities for black groups at Catoctin in this period.[51]

With Nazi Germany expanding its control over Europe and its submarines striking across the Atlantic, President Roosevelt proclaimed an unlimited national emergency in the spring of 1941. Later that summer, even before the United States entered the war some of the Army's young, draftees were sent to Catoctin, where tents were set up next to the CCC barracks at Round Meadow, and the new recruits were given military training.[52]

In a gesture of goodwill towards Great Britain, half a dozen Recreational Demonstration Areas were made available for British sailors to relax while their warships were being repaired at shipyards like Baltimore's, and between June and November 1941, Catoctin RDA was host to 630 British sailors. They and their officers came in groups of 60 to 100 to spend one or two weeks in the facilities of the park.[53] Mike Williams reported that the British sailors behaved themselves in a gentlemanly manner and kept the grounds and buildings in excellent order, although a sizable telephone bill arrived after they had gone.[54]

The National Park Service and World War II

The National Park Service was put in a difficult situation by the country's mobilization during World War II. Although its original mandate was to hold the resources of the national parks in trust for the American people, during the wartime emergency, it was often asked to place those resources at the disposal of the armed forces and war production agencies. The new director of the National Park Service, appointed in 1940, was West Coast preservationist Newton B. Drury.[55] A dedicated Republican with a suspicion of bureaucracy, Drury disagreed with the NPS's dramatically expanded missions under New Deal, believing instead that the agency should have a more passive, caretaker role.[56]

Given the shift of government priorities to national defense, the National Park Service in World War II was reduced, as Drury later explained, to simply a "protection and maintenance basis."[57] Like other domestic federal agencies, NPS saw its appropriations and personnel drastically reduced during the war.[58] But at the same time, the war confronted the National Park Service with unusual demands. The War and Navy Departments sought to use the national parks in many ways. Rest camps for military personnel were established in Grand Canyon, Carlsbad Caverns, and Sequoia national parks as well as many other NPS facilities, including a Marine Corps rest camp at Catoctin Recreational Demonstration Area in 1945. Many of these cases, service personnel were housed in cabin camps, lodges, or even refurbished old CCC work camps on the park grounds. Hospitalization and rehabilitation was provided at Yosemite and Lava Beds national parks. The military held overnight bivouacs in a number of parks and extended maneuvers in Hawaii National Park and

Mt. McKinley National Park in Alaska. The War Department also obtained permission to conduct extensive training in national parks such as in Yosemite, Shenandoah, Yellowstone, Isle Royale, and Death Valley as well, of course, as in Catoctin and Chopawamsic (Prince William Forest) Recreational Demonstration Areas. Some 2,600 permits and authorizations for use of park facilities and resources were granted for defense-related purposes.[59] Most of the parks and recreational demonstration areas were also made available for use by military as well as civilian visitors, but with wartime rationing of gasoline and rubber, the number of visitors plummeted to one-third by 1943. Still some 35 million visitors, including eight million men and women in uniform, visited the national parks during the war putting an additional burden on the severely reduced park staffs.[60]

Faced with wartime demands, the National Park Service concluded that the good of the nation lay in conserving the parks intact for future generations. As Drury declared during the war, while the National Park Service recognized its obligation to help the war effort, it needed "to hold intact those things entrusted to it—the properties themselves, the basic organization to perform its tasks, and, most important of all, the uniquely American concept under which the national parks are preserved inviolate for the present and future benefit of all of our people."[61] As early as November 1940, with the beginning of the defense mobilization before the United States entered the war, Drury and his staff had formulated criteria for evaluating possible war uses of the parks. The criteria emphasized the need to determine that the request for usage came from a defense agency, that the use of park land was necessary for defense and that it would not cause irreparable damage to the park. NPS also wanted to ensure that alternatives to using park land had been considered and exhausted. If it was agreed that the military's use was essential, then the use permit should include specific conditions to protect the park, and there should be provisions that when the military usage was terminated, the military repair any damages and restore the property to its previous condition.[62] Drury later concluded that NPS had been surprisingly successful in meeting its criteria and preventing irreparable harm to the parks.[63]

Obtaining Permits for Military Training Camps

At Donovan's headquarters in Washington, D.C., Colonel M. Preston Goodfellow and his executive officer, Lieutenant Colonel Garland H. Williams, had determined early in 1942 that Catoctin and Chopawamsic (Prince William Forest Park) would make valuable sites for the paramilitary training camps. Negotiations for military use of these two NPS parks proceeded at the departmental level between the War Department and the Department of the Interior.[64] On March 24, Secretary of War Henry L. Stimson wrote to Secretary of the Interior Harold L. Ickes that "it has been determined that a military necessity exists for the acquisition by purchase and by transfer" of certain government lands, buildings, and other improvements under control of the National Park Service in Catoctin and Chopawamsic Recreational Demonstration Areas. Stimson asked that these be transferred by executive order to the War Department for military purposes. The matter was urgent, he wrote, as troops were scheduled to arrive at those locations that week, and he asked for immediate permission to occupy and use those NPS sites, pending their formal transfer to the War Department. He noted that the War Department planned to purchase tracts of land within the training area that remained in private hands. Consciously avoiding any mention of Donovan's organization, Stimson cautioned that "The War program demands that the utmost secrecy be maintained regarding the military use of this property. That fact, or the arm of the service so using this property, should not appear in published Federal Records."[65]

Neither Secretary Ickes nor NPS Director Drury wanted to have the parks turned into military training camps and certainly not for the duration of the war as Stimson had requested. On 25 March 1942, the Secretary of the Interior wrote to Stimson, reminding him that these RDAs had been developed for work "of vital importance, "primarily for rebuilding the health" of children and adults from lower income groups in Baltimore and Washington. He directed that the War Department be granted a use permit but only until 1 June 1942—just over two months—with the understanding that the NPS would help the War Department locate an alternative site and bring in portable CCC barracks for housing. If an acceptable alternative could not be found, Ickes agreed to grant a special-use permit but only on a year-to-year basis. In keeping with Drury's criteria, Ickes insisted that the military was not allowed to make any changes in the park without prior review and concurrence by the National Park Service. "The [NPS] managers who are now on the areas [Catoctin and Chopawamsic] will remain there, will be of all possible assistance to the Army, and also will aid in protection of the areas from fire and abuse."[66]

Donovan's organization was not to be deterred, however, and on Saturday, March 28, 1942, three days after Ickes' temporary, short-term authorization, two military officers representing Lt. Col. Garland H. Williams' office arrived unannounced at Chopawamsic (Prince William Forest Park). They introduced themselves to NPS Project Manager Ira B. Lykes and requested permission to inspect Cabin Camps 2 and 5 with him to determine their housing capacities. Afterward at Lykes' office, they took notes on the layout plans for all five cabin camps. Lykes handed them copies of the Rules and Regulations for the use of the facilities issued by the Secretary of the Interior in 1937. Designed to preserve public property, the regulations protected structures, trees, flowers, rocks, wildlife and relics, prohibited camping outside designated areas, and banned firearms and explosives. They last seemed particularly quaint, given what Special Operations plan to do in the two parks, but it did put Donovan's organization on notice that the National Park Service intended to continue the preservation of the property entrusted to its care. The two officers told Lykes that their men would start arriving to occupy Cabin Camp 5 in four days on Wednesday, April 1st.[67] The same military takeover date, 1 April 1942, was set for Catoctin RDA.[68]

Rumors immediately flew thick and fast around both Chopwamsic and Catoctin about what was happening to the parks. On 5 April 1942, an outdoor sports columnist reported in the Baltimore American that Catoctin RDA had been "taken over for use in the present war effort" and was closed to the public.[69] The head of

Maryland's State Parks wrote to NPS concerning rumors that the Army was going to establish an Officer's Candidate School there and conflicting rumors that the Army would take all of the park or simply the half north of the Thurmont-Foxville Road (today Route 77). Conrad Wirth responded that the Army wanted only the area north of the road and that a use permit had been granted but only until 1 June. It was hoped, he wrote, that the Army would vacate by then, but "it may be necessary to extend this permit for a much longer period."[70] On 10 April, the *Catoctin Enterprise*, a weekly newspaper from Thurmont, reported that although no announcement was made as to how the area would be used, the government was taking over Catoctin "for use in the war effort" and "the public will be barred from entry into the area proper."[71] Indeed, during the war, public recreational use of the Catoctin RDA was limited to picnicking in the West Picnic Area and trout fishing on Big Hunting Creek, both of which were south of the Thurmont-Foxville Road and therefore outside the military area of the park.[72] During the war, although local residents around Catoctin—as well as around Chopawamsic RDA down near Quantico—realized that the Army had men in the closed off parks, they never knew until recent years that these were special soldiers in "Wild Bill" Donovan's clandestine organization.[73]

When Donovan's organization took over Catoctin and Chopawamsic Recreational Demonstration Areas in April 1942, the public and any unauthorized visitors were barred from entry, warning signs were posted, sentries were stationed at the entrances and armed guards periodically patrolled the perimeter. It soon became clear to the National Park Service that the military intended to stay. Years later, Ira Lykes recalled that in early spring 1942, NPS Director Newton Drury and his staff had advised him that "Colonel Donovan and his cloak and dagger boys" were looking for a place and had decided that Chopawamic was ideal for their purposes. "We would have to cancel all the leases on camp grounds, and they would move the OSS in there. I was to remain as the liaison between the National Park Service [and OSS] and help them in any development they wanted done in the war effort."[74] NPS did notify the charitable and other social agencies that had been issued permits for the summer 1942 camping season at Catoctin and Chopawamsic RDAs that because of the military, they could not use the park. It offered them alternative facilities at abandoned CCC camps in smaller RDAs in Virginia and Pennsylvania.[75] Faced with the upcoming expiration on 1 June of the two-month temporary permit, Secretary of War formally requested permission to continue to "occupy" the two parks for "the duration of the present emergency."[76]

The Department of the Interior accepted the fact that the military would use the parks, but it still sought to control the duration and the terms of that usage. Consequently on 16 May 1942, the Secretary of the Interior issued special use permits, retroactive since March 25th, authorizing the War Department to use certain lands and facilities within Chopawamsic and Catoctin RDAs. The permit did not provide an expiration date but indicated that it was "revocable at will by the Secretary of the Interior."[77] Seeking to preserve the parks during their use for commando-style training, the permit had a number of protective provisions. It excluded from the NPS administrative and staff facilities from military control. To protect the environment, the permit required "that precaution shall be taken to preserve and protect all objects of a geological and historical nature....that wherever possible, structures, roads, as well as trees, shrubs and other natural terrain features, shall remain unmolested...that every precaution shall be taken to protect the Area from fire and vandalism" and that the military would help fight forest fires. The normal prohibition against weapons or explosives in the park was omitted in this special permit for the military, but the permit did require NPS approval for any new structures the military would build and required that their location was to be determined through consultation between the military and the NPS park manager. Upon termination of military usage, all structures erected by the Army were to be transferred to the Department of the Interior or removed by the War Department, "and the site restored as nearly as possible to its condition at the time of the issuance of this permit, at the option of the Secretary of the Interior." An accompanying letter from the Secretary of the Interior stated, as the Army had agreed, that private lands purchased by the War Department within Catoctin and Chopawamsic RDA would be transferred to the Department of the Interior to inclusion in the parks when the need for them for national defense was ended.

Acquiring Additional Land for the Camps

The War Department moved quickly to facilitate the training camps for Donovan's organization. On 30 March 1942, a representative of the Army's land acquisition office arrived at Chopawamsic RDA (Prince William Forest Park) with the intention of acquiring privately held tracts within the area assigned to the military as quickly as possible.[78] Lykes and John E. Gum, Acting Superintendent of one of the CCC work camps in the park, spent the morning with him going over the acquisition status of the remaining private properties lying north of Joplin Road (today Route 619).[79] Lykes urged the War Department to acquire the ramshackle, wooded area on the eastern edge of the park known as "Batestown," that the park manager denounced as "a hot-bed of bootleggers, thieves and vice." He warned that it would be "a potential health hazard to the adjoining Camps 1 & 4 and the Quantico Creek Watershed." For years, the park manager and the local sheriff had tried to catch the moonshiners, but they never did.

Now, as Lykes wrote to Conrad Wirth, "I believe every effort should be made to induce them [the War Department] to acquire these properties as well as other interspersed tracts throughout the [Chopawamsic] Area."[80]

At that time, Chopawamsic RDA (Prince William Forest Park) was composed of 14,446 acres obtained by NPS between 1935 and 1941.[81] During the next year, the War Department acquired an additional 1,000 to 1,500 acres of isolated, privately held tracts scattered within the park, which were acquired for security purposes.[82] The "Batestown" area was not among the parcels acquired during the war. It still exists today, although the park acquired some of it in the first decade of the twenty-first century.

Similarly, in Maryland, the War Department gave priority to acquire the last in-park private properties at Catoctin RDA (now Catoctin Mountain Park). Starting the first week in April, the Real Estate Branch of the Corps of Engineers mapped the area. The War Department soon leased 20 acres from the Church of the Brethren. By the end of 1942, it had by various means, optioned and then purchased from private owners 225 acres of land within the park north of current Route 77.[83] The next year, it obtained the last fifty acres of private property within the areas of Catoctin park used by the military, bringing the total War Department acquisition at Catoctin to 275 acres.[84]

As Colonel Preston Goodfellow recalled proudly in 1945: "We got three camps [OSS Areas A, B, and C] from the Department of Interior for a dollar a year, provided we'd clean up [acquisition of] the farm areas which they contained. We had the War Department real estate people condemn them, then bought them up."[85]

CHAPTER FOUR

CONVERTING CATOCTIN MOUNTAIN PARK INTO MILITARY CAMPS

Within days after the Secretary of the Interior had granted a permit for Catoctin Recreational Demonstration Area to be used for military training, officers and enlisted men from Donovan's organization began to arrive on 1 April 1942, and started to convert their part of the park into a basic paramilitary training school.[1] Major Ainsworth Blogg, commanding officer of Training Area B, was the first to arrive. A 43-year old reserve officer, Blogg had been a personnel manager for an insurance company in Seattle, when he was called to active duty in March 1941. As a captain, he commanded a military police unit and then served with the headquarters detachment of the IX Army Corps based at Fort Lewis near Tacoma, Washington, until he was recruited by Donovan's organization.[2] The second to arrive was 1st Lieutenant Charles M. Parkin from Lancaster, Pennsylvania, an able young officer in the Corps of Engineers, who was assigned as executive officer and chief instructor.[3] Other staff members soon got there and began to set up firing ranges and obstacle courses.

The first group of trainees, a dozen or so officers and enlisted men destined to be sent to India and the jungles of northern Burma by the end of the year to begin to organize guerilla teams behind Japanese lines, disembarked from the back of canvas covered Army trucks by late April. They were half of the original contingent of OSS Detachment 101—the other half, as noted earlier, were training at Camp X in Canada—which would become one of the most famous and successful of Donovan's guerrilla organizations, credited with mobilizing 11,000 Kachin natives and providing strategic support for regular combat operations that eventually defeated the Japanese in Burma.[4]

Although the area at Catoctin available for field exercises and maneuvers included several thousand acres in the northwestern part of the park, the center of the Training Area B was Cabin Camp 2 (Greentop), which Blogg designated B-2. That camp initially housed the entire COI/OSS complement; however, in September 1942, the permanent cadre, that part of OSS Detachment B responsible for maintaining the operation, moved down the slope to the former CCC Camp (near today's Round Meadow).[5] Designated B-5, the area became the OSS Special Operations camp's headquarters with offices, cadre accommodations, motor pool, and storage facility.

The basic training facility for the students and instructional staff remained up the hill at B-2. With the park closed to the public, the National Park Service personnel were reduced to three: Park Manager Garland B. ("Mike") Williams, a clerk, and a handyman-mechanic.[6]

Although Donovan's organization controlled the northern half of the park, the National Park Service retained several facilities within that area: the park office, located on the site of what is today the Visitor Center; the Park Manager's residence, then located behind the office; and the park maintenance facility at what is today Round Meadow.[7]

Between the efforts of the Special Operations cadre and staff at Area B and the Army Corps of Engineers, the area around B-2 was soon converted into a commando-style training camp. For that purpose, the camp needed obstacle courses and physical training areas, target ranges for various types of small arms, grenades, and mortars, demolitions areas for instruction in the use of explosives, storage facilities for guns, ammunition and other munitions, and explosives, classrooms and eventually a full-scale mockup of a house to use in urban fighting. In addition, since the cabins, washrooms, dining and recreation halls and other facilities had been designed by the National Park Service only for summertime use, they would have to be winterized. The training schools of the OSS were expected to operate all year long for the duration of the war.

The Army moved fast to prepare Training Area B. The construction by the Corps of Engineers with their heavy equipment put considerable strain on the roadways, and a visitor from OSS headquarters in June 1942 noted that the winding macadam road from Thurmont along the base of the mountain was in "a rather poor state of repair."[8] Since much of the field exercises, particularly with mortars and other weapons, took place in the northwest part of the park, the Foxville-Deerfield Road, which ran through the middle of that area, and led from Foxville around to Manahan Road, was closed to the general public due to the hazardous operations at the paramilitary training camp.[9]

The main entrance to the park in the 1930s and 1940s, and thus to the OSS camp in World War II, was on the opposite side of the mountain than it is today. When the old gravel east-west road along the south side of Catoctin Mountain was replaced after the war with a modern, paved east-west road (Route 77) between Thurmont and Hagerstown, the main entrance to the park was re-located along it on the south side of the mountain. But before and during the war, the entrance to the park had been from the north side of the mountain. In those days, visitors drove north about five miles from Thurmont on the Sabillasville Road (today Route 550) to the bottom of the northern slope of Catoctin Mountain. At the village of Lantz, where there was a railroad stop, they would turn west to the even smaller village of Deerfield, then drive south up the dirt and later gravel Manahan Road through what was then the main entrance to the park. Eventually they would arrive south of what is today the paved Park Central Road, which was built after World War II. At that time there was a dirt or gravel road father south and west that connected the CCC camp, NPS maintenance area and Cabin Camp 2. A turn to the east and a mile and a half dive up the hill brought them to the cabin camp for polio-crippled children, which was converted into Special Operations Training Area B-2.[10]

One of the first instructors to arrive at Area B-2, 1st Lieutenant Jerry Sage, who got there at the beginning of April, recalled that his group of men in their army truck encountered guards and barbed wire on the little road up through the hills, and they were stopped several times along the way. When they arrived at B-2, it had "some log cabins, a big mess hall and kitchen and a headquarters building surrounded by trees. There was a swimming pool and a large green field, possibly for football or soccer."[11] Apparently, the swimming pool was subsequently filled in and covered over during war.[12] Two months later, a visitor from the Secret Intelligence training office provided a more detailed description of the camp. Arriving at the main gate down toward Lantz, he now found the road blocked off by a wooden obstruction as well as a sentry on guard. Having passed inspection, he reported that he drove a mile or more (probably closer to three miles) up a mountain road through thick, second-growth forest to the center of the training camp. There, at Area B-2, he saw a large, T-shaped mess hall, and many roughly built cabins scattered throughout the trees and providing housing for the trainees. Each of the four groups of cabins had a latrine with toilets and washbowls but, initially at least, no hot water. Two buildings that served as separate quarters for the officers of the cadre and instructional staff had hot water and bathtubs. Two recreational buildings were used for multiple purposes: a classroom, a theater, and a reading room with desks and chairs and supplied with books, magazines, and other literature.[13] More than a year later, in October 1943, B-2 was reported to have a capacity of 149, including 20 officers and 129 enlisted men, although this could be expanded by converting the two recreational buildings into barracks lodging 15 enlisted men each, for a total camp complement of 179 men if necessary.[14]

Altering a Summer Camp for Year Round Training

A number of alterations were made to transform a summer cabin camp into a year round paramilitary training facility, much of the work being done by the Corps of Engineers. The instructional staff, and before they moved down to B-5 in September 1942, the camp cadre as well, lived in cabins in the woods just to the west and south of the main buildings. But the cabins, which had four cots for the young campers, were turned into two-man cabins, at least for officers, and in the officers' cabins, indoor toilets, chemical or flush, were installed. The Army Engineers installed insulation and wood-burning stoves to heat the cabins.[15] Some of the officers shared rooms in larger cabins or lodge-type buildings. The latter were composed of four, two-bed cubicles, each enclosed by high partitions that left an airspace for ventilation. At the far end of these officers' sleeping quarters, there was a lavatory with a flush toilet and a common room.[16]

For the trainees, particularly when large numbers began to arrive in 1943 as the Operational Groups were being prepared for duty in 1943 and 1944, there were temporary barracks, comparatively inexpensive, rectangular buildings each housing about ten men. These were classic Army-style open barracks with double rows of bunks. A potbellied stove was supposed to provide enough heating, but the cold on the mountain became so intense while OGs and others were training there in January and February, 1944 that the barracks were evacuated and the trainees put in groups of four into heated, cabins nearby. The barracks were located between the cabins that housed the cadre and the mess hall.[17] There was also a bath or wash house with running water in sinks and shower stalls; the Army Engineers built a frame addition to the bathhouse and installed a barracks heater, electric force hot air blower type, and galvanized steel hot air ducts to the various parts of the bath house.[18] In preparation for the large numbers of Operational Groups being trained for the invasion of France, the Army Engineers installed sanitary toilets to replace Vogel toilets and renewed practically all of the electrical wiring in early 1944.[19]

The mess hall was on the same site of the current dining hall at B-2, but the original dining hall was destroyed by a fire in the 1950s.[20] The OSS mess hall was constructed as a long, tarpaper covered building with two or three lengthy tables and with the cooking facilities at the other end. "We would sit down and eat in whatever we were dressed in," recalled Reginald ("Reg") Spear, who was there in early 1944 as Special Operations trainee preparing to be sent to some of the islands in the Pacific.[21] Near the mess hall was the recreation building. It was a different structure than the present one. OSS veterans who toured the park in May 2005 said that their recreation hall was smaller and was a temporary building of wood frame and tarpaper with a foundation of black and gray crushed rock.[22]

There were classrooms behind the camp headquarters at B-2, which was in a cabin which today has a totem pole by it, before the headquarters was moved down to B-5 in the end of the first summer. The classrooms were not very big and held about 15 to 20 persons.[23] They may have been NPS cabins but with the interiors converted to serve as classrooms.[24] The motor pool, where the OSS staff stored and maintained the Army vehicles, was located down the hill in the former CCC camp in what is today the parking lot for the gymnasium.[25] The area that once held the CCC camp and then Area B's headquarters and motor pool, has been radically altered since those days. Most of the old buildings were torn down and the area re-graded and rebuilt in 1965 when this was the site of the first Job Corps camp in the United States.[26]

At the transferred Civilian Conservation Corps (CCC) camp down the slope about a mile to the west of B-2, the Army Engineers made a number of improvements, so that the officers and enlisted men of the permanent staff could occupy that facility when the headquarters and cadre moved down from B-2 in September 1942.[27] The facility was designated Area B-5 and was located adjacent to the NPS Service Area, today called Round Meadow. The CCC camp already contained an office, officers' quarters, technical quarters, recreational building, educational building, dispensary, mess hall, and five tarpaper, wood striped, barracks. Those barracks, now gone, were located to the south, behind what is today the dining hall at Round Meadow.[28] The Army Engineers improved the water supply and waste disposal systems, building a 25,000-gallon water reservoir of reinforced concrete with a wooden roof and a pump house, and laying 1,500 feet of water pipe, as well as 7,600 feet of waste pipe, and constructing a concrete septic tank. In a number of buildings, the Army Engineers not only provided flush toilets but hot water tanks for showers. They put partitions in one of the barracks to provide some privacy for the enlisted men of the station complement, and they added a walk-in refrigerator to the mess hall.[29] Area B-5, at least in October 1943, had a capacity to hold a station complement of 25 officers and 40 enlisted men, plus up to 200 trainees, a total of 265 OSS personnel.[30]

As frost arrived on the mountain in the fall of 1942, the Park Manager Mike Williams knew that the camp would soon face a major test, operating amidst sub-freezing temperatures. By November, the Army Engineers had painted buildings inside and out, laid new roofing tarpaper down, and put in additional heating and plumbing facilities. Insulation board was ready to be installed on ceiling and sidewalls of all buildings except the garages. This would be the park's first experience of trying to operate the cabin camps throughout the winter, and Williams reported that "as the plumbing and water systems were not designed or constructed for winter use, there was considerable work connected with protecting them."[31]

"Cold as the devil this morning, nearly froze shaving and everyone had long underwear but me so I beat it to the supply [room] and got a pair," wrote a new arrival, Private Albert R. ("Al") Guay in his diary in late October. A member of the cadre, Guay was appointed assistant company clerk of Detachment B. Years later he recalled "I can remember, it was probably the previous week, the first time I went in to take a shower, there was ice on the floor. There was hot water. I believe we had a coal-fired boiler. The barracks were just board shacks with no insulation in them. I don't think there was any heat in them. There might have been a pot-bellied stove in there, but I don't remember it now. I do remember it being cold!"[32]

The largest building at B-5 was the mess hall It was insulated and heated, and it held perhaps fifty enlisted men (the officers ate in a different location).[33] It also served as a theater where training films and sometimes Hollywood movies were shown. Across from the mess hall was tiny wood building about 12 by 12 feet, which the cadre used as a kind of PX or post exchange (today it is the nurse's station). One side had been cut in half horizontally, like a Dutch door, and the top half would be swung up like an awning, the lower half had a counter inside the small, room-like building. At this little PX, the cadre could purchase toiletries, aspirin, snacks, soda, and various sundries. "It looked like a hot-dog stand at a small baseball field," Guay remembered. "I would usually go over there almost every day and get a can of sardines and a small box of crackers. I would go and sit down somewhere, maybe lean against the PX and eat it."[34] In addition to that former PX, the only other buildings currently at Round Meadow that date back to the CCC period, according to Park Superintendent Mel Poole, are the original blacksmith shop and the original project office and then the first park headquarters, which is today the Catoctin Research Center. These two buildings would have been used by manager Mike Williams and his crew in running the park in the 1930s and early 1940s.[35]

Preparing for Combat Training and the "House of Horrors" at the Park

Commando-style training was taught up the hill at B-2, and it involved learning to handle various weapons and munitions, developing knife-fighting and other close-combat techniques. It also meant learning about explosives for sabotage. It meant becoming familiar with guerrilla field operations. And perhaps above all, it required developing and maintaining top physical and mental condition and a predominant attitude of self-confidence, daring, and initiative. The basic Special Operations course at B-2, therefore, required a number of specialized facilities not found in a park.

Among the first to be constructed was a pistol range for target practice with the U.S. Army's standard sidearm, the .45 Caliber Colt automatic pistol, as well as handguns from Allied and enemy countries.[36] It was a quickly made 10-to 20-yard range. "It was nothing fancy," instructor Frank A. Gleason, recalled, "there was just dirt behind the targets, no concrete butts."[37] However, as a substitute for the standard concentric circle, bull's eye targets used by the Army for marksmanship training, the special operations instructors put up waist-high, frontal silhouettes of enemy soldiers In another departure from standard Army military training, instead of raising the pistol to eye level and taking careful aim, the students were taught "instinctive shooting" or "point and shoot," a technique of quick firing from the hip. Within two months, the stationery targets were augmented for advanced students with targets that popped up from behind shrubs or other types of concealment. A visitor to B-2 described the pistol range in early June 1942: "In front of the administration building is a field at the end of which are targets for shooting practice. An apparatus is rigged up for providing a moving target to operate behind the frame of a window and door."[38]

Even this was not realistic enough, and a complex "house of horrors" was soon built to help reproduce at least a sense of the disorientation, confusion, stress and fear of actual combat. At B-2 on the edge of the training field, approximately where the stable is today, the Donovan's organization built what the official records refer to as a "pistol house" or simply a "training building," but which was known by the trainees and staff as the "mystery house," "haunted house," or "house of horrors." It was modeled after training structures used by the British Special Operations Executive in the United Kingdom and Canada.[39] A small house with a rectangular floor plan, 40 by 78 feet, it was complex and also costly; indeed, it was the most expensive structure built by the OSS at Area B.[40] It was still under construction in early June 1942 when a visitor from the Secret Intelligence Branch reported that "there is also a house which they had been working on recently and which was apparently not completed. It is fitted with wires operating hidden targets and with sandbagged partitions. One end of it, behind the targets, had been heaped up with turf and mud to provide a non-ricocheting backstop for bullets. The purpose of the house is to provide close-in shooting practice under realistic conditions."[41]

To achieve those "realistic conditions," trainees were awakened in the middle of the night, given a .45 caliber pistol and two clips of live ammunition, and sent into the house where they were told to expect Nazi guards. Sometimes, they were told to kick open the front door and be ready to shoot at once. As they moved through the darkened corridors and rooms, cardboard cutouts of armed enemy soldiers or papier-mâché Nazis with pistols would suddenly pop up to test the trainee's instinctive firing abilities. Years later, Richard Dunlop, a member of Detachment 101 in Burma, remembered what had been required. "Trainees... walked through a darkened hall about four feet wide, when suddenly the floor dropped some six to eight inches and a papier-mâché enemy appeared. Even though they were thrown off balance, they had to fire from the hip and hit their hated adversary in the head. In the next instant the head popped up, and they had to fire again."[42]

Two former OSS agents reiterated the experience of the trainee: "With a masked instructor at his elbow, he would move with drawn revolver through a cleverly designed Scare House that rivaled any Coney Island Chamber of Horrors in one-a-minute thrills. Boards would teeter realistically underfoot as he felt his way along a dark hall, footsteps would echo mysteriously ahead of him, a concealed phonograph would grind out the rumble of guttural German voices around a poker table, the clink of glasses and slap of cards. A turn of the corridor would reveal a suddenly lighted dummy dressed in the uniform of a Nazi Storm Trooper, confronting him. Whirl, fire. Careful, now, there's someone in the room just ahead. Reload, safety off, hammer cocked. Burst open the door, fire; fire again."[43]

Initially, for rifle practice as well as training in the use of various submachine guns, the students were sent by truck to the extensive firing ranges at nearby Camp Ritchie, headquarters for the Maryland National Guard and also the training facility for the U.S. Army's Military Intelligence School. However, by the end of 1943, the Army Engineers had constructed a comparatively small rifle range near Area B-5, located about 500 yards west of the NPS service maintenance area at Round Meadow. It was a standard U.S. Army rifle range with manually operated targets manipulated by ropes by enlisted men down in pits dug beneath them and earth and concrete butts behind the targets. In this case, it was only a 300-yard range, considerably shorter than the 600-or 1,000-yard rifle ranges the Army used for standard infantry training.[44] But the Special Operations instructors were less interested in long-range, battlefield firing then in training guerrilla leaders and saboteurs for staging close-range ambushes or fighting their way out of a trap. As with the "instinctive fire" for pistols, so the training here emphasized laying down a quick and rapid fire rather than sniper-like marksmanship. The OSS men were trained with American weapons—rifles, carbines, and submachine guns—as well as similar small arms from various Allied and Axis countries. The rifle range at B-5 could handle only a comparatively small number of shooters. When larger numbers were involved, such as Operational Groups being prepared for duty in Europe, the trainees were trucked up to Camp Ritchie. Numbers were not the only reason for that move, however, the extensive use of automatic weapons was another. "It is dangerous to have [firing] ranges too close to trees and brush," noted Reginald ("Reg") Spear, who trained at B-2 in early 1944. "Bullets get hot. They can and do cause fires, tracer bullets especially." Training sizable groups to fire submachine guns also often involved a trip to Camp Ritchie, for as Spear recalled with gusto, "behind Ritchie for miles there was nothing, and we would do things like let a balloon go and take a grease gun [a compact, rapid-firing submachine gun] and shoot it down."[45]

For training and practice in the use of explosives as well as throwing hand grenades and firing rifle-propelled grenades, and sometimes the firing of regular rifles before the range at B-5 was constructed, the OSS instructors used what they called a "demolition area." It comprised about 15 acres. The location is difficult to determine today. It may have been situated about 1,000 feet east of the old CCC camp, now Round Meadow, in an area currently in the vicinity of the Fire Cache, where much of the park's fire-fighting and other emergency equipment is stored today, or it may have been in an area 8,000 feet north, northeast of Round Meadow.[46] Within the demolition area, Army Engineers bulldozed part of the woods there to make a level area 100 by 100 feet square. The approximately one-thousand cubic yards of material from making that cut was then pushed by the bulldozers into making an embankment as a backstop along the upper edge of the demolition and firing area. The earth and timber embankment was 10 to 12 feet high and 300 feet long. In the slope below the square, 15 observation pits were dug to provide safe viewing of the use of demolitions, grenades, and mortar fire.[47] The demolition area was apparently located in what is today the Chestnut Picnic Area on the north side of Park Central Road between Greentop and Round Meadow.[48]

To house the munitions and explosives, the first arrivals on the staff of Area B in April 1942 simply dug caves in the side of the hill and stored them inside. Charles Parkin, the first chief instructor, said that in case of an accident explosion inside, the substantial weight of the earth above it would have contained any blast.[49] Fortunately, there was no such explosion in the storage facilities at Area B. When the cadre moved down to B-5 in the fall of 1942, much of the ammunition was stored temporarily in a small old wooden building, about twenty feet square, called the arsenal. Private Albert Guay, the assistant company clerk who described it, noted in his diary in November 1942: "Helped take inventory in the arsenal today. Saw thousands of rounds of ammunition, mortar shells, grenades, flares, etc." An avid hunter, the young 22-year-old added: "Swiped 5 tracer bullets and a box of .22 [caliber] long rifles [cartridges]."[50]

Later, in 1943, the Corps of Engineers built one of the Army's standard magazines along with a sentry box for guarding it, about 600 yards north of the old CCC Camp.[51] Designed to store small arms ammunitions, grenades, and mortar rounds, plus TNT and other types of explosives, the standard Army magazine could be either a small one, 10 by 12 feet, or a large magazine, 20 by 24 feet. A four-walled windowless structure, it was made of cinder blocks with a locked steel door and often with steel plates on the walls inside to channel any blast upwards. The flat slightly sloped roof was generally of saturated rag felt and asphalt or of cork and plywood covered with tarpaper.[52] The structure was designed so that in case of explosion, the force of the blast would be hurled up through the comparatively flimsy roof rather than sideways which would have sent shards of concrete cinder block flying with lethal force across the area.

At Area B-2, the weaponry for special operations training was, at least in the spring of 1942, kept in an arsenal in the general area of the field in front of the current dining hall. It included rows of M-1 Garand semi-automatic rifles, Springfield 1903 bolt-action rifles, Thompson submachine guns, .45 caliber automatic pistols, .45 caliber revolvers, and .33 caliber automatic pistols. It also contained hand grenades and 60 millimeter trench mortars.[53]

The use of explosives or live ammunition, which were so much a part of the training at Area B during World War II, were prohibited whenever President Franklin D. Roosevelt came up to Catoctin Mountain. The Secret Service wanted no untoward accidents happening to the Commander-in-Chief. Roosevelt frequently came to the mountain park in the summer, because in 1942, he chose it as the site of his wartime Presidential Retreat.

Physical Training at Area B

Getting the trainees into top condition to build up their physical abilities and mental and emotional self-confidence was a priority as important to COI/OSS as teaching them particular skills. In the field in front of current dining hall at Cabin Camp 2 (Greentop), the instructional staff constructed at B-2 several pieces of equipment designed to enhance physical conditioning. There were ropes for hand-over-hand climbing, a football tacklers' dummy that was used to simulate an enemy in jiu-jitsu or similar instruction. There was a wooden platform erected over a sand pit. It was to facilitate practice in jumping and tumbling, its eight-foot height providing about the same shock that a parachutist felt upon landing.[54] Similarly, there were several large, 30-30-foot square, but shallow open pits dug in the training field in front of the mess hall. They were mainly full of sawdust and sand. The pits were used for ju-jitsu, wrestling, knife-fighting and other close-combat exercises that involved throwing opponents over and down.[55]

Going beyond such familiar exercises, the instructional staff at B-2 developed a new piece of training equipment on their own, which they called the "Trainazium." It was located in the field in front of the mess hall.[56] "We cut a bunch of fine oak trees, 15" to 18" in diameter," Frank Gleason remembered. "We dragged them back with ropes. The park ranger ["Mike" Williams] was mad as hell at us."[57] Parkin had designed it and Lazarsky and some other enlisted men built it. They trimmed off the branches, leaving long, heavy logs. Burying the lower part of each log securely in the ground, the men arranged them as tall posts, like telephone poles, in two parallel rows of three poles each. Nearly two stories above the ground, each of the six vertical posts was connected horizontally by narrower rounded poles. These formed a wooden grid both along and across the posts but have above the ground. The entire rectangular structure of the Trainazium measured about 20 by 20 feet for its footprint and it soared almost 18 feet into the air.[58]

"It was designed to build the men's self-confidence, to build up their physical strength and dexterity, using their legs and arms and upper body to maneuver around in tight, narrow places, and to be agile on narrow high places," Gleason declared.[59] "We could train our agents to walk along narrow places and to climb up and chin up and do that [kind of thing]." "These were big [tall] logs and soft down below in case you fell off."⁶⁰ Charles Parkin, who designed the structure, had put safety nets below. "We had these landing craft [embarkation] nets that you see on ships. They hung down. And we would jump from one to the other like a group of crazy monkeys."⁶¹ As one of the students remembered it, "Trainees ran along narrow boards about fifty feet off the ground to get them used to running over housetops. If they slipped, they plummeted into a net."⁶²

An Accident on the Demolition Trail

Even more demanding and dangerous were an obstacle course and a "demolition trail" built in different parts of the larger Area B. The first obstacle course erected in early 1942 involved a tricky passage across Owens Creek. A thick piece of wire was tied to trees on opposite banks and stretched taut across the rushing stream; a rope line was run parallel to the wire but several feet above it. Trainees, sometimes wearing full field packs, were required to make their way along the swaying wire while holding on to the rope, successfully crossing over the stream or falling into the cascading, often freezing waters below.⁶³ More intricate and even more hazardous was a "demolition trail" located in the forest down the hill between Areas B-2 and B-5. Trainees were ordered to move stealthily and carefully along the trail in the woods, keeping their heads down and eyes open, looking for booby-traps along the way. The traps were trip wires set up by 2nd Lt. Frank A. Gleason, a recent Penn State University engineering graduate, and Sergeant Joseph Lazarsky, a fellow Pennsylvanian, both members of the Corps of Engineers. The two of them tied small charges of TNT to tree branches away from the trail. A trip of the wire would set off a small, if real, explosion nearby. The trainees were told to keep their heads down, stay on the trail, and move forward in a crawl or crouching position.

Unfortunately, in one exercise, one of the students, a 30-year-old New York lawyer who was in training for the Secret Intelligence Branch, failed to crouch down, and when he blundered somewhat off the trail, snagged a trip wire and set off a charge of TNT. The blast sent a chunk of branch flying through the air. It caught him right in the side of the face and broke his jaw. He had to be taken to the nearest hospital and his jaw reset and held in place with wires until it healed. It was the only training accident at B-2 while Frank Gleason was there between June 1942 and March 1943, and he felt terrible about it. The injured trainee, like the other agents in training, was known to the staff only by a fictitious code name in order to maintain the cover of future secret agents and saboteurs. It was only much later, that Gleason and Lazarsky learned the real identity of the careless trainee whose jaw had been broken in that training accident on Catoctin Mountain. His name was William J. ("Bill") Casey, a Navy lieutenant and espionage trainee, who would by the end of the war be put in charge of OSS Secret Intelligence Operations in Germany, and who forty years later would be appointed Director of Central Intelligence by President Ronald Reagan.⁶⁴

Roosevelt Comes to Catoctin and the Marines Set Up Camp

For respites from the heat and humidity of Washington in the summer, President Franklin D. Roosevelt had long used his family estate on the Hudson River at Hyde Park, New York, or the camp for polio victims, like himself, at Warm Springs in the mountains of western Georgia.[65] For relaxation closer to the nation's capital, he had favored cruises on the President yacht, *U.S.S. Potomac* in Chesapeake Bay. With the nation at war, however, and with German submarines sinking shipping off the entrance to the bay, the White House began to look for a more secure and secret alternative for summer weekend retreats. Roosevelt's Republican predecessor, President Herbert Hoover, an avid fly fisherman had established a fishing camp on the Rapidan River in Shenandoah National Park for weekend escapes, but the Democratic Roosevelt chose to ignore the camp of the man he had defeated in 1932.[66] Roosevelt was not a fly fisherman, and the damp atmosphere of the river bottom land aggravated his asthmatic and sinus conditions. His personal physician recommended the cool, fresh air of the mountains. Consequently, in late March 1942, Roosevelt asked National Park Service Director Newton B. ("Newt") Drury to find a mountain cabin or small lodge fairly close to Washington for his occasional use. Drury gave the assignment to Conrad L. Wirth, his assistant in charge of recreation and land use planning, telling Wirth that Roosevelt wanted a summer retreat in a secluded mountain location within about 50 miles or a ninety minute drive from the nation's capital.[67]

After NPS had checked on seven or eight mountainous sites in Maryland and Virginia with utilities and adequate roads and within the prescribed radius, Drury wrote to Secretary of the Interior Ickes with three suggestions, one of which was Catoctin Mountain.[68] Within two weeks, the head of the President's Secret Service Detail, Michael F. ("Mike") Reilly, drove up and made a preliminary inspection of the facilities at Catoctin Recreational Demonstration Area. As he walked around the camp grounds, he was accompanied by Connie Wirth and a few other NPS men, guided by Park Manager Mike Williams. The next day, Wirth reported to Drury that they had first looked over the site that NPS had proposed. Wirth did not bother to name it, but was probably the site NPS had planned for the yet unbuilt Camp 4 or another undeveloped site nearby.[69] Then, Reilly and Wirth walked over to Cabin Camp 2 (Greentop), the facility for crippled children, which Donovan's organization was currently occupying as Training Area B-2. At that location, Wirth reported that Reilly had declared that "a similar layout is exactly what is desired by the White House. With this in mind, we investigated the Hi-Catoctin camp which is on a side road and has a much better distance view." Reilly was extremely pleased with Hi-Catoctin, Cabin Camp 3, which had been used by federal employees and the Boy Scouts. Part of Hi-Catoctin had been designated for use by Donovan's organization, but Reilly and Wirth picked out one of the three subunits of Cabin Camp 3, a cluster of cabins around a lodge, for the President. Reilly suggested a few changes. He also said the President wanted to drive up for a visit, so Wirth told Park Manager Mike Williams to clear out some of the underbrush for better access by the Presidential limousine. If the President liked what he saw, Wirth said that the already developed

Hi-Catoctin facility could be ready for Roosevelt's immediate use and that if the President authorized renovation right away, the alterations could be completed by the first of June 1942.[70]

The President was driven up to Catoctin on 22 April, and in the words of NPS Director Newton Drury, who accompanied him, "he was very much pleased with the area and asked us to proceed immediately with plans and estimates." Because cabins, lodges, and even a swimming pool already existed there, the Presidential Retreat could be created at a comparatively minimal cost and within a short time. Drury submitted to the Secretary of the Interior a map showing the location the President selected and also a plan showing a suggested new floor plan for converting of the recreational lodge at one of the units at Camp 3 into the Presidential lodge.[71]

Roosevelt had made his choice. The appeal was clear: Catoctin offered seclusion and security in what was during wartime a secret, guarded military reservation for the OSS. Atop the mountain, the President could enjoy fresh air, temperatures five to ten degrees lower than Washington, and spectacular views, including the Monocacy River in the valley below and mountain ranges in the distance. Roosevelt allegedly took one look at the view and declared, "This is Shangri-La!"[72] He was referring to the wondrous, enervating and hidden mountainous world depicted in the 1937 film *Lost Horizon* based upon James Hilton's novel.[73] The Presidential Retreat would be known as Shangri-la until the 1950s when President Dwight D. Eisenhower renamed it Camp David, after his grandson.[74]

Cabin Camp 3 contained three sub-units—each a group of cabins around a larger building. The Presidential Retreat would encompass one of them. The OSS planned to occupy the other two sub-units as part of the spillover from its occupation of Cabin Camp 2, re-designated Training Area B-2, and the OSS training camp during the war did extend up into what is now part of Camp David.[75] Both National Park Service and the White House acted quickly. Conrad Wirth submitted proposals and cost estimates based on "what we interpreted to be the President's wishes." Existing or remodeled buildings would provide accommodations for three dozens persons, both guests and staff. The OSS could lend Army tents if more accommodations were needed. Wirth noted, however, that there was only one washhouse with showers for the entire Cabin Camp 3, and Training Area B-2 would need those existing showers, since Donovan's trainees would continue to use two-thirds of the camp. Consequently, Wirth allocated $600 to put in new shower rooms next to existing toilet facilities so as not to interfere with the use of the rest of the camp by the paramilitary trainees of COI/OSS.[76] Including all proposed changes to Hi-Catoctin, Wirth estimated at first that modifications necessary for the Presidential Retreat could be made for under $18,000.[77] That initial estimate quickly became outdated, however, and as Presidential plans for the facility expanded, Wirth revised and revised, finally providing an estimate of $150,000. But as he later recalled, the President had said laughingly that Congress would never give him more than $15,000 for it.[78]

The President gave his final approval to the project on 30 April 1942, and a week later, Catoctin Project Manager Mike Williams was authorized to use any resource needed to complete the work as expeditiously as possible.[79] Two of the OSS demolitions instructors from the Corps of Engineers, Capt. Charles M. Parkin, Jr., and Lt. Frank A. Gleason, were called upon to help out with a difficult problem. A special small room was added onto the lodge for the President. After it was constructed, Gleason said, the contractor found that there was solid rock under the soil in the area where sewage pipes were to go. He asked the two experts in explosives to split the rock so the pipes could be installed. As the rock was right next to the cabin, it was a tricky task; they had to blow the one without blowing up the other. "We used a powder explosive, nitrostarch," Gleason said. "It's less powerful than dynamite. We drilled holes in that rock, filled the holes with nitrostarch and then covered it [the rock] with sandbags, and we set it off. We cracked the rock, and they put the sewer line in."[80]

Heavy rains in May and June plus wartime shortages of labor and materials delayed progress somewhat, but Mike Williams kept close to schedule on preparing the expanded facility.[81] With NPS employees, skilled craftsmen, WPA workers, U.S. Marines and forty soldiers from the Corps of Engineers all helping, Shangri-La was finished enough for the President, joined by his private secretary and four guests, to inaugurate the Presidential Retreat officially on a daylong visit on 5 July 1942. "U.S.S. Shangri-La—Launched at Catoctin July 5, 1942" Roosevelt wrote in a Navy blue log book as he whimsically declared that the Presidential yacht, the *U.S.S. Potomac*, had been replaced by a new, secret "yacht" for the duration of the war. In his jocular manner, Roosevelt always asked his guests at Catoctin to sign aboard the "U.S.S. Shangri-La" for visits he listed as "cruises."[82]

The President and Guests at Shangri-La

Anticipating the President's visit on the 5th of July, three of the OSS instructors, engineers who were experts in explosives, planned to make a gift for Roosevelt when he arrived and to do it right in front of him. Using the new flexible, "plastic" explosive, Compound C, 1st Lieutenant Charles M. Parkin, 2nd Lieutenant Frank A. Gleason and Sergeant Joseph Lazarsky, were going to burn the President's initials into a steel plate nearly two inches thick. All of this was going to be "done in the presence of the President," Parkin recalled. "But the Secret Service would not allow us to do it."[83] They did not want any explosives near the President. So Parkin decided to create the gift in the demolition area and give it to the President after it was finished. One Saturday not long afterward, Parkin was in the demolition area practicing making Roosevelt's initials with explosives. "The red [warning] flag was up," he said, "so the President's car should not have driven into the explosives area, but it did. There was just the President himself and his driver. He introduced himself. But, of course, I knew who he was. He talked to me—imagine the President of the United States, and I was only a first lieutenant or recently promoted to captain, I don't remember which—but here was the President of the United States apologizing to me that he was sorry [for coming into the demolition area when the warning flag was up]. It shows what a big man he was. And I'm a Republican."[84] Parkin, Gleason and Lazarsky later completed the sign in the demolition area and presented the inscribed steel plate to the President. Using plastic explosives, they had blasted letters six inches high, spelling "F D R" into a rectangular plate of steel two and a half inches thick, measuring about three feet by four feet, and weighing two to three hundred pounds. It took six enlisted men to pick it up.[85] The President expressed his amazement and appreciation. What the White House did with the massive steel plate is not known.

Roosevelt and his guests were frequent weekend visitors at Catoctin, going up to the Presidential Retreat nineteen times in the summers of 1942 and 1943. A busy schedule in the Presidential election year of 1944, precluded Roosevelt from making more than four visits to Catoctin that year; and he made no visit to it the following year before his death in April 1945. During the war, the President spent a total of 64 days at his retreat at Catoctin.[86] On some of those Presidential visits in addition to Roosevelt and his guests and there Secret Service agents, there were sometimes as many as a hundred persons there, most of them staff members who came to help run the facility while the President and his guests were in residence. Because of the cold, the President's camp was basically closed down in the winter from December through May, but many of the staff who accompanied Roosevelt there in late fall, remembered how bitterly cold the nights could get in October and November atop the mountain.[87]

OSS Director William J. Donovan was one of Roosevelt's guests during the President's first overnight visit to Catoctin on the long weekend of 18-21 July 1942. There were a number of guests, including Donovan and former Senator James F. ("Jimmy") Byrnes from South Carolina, a politico who was currently on the Supreme Court but would soon to be appointed to head major wartime mobilization agencies. At one point during the four-day visit, Donovan, Byrnes and his wife, walked over to the closest part of paramilitary training camp only a few hundred yards away. There the instructors at B-2 put on a show for the distinguished guests. Captain William ("Dan") Fairbairn, a lanky, taciturn, bespectacled Scot with graying hair and looked like a harmless old man. But he was a renowned authority on martial arts. Known as "Dangerous Dan" and the "Shanghai Buster," he had been assistant commissioner for the international police force in Shanghai before the war and organized its tough riot-control squad. Since 1940, as the top close-combat instructor for Britain's Special Operations Executive, Fairbairn had taught martial arts at training camps in the United Kingdom and then Camp X in Canada, and in 1942, SOE loaned him to Donovan's organization.[88] He brought the idea of the "mystery house" or "house of horrors" to replicate more effectively the unexpected conditions of combat.[89]

Now Donovan and the Byrnes watched as the 57-year-old Fairbairn, flipped 20-year-old trainees over on the ground with their arm twisted behind their back. He showed the erstwhile spies and saboteurs how to disable or kill the enemy, even armed guards, using only their hands and feet. While the guests watched, American instructors showed the trainees how to leap into a dangerous situation, firing off groups of two pistols shots from the hip quickly and accurately. One of their targets that day was a papier-mâché dummy painted to look like the Japanese premier, General Hideki Tojo. Writing later in his memoirs, Byrnes said he considered it "circus stuff." But both he and Donovan knew that Roosevelt loved a good show, and they took the British martial arts expert back up the hill to entertain the President with demonstrations of jujitsu, specialized weapons, and stories of curbing battling street gangs in prewar Shanghai.[90]

British Prime Minister Winston Churchill, the most famous guest Roosevelt had at Shangri-La, first visited Catoctin on May 15, 1943. The two leaders drove there from Washington during a break in the Trident Conference on Anglo-American strategic planning for 1943-44.[91] In addition to discussing military and diplomatic matters, the pair took time out to try some fishing in a nearby creek. A couple of photographs were taken showing the two sitting by the water with their fishing poles. Sending copies of the pictures to Churchill later that summer, Roosevelt wrote that the photos did not turn out well but they "prove we tried to catch fish."[92] At least one member of the OSS cadre at Area B reported meeting Churchill on a couple of evenings during that visit. Joseph Lazarsky, from Hazelton in the coal-mining region of northeastern Pennsylvania, was serving as sergeant of the guard. "It was about 11 o'clock at night, and we had a couple of people of ours to do some guard duty up here [at the Presidential Retreat]," Lazarsky remembered. "I was the sergeant on duty. I went up to check on how they were doing. Churchill comes walking down alone. There wasn't much light, but you could see him. He had his cigar, and in one hand, he had a drink in his hand. He introduced himself. I knew what he looked like. I had never met him before. We chatted about two or three minutes, an exchange of greetings. He asked me if I wanted a drink. I said, 'I don't drink.' If I did drink, I wouldn't have taken one, because I was on duty. I told him I had to check the guards." On another night, Lazarsky met the Prime Minister once again, out for a similar stroll. Checking on the guards, the sergeant ran into him once more. "There wasn't much of an exchange. 'It's good to see you again.' That's what he said."

Subsequently, it was determined that more protected room was needed for such strolls by Roosevelt's visitors, because a few months later, Lazarsky recalled, the Marine Corps "put the guards and guard posts down lower, before you could make your way up the mountain."[93]

At least one OSS trainee met the President at Catoctin. Reginald ("Reg") G. Spear, a highly talented young American from Pasadena, California, was descended from a distinguished British and Canadian family. His paternal grandfather had been knighted, his father was a highly decorated Canadian officer in World War I, and his aunt, Mary Ellen Smith from British Columbia, had became Canada's first woman cabinet minister and the first woman Speaker in the entire Commonwealth.[94] In December 1942, young Spear had enlisted in the Army in Los Angeles, went to OCS at the Aberdeen Proving Ground in Maryland, and became a second lieutenant in the Army's Ordnance Department. Spear was recruited for the OSS and sent to Training Area B in the winter of 1943-44. One day in mid-February 1944, according to Spear, the camp commander, Colonel Blogg, called him out of ranks and told him that the President wanted to see him. Roosevelt may have known about Spear's family from Churchill. Up at Shangri-La, Spear gave his snappiest salute and then listened as the Roosevelt chatted with him and in the first of a series of meetings, Roosevelt unveiled the nature of a special assignment he had for him.

The President pledged him to secrecy.[95] He was going to send Spear into the Southwest Pacific Theater of Operations where General Douglas MacArthur had refused to allow any of Donovan's OSS men. When his training was completed, Spear was to report to Admiral Chester Nimitz, commander of U.S. naval forces in the Pacific. Although an Army officer, Spear would wear a naval officer's uniform. As will be seen in a subsequent chapter on the OSS in the Far East and the Pacific, Spear would be sent on a number of highly dangerous missions.[96]

Protecting the President and the Relationship to Area B

At Shangri-La, security for the President was, of course, a major concern, even more so in wartime. On May 14, 1942, Mike Reilly, head of the Secret Service's Presidential protection unit, had inspected the site with Conrad Wirth, and subsequently for three days at the end of May, Reilly met there with Lt. Col. Charles Brooks, USMC, since a unit of Marines was assigned to guard the Presidential Retreat. Before the Marines arrived, the OSS cadre had helped guard the Presidential Retreat in addition to their responsibility for guarding the training camp itself. In addition, Colonel Blogg, the commander of Area B, had agreed to allow Secret Service agents to practice at the pistol range at B-2 and offered the services of British instructor William ("Dan") Fairbairn to instruct agents in jujitsu.[97] But by the end of August 1942, when William D. Hassett, a former newspaperman now an aide to the President, first accompanied Roosevelt to Shangri-La, he confided in his diary that the nearby group cabin camp was "thrice hush-hush" and was "referred to mysteriously as a camp for training commandos." Soon Hassett found out that it was actually a camp established by Donovan for training saboteurs. In the interest of security, the Secret Service apparently sought by then to minimize interaction between the two facilities. In his diary, Hassett noted that Mike Reilly, had indicated that both Donovan's and the President's camps sought to avoid strangers and to maintain their secrecy, and in regard to the OSS camp, Reilly had said. "We haven't called [on them], because we don't want them to call on us."[98]

The establishment of the Presidential Retreat meant additional secrecy and security for the mountain park. For wartime security purposes, the President's staff, the National Park Service, Secret Service and the Marine guards sought to avoid direct references to the location or name of the Presidential Retreat. Instead, they employed such euphemisms as the "Marines on the hill," or that the President was going to "the camp."[99] Few of the OSS trainees and only a handful of the enlisted members of the cadre at Training Area B knew who was behind the barbed wire fence in the woods only a few hundred yards away. Sometimes the officers on the cadre mentioned that "Mr. Jones" was coming that weekend, but very few of the enlisted members of the cadre, and even fewer of the trainees, knew that "Mr. Jones" was the President or that there was the Presidential Retreat was in the woods. The enclosed compound was off limits, and OSS personnel who happened to wander through the forest and up to the fence were confronted by armed Marine guards, held for identification, then given a warning and sternly escorted back to the camp at Area B.[100]

Local people in Thurmont, the closest town, knew when the President arrived. They would see the Chief Executive come through the little town on his way up the mountain in his special limousine. When the United States went to war in December 1941, the only armored motor car the U.S. government had that could be used for protection of the President was the one the FBI had seized from Al Capone, when the Chicago gangster went to jail in 1932 for income tax evasion. It was used by Franklin D. Roosevelt from December 1941 until a new armored limousine could be specially made for him in Detroit.[101] Longtime Thurmont resident William A. Wilhide was 16 in 1942, and remembers the Presidential motorcade. "He was escorted by the state police. It was not an open car, you couldn't see him. But I knew who it was," Wilhide recalled. "We knew the President was up there [on the mountain]."[102] Another old-timer, said that a busload of Marines was the tip off. "When the President was going to be up there," James H. Mackley said, "they would drop off guards under bridges going up the mountain road.... to guard them as the President went over the bridges. On one occasion, I was at the square corner... in the center of town. I saw them standing guard there. People in town knew when the President was coming here when they saw the bus coming and dropping off the Marine guards." Mackley recalled impishly that "after the war, Harry Truman went up there in a convertible—an open convertible. He slipped away from the Secret Service and went up there and drove around himself with the top down."[103] Catoctin area residents, however, kept information about the Presidential Retreat to themselves. When newspapers began to print rumors of its location in October 1943, one longtime resident blamed the publicity on outsiders and wrote to the Democratic President that "Republican Thurmont had been ... pleased and silent" about Shangri-La.[104]

Marines at Misty Mount

The basic security for the Presidential Retreat was provided by the U.S. Marine Corps. In the summer of 1942, Cabin Camp1 (Misty Mount), a WPA-built camp on a steep, boulder strewn slope on the east side of Catoctin Mountain, was taken over by the Marine Corps to guard Shangri-La. The Marine detachment at Catoctin did not simply patrol the area to protect the President. As active duty Marines, they engaged in drilling and training exercises when the President was not in residence.

The Marine camp at Misty Mount was nearly one and a half miles away through the woods away from the OSS camp (B-2) at Greentop and almost two and a half miles from the OSS's cadre headquarters (B-5) and the NPS maintenance facility at Round Meadow. There was little contact between the Marines and the OSS unit, each of which sought to maintain the secrecy of its operation. "We never had any interface with the Marines," said OSS demolitions instructor Frank Gleason, who was at B-2 most of the time from June 1942 to March 1943. "Their job was to look after the [Presidential] facilities there. That's all. The Marines never were up in our area. Never, while I was there, did we have any relationship with them."[105] Neither Gleason nor his superior in the Special Operations Branch, Captain Charles Parkin, who were instructors at B-2 from 1942 to early 1943, ever visited the Marine camp at Misty Mount.[106]

In contrast, while OSS Operational Groups, many of them with paratrooper officers, were training at Area B in 1943 and 1944, some of the more rambunctious OG members found and probed the Marine-guarded compound. "We shadowed Marine patrols during their training," recalled Otto N. Feher, from Cleveland, the son of Hungarian immigrants, and a member of one of the Yugoslavian Operational Groups. "They said if you ever get caught, get ready to get your ass kicked by the Marines."[107]

Albert R. Materazzi, a member of the first Italian OG, who was there in July 1943, said "We sought to infiltrate Shangri-La. I don't know whether it was a prank, boredom or just testing ourselves, or whether we just wanted to know whether we could find a weak spot where we could get over the fence. We never succeeded."[108] Other OGs jousted with the Marines. Roger Hall, who was briefly at Area B as an instructor that winter, wrote in a jocular, iconoclastic, if not always accurate, account of the OSS, that "trigger-happy Marines" were continually firing at OSS personnel "whenever we got within an extremely variable distance they called 'range,' and the fact that no one was sent home in a box bordered on the miraculous."[109] In reality, the situation was probably not have been as dangerous as Hall made it seem. "One of our training assignments was to get as close to the Presidential Retreat as we could without endangering lives," recalled former 1st Lieutenant Robert E. Carter an officer in the OSS German Operational Group. "We knew there were armed Marines there, so we didn't approach to close. Never close enough to raise an alarm or problem for ourselves. Never close enough to be fired on."[110]

But some groups apparently took a more reckless approach. "We tried to penetrate the Marine guards surrounding Shangri-La," recalled Caesar J. Civitella, a trainee in the second Italian OG, about a night infiltration exercise that took place in November or December 1943. "They were alerted. We'd try to get close. We'd get exposed. They fired blanks at us. It was not live ammunition. They must have been in on it."[111]

For Catoctin RDA Manager Mike Williams and the National Park Service, there were increased responsibilities in having the Presidential Retreat at the park. Preparation of the facility for the use of the Commander-in-Chief that summer had been a major job. Preparation for the winter brought added tasks. "This has been our first experience in attempting to operate the camps through the winter," Williams reported to Conrad Wirth in November 1942, "and as the plumbing and water systems were not designed or constructed for winter use there was considerable work connected with protecting them."[112] With Marines and OSS personnel at Catoctin all year round, there was also a significant increase in trash and sewage. Garbage and trash at a large garbage dump used by the two military outfits required periodic burning. While the Army had its own trucks, the Marines used NPS vehicles to haul sewage, which had to be taken four and three-quarters miles from camps in order to protect their water supply. Because of the war, gasoline was rationed and in short supply in America. In March 1943, Mike Williams had to warn the Marines that unless the Marines Corps provided gasoline for their own use as well as reimbursing the National Park Service the 169 gallons of fuel he had loaned them, the park facilities would run out of gasoline within 24 hours.[113] Presumably the Marines responded from military stockpiles as the park and the two military camps in it--both OSS and the Marine Corps—continued to operate.

End of OSS Use of Area B at Catoctin Mountain Park

As early as March 1943, the OSS Director of Training, citing a number of reasons, had recommended that Area B be abandoned. The creation of the Presidential Retreat had caused the restriction and sometimes complete stoppage of OSS training while the Chief Executive was in residence. The site had been deemed not suitable for trainees of Morale Operations Branch, which was preparing for new disinformation campaigns and who included women as well as men. There was only limited manpower within Schools and Training Branch to staff Area B. Finally, the Maryland state health department had reported that the sewage disposal system was inadequate and that a new system would have to be installed.[114] Although Donovan tentatively approved the decision to abandon Area B, it was not implemented at that time, because it was decided that the area could be useful in handling the overflow of Operational Groups from Training Area F, the former Congressional Country Club.[115]

There had been thoughts of closing OSS Area B as early as the winter of 1942-43, when OSS training began to be centralized,[116] but the facility was maintained through early 1944 before it was finally abandoned by Donovan's organization. The station complement, Detachment B, had remained throughout the period, 1942-1944, at Area B-5, but the training camp at B-2 was used sporadically during those years of operation. Area B played a major role in the initial training of SO and SI personnel in the spring and summer of 1942, and then again performed an important function in the preparation of Operational Groups during the summer of 1943 and the winter of 1943-1944.

Early in 1944 with the OSS Operational Groups and Special Operations teams already trained and in place for the invasion of France scheduled for that summer, Donovan's organization decided to end its use of Area B at Catoctin. OSS was shifting its training focus to the Far East and building up training camps on the West Coast. Thus, OSS would abandon, at least temporarily, the use of Training Areas B-2 ("Greentop") and B-5 (the abandoned CCC Camp MD-NP-3), as well as the approximately 2,000 acres in the northwesterly undeveloped portion of Catoctin RDA, which had been used for field exercises.[117] The plan was to discontinue the training camps at Area B as soon the last Operational Group completed its training. Schools and Training Branch had rejected the suggestion Capt. Montague Mead, successor to Ainsworth Blogg, and the final commanding officer of Detachment B, who had recommended additional improvements to Area B. Some OSS training of Operational Groups continued at Area B-2 until the end of April 1944.[118] By the end of May 1944, the camp was evacuated, except for a handful of fire guards composed of soldiers from nearby Camp Ritchie. The OSS instructors had left from B-2 and most of the cadre of Detachment B based at B-5 had been reassigned. In mid-May, headquarters notified Mead that all OSS training at Area B would be discontinued as of 1 June 1944. Mead was directed to close out the post exchange, transfer all OSS equipment and other property at Area B to other OSS training areas, and turn over use of the facilities to the commanding officer of nearby Camp Ritchie.[119]

Catoctin Park, NPS and Military Intelligence at Camp Ritchie

From Catoctin Mountain Park, both the OSS and the NPS had a sporadic and congenial relationship with Camp Ritchie located at Cascade, Maryland, only a couple of miles north of the park. In 1926, the State of Maryland had purchased extensive land for the camp for the Maryland National Guard and named it in honor of Governor Albert C. Ritchie. In the spring of 1942, the U.S. War Department took over the camp, and it served as the Army's Military Intelligence Training Center until it was deactivated in 1945.[120] When it became the military intelligence training school, Camp Ritchie disappeared from the public view. As with Catoctin RDA, an air of secrecy cloaked not only the activities going on at Camp Ritchie but all the Army personnel sent there. The military trainees at the Army's Military Intelligence School arrived individually or in small groups under classified orders. They were told not to reveal their identity to other trainees and not to disclose to anyone including their families that they were in military intelligence. More than 19,000 combat intelligence specialists graduated from the Army's Military Intelligence School at Camp Ritchie during the war and went to serve in the Army or Army Air Forces as interpreters, interrogators of prisoners of war, specialists in enemy units and equipment, and aerial photograph analysts. Many of them were sent on field exercises into the northwestern part of Catoctin Mountain Park.[121]

In April 1942, at the same time that the OSS took over Catoctin RDA, the commanding officer of the Military Intelligence Training Center and Camp Ritchie, Colonel Charles Y. Banfill,[122] requested permission from the National Park Service to use the northern portion of Catoctin RDA for field exercises under the permit granted to the War Department by the Department of the Interior. Banfill informed Park Manager Mike Williams that the operations would include short field maneuvers, the construction of gun emplacements, slit trenches, camouflaged installations, as well as the placing of booby traps and barbed wire entanglements. They would involve cross-country patrols day and night, the movement of cavalry patrols and horse drawn artillery. Tanks and artillery would be simulated by light vehicles and even plywood mockups. There would be no firing of guns or use of high explosives, but there would be pyrotechnics and smoke, which would be very closely supervised to insure maximum safety. Booby traps would be electrically detonated by engineer troops and removed immediately after the field exercise. No booby trap would be left unguarded at any time. After each operation, the area would be cleaned up to the satisfaction of the Park Manager. Given such assurances, Williams recommended the permit be issued as the Center's activity was similar to that being conducted by the OSS. The permit was issued in May 1942.

Mike Williams had visited the training exercises on several nights to ensure that they complied with the permit and did not damage the park property. He reported that the vehicles used were mainly lightweight, both motorized and horse drawn, and that they were confined to existing roads in the area. After observing the night operations, he concluded that the park would not suffer serious damage. NPS policy during the war to allow short-term military use of the national parks for bivouacking and maneuvers as long as it did not cause serious damage. During the summer of 1943, the Military Intelligence Training Center conducted field exercises and problems about three times a week, mainly at night in the northwesterly 1,800 acres of the park, an area west of the Foxville-Deerfield Valley Road, that had not been developed by the RDA. On average, these involved some 300 soldiers and 50 vehicles. So scrupulous were the soldiers from Camp Ritchie in protecting the park environment, that at the end of the summer, Williams wrote personally to Colonel Banfill congratulating him and his men for taking all precautions to preserve the natural growth in the area and for having removed all the debris resulting from the maneuvers.[123]

The Military Intelligence Training Center at Camp Ritchie used the northwestern area of the park again in the summer of 1943, but NPS had to cancel the permit in October 1943 because the Capt. Montague Mead, the commanding officer of OSS Training Detachment B, had informed Mike Williams that OSS use of Area B would be increased dramatically.[124] This was due to the expansion involved the massive preparation of Operational Groups (OGs) for use in the Allied invasion of France in 1944.[125] For those purposes, Training Area B had to use all its facilities. It was packed to overflowing (260 men when the maximum capacity was 251) with OG trainees and holdees; they occupied both Area B-2 and B-5 and conducted extensive field maneuvers in various areas of the park.[126]

The Operational Groups—French, Italian, Norwegian, Yugoslav, and Greek—were trained at Areas F and B and then sent to Europe in the summer of 1943 and the winter of 1943-1944, and the OSS role in Catoctin Mountain Park came to a close in mid-1944. Although the OSS was leaving, the War Department had other uses for the property at Catoctin Mountain Park. It intended to turn its area of the park over to the commanding general at the Military Intelligence Training Center at Camp Ritchie.[127] After acknowledging the May 1944 notice from Secretary of War Henry Stimson that the OSS no longer needed the facilities at Catoctin, Acting Secretary of the Interior Abe Fortas issued a new permit effective 1 June 1944. This permit allowed the use "for war purposes" of Cabin Camp 1 and Cabin Camp 2 as well as the old CCC camp, plus 1,800 acres in the northwest undeveloped portion of the park, which had been used for field exercises by the OSS and by the trainees from Camp Ritchie. The idea was that the intelligence and counter-intelligence trainees of the Army and Army Air Forces from currently overcrowded Camp Ritchie would be housed in what had been OSS Areas B-2 (Greentop) and B-5 (the old CCC camp) and they would continue to hold field exercises in the northwest quadrant of the park.[128] Banfill, now a brigadier general, would as the commandant at Camp Ritchie, replace the OSS commander of Area B as the officer responsible for War Department usage of the formerly OSS area of the park.[129]

The National Park Service took the opportunity of a change in military jurisdiction within the park to implement some of the lessons it had learned during the previous two years of military use of the park. The new permit issued in May 1944 put some restrictions on the use of weapons in the park and also allocated some of the additional costs to the military.[130] The Army's plan for using the park to reduce the overcrowding at Camp Ritchie was eventually deemed unnecessary by General Banfill. Consequently the trainees and instructors from Camp Ritchie did not take up residence at Areas B-2 or B-5. The Marines, however, did continue to reside in Cabin Camp 1 (Misty Mount).

The Marine Corps suffered extremely heavy casualties in the capture of Japanese-fortified islands of Iwo Jima and Okinawa in 1945, and in July, the Corps requested the use of Cabin Camp 2 (Greentop) for the "physical rehabilitation" of its troops returning from those battles in the Pacific. The overcrowding at Camp Ritchie had ended and the Military Intelligence trainees had never taken up residence in Cabin Camp 2.[131] Because Camp 2 was available, the Marines would, in fact, move from Cabin Camp 2 up into it in January 1946. In fact, Greentop, OSS Training Area B-2, which had been vacated by the OSS in the spring of 1944, would be occupied by the Marine Corps from January 1946 until March 1947.[132]

CHAPTER 5

TRANSFORMING PRINCE WILLIAM FOREST PARK INTO MILITARY CAMPS

The hilly, forested lands in the large northern part of Prince William Forest Park near Quantico, Virginia, became the site of training camps for the OSS Special Operations and the Communications Branches in 1942. In a 5,000 acre section of the park, Donovan's organization set up Training Area A for advanced paramilitary work in sabotage and guerrilla activity and other aspects of unconventional warfare behind enemy lines. Additional acreage in the northern part of the park was subsequently designated Training Area C and was assigned as the radio operators school. Within Areas A and C, the OSS converted the five National Park Service summer cabin camps and one of the Civilian Conservation Corps work camps into sub-camps capable of training several hundred men. The U.S. Marine Corps base at Quantico was located adjacent to the park, and the Marines used the acreage in the southern part of the park, where there were no summer cabin camps, for their own type of field training. The wartime military facilities in the 14,000 acre park played an important part in OSS and Marine Corps training. They also led to important changes at the park.

Area A (Advanced Special Operations School)

Three officers from Donovan's headquarters were the first to arrive at the park, which was then known at Chopawamsic Recreational Demonstration Area. On the last day of March 1942, they drove down U.S. Route 1 about 35 miles south of Washington to meet Park Manager Ira Lykes and to inspect the facilities. Lieutenant Colonel Garland H. Williams, executive officer of the Special Operations Branch, the man in charge of SO training, headed the group. Accompanying him were Captain William J. Hixson, an engineer, who would become the first commanding officer at Training Area A, and Lieutenant Rex Applegate, a military police officer soon to become a noted instructor in combat pistol shooting, first at Area B in Catoctin and later at other OSS stateside training camps. Accompanied by Lykes, the three officers inspected Cabin Camps 2 (Mawavi) and 5 (Happy Land) in the western section of the park. Williams told Lykes that the organization would begin occupation the next day, April 1, 1942. As commanding officer of Detachment A, Hixson would establish his headquarters in Cabin Camp 5, which was renamed Area A-5.¹

Williams emphasized that as a military facility, the park would be closed to the public. The War Department planned to use the large part of the park that lay north of Joplin Road (today County Route 619). That road, which ran from the towns of Quantico and Triangle northwest toward Manassas, could remain open, but between it and Dumfries Road (today Route 234) the secret training camp would be guarded from intruders by sentries and military police patrolling the perimeter.² Lykes later recalled that everything the OSS did was covered with the "greatest secrecy." Even Lykes had to pass through military checkpoints to enter and leave the park.³ Guard houses were built at the two main gates, the one for Area A on Joplin Road and the one for Area C on Dumfries Road. In addition, as sub-camps were created in other cabin camps, they often had guard houses at their entrance.⁴ The area was patrolled by mounted soldiers and armed guards with dogs, and barbed wire fences may have been erected in some sectors of the park, perhaps toward Batestown.⁵ In converting the facilities to military use, the colonel told the park manager that the War Department would provide the needed workers—carpenters, plumbers, electricians— to recondition the camp, and that in accordance with the use permit, Lykes would be consulted about the work.

Within two weeks after that visit, the Army Engineers had come up with a construction plan for altering Chopawamsic to the needs of Donovan's organization. The Civilian Conservation Corps and its youthful workers would do much of the construction as a national defense project. The National Park Service concurred and recommended that a CCC unit in the park which had been scheduled for termination, be maintained and assigned the job. Its personnel had been working on projects in the park for some time and were quite familiar with the area and local resources.⁶ The Army's new projects included construction of roads and trails, extension and correction of sanitary systems and water supply systems, grading and landscaping for unidentified military purposes, obliteration of some buildings and roads, and other incidental work necessary for the maximum utilization of the area by the military.⁷ "It is estimated," Lykes' supervisor wrote in late April 1942, "that it will take a full 200-man company [of the CCC] approximately 72,000 man-days to complete the work outlined," in other words, nearly a year.⁸ The CCC, however, was terminated by Congress in 1942, and subsequently, most of the construction work done for OSS in Areas A and C of Prince William Forest Park was performed by the larger workforce of a private firm, Kent and Company of Richmond, Virginia.⁹

Simultaneously, the War Department moved quickly to acquire land within the park by the OSS training area that was still in private hands. A representative of the Corps of Engineers had arrived at Chopawamsic at the end of March with copies of the Army's property acquisition plan. Ira Lykes spent the morning with him going over the acquisition status for the private properties. The park manager called in John T. Gum, the NPS representative and acting manager for all the CCC facilities in the Chopawamsic RDA, both those still existent and the CCC work camps that had already been abandoned.¹⁰ Gum lived in the nearby town of Dumfries with his wife and three children.[11] Having worked as liaison with CCC and the Resettlement Administration for several years before Lykes arrived, Gum had an intimate knowledge of the land owners and land prices in the Chopawamsic RDA, and he shared that personal information with the Army's representative. Because the OSS project remained secret, Lykes did not at that time inform Gum about the specific nature of the planned military use of the park. But John Gum, in his dual capacity of representative of CCC and NPS, would soon learn the details. In fact, Gum would supervise most if not all of the construction work performed by the CCC work crews for the OSS. Subsequently, John T. Gum was recruited into the OSS and given the rank of Army Technical Sergeant. He remained assigned to OSS Detachment A until it was closed down and was held in stand-by status from November 1944 through July 1945. At that point, Gum was transferred to OSS Area C in the eastern part of the park.[12] In the meeting of 30 March 1942, with Lykes and Gum, the representative of the Army Engineers told them that he planned to give full attention to acquiring the privately held tracts of land located within the area.[13] Lykes had emphasized to the importance of including Batestown in the Army's acquisition plans. "This so-called community is a hot-bed of bootleggers, thieves and vice," Lykes said he told the Army "and is likewise a potential health hazard to the adjoining Camps 1 & 4 and the Quantico Creek Watershed."[14]

Although the Roosevelt Administration had insisted that the people in the Chopawamsic RDA needed the help of the U.S. government to escape poverty, many of the local people did not consider themselves to be poor and did not want to be forced out of their homes. These people, whites and blacks, resided in small houses scattered throughout the wooded area or in a few tiny settlements. One of them, Batestown, a scattering of houses, a country store and a church, just east of the park boundary, was a largely black community. It had been founded after the Civil War by ex-slaves and free blacks who were members of the family of Betsy Bates, the matriarch. By the 1930s, Batestown had a population of around a hundred persons.[15]

Under the New Deal programs, the Department of the Interior had obtained title through purchase or condemnation of most of the lands within the new park. The developing agencies had been gradually moving the residents from these lands out of the park. But U.S. entry into the war and the designation of Chopawamsic RDA as a military facility dramatically altered the situation. With the secret training camp going into operation in April 1942, the War Department moved swiftly to remove remaining residents and acquire additional in-park tracts. Residents within the park area who had not moved were now quickly evicted. Some of them moved just outside the park; others moved farther away. Many of the longtime residents did not want to sell their land, and the government's negotiations were often acrimonious.[16] The village of Hickory Ridge with its houses, church and country store, was abandoned, and during the war, its empty buildings were destroyed in OSS field exercises including demolition practice. Today the site of Hickory Ridge village is marked only by the Virginia pine trees near the park's Parking Lot D and Pyrite Mine Road, and by the nearby small cemeteries containing the remains of generations of Hickory Ridge residents who died before the park was created.

The War Department did not purchase the Batesville land outside the park, but incursion into the eastern sector of the park was impeded by the security that the military established. Batestown continues to the present day, although in recent years, the park has acquired some of its properties. In contrast, in the southern part of the park, the village of Joplin, which consisted of a number of houses, a grocery store, school, and a church at Joplin Road around the intersection with Breckenridge Road, no longer exists.[17]

When the United States entered the war, Chopawamsic Recreational Demonstration Area included 14,466 acres, both north and south of Joplin Road, but the government did not yet have complete ownership or possession of all that property.[18] There were leases, options to purchase, and deeds not yet conveyed. Many residents were still in their homes within what would become the military training camp. When Donovan's organization took over the park north of Joplin Road, however, the process of acquisition speeded up. During the war, indeed mostly by the end of 1942, the War Department purchased 44 additional tracts of land within the park north of Joplin Road. They comprised a total of 1,139 acres. The price paid was $54,890, an average of about $48 an acre. They were privately-owned tracts that were acquired for security and demolition purposes. Two munitions storage magazines and three demolition areas were later constructed on them. After the war, in accordance with the usage permit, these tracts, plus the CCC camp lands, were turned over to the Department of the Interior and made part of the Prince William Forest Park.[19]

Transformation at Area A

Even as workers began to convert the northern part of the park into a Special Operations school, the first advanced training class began in the early summer of 1942. Members of one of the initial classes at the Basic Special Operations course at Catoctin arrived for the Advanced Course at Area A in Chopawamsic RDA the second week in May. Private Edgar Prichard, a journalist from Tulsa, Oklahoma, was one of them. He had joined the Army and become bored as a company clerk before being recruited by Donovan's organization. "We arrived there in later afternoon, but it was still broad daylight," Prichard, now a Virginia lawyer, recalled fifty years later. "The camp we were taken to appeared to have been a scout camp. There was a cluster of cabins beside a lake and a Mess Hall. The cabins had no window sashes, just screened openings, but there were Army cots in place. We were told to pick out a place in one of the cabins. I walked into a cabin which hadn't been inhabited for a while and a big rat ran across the floor. I was carrying a .38 [caliber pistol] in a shoulder holster. I fired at the rat. I think I wounded him. Anyway, he took off squealing. And about that time, the Captain I had just met came storming down the hill and confiscated my .38."[20]

That facility was either Area A-2 or A-5, the first two training camps established by the OSS at Area A. The captain, who confiscated the pistol from Pritchard, may have the commanding officer, Captain Hixson, or the chief instructor, George H.

White. Like Garland H. Williams, White was a veteran agent in the Federal Bureau of Narcotics [FBN], who had been recruited by Donovan's organization. In March, White had been sent Canada to attend the British SOE school at Camp X. After graduating from what he called the "Oshawa School of Mayhem and Murder," White helped teach the first Basic Special Operations class at Area B-2 in Catoctin in April He was then assigned as chief instructor at A-5 at Chopawamsic RDA.[21] White was an inspiring instructor, but his countenance belied his lethal skills. Five feet, seven inches tall and weighing 179 pounds, the roly-poly agent reminded his boss of a Buddha statue. But behind that innocent, round face with the disarming smile, the short, plump White was, as one OSS official recalled, "the most deadly and dedicated public servant, I have ever met."[22] By late summer 1942, White had transferred to OSS counter-intelligence first as an instructor and subsequently in the field overseas, achieving considerably notoriety, in one case for personally identifying and assassinating the head of Japanese espionage system in Calcutta, India. In 1943, the former narcotics agent collaborated with Stanley Lovell, OSS director of Research and Development, in an unsuccessful search for a "truth drug" that would lead enemy agents or prisoners to disclose secret information. White was, according to one account, "probably the most dynamic, flamboyant and prolific agent in the FBN's history."[23]

From Prichard's description of his cabin, the work on preparing the military camps had just begun. His hut had not yet been fitted with window sashes nor had it been winterized as it would be later, but the Army cots had been delivered. During the war, the OSS did not remove or destroy any of the buildings in the NPS summer cabin camps, but it did make significant changes to some of them, and it erected additional structures and training facilities related to its military mission. Most importantly on a permanent basis, key buildings in all the cabin camps in the park were made suitable for operation during winter. The NPS camps had been designed for occupancy only during the summer months. The OSS training camps would operate year round. The occupants found them cold and damp in the winter months. To make them habitable at that time, the OSS installed pot-bellied, cast-iron Franklin stoves, and hot water tanks first and subsequently put in insulating wall board inside and creosoting outside many of the buildings. The pit-latrines were augmented or replaced by Army latrine/washhouses often with hot water showers. The kitchens in all of the mess halls were upgraded with new wiring and the addition of an array of new electrical equipment for food preparation and storage, including standard Army gas ranges, electric potato peelers, deep fryers, and dishwashers, plus large refrigerators.[24]

In general, in addition to winterizing and upgrading the NPS summer camps at Prince William Forest Park, the OSS made a number of changes it deemed necessary for the operation of the areas as military training camps. As on any military base, that meant the allocation of more individual accommodations for officers than enlisted men on the staff (in contrast, the trainees were all treated alike regardless of their rank, which was unknown because the students dressed alike and had fictitious names). It also meant the creation of post exchanges, or PX, that is shops on a military post selling items of food, clothing and sundries, guard houses and sentry boxes, armories for weapons and ammunition, and magazines for storing explosives. It meant the creation of firing ranges, indoor or outdoor, for various types of weapons: pistols, rifles, submachine guns, which were a favorite among OSS Special Forces and Operational Groups, as well as areas to practice throwing hand grenades. There were demolition areas for instruction in the use of various types of explosives and fuses for blowing up samples of wood, steel, or iron, to simulate bridges, buildings, or railroad tracks. Extensive areas of woods and fields were designated for day and night exercises in map reading and deployment and for practice in field craft, that is the ability to operate stealthily and effectively regardless of weather or terrain, to "see without being seen," either for attack or for evasion or escape from the enemy.

Detachment A was ultimately divided among four sub-areas in the park. Three of them, Areas A-2, A-3, and A-5, had been NPS cabin camps previously used for summer camping by various charitable or social service youth groups. They had a common pattern. Each camp consisted of several circles of huts, each cluster grouped around a recreation all. In the center of each camp, for use by the occupants of the various clusters, was a large central dining hall as well as some smaller structures: an administrative office, arts and crafts building, and an infirmary. Unlike these three former summer camp areas, OSS Area A-4 had been a year-round work camp for the Civilian Conservation Corps. With its already winterized facilities for a headquarters, barracks, motor pool, workshops, and communications building, the former CCC work camp fit the OSS needs for an administrative headquarters, supply facilities and motor pool for the entire military training camp at Area A.[25] Area A-4 served primarily as the headquarters of Detachment A, but because of the CCC barracks, it also served as a training area. When the military conversion of the entire Area A was complete, each of the sub-camps had its own small headquarters. At each sub-camp, an executive officer was responsible for the administrative staff and the running of the facility; a chief instructor had authority over the instructional staff, the curriculum and the training of the students. The entire Area A was the responsibility of the central headquarters of Detachment A located at A-4. That former CCC work camp housed the commanding officer and his staff as well as the motor pool and other facilities for furnishing supplies and services to all the sub-camps. It was also the location for the chief instructor for Area A and his assistants. The total staff at all four sub-camps plus Detachment A headquarters numbered approximately 280 officers and enlisted men. At capacity, Area A could hold some 600 trainees, giving a total of nearly 900 personnel that could be handled by OSS Training Area A.[26]

Area A-5: First Headquarters, and SO Waterborne Infiltration School

As the first commander of Area A, Captain Walter J. Hixson had indicated to Ira Lykes on 30 March 1942 that he intended to begin by occupying Cabin Camp 5 in the western end of the park, which he did within the next few days.[27] Hixson set up the headquarters for OSS Detachment A there until the main headquarters was moved to the old CCC camp at what became A-4 in November 1942. Before the war, the NPS summer camp, known as "Happy Land," had been used by needy children from Washington, D.C., brought to the park and supervised by the Salvation Army. Now it was converted to much different purposes. Beginning in mid-May 1942, the first class of trainees for Advanced Special Operations, arrived to learn more about sabotage, guerrilla leadership, armed and unarmed close combat technique, including silent killing, and other aspects of unconventional warfare.[28] These first groups of trainees were housed at A-5 and A-2. Later, between 1943 and 1944, Area A-5 served as a "finishing school," basically a holding area for OSS personnel finished with their training and awaiting assignment overseas. The purpose of the finishing school was largely to keep the graduates in top physical and mental condition. Consequently, much of the activity was in physical exercise, weapons use, and field exercises.[29]

Smaller than many of the camps, Cabin Camp 5 included only two subunits, A and B, each a cluster of cabins and a few other buildings. Depending on their size, the cabins slept four or eight trainees. OSS used two buildings as classrooms, one capable of holding 70 men for lectures or training films, the other accommodating 40 students for Morse code practice, map exercises, or other interactive training. An outdoor demonstration area seated 50. Later more classroom space was added through temporary buildings. Existing structures were converted to create officers' quarters, a cabin for staff enlisted men, an armory, and a combination recreation hall and post exchange. Beginning in the winter of 1943-1944, key buildings in A-5 were winterized.[30] Area A-5 had a capacity to hold 8 officers and 10 enlisted men in the station complement, and a maximum of 112 trainees in double-decked bunks, a total capacity of 130 men.[31] Normal housing, however, was for about 60 rather than 112 trainees.[32]

A number of military facilities were constructed near Area A-5. A 275-yard firing range with 20 silhouette targets of the advanced, pop-up type was constructed for practice with rifles and submachine guns, the latter a favorite with the OSS. Farther away were demolition ranges for testing explosives. For outdoor and indoor pistol practice, students from A-5 had access to the pistol house and outdoor firing ranges erected at A-2 only half a mile away. To the east of A-5, an extensive area was set aside for practical exercises in map reading and in field craft.[33]

For maritime use, a small boat house and dock were constructed near Area A-5 on a lake on the South Fork of Quantico Creek. Arthur F. ("Art") Reinhardt, whose group of radio operator trainees was kept briefly at Area A in 1944 until space became available for them at the Communications School at Area C, recalled that there was also a pylon type structure with rope ladders. "The idea was to practice jumping off— supposedly in full battle gear—and tread water in case, if you went by ship, it was torpedoed, you would be able to tread water and survive."[34] For the Special Operations trainees, the lake was used for practice in clandestine seaborne landings or river crossings.[35] It was the initial site for instruction in OSS waterborne operations before responsibility was transferred from Special Operations to the new Maritime Unit (MU) and instruction was moved to Area D on the Potomac River. So the lake near

A-5 served a variety of purposes including amphibious and aquatic training in numerous exercises from evacuating from sinking ships to clandestine waterborne landings in kayaks or rubber rafts.[36] It may also have served for recreation for the OSS personnel if the photos taken of Technical Sergeant Louis J. ("Luz") Gonzalez and his fiends horsing around at the lakeside dock in July 1942 are indicative.[37]

Area A-2: First SO Training School

One of the first two camps opened by the OSS in Area A was sub-camp A-2, located only half a mile from A-5 where Hixson had set up his headquarters. Formerly NPS Cabin Camp 2, known as Mawavi, it had been used by the Girl Scouts before the OSS took it over and employed it first for an Advanced Special Operations course. Subsequently, A-2 was used for a variety of wartime purposes: a holding area for incoming personnel awaiting security clearance, a training area for Operational Groups in 1943-1944, and a holding area for already trained OG or SO personnel awaiting further assignment. Beginning in the spring of 1944, it offered basic military training for OSS personnel being sent overseas who had for some reason not already undergone such training.[38] The OSS retained all the NPS buildings in Cabin Camp 2. It was one of the larger camps and included four subunits: cabin camps A, B, C, and D, each clustered around a recreational hall. The OSS put four to eight enlisted men or trainees in each cabin and 12 men in each of three recreational halls. In the center of A-2 was a dinning hall, infirmary, a central shower building and an office. In addition to upgrading these, OSS converted two other buildings into classrooms for lectures and code work. For military purposes, the OSS built a small guard house, an armory and an ammunition storage building. It also erected a "pistol house," actually a "mystery house" or "house of horrors" like the one created at Area B-2 at Catoctin under the direction of William ("Dan") Fairbairn.. Outside in a neighboring area of A-2, the Army Engineers constructed a pistol range, a submachine gun range, and a jogging-type obstacle course that extended 150 yards. A map-reading and field craft exercise area were nearby.[39]

Beginning in late fall 1943, the cabins and other accommodations at A-2 were winterized because of the heavy usage by the Operational Groups. In the spring of the following year, the basic training course had filled the camp to such an overcapacity that a new sewer field had to be installed.[40] Area A-2 was able to hold 44 officers, a large number probably because for the first six months, it served as the headquarters for Detachment A. In addition to the staff officers, it had the ability to accommodate 22 enlisted men in the station complement, and in an emergency, 206 trainees or holdees, in which case many of them were bunked in the mess hall, a total capacity of 274 men.[41] Under normal conditions, however, A-2 held a much smaller staff and housed 78 students or 98 holdees.[42]

Area A-3: Parachuting and Training

In the southeastern corner of the park, Cabin Camp 3 became OSS Area A-3. The NPS camp known first as Goodwill and later as Orenda had been used before the war by several different social service organizations. When Donovan's organization first took it over, A-3 became the OSS Parachute School. Lieutenant Colonel Garland H. Williams had planned that after basic and advanced Special Operations training, students would then learn infiltration by sea or by air. The lake and then Area D were where waterborne infiltration was taught. Most of the OSS Parachute training was provided at the U.S. Army's new Airborne School at Fort Benning, Georgia, but during a brief period in 1942 the OSS had a Parachute School in at Area A-2 in Prince William Forest Park. Captain Lucius O. Rucker, Jr., a dedicated career soldier and an avid paratrooper, from Jackson, Mississippi, was in charge. He brought a group of personnel qualified in parachute rigging and jumping from Fort Benning to the OSS camp in Virginia in June 1942 to establish the OSS Parachute School. The training equipment was primitive, but it did provide for landing training.[43] According to friends like SO instructor Captain Jerry Sage, "Ruck's" slow, southern drawl contrasted with the speed with which he volunteered to jump out of airplanes. He had done well over three dozen jumps by November 1942. "We always said he could jump out of a plane faster than he could talk," Sage recalled.[44] Rucker was often assisted in instruction by Sergeant John Swetish, who once set a record for the most jumps in twenty-four hours.[45]

Air travel was not common until after World War II, and in the early 1940s few people had even ridden in an airplane, let alone jump out of one. Parachuting, however, was the fastest and most efficient way to infiltrate behind enemy lines, and paratrooper training also gave agents a further sense of self-confidence, courage, and pride and a feeling of belonging to an important, elite group. At parachute school, whether at Area A or at Fort Benning, trainees spent the first two days at Parachute School doing ground exercises of body control—jumping from a platform and learning how to tumble, practicing in a suspended harness—and also how to control the open parachute upon landing. They also learned how to pack their parachute and containers and how to dispose of both afterward so they would not be discovered. Finally, trainees had to make a total of five jumps in order for certification and award of the winged parachute insignia.[46]

At the OSS Parachute School at A-3, candidates would board a plane at the nearby U.S. Marine Base at Quantico, fly low over Prince William Forest Park and parachute into clearings, simulating the manner in which agents and equipment would be airdropped into partisan areas behind enemy lines. They would walk back to the Marine base and jump again. Among SO instructors and trainees a tradition emerged by late 1942 of earning their winged paratrooper insignia by doing the ground preparation and all five required jumps in a single day.[47] In December 1942, OSS headquarters decided to transfer Rucker and the parachute school to North Africa, and, thereafter, if OSS trainees received parachute training in the United States, most of them did it at Fort Benning, Georgia. When the OSS went overseas, Rucker established and ran the OSS parachute training school in Algiers from

December 1942 through August 1944, after which he was sent to China to help prepare the first Chinese paratroopers being trained and led by the OSS, including former Area B instructor Charles M. Parkin, Jr. Rucker left behind in Algeria some inadequate financial records, which the head of OSS finances subsequently excused, noting "Lt. Col. Rucker with the other officers that worked with him are typical parachute men, quite casual about everything except jumping. This is to be expected."[48]

Beginning in October 1943, A-3 became the site for advanced training for members of Special Operations and Morale Operations (psychological warfare). An "E-type" course developed at Area E, the basic SI school north of Baltimore, the course was designed to give an overall idea of all the various tasks by OSS personnel in the field so they could better coordinate with others and even do the others' work if required. At the end of the two-week course, three days were allotted when students could be sent out of Area A on practical problems related to their specialization, mock exercises of espionage, sabotage, or the spreading of disinformation. There are reports that at various times before the OSS formal assessment program began in 1944, A-3 had been used for weeding out the unfit or unsuitable and determining those who would be sent on for advanced training for missions behind enemy lines.[49]

OSS took over the NPS Cabin Camp 3 pretty much as it was. In three subunits, A, B, and C, it had clusters of cabins around recreation halls, several of which were turned into classrooms holding 40 men. Enlisted men and trainees slept six to a cabin and 12 in each barrack. One building was retained as a combination recreational hall and post exchange; the dining hall and infirmary were maintained but improved, as were the central showers. One small building was converted to a code room, a larger structure was retained as the headquarters for A-3. Of course, like the others, this cabin camp had to be winterized with insulation and hot water as well as enhanced utilities, water and sewage system. That did not occur, however, until well into the winter of 1943-1944. Meanwhile, pot-bellied, cast-iron heating stoves were installed to provide warmth. At A-3, a 25-man latrine/washroom was the only new building constructed by the Army.[50] Outdoors, the Army Engineers created firing ranges for rifle and submachine gun practice and a Demolition Area for using explosives in sabotage training.[51] Area A-3 had a capacity to hold 6 officers and 20 enlisted men in the station complement, and 162 trainees, for a maximum capacity of 188 men.[52]

Area A-4: Permanent Headquarters and Basic SO and OG Training

Although sub-camp A-4 replaced A-5 as the headquarters for all of Detachment A in November 1942, it also sometimes served as a training camp for basic or advanced Special Operations and Operational Group training. One of the new trainees there in 1942 was Prince Serge Obolensky, a royal Russian émigré and New York socialite, who had been an officer in the Czarist Army and leader of White

Russian guerrillas against the Reds after the Revolution. After Pearl Harbor, Obolensky enlisted in the American Army, was commissioned a major and was soon ordered to report to Donovan's headquarters in Washington. Donovan assigned him to Col. M. Preston Goodfellow, head of Special Operations for training of commando units to jump behind enemy lines. This was part of Goodfellow's plan for a Strategic Services Command, an idea put forward in the fall of 1942, which was rejected by the Joint Chiefs of Staff but which ultimately was adopted in a smaller scale in the Operational Groups. Obolensky trained first at Area B.[53] Then as he later recalled, "I was sent immediately to a commando school in Virginia [Area A]. None of us had names. We were all given nicknames, which were printed on badges. Nobody knew who we were. My name was Sky. This was a physical toughener as well. We got up early and went to bed late, training in special tactics, shooting at sounds at night, guerrilla tactics, memory tests of small and configurations of places in total darkness, and especially demolitions work."[54] After graduation, Obolensky reported to Goodfellow's office to help write a proposed training schedule for the groups of foreign nationals and first generation American citizens into military units in the proposed Strategic Services Command. A variation of these would become the OSS Operational Groups (OGs).[55] Subsequently, as will be shown in a later chapter, Oblensky would use his skills and training effectively behind enemy lines in Italy and France.

Subsequently, a course was developed called the A-4 course because it was given over such a long period at that facility. It was, in fact, SO basic training, designed for men who were going to perform the rougher kinds of missions overseas. Many SO trainees went first to Area B-2 or sometimes Area E for their basic training and then to A-4, but many reversed that process and went first to A-4. The course lasted about three weeks and comprised such subjects as field craft, map reading, demolitions, weapons, Morse code, close combat and physical training. Students were taught how to make a few basic types of demolition charges adequate for the sabotage of industrial establishments, rail lines, and bridges. They were taught how to use small arms most effectively for rapid firing in daylight or at night. Practical problems assigned to the students at the end of the course included night map and compass problems, problems in reconnaissance, and problems in the placing of demolition charges on mock targets. A-4 courses were given not just at Area A, but also at various times at Areas B, D, F, and on the West Coast on Catalina Island.[56]

Because it had been a CCC work camp rather than an NPS summer cabin camp, A-4 in the southeastern part of the park had a rather tenuous relationship during the war with the National Park Service. It was located not far from the park headquarters, but the CCC camps did not become the property of the NPS until after the war. Consequently, OSS concluded that the National Park Service "had nothing to do with the buildings and equipment" there. Indeed, John T. Gum, who worked there as liaison between the Civilian Conservation Corps and NPS, served as the CCC liaison with OSS in the park until CCC was terminated in the winter of 1942-1943. Thereafter, Gum worked as liaison between the NPS liaison and the OSS. Later, Gum recalled that in the prewar period the National Park Service had no definite plans for using the CCC work camp site as a summer camp.[57]

Twenty-four buildings at the work camp of the Civilian Conservation Corps were taken over by the OSS. There, as at the NPS cabin camps, the use made of the structures by the two organizations was often similar. The CCC headquarters became the OSS Area A headquarters, five CCC barracks holding 80 young men each in double-deck bunks, remained barracks for the OSS enlisted men of the cadre or for trainees. The CCC supply warehouses, gas house, storage shed, and water tower continued to function in the same manner for their new tenant. Sometimes, however, buildings were put to new use: The CCC infirmary became the OSS Guard House, a CCC structure listed only as "Building," became the OSS Post Exchange (PX) and Recreation Hall. Two-thirds of one CCC barrack became a gymnasium complete with parallel bars, punching bags, rowing machines, and place for instruction and practice in knife fighting and unarmed, close combat. The dining hall was retained as the mess hall, but the kitchen was completely remodeled and upgraded.

Nearly a dozen new buildings were constructed for OSS and remained in Area A-4 at least until the end of the war. They included an auto and truck washing building, carpenter shop, commissary, armory, code room, storage shed, a small guard house and a sentry box. Most expensive was a ten-room, bachelor officers' quarters (BOQ), which cost $13,000. A-4 also had two large classrooms, each of which could accommodate 50 students, where the trainees heard lectures and watched films about field craft, weaponry, and demolitions, plus a smaller classroom where code work was taught and practiced.[58] Area A-4 had a capacity for 15 officers and 128 enlisted men as part of the cadre and instructional staff, plus the ability to hold 76 trainees if necessary, a total capacity of 219 men.[59]

For use in the military training courses at A-4, the Army Engineers built three indoor firing ranges, one each for pistols, .22 caliber rifles, and submachine guns. Outdoors in a neighboring area, they installed another pistol and submachine gun range. About a mile west-northwest of A-4, an M-1 rifle range with ten standard target frames was created. Not far away, in the same general area, engineers constructed two standard cinder block magazines for storing ammunition or up to 50 tons of explosives between them.[60] For physical training, the engineers erected an obstacle course as long as a football field and which included various difficult, optional objects to be overcome. Farther from the camp, there were practice areas for field craft and stalking the enemy, and a demolition area for practicing with various types of explosives on the kinds of wood, iron and steel fixtures used in bridges, railroads, and buildings.[61]

The capacity of Area A overall, including all four sub-camps when they all had been winterized, equipped with extra temporary classrooms, and completely operational in early spring 1944 was 550 men—trainees or holdees. This was in addition to the 130 men assigned as the station complement for Detachment A.[62]

Teaching Effective Use of Violence at Area A, including the "Mystery House"

Overall at Area A, a variety of firing ranges, demolition areas, pistol houses, and other facilities were installed by the military for instruction in the effective use of violence in pursuit of OSS objectives. A list of what was available in Area A and environs in the spring of 1944 began with ten target ranges for practicing with pistols, carbines, rifles, and submachine guns, including a 200-foot rifle range with ten, 6 by 6 frame sliding targets, a 400-yard mortar range, a range for the use of hand grenades, rifle-propelled grenades and M-1 rocket launchers. Also listed were two booby trap areas and tactical problem houses; four objective tactical areas, a proposed infiltration course, three map and reconnaissance areas, a marine dock at Quantico for use of amphibious tactics on the Potomac River, and fifteen thousand acres of all various types of terrain for tactical problems. Plus outside the area, the commanding officer of Area A reported, there were numerous civilian facilities available for exercises involving simulated capture or demolition—from the Manassas Power Dam and Control Tower to the bridge at Woodbridge, Virginia for the Richmond, Fredericksburg and Potomac Railroad.[63]

Demolition areas, where explosives were used against wooden, iron, or steel objects, mock bridges, towers, buildings, or railroads, were extremely hazardous areas. There were three of them, ranging from one to four acres, and two magazines for storing explosives and other munitions had been erected on lands acquired by the War Department in 1942 for that purpose.[64] Because of the danger from the use of explosives, the Army convinced Prince William County officials in April 1944 to close old Route 644 which ran through the park near the demolition areas and to abandon several sections of secondary roads in and around Chopawamsic RDA until six months after the end of the war.[65] A number of the OSS field exercises involving weapons and live ammunition and explosives took place in a large corner area of the park formed by old Route 644 and the Joplin Road.[66]

Within Area A, there were several specialized houses built for OSS. They included two or three "Problem Houses" and a "Mystery House." The Problem Houses involved field exercises. Trainees might be required to recognize and dismantle booby traps or making a nighttime rendezvous, or message drop, at a "safe" house, or they might assault it as part of an enemy outpost. What Fairbairn called the "Mystery House" at Area A, like the similar structure at Area B at Catoctin Mountain Park, was created for indoor weapons training and testing reaction time.[67] It contained sounds of gunfire and visual props such as pop-up targets, resembling armed enemies soldiers to test the trainee's nerves and skill in instinctive pistol firing as he proceeded, gun drawn, through the rooms and corridors of what many students called the "House of Horrors."

At Area A, like Area B, there was a Mystery House or House of Horrors. It was constructed by the CCC work crews under the supervision of the OSS and the NPS Park Manager Ira Lykes. Years later, Lykes, after a career of preserving the forest and its wildlife while fostering healthy recreation in nature, recalled, still with some concern, the brutal elements involved. "I remember we built a little place we called Little Tokyo[68] in which we had one building, and [they] had to train the men to get through the door of that thing and walk down a labyrinth in a blue light, a very, very dim light while they were being fired at [and were firing back] in order to protect themselves. They taught them such delicate things as when they burst into a room of German staff officers which one to shoot first. The one that stood up was the one that got shot first and the one that had the wild look and stare on his face got shot last because he hadn't collected himself yet and he didn't know what to do. This was the fine shadings in murder. Of course, we had to do it because we had to fight fire with fire and it was a really justified program."[69]

Normal ethics were put aside and the idea of a fair fight rejected in the OSS training program that emphasized the need for an agent behind enemy lines to use any effective means available to accomplish their mission and avoid capture. As an example of such training, two former OSS Special Operations officers, writing immediately after the war, informed the public with barely concealed relish about what a typical recruit would learn at an OSS training camp. "Here he would be taught to meet the enemy on the enemy's terms. He would park all idea of clean sportsmanship along with his name, and learn to fight barehanded, gutter-style, with all the dirty tactics of gutter fighting: a knee in the groin, a savage slash with the side of the hand across an Adam's apple, a jab at the eyes with fingers stiffly hooked in a tiger's claw."[70]

Destruction in the Park

During the OSS training exercises a number of buildings within the park were destroyed. Most were old structures—farmhouses, barns, outbuildings, sheds and shacks—which had been abandoned or whose owners had been evicted. They were destroyed in exercises involving booby-traps, explosives, or mortar shells. During a demolition exercise in March 1943, the OSS blew up an uninhabited, dilapidated old house in the western border of the park. But a few weeks later, they learned that the house and its two-acre site on County Route 619 were not owned by the government, but had remained the property of the Stark family, which complained about their destruction. As Captain Eliot N. Vestner, current chief instructor, reported to OSS headquarters, he and the then chief instructor, Captain Arden Dow, were operating on the basis of a map showing a training area that included the property and the unoccupied house. "We have been accustomed to maneuver all through this [park] land and have likewise demolished other houses in the immediate vicinity," Vestner reported. It was only on investigation, presumably by conferring with Park Manager Ira Lykes, that Vestner learned three months later that the two-acre plot was only under option to buy and not yet been purchased by the War Department. The National Park Service had been anxious to acquire the property for the park, and the government had originally offered $1,000 for the property, which they, but apparently not the owner, considered a fair price.[71] The entire Stark property, all ten acres, on Route 619 was eventually purchased for the park.

Building and Maintaining Roads

The National Park Service had established five cabin camps in Chopawamsic RDA, but the camps were not connected by any system of internal roads. They had been used by separate civilian, non-profit organizations, and the Park Service did not have the funds for the expensive work of constructing an interconnected road network through the forest. This meant that one had to drive around the periphery, nine miles over state roads, for example, to get from Cabin Camps Numbers 1 and 4, in the northeast section to Cabin Camp Number 3 near park headquarters in the southeastern sector of the park. In 1942, Superintendent Ira Lykes hoped that the "Army occupation" of the park would provide a "splendid opportunity" to build "at least temporary connecting roads between the organized camps, particularly camps one and four and the central road."[72] But Lykes saw little chance that this would happen, because during the first few months after the military had occupied the park in April 1942, he had run into major difficulties trying to get "the Army to stop unnecessary auto traffic," the volume of which Lykes believed was counter to sound conservation and indeed ultimately harmful to the preservation of the park.[73]

The effects of the heavy military vehicles on the roads within the OSS training area, plus the deterioration accelerated by the weather during the winter of 1943-44, led Corps of Engineers to agree that it was imperative that the roads in Training Area A be repaired. Furthermore, the engineers recommended that culverts be installed and other drainage be put into place before the fall of 1944 and even greater damage created in the following winter. The Army Engineers furnished the labor, but the material would have to be purchased. The commanding officers of both Areas A and C agreed that the first priority was for the roads leading to the OSS constructed magazine for storage of ammunition and explosives, as the dirt road had become almost impassible in the winter. Second priority went to repairing roads which connected existing camp areas. The third priority was for repairing roads classified as secondary roads within the training area and which were used mainly for training purposes and in connection with field exercises. The roads were in such poor condition that they were inflicting undue wear and tear upon military motor vehicles, which consequently were frequently in need of repair. OSS concluded that under the agreement between the War Department and the Department of the Interior, "we are required to maintain existing roads and, of course, maintain any roads we ourselves constructed."[74] By 1945, Superintendent Lykes could report that the Army did maintain "certain roads in good condition."[75]

Closure of Area A

Recruiting and training of paramilitary personnel in the United States for the European Theater of Operations ended in the summer of 1944, and consequently Area A was closed down as a training area in November 1944. Most of the OSS training in 1945 for the war against Japan was done in West Coast camps. Area A was declared officially closed effective 11 January 1945 with the exception of normal maintenance and safeguarding of property, and was held in stand-by status until it was permanently closed in July 1945. When the OSS personnel vacated Area A, responsibility for the facility was transferred to the commanding officer of Area C, Major Albert Jenkins.[76] Of course, the two NPS officials at Chopawamsic RDA, Ira Lykes, the park manager, and John T. Gum, the NPS liaison person, first with CCC and then with OSS, also maintained responsibilities for the preservation of Area A as well as for all of the park. Area C remained in operation until October 1945.

OSS Maritime Training (at Areas A and D)

Special Operations had initial responsibility for training OSS agents, especially from SO and SI, in techniques of clandestine infiltration by water—from the sea or across a lake or river into enemy territory. The original training for such waterborne landings began in May 1942 using a boat house on a lake near Area A-5 and sometimes a dock at the nearby Marine Base. About the same time, Special Operations had begun that same month to create a maritime training camp on the east bank of the Potomac River, a few miles from Quantico. It was subsequently designated OSS Training Area D.[77] The precise location of Area D is still the subject of debate among some veterans, but the CIA records indicate that it was at Smith's Point, in Charles County, Maryland.[78] That would place it just below what is today Purse State Park on Wade's Bay, a half a dozen miles down river from Quantico, and about three miles west of the village of Grayton, Maryland. OSS. The temporarily obtained property included a long stretch of river frontage and 1,400 acres behind it.[79] After a few yards of river bank, the area was covered with light woods interspersed with open fields. The camp site was about 50 feet from the river. Three or four miles away there were some scattered dwellings, including a tavern, sometimes frequented by the OSS staff.[80] Temporary prefabricated buildings, possibly CCC-type portable structures, had been secured from the Army and used for housing the students and the equipment.[81] When expansion of Area D was completed by the summer of 1943, and by that time was used mainly by Special Operations, it had a station complement of 75 officers and enlisted men and could house an additional 180 trainees, for a total capacity of 256 OSS personnel.[82]

The initial development of the maritime infiltration school at Area D was by OSS personnel from the U.S. Navy under the supervision of a commando experienced Royal Navy officer, Lieutenant Commander H. G. Woolley of British Combined Operations. Among the American naval officers assisting him were Lieutenant Jack Duncan, who had trained with the U.S. Marines' Raider Battalion and with the British commandos in England, and Lieutenant John ("Jack") Shaheen, later chief of special projects for Donovan, and who the following year would run a small boat mission, the McGregor mission, to obtain information about Nazi weapons development from Italian naval officers. Instruction in clandestine waterborne landings had begun at Area D for SO and SI agents in August 1942. In October, OSS planned to expand use of the area to include not just for small craft amphibious landings but full blown field exercises, including airborne infiltration by parachute drops and small planes from the Marines' airstrip at Quantico.[83] Army Lieutenant Jerry Sage, an instructor from Area B-2 who had just returned from SOE training in Britain, arrived in the late aumn of 1942.[84] He became executive officer for Major Albert B. Seitz, commanding the SO training exercises. (Within a year, Sage would be trying to escape from a German prison camp after having been captured in North

Africa, and Seitz would be parachuting into Yugoslavia as a member of the first U.S. military mission to the Chetnik partisans fighting under General Draza Mihailovich.)

In the autumn of 1942 at Area D, however, Sage's assignment was to build an earthen landing strip at the camp in woods to experiment with short takeoffs and landings by light planes. Obtaining a bulldozer from the Corps of Engineers, Sage used it to tear up trees and stumps and to scrape an even landing surface. The first test landing went off perfectly, but when the pilot tried to take off, the unencumbered strip was not quite long enough. Suddenly confronted with the camp flagpole, the pilot banked sharply, lost control of the plane and crashed, fortunately without sustaining serious injury.[85] Another near fatal mishap occurred during a major field exercise in November, Sage was leading the simulated attacking force coming down a road in Army trucks. Lieutenant Frank Gleason, Corps of Engineers, a fellow instructor from B-2, was leading the simulated guerrilla force setting up an ambush including explosives. Someone else had set a booby trap with a wire, pull switch, and two blocks of TNT, about half a pound of explosive. When Gleason pulled the pin, the trip wire had been wound too tight and instead of triggering the time fuse, it set off the explosive, which was only four or five feet away from him. "I almost bought the farm that day," he recalled. "Fortunately I had a helmet and heavy back pack on. The blast knocked me off my feet and onto the ground. When I came to, I couldn't hear. I couldn't hear for a week."[86]

In one of the waterborne exercises in the Potomac River off Area D in 1942, one of the trainees accidentally drowned. earlier that year. Seaman John Pitts Spence, an instructor who became the Navy's first frogman, remembered that during the autumn while the weather was still mild, the men trained in underwater demolition and the use of small rubber boats which were launched from a converted yacht, the Marsyl. It was during one of those exercises that one of the trainees in underwater demolition apparently became disoriented in the murky waters or was trapped in debris, was carried downstream by the current and drowned.[87]

Underwater swimmers from Italy had been sinking British ships, and Donovan's organization decided to initiate a group of underwater warfare swimmers (later called combat swimmers or "frogmen). Spence, was one of the first chosen. Beginning in November 1942, he tested a new self-contained underwater breathing device that re-circulated the oxygen and left no telltale bubbles on the surface.[88] This was the Amphibian Breathing Apparatus or Lambertsen Respiratory Unit, named after Christian J. Lambertsen, a physician, who invented it in 1940. The underwater combat swimming groups began training in May 1943 at Annapolis. Meanwhile, in January 1943, OSS designated maritime activities the Marine Section of Special Operations Branch, and in March 1943, that section assumed complete control of Area D. The emphasis was still on waterborne infiltration rather than underwater sabotage, but the combat swimming groups did proceed from Annapolis to Area D for training in small boats and the use of limpet explosive devices for use against ship's hulls. The Marine Unit (MU) of OSS was established as a separate branch on 9 June 1943, and it did emphasize the use of "frogmen" for underwater demolition work. Polluted water in the Potomac and ice in winter prevented effective underwater combat swimming training, and as a result, training for the "frogmen" was shifted to Silver Springs, Florida, and in 1944 to Nassau in the British Bahamas as well as Catalina Island and the Camp Pendleton Marine Base near San Diego, California. Dr. Lambertsen transferred from the Army Medical Corps to the OSS Maritime Unit in 1943 and remained with it until the end of the war.[89]

When the Maritime Branch was created in June 1943 and vacated its training camp on the Potomac for bases in the Caribbean and on the Pacific Coast, Area D, still under control of Special Operations Branch became available as an adjunct to SO's Area A. In October 1943, for example, the overflow of students at A-4 basic SO training was briefly sent to Area D for Special Operations training there. Area D also served as a holding area for already trained SO and OG personnel awaiting assignment overseas. The SO role continued until April 1944, when Area D was closed, its staff moved to Area A-5 at Prince William Forest Park, and the property in Charles County, Maryland, returned to its owner.[90]

Area C: the Communications School

In the hilly, wooded northeastern corner of Prince William Forest Park, Donovan's organization established Area C as a site for training in clandestine radio operations and equipment, probably in June 1942.[91] At that time, communications was part of the Special Operations Branch, and the initial purpose of Area C, then called the SO-SI Communications School primarily to train agent-operators for the Special Operations and Secret Intelligence Branches. They were taught a smattering of International Morse code, secret cipher and radio techniques and equipment operation and repair.[92] It was initially a very small facility, part of one of the two National Park Service summer cabin camps there.

To create a secure, rapid, global communications system connecting agents in the field with regional bases and ultimately with the OSS message center in Washington, Donovan established a new and separate Communications Branch (CB) of OSS on 22 September 1942. It replaced the separate communications systems that SO and SI had been creating, at least at the higher, central level.[93] The Communications Branch was headed by Lawrence ("Larry") Wise Lowman, longtime vice President in charge of operations at the Columbia Broadcasting System in New York, who had been recruited despite the opposition of CBS chief William S. Paley. Lowman soon became one of the few full colonels in the OSS.[94]

The Communications Branch combined all previous OSS signal and traffic facilities, was charged with developing a global communications network, and most relevant to this study, in October 1942, it was given responsibility for all communications training in OSS. The CB or "Commo" Branch, as it was informally called, had responsibility for recruiting and training staff and field personnel, both civilian and military, and providing instructors for communications courses in all the OSS training camps, not just its own communications school at Area C but at the other camps as well.[95] Despite an attempt in November 1942 by the newly centralized Training Directorate of Kenneth Baker, Lawrence Lowman, and British

SOE advisor Bill Brooker to move the communications school from Area C to a country house near Area F, the former Congressional Country Club, the main CB training center remained at Area C for the duration of the war.[96] Although all training at Area C, including the selection of the trainees and the instructors and the course of instruction, was the responsibility of the Communications Branch, the camp itself and the cadre who ran it remained for administrative purposes in the Special Operations Branch. Schools and Training Branch never obtained much authority in regard to Area C, which remained primarily the responsibility of the Communications Branch. More than 1,500 communications personnel were trained at Area C between 1942 and 1945 in highly technical courses that ran from nine to thirteen weeks and included telegraphy, short-wave radio, codes and ciphers, OSS equipment operation and maintenance, weaponry, and demolition work, as well as physical and mental conditioning, field craft, and close combat skills.[97]

Around November 1942, when Lowman was considering moving the Communications School closer to Washington, he sent one of his staff to spend a week at Area C and report on conditions there. R. Dean Cortright, a civilian engineer for the Navy and like many of the "Commo" people an amateur, short-wave radio operator, a "HAM," as they were called, had joined OSS in October 1942, becoming perhaps the third person to be assigned to the new Communications Branch.[98] Years later, he recalled his visit to the Communications School at Area C. "Upon arrival, I found that there was a building housing the code room. There were a few 'agent operators' in training; from Thailand, I recall.[99] There was a housekeeping group [cadre] in place, with supply arrangements capable of handling a few operators. There was no commo [communications] equipment for practicing, other than a couple of British agent units. I can't remember the officer in charge at the time, but later on it was Maj. Albert Jenkins."[100]

After the decision was made in November 1942 to retain Area C as the main school for the new Communications Branch, Cortright was sent back there with orders to redesign the technical program and establish the kind of training necessary to produce the large numbers of communications personnel that would be needed for OSS's global operations. Winter was imminent, and the first priority was to get the NPS summer camp cabins and dining halls winterized. They were soon provided with insulation and cast-iron stoves that could burn local coal or wood. As in Area A, the mess hall was equipped with improved kitchen facilities. Major Jenkins arrived in December and quickly set up areas for physical training, an obstacle course, and pistol, rifle, hand grenade ranges. Jenkins and Cortright obtained vehicles, set up a motor pool, and also made arrangements to purchase food supplies from the nearby Marine Corps base at Quantico. Area C was soon in full operation and would continue so until the end of the war.[101]

Major Albert H. Jenkins remained commanding officer of Detachment C from December 1942 until February 1945. A reserve Marine captain who served in World War I, Jenkins and had been recalled to active duty in World War II, he had been recruited by the OSS, in June 1942 to be an instructor on small arms, mapping and field craft at the Secret Intelligence school at RTU-11 ("The Farm"). In July, he was promoted to major and made executive officer of that training facility.[102] Five months later, when the new Communications Branch began its school at Chopawamsic RDA, Jenkins began the commanding officer there. He was "quick and snappy, not a man for small talk," recalled Army Lieutenant Allen R. Richter, a Commo veteran from Detachment 101 in the China-Burma-India Theater. "If you met him on the street, you would know he was a Marine, the way he looked, walked, and talked." Jenkins was approximately 6'1" tall, slender, and although slightly balding, he was fit at about 150 pounds.[103]

Like most of the other officers, Lieutenant James Ranney, a radio instructor at Area C from August 1943 to January 1944, remembered Jenkins fondly. "Major Jenkins was in charge. He was a very fine gentleman. Everybody liked him…. He was a Marine reservist. I don't think he had anything to do with radio or communications. He was a nice guy, easy-going," One incident showed how imperturbable Jenkins was. "I remember while we were there," Ranney said. "they issued us our side arms, our .45s, and we were all looking at them….and wiping them off and checking the trigger pull. And the Officer of the Day wandered in, and he was carrying a side arm. He pulled his side arm out without remembering it was loaded, and he said: "This has a very good trigger pull." And BANG, he shot a big hole in the floor. Major Jenkins came rushing in. He said: "Anybody hurt?" They said, "No, just the floor." He said, "O.K., carry on." And that was that."[104]

While members of the cadre were familiar with the commanding officer, the trainees had little to do with him. They saw mainly the instructional staff: commissioned officers and NCOs. Private Marvin S. Flisser from the Bronx spent three months in the winter of 1943-1944 at Area C as an OSS Commo trainee, and was struck by how informal things in the OSS. "We hardly ever saw him [Major Jenkins]," Flisser recalled. "It was as if there was nobody there. There was no saluting or anything along that line. It was very informal. There was a basketball court, and we were all playing basketball, officers and men, all together, first name [code name] basis. Once in a while, Major Jenkins would come out to play basketball….When he came out to play, he wanted to win. So we always let him win."[105] Timothy Marsh, a civilian instructor in Morse code there from December 1943 to December 1944, remembered Jenkins as the senior officer and camp commander but had little interaction with him. Like other members of the instructional staff, Marsh dealt mainly the chief instructor, then Army Air Corps Captain Paul M. McCallen, who had his own office across from the mess hall and who, Marsh said, "more or less ran the Communications School."[106]

To many of the Commo men who came from cities or suburbs or directly from a standard Army base, Area C at Prince William Forest Park was a blissful, wilderness setting. Lawrence L. ("Laurie") Hollander, was a Chicago lawyer and amateur "HAM" operator, when he was recruited by OSS, commissioned a captain and sent to Area C in early 1943 to organize instruction in International Morse Code.[107] Years later, he recalled the natural beauty at the camp in the Virginia woods. "The complete area was about three quarters of a mile square. It is, and was, a beautiful area with its many shade-tolerant oak, hickory and pine trees. It also had its share of snakes and small animals. Located about 35 miles from Washington, with its entrance off

Highway No. 1 on a dirt side road, it became a top secret installation, its location and purpose known only to those with a need to know."[108]

OSS Training Area C was composed of two cabin camps some 400 yards apart in the northeast section of park between Quantico Creek and the Dumfries-Manassas Road (Route 234). The main gate was off the Dumfries Road, and no one was allowed to enter without proper authorization. When taxicabs from Triangle, the nearest bus stop on Route 1, when they brought trainees or staff back to the camp, they were not allowed to go past the main gave, and most of them did not even want to drive into the park up to the armed guards at the gate of the secret camp.[109] Roger Belanger, a trainee from Portland, Maine, who returned from England after D-Day as an experienced instructor, recalled that 'if we took a cab, the driver would let us out at the little white church on the Manassas Road. The cab drivers were reluctant to take us any closer."[110]

As at Area A, the OSS made alternations in many of the NPS summer camp buildings for year round operation and military training purposes. The layout of the camps at Area C was similar to those in the western park of the park, clusters of huts in sub-groups and the whole camp organized around a lodge, a dining hall, an office and other buildings. Most of the accommodations were quickly provided with insulation, heating stoves, and hot water facilities. Most dramatically, in 1943, the OSS upgraded the kitchens, adding a 500-gallon hot water tank and heater, gas ranges, a heavy-duty, electric baking oven, gas-heated, hotel-type steam tables, as well as electric coffee grinders, gas-heated coffee urns, plus ice boxes, refrigeration units, and an electric meat slicer, electric food mixer, and an electric-powered dishwashing machine.[111] Following the termination of the Civilian Conservation Corps in 1942, most of the work was done, as in Area A, by Kent and Company of Richmond, Virginia.[112] But the existing cabins did not prove adequate for the rapidly expanding numbers of radio operators being trained, and it became necessary to set up several dozen tent-roofed wooden huts to handle the overflow, particularly among the trainees. That gave Area C a housing capacity of 26 officers, 84 enlisted men in the cadre and instructional staff, plus 247 trainees, with 125 of the trainees living in temporary tent-roofed cabins, for a total occupancy of 357.[113] In the summer of 1944, the OSS also approved the construction of a swimming pool at Area C, if it were required for water-borne instruction and practice for Asian Special Operations agents, Communications Branch radio operators, and Operational Group members scheduled for training in Area C in 1945 as part of OSS's expanding mission in the Far East.[114]

"Commo" Camp Headquarters: Area C-1

Before the war, National Park Service Cabin Camp 1 (Goodwill) had been a boys' camp used by African-American youths from what was then called the Negro YMCA in Washington. It had four clusters of huts, labeled A, B, C, and D, plus the central dining and administrative area. During the war, the OSS designated it Area C-1, and it served primarily as the headquarters, accommodations, and maintenance facilities for Major Jenkins and the cadre of officers and enlisted men who operated Training Area C. When necessary, however, C-1, which was separated from the main training camp at Area C-4 by nearly a mile, was also used for a variety of training purposes. These included brief communications training for agents in operational branches such as SO, SI, or MO, or basic Army training for OSS personnel going overseas who had not yet fulfilled that requirement. At other times, it was used as a holding area for men awaiting further assignment, including graduates awaiting shipment overseas or veterans returning from abroad. In 1945, C-1 served as an area for training Koreans and perhaps other Asians as well for OSS operations in the final offensives planned for 1946 against the Japanese Army in China, Korea, and Japan itself.

In addition to winterization and other alterations in existing buildings at the cabin camp, the OSS erected several new wooden structures at C-1. A radio repair shop was the largest, a portable, plywood building measured 16 by 16 feet and may have served as a classroom. Smaller new structures included a radio transmitter building and two guard houses, one at the main gate and one at the northern end of Area C-1.[115] As completed for use by OSS's Communications Branch, Area C-1 included an administration building, two buildings serving as quarters for 16 officers, a drivers' quarters holding six enlisted men, a garage, a work shop, five washroom/latrines, a mess hall, a lodge, and a series of huts as well as two, 12-man barracks for enlisted men on the staff. In addition, there was a radio transmitter station in a building in the D cluster of cabins There was also another transmitter station, perhaps the main radio transmitter, in what was identified as an old CCC camp (probably the abandoned CCC work camp SP-22) across the road and a little more than 100 yards east of administration building at the entrance to Area C-1.[116] Both the transmitters and the receiver stations were powered by electrical generators and were equipped with sizable antennas, long poles with radio antenna aerials on them.[117]

Communications School: Area C-4

Nearly a mile south of C-1, the main Communications School of the OSS was located at Area C-4. In the years before the war, it had been NPS Cabin Camp 4, known as Camp Pleasant, and it provided a healthy respite of wholesome outdoor summer recreation for African-American girls and young women from Washington, D.C. Now it was the primary training facility of the OSS Communications Branch. Beginning in the winter of 1942-1943, it was transformed into an intensive training center where young men in Army fatigues spent two to three months learning to be clandestine radio operators behind enemy lines or more often operators and other technical personnel at OSS regional base stations in war zones and theater headquarters around the world. They learned and practiced repair and maintenance of the OSS's specialized radio and wireless telegraphy equipment as well as how to code and decode and send and receive telegraphed messages rapidly and accurately.

Area C-4 was larger than C-1. The NPS Cabin Camp had five sub-clusters of huts, A, B, C, D, and E, each with a lodge, plus there was a central dining hall and administrative area for the entire camp. Winterization of the entire facility was accomplished during the winter of 1942-1943, and the kitchens at the mess hall were modernized, as at the other cabin camps. Many of the NPS buildings continued to be used in the same manner, but some, such the nursery, today Building 78, was the OSS code room, where students learned and practiced becoming faster and more accurate in Morse code.[118] Perhaps because of the demand for communications personnel and the comparatively long time it took to train them in their highly technical skills, new construction at C-4 was much more extensive than at C-1, or Areas A or B. In the summer of 1943, Donovan personally approved the construction of a new multi-purpose building, which at $24,000 was the most expensive structure built by the OSS at any of East Coast training camps.[119] The cavernous building, today the Theater/Recreation Hall, Building 250 at Prince William Forest Park, was used by OSS as a theater to show training and entertainment films, for indoor assemblies by the entire camp such as when General Donovan came to speak, and for basketball and other spots during the winter months.[120] Other new structures built at C-4 included a Bachelor Officers Quarters, a 100-man latrine/washroom, three portable plywood buildings, probably used as classrooms, an addition to the mess hall, a guard house, and a metal Quonset Hut.[121]

By mid-1943, the Communications School had grown to 250 trainees, which meant that more housing had to be provided.[122] To handle the overflow, OSS created a "tent camp" at C-4. There were more than two dozen "tents" or more accurately, "cabin-tents," since each of them combined a canvas peaked roof with a plywood floor and four wooden sides, one with an entrance, the other three with screened openings and hinged plywood flaps. They measured 16 by 16 feet, and most held four trainees. Although primarily for summer use, cabin-tents were eventually fitted with cast-iron heating stoves that provided warmth in winter. The tent camp, which held up to 125 trainees, was erected at a cost of $5,500.[123] Various groups of these cabin-tents, often in sets of six or more were located between NPS cabin sub-clusters A and C in Area C-4, but the cabin-tents were only occupied when the regular cabins in the camp were filled.[124]

Each of the sub-camps, A, B, C, D, and E in C-4 had a wood cabins sleeping four and some of the sub-camps had wooden barracks for ten men each. The barracks were divided into three rooms with wooden bunks, an Army field telephone in a leather case hung from the wall beside the door. If it rang, you cranked it up to hear; it was for communication only within the base. The washroom/latrines were outdoors and trainees remembered that it was cold going there on winter nights. Most of the sub-camps also had a classroom building.[125] Often the lodges were converted into a post-exchange and a "Day Room" for relaxation and recreation. In sub-camp C, the "Day Room," was in what is today Building 12, the lodge. There was a ping-pong table and a big table where six or eight men played cards. Poker was the main game and the men played openly for money, while others watched. Other men read or listened to the radio.[126]

For training in weapons and hand grenades for its communications personnel, OSS built target ranges about 200 yards east of sub-camp C of Area C-4. A pistol range with six silhouette stationery targets, was constructed first, then relocated when it proved to be in the way of the rifle range with eight targets.[127] Both were pointed away from the camp, with shooters firing at targets mounted in front of an earthen embankment across Quantico Creek. A "utility range," apparently the site of the grenade range and perhaps demolitions as well, was over a ridge and in a valley formed by Quantico Creek about another 200 yards southeast of the rifle range.[128] Communications veterans Arthur Reinhardt from near Buffalo, New York, who spent ten weeks training at C-4 from April to early July 1944, before being sent to China, identified the old pistol range on a return visit to the area sixty years later. "We were encouraged to come down here and shoot our [Colt] .45s," he said. "I used to come down virtually every day and shoot."[129]

Harry M. Neben, an amateur radio operator ("HAM") from Illinois, had been recruited to help with install radio equipment at Area C. He arrived in December 1942 and helped clear trees and brush for the firing ranges and an obstacle course as well as assisting in electronic installation for radio training. In addition to the obstacle course, he also remembered that trainees at Area C-4 "practiced on a landing net on a small lake near [Sub-]Area B."[130] This would be a lake on the north fork of Quantico Creek a quarter of a mile from C-4 and presumably involved some of the same kind of aquatic activities that trainees at Area A remembered undergoing at a similar lake across the park near OSS Area A-5.

Unlike the Special Operations paramilitary field exercises with firearms, mortars, bazookas and explosives in Area A, the OSS Communications Branch focused on radio training with merely some marksmanship training at firing ranges. Consequently, none of the deserted houses or other structures was deliberately demolished in Area C. However, one NPS cabin was inadvertently destroyed. The cause was a fire in a cast-iron heating stove with a loose ash dump door. When the trainees returned from an afternoon of field radio training away from the camp on a winter day, the cabin had caught fire and burned to the ground.[131] In a separate incident involving fire, one part of the former NPS building used by OSS for the motor pool burned down. In both cases, OSS rebuilt the structure.[132] These were apparently, the only NPS buildings destroyed by the OSS during the occupancy of Prince William Forest Park.

For any kind of maintenance work at the camp, the OSS cadre turned to John T. Gum, the National Park Service employee who previously had been the NPS liaison with the Civilian Conservation Corps that had worked in Prince William Forest Park. During the war, Gum had been drafted into the Army, and was recruited by the OSS. He held the rank of Army technical sergeant, and a roster listed him as post engineer and the sergeant in charge of utilities at both Area A and Area C.[133] The friendly, six-foot-tall Gum visited the camp every day, always accompanied by his big black dog. Corporal Joseph Tully, a member of the cadre at Area C from November 1944 through October 1945, remembered Gum well. "He was in charge, the enlisted man who ran the entire camp. There were guys who rated higher than him. But if you wanted any kind of maintenance, he was the one you went to."[134]

Area M in Illinois and W-A in California: Adjuncts to Area C

So large were the numbers of communications trainees in the winter of 1943-1944—the Communications Branch estimated that 430 Commo personnel would be trained by December 1943, exclusive of SO, SI and other agents being trained in communications[135]—that even with the cabin-tents Area C could not handle them all. Consequently, OSS acquired an adjunct Communications Training Area near Napierville, Illinois in October 1943.[136] It was designated Area M because it was located at Camp McDowell, which had previous served as a radar training school for the Army Signal Corps. Captain Lawrence L. "(Laurie") Hollander, a lawyer and amateur radio operator from Chicago, headed Area M, which was used, during its brief existence in 1943-1944 for training recruits from the Midwest and Far West, including a number of Japanese Americans who volunteered for hazardous duty as radio operators with Donovan's organization.[137] Graduates probably served with the OSS against the Imperial Japanese Army in China.[138] Instructors at Area M offered a more elementary and simplified course than at Area C.[139]

In January 1944, when OSS turned increased attention to preparing increased numbers of personnel to be sent to the Far East, a West Coast Training Center, Area W-A, was established on Catalina Island off the California coast not far from Los Angeles. There the Communications Branch coordinated Commo instruction with training by the SO, SI, and MU Branches. By mid-1944, almost most of the OSS training in the United States for SO, SI, and MU had shifted to the West Coast, Area C continued to be the main training center for the Communications Branch until the end of the war.[140]

The Marine Corps and Prince William Forest Park

From the inception of Prince William Forest Park in 1935, much of its southern border adjoined the U.S. Marine Corps Base at Quantico, Virginia. The original plan of the National Park Service was to acquire much of the land west of U.S. Route 1 that was in the drainage areas of Quantico Creek north of Joplin Road (today County Route 619), and Chopawamsic Creek to the south of it, that were not already held by the Department of the Navy for the Marine Corps.[141] But lack of adequate funding prevented the purchase of the Chopawamsic watershed property, and it also limited recreational development there. In practice, the park developed cabin camps and other organized recreational activity only north of Route 619 in the Quantico Creek watershed. The area south of that road, the area of the Chopawamsic Creek watershed, was preserved solely as a wilderness area with no organized recreational facilities. In that southern sector, NPS maintained only a temporary, one-room, park office and an undeveloped maintenance area just south of Joplin Road.[142]

Given its proximity to the U.S. Marine Corps Base in Quantico, Virginia, Prince William Forest Park has had a continuing relationship with the Marine Corps and its parent organization, the Department of the Navy. The Marine Base dated back to World War I.[143] When the Department of the Interior acquired the adjoining lands in the mid-1930s, the Marines expressed interest from the beginning in finding ways that were "mutually advantageous" for using what they called the "contiguous areas under Federal control." In the park area south of Joplin Road, the Corps wanted to build a dam and water reservoir on Chopawamsic Creek to augment the water supply for the base. They also wanted to clear areas within the southern section of the park for field exercises in which "weapons may be used in field firing and musketry training," and they sought to use the dirt roads and abandoned buildings there for military training purposes.[144] In 1938, the Department of the Interior granted the Department of the Navy the right to build the dam and what became known as Breckenridge Reservoir.[145] Beginning in 1938, the National Park Service began to grant use permits to Marine commanders to conduct specific field maneuvers in the southern portion of the park.[146]

After the declaration of war in December 1941, Lykes was as diplomatic as possible in dealing with the Marines. He complained to his superiors that the Marines' field maneuvers in the southern park of the park had become so frequent that they "have assumed the right to enter upon the area without advising or even consulting this office."[147] But Lykes continued to maintain a cooperative relationship with the personnel of the Marine base, emphasizing to the Marines the park's recreational value to military personnel, including fishing in its well stocked streams, ponds and lakes.[148] In the hot, dry summer of 1942, Lykes voluntarily taught Marines the latest techniques in fighting forest fires so they could more effectively control fires when they began in the woodlands of their reservation.[149]

The following year, the commander of the Marine base, Major General Philip H. Torrey, offered, and Lykes accepted, a commission as a first lieutenant and a position at the base.[150] During his active duty in the Marines, from April 1943 to December 1945, Lykes was assigned as a security officer, particularly responsible for the forestry program and the prevention and control of fires on some 94,000 wooded acres of the expanded wartime Marine Base.[151] That included some 5,000 acres of the park land south of Joplin Road that the Marines used as an extension of their base. That included, Lykes later recalled, what the Marines called the "Guadalcanal Area," a reference to the Pacific island which was the site of the first jungle fighting between the Marines and the Japanese. In that area, the Marines erected 13 firing ranges as well as pistol houses, mockup villages, and other installations for field exercises using live ammunition and other munitions.[152] The little village of Joplin, a hamlet with a general store, a schoolhouse, and a church, located near the intersection of Joplin and Breckenridge Roads, was demolished, only a concrete slab foundation of the old school house remaining.[153] Lykes reported directly to the commanding general of the Marine base. The National Park Service officially put Lykes on furlough due to his military service.[154] But because the base and the park were adjacent, Lykes, as he later recalled, could and did "serve two masters."[155]

Lykes remained at his new duties with the Marines on weekdays during the war, but on weekends, he continued his role as manager of the park. Lykes remained in the park manager's house just south of Joplin Road for the duration of the war, and he maintained social and professional relations with the commanding officers of the OSS training camps, such as Captain Walter Hixson and his successors at Area A and Major Albert Jenkins at Area C. Lykes kept an eye on the park property being occupied the the OSS north of Joplin Road and being used by the Marine Corps south of there. Weekdays, the only administrative employee of the park on duty was Lykes's secretary, Thelma Williams. She worked in a one-room temporary headquarters for the park off Joplin Road. Sharing the small office with her was a clerical employee of the OSS, a secretive man, who Lykes reported was given to drink. Late in the day on Fridays, when the Lykes came by the park office, Thelma Williams would fill him in on developments in the park during the past week.[156]

Complicated Relationships: USMC, OSS, and NPS

The Relationship among the U.S. Marine Corps, the OSS and the National Park Service at Prince William Forest Park during the war was often intricately interrelated and sometimes complex. The Marine base provided many services for the OSS. Area C, and probably Area A as well, purchased much of its fresh food supplies from the Marine commissary at Quantico.[157] The Maritime Unit at Area D sent a boat every Sunday to get a variety of supplies.[158] Representatives at Areas A and C drove over for similar purposes. The airfield at the Marine base was utilized by Donovan's organization. The OSS Parachute School used it for two-engine planes for parachute jumps at Area A in Prince William Forest Park and Area D across the river. Practice runs by the single-engine light planes that the OSS also employed for inserting or removing agents and carrying messages also flew in and out of the Marine airstrip when necessary.[159] The relationship between the National Park Service and the Marine Corps could be especially complicated.

Park Manager Ira Lykes sought the Corps' aid in obtaining 1,900 acres west of the park that would secure the Quantico Creek watershed and help round out the borders of the park. In 1941, Lykes reported that Major General J. McCarthy Little, the base commander, had suggested that the Navy might be able to purchase those that acreage with the idea of transferring them to the Park Service in exchange for a comparable amount of land on the Chopawamsic watershed in the southern part of the park that could be transferred to the Marine Reservation. The Secretary of the Interior followed up with a request to Secretary of the Navy proposing that the two agencies draft legislation for such a transfer.[160] On its own, the Navy announced in October 1942 that it was going to expand the 6,800-acre Marine Base at Quantico by adding 50,000 acres west of the base.[161] The Marines included the southern part of the park as well as private holdings in that plan.[162] In conversations with Major General Philip Torrey, the new commanding officer at the base, Lykes expressed concern about the impact of the Marines' acquisition on the western part of Quantico Creek upon the quality of water flowing into the cabin camps in the northern part of the park and that the park lands "not be used for maneuver purposes when the Marines Acquisition Program is completed."[163] Lykes later said that he had a very friendly relationship with General Torrey, who had proven helpful and cooperative. When the OSS occupied the park north of Joplin Road, Lykes and his family moved from the park manager's residence (today the park headquarters) into a park building on the south side of Joplin Road. Lykes used part of this residence as an office. Although the building was in the southern sector used by the Marines during the war, General Torrey not only allowed Lykes to live there but the Marine base winterized th dwelling, furnished the heating oil, and paid the electric bill. General Torrey may have had other ideas for its eventual use. One day while touring the area, Torrey walked into the building and looked around. "The first thing he said was, what a damn good officers' club this would make," Lykes recalled. "I didn't know how to respond to that so I didn't respond." The park manager's coolness put the damper on the general's enthusiasm. "They never did turn it into an officers' club," Lykes said.[164]

With the Marines already using the southern part of the park due to wartime needs and in anticipation of an exchange of land and permanent agreement, the Secretary of the Interior issued a temporary permit in June 1943 for the Department of the Navy (the Marine Corps) to use the 4,862 acres of park land on Chopawamsic Creek south of Joplin Road for the duration of the war and six months afterward.[165] So while the OSS Special Operations and Communications Branches were doing their training in the northern part of the park, the Marines conducted their field exercises in the southern part.[166]

What Lykes believed to have been a consensus on transferring Navy purchased lands on the western border of the park to the National Park Service in exchange for park lands south of Joplin Road to be transferred to the Marine Reservation broke down between 1943 and 1946. The joint Navy-Interior committees formed to draft legislation for the permanent transfer could not reach agreement. In Lykes' opinion, he had a verbal commitment from Marine leaders at the base for the Navy to purchase 1,900 acres on the west in exchange for nearly 5,000 acres of the park in the south. It was with such an understanding, Lykes said, that the Marines had been allowed to use the land south of Joplin Road for major field maneuvers, including changes in the landscape for their training purposes that in the park manager's words had left the area "no longer... in any way suitable for recreational development."[167] By 1944 with Allied victory only a matter of time, NPS was concerned cutbacks in defense spending would eliminate the funds necessary for the Navy Department's purchase of the land and the transfer. Attempts to get a transfer agreement were blocked by the Navy which wanted to wait for a determination of the peacetime size of the Marine Corps which had been increased from 19,000 Marines in 1939 to 475,000 during the war.[168] The matter of the transfer of lands and the status of the Marine usage of the southern part of Prince William Forest Park remained unresolved when the war ended in September 1945. Although the 1943 temporary use permit for the Marines was technically terminated in 1946, the entire southern tract of nearly 5,000 acres appeared to have become a fixture of the Marine Corps base. That expansion led to a longstanding controversy about control over the southern area of the park that would continue until its final resolution in 2003.[169]

CHAPTER 6

INSTRUCTING FOR DANGEROUS MISSIONS

Physically establishing the training camps was comparatively easy because of the pre-existence of the cabin camps in the two national parks. More difficult was creating the training process. To prepare spies, saboteurs, guerrilla leaders, radio operators, psychological warfare specialists and commando teams for their clandestine missions, Donovan's organization had to obtain instructors, prepare a curriculum, develop courses, and devise practical exercises. Ultimately, the effectiveness of the training would be judged by the success of the OSS.

Although OSS was deliberately lax in its approach to Army protocol and procedures, its paramilitary training camps were at least organized like military bases. There may not have been any saluting, close-order drill, bugle calls, or marching to class, but each training area had a commanding officer, executive officer and a cadre of officers and enlisted men who kept the camp running. Because the camp was also a school, there was an instructional staff. Headed by a chief instructor, it included officers and enlisted men and sometimes some civilian instructors. Few of the cadre or the instructors had been professional soldiers, although many had been in the reserves and most had enlisted voluntarily, first in the armed services and then in the OSS.

While the cadre kept the camp going with food, fuel, and other supplies, the instructional staff was faced with diverse groups of students recruited for unconventional warfare. Among the first to take basic Special Operations training at Area B at Catoctin Mountain Park in the spring of 1942 was a group of two dozen young men from Thailand, who had been students in American colleges and universities. They were succeeded by a group that included a number of Yugoslavian guerrillas. Other foreign nationals and American ethnic groups followed. Some classes were a mélange of men whose only commonality appeared to be their desire to fight against the Nazis. Edgar Prichard, who had been a journalist from Oklahoma and a company clerk in the Army before Donovan's organization recruited him, recalled years later that his class of sixteen men at Area B-2 in Catoctin Mountain Park in 1942 had been an "interesting lot:"

"We had a Finnish fellow named Aalto who had fought on the Republican side in the Spanish Civil War and another fellow named Doster who had fought on the Fascist side in the same war. They weren't overly friendly with each other. We had a Zionist fellow who had fought with the Haganah [a Jewish paramilitary group] in Palestine and a Lieutenant in the Norwegian Army who had escaped from Norway with the British after Narvick [a key shipping port in German occupied Norway, raided by British commandos in April 1940]. We had a couple of Frenchmen, one of whom had escaped from Occupied France. We had a Polish fellow who had escaped from a German prison camp after the Polish defeat and made his way to Spain. We had a young Greek boy from the island of Samos. We had a football player named Jim Goodwin whom we later [air] dropped into Tito's headquarters. We had a Brazilian whose mother was an American. We had a Scot who had grown up in Peru, an Italian fellow who had been a Treasury Agent. We had a French Canadian who had washed out of the RAF [British Royal Air Force]. We had a young Foreign Service officer from Chicago, and we had a couple of central European types."[1]

Cadre and Instructional Staff at Area B, Catoctin Mountain Park

OSS Area B opened in April 1942, offering four-week courses to prepare recruits for hazardous missions of espionage, sabotage, and guerrilla leadership behind enemy lines. The first arrival was the commanding officer, Major Ainsworth Blogg, a tall, gentle man, who had been an insurance company executive in Seattle and a reserve infantry officer, assigned to military police when the Army called him to active duty and he was subsequently recruited by Donovan's organization. Blogg established his office at Catoctin and set up his residence with his wife, Jane, in officers' family quarters at Camp Ritchie half an hour away.[2] In administering Area B, Blogg was assisted Lieutenant James Johnson, a Midwesterner, as his executive officer, and Captain Louis Lostfogel, a short, friendly, pipe-smoking physician originally from Philadelphia, who served as medical officer.[3]

A former Civilian Conservation Corps work camp, located near an NPS maintenance facility that today is called Round Meadow, was designated Area B-5 by Major Blogg, and it became the headquarters of Detachment B. At B-5 were located the offices, mess hall, motor pool, and accommodations for most of the cadre. There were several commissioned officers, but the majority of the cadre were enlisted men, perhaps most prominent among them, the company clerk, an enlisted man, Corporal Georges George. Despite the variation in spelling, his first and last names were both pronounced "George." An Arab American from Lebanon, Corporal George was a handsome, dark-haired man with a somewhat haughty attitude, who, according to his assistant, would boast that he was "probably the only man in the OSS who speaks and really knows Arabic."[4]

In the fall of 1942, the assistant company clerk was Private Albert R. ("Al") Guay, a recent graduate of American University in Washington, D.C., who had majored in personnel management in college, jotted down in his diary brief assessments of some other enlisted men on the camp staff.[5] "Milton Giffith is first man I knew here. Swell personality, from Pittsburgh. Bus driver, not college man, but damn good company. Likes liquor & women, not offensively. Is married." "Benton E. Bickham from Louisiana. Real drawl, black, curly hair. Quiet but a real friend and fun. I would trust him & Griffith with anything. Don't know his background at all." "Deisher—don't know his first name. Religious fanatic. Listens to prayers over radio constantly. Got outside & prayed for whole camp one night. Won't swear & is pained by dirty jokes." "Corporal Ressler has let his stripes go to his head. A smart guy and an agnostic. Thinks he can do anything. Always talking about himself or what he has done. He has poor taste in women. Wild ones." "Velleman is a Dutch Jew raised in Antwerp. He isn't too bad, but he thinks he is superior to us; has very definite ideas about class. Orders people around as if he were a general." Guay later recalled that Velleman, who spoke several languages, had in the French Army, and after France's defeat in 1940, he had come to the United States, enlisted in the U.S. Army and been recruited by the OSS. Winding up Guay's list was "Private Baker, an 18-year-old kid, a good soldier, conscientious and well liked by all. Homesick too."[6]

Working in the company clerk's office, Guay was responsible for keeping the Morning Report, a record of the status of every member of the company each morning, whether they were present for duty, on sick call, on leave, or absent without leave. He seldom saw the commanding officer (C.O) of the camp, but he certainly remembered the day that Major Blogg was promoted to lieutenant colonel, because, as he noted in his diary on Saturday, 5 December 1942, the jubilant Blogg let the staff off early, at 2 o'clock in the afternoon and let them spend the rest of the day and evening in nearby Hagerstown on a one-day pass.[7]

Lieutenant Colonel Blogg served as C.O. at Area B through June 1943, when he was succeeded by Lieutenant Montague Meade. Blogg soon went to OSS headquarters in Washington where he served as executive officer of the Schools and Training Branch, supervising the administration of all the OSS training camps in the United States, a position he held until the end of the war, when he returned to the insurance business in Seattle.[8]

The Special Operations School itself, where the students and their instructors were quartered and the classrooms were located, was established up the hill at the former NPS Cabin Camp 2, Greentop, designated during the war as Area B-2. Finding suitable instructors for paramilitary training in unconventional warfare posed a major challenge to Donovan's organization. The leaders of the Special Operations Branch, Lieutenant Colonel M. Preston Goodfellow and Major Garland H. Williams, faced the problem of finding and developing men who could instruct in a multitude of subjects for which many of them had little or no background or experience, at least not in combat. The very fact that the United States had just entered the war and that the OSS was a new organization and was striking out in new directions made it virtually impossible to recruit men experienced in unconventional warfare. The operational branches—Special Operations (SO), Secret Intelligence (SI), Morale Operations (MO), Maritime Unit (MU), and later Counter-Intelligence(X-2) and Operational Groups(OGs), plus the Communications Branch (CB or "Commo")—were themselves almost totally inexperienced in regard to their missions and, therefore, could initially offer little practical direction about the kind of training needed. Consequently in the early period, Donovan's organization relied heavily on the British Special Operations Executive and Secret Intelligence Service for assistance. It sent some of the new instructors to British schools in Canada and Great Britain. It borrowed British instructors, and it initially drew upon British curricula, teaching materials, and even some of its weaponry and explosives.[9] Yet the Americans also drew upon various specializations and connections within U.S. institutions as well.

Lieutenant Louis D. Cohen served as the first chief instructor at Area B. He had been one of the Americans trained at British SOE's Camp X in Canada earlier in 1942. At B-2, he headed the instructional staff from May to August 1942, when he was succeeded by Captain Morris M. Kessler. A psychiatrist, Kessler had succeeded Louis Lostfogel as medical officer. But Kessler also succeeded Louis Cohen as chief instructor and remained in charge of instruction until the school closed in June 1944.[10]

As instructors, Major Blogg brought in a number of reserve Army officers from the state of Washington that he knew and trusted, many were from Washington State University, the state's land-grant college and host to a sizable Reserve Officers' Training Corps (ROTC) program. At least half of the first ten officer instructors at B-2 had been athletes and ROTC cadets there.[11] Several of them had been in the Military Police where Major Blogg had served at Fort Lewis, near Tacoma. One of them, Jerry Sage, 24, would later become one of the most famous members of the OSS agents because of the number of times he escaped from German prisoner of war camps, including his early participation in the mass escape planned at *Stalag Luft III*, dramatized in the 1963 film, *The Great Escape*, only to be recaptured, until his final successful breakout in January 1945.[12] In the fall of 1941, Sage had been a salesman for Procter and Gamble marketing its products on the West Coast. He was also a 1st lieutenant in the Army reserves, having gone though ROTC, while at the same time playing end on the Washington State football team and becoming Phi Beta Kappa for his academic performance. In September 1941, stirred by radio reports of the war in Europe, he requested active duty and was assigned to Fort Lewis. He proved so proficient at training his unit, that in February 1941, the camp commander put him in charge of training large numbers of reservists. In March 1942, Sage received a telegram from the "Coordinator of Inf." (Sage thought it meant the Coordinator of Infantry) directing him to report to Washington, D.C. There he was interviewed by William J. Donovan himself. As Sage later described the scene, "Seated behind the desk was an imposing, somewhat heavyset man—not as tall as I, but with good shoulders. He had piercing Delft-blue eyes. His gaze did not waiver for an instant."[13]

"The reason you're here is that you've been hand-picked," he said, "We have information from your friends at Fort Lewis, and the FBI also mentioned you. Apparently you're not afraid to mix it up. You're not afraid to use your fists, you're a good athlete, a football letterman, and you learn fast. My guess is that you can learn rapidly enough to keep up with the people who do what we do."

"I was about to ask what they *did* do, but he quickly continued. "This is a completely volunteer outfit. I want to tell you what it's like. If you join us here, you will be working on the most dangerous assignments in the military. We'll be seeking a payoff from your work. You'll be an agent, a saboteur, maybe an assassin, certainly a guerrilla fighter....." (With each word, my grin got a little bit bigger.) "Your folks will never know where you are. They'll communicate with you only through a post office box in Washington, D.C. You won't ever earn any medals. Nobody will know anything about what you do. But it will be a great service to your country. We can really hit 'em where it hurts most, behind their lines, and we can bring out the intelligence that we must have to win this war."[14]

Sage accepted eagerly and was sent directly to Area B-2 to meet the other instructors. He taught there from April to July 1942. "My duties were to indoctrinate the new recruits—who were everything from jailbirds to college professors—and teach them the fundamentals of soldiering and physical conditioning."[15] According to Sage, the initial instructors tried to make the early trainees, many of whom were civilians, look and act like soldiers. Because all countries had soldiers, that was the simplest cover for an agent. "We taught them how to march and react like military men. Then we turned our emphasis on tough training—physical conditioning, explosives work, hand-to-hand combat and knife fighting." Each man who completed the training had learned how to survive behind enemy lines, to gather intelligence, get it out, and how to conduct sabotage and guerrilla operations.[16]

The little group of Washington State alumni also included three other able instructors. Army Captain Joseph H. ("Joe") Collart was an agile gymnast, tumbler, and diver in college. Now he taught physical training, compass and map reading.[17] Marine Lieutenant Elmer Harris, nicknamed "Pinkie" because of his red hair and pale complexion, came from Ketchikan, Alaska and after graduation had been a marine use salesman for General Petroleum Company. In the spring of 1942, following three months at Marine Corps School, he was assigned to OSS Area B where he taught field craft.[18] Field craft are the skills used in operating stealthily day or night in open or closed areas, using camouflage or natural concealment, moving undetected across open ground or across obstacles, selecting effective firing and resting positions, and conversely being able to detect camouflaged enemy positions. The motto is "to see without being seen."

Army Lieutenant Arden W. ("Art") Dow had recommended the other instructors to Blogg. A tall, ramrod-straight, 21-year-old ROTC graduate from Washington State, Dow had served under Blogg in the Military Police unit at Fort Lewis in 1941. After the major was summoned by Donovan's organization to command Area B, he sent for Dow and the Washington State comrades Dow recommended. The spring of 1942 was certainly a momentous time for Dow. In March he married Dorothea ("Dodie") Dean, a recent graduate of the University of Washington, and when they returned from their week-long honeymoon, orders were waiting for Dow to report to Area B as an instructor. The two newlyweds drove across the country, accompanied by another Washington State football star and friend of Dow's, Lieutenant Ira Christopher ("Chris") Rumburg, center and team captain. Rumburg had also been recruited as an instructor at Area B by the OSS. After reporting for duty at Catoctin, Dow put his wife up at an apartment in Frederick, Maryland, visiting her on weekends. He was not permitted to tell his wife, or anyone one else, what he was doing up at the mountain camp.[19]

Tall and husky, Lieutenant Rex Applegate hailed from Oregon and had played football for Oregon State, but he became acquainted with Blogg in the Military Police Unit at Fort Lewis. Blogg brought him to Area B where Applegate was one of the instructors in close-combat and pistol shooting techniques.[20] He popularized a method of pistol shooting, known as "instinctive firing" or "point shooting" that emphasized training for close range, fast response shooting. Rejecting traditionally careful sighting of the target with the pistol raised to eye level, Applegate argued that speed even more than accuracy was essential. He advocated quickly crouching, feet squarely apart, both eyes on the enemy, and simultaneously, swiftly whipping up the pistol to hip level with the arm fully extended as though one were pointing his finger at the enemy, and immediately and instinctively firing off two shots, followed by another two if necessary. "You hit where you look," he said, and his system was adopted throughout the OSS.[21] In addition to Area B, he taught at most other OSS training camps in the United States during the war. In 1943, he wrote an instructional manual for hand-to-hand combat and combat pistol shooting, entitled *Kill or Get Killed*.[22] Applegate become famous for this pistol firing technique which he taught after the war to the military and to law enforcement officers.[23]

Most of the contingent of officer instructors from the Northwest—Applegate, Collart, Dow, and Harris—as well as executive officer James Johnson from the Midwest, were newlyweds, and around June 1942, they rented a big old house just across the state line in Blue Ridge Summit, Pennsylvania, that could serve as quarters for the five young couples. It was closer to Catoctin Mountain Park than Frederick, Maryland had been and was also more secluded. The married officers drove home together each night in an Army truck and returned to the camp every morning, except Sunday. Each couple had a private room, and the women took turns cooking. "We were all young brides," Dorothea ("Dodie") Dow recalled. "We didn't know anything about housekeeping, cooking or anything.... We had a lot of hot dogs and pineapple and cottage cheese salad. I think finally the men complained about the hot dogs." For six weeks in the fall, most of the men were sent to England for SOE instruction. For protection for the wives due to the secluded area, they gave pistols to each one and had target practice in the basement. The women were not enthusiastic. "We were afraid to get up at night and go to the bathroom," Dodie Dow said, "for fear that somebody, one of the others, would shoot you."[24] The women lived there until October 1942, when most of the men were transferred to new assignments.

In addition to the athletes and military police officers that Blogg and Dow brought from the Northwest, the officer instructors at Area B included one agent from the Federal Bureau of Narcotics and two demolitions experts from the Army Corps of Engineers. Captain George H. White was one of the top investigative agents from the Federal Narcotics Bureau, who had attended the SOE's school at Camp X in Canada.[25] Two of White's early students, Lieutenants William R. ("Ray") Peers and Nicol Smith, both of whom served in the China-Burma-India Theater, recalled that White had taught undercover work. He gave practical insights into how to trail people without being noticed, how to use recording devices and tiny cameras, how to detect if someone had entered your room (by stretching a hair across the keyhole and attaching it with tiny specks of gum), how to interrogate or how to resist interrogation.[26]

Explosives were taught by two young, engineering officers, both Pennsylvanians and fraternity brothers from Penn State University, 1st Lieutenant Charles M. Parkin and 2nd Lieutenant Frank A. Gleason. Born in Pittsburgh and raised in Lancaster, Parkin was an "Army brat," the son and grandson of regular Army soldiers. After graduating with honors in engineering from Penn State and the ROTC, he spent a year in training with reserve officers, graduated at the head of his class, and received a commission in the regular Army. He spent the next year, as an instructor at the Corps of Engineers School at Fort Belvoir, Virginia, teaching, as he later said, everything the engineers did, from how to build bridges to how to blow them up.[27] He actually did blow up a number of bridges in the winter of 1941-1942, that were going to be abandoned and buried beneath reservoirs being constructed by the Army Engineers for the Tennessee Valley Authority. In testing explosives, particularly in the destruction of bridges with stone or concrete pillars, Parkin discovered that the Engineers Field Manual written during World War I greatly underestimated the amount of explosive charges needed for destroying such bridge supports. As a result of his findings, the Field Manual was rewritten.

On the way back to Belvoir, Parkin, on his own initiative, took a platoon of engineering troops out at night and secretly placed dummy charges around the pillars on the U.S. Route 1 Bridge at Quantico, demonstrating how inadequate the security was with just one night watchman on the main southern route to Washington, D.C. Lieutenant Parkin wrote up a report on the exercise and turned it in to his commanding officer Monday morning. He soon received a call from the commander of Fort Belvoir, who told him "I think you are either going to be court-martialed or honored for this." A little later, Parkin received a summons from Goodfellow and Williams in Donovan's office, recruiting him as an instructor in demolitions and night field exercises at Area B.[28]

Enthusiastic, energetic and impetuous, Frank Gleason was, at 21, the youngest officer at Area B. Boyish and engaging, the red-haired Gleason was friendly and well-liked by all, regardless of rank.[29] He came from the suburbs of Wilkes-Barre in the coal mining region of northeast Pennsylvania, and one of his colleagues recalled that "we used to call him the 'Wilkes-Barre Kid.'"[30] Because his relatives had worked in the mines, Gleason was already familiar with explosives. At Penn State, he graduated with a degree in chemical engineering and an ROTC commission. Called to active duty in the Corps of Engineers in April 1942, Parkin got him assigned to the COI and Area B.[31] Parkin also brought up two sergeants from Fort Belvoir, Joseph Lazarsky and Leopold Karwaski, who, like Gleason, were from the coal regions of northeastern Pennsylvania. The three—Lazarsky, Karwaski, and Gleason—were soon nicknamed, "the Three Skis."[32] Gleason had an infectious enthusiasm. "He loved to blow up simulated enemy targets," Jerry Sage remembered. He and the other engineers had a range of explosives to work with: dynamite, TNT, the new more stable and moldable "plastic" explosive, Composition C, and a variety of igniters: fuses, blasting caps, time-pencils, and delaying devices.[33] "With a wide grin, he would discuss our upcoming demolition blasts, calling them big booms."[34] Later, as will be detailed in a later chapter, Gleason would get a chance to set off some of the largest "booms" in China.

At Area B in 1942, the instructional staff from the Army Engineers, Lieutenants Parkin and Gleason and Sergeants Lazarsky and Karwaski, led the trainees in studying and discussing British commando raids on Norway, France, and North Africa. Although these young Americans had no personal experience with such raids or guerrilla warfare, they learned from the lectures and reports from British and American sources. Between March and June, several Special Operations Branch officers had attended the SOE paramilitary school near Toronto.[35] In the late summer of 1942, during a lull at B-2, Gleason and half a dozen other instructors from Area B, were sent to SOE schools in Britain for six weeks and trained there for raids and industrial sabotage with people from Norway, France, Czechoslovakia, and other German occupied countries.[36] From all these sources they passed on as much information as they could obtain to the students.

One of the instructors at Area B had actual combat experience in a commando raid. Edward ("Ed") Stromholt, a lieutenant in the Norwegian Army, had helped British commandos in their raid on the Lofoten Islands in German-occupied Norway in March 1941. That raid destroyed factories producing glycerine for explosives and more importantly, captured the current settings for the German Enigma coding machine, which enabled the British Code and Cypher School at Bletchley Park to break the enemy's naval coded radio traffic for the previous month, a major intelligence coup. After the raid, Stromholt taught raiding techniques at commando schools in England and then at OSS schools in the United States. At Area B in 1942, he gave instructions in compass and map reading and other forms of scouting, which he called "orienteering."[37]

Unexpectedly, Stromholt became the center of attention at Area B one summer afternoon when Franklin D. Roosevelt arrived for a visit to his new Presidential retreat. Riding with the President in the back of his open-top car was a beautiful Norwegian princess, Princess Märtha, who was in the United States to accept two U.S. Navy destroyers for the Norwegian Navy. As the Presidential automobile stopped by a group of OSS instructors and trainees, the princess suddenly stood up, pointed a finger at one of officers and exclaimed loudly in perfect English, "Stromholt! What in the hell are you doing here?" As instructor Charles Parkin said later, Stromholt probably thought, "Well, what in the hell is she doing here?" A tall, handsome man, he later married an American movie actress, but not before continuing his commando work, including a parachute mission back to Norway.[38]

"Dangerous Dan" Fairbairn, the "Shanghai Buster"

The man who would become famous among the instructors at the OSS Training Camps was British Army Captain William Ewart ("Dan") Fairbairn. He was one of the world's foremost experts on close combat techniques, and none of the American trainees who saw him in action would ever forget him. After teaching at British SOE schools in Britain and Canada, Fairbairn had been lent to the OSS in April 1942. He remained for the duration of the war, taught at all the OSS camps, and rose to the rank of lieutenant colonel. As a youth, he had left Scotland, enlisted in the Royal Marines, served in the Far East, and then joined the British-led Shanghai Municipal Police Force. In response to a beating by Chinese gangs, he became a master of jujitsu, a Japanese Samurai system of unarmed fighting, as well as Chinese boxing and other martial arts. He learned to disable or kill an attacker with his hands and feet, a knife, or any instrument at hand, and reportedly engaged in hundreds of street fights in his twenty year career in Shanghai, where he organized and headed a special anti-riot squad.[39] Much of his body, arms, legs, torso, even the palms of his hands, was covered with scars from knife wounds from those fights, recalled an American officer who roomed with him at Area B.[40] Fairbairn had retired from the Shanghai police at 55 and returned to Britain in 1940 to help train commandos and SOE agents in his own mélange of martial arts, he called Defendu, and which he described in a book entitled, *Get Tough*.[41]

While training British commandos and SOE agents in England, Fairbairn and British Captain Eric Anthony Sykes, a colleague from Shanghai, also developed innovative and effective techniques for combat pistol shooting,[42] and they designed the famous Fairbairn-Sykes fighting knife. A razor-sharp stiletto dagger, it was carried by British special forces and initially by some OSS agents before the OSS replaced it in 1944 with the less brittle, M3 trench knife.[43] In 1942, Fairbairn was transferred first to Canada and then the United States, where he taught his techniques to instructors and recruits in Donovan's organization.[44]

Fairbairn's basic message to the young recruits—and it was a message that pervaded the training at Donovan's organization—was to forget any idea of gentlemanly conduct or fighting fair. Get tough, get down in the gutter, win at all costs was his mantra. "I teach what is called 'Gutter Fighting.' There's no fair play; no rules except one: kill or be killed," he declared. The average American's sense of sportsmanship and repugnance to such fighting presented a handicap, Fairbairn admitted, but he emphasized, "when he [the student] realizes that the enemy will show him no mercy, and that the methods he is learning work, he soon overcomes it."[45] Fairbairn taught the techniques not only as survival skills, but to build up a strong self-confidence that would enable agents to overcome their fears and be more effective in accomplishing their missions.[46]

Because the thin, gray-haired, bespectacled Fairbairn looked more like an accountant or a mild-looking bureaucrat than an expert killer, the overconfident, young trainees usually misjudged him at first. But they were soon impressed with what they saw. "He didn't seem like anything special," was the first impression of communications trainee Marvin S. Flisser, when he saw Fairbairn at Area C in late 1943, but he quickly changed his mind. Often the Scot would call out one of the largest students first and ask the young giant to throw a punch at him. If the youth refused to hit the older man, Fairbairn would kick him in the groin. The infuriated student would lunge, and Fairbairn would flip him onto the ground, face down, with his arm twisted behind his back. "He was really something else, I'll tell you. Quite a guy," Flisser recalled. "Everybody looked up to him. We didn't play around with him. He was very serious. He was all training. He wanted to make sure that we stayed alive! He taught us all the holds and moves. As a civilian afterwards, I was very careful not to get into fights or anything, because by this time I was sort of a –well, if somebody hit me, I knew what I was going to do! I thought it was dangerous."47

In addition to showing instructors and students how they could overcome their enemy whether unarmed or armed with a knife or a pistol, Fairbairn had brought the idea of a realistic testing facility, a "house of horrors" or what he called an "indoor mystery range," to the OSS. He had started these in Shanghai, then brought the idea to Britain and the United States. "In the Mystery Ranges," Fairbairn explained, "a simulation was affected of actual battle noises, conditions under which shooting affrays occur, especially in house to house combat. The training included methods of entering closed and locked doors, methods of bursting open such doors, methods of using trap doors, methods of roof top fighting and firing on moving and possibly concealed targets. Under varying degrees of light, darkness and shadows, plus the introduction of sound effects, moving objects and various alarming surprises, an opportunity is afforded to test the moral fibre of the student and to develop his courage and capacity for self control."48 For Donovan's organization, the first such "indoor mystery range," was built at Area B-2, then at A-2 (or possibly A-3), and finally at Areas E and F.49

When first assigned to the OSS in April 1942, Fairbairn trained instructors and students at Special Operations camps at Area B and Area A. Upon the formation of the first Operational Group (OG) units in spring 1943, Fairbairn was based at Area F, but this did not keep him from continuing his rounds of instruction and demonstrations on a routine basis at the other OSS training camps, Areas A, B, C, D, E, F, and RTU-11 ("The Farm").50 The official history of the Schools and Training Branch of the OSS, written at the end of the war, credited not simply Fairbairn's unquestioned knowledge and skill in close combat but also the strength of his personality in building a cadre of able American paramilitary instructors and raising their confidence and morale to a very high level.51

Hollywood Director John Ford and OSS Training Films

From the beginning, Major Garland H. Williams had insisted that instruction be varied and go beyond classroom lectures to include field exercises and training films. Donovan too wanted OSS to have photographic and motion picture capabilities not just for training but to record visually developments in modern warfare and the contributions of the OSS and to show them to the President and to the public. Thus, he set up Field Photographic Branch under Hollywood director John Ford and subsequently a Visual Presentation Branch. In addition to making documentary films in the war zones, Ford, his cameramen, and screenwriters like Bud Schulberg, produced training films.52 "OSS Training Group," for example, which shows students, Fairbairn, Applegate and other instructors, all with masks to conceal their identity from the audience, engaged in hand-to-hand combat, pistol shooting, demolition, and field exercises, was filmed under John Ford's direction at Area B in 1942.53 Trainees and instructors were forbidden to take photographs at the secret training camps, but OSS photographers and cameramen took pictures of training area Areas A, C, and other OSS schools.[54] The Visual Presentation Branch praised the cooperation they received from instructions at Area A while making a film there on Special Operations and demolitions but complained about difficulties of getting the approval from suspicious, mid-level officials in some other branches, such as the Communications Branch's Peter Mero, who was reluctant to let them film at Area C.[55] This situation was certainly rectified, because in the spring of 1944, Commo chief Lawrence Lowman praised the Field Photographic Unit for the "very valuable and excellent work' done in the "production of the rather complicated and highly technical film of the SSTR-1 [OSS's "suitcase" radio transmitter/receiver] set."[56]

Practicing Espionage and Sabotage Techniques in American Cities

Sometimes following completion of their courses, new SO and SI graduates were sent on local undercover operations, including mock industrial espionage and sabotage exercises in industrial cities. Lieutenants William R. ("Ray") Peers and Nicol Smith were in one of the first classes at Area B-2. Their final assignment was to infiltrate the Fairchild Aircraft Plant in nearby Hagerstown, Maryland, which they did, returning with a detailed plan of how to sabotage the facility.[57] Such realistic undercover exercises proved appealing to the OSS hierarchy, which in late 1942 approved a plan—they called it a "scheme"—to have SO students about to graduate at Area A in Prince William Forest Park try to penetrate war production plants in Pittsburgh. Their assignment was to gain employment at one of twenty selected plants using cover stories, fictitious names and forged credentials, to work there for a day, and then develop a theoretical plan for sabotaging that plant. Only the chief executive officers of the plants had been alerted. If any of the trainees were arrested, they were to be turned over to the FBI, with the head of the local office informed of the operation. The men were to be released after being interrogated. OSS considered Pittsburgh an ideal city as a testing ground for sabotage training, not only due to the extensive number of industrial plants, but also due to the large number of central European nationalities working there, with whom some of OSS's ethnic and bilingual students could operate with their undercover plans.[58]

Changing Uses of Area B

After sixteen classes had graduated at B-2, the training school there was closed temporarily in the late summer of 1942 because the War Department was slow in approving positions and assignments overseas for OSS graduates. While most of the headquarters cadre remained at B-5, half a dozen members of the command and instructional staffs spent six weeks studying at SOE schools in Canada or Great Britain.[59] Louis D. Cohen, who had gone to Camp X in Canada, went to Area A-2 as chief instructor in late summer 1942, and then to the new Area E north of Baltimore in November 1942. Elmer ("Pinky") Harris went to Camp X in August, then was sent to Area A. Area B-2 would remain largely inactive until the inflow of Operational Groups began in the spring of 1943. Meanwhile, among those who had been sent to SOE schools in England, only Blogg and Gleason returned to Area B. Sage went to Area D; Dow, Parkin and Collart were sent to Area A. Later Sage and Harris would leave for North Africa. Collart would join the airborne engineers and head for Europe. Dow, Parkin, and Gleason would go to China in spring 1943 to train Chinese guerrillas and to conduct Special Operations.

Training Operational Groups at Areas F, B, and A

Donovan had from the outset wanted to infiltrate ethnic Americans and foreign nationals into German occupied lands to encourage and lead guerrilla actions. Planning the training program in early 1942, Major Garland H. Williams wrote that "The special courses for racial or national groups will consist of the same instructional material as is given in the other courses, and the only changes will be those made necessary by the character of the group and the special area in which they are to operate. Special provisions must be made for segregation of these groups, and special instructors and instructional material will have to be provided."[60] Even before the term Operational Group was adopted in 1943, these kinds of groups trained at Area B in the spring and summer of 1942 and were composed mostly of foreign nationals. Demolitions instructor Lieutenant Frank A. Gleason remembered that in addition to Americans training separately at the Special Operations camp, "we trained groups of Thais, Norwegians, Frenchmen, and Yugoslavs, members of Mihailović's organization [Chetniks who fought against both the Nazis and the Communists]." "We even trained a group from Spain, or rather Americans who had fought in the Abraham Lincoln Brigade in the Spanish Civil War [in the 1930s], some of them Communists," Gleason added. "Donovan had gotten them into the OSS, and they were going to be infiltrated into Spain, because they knew the country so well. We trained them in sabotage and that stuff."[61]

In December 1942, the Joint Chiefs of Staff approved OSS's request to form Operational Groups (OGs). Foreign-language speaking Americans would be recruited from the armed forces and trained by OSS to operate as military units behind enemy lines in coordination with resistance groups.[62] Ability to speak a foreign language was a requirement. Unlike the smaller Special Operations teams, the OGs were military units and always served in uniform. Officially, each county-specific unit had four officers and 30 enlisted men, although the sections sent behind enemy lines frequently numbered 15 or less. Most of the Operational Groups were trained during the year between the spring of 1943 and the spring of 1944. There were French, German, Greek, Italian, Yugoslavian, and Norwegian Operational Groups. In 1945, the OSS recruited and organized Chinese Operational Groups in China; however, the Chinese OGs were composed of indigenous recruits, not Chinese Americans.[63] (Chinese Americans, Korean Americans and Japanese Americans, the third group despite the fact that many of their relatives were incarcerated in U.S. internment camps, volunteered to serve as interpreters with the

OSS in China and the Pacific.) Unlike Special Operations agents who were trained to work as individuals or in two-or three-person teams. Operational Groups were designed to operate as larger groups of uniformed combatants, sections of a dozen men, and their officers participated in their training. Much of the OGs' success would depend upon a cooperative spirit and teamwork within the groups.

French-educated Jacques F. Snyder, a member of one of the French OGs, being trained in Fall 1943, recalled that at that time, before the Americans had much experience with the French resistance groups, the *maquis*, the initial French OGs were told to operate on their own behind enemy lines and avoid local contact. This he said was because the average member of the French OG came from northern New England or Louisiana and their French Canadian accent was so strong that they could never pass as native Frenchmen. Agents speaking such a patois would quickly be identified, and because of the profusion of collaborators and would-be informants in France in 1943, they might be quickly betrayed. Consequently, the initial plans were for units to be infiltrated, hide in the woods, make attacks, and be quickly gotten out of the country. Only later, when the importance, real value, and trustworthiness of the French *maquis* became known did the policy change to encourage the OGs to make full use of the local underground for liaison and support.[64]

Preparation of Operational Groups in significant numbers began in the spring of 1943. They were to receive six weeks training in the United States followed by additional training overseas. Most of the OGs spent two to three weeks in preliminary training at Area F, the former Congressional Country Club, in Bethesda, Maryland. There, they underwent physical conditioning, training in various weapons and munitions, and field exercises. Afterward most of the OGs went to Area B for two to three weeks of intensive training especially in weaponry, field maneuvers, and demolitions. So many OGs were at Area B—more than 300 men in some instances—that they filled both B-2 and B-5 to capacity. Area A, specifically A-2 and A-5, also provided training sites for OGs, and some of the overflow went to Area D and Area C.[65] Many of the OG units and SO teams from Europe and the Mediterranean returned to the USA in late 1944 and early 1945, and a number of the Training Areas, particularly F, A and C were used as holding areas for them, while some of them were prepared for assignment to the Far East.[66]

Curriculum for Operational Groups

A written guide for instructors training Operational Groups emphasized the need for the trainees to achieve proficiency, self-confidence and determination and to recognize that unconventional warfare behind enemy lines was a hazardous undertaking and required not only much skill but a certain degree of ruthlessness. Students were required to master at least ten of twelve different techniques in martial arts, plus the use knives, to disable or dispatch an opponent in a course that was entitled "Silent Killing." Instructors were told to get the trainees away from ingrained ideas of fighting fairly. The use of normally unacceptable methods, the manual indicated, was to be justified to the students. "Naturally under such conditions as exist in present day warfare," the students were to be told, "one must have no scruples as to the way and manner in which it is intended to fight an opponent. The justification for using any foul method in accomplishing this end does not, when one stops to think what has already been used against us, have to be considered. It simply resolves itself into a matter of two things—kill or be killed."[67] The syllabus noted that since the time available to the students was limited, "it is essential, therefore, to confine the teaching to what is simple, easily learned and deadly."[68] Hanging dummies, some filled with straw were used, but for noiselessly attacking a sentry from behind, an individual with a German helmet helped in the exercise. In that case instead of a knife, a handle with the "blade" of thick rope was used. If anyone expressed qualms about "foul play," the instructor was to inform them that, "This is WAR, not sport. Your aim is to kill your opponent as quickly as possible. A prisoner is generally a handicap and a source of danger, particularly if you are without weapons.... 'Foul methods' so-called, help you to kill quickly. Attack your opponent's weakest points, therefore. He will attack yours if he gets a chance." Thus, students were instructed how to kill, temporarily paralyze or badly hurt their opponent by striking or jabbing vulnerable parts of the body. Stressing "attack mindedness," the instructor was to encourage students to attack and keep attacking until the enemy was dead. The syllabus declared bluntly, "this course of instruction is meant to teach you to kill."[69]

In the curriculum for OGs, the Preliminary Course, usually taught at Area F, but sometimes at Area A or B, began with an hour of Introduction and Training Objectives. Over the next few weeks, it would include 22 hours of Map Reading, Sketching, and Compass Work, both theoretical and field problems; 20 hours of Scouting and Patrolling; 14 hours of Physical Training; 7 hours of Camouflage and Field craft; 4 hours of Close Combat and Knife Fighting; 6 hours training on the Obstacle Course; 4 hours instruction on the .45 caliber pistol; and 4 hours on the submachine gun. There would be 7 hours of Training Films. The longest amount of time, 57 hours, was devoted to Tactics. That included compass runs, target approach, and day and nighttime field problems. Finally 2 hours were devoted to Hygiene and Camp Sanitation; and 4 hours went for Special Subjects: enemy organization, communications, security, and current events. Total OG preliminary instruction and training was 152 hours.

Then the OG team moved to Areas B or A, where the Final Course involved 8 hours of Physical Training, 22 hours of Demolitions, 40 hours of Weapons training, which included 2 to 3 hours each on the mechanics and firing of the M1 rifle, carbine, light machine gun, Browning Automatic Rifle, Colt .45 automatic pistol, British Sten gun, Thompson submachine gun, Marlin submachine gun; M1 and AT rocket launcher, 60 mm mortar, 81 mm mortar, and the .50 caliber machine gun. There was also a bit of hand grenade and anti-tank training. Ellworth ("Al") Johnson, remembers firing a bazooka at Area B, "just to get the feel of how it worked."[70] Thereafter students went through 4 hours on care of clothing and equipment, 4 hours on hygiene and camp sanitation, and 8 hours of training films. Finally, there was ground training for parachute jumps that would be made at Fort Benning or overseas. Total OG advanced training was 106 hours; grand total 250 hours.[71] In large part because of their success in World War II, the OSS Operational Groups became the predecessors of the U.S. Army's Special Forces.[72]

Instructors at Area B, during OG training, 1943-1944

The Operational Groups that filled both B-2 and B-5 during much of 1943-1944, trained under their own officers, but there were also some specialist instructors. Rex Applegate continued to teach instinctive firing techniques of pistol shooting. William ("Dan") Fairbairn continued to make his rounds of the training camps teaching his own methods of close combat. He was joined, beginning in February 1943, by an old friend Shanghai, Hans V. Tofte. A fine athlete and an exceptionally able student, the Danish born Tofte had mastered the art of silent killing under Fairbairn's guidance in Shanghai. After the Germans occupied Denmark, he had joined British SOE and led teams taking weapons to Chinese guerrillas through Burma. In the autumn of 1942, Tofte came to the United States and enlisted in the Army. Fairbairn had his former student transferred to the OSS as a special assistant.[73] Tofte worked out of Area F, but he spent considerable time at Areas A and B.[74] In addition to silent killing, the long-legged Tofte also tried to teach the recruits a new style of cross-country running he called the "elastic stride," in which instead of jogging, he bounded along, stretching and leaping, like an antelope.[75] After more than six months as an instructor, Captain Tofte was sent in August 1943 to SO in Cairo, Egypt and then Bari, Italy. At the latter, he headed Operation Audrey, supporting Tito's partisans in Yugoslavia.[76]

Among the American officer instructors at Area B were Lieutenants Paul A. Swank and John K. Singlaub. Both taught demolitions at B-2 in the fall of 1943. Singlaub was a graduate of the University of California at Los Angeles, and had joined the Army with a commission through ROTC there. He joined the paratroopers and was selected by SO at Fort Benning. Swank was a West Point graduate and a member of the Corps of Engineers. He was the demolitions instructor at B-2. Singlaub, who arrived as an SO student from Area F, the Congressional Country Club, broke his leg, and was then assigned to assist Swank, since Singlaub had learned American and British explosive at parachute demolitions school.[77] As will be described in a subsequent chapter, Singlaub and Swank later parachuted into different areas of occupied France in the summer of 1944 to lead French guerrillas in hindering German reinforcements against the Allied landings in Normandy and the Rivera. Singlaub returned to head a successful mission in China. Swank was killed in France.[78]

Cadre and Instructional Staff in Prince William Forest Park, Area A

Since OSS Area A was planned as the Advanced School for Special Operations, its instructional cycle was to follow from the basic SO training given at Area B. The first Commanding officer of Detachment A, Captain William J. Hixson, had arrived in April 1942 and established his headquarters at A-5. In November the headquarters was moved to A-4. Captain Harold Rossmiller of the Medical Corps arrived in the summer of 1942 to serve as medical officer for both Areas A and C in Prince William Forest Park. Rossmiller was sent to Algiers as the OSS prepared for the American invasion of Vichy French North Africa, and he was succeeded as medical officer at A and C by Captain Louis Lostfogel, who was transferred from Area B. Lostfogel remained at Areas A and C in Prince William Forest Park until the closure of Area A in early 1945.[79]

The first groups of advanced Special Operations students arrived at Area A in late May 1942. Captain George H. White, the former narcotics agent, who had been instructing in undercover work at B-2, was transferred down to A at the end of May to serve as the first chief instructor.[80] In November 1942, he was succeeded by Captain Arden W. Dow, one of the original instructors at B-2, who had just returned from SOE training in Britain. Under Dow were several friends and colleagues from Area B and Washington State University, among them Joseph H. Collart and Elmer ("Pinkie") Harris, now joined by former WSU football team captain Chris Rumburg. Outside the circle of Washingtonians were at least two men from Area B who came to Area A in late 1942: Charles Parkin, a Pennsylvanian who had taught demolitions, and James Goodwin a football player from upstate New York who had been an early student at Area B and then had attended SOE school in Britain. At Area A, the married officer instructors rented houses in the nearby towns of Triangle and Dumfries, brought their wives down and went home each night.[81] Unmarried officers, like Parkin and Rumberg, lived in accommodations in the park, as did most of the officers of the base cadre and all of the enlisted men, as well as the trainees and those who were being held there awaiting assignment or transportation overseas. Parkin recalled that during the winter of 1942-1943, "we cleaned up the old CCC camp [A-4, the new headquarters] there and make it livable [winterized it]. We mainly put the men through exercises, physical exercises. It was then being used mainly as a holding area."[82] In the spring of 1943, many of the instructors from Area A as well as B would be sent to active duty overseas: Dow, Gleason and Parkin to China; Collart, Goodwin, Harris, Sage, and Rumburg to the Mediterranean or European Theaters of Operation. Only Rumburg would never return.

Arden Dow, who left for China in March 1943, would remain there until May 1945, serving first as commander of an OSS training camp for Chinese guerrillas located just south of Loyang on the Yellow River. Frank A. Gleason from Area B was his deputy at the camp.[83] The school they established in China was modeled after the SO schools at Areas B and A. At the camp, nicknamed "Camp Rowdy," despite being located by an old Buddhist temple, twenty American instructors trained hundreds of Nationalist Chinese guerrillas to conduct raids and sabotage behind Japanese lines.[84] Later, Parkin personally commanded the first airborne unit mission behind Japanese lines in China (Operation Akron), a feat that earned him the Legion of Merit and a Bronze Star Medal.[85]

Four Sub-camps in Area A, Prince William Forest Park

Detachment A at Prince William Forest Park ultimately included four separate sub-camps: Areas A-2, A-3, A-4, and A-5. Their main role was training Special Operations agents, but they often performed other functions as well. The nature of the regular courses at the sub-camps changed over time in response to lessons learned at other camps or overseas. Courses varied in length, but they all assumed 40 to 48 hours of training per week.[86]

Area A-2, the former NPS Cabin Camp 2, Mawavi, was the first OSS camp opened by Detachment A. It began instruction in May 1942 as Advanced Training in Special Operations, and the emphasis was on organization, recruiting and handling indigenous resistance forces and the conduct of sabotage activities. The faculty included Captain George White, as chief instructor, later succeeded by Captain Arden Dow, plus Lieutenant Christopher Rumburg and two sergeants. An inspector from headquarters reported in early July 1942 that students attended lectures and also engaged in target practice, each given a quota of 100 rounds of .45 caliber night pistol firing at bobbing targets, and 15 rounds of submachine gun practice. "Advance demolition practice is given with actual targets such as concrete retaining walls and abandoned houses. Advanced booby trap training is given with electrical and time fuse connections." The inspector also reported that "excellent meals are served, considerably better than those are regular Army camps."[87] The following year, A-2 was used to train or hold Operational Groups waiting their combat missions abroad. Then, beginning in the spring of 1944, it was used to give basic military training to OSS personnel, when the Army required this for everyone being sent overseas.[88]

Located near A-2, Area A-5, which had been NPS Cabin Camp 5, Happy Land, served briefly as the first headquarters for OSS Detachment A before the command center was transferred to A-4 in November 1942. Thereafter, for much of 1943, A-5 was used as a holding area for individuals or Operational Groups. In late 1943, it initiated a 118-hour course with particular emphasis on physical training, weaponry, and demolitions. In spring 1944, A-5 was used as a holding area or finishing school for SO personnel who had completed all their other courses.

Area A-3, which under NPS had been Cabin Camp 3, Orenda, served temporarily in the summer of 1942 under Captain Louis O. Rucker as the OSS Parachute School in connection with the airstrip at the Quantico Marine Base. In the fall of 1942, Rucker was joined by Marine Lieutenant Elmer ("Pinky") Harris, just returned from Camp X training in Canada, as an instructor in the OSS parachute training unit.[89] After Rucker and Harris were sent to North Africa, A-3 seems not to have been much used by OSS until October 1943, when it was designated to provide SO and MO students with "E-type" training. The "E-type" course, named for Area E where it originated, gave an overall picture of the work of the operational branches of OSS in order to improve their coordination in the field. The 156-hour course included a little of everything: undercover techniques, intelligence gathering and reporting of SI, the counter-espionage of X-2, the sabotage, weapons, demolitions, close combat techniques of SO and the spreading of rumors and other disinformation of MO. Like the early SO and SI training at Areas B and A, it concluded with a three-day undercover penetration by Special Operations and Secret Intelligence trainees of defense plants, usually in Baltimore, Philadelphia or Pittsburgh. The E-type course was given at A-3 from late 1943 to November 1944. A-3 also served at different times as a holding area for men awaiting training or assignment elsewhere. In actual practice, however, the SO Branch during 1943 came to see A-4 as the basic SO training area (as SI came to see Area E as its basic training school).[90]

The use of the "E-type" general, introductory course at Area A-3 came about because of the dissatisfaction of SO and MO with the SI orientation at Area E. In October 1943, Anthony ("Tony") Kloman, a civilian instructor at Area E was assigned to replicate the course as chief instructor at A-3.[91] When the first class began late in the fall, the cabins and other buildings at A-3 remained fit for summer use only. Consequently, Kloman and his instructional team had to teach the classes while the buildings were being winterized. In February 1944, A-3 had seven civilian instructors, who taught police work, field craft, knife and close combat, SO material or MO material, and seven military instructors, who taught weaponry, demolitions, maps and ciphers.[92] By spring 1944, fewer SO agents were needed in Europe and enrollment at A-3 declined as did the instructional staff headed by a civilian, H.B. Cannon from Area E. The course shut down from March to June 1944. It reopened under Captain Eldon Nehring from Area E, who served as chief instructor at A-3 from July to November 1944, when he left to become chief instructor at an OSS training school in China. A-3 closed in December 1944.[93]

In early 1944, while A-3 was simply a holding area, several recruits for the Communications Branch, awaiting openings at Area C, were temporarily housed there. Perhaps they were bored or maybe they just became friends; whatever the reason, and fortunately for us, they told each other their names and provided information about their backgrounds, despite OSS rules against it. They represented some of the diversity in the OSS. David Kenney, 19, had grown up in a hamlet of 400 people on the windswept grasslands of southern Wyoming. When he graduated from high school, he was drafted into the Army and trained to be a radio operator. Young and eager, he volunteered for hazardous duty and wound up in the OSS. Jerry Codekas was a Greek-Cypriot American, who later became part of one of the Greek Operational Groups. After the war, Codekas had a date growing farm near Palm Springs, California, and Kenney who also settled in California, renewed their acquaintance. Leonard Iron Moccasin was a Sioux/Lakota Indian from South Dakota. "I asked him, what was an iron moccasin?" Kenney recalled. "He said, it was a horseshoe....He lived on a reservation and was a high school teacher there. I don't know what he taught. He was very talented. He was good with electronics. He and Jerry [Codekas] disappeared from me in England [where they were sent in July 1944]. I don't know what happened to Leonard [Iron Moccasin]."[94]

Area A-4, a former CCC work camp, had been opened as an OSS facility in November 1942 as the new headquarters of Detachment A as well as a training camp. During the winter of 1942-43, Area A-4 replaced B-2 as the preliminary training area for Special Operations (Area B-2 would resume in spring 1943 as the advanced OG training camp). Under Schools and Training's attempt at centralization in the winter of 1942-1943, the two-month, S&T program of training each OSS operational recruit, whether SO, SI, MO, or X-2, began with two weeks preliminary training at A-4 (although many of the older men in SI skipped it as too strenuous). The "A-4" course, as it came to be called, was an intense, basic paramilitary program that could be given to people from any branch. Initially two weeks, its 130 hours of instruction included 27 hours of field craft, 23 hours of demolitions and explosives, 23 hours of weaponry, 12 hours of code and ciphers for wireless radio messages, 10 hours of map reading, sketching, and compass, and then assorted other instructions. Each class consisted of fifteen to twenty men and two classes generally running concurrently on a staggered basis. A constantly changing roster of some twenty to thirty instructors maintained the course.

One of those instructors at A-4 in 1943 was a future "Jedburgh," Captain Francis L. ("Frank") Coolidge, who had gone through SO training at Area B and then Area E. It was probably during that instruction at Area A that he met Marine Major Peter Julien Ortiz who underwent more OSS training after recuperating from a wound suffered in Tunisia. Ortiz, the son of a French-Spanish father and an American mother, would receive the Navy Cross, the second highest military decoration of the U.S. Navy (after the Medal of Honor) for his first parachute mission into France in 1944. As will be seen in a subsequent chapter, his second, in August 1944, in mission that included Frank Coolidge, was not so fortunate. One team member was killed in the parachute drop. Later the team was trapped in a village by the Germans. Ortiz and two sergeants surrendered to save the villagers from Nazi reprisals. Coolidge, although wounded in the leg, and another sergeant were able to escape unnoticed by the Germans. Freed from a German POW camp in April 1945, Ortiz was awarded a second Navy Cross, and with his many other medals became one of the most decorated U.S. Marines in World War II.[95]

Under the centralized plan of Schools and Training Branch, the preliminary paramilitary training at Area A that began in 1943 was followed by two more weeks of advanced agent training at Area E, the newly established training school in country estates near Towson and Glencoe, Maryland. Then there would be two more weeks at the specialist and finishing school at Area RTU-11 ("the Farm") near Clinton, Maryland (although few SO agents went there in practice), and additional weeks in parachute training at Fort Benning, Georgia, or in advance SO and sabotage and guerrilla techniques at Areas A or E. Ultimately, however, the two-week course at Area E evolved into a basic OSS agent course, and the SO and SI students were given advanced and finishing training in their own separate areas under their own training staffs.[96] Beginning in February 1944, as a result of requests from several OSS branches, the training course at A-4 was expanded to three weeks, increasing the number of hours from 136 to 194, and including more instruction on techniques of operating in the field.[97] In July 1944, after the bulk of SO and OG personnel had been deployed following the Normandy invasion, A-4's role was reduced, the course was cut back to 132 hours, and it became simply another site for advanced SO training.[98] Still, other uses of A-4 were training for some of the Operational Groups and also for communications training for SO or OG radio operators members who needed it for work behind enemy lines.[99]

Although Area A began as the Advanced SO camp, it functioned as the main training center for Special Operations agents from late 1942 through early 1945. Its instructors trained individual agents or two-or three-man teams who would conduct sabotage, espionage, guerrilla activity, or hit-and-run raids. They also trained Operational Groups and some Morale Operations personnel. The various subcamps—A-2, A-3, A-5, and most especially and continuously A-4—offered basic or advanced courses ranging from two to four weeks in length. As the official history of the Schools and Training Branch indicated, "the A training areas during their period of operation built up a creditable and enviable tradition. No only did students see action and engage in direct operations against the enemy, but many of the instructors, too, were assigned overseas in active theatres on operations—where they really wanted to be anyway."[100] As will be seen in subsequent chapters, several of them won medals there. For some of them, like A-4 instructor Lieutenant J. Holt Green from an old Charleston family who subsequently led missions into Yugoslavia and then Czechoslovakia, the award was posthumous.[101]

SO Trainee George Wuchinich, Serbian American from Pittsburgh

The anonymous student identified only as "George" in the history that Schools and Training Branch wrote at the end of the war was, in reality, Lieutenant George Wuchinich. In many ways, he had a typical student experience in the early SO training program at Areas B, A, and C. [102] Like many so many members of the OSS, however, he was also an extraordinary individual. A second-generation American, whose parents were Eastern Orthodox Serbs from Slovenia, Wuchinich had been born in Pittsburgh, Pennsylvania. As a youth, he worked and saved enough to attend Carnegie Tech but became jobless during the Great Depression and hitchhiked across the country looking for work. In 1936, he considered enlisting in the Abraham Lincoln Brigade to fight against Franco in the Spanish Civil War but was deterred by pleas from his mother and sister. He did find employment in Pittsburgh, and the big, energetic, ebullient Wuchinich gradually worked his way to become a top salesman of major industrial steel products from locomotives to tugboats.

Hitler's invasion of Yugoslavia in 1941 led Wuchinich to enlist in the U.S. Army, and he was soon recruited into Donovan's organization.[103] Wuchinich was a member of the fourth class (Class G-4) to go through the OSS SO training process in spring 1942. The first week of training at Area B in Catoctin Mountain Park consisted of an introduction to military life, since unlike Wuchinich many of the recruits had come directly from civilian life. The second week introduced the students to map-reading, field-craft, protection against poison gas, familiarity with weaponry, including small arms and also a 60 mm mortar. The final two weeks involved lectures and demonstrations of close-combat techniques, house-to-house fighting, cryptography, and the nature and ideology of the Nazi regime, German and Japanese unit identification and order of battle, surveillance, photography, sketching, interrogation, and some basic demolition work. In the final field exercise, the graduating class had to penetrate an "enemy" camp, overpower the guards, generally obtained from enlisted men on the training staff, but sometimes played by other students. Accomplishing that objective and passing all written tests and personal evaluations, the class graduated in June 1942.

Following graduation from Area B and a brief leave, Wuchinich and his classmates were taken to the Advanced SO School at Area A in Prince William Forest Park. The emphasis there was on direct action, and the three main subject areas included Organization; Recruiting and Handling of Indigenous Agents; and Selection and Execution of Sabotage Activities. At Area A in the summer of 1942, Wuchinich found no emphasis on military routine. Instead there was an informal atmosphere, much different from a regular Army facility. Still the training was rigorous, from 7 a.m. when training began with calisthenics and swimming until 5 p.m. when it ended with close combat, although in some evenings, instruction continued with night field problems. In this advanced course, the trainees learned the finer points of agent operation. That included actual practice in listening devices, the use of secret inks for passing messages (a course later dropped by SO but maintained for SI agents), propaganda and influence and control of civilian populations. Wuchinich and the other students were given problems in reconnaissance as well as concealment of weapons, radios, and other equipment. The students became proficient and self-confident in close-combat, the use of American and foreign weapons, and most importantly, in sabotage—the employment of various types of explosives to destroy bridges, road lines, industrial plants or warehouses, or even ships in harbors. As a graduation exercise, individuals or teams of students were required to penetrate an industrial plant in a nearby city and return with information about its operation or how it could be disabled by explosives.[104]

The final part of the Special Operations training cycle for Wuchinich and his classmates involved three more specialized qualifications. At Maritime School, either at Area A or Area D, they spent two weeks learning about infiltration by water into hostile areas. Then for two weeks at Communications School in Area C they went through a crash course in ciphers, telegraphy and wireless radio operation. Finally, they went to Parachute School either at Area A or at Fort Benning, Georgia, and earned their paratrooper's wings after completing the required five parachute jumps.[105]

After completing his training by the end of 1942, SO agent Wuchinich was ready to be sent behind enemy lines in German-occupied Yugoslavia. But his and others' dispatch was held up for several months. The reasons were complex, involving an initial reluctance by the Army and particularly its theater commanders to authorize significant numbers of positions for OSS personnel, delays in political clearance by the State Department, disagreement within the OSS, and, in this case, problems caused by the internecine fighting within the Yugoslavian resistance movements, between royalist Chetniks under Mihailović and communist-led partisans under Tito. With a number of proposed OSS missions cancelled, the Yugoslavian-American trainees like Wuchinich, who had been brought to a peak of eagerness and self-confidence, were suddenly left with nothing to do but wait.[106]

Lieutenant Colonel Garland H. Williams, who had been in charge of the SO training program through August 1942 had been aware of this type of problem for recent graduate and had insisted that there must be "a planned course of instruction which will not permit a lessening of his enthusiasm or desire to active work."[107] Wuchinich did take some "refresher" courses, but as he later complained: "Delay is the agent's greatest frustration, hanging around holding areas waiting to get overseas, hanging around field bases waiting to get a plane, hanging around in a plane searching for signals to make a pinpoint landing, then losing it entirely and being forced to come back to base. All this gets on a man's nerves. And there's nothing that can be done about it!"[108] Finally, in February 1943, after near a year in training and holding camps, Wuchinich finally found himself on his way to Yugoslavia. Despite further delays, he eventually led the "Alum" team that parachuted into Tito's camp. He would spend considerable time working with the partisans in Yugoslavia and engaged in a number of daring operations, as will be described in a subsequent chapter. George Wuchinich came home with many stories and a Distinguished Service Cross.[109]

Friction and Demoralization among the Staff at Area A

Although many of the original C.O.s and instructors at Area A left in the spring of 1943 for assignment overseas in the European, Mediterranean or China-Burma-India Theaters of Operation, new officers and civilians took their places. Sometimes there were personality problems. Captain Joseph J. Grant, Jr., became commanding officer of Detachment A in 1943. He had been a member of the first group of Americans to attend SOE training at Camp X near Toronto in early 1942.[110] Unlike most OSS officers, Grant was punctilious, issuing dozens of regulations and alienated a number of his subordinates. The record reveals increasing friction between the testy C.O. and his chief instructor, Captain Eliot N. Vestner. Admittedly, the chain of command between the commanding officer and the chief instructor of the OSS training camps had never been entirely clear, and they each reported to different superiors at OSS headquarters. But Grant's abrasive style strained that inherently ambiguous relationship.

Soon after taking command, Grant launched a series of caustic memoranda. Chief Instructor Vestner became the target of his ire. On 10 May, Grant dispatched a memorandum demanding that Vestner explain why fire guards and firefighting equipment had not been posted during mortar target practice when, the target, an abandoned house had caught fire and burned to the ground, "seriously endangering the countryside." Two days later Grant demanded to know why Vestner had not responded, and fired off a copy to his superiors in Washington.[111] Vestner answered the next day, stating that he was responding within forty-eight hours and had had to wait until the instructor involved had free time to write a report. The fire had never endangered the countryside, Vestner insisted; it had been prevented from spreading to the grass around the house. It was a small, dilapidated house, he indicated, that had been gradually being destroyed over the past five months as a target for mortar practice. Discussions with the National Park Service about fire hazards, Vestner wrote, had concerned only demolitions using explosives, not mortar rounds which flare upward on contact. The vexed chief instructor added a personal note to Grant: "Living as we are in close association with one another, it seems absurd that you feel it necessary to waste time and correspondence when the same ends could be met by friendly contact."[112]

Within the next few weeks, Grant issued a plethora of regulations for Area A, each with penalties for infringement. He imposed new speed limits, requirements to read bulletin boards, prohibition of swimming in park waters without a lifeguard, mandates about taking bottles away from the PX, restrictions on late night snacks at the mess halls, and requirements that beds be made up according to Army regulations. Violations would result in "appropriate disciplinary action."[113] When there were complaints about Grant's latest set of rules, and perhaps Captain Queeg-like behavior, Grant's superior in Washington required an explanation. Grant's justifications only confirmed reports of their demoralizing impact, and the head of the Schools and Training Branch required Grant to send all future camp regulations he wished to promulgate to him for review.[114]

Most embarrassingly in the midst of this exchange, Grant humiliated two new civilian instructors at Area A in mid-June 1943. Chief instructor Vestner had invited the new instructors to eat with the officers, but when he learned of this, Grant immediately issued an order forbidding these civilians to use the officers' dining facilities or sleeping quarters. Not being officers, they were relegated to using the eating and sleeping accommodations of the enlisted men or the students. When OSS headquarters learned of this insulting and demoralizing edict, they quickly reversed it. They also recommended an apology, which Grant declined to make.[115] An inspector from headquarters concluded that Grant was in the wrong job: "You can't make a silk purse out of a sow's ear."[116]

Grant was later transferred. He was succeeded as commanding officer by Captain Don R. Callahan, his adjutant, who remained C.O. at Area A for a year. Chief Inspector Vestner left in the fall of 1943 and was succeeded by a series of officers until March 1944 when Captain Stephen W. Karr, became the last chief instructor, serving until September 1944, when training essentially ended at Area A. Although the instructors and trainees left, a reduced base cadre remained, commanded by Captain James E. Rodgers, former supply officer at the camp, who served as the last C.O. of Detachment A. The unit was shut down in January 1945, and responsibility for it was shifted to the commanding officer of Area C in the eastern section of Prince William Forest Park.[117]

Problems with the British SOE Training Model

Admitting that SOE's assistance had been "indispensable" at the beginning of SO and SI training, OSS Schools and Training's official history declared at the end of the war that "a number of the original tenets laid down later either proved unrealistic or were outmoded…."[118] Some problems proved difficult to eliminate. By necessity the initial training had been based on the experiences of SOE, which had been in the war for two years before the Americans entered. But the British organization had only been created in 1940 and did not have the maturity and solid experience that its officers often led others to suppose. The British instructors at Camp X in Canada and then in the United States—men like Major Brooker, Lieutenant Colonel Skilbeck, and Major Dehn, who lectured at OSS training camps in 1942 and 1943—had little or no personal commando experience and were a long way in time and distance from the active theaters of the war in Europe, the Mediterranean, and the Far East and were thus removed from current developments in the theaters effecting the training goals. Finally, because of the bureaucratic division in the British system, SOE's functions were primarily special operations and psychological warfare, not foreign espionage and intelligence which were assigned to the entirely separate Secret Intelligence Service.[119]

British SOE courses and instructors stressed the secret means by which agents achieved their purpose in special operations, disinformation distribution or intelligence gathering and reporting. Their specialty was projects for Special Operations or Morale Operations Branches—hit-and-run raids, sabotage, black propaganda efforts—rather than Secret Intelligence. They emphasized agent techniques and provided examples of successful and unsuccessful undercover operations by British or German agents. In doing so, the focus was particularly on secrecy and the need for cover. This "cloak and dagger" approach, as it was called, permeated the entire recruitment and training process. During the course of the war, OSS Schools and Training Branch came to conclude that this extreme emphasis on cover and on means rather than goals was misdirected and problematic. It was not until 1944, asserted the official history of S&T, that OSS training "finally achieved emancipation, largely discarded the 'cloak and dagger' approach, and got down to brass tacks on such questions as 'What is intelligence?', 'How do you get intelligence?', How do you report intelligence?', 'Who is a customer for what kinds of intelligence?', 'Exactly how may a power plant be most effectively sabotaged?'"[120]

This was the result of American pragmatism: the test of how things actually worked. As S&T reported: "This change came about when OSS experience in the field demonstrated that 'intelligence's the thing', and many amateur but aggressive operators who possessed none of the finesse of the classical figures [in espionage and sabotage], oftentimes produced just as good results in the midst of the hurry and confusion of the war by keeping their eye on the objective rather than the means."[121]

Operational Group (OG) Training and Field Exercises

A guide for training the Operational Groups at Areas F, B, and A was prepared by Lieutenant Colonel Serge Obolensky, the former Czarist Army officer and New York socialite, with the assistance of Captain Joseph E. Alderdice. "The greatest operational value of the groups," the training manual indicted, "is the breaking down of enemy communications and supply lines and the uplifting of the morale of the subjugated peoples in occupied areas, who need only be organized, supplied and led to offer effective resistance to the enemy."[122] In a lecture on guerrilla warfare, the instructor was told to emphasize that the main objects are "to inflict the maximum of damage to the enemy and to force him to tie up regular troops who might otherwise be engaged in the major effort."[123]

The two-week specialized training course for OGs at Area A-2 in August 1943, for example, included 12 hours of physical training and 43 hours of instruction in the use of various forms of explosives—TNT, nitrostarch, dynamite, nitramon, composition C and C-2—plus various forms of blasting caps, fuses, and primacord. These were for demolishing wooden, steel or concrete structures—bridges, buildings, railroads, ships, oil tanks and refineries, ammunition dumps, dams and powerhouses. The trainees also learned how to use anti-tank mines, anti-personnel mines, and booby-traps. They spent 23 hours learning tactics, including both day and night field problems. They were given 33 hours of instruction and practice with a dozen different types of weapons. They also spent three hours watching training films. The total was 114 hours of instruction. Training lasted every day, from 6:15 a.m. to 5:30 p.m., plus some night exercises.[124]

One of the original Operational Group officers, Lieutenant Aaron Bank, who later would become known as a founder of the U.S. Army's Special Forces, contrasted the situations at Areas F and A as they received their OG training in 1943. At the Congressional Country Club and its vicinity, during the tactical field problems in the area, the trainees were often frustrated when after carefully creeping and crawling through the underbrush toward their targets such as guarded road bridges and culverts, their silent progress would be interrupted with shouts of "There they are!" from neighborhood youngsters, who made a game out of spotting the stealthy trainees. But this kind of harassment was absent when the OGs moved to Area A at Prince William Forest Park. It was completely isolated. Bank also thought the training was better there. "Here [at Area A] our exercises lasted longer and were more vigorous and we did perfect ourselves as commandos with a guerrilla flare… Physically, the unit was in really good shape: no calisthenics, but good hard rope climbing, chinning, pushups with a knapsack on our back, and crossing streams on ropes slung horizontally between trees, topped off with five-plus miles of daily running. We had a good martial arts instructor and we specialized in knife fighting and throwing, a silent form of killing."[125]

Field Exercises in and around Prince William Forest Park

There were thousands of acres of varying types of terrain in Area A in Prince William Forest Park, and the OSS used them as well as areas outside the park for its training exercises and tactical problems. For field exercises outside the park, the Occoquan Bridge and Power House and the Manassas Power Dam and Control Tower were available; so were the Richmond, Fredericksburg and Potomac Railroad Bridge at Woodbridge, and the Fredericksburg Highway Bridge.[126] In one exercise, for example, after studying explosive charges and their placement in the classroom using drawings and models, the students were then were called upon for a field test. The problem site was eleven and a half miles north of Area A-2. The mission was to stage a night sabotage raid that would in theory drop the steel-trussed Bland's Ford Bridge into the creek, destroy the locks of the adjoining dam, and blow up the turbine generators in the power house. After midnight, the class, along with the instructor, was trucked to the target area, where the students would make a reconnaissance through the woods and the instructor would show them where to place the demolition charges and how much explosive to use. Each student was told to assume that he was the leader of a 16-man OG section. After the 2:20 a.m. attack, the group was to rendezvous at a friendly farmer's house as quickly as possible. In this tactical problem, instead of actually re-enacting the attack, each student would plan and write out the calculations for obtaining each objective—who would do what, when, where and how, including a diagram of the placement of each charge—from the initial assembly point in the woods to the placement of the charges and then the final rendezvous. Afterwards, there would be a critique and a discussion of the problems. The entire night exercise was scheduled to take four hours, from the time the class left Area A-2 for Bland's Ford Bridge, to the time it returned to the training camp in Area A.[127]

In some field problems, there were more realistic exercises involving actual rather than theoretical maneuvering, but these were only held within the OSS training area itself. In such cases, some members of the cadre were dressed as enemy soldiers, and staff officers served as umpires for these small unit "war games." In one field problem labeled "Clearing Buildings of Personnel," the target was an enemy command headquarters, located some seven hundred yards north of Joplin Road, which was patrolled by motor vehicle every hour on the hour. Allegedly the target was the headquarters for an enemy regiment, bivouacked at Triangle, Virginia. The mission of the OG team was "to attack enemy command headquarters, located at (Point X on sketch) and eliminate all personnel quartered there, and secure all the information possible."[128]

Such exercises proved quite similar to much of what the SO teams and OG units did when sent into action overseas—blowing up bridges, tunnels, warehouses, power stations—to impede the enemy. SO and OG operations were conducted in a dozen or more countries in all major theaters of operation. Probably the largest effort was put into operations in support of the invasion of France in 1944. There 85 officers and radio operators in American SO teams and 83 Americans in multinational "Jedburgh" teams parachuted mainly from Britain into northern and central France between June and September 1944. The Jedburgh teams later received considerable publicity for their successes in blocking German reinforcements, but the 85 members of regular SO teams and the 355 members of 22 Operational Group sections, although receiving less publicity, were no less successful overall. In support of the Allied invasion of Normandy and of the Rivera, the OGs were parachuted into southern and central France by planes from Algiers. Like the SO, the OG units had to depend on the protection of resistance groups and their own ingenuity to avoid capture since they were in enemy occupied territory. Of the 523 American SO and OG personnel, who worked behind enemy lines in France in 1944, 18 were killed, 17 missing or captured, and 51 were wounded in action. Total American SO/OG casualties: 86 or nearly 17 per cent of those involved.[129]

A Faltering Beginning at Area C

In the northeastern part of Prince William Forest Park, the OSS had established Training Area C in the early summer of 1942. Initially run by the Special Operations Branch to provide SO and SI agents with a smattering of knowledge about OSS communications and equipment, the so-called SO-SI School of Communications had not been very successful that summer. Few of the SO and SI trainees had any previous telegraphy or short-wave radio experience, but had simply had shown some aptitude for it. After the excitement of the regular SO or SI training camps, students found themselves in a rather makeshift and demoralizing situation. An anonymous student reported in August 1942 that the camp was seriously pervaded by a "lack of morale." In contrast to the exciting paramilitary training at Area B and A, he complained that "radio work is thankless and unrewarding. None of the students had entered this course from any real love of the work. Not a single thing was done at this camp to keep the spirits of the students up....the morale of the complement [the cadre] was poor also. The blame for all of this should be laid directly at the feet of the officers who either do not remember or have never been told that their function is more than the simply mechanical administration of the camp plus some radio repair work. The rest of the course itself is quite interesting for those who like radio work. For the others it is dull and uninspiring. These others should be told clearly and explicitly why they are here. Otherwise they are bound to become very depressed."[130] Not only were the instructors lackluster, but apparently the initial commander was a martinet. Two years later, an S&T report on an interview with returning veteran George Wuchinich, who had between a student at Areas B, A, and C, stated that in the summer of 1942, Area C was "a tough camp for student-agents. It was the only one of our schools at the time to subject them to some kind of GI regulation such as K.P., bed-making, and restricted liberty, etc... certain of our more sensitive students had objected violently to this kind of treatment."[131]

Creation of the Communications Branch School

When the new Communications Branch (CB) took over at Area C at the end of 1942, it transformed the Communications School. As indicated in the previous chapter, when Marine Major Albert H. Jenkins arrived from his previous position as an instructor and executive officer at SI's RTU-11 ("The Farm"), in December 1942, he oversaw the arrival of new equipment and new instructors, new staff, and the creation of more comfortable year-round accommodations. Included in his improvement plan and in keeping with the OSS policy, were plenty of physical exercise, meals that were far superior to regular Army bases and an overall atmosphere that was informal rather strict military. Morale was also improved by the fact that most of the students at Area C were recruits for the Communications Branch, learning to be operators and technicians at the regional base stations that the Commo branch was establishing around the world. Some of them would also be radio operators in the field. Most of the recruits for SO, SI, MO or other operational branches who would be given brief training in wireless telegraphy, codes, and radio operation, would thereafter generally receive their instruction from Commo instructors at their branch training camps, from Area A to Area F. At Area C, the Cadre lived primarily in Area C-1, the former NPS Cabin Camp 1, called Goodwill. The Communications School itself, with accommodations for students and most instructors, was half a mile away in Area C-4, the former NPS Cabin Camp 4, called Pleasant.

Major Jenkins, commanding officer of Detachment C, may have been relaxed in regard to military protocol, but he had a serious, no-nonsense attitude toward the essentials of training men for their missions overseas. As Lawrence ("Laurie") Hollander, one instructor remembered, Jenkins "was obsessed with the thought that everyone in the outfit, military or civilian, should be physically fit and be qualified in the use of weaponry. Classes in these categories were compulsory."[132] He often gave instruction personally in the use of the Colt .45 automatic pistol. But on the whole the paramilitary training in weaponry, demolitions, close combat, and field craft, given to Communications Branch trainees at Area C was provided mainly by SO instructors from Area A, augmented by the peripatetic William E. Fairbairn.[133]

Every new arrival at Area C, beginning in August 1943, received a typed, four-page copy of "Information and Regulations for All Personnel at Detachment 'C'" signed by Major Jenkins. It was straightforward and precise, written in easily understood language, not legalistic bureaucratize, and it typified Jenkins view that these OSS trainees should be treated as adults with their own responsibility for following rules which were explained and justified not simply promulgated. He relied on the self-discipline of individuals and their own motivation. The instructions were simple and direct.[134]

In February 1945, Major Jenkins retired as commanding officer at Detachment C. "He was a tough nut, an old Marine.... You did not want to cross him. A tough little guy. He let you know that he was in command and you would do what he said. He was a disciplinarian, not unfair, but tough," recalled Joseph J. Tully, an enlisted member of the cadre. "When he retired, we threw him a hell of a party."[135]

Jenkins was succeeded as C.O. by Captain Howard E. Manning. Veterans remember Manning as a fine commander, a pleasant, serious officer, who, in keeping the OSS tradition did not insist on strict military protocol.[136] Manning came from Chapel Hill, North Carolina, where his father had been dean of the medical school and his grandfather, dean of the law school. Howard Manning was graduated from the University of North Carolina and then Harvard Law School. He returned home to become a practicing attorney, and an amateur ham radio operator.[137] In January 1941, at age 26, he enlisted in the Army as a private but within a year had risen to sergeant-major and after OCS was commissioned a 2nd lieutenant in 1942. Assigned as an instructor at the Field Artillery School at Fort Sill, Oklahoma and then Camp Roberts, California, Manning earned excellent reviews and promotions.[138]

Jenkins' replacement was personally selected by Colonel Henson L. Robinson, head of the Schools and Training Branch. He picked Manning because, Robinson thought the was "ideally suited by civilian background and military experience to do a superior job" as commanding officer at Area C, "where a knowledge of radio communications, ability to command troops, and supervise and control the varied and complex administrative functions incidental to camp operation" was required.[139] The tall, imposing 30-year-old Manning replaced the small, wiry, and aging Jenkins. Manning brought a new burst of energy to the command, and Donovan himself came down to Area C shortly after Manning had assumed command.[140] The two COs at Area C had different leadership styles. Jenkins was commanding but Manning was persuasive.[141] Effective accomplishment of mission was a strong point with Manning.[142] "He was an aggressive man and a tough man. He could control other men," his widow recalled. He was also imaginative, she said. "He told me one time they needed a building, and he made a 'midnight requisition,' that is, he stole an entire small building and bought it to the camp."[143]

In addition to the commanding officers, the cadre of Detachment C included an adjutant, supply and mess officer, and an officer in charge of transportation, the post exchange, and the theater, plus a medical officer. From late 1942 through 1945, the medical officer for both Areas A and C was Captain Louis Lostfogel of the Medical Corps. His infirmary at C was like a doctor's office; it had no beds. If someone was sick enough to stay overnight, they were sent to the medical facility at the Corps of Engineers School at Fort Belvoir.[144] One of the officers at Area C developed spotted fever after being bitten by a tick and died in June 1945.[145] In addition to the officers, the cadre at Area C included 25 sergeants and 32 privates, under the command of a First Sergeant, who, at least in 1944-1945, was Leonard ("Len") Putnam from Manassas.[146] They included clerks, cooks, mechanics, drivers, firemen, projectionists, even a camp barber.

Perhaps the most popular officer in the camp was Captain John F. Navarro, the supply and mess officer. Navarro had been a restaurateur in New England, and he ran the mess hall, which was open to officers and enlisted men, like his family restaurant in Boston. He brought in supplies of fresh food from the Marine base at Quantico.[147] All the veterans of Area C agreed: "We ate very well," as staffer Joseph Tully remembered. "We had two of the best cooks in the Army…our own bakery, etc., fresh donuts, fresh Danish every morning."[148] "The food was out of this world," said trainee Marvin S. Flisser. "It was the best food you could get….We got everything fresh. We didn't get powdered eggs. We got everything we wanted like in the best restaurant. They had all kinds of chops and steaks. On top of which, there was one area, one corner of the mess hall, where there were refrigerators, big, huge refrigerators, complete with all kinds of food for us to use…. When we came back at night [from field exercises], we were free to go into that kitchen and help ourselves. And we all did, making salami and all kinds of sandwiches and coffee. They had it all set out. There were people doing guard duty, and there was food for them [too]. The only thing we had to do was clean up after ourselves. The word from the captain was, he never wanted to hear anybody say that they were hungry….This was not regular Army fare. We used to call it the 'silver foxhole.' We all put on a lot of weight."[149]

Civilian and Military Instructors Teaching Commo Recruits

The instructional staff at Area C was divided into two groups. The communications instructors were initially mainly civilians, hired to teach Morse Code, OSS ciphers, and the operation and maintenance of transmitting and receiving equipment. Numbers varied, but at one point there were nine civilian instructors, three officers, and five enlisted men. The paramilitary instructors, were all uniformed personnel, half a dozen officers and a similar number of enlisted men, assigned to teach weaponry, close-combat, and field craft.[150] By August 1943, the beginning of a peak period of training when there were 210 students at Area C, the cadre numbered 84 to keep the camp running and the instructional staff numbered 27 civilian and military men.[151]

Captain Paul M. McClellan was the Chief Instructor, who administered the instructional program, at least in the latter part of 1943 and all of 1944. The instructors dealt with him, not Major Jenkins. A married officer, McClellan lived in Manassas, but he had his own office across from the mess hall.[152] A Lieutenant Lethgo, who called himself the oldest lieutenant in the U.S. Army, was one of the few military officers who taught communications. Several enlisted men helped out.[153] The majority of instructors in radio operation and maintenance and International Morse Code work were civilians not military personnel. They were contractual employees of the Communications Branch of the OSS.

John Balsamo a civilian instructor there for much of the war, became a legend among Commo recruits. He was extremely fast and accurate at Morse Code. He could send or receive 40 words a minute. For comparison, the Communications Branch tried to get members of other branches up to six words per minute, and its own operators to around 15 or 20 words per minute. Balsamo had worked as a telegrapher on Wall Street and, as one CB trainee recalled, "He emphasized, that if he had made one mistake, it might have cost millions of dollars. So he really stressed accuracy in our transmission and reception."[154] The code room building was where the Morse Code training took place indoors, there were also field exercises outdoors. The trainees would file into the room. Balsamo liked to take roll call at extraordinarily high speed, say up to 40 words a minute, Arthur ("Art') F. Reinhardt, one of the trainees, remembered. "So you had to know your code and acknowledge it. He was going so fast that he couldn't catch everybody raising their hands. That was a little game he played."[155]

The Communications Branch had sought experienced, federally-certified radio operators from radio schools and businesses, and hired them to teach at Area C. Many of them were also "ham" radio operators. A group of the civilian instructors at Area C came from Coyne Radio School in Chicago. Timothy Marsh was one of them. "Like a lot of rural, southern boys, I went north during the war," Marsh recalled. In 1941, 20 years old and newly married, he moved to Chicago from rural Lincoln County in southern Tennessee. After graduating from Coyne, he worked as a civilian radio operator for the Army Signal Corps. In November 1943, his old instructor from Coyne, William Barlow, who was working as a civilian for the OSS Communications Branch, recruited Marsh for the CB's new Training Area M at

Camp McDowell near Napierville, Illinois. A month later, Marsh and a dozen other civilians and one career military man, Corporal Ray Cook, were reassigned to Area C. Barlow was already there. At Area C, the civilian instructors were bunked together six to a cabin; the military instructors as well as the cadre had their own facilities. Marsh, who spent a year between December 1943 and December 1944 teaching radio operation at Area C, recalled that "the military [the cadre] as a whole did not mix with us. Of course, we all shared the mess hall."[156] The civilians worked as instructors and so did Corporal Cook, who also served as secretary to the director of the Communications School, Captain McClellan. Marsh, who could by that time send and receive more than 20 to 25 words a minute, worked alongside John Balsamo, helping to get students up to speed. That was in the morning, in the afternoon he worked in the "radio shop," one of the old barracks, fixing radios that needed repair. Having had training in transmitters, Marsh also filled in part-time with the transmitters, including the main transmitter for Area C. It had overseas circuits and a huge diamond-shaped antenna reaching far up into the air and aimed at Europe.[157]

Among the military instructors for communications was 1st Lieutenant James F. Ranney from Akron, Ohio. After two years of college, Ranney, an amateur HAM radio operator, became an engineer at a radio station in Youngstown. When he was drafted, the Army sent him to Officers' Candidate School at the Signal Corps School, Fort Monmouth, New Jersey. In the spring of 1943, when he graduated as a second lieutenant, he was recruited by the OSS, and he served as an instructor at Area C from August 1943 to January 1944. He emphasized the demanding nature of clandestine traffic. The dots and dashes of Morse code came in using OSS's secret code, in five-letter groups. "It was in letters or numbers without obvious meaning," he recalled. "That puts an extra strain on the operator, because he can't make any guesses. If it was in plain language and you miss a letter, you can guess it from the context, but you have to be perfect in the code groups."[158] In 1944, Ranney was sent to OSS base stations in Cairo, Egypt, then Bari and Caserta in Italy. He was part of an OSS communications team at those base stations that helped to maintain agent communications circuits to missions in Northern Italy, Yugoslavia and Albania. When the European war ended, he volunteered for the Far East and served at an OSS base Chihkiang, China until the Japanese surrender.[159]

Recruiting Radio Operators for the OSS

OSS faced a formidable challenge in obtaining and training hundreds of radio operators to maintain the clandestine communications network that Donovan's agency was establishing around the world. Commercial radio schools took an academic year or more to prepare operators. Major Garland H. Williams at SO had recognized the problem early in 1942 of how to get large numbers of radio operators in the OSS without spending many months, even a year, training them. The answer he said was to recruit people who already had extensive experience as a radio operator. "In such cases the training will consist of special instruction in the use of the equipment used in this service. The training period in such instances may well be of [comparatively] short duration."[160]

The new Communications Branch agreed. It sought individuals who already had experience in the operation and maintenance of short-wave radios.[161] It would be even more preferable if they were also familiar with wireless telegraphy—the sending and receiving of messages in the dots and dashes of International Morse code. One source of such individuals was amateur radio operators. These "hams" were licensed by the Federal Communications Commission (FCC), which regulated the airwaves in the public interest and which certified such operators and provided each of them with a call sign, a brief combination of letters and numbers that identified them, as they sent and received messages on the Citizens Band spectrum of radio frequencies. There were thousands of licensed amateur radio operators in the United States. Another source were the tens of thousands of others who had varying degrees of experience with radio operation and maintenance, and sometimes Morse code telegraphy, from training in the Boy Scouts, amateur clubs, commercial radio schools as well as the telegraph and radio industries. Another source was the military, trained radio operators in the Signal Corps of the U.S. Army or Army Air Forces, or as radiomen in the U.S. Navy or Coast Guard.[162]

Beginning in January 1943, the Communications Branch made recruiting capable instructors and interested and able students a major priority. Competition for trained radio and electronics personnel was intense in the wartime buildup among the armed forces, the OSS, and commercial agencies, but the OSS proved quite successful in obtaining men with experience or proclivities for radio and code work from civilian and military sources. Volunteers with experience were interviewed, many at the Signal Corps School at Fort Monmouth, New Jersey, but less than half of those interviewed were accepted for a task which often demanded not only skill but ingenuity, stability and courage.[163] The Communications Branch sought not just men with radio and Morse code experience, but individuals who were familiar with one of 22 foreign languages from Arabic to Swedish, and also had demonstrated particular initiative. As Donovan explained to the General Staff, "OSS communications men in the field require a higher degree of self-reliance than those assigned to a normal military operation, due to the autonomous nature of each mission."[164] Nearly 700 officers and enlisted men were recruited in 1943, and 220 of those enlisted men were familiar with a foreign language, and thus might become radio operators in SO or OG teams used behind enemy lines.

Area C was already overcrowded by August 1943, and future requirements for additional communications personnel had expanded beyond its capacity.[165] Headquarters recognized that the OSS role overseas was accelerating rapidly and would require a dramatic increase in trained communications personnel. Over the winter of 1943-1944, a total of 400 communications operators and technicians would need to be trained for base stations and mobile unit, a figure wholly aside from the training of agent operators and Direction-Finding teams. Consequently, the branch increased its recruiting drive, acquired another training facility, Area M at Camp McDowell, Illinois, which would operate over the winter, and expanded communications training schools overseas.[166]

One of the most successful recruiters was Major Peter G.S. Mero, a former investment executive from Chicago. Energetic, engaging, handsome, fluent in several languages, but all inflected by his strong Hungarian accent, Mero helped design the training program at Area C and also obtain recruits for it. Later he was dispatched overseas, and from late 1943 to 1945, he was the chief communications officer first for North Africa and then for Italy. He was responsible for communications personnel and equipment as well as preparing radio, code plans, and frequency allotments for all OSS missions in the Mediterranean Theater of Operations.[167] More than one OSS Communications veteran (or CommVet, as they called themselves) remembered being recruited by the dynamic Major Mero. W. Scudder Georgia, Jr., who along with Frank V. Huston and James Herbert, were new naval officers and radio instructors in Chicago in 1943, when Mero gave them a pep talk about joining the OSS. Georgia remembered Mero's "dedicated and passionate eloquence as he described the wonders which would befall us 'out there' if we joined up." Georgia began to daydream about a glamorous Hollywood version of the Far East. "I heard the words 'OSS' reverberating like a distantly struck gong. After that the room became silent and I sort of remember turning to Frank and saying with eyelids still at half mast: 'OSS? Why not?' Upon hearing that innocent question, the room reverberated as Major Mero slapped the conference table with the flat of his palm and sternly announced: 'Ah, vot I want is man vid gutts.' Frank, never one to falter at drawing to an inside straight, nodded as he also said: 'Why not?' The next thing we knew, we had signed a bunch of papers and were shaking hands all 'round. We had joined up!"[168] After a stint at Area C, the three friends found themselves in the Mediterranean Theater: Frank Huston working out of Bari, Italy, managing training and the message center there as well as helping to run guns to Tito's partisans in Yugoslavia, and Georgia Scudder, eventually also working out of Bari and then on a small island near Corfu, where he trained agent-operators for German occupied Greece and helped keep their equipment repaired.[169]

Commo recruits came to OSS from a variety of backgrounds and through various means. Frank Huston had learned radio operation in the Coast Guard in the 1930s and was recruited from the Navy by Peter Mero. Roger L. Belanger came from a small town near South Portland, Maine, joined the Army at 18 in February 1943 and was trained as a radio operator by the Army Air Corps, from which the OSS recruited him because of his skill with radios and his fluency in French. After training at Area C from September to November 1943, he was sent to England on the *Queen Elizabeth* and worked outside of London at an OSS base station that handled radio traffic for the D-Day invasion. He returned to the United States and then in January 1945 to Area C, where he trained operators for duty in the Far East.[170] Edward E. Nicholas, Jr. from Rock Island, Illinois, was studying electrical engineering at the University of Illinois when he joined an Army Signal Corps program that sent him to the University of Chicago to study advanced microwave design. Recruited by OSS, Corporal Nicholas spent nearly three months training at Area C in the summer of 1943. Shipped to the Mediterranean he worked a transmitter in Algeria and then in Corsica and Italy before volunteering in January 1945 for a sub-base assignment, which turned out to be Tirana, Albania, where with a little 50-watt transmitter he communicated between field agents and the Bari base station.[171]

Vincent L. Gonzalez, Jr. was born in Havanna, Cuba in 1918 and came to New York City with his family around 1930. After high school, he worked as a runner, got married, and when classified 1A by his draft board, joined the merchant marine, where he obtained his commercial telegraph license from the FCC in December 1943. Soon he was recruited by the OSS, given basic training, and as a telegrapher also fluent in Spanish was assigned as a clerk to the American Embassy in Spain a neutral nation but filled with intrigue and also an escape route for downed fliers. That was his cover for espionage while in Madrid, where he remained until the war ended.[172]

OSS Communications Branch recruited Spiro Cappony from U.S. Navy. The son of Greek immigrants, Cappony was born and raised in Gary, Indiana. He attended Michigan State College but dropped out in 1943 at age 20 to join the Navy. After boot camp, he was sent to naval radio communications school being held at Miami University in Ohio. Skilled in telegraphy and fluent in Greek, Seaman Cappony was recruited by OSS to serve as a radioman in Special Operations missions in Greece. Cappony learned OSS communications equipment and procedures at Area C. Then underwent paramilitary training at Area B. After additional training in North Africa and the Middle East, Cappony, as will be seen in a subsequent chapter, accompanied a team of Greek Americans to destroy bridges and railroads in Greece in spring and summer 1944 to prevent German troops there from reaching France to opposed the Allied landings.[173]

Raised on a family farm near Buffalo, Arthur ("Art") Reinhardt had learned Morse code as a Boy Scout. After graduating from high school in 1943 and joining the Army Air Corps, he was sent to communications school in Scott Field, Illinois. When he finished second or third in a class of 1,500, the OSS recruited him for the Communications Branch, and from June to August 1944, Reinhardt spent ten weeks training at Area C. After graduation, he and his 23 classmates were sent to China. Serving as a radio operator in forward substations, he received, transmitted and decoded radio messages to and from OSS Secret Intelligence teams and Special Operations teams behind Japanese lines to the OSS main base station in Kunming. His substation in Suich'uan survived regular bombing attacks by enemy planes and was eventually overrun by the Japanese Army in December 1944. Reinhardt was then assigned as a radio operator/cryptographer for a field intelligence team, communicating from the field to a sub-base at Ch'angt'ing. Later, in June 1945 he was sent to a sub-base south of Shanghai, which received information about Japanese ship movements from a host of coast-watchers and transmitted it to Admiral William F. ("Bull") Halsey's fleet in the western Pacific. The information proved extremely valuable, enabling the U.S. Navy, for example, to locate and destroy all 26 Japanese ships in one major convoy.[174]

Curriculum at Communications School

Communications School at Area C involved ten to twelve weeks of course work for Commo Branch trainees, depending on their degree of skill and previous experience. A ten-week course included 490 hours of instruction and practice, most of it in communications but with some paramilitary training also included.[175] Most of the trainees at Area C were already quite familiar with radio work and Morse code. There are two radio methods of transmitting energy or electronic messages—one is by voice, the other by telegraphic signals referred to as CW, for the continuous wave generated in those days by a vacuum tube. The main training at the Communications School involved the quick and accurate use of wireless telegraphy, International Morse (CW) Code. Instructors sought to get the trainees up to around 20 words per minute, accurately sending and receiving those dots and dashes. In the Code Room building, students would practice sending and receiving messages from each other or the instructor. They would also transcribe messages from taped transmissions operating at faster and faster speeds.

OSS had a variety of radio equipment, and students had to learn how to use and maintain these transmitters and receivers, their power sources, and their antennae. OSS special equipment included the SSTR-1, a small portable field wireless set that could fit into a "suitcase" with AC generated by batteries or a hand crank; the SCR299/399, a larger mobile unit mounted in the back of an Army truck with a gasoline-powered AC generator in a small trailer, and major base station transmitters and receivers with their independence power sources and enormous antennae.[176]

Most of the operating practice was done off the air, simply in the classroom, but to achieve greater realism, the Communications School wanted to have its trainees practice on the air, that is communicating with wireless radio telegraphy between the field and the base. Peter Mero, one of the most energetic recruiters and entrepreneurs in the Communications Branch, obtained from the Federal Communications Commission (FCC) specific frequencies for use in OSS training, as well as a call sign, XBLCD, so that FCC monitors could recognize traffic as being from the OSS Communications School.[177]

Instructors like James Ranney and Timothy Marsh took the trainees out into isolated rural areas so that they could practice communicating from the field with the base station at Area C. Ranney would go the motor pool at Area C, and gas up a customized version of the standard 2 ½ ton ("deuce and a half") Army truck. Inside the back was a 500 to 600 watt transmitter and two or three operators' positions, each with a sending key and receiver. An antenna emerged above the truck's cab, and a small trailer containing a five-or ten-kilowatt, 120-volt portable generator was towed behind. "We would go out usually in the vicinity of Fredericksburg, twenty-five miles or so from the camp, and find a good place to park," Ranney said. "We would send the [signal] traffic back and forth between the base and our unit." "Generally we were stopped when we were transmitting, but sometimes, just for fun, we would fire up the generator on the run, and they would send back to camp while we were on the road. That was good practice, because then you would run into fading and dead spots and so forth."[178]

Codes and Ciphers

Codes and ciphers were essential to maintaining the security of OSS communications, and trainees had to learn and practice them. The OSS defined a code as a method of concealing a message in such an agreed upon way as to make it appear innocent. For example: "No news received for ages, are you well?" might be agreed upon to mean "keep on with the plan as arranged." A cipher was a way of converting a message into symbols which did not appear innocent but which had no meaning to a person not possessing the key. As an example, the "playfair" substitution cipher, first used in World War I, combined relative security with great simplicity. It was based on a grid 5 by 5 containing 25 cells. A key word containing at least eight different letters was memorized and then used as the key to unscrambling the letters in the boxes.[179]

The Communications School spent 22 hours on cryptography out of the 490 hours of instruction in the ten-week course at Area C for radio operators and technicians of the Communications Branch. OSS also trained some men and women specifically as cryptologists, code and cipher clerks they were called, who would work at base stations. The crypologists, men and women, received their training at the Signal Center at OSS Headquarters in Washington. Male cipher clerks were subsequently sent to Area C for weapons and other paramilitary training before being sent overseas. One of those male cryptologists was Gail F. Donnalley, who had earned a merit badge in radio communications as a Boy Scout in Lisbon, Ohio. In his sophomore year at Ohio Wesleyan University, he was recruited by OSS and trained in ciphers in Washington He spent ten days receiving paramilitary training at Area C in October 1943. Sent abroad, Technical Sergeant Donnalley worked at cryptology at OSS bases stations in Cairo, Egypt and then Bari and Caserta, Italy between January 1944 and July 1945.[180]

Another cryptologist who took paramilitary training at Area C was John W. Brunner from Philadelphia, Pennsylvania. Brunner had a gift for languages. As a child, he was already reading well before he went to kindergarten. Later, he was studying classical Greek and Latin at College, when he was drafted at 19 in 1943. The Army sent him to study Chinese at Berkeley, but less than a year later, he and the other top student in the class were recruited by the OSS and trained as cryptologists ."They decided that people skilled in languages could learn coding most quickly," Brunner said. After cipher training, he was sent to Area C for a week of weapons instruction in October 1944. Sent to India by ship, and flown across the Himalaya Mountains to China, Brunner served in Kunming, working in the main OSS base station message center, coding and decoding, messages to and from Washington. The code room was considered so secret, that only a few authorized personnel were allowed to enter. "We were told that even if General Donovan arrived, he would have the permission of the officer in charge to enter. We should not let him in without the captain's permission." After Japan's surrender, he worked in Shanghai and then Tientsin doing interpretation as a linguist in Chinese for counter-intelligence and secret intelligence branches until his discharge in March 1946.[181]

OSS's code and cipher systems evolved over time, and they were taught on a more elementary basis to all the Communications Branch trainees studying at Area C. The cryptography course at Area C, as one student radio/base station operator remembered "was two fold. One was the encryption and decryption, [the other was] how you set the message up, so the recipient would know how to break it. In other words, you have to encrypt it and send it, and the recipient has to receive and decode it. It is very, very procedurally oriented. You have to do it in a very precise way."[182] In early 1944, the students at Area C were learning what was called "double-transposition." You chose a word, a code word. Using that word and depending on how the letters of the alphabet showed up, that would be number one. You made a matrix, lining them up one way and then the other way. The recipient on the other end, who knew what the code word was, would go through the reverse process to decipher it. It was a very laborious process. But the double-transposition system was cumbersome and also not very difficult for experts to break.

Consequently, in late 1944, OSS switched to the One-Time-Pad (OTP) as the main coding system in the field. It was based on a random key text and the Vigenére cipher, which encrypted an alphabetic text based a triad.[183] The triad included a key letter, the plain text letter and the cipher text letter. The operators learned to memorize 676 combinations.[184] Using a conversion chart on the back of a One-Time-Pad codebook, the key letter was read along one margin while the clear/plain text letter was read on the intersecting margin. The cipher text letter was where the two letters intersected. In the One-Time-Pad, each page contained a key which was used only once and then destroyed. The technique was comparatively quick, simple, and practically unbreakable because of the one-time use feature. According to OSS cryptographer John W. Brunner, a skilled cryptographer using a One-Time-Pad could encode or decode brief messages more quickly than on the electric code a skilled cryptographer using a One-Time-Pad could encode or decode brief messages more quickly than on the electric code machines.[185]

During his training in encoding as well as code-breaking before he went to China, Brunner had sent to the Message Center directly under Donovan's office in October 1944. It was an exciting place, he recalled. "They showed us the famous machine that made the One-Time-Pad. It was the only machine at that time that could do that. It was impossible [for a person] to make the key to the code books. The key had to be a completely random set of letters. But the only way to make them was to hire a couple of hundred women and set them down at typewriters and have them type randomly. There was no way that they could produce enough code books for general use. But then a guy in OSS invented a way to produce random letters. With this machine, they could produce enormous numbers of code books. They did, and they distributed them to all the OSS bases around the world. It was an unbreakable code system. We were trained in it. All the field offices got code books and stopped using the old double-transposition system, which was easy to break, and switched to the new one."[186]

Long Distance Final Field Exercise

Radio operator training at Area C concluded with a field exercise in which trainees spent a week communicating with the base station in Prince William Forest Park from a distant, remote location under adverse conditions.[187] The ubiquitous Peter Mero had obtained the remote facility, located nearly 300 miles away on an isolated coast near Wilmington, North Carolina.[188] Marvin S. Flisser, a Brooklyn College graduate, who trained at Area C during the winter of 1943-1944, remembered the week-long trip to North Carolina. They took their food and other supplies with them. After setting up the equipment, they sought to make the long-distance contact with Area C the first night. "Nighttime is a very difficult time to get message through," Flisser explained. "Up in the sky there is something called the heavy-side layer, which billows. It goes up and down and sideways. You have to be able to catch it at the proper time, so that the skip of your message would hit it and bounce down and get to where you want. You had to position yourself to get your message through, and I always got my message through."[189]

Getting the messages through—whether from agents behind enemy lines or to headquarters in Washington—was what the "Commo" operators had to learn how to do, even under the most difficult circumstances. Some 1,500 communications personnel received their training at Area C during the war. Although many of them worked as fixed station operators at base or sub-base stations, many served as operators in the field with Special Operations, Operational Groups, or Secret Intelligence teams. As one Communications veteran later wrote, "Commo underpinned everything OSS did."[190]

Other Uses of Area C

Area C was sometimes used by the Communications Branch or other OSS branches in special ways. When these had little to or nothing to do with communications, it was generally in the final year of the war when other areas, like A, B, D, and E, had been closed and F and RTU-11 ("The Farm") were unavailable. Since the Army required basic military training, including weapons training, for any military personnel sent overseas, the OSS in 1944 created a four-week basic training course to satisfy that requirement. It was taught mainly at Area A-2, but a similar basic training course was offered at Area C in 1944.[191]

Although Camp C-4 was the main training area for the Communications School, an Incoming Holding and Training School at Area C was in Camp C-1. Beginning in April 1945, a preliminary training program for SO and OG personnel going to the Far East was established there.[192] Captain William ("Bill") McCarthy was in charge of C-1 at the time. A wiry, rugged, friendly officer from Newark, New Jersey and a graduate of Fordham University, McCarthy was, according to his company clerk, Corporal Joseph J. Tully, from Philadelphia, easy to get along with and also a good baseball player.[193]

Area C-1 in the spring of 1945 was being used as a facility for housing returning veterans from Europe and the Mediterranean and also for training OSS troops who were going to the Far East. Ninety cocky new paratroop officers who had been recruited by OSS to lead Chinese Operational Groups being trained in China arrived at C-1 in June. "We tried to keep all the paratroop officers from destroying the place," Tully remembered. "They horsed around a lot. They came out of [parachute school at Fort] Benning, and a lot of guys were still sowing their oats."[194]

In a number of instances, Area C was used to train foreigners, primarily in radio and code work, although sometimes in paramilitary skills as well. In the fall of 1942, more than a dozen Thai students at American universities who had volunteered with Donovan's organization and already undergone paramilitary training at Areas B and A, were given training in radio operation and code work at Area C-4.[195] Radio training for some foreign recruits continued during the war at Area C. One of the instructors, Robert L. ("Bob") Scriven, who had worked for an Iowa radio company before being trained as a radioman first by the Navy and then by the OSS, was frustrated that most of the foreign students failed the examination on the OSS "suitcase" radio transmitter/receiver, the SSTR-1. Consequently, he suggested that a training film be made to explain the equipment visually. His idea was accepted, and he participated in front and behind the camera in preparing the film, which was given high praise in 1944 by Communications Branch chief Lawrence W. Lowman, who had been a vice President at CBS.[196]

For the Far East, OSS trained a number of Chinese Americans, Japanese Americans (the Nisei), Korean and Thai Americans. Of course, overseas, OSS recruited a number of indigenous agents—from India, Burma, Thailand, Indochina and Chinese an Koreans from China. John W. Brunner, who served as a cryptologist in China, recalled that "we used a lot of Chinese, Thais and Japanese Americans, Nisei. We used a lot of Nisei in OSS, but I never saw any of them in training. The radio [operations] people did use Asians: Thais and Chinese. But the Nisei, Japanese speakers, were too valuable to use for radio [operations] work. They were used as translators."[197]

Communications instructor Timothy Marsh recalled training foreigners at Area C in wireless telegraph operation. "We did train them in how to use the little 'suitcase' transmitter/receiver. Most of them were Asians. We knew what was going on. They were being trained to be dropped behind enemy lines."[198] A number of Koreans were trained at Area C, particularly in 1945, as the United States began to prepare for the final resistance of the Japanese armed forces in the Far East. Roger L. Balenger returned to the Communications School after serving in England. His main assignment was to train Koreans for radio and coding/decoding operations. He spent six months with them. They studied radio operations in groups of five. Usually only one of them spoke any English, and he translated for the rest of them. Balenger recalled a near disaster when they took a paramilitary course at Area C. During hand grenade practice, one of them threw a grenade, but it bounced off a nearby tree and back into the trench. Fortunately, one of the other Koreans threw it out before it exploded.[199]

In July 1945, a special group of Koreans for Secret Intelligence Branch were trained at C-1.[200] The contingent included about 30 to 50 Korean officers—captains, majors, colonels. They were known only by their Anglicized code names, such as

Peter, Sam, and Joe. "Half of them could not speak English," Joseph Tully remembered. "We got those who could, and they explained to the others." The Koreans were trained in close order drill, pistol and rifle firing, the use of the bayonet, close combat, radio operation, and field exercises. They were at C-1 for about two months.[201] Years later, the last C.O. at Area C, Howard Manning, then a successful lawyer in Raleigh, North Carolina, told his wife that among one of the groups at Area C that had been trained to be infiltrated behind enemy lines was Korean. Seven of those trainees, he said, later became members of the cabinet of the Republic of Korea.[202]

Both Donovan's new organization and the training process evolved along with the U.S. participation in the war. To be effective, the training system would also need constant tuning to the changing needs and circumstances occurring in the field overseas. The great difficulty was that Donovan's organization had little or no experience at what it was doing at the same time that it was obtaining recruits and instructors and building its training system. The recruits, at least initially, were being trained without their, or often their branches, knowing yet what their mission would be. Later, as it gained experience in the field, the OSS had difficulty in bringing knowledge from the war zones back to the training program in a timely and integrated manner.[203] The OSS was not alone in this; the U.S. Army faced similar difficulties in bringing its training programs in line with experience gained in the battlefield.[204]

In contrast to OSS, British SOE by 1943 had three years experience in training and in putting agents behind enemy lines. One of the American SO training officers was dispatched that year to evaluate instruction at SOE schools in Britain. He was so greatly impressed that he reported that "The training any prospective SO agent has received in our Washington [area] schools prior to his arrival in this theater is entirely inadequate and no trainees should be considered for field operations until they have had further training in this theater, which in many cases will involve a period of three months."[205] Special Operations Branch accepted his recommendation, but concluded that "a short training course in the United States would, at the very least, serve as a process for weeding out undesirable personnel."[206] Years later, comparing the training he received at OSS Areas F and B in 1943 and in Britain's Brock Hall in early 1944, before being sent behind enemy lines in France and then China, OG Sergeant Ellsworth ("Al") Johnson made two important observations. One was that the British instruction was almost one on one, one instructor giving personal attention to no more than a couple of students, whereas the Americans were trained in groups, whether in the classroom or outdoors. His other point was that the British instructors, unlike the Americans at least in 1943, had actual experience. "The American instructors would suggest to you what might happen. The British knew what might happen and told you. They had experience. They knew what they were doing."[207]

OSS operational training continued through the end of the war to include assessment and preliminary training in the camps in the United States, but with additional and intensified training in OSS training camps established in overseas theaters around the world. Even then, the OSS training system was a process of trial and error, much uncertainty and considerable bureaucratic infighting between the operational branches, such as Special Operations, Secret Intelligence, and Maritime Unit that ran their own schools, and the Schools and Training Branch that sought to coordinate and centralize the training program of the OSS.

CHAPTER 7

DAILY LIFE IN CAMP, PARK, AND TOWN

Since they might well serve in sabotage teams or guerilla units behind enemy lines, the typical recruit for Special Operations, Operational Groups, or the Communications Branch, was generally obtained from the armed forces. Most were former civilians, not career military, men who had simply enlisted or been drafted into the armed forces because of the war. For the agents who would operate behind enemy lines, OSS was especially looking for a combination of intelligence, imagination, courage and, if necessary, ruthlessness. OSS recruiters visited military training centers looking for high aptitude individuals with demonstrated initiative and ability and where possible fluency in an appropriate foreign language. They accepted only volunteers for possible hazardous duty. Most of the young recruits craved the excitement and challenge of a special overseas assignment. A college graduate seeking a different lifestyle, Richard P. ("Scotty") Scott was drifting in 1942 when he was drafted into the Army. Despite his business degree, he was working as an electrician on a 247-foot diesel yacht in the Caribbean. The Army sent him to Officers' Candidate School (OCS), but after becoming a 2nd lieutenant, he spent two years behind a desk in the Signal Corps. He yearned to get overseas. He was, as he recalled, rescued by an OSS recruiter, "a dashing Major in paratroop boots," who interviewed him and got him into the OSS Communications Branch. After training at Area C, he was sent to the Far East. It changed his entire life. Two decades later, he became Assistant Secretary of State for Communications and then Director of Communications for the Central Intelligence Agency.[1]

"In September 1943, I was in paratroop school at Fort Benning, Georgia," recalled Caesar J. Civitella, a private from Philadelphia, "and then at Camp Mackall near Fort Bragg, North Carolina, for airborne and gliders. A guy from the OSS came and interviewed me. I was 20 years old. He asked me did I speak Italian. I did. He asked me if I as sent to Italy and had relatives there and they were working with the Nazis would I shoot them. I didn't know if I had any relatives in Italy. So I said, 'Of course.' He said, 'You'll hear from us.'"[2] A month later, Civitella reported for training with an Italian Operational Group at Area F and then Area B before being sent to North Africa, France, and Italy. He spent the next thirty years in the special operations: OSS, Army Special Forces, and CIA.

Asked about his fellow Special Operations trainees at Area B in the fall of 1943—many of them brash young paratroopers with parachute wings on their blouses and trouser cuffs jammed into their jump boots—one of them, future Jedburgh John K. Singlaub, who would serve in France and China, and end his Army career as a major general, recalled that "they were all independent thinkers. I mean I had this impression. They were all very active type people who were willing to do things rather than to sit back and wait. They quickly took charge when they were given a mission. They all wanted to be there and to succeed."[3]

William ("Bill") B. Dreux, a young lawyer from New Orleans, was an infantry lieutenant not a paratrooper and was therefore more typical of the Special Operations volunteer. He had enlisted in the Army, done his basic training, and completed OCS at Fort Benning, Georgia. It was there in June 1943, that he was interviewed by a visiting lieutenant colonel. After learning that Dreux spoke French fluently, the officer asked if he would be interested in a mission behind enemy lines in German-occupied France, working with the *Maquis* in ambushes and sabotage. "We are asking for volunteers," the colonel said, "because these will be hazardous missions. We expect maximum casualties."[4] Elsewhere, Lieutenant Rafael Hirtz was told that operatives behind enemy lines had only a 50-50 chance of survival. Born in Buenos Aires, Argentina to a wealthy French-Spanish family, Hirtz had grown up in Hollywood in the 1930s. After the U.S. entered the war, he became an Army officer and was subsequently approached by the OSS to train for operations in France.[5] While such warnings deterred some potential recruits, Dreux and Hirtz had both volunteered. Both trained first at Area F, then Hirtz went to Area B and Dreux to Area A. Both were later parachuted into Brittany, France in 1944 in support of the Allied invasion. After the war, Dreux returned to New Orleans to become a highly successful lawyer. Hirtz finished college and spent the next forty years working international finance, grain, and trade.

The son of an American doughboy who married a French woman after World War I, Jacques F. Snyder was born in New York City but raised in France. His father was killed in an automobile accident. When the Germans occupied the country, 19-year-old Jacques Snyder returned to the United States to fight the Germans. He joined the Canadian armed forces, and when America entered the war, he enlisted in the U.S. Army Air Corps. He wanted to be a pilot but could not, and because of his fluent French, he was put in a language training school in Tennessee. On off-duty evenings, he played clarinet and saxophone in a swing band. Returning home late one night, he was hit by a car and suffered several broken ribs. An OSS recruiter visited him in the hospital and asked if he would volunteer for hazardous duty. Snyder's son said his father told him that "he thought it couldn't be more dangerous than playing in the swing band" and he volunteered. After training in Areas F and A, he was sent to North Africa and then parachuted into France and later into Indochina. After the war, he worked for military intelligence, retiring from the Defense Intelligence Agency in 1991.[6]

Bored as a company clerk in the Army in spring 1942, former Tulsa journalist Edgar Prichard was interviewed by the base intelligence officer, a major, who after reviewing his record and learning that he could speak Spanish and French, asked

Prichard if he would like to volunteer for some "interesting duty." "The only thing he told me was that if I agreed I would be placed on the detached enlisted men's list," Prichard recalled years later. "Of course, I had never heard of it, but the first sergeant of the company I was in told me it [detached service] was great duty, so I volunteered." Under orders marked "Secret," he arrived in Washington a week later "I learned there that I was in the office of the Coordinator of Information and that I was assigned to the S.O. Branch and that I was going to be sent to a secret camp. I was told to write any letters I wanted to write because I would be out of communication with my family and friends for a time. I was admonished not to tell anyone about what organization I had just joined or where I would be."[7] After training at Areas B and A, Pritchard was sent to North Africa. After the war, he became a lawyer in Arlington, Virginia.

The secret nature of the organization and the security background checks on recruits caused some difficulties as well as considerable anxiety for the recruits and the folks back home. Initially, Donovan's organization used Dun and Bradstreet, the commercial, credit-rating firm,[8] but usually the background checks for the OSS were conducted by the Military Intelligence, the FBI, Treasury agents, or other federal law enforcement agencies. With agents asking questions of former employers and neighbors, parents knew that their youngster had gotten involved in something, but they did not know what. Some thought their son had gotten into trouble—desertion, being absent without leave (AWOL), or some such military offense—and not told them. "When I got home [on leave], everybody wanted to know what I had done because the Treasury Department had been around checking my references," said Lieutenant James Ranney, of Akron, Ohio, who had been recruited for the Communications Branch. "I said that I was going into a job that required security clearance."[9] "We were investigated all the way back to high school," said Albert ("Al") Materazzi, a young engineering officer from the small, candy manufacturing town of Hershey, Pennsylvania, who had volunteered for the Italian Operational Group. "I remember a frantic call from my mother asking what had I done, the FBI was in Hershey asking all sorts of questions. I was able to calm her down by saying I was being considered for a government job and such investigations were routine."[10] Materazzi trained at Areas A, F, and B and then served in Italy. With his postgraduate work on graphic arts, Materazzi worked after the war as vice President of a printing and lithography company and subsequently as director of research and quality control at the U.S. Government Printing Office.[11]

OSS recruits were prohibited from disclosing their new affiliation or describing or identifying the location of their training camp, even to their parents. 18-year-old Art Reinhardt from near Buffalo, was recruited for the Communications Branch from an Air Corps' radio school. The OSS officer emphasized it was an elite group, spoke of an intriguing if vague assignment and mentioned that "there's some danger involved." Reinhardt readily volunteered. "I wanted to get involved in the war. That was my objective. When you are that old, you are adventurous. You don't really concern yourself with the adverse consequences."[12] On detached service from the Army Air Corps, he was sent to OSS headquarters and then to Area F, Area A, while he was cleared for security purposes, and finally to the communications school at

Area C. Not allowed to provide any details to his parents, he simply wrote that he was in Washington undergoing training and gave them, as he was told, a postal address: P.O. Box, 1925, Washington, D.C.[13]

Many of the men recruited from Army bases for the OSS did not know the name of the organization recruiting them or, if they learned its initials, did not know what OSS stood for. Captain Francis ("Frank") Mills, an artillery officer from Oklahoma, recruited for Special Operations, was told that O.S.S. stood for "Overseas Supply Service."[14] Information about the organization was withheld from recruits until they were cleared for service. The OSS buildings in Washington, D.C. at 2430 E Street, N.W., were supposed to be top secret, but as Ralph Tibbetts, a young recruit for the Commo Branch, found out from the waitress at a nearby delicatessen when he arrived too early, the neighbors correctly suspected that the buildings were filled with "a bunch of spies."[15] The entrance was protected by an armed solider and a guard dog, and no one was allowed in without an OSS pass, identification or an escort.[16] Inside when William Dreux reported for duty, he found the long corridors were filled with men and women hurrying back and forth. Some were civilians, but most were military officers. The majority wore Army or Army Air Corps uniforms but some were from the Navy or Marines. Most strikingly were the foreign officers: British officers with Sam Brown leather shoulder belts and riding boots; a Scot in kilts, and a Free French officer in light blue uniform and kepi cap. Dreux blinked when a major told him that his first OSS training would be at the Congressional Country Club. "Did you say Congressional Country Club, Sir?" he asked. "That's right, Lieutenant. Very plush place. Herbert Hoover was one of the founders. It's about six miles out of town. We've taken over the whole club for training, golf course and all."[17] Dreux and the French OGs were there for a few weeks before being transferred to Area A where they would undergo intensive guerrilla training for the next two months.[18]

The Secret Intelligence Branch also recruited in various ways and from diverse groups. William J. Casey, born and raised in a Roman Catholic family in the Borough of Queens in New York City and attended Fordham University and St. John's Law School. An able entrepreneur, Casey went to Washington as head of a private research bureau in 1941 providing information to businesses on forthcoming mobilization and war orders. After Pearl Harbor, he worked as a consultant for the Bureau of Economic Warfare, helping the Navy get through bureaucratic tangles to increase production of landing craft. Casey was commissioned a naval lieutenant, but as he later wrote, "sailing a desk in the old Navy building on Constitution Avenue was not my idea of helping to win the war. I itched for action and the OSS seemed the best place to get it." Consequently, he looked up an old friend with whom he had parked cars at Jones Beach as a teenager and who was now a member of Donovan's law firm, and this led to introductions to Otto Doering and Ned Putzell, partners in the firm who had followed Donovan to Washington and become senior officials in the OSS. The 30-year-old Casey also impressed Donovan himself, who 'poached" the talented young officer for detached duty from the Navy. After four months of training and apprenticeship in Washington, including a hazardous paramilitary course at Area B, Casey was dispatched to London. Casey's eyesight was too poor for working in the field, which he longed to do, but from his office, he ran a number of espionage operations, including efforts within Germany. By the end of the war, Casey briefly headed OSS intelligence operations for all of Europe. After the war, returning to become a Wall Street lawyer and active Republican, he eventually served as Ronald Reagan's Director of Central Intelligence.[19]

Like Casey, Richard Helms, also worked in the Secret Intelligence Branch and subsequently served as Director of Central Intelligence (DCI), but via a different route. Helms came from a prominent family on Philadelphia's Main Line, and he was educated at private schools in Germany and Switzerland. Graduating from Williams College in 1935, he worked as a journalist in Europe and the United States. When the U.S. entered the war, Helms enlisted in the Navy as a lieutenant and was assigned to the anti-submarine warfare operations center in New York City. In August 1943, the OSS recruited him, without his seeking it, because of his experience and connections in Europe and his fluency in French and German in Europe. He worked first on the SI Planning Staff, and then from 1944 to 1945, he served under Ferdinand Meyer and then Allen Dulles conducting espionage operations from posts in England, France, and Sweden.[20] After the war, he worked his way up in the CIA, serving as DCI under Presidents Lyndon Johnson and Richard Nixon. Most of the SI training took place at Area RTU-11 ("The Farm") an estate a dozen miles south of Washington or in Area E, estates north of Baltimore, but a number of SI recruits, like William S. Casey, received paramilitary or other Special Operations training at Areas A or B, and some of them received some communications training at Area C.

For the Communications Branch, most of the radio-operators for OSS's clandestine field or regional base communications network were obtained from the radio, radar, or other electronic schools of the Army, Army Air Force, and Navy. They were generally enlisted men, who a short time earlier had been civilians. David Kenney from the small town of Encampment, Wyoming, turned 18 in 1943 and was drafted into the Army. He scored high on the aptitude tests and was also an amateur musician. He believes the Army thought musicians would be good at radio, because they sent him to a Signal Corps' radio school at Camp Crowder, Missouri. "Then a friend of mine saw a notice on a bulletin board: 'Forty volunteers needed for dangerous assignment,'" he recalled. "My buddy volunteered us. I talked to on OSS captain. I didn't know he was from the OSS or a recruiter. I said that I was worried about the dangerous part. He said, 'everything overseas is dangerous, but you'll never regret doing this.' He was right. I never regretted it. It was a wonderful experience."[21]

Many of the "Commo" recruits had been amateur short-wave, radio operators ("hams") before the war or had been trained or further trained in International Morse Code and telegraphy by commercial or military radio schools. In the first several decades of the twentieth century, when radio was in its infancy, amateur experimentation was an exotic hobby. Some individuals with scientific or technical inclinations and a desire to do things themselves, would tinker with equipment, constructing radio crystal sets, with voice or code wireless communication. They learned that they could communicate through the airwaves from their own home and reach out, depending on atmospheric conditions, to places around the country and even around the world. OSS CommVet Art Reinhardt, who continued his work in clandestine communications for the CIA for thirty years after the war, estimated that probably 30 percent or more of the OSS technical personnel had been hams.[22] The importance of these amateur radio hobbyists, he said, cannot be overstated. "The hams were the real backbone of the OSS. They were especially important at the base station receivers. Because out in the field you had agents [especially the indigenous agents] who were poorly trained, who were operating under adverse physical, security, and technological problems against interference.... So, to counter all that, you had more powerful transmitters and receivers at the base stations, you had better antennas, and you had these wonderful people with good, selective ears that could listen and sort of drag out these weak signals. These ham operators from the 1930s who had learned how to seek out the exotic signal, they are the guys who were really, really good, and they were the backbone of the OSS communications system."[23]

Most officers recruited for the operational branches of the OSS met Donovan personally. He interviewed them briefly in his office on the second floor of the headquarters building in Washington. Lieutenant Jack Singlaub was partway through his training for Special Operations at Areas F and B when he was summoned to the director's office. He did not know then that he was being selected for one of the elite, multi-national "Jedburgh" teams that would be parachuted deep behind German lines in support of the Allied invasion of France. Standing at attention in front of Donovan's desk while a colonel read his military record, Singlaub eyed the rows of ribbons on the general's uniform, particularly the light blue, star-spangled ribbon of the Medal of Honor he had earned in World War I. The maps on the wall behind him were marked "Secret." Donovan seemed pleased, smiled and said "Lieutenant, you have an excellent training record. And you know how I feel about thorough training." "Yes, sir," Singlaub replied. "Well," he said, and his bright, probing blue eyes fixed upon the young officer to reinforce the point, "I just want you to know that the kind of combat we'll be in is a lot rougher than any training."[24]

Arrival at Training Camp

OSS recruits were driven to the secret training camps in standard, 2-½ -ton Army trucks, the back enclosed in olive-drab canvas with the rear flap down.[25] They were told not to open it, because their destination was classified information. Sometimes the drivers would talk with there recruits. As with the other OSS men dressed alike in Army fatigues, their background sometimes belied their appearance. The OSS Army private driving Edgar Prichard and his classmates to Area A in May 1942 turned out to have recently been a professor of English at Williams College.[26] Many recruits did not know that these had been government parks. Robert R. Kehoe from New Jersey, a 21-year-old recruit from the Signal Corps, thought Area B-2 on Catoctin Mountain had been a "private hunting lodge in pre-war days."[27] Marvin Flisser, a communications recruit from the Bronx, was told that Area C was originally a private camp, perhaps a hunting camp.[28]

Commo student, David Kenney from Wyoming, remembered the long trip from the Army radio school at Camp Crowder. "They put about forty of us on a train from Missouri to Washington, D.C. We didn't have an inkling of what was up. Then they put a bunch of us in a canvas covered truck with the flaps down, and we headed south from D.C. Finally we could see the entrance to the Marine base at Quantico from our back flap, and we cheered when the truck turned away from it. We didn't want to be there. We turned off to the west and up a dirt road into a forested area." At Prince William Forest Park, they were taken temporarily to Area A-3 for security clearance. It was, Kenney recalls, "a bucolic campground consisting of a number of small, rustic cabins and several larger structures. It was my understanding that it was a former Campfire Girls facility.... We poured out of the trucks and into a conference room. They said that we are in the OSS, but they did not even tell us what OSS stood for. They said it is a secret thing. We were not to tell anyone about it. Someone [in the group] had picked up a newspaper, probably the Washington Post. It had a column by Drew Pearson about the OSS. Pearson called it 'Oh, So Social' or "Oh, So Secret." He didn't like it.... Someone had that paper... when we arrived at Area A, and from that and what they told us, we figured it out that we were in the Office of Strategic Services, a secret War Department operation."[29] "We were told not to discuss it [the OSS] with anyone. If an MP [Military Police Officer] stopped us and asked for identification, we were to tell him that it was a 'supply outfit.' We were also required to supply about ten references for an FBI clearance.... We spent about a week there, mostly undergoing indoctrination. We knew Morse code, but their equipment was different than the Signal Corps'. I don't recall receiving any training there but, during this period, several in our group simply disappeared, apparently after failing to meet security or other OSS-specific requirements"[30]

OSS trainees were given fictitious code names in order to keep their identity secret. The code names (or "school names" as OSS Schools and Training called them) were generally common American first names such as "Harry," "Ed," "Jack," "Sam," "Bill," and "Pete," but sometimes they were closer to their ethnic background such as "Maurice," "Leif," "Ivan," "Spiro," "Gino," or "Bruno."[31] Sometimes it was their real first name but with letters and numbers were added. New arrivals were required to turn in all their personal possessions, even their driver's license as well as their uniform. All the trainees received the same kind of Army uniform along with their "school name."[32] Cameras were confiscated, because neither the trainees nor the staff was allowed to take pictures, as they might lead to future identification of secret agents or the secret camp.[33] The cadre and instructors knew each other's real names, but in regard to the students, the staff was given information only on a "need to know" basis, and asking too many questions was frowned upon.[34]

Herbert R. Brucker, who would later become one of the founders of Army Special Forces, was born in Newark, New Jersey to a German-American mother and a French-born, American father who served in the U.S. Army, Brucker grew up in Alsace, France, raised by various relatives, before returning to the United States with his father in 1938. At age 19, he joined the U.S. Army in 1940 as a private and became a wireless radio operator soon sending and receiving twenty words a minute. He was so proficient that in 1942, he was made a Morse Code trainer at Fort Meade and promoted to sergeant. But he wanted to see combat, and so in August 1943, he volunteered when he learned about the OSS. Brucker was not only an excellent radio operator but was fluent in both French and German, and the OSS snapped him up.

After getting through the OSS Assessment program, Brucker and thirty other men were trucked up to Area B. Brucker remembered what looked like a CCC camp, eating at the mess hall and being put on guard duty the first night. The next morning when the new men were lined up in three ranks, Brucker was surprised to see a group of Norwegians carrying M-1 Garand rifles jogging by them at double-time. Just then a runner came looking for him. Trainee "Herbert E-54" (Brucker) had been sent to the wrong camp, presumably because Area B was being used mainly for OG training at the time. He returned to Washington and subsequently directed by the Special Operations desk to the Phoenix, Maryland railroad station and taken to OSS Training Area E-3, the Nolting Estate, a mile east of Glencoe, in a rural area north of Baltimore. He went through SO training there in August and September 1943, including infiltration into an Baltimore shipyard as a graduation project.[35]

Trainees, particularly those who were training for individual infiltration behind enemy lines, like SO and SI agents, and who might be captured were not encouraged to make friendships with their fellow trainees. Plus, as the course progressed, trainees were weeded out for any number of reasons, such as failing written or performance tests, being unable to function under stress, deemed too loquacious for a secret organization, or just giving up and voluntarily requesting return to their previous organizations. "We would sit at a round table at night... and one given day, I would see a face was missing. It wasn't there anymore," said Sergeant Stephen J. Capestro ("Steve A3"), a former football player from Rutgers College, who trained at Area A. "So, you'd ask, 'What happened to "George A2?"' He [the instructor] would say, 'Well, he got caught divulging who he was or what he was doing.'"[36] "You didn't make lifelong friendships," said Reginald ("Reg") Spear from Los Angeles, who trained at Area B, "because you never saw the guy again, and you didn't know his name."[37] James L. Boals III, a trainee at Area C in 1944, remembered someone outside their group being called out at roll call as "Jack-S1." "At the time, we were told that a Nazi had gone through the school with our [OSS headquarters'] knowledge and was not arrested until graduation."[38]

Daily Routine

As at any military facility, each day was tightly organized, but at OSS facilities, in keeping with the informal, quasi-civilian nature of Donovan's organization, the emphasis was on individual initiative and responsibility rather than regimentation. The basic schedule, particularly at paramilitary camps like A, B, C, D and F, was similar to that at a military training base. Each day began with a wakeup at 6:15 a.m., the breakfast at 7:00 a.m., classes in the morning, lunch at noon, more classes in the afternoon, supper at 6:00 p.m. and lights out at 11:00 p.m.[39] Unlike regular Army bases, however, the OSS paramilitary training camps did not use bugle calls, no bugler trumpeting out the staccato notes of reveille or the slow, mournful echoing of taps. Instead, the men were awakened by the playing popular music over the public address system. Most of the camp commanders allowed the company clerk's office to play almost any record they chose. At Area C, for example, contemporary musical hits were a favorite for reveille. Terry Samaras, remembered waking up to "Holiday for Strings." Others recalled "Oh, What a Beautiful Morning," and the country music hit "There's a Star Spangled Banner Waving Somewhere."[40]

"Everything that I experienced was very informal," said Private John W. Brunner a cryptologist from Philadelphia training at Area C. "There was no military routine at all. No marching or parading. We simply got up when we were told. An enlisted man would come down to our hut to wake us up. We went to the mess hall. Afterwards, we went out to our assignment for the day. There was no marching there. We just walked."[41] "They treated you like a man," recalled another trainee at Area C, Private Art Reinhardt. "There was very little discipline. He [Major Albert Jenkins, the C.O.] was relying on the [self] discipline of the individuals and the motivation of the people."[42] To keep the men in shape, there were calisthenics every morning. Instructors and staff generally participated as well as the trainees. All would assemble individually, then line up and follow the directions of a physical training instructor on an elevated platform. It was held near the mess hall. Afterwards, they would walk into the mess hall for breakfast.

OSS training in the National Parks began in April 1942, when the first basic special operations course held in the United States was conducted at Area B-2 in Catoctin Mountain Park. The basic or preliminary SO course in 1942 provided physical conditioning as well as a variety of skills, which, as Lieutenant William R. ("Ray") Peers, who would later command Detachment 101 in Burma, remembered it the students would later find very useful in their subversive activities. The included "methods of agent operations, secret writing, resisting an interrogator, searches for downed aircrews, cryptography, experiments with a variety of high explosives, learning the difference between blowing a stone or steel bridge." Along with their other studies, they read and discussed British commando raids in Norway, France, and North Africa. The basic course at Area B in spring 1942 lasted only two weeks, but it was intensive. "It was," Peers later wrote, "abnormal to get more than six hours sleep a day."[43]

An inspector from headquarters visited Area B-2 in June 1942, by which time the basic SO course had been expanded to four weeks.[44] The inspector, who was from the Secret Intelligence Branch, talked to one of the SI trainees taking the paramilitary course, John R. Brown, whose code name was "Bill." Later that student provided the SI office with a chronology of the four-week basic SO course at Area B.

FIRST WEEK
Military Drill and Manual of Arms
Interior Guard Duty explained; Articles of War read
Elementary First Aid course
Rifle instruction and practice (M-1 Garand and Springfield '03)
Elementary principles of explosives (with the use of dynamite)
Guard problem opposed to raiding senior class
Exercises and sports twice daily.

SECOND WEEK

Gas training (different types, precautions against, use of different types of gas masks)
Elementary explosives and demolition
Practice throwing hand grenades and firing rifle-propelled grenades
Study of pistols and Thompson submachine guns (shooting from the hip)
Rifle practice on moving targets with Garand and Springfield '03 marksman rifle
Field patrol training how to guard against raids
Study and firing of flares and of .30 caliber Browning machine gun
Study and practice firing 60 mm trench mortar
Elementary compass instruction, map reading, and field compass problems
Guard problem against raiding senior class
Exercise once daily

THIRD WEEK

Elementary Cryptography and codes
Practice firing pistols and submachine guns
Advanced explosives and demolition
Jiu Jitsu and close combat training without weapons
Compass problems in field scouting
Political studies of Nazi party, secret police and regular German police.
Elementary photography
Raid problem against guarding junior class
Exercise once daily.

FOURTH WEEK

Extensive pistol practice at stationary and bobbing targets
Advanced demolition and incendiaries
Raid problem against guarding junior class
Elementary studies of German and Japanese Armies
Close combat and unarmed defense
Raid problem in mined field ("booby traps")
Lectures on street and house to house fighting
Lecture on surveillance and criminal investigation
Exercise once daily.[45]

Recreational Activities

Although the class schedule was full at the training areas, there was some free time and some places to enjoy it. At every camp, there was a building large enough to be used as a recreation center at least part of the time. Movies were shown there, not just training films and footage of devastation caused by German and Japanese armies and air forces but occasionally Hollywood movies as well. In winter, the men could shoot hoops and play basketball there. Most camps also had some kind of common room, often called the "Day Room," in which the men could relax in the evenings, listening to the radio, reading, or playing cards. Some, like the one at C-4, included a ping-pong table.[46] In the summer, the men at Area C played baseball in a large parking area.[47] At Area B-5, assistant company clerk Albert Guay played 78 rpm records on the public address system in the late afternoons after duty time and early in the evening. At the little PX, the men could buy candy, soda pop, snacks, toiletries, and perhaps a bottle of beer. Guay would purchase and eat crackers and sardines there. His diary in late 1942 indicated that he saw a movie, a training film, in the mess hall, and that in the recreation room, he watched other enlisted men on the staff playing darts and checkers, and gambling at poker. In the barracks, those enlisted men, he noted, wrote letters and did a lot of talking and joking. Some also took walks in the woods, and a few, like Guay, went hunting there. But many of the men got bored sitting around at night, and Guay wished they had some recreational equipment.[48]

Most camp commanders encouraged the healthy use of leisure time after a day of intensive coursework. The second C.O. at Area B, Lieutenant Montague Mead, did purchase recreational equipment. He also encouraged a healthy competition between the men at Area B-2 and those down the hill at B-5, half a mile or so away. By the summer of 1943, an inspector from headquarters reported that the practice had developed there of playing games, such as soccer, between the two Training Areas.[49] At Area A in the spring of 1944, a group of Commo recruits had very little to do while they waited at Area for clearance to proceed to communications training at Area C. There were about twenty of them with no recreational facilities. "So one day, a gentleman came by," recalled Art Reinhardt. "It was one those two handsome movie actors who were in OSS—Sterling Hayden or Douglass Fairbanks, Jr.—I'm pretty sure it was Sterling Hayden. He saw us sitting around and asked: 'Do you boys need anything?' We said, 'We'd like to have some athletic equipment.' The next day, an open-stake truck came, and it was full of athletic equipment: basketballs, gloves, baseballs, softballs, anything you can imagine."[50]

Special Operations Training Experiences

"Our training consisted mostly of sabotage operations and the conduct of guerrilla actions ranging from raids, ambushes, and assassinations," Jacques L. Snyder recalled. "Emphasis was placed on stealth, approaching the target without raising any suspicion, crawling when necessary for long distances, using maximum cover and concealment as camouflage when needed. Another important point was the total ruthlessness and efficiency in dispatching guards and sentries in the target area.... The overriding concern was silence and speed in execution."[51]

Everyone who went through OSS paramilitary training remembered the British close-combat instructor, William ("Dan") Fairbairn. Most were shocked and some appalled by his "gutter fighting" techniques. But he argued that they had a job to do to defeat an evil enemy and they had a duty to learn to survive to do that job effectively. Memoirs that cover OSS training almost always mention Dan Fairbairn, the "Shanghai Buster."[52] "We called him delicate Dan the deacon," Edgar Pritchard remembered. "He knew at least 100 ways to kill people without shooting them. He was a knife fighter and an alley fighter. It was said that he had killed dozens of people in hand to hand combat....Fairbairn had us throwing each other all over the place. Fortunately none of us killed any of our classmates. We all survived."[53] Most memoirs reveal a few of the surprising methods he taught them about hand-to-hand combat. How to fight barehanded (or with anything that came to hand), with all the dirty tricks of gutter fighting: "a knee in the groin, a savage slash with the side of the hand across an Adam's apple, a jab at the eyes with fingers stiffly hooked in a tiger's claw."[54] How to fall, roll, and come up on the offensive. How to kick, jab, punch– not with a fist but with a rigid open hand. How to dislocate an enemy's shoulder or break an arm or leg; how to break a sentry's neck from behind. He showed them how to use knives, hatchets, and other cutting or striking instruments. "Fairbairn taught us how to get rid of an opponent using normal everyday items such as a rolled-up magazine or newspaper, a matchbox, a pen or a pencil," recalled Lieutenant Jerry Sage, who taught at Areas A and D. He showed us the pressure points on crucial arteries and demonstrated the sensitive nerve areas where a single blow could immobilize an enemy. With every tactic we also learned the counter-measures to protect ourselves in a similar attack."[55] Army Corporal Edward E. Nicholas from Rock Island, Illinois, who was trained as an OSS radio operator at Area C in the fall of 1943, said the students there learned "'dirty fighting' or pseudo judo as we called it."[56]

Since few of the students, even those with military training, had any experience with explosives, the instruction in demolition work was generally new to them. It was dangerous work, and it was taught at Area B in 1942 by knowledgeable instructors from the Corps of Engineers, like Charles Parkin, Frank Gleason, Leo Karwaski, and Joe Lazarsky. They taught how to blow up railroad lines, bridges, and buildings and how to make booby traps. They would first lecture in class and then go outside, down the hill and have the students set small charges and explode them.[57]

Plastic explosives, called P.E. or Composition C, had only recently been developed. At Area A in 1943, Lieutenant William Dreux and his fellow French OGs, watched intrigued as their demolitions instructor, a tough, grizzled engineering sergeant named Bolinsky, who had attended a British commando school, showed them how it worked. Unlike dynamite, which is highly sensitive and volatile, plastic explosive is extraordinarily stable, and unlike rigid dynamite sticks, P.E. is soft and malleable like a piece of clay. It was a highly effective explosive when a primer and detonator were inserted and a connecting fuse lit. The trainees practiced with various amounts of plastic explosive on different kinds of material used particularly in bridge construction, blowing up pieces of wood, concrete, and steel. In addition, in order to hamper enemy truck convoys, the students learned how to explode craters in roads. Sometimes, despite the National Park Service's policies, they blasted off stout tree branches or even blew up trees in demonstrating the explosives.[58]

Seaman Spiro Cappony, a Greek-American radioman from Indiana, receiving SO training for infiltration into German-occupied Greece, recalled that at Area B in 1943 the instructors there also showed them the new Composition C "They showed us how to put fuses into it. They would also take a line of fuse cord and wrap it around a tree and show us how fast it traveled [burned]. So we became familiar with that. In Greece, [OSS] would airdrop it from Bari, Italy. They would airdrop it, but it would not explode. You had to put the fuses into it. We taught the [Greek] guerrillas how to do that."[59]

In addition to using plastic explosives, instructors sometimes taught the trainees how to rig improvised explosive devices from artillery rounds. At Area B, Reginald Spear remembers that they blew up a 155mm. artillery shell as an explosive device along a fire trail. "You used [a particular chemical] to explode the 155 mm round. That gave you a few minutes to get away. But the enormous explosion could kill you. We were very respectful of those rounds. We wore steel helmets."[60]

What most excited many of the trainees, the majority of them young men not long ago civilians, was learning and firing so many weapons. "We fired the Thompson submachine gun and the M-3 submachine gun, a little 'burp' gun. I wish I had more time for firing those," John Brunner noted in his diary at Area C on an October day in 1944, "It was fun."[61] The OSS emphasized fast combat shooting rather than careful marksmanship and familiarity not just with the standard U.S. Army small arms, the .45 caliber automatic pistol and the M1 Garand semi-automatic rifle, but, since the men were being trained for service behind enemy lines, with a wide variety of Allied and Axis weaponry. They fired many different pistols including German Lugers and Walthers and Japanese weapons. The Colt .45 was the standard U.S. sidearm. Brunner called it "a sweetheart." "It is superbly accurate at even 100 yards. It puts a man down. A 9mm. [pistol] causes hemorrhaging and death more effectively. You might survive more easily with a .45 wound, but you will be down and in pain and no longer a threat. And it will take several of your buddies to carry you back."[62] The .45 was big and heavy, consequently, Reginald Spear recalled, a number of OSSrs carried a 9mm or .380 caliber pistol, because they were smaller and lighter.[63]

OSS training focused on rapid combat shooting rather than deliberate marksmanship. The primary instructors in pistol shooting at OSS training camps in the United States throughout the war were Rex Applegate and Dan Fairbairn. They taught their new method of couched, two-shot "instinctive firing" or what Applegate called "point-shoot." With bursts of two shots fired within seconds, the first would unnerve the enemy even if it did not hit him and the second could kill him. "You chaps have been trained to aim a weapon," Fairbairn told Jack Singlaub and the other SO trainees at their first session with him at Area B in fall 1943. "That's all well and good on the range. The only problem is that your average Hun or Nip rarely stands still with a bull's eye on his nose." With that, three silhouettes of helmeted Germans suddenly sprang up in the bushes about 60 feet away. Fairbairn twirled around, pulling his pistol out of his belt, crouched and fired off three bursts of double shots, each pair right into the center of the paper targets. "The man and the pistol had become a single weapon."[64] At Area A, Jacques Snyder remembered that Fairbairn also taught the students how to use a submachine gun, the Thompson was his favorite, "with maximum efficiency, firing only three rounds at a time."[65]

John Brunner remembers that the day after his group arrived at Area C-4 in October 1944, Fairbairn instructed them in knife and pistol work. "The first thing he told us was that if someone comes at you with a knife, pull out your pistol and shoot him! None of this fair fighting, Marquis of Queensbury rules nonsense. If you do not have a pistol, turn and run away. Do not attempt to fight him. If you cannot run, then and only then, here is what you do. Here is how you disarm a man and incapacitate him. He repeated, however, never enter a knife fight if you can avoid it. You can never predict how good the other guy is with the knife." "From then on, Brunner said, "we spent little time with the knife. We trained with the pistol. Don't ever be without your pistol [in a war zone]," Fairbairn said. "Even take it into the shower with you [hung up at arm's reach]. I did [in China], even into the shower."[66]

Applegate and Fairbairn were sticklers for realism and detail. They required students to be able blindfolded, in simulated night conditions, to disassemble and reassemble not only American but German and Japanese weapons as well. To get a sense of what it felt like to be shot at, they actually shot at—or rather near—the students. "I remember Delicate Dan the Deacon having us stand in front of a target and then firing a .45 so that it just missed us and we could feel the muzzle blast," Edgar Prichard recalled vividly.[67] The former Tulsa journalist remembered the practical final examination held at a pistol house, what Fairbairn called a "mystery house," built to test the trainee under stress as well as to replicate the dangerous reality that an agent might face in attacking an enemy-occupied house or command center. "Each of us over a period of a couple of days would be awakened in the middle of the night and hauled off to carry out a special mission. When it came my time I was told that there was a Nazi soldier holed up in a building and that it was my job to go in and kill him. I was given a .45 and two clips [of ammunition]. The house I was sent into was a log house with long corridors and stairways. I wasn't sure whether there really was a Nazi soldier there or not. I kicked a door open with my gun at ready. Paper targets with photographs of uniformed German soldiers jumped out at me from every corner and every window and doorway. We had been taught to always fire two shots at the target. There must have been six targets because I got two bullets in each one. The last one was a dummy sitting in a chair with a lighted cigarette in his hand. If you didn't shoot him you failed the test."[68]

Field Maneuvers

Training exercises became increasingly demanding as the curriculum progressed. Sometimes in the hot, summer field exercises, the trainees found themselves plagued by insects. After a Sunday morning field problem at Area B in August 1943, Private First Class Arne I. Herstad, a member of the Norwegian Operational Group recently arrived from Colorado, wrote to his finance, "I have never seen so many bugs, flies and pests anyplace as there are here. I even walked past a snake hanging in a tree. Don't like them."[69] Jack Singlaub recalled more dangerous and demanding aspects of the training. "By the end of November, our training at Area B... had become a grueling marathon," he wrote. "We fired American, British, and German weapons almost every day. We crawled through rain-soaked oak forests at night to plant live demolition charges on floodlit sheds. We were introduced to clandestine radio procedure and practiced typing out code and encrypting messages in our few spare moments. Many mornings began with a run, followed by a passage of an increasingly sophisticated and dangerous obstacle course. The explosive charges under the rope bridges and wire catwalks no longer exploded to one side as exciting stage effects. Now they blasted directly below, a moment before or after we had passed."[70]

Throwing hand grenades is a dangerous business for the jagged steel fragments fly in every direction. In every camp, the grenade range was positioned some distance away and a trench was dug for the trainees who would pull the pin and hurl the grenade. "It's not like what you see in the movies," Ellsworth ("Al") Johnson, a medic from Michigan who trained with the French OGs at Area F and Area B in 1943. "You don't pull the pin with your teeth or you loose them; and the grenade doesn't blow up a building or a truck. It's an anti-personnel weapon."[71] In Area C, the grenade range began with the trench that had been dug at the top of a big ravine in the woods. "We were in the trench," John Brunner recalled. "We raised up and threw live grenades into the ravine so the fragments would not go all over the countryside. We ducked back down fast. The grenades exploded, and you could hear the shrapnel whipping through the trees above your head. You learned to respect those little things, those grenades. We also fired 60 mm mortars. You had to make sure they were firmly anchored. Then you just dropped the shell down the tube."[72]

Realism was what the OSS instructors were after. At Area A at least, exploding dynamite caps—not dynamite sticks—were used to simulate the sounds and dangers of battle. "One of our [field] problems involved dividing into two groups with one group being assigned to protect another summer camp area and the second group to try to penetrate the area. I was in the defending group," wrote SO trainee Edgar Prichard about the summer of 1942. "We spread out in the woods and set booby traps all over the place. They were not armed with anything except dynamite caps, but it was a wonder someone didn't get his eye put out from flying fragments. I recall one person got a piece in his abdomen, but it didn't go in deeply."[73] The following summer, Norwegian OG, Arne I. Herstad wrote to his fiancé about a night field exercise his group had at the OSS training camp in Virginia, which produced "a little bit of excitement." The instructors had set out some "booby traps," and Herstad managed to set off four of them. "The last one was set very close to us, and a few of the boys got hit with a few pieces of copper. These little pieces go in about 1/8th of an inch, and if there are enough of them in one place, it makes your arm, or whatever it hits, a little numb. The traps were dynamite caps, and they are pretty noisy if they go off beside your ear."[74]

Field exercises sometime used live ammunition, sometime blanks. In either case, it was rigorous and intense. SO trainee Erasmus ("Ras") Kloman, a Princeton graduate from Baltimore, said that the obstacle course at Area A-4 was "designed for the toughest and most combat-ready personnel. I remember keeping my head and the rest of my body down at snake level as we went through the exercise just as if live [and not blank] ammunition were being fired."[75] Private Albert R. Guay had never had Army basic training, so the field exercise he participated in at Area B in December 1942 was a new and sobering experience for the 21-year-old graduate of

American University. The winter had set in and it was cold and the ground was covered with ice and snow on Catoctin Mountain. "Had my first combat training today," he wrote in his diary that night. "Advancing over open field under fire. Ground frozen & awful hard. Slipped once & banged my knee cap hard. Probably be sore tomorrow. Enjoyed it very much, but it showed me how tough it would be to actually advance under fire. Casualties would be high. We are expendable."[76]

Spiro Cappony was a 19-year-old sailor recruited by the OSS as a radio operator, but his naval training had not included ground combat exercises. At Training Area B, he remembered, he underwent field exercises amidst live ammunition. "We had ground fire, where they fired machine guns at us. We're on our bellies crawling with live ammunition going over our heads." This was in late summer 1943. Within six months, he was under similar fire, this time from the enemy, in northern Greece. "In fact that happened to me and Johnny—Johnny Athens [an OG lieutenant]—in Greece, where the Germans were firing at us. Our horses stumbled, and we went down a cliff, and we ended up in the vines, grapevines, and the Germans shooting at us. We were lying on our backs. The bullets were flying over our heads. And there we were. It seemed like we had trained for that. There I was doing the same thing. We crawled out of there with the live ammunition going over our heads." Asked in 2006 to what degree the OSS training had prepared him and proven effective, Cappony responded without hesitation: "One thousand percent! Without that training, I wouldn't have made it. It saved my life!"[77]

As a member of a French Operational Group, Jacques L. Snyder's OSS training began in the fall of 1943 at Area F, the former Congressional Country Club. Part of it was learning how to live off the land, surviving on what they could find in our near the woods. "Some of our people became very adept at stealing pigs from local sharecroppers and organizing some fabulous barbecues in the boondocks. This activity reached such proportions that the locals living in the sticks were terrified of the 'night people' and no longer dared to get out at night, keeping a loaded shotgun behind the door. This... training included the quick and silent disposition of rabbits and chickens without raising the alarm, and the different ways to prepare them to make them edible. This talent came in handy when I had to sustain my group while operating behind German lines in Alsace [northeast France] and had to raid farms to steal rabbits, chickens, and eggs."[78]

Since Jacques Synder had grown up in France and was entirely fluent in the language and customs, OSS decided to shift him to Special Operations for solo infiltration as a spy and saboteur. For that training, he was sent to Area A. "The accent was on sabotage, learning more sophisticated techniques and acting as an individual agent…. The course was even more rugged than the previous one and had to meet all the requisites of security, using false identity and cover stories." He learned radio transmission and codes and other clandestine tradecraft with dead letter drops, cutouts, concealing devices, recognition signals, and the forging of identity documents. In addition, "we were conducting night operations, being dropped in an unknown territory with just a map and a compass and had to rendezvous with other groups at a certain time, generally in secluded wooded areas, using prearranged recognition signals. We had a very extensive, in depth sabotage training, which included all the latest methods of destruction, by fire, by explosives, by chemicals, to attack any targets of the opposition.... We were also taught how to manufacture our own incendiary mixtures and explosives out of ordinary chemicals."[79]

Espionage and Sabotage Exercises in the Region

For many future spies, saboteurs or guerrilla leaders, OSS field training exercises culminated in real-life, if mock espionage or sabotage missions at nearby facilities or cities. Bridges and dams were ready made targets. Lieutenant Al Materazzi from Hershey, Pennsylvania, along with a dozen other officers of the first Italian Operational Group were sent to Areas A and C in the spring of 1943 before meeting and training the enlisted men of their OG. At A-4, they were given some night problems. One was to sabotage a major iron bridge carrying U.S. Route One across the Rappahannock River. "At night, we clambered out and placed charges strategically—dummy charges, of course," Materazzi said. "Another exercise was when we were to sabotage a dam on the Rappahannock River. That one required descending into the gorge and climbing up the other side and—ah—taking care of any guards that might be there and sabotage it. The only one that was there was the poor night watchman. We scared him half to death. Nobody had told him."[80]

Fredericksburg, not far from Prince William Forest Park, was chosen for a training problem for one of the early classes to graduate from Advanced SO school at Area A in June 1942. That class of more than a dozen men included, among others, Jim Goodwin, a football player from upstate New York; Ben Welles, whose father, Sumner Welles, the Assistant Secretary of State; Robert Schlangen from the State Department,; Jack Okie, from Marshall, Virginia; Roger Prevost, a lieutenant in the Free French Army, who had connections at the White House; and Edgar Prichard. The entire group was taken to Fredericksburg and told to find out everything they could about the city that would be useful if they were a sabotage team in enemy territory. With a forged letter on White House stationery which requested assistance for Lieutenant Prevost in establishing a supply base for Charles de Gaulle's French forces, three of them visited the City Manager and came away with complete information on Fredericksburg, the water supply, the electric supply, the telephone system, the labor force and all of the factories then in Fredericksburg. Meanwhile, Prichard and Scotty Lockwood posing as fishermen pretended to fish on the city reservoir. They crawled through the inside of the dam into the hydroelectric plant from below, located the generators and marked them "destroyed" with chalk. "There wasn't much we didn't know about Fredericksburg by the end of the day," Prichard said. But he added, "I do know that later on the same type of reconnaissance program we carried out in Fredericksburg was repeated and that some of the students ended up in the clink."[81]

The practice continued throughout the war, particularly using larger industrial cities such as Philadelphia, Baltimore, and Pittsburgh. In the winter of 1943-1944, a four-man team of SO trainees from Area A was assigned to breach the security of steel plants in Pittsburgh. By that time these OSS industrial field problems—"schemes" they were called—were run in cooperation with the Civil Defense Agency, which assigned the targets in order to check their security. In each city, OSS assigned a senior person to watch over the mission, obtain release of the students if they were arrested, and prevent publicity. This particular team included Ras Kloman, 23, a Princeton graduate; William Woolverton, 23, a Yale graduate from a socially prominent Manhattan family; William Underwood, 32, from an affluent Connecticut family, who left the prominent Wall Street firm of Sullivan and Cromwell for the OSS; and Norman Randolph Turbin, from Richmond, Virginia, who had become a paratrooper at Fort Benning, and who would later, after behind the lines operations in Italy, would be killed in action in the Philippines.[82]

At Area A, the team Kloman was on worked out cover names and stories and prepared false identification papers. Their assignment was to penetrate US Steel and National Pipe and Tube, both which produced gun barrels and other military ordnance. They rented separate rooms in Pittsburgh and met regularly in working-class bars. The radio operator assigned to them from Area C relayed their progress. Woolverton got a job at U.S. Steel by pretending to be a metallurgist. Underwood posed as a magazine reporter got some information and had interviews lined up with senior management, until the company checked with the magazine which had never heard of him. Unsuccessful at getting a job at National Pipe and Tube, Kloman became discouraged and decided on the last night of the mission to break in at night. Risking being shot by a guard, he squeezed through a space in the chain-link fence. The plant was working round the clock, and no one seemed to notice anything wrong as Kloman walked around making observations and gathering up a number of documents, including plans of the plan. Kloman later learned that he had been the first OSS trainee to actually break into a plant rather than gaining access through a cover story. It was a bold and risky gamble, and he gained some acclaim for it. Richard Helms on SI's European desk, later asked him, probably facetiously, whether he would like an assignment parachuting into Berlin. Kloman did not go to Berlin, but he did go to Egypt, Algeria, and Italy. At 24 years old, he became Acting Chief of Operations in the Mediterranean Theater, and during the war, he helped organize and coordinate Special Operations teams that infiltrated Italy and France on sabotage missions.[83]

Communications Training Experiences

Trainees at the Communications School at Area C were learning primarily how to be OSS base station operators overseas; they also learned how to be radio operators in the field if needed. They had to know electrical theory, and they all had to know how to communicate between the base stations and the operators in the field, who were equipped with the mainstay of the agents, the small SSTR-1, a suitcase sized clandestine wireless telegraphy sender/receiver.[84] "The transmitter was absolutely exceptional. The receiver was very difficult, especially in the lower frequency ranges, of course, because the spectrum was noisier, and there was more interference…,"

Art Reinhardt, who used the SSTR-1 in China explained. "You did not have very much band spread, so your selectivity was very critical. So the operators relied on their expertise and also the phenomenal ability of the communicators [the operators at the base stations], digging out signals from the interference, static, and what have you." "The major problem sets had was when they were airdropped to agents in the field. They were not designed to handle the shock, and many of them broke on impact.[85]

Agents from OSS branches and operators from the Communications Branch operators both learned how to use the SSTRs out in the woods. Mostly this was done in the woods of Prince William Forest Park, but often this was done in the Virginia countryside. Directing the antenna was crucial in either case. "We would have to do it all in a quick period of time, because if we were really in the field," trainee Marvin Flisser recalled, "we would be triangulated by the enemy. We would have to get that antenna up, send out messages, and get the antenna down and out of there before the enemy gets you. Within a half hour we had done everything."[86] Reinhardt had similar recollections: We were sent out on live field exercises, where we took our SSTR-1s, and they drove us out in trucks to areas in Virginia somewhere, to abandoned farms. They would give us a lunch. We would encrypt our message, and we would contact the base station back at C. We would create simulated field operations.... Of course, as part of our training, Area C had a base station, which had the more powerful transmitters that you would find at an overseas base station. We were taught how to tune them and be familiar with them. Because, once you went overseas, you didn't know whether you would end up at a base station or whether you would be an operator at a field station, or what you would be. It depended on the need."[87]

During classroom instruction, students would sit at telegraph keys with earphones on their heads, practicing sending or receiving messages in code. For non-Communications Branch student agents the minimal goal was at least six words, or more precisely six five-letter code groups per minute. But for Communications Branch personnel the minimum as Private Harry M. Neben from Illinois remembered, it was 25 words per minute[88] In the code training room everyone would be wearing earphones, and civilian instructors like John Balsamo and Timothy Marsh would supervise as a dozen or so trainees. The students would first work with simple telegraph keys, but as they became more proficient, they would graduate to using superfast, double actions, fiberplex "bugs," as these semi-automatic telegraph keys were called. The class would not all be doing the same thing, it would depend on the individual's progress and also the instructor's decisions. "You know we all came in from all walks of life. We were all used to operating with different procedures and protocols," Reinhardt recalled. "They were trying to standardize somewhat. But that changed, because depending upon what field or theater you went in, because sometimes they might want you to simulate the German field operators, the Italians, or whatever. So it varied from place to place."[89]

Sometimes, the instructor would simply play audiotapes of someone sending Morse code, and the students would copy it, but while that improved speed, it was not realistic. "The stuff on the tapes is all perfect sending, and they weren't going to hear much of that [in the field]," Lieutenant James Ranney said. "So I would do it by hand, so they could get used to that. It would be imperfect. Everybody has a different—what they call a 'fist,' and you can actually identify someone on the key by his 'fist.' You can have two people send, and they won't sound anything alike. It is all in the international Morse [code], but it will be different in the spacing and in how careful they were and how many mistakes they made and so forth."[90]

Area C Relationship with Area A

Most of the students at Area C knew little or nothing about Area A, the secret training area for saboteurs and spies, even though it was only three or four miles on the other side of Prince William Forest Park. Because of the widespread field exercises with live munitions, it was off limits. Area C had its own pistol and rifle ranges, grenade trench and ravine, and even its own obstacle course, as well as a meadow in which trainees crawled along the ground with rifles under barbed wire.[91] Morse code instructor Timothy Marsh remembered that Area A was a place where OSS was using explosives and other munitions and that "Commo" instructors got to meet some of the agents from there.[92] "We had a vague notion that it was a big wooded area and that there were other people in other areas around us," trainee John Brunner said. "We did not know the nearest town. The truck did not open the flaps until well into the woods. We did not know that Quantico Marine base was nearby."[93] Wandering around the forested area was discouraged, and the men were directed to stay in camp. Art Reinhardt, a student at C-4 for three months in the summer of 1944, recalled that he stayed at the training camp, except to go on leave. He did not explore the forest park.[94]

Communications instructor James Ranney got into trouble when some of his students wandered into Area A by mistake. He assisted in planning a nighttime field exercise for SO or SI agents that was supposed to be held within Area C, which was unfamiliar to them. "They wanted me to make up a set of instructions guiding them to a certain point, where they would then set up their radio and talk back to the base station. Either my directions were not too clear, or they didn't read them right, or they just got lost. But they—at least some of them—mistakenly ended up in Area A, where there were all kinds of unexploded mortar shells and everything."[95] They were not the only ones who got lost in the forest during the war. Ranney himself, an instructor at Area C for six months in 1943, said "I never was really sure where I was in that camp and forest. It was a forest, and we routinely got lost in the place."[96]

Accidents

Accidents do occur, particularly with weapons and explosives around, and there were injuries and a few fatalities at OSS training camps. There is little record of minor injuries, but veterans do remember some of the more dramatic ones. Instructor Frank Gleason certainly recalled being blown off his feet and knocked unconscious by a demolitions accident at Area D on the Potomac when someone had improperly wired the detonator.[97] A communications student from Area C, while helping set up a remote transmitting site near Richmond picked up an antenna wire already connected to the power source and it knocked him about fifteen feet. The student was sore but otherwise unhurt.[98] Another student received a wound, albeit not a deep one, in the abdomen from debris from one of the dynamite caps being used to simulate the noise of battlefield explosions at a field training exercise in Area A in the summer of 1942.[99] William J. Casey, a future Director of Central Intelligence, but an SI student in 1942 had his jaw broken by a flying chuck of tree branch when he triggered a trip wire on the demolition trail at Area B and was down crouched down far enough.[100] "I was just sick about it," Instructor Frank Gleason said. "It was the only accident we had that I know of.... We were very, very careful that we did not hurt anybody."[101]

Although the final war report of the OSS did not list them, there were several fatalities during paramilitary training. At least two occurred on the obstacle course at Area F, the former Congressional Country Club, where, as at the other paramilitary training camps, in order to simulate battle conditions, students were required to crawl with their weapon along a trail while machine guns fired live ammunition over their heads. Two students were killed when they rose up into the machine gun fire, one of them because he unexpectedly came face to face with a snake.[102] In another instance, this one at the Maritime Unit School at Area D in 1942, a trainee accidentally drowned in one of the underwater exercises on the Potomac River.[103]

A bizarre tragedy occurred during a nighttime field exercise at Area F in 1943. It was planned as a standard exercise, very similar to those done at Area A or B, and it involved a group of OG trainees assigned to make a mock attack on a bridge nearby. The plan was to overcome the guards, whose part was played by enlisted men on the camp cadre, and then set up mock explosive charges to destroy the bridge. Lieutenant Al Materazzi was an observer. Years later, he recalled the events of that summer night. "In the morning, we had received our knives, the famous Fairbairn knife [a razor sharp stiletto]. We were told how to place it, where to carry it. We were given strict instructions not to remove it from its scabbard for the exercise. The exercise was that the officer [Materazzi] would be an observer, and the enlisted men would do the mission, which consisted of going, as I remember it, to a bar along the Cabin John Creek, rendezvous with the others and attack the place with another group. Opposing us was the station complement [enlisted members of the cadre at Detachment F]. The patrol moved in like a classic infantry patrol. There were two points [two members out in front], and as they reached the bridge, they were captured. Now, nothing happened, and one of them [one of the remaining men in the patrol] came to me and said, 'Lieutenant, I can get up there and do it.' And what did he do? He put the knife between his teeth, did a magnificent job of crawling over the parapet, waited until the guard turned his back, and then put the knife loosely on the man's back, and then said, 'All right, you're captured. Now, we can go on.' The guard wore glasses thick as bottles, you might say, and he turned and walked into it [into the knife]. He was killed immediately. Then they started to scream for me...I broke cover. Found out what it was, and arrested him on the spot. I sent him [back] to the F Area. I then took over the platoon to get to where we were supposed to go, which I did."[104] The man went before a court-martial. He was found not guilty but was transferred out of the OSS.[105]

Serious accidents were indeed rare in OSS training in the United States. Instructors emphasized safety measures with the weapons and took great precaution with the explosives. Spiro Cappony, a sailor and a radio operator, had never had any paramilitary training until the OSS recruited him and send him to Area B in 1943. He remembered that a big, red-haired instructor was particularly good. "He taught you to be confident in yourself. Not to be afraid to do things. Learning about your weapons. The weapon is your friend, but it is also your enemy. Because you don't want to make a mistake with it either. You had to be very careful how you used your weapons. How to store them, how to carry them, and all of that. You were taught not to be careless, but to be careful with them and to respect their firepower.[106] Learning and practicing with hand grenades and mortars was, Cappony, recalled, "really scary. Because you knew it could explode. That was live stuff, not play stuff. They were very, very strict about training us with that. They were really on top of us. They made us appreciate the danger...to be very, very careful. There was very close supervision because they didn't want to see any accidents. Thank God, we never had any accidents."[107]

Celebrities in Disguise at OSS Training Camps

OSS training camps had a few nationally known celebrities as trainees. Among those students were professional wrestler "Jumping Joe" Savoldi; professional baseball player "Moe" Berg; New York socialite and former Russian prince, Serge Obolensky; and movie star Sterling Hayden.

Born in Italy but growing up in Michigan, Joseph Savoldi first became famous as an All-American fullback on Knute Rockne's football team at Notre Dame University in the late 1920s. The Chicago Bears wanted him as a halfback, but Savoldi decided to become a professional wrestler, and in 1933, he won the world wrestling championship by defeating Jim Londos at Chicago stadium. His winning streak continued, and in 1938, he garnered the European wrestling championship.[108] In 1942, the 34-year-old Savoldi was recruited by Donovan's organization, and he was a member of one of the first classes trained at Area B in Catoctin Mountain Park.[109] During close-combat practice one day, instructor Jerry Sage, who knew Savoldi only by his code name "Vic" was taken aback when "Vic" leaped at him feet first, caught him around the legs, toppled him and pinned him to the ground. "Where'd you learn that?" Sage asked, spitting grass. "That's my specialty in the ring," Vic replied, "It's called the flying scissors." He then acknowledged being a professional wrestler.[110] After his SO training, Savoldi was selected in 1943 for the "McGregor" Mission to negotiate with Italian admirals for the surrender of their fleet. Michael Burke, all-American football player for Cornell and later president of the New York Yankees, a member of the mission recalled that Savoldi "was built like a gorilla and moved as lightly as a leopard. His wrestler's face had been mashed against the ring canvas a thousand times. He was enthusiastic; I thought he would be perfect. He would terrify Girosi [Marcello Girosi, New York brother of the Italian admiral they were trying to contact] and maybe the entire Italian fleet."[111] Savoldi would be Girosi's bodyguard. Before the group left, however, Mussolini's fascist government had been toppled, the new Italian government had joined the Allies, and the OSS had already smuggled out of German-occupied Italy, a dozen Italian naval officers and scientists with crucial information about the new radio-guided bombs, high-speed "commando" submarines, and special torpedoes the Axis had developed and were using against Allied forces.[112]

Morris ("Moe") Berg was an extraordinary individual, an intellectual, linguist, spy, and most famously a professional baseball player. Growing up in Newark, New Jersey, the child of Russian Jewish immigrants, he starred academically and athletically in high school and at Princeton University, graduating Phi Beta Kappa and captain of the baseball team in 1923. He spent the next 16 years as a catcher for the Chicago White Sox, Boston Red Sox and a couple of other teams, while also earning a law degree at Columbia and taking classes at the Sorbonne. Sportswriters called Berg, "the brainiest guy in baseball."[113] Berg liked being a hero and the excitement of taking risks. That and his intellect, self-confidence, charm, and his iron nerves made him a perfect OSS agent. In August 1943, he was hired as a civilian by the OSS.[114] Apparently, Berg received some OSS training at Area B at Catoctin Mountain Park.[115] Subsequently, he was assigned first to the Balkan desk until May 1944, when he left for England and worked on special missions throughout Europe.[116]

Berg directed a series of daring field operations, in many of which he played the primary role himself. He parachuted into Yugoslavia, and his report supported Tito's efforts against the Germans. Later, he was flown into Norway, where he obtained more information about the Germans' nuclear-related heavy water plant, subsequently destroyed by Allied bombing. But his most extraordinary mission was in 1944, when because of his knowledge of physics, Berg was sent to a conference in neutral Switzerland to listen to a lecture by Werner Heisenberg, the leading nuclear physicist in Germany. If Berg judged from Heisenberg's remarks that a German atomic bomb was imminent, he was to assassinate Heisenberg. Listen to the lecture, Berg correctly concluded that the Germans had not made significant progress toward an atomic bomb. Later, Berg was awarded the Presidential Medal of Freedom for his accomplishments with the OSS.[117]

Prince Serge Obolensky was born in St. Petersburg in Czarist Russia in 1890 to a family that traced its ancestry back to Igor, the Grand Duke of Kiev in the tenth century. Although not a member of the Romanov family, the Obolenskys won the right to use the title prince in the reign of Ivan the Terrible. Serge's father was a general and his mother was from one of the wealthiest families in Russia. Serge studied at St. Petersburg and Oxford universities. When the First World War broke out, he left college and joined an Imperial Guards cavalry regiment. He fought against the Germans in World War I and against the Red Army during the Russian Revolution. After the communist victory in Russia, Obolensky became one of the many white Russian émigrés, settling first in London and then in New York.

Working in banking and real estate in New York in the 1920s, he married into the socially prominent and extremely wealthy Astor family. He became a US citizen in 1931, and in 1935 having lived in many of the best hotels in Europe and America, was given the job by Vincent Astor of consulting and promoting the renovated St. Regis Hotel in New York. This led him into the hotel business, and he later became vice President of the Hilton International hotel chain.

In 1940, when the German Army overran Western Europe, Obolensky tried to enlist in the U.S. Army, but was rejected because he was 49. Instead, he joined the New York National Guard as a private and was soon promoted to captain. After Pearl Harbor, he sought out William Donovan and asked to join as a commando. Donovan obtained an Army commission for him as a major. The 51-year-old Obolensky reported to Colonel M. Preston Goodfellow in Donovan's office in Washington in early 1942 and was sent immediately to Area B-2 for Special Operations training.[118]

At B-2, Obolensky worked hard. He recognized that the course was designed to toughen the men as well as provide them with commando skills. "We got up early and went to bed late, training in special tactics, shooting at sounds at night, guerrilla tactics, memory tests of smell and configurations of places in total darkness, and especially demolition work."[119] Obolensky's code name was "Sky," it was written on an identification tag he wore. Lt. Jerry Sage, then an instructor at B-2, remembered "Sky, the tall gentleman... then about fifty-four years old... He was a real man—tough, resilient and good-humored. He trained hard, and despite his greater age, he kept up with all the younger fellows and did everything we did."[120] Sage recalled that despite his age, Sky (Obolensky) insisted on going along with the OSS tradition of making all five qualifying parachute jumps in one day. When the group had completed the first two jumps, Sky's legs began to hurt. He had the medics wrap them in tape, but the tape broke when he hit the ground again—on both the third and fourth landings. The jumpmaster tried to get him to stop, but he insisted on completing the fifth jump that day. Up in the plane, his legs hurt so badly he could hardly walk and the jumpmaster tried to block his way. But Obolensky yelled out, "Throw me out of the plane, damn it!" They did, and he received his paratrooper's insignia that day.[121]

Back at OSS headquarters, Goodfellow put Obolensky to work in the summer of 1942 helping to prepare tables of organization and manuals for guerrilla units, what would later become known as Operational Groups. Obolensky was told that the Joint Chiefs of Staff only reluctantly approved Donovan's idea for these uniformed, military units and that the U.S. Army had feared a repetition of the nineteenth century's U.S. Volunteers. Those had been ad hoc units of citizen soldiers led by amateur officers such as in the Civil War and most famously in the Spanish-American War, Theodore Roosevelt and his "Rough Riders." The Army, Obolensky claimed, would not even allow the OSS units to use the standard military nomenclature—platoons, companies, battalions. Nevertheless, Obolensky declared, eventually "we got everything we wanted, the newest and most secret weapons. We had the bazooka long before anyone else, high explosives, pliable plastics, quite safe with detonators."[122]

After being dispatched on a tour of Army training camps by Goodfellow, Obolensky returned to Washington and was sent to OSS Area A in Prince William Forest Park, Virginia. Colonel Russell ("Russ") Livermore was there in the spring of 1943 and Obolensky says that he worked with Livermore on the development of the organizational and training program for the OGs. It was while he was at Area A that he made the five jumps in one day and earned his paratrooper's winged insignia. Afterwards, Obolensky, later wrote, he was assigned to prepare a camp for guerrilla warfare training of the OGs, or what he called "our OSS commando combat teams." The facility was Area F, the former Congressional Country Club.[123]

Obolensky obtained two young instructors in guerrilla warfare at the infantry school at Fort Benning who would play key roles in the Operational Groups: Captains Joseph Alderdice and Alfred T. Cox. While the men for the OGs were being recruited, Obolensky and Alderice wrote the training curriculum, lectures, and manual for the OGs. They drew upon British commando material, Fairbairn's methods, insights that Obolensky had gathered in his career and his tour of American bases, and in addition, the translation of a smuggled copy of the *Soviet Russian Guerrilla Manual*. The manual they wrote, *The Operational Group Manual*, eventually became the *United States Army Guerrilla*.[124] The brilliant Albert ("Al") Cox, who had been a civil engineering student at Lehigh University in Pennsylvania, President of his class, captain of the football and baseball teams, and Phi Beta Kappa, also played a significant role in organizing and training Operational Groups at Area F, and later in leading and directing them in action. In the Mediterranean Theater, Cox led the first OG teams sent to Corsica from Algiers to stage raids along the Italian coastline, and he also led the final OG team into southern France to coordinate all the OG sections in the Rhone River Valley behind German lines.[125]

In late August 1943, Obolensky, by then a lieutenant colonel, was sent overseas to the Mediterranean. In Algiers, Donovan himself explained to Obolensky that the Allies were soon going to invade the Italian mainland south of Naples from Sicily. General Dwight Eisenhower, the theater commander, wanted a prestigious person—in this case Obolensky—to neutralize the Axis forces on Sardinia by delivering special letters to the Italian commander on Sardinia from the General Eisenhower and by the head of the new Italian government, Marshal Pietro Badoglio, ordering him to surrender. The four-man OSS team, including Obolensky, an assistant, Lieutenant James W. Russell, a British sergeant as a radio operator, and one of the Italian American OG officers, Lieutenant Michael Formicelli, from Areas A, F, and B, as an interpreter, parachuted into mountains. At 52, Obolensky thus became the oldest combat paratrooper in the Army. After leaving two members of the team and the radio in the valley, Obolensky and Formicelli made contact with the Italian commanding general, who agreed to surrender. Although Obolenksy could not convince the general to use his troops to try to neutralize the 19,000 Germans already leaving the island, his mission to Sardinia proved successful in achieving Allied control of Sardinia and its Italian garrison of 270,000 troops without a fight.[126]

The best known celebrity among the OSS Special Operations agents who trained in the parks was Hollywood star, Sterling Hayden. He had dropped out of high school at 17 and run away to sea, and the 6'5", 230-pound, ruggedly handsome, blonde adventurer had been picked by Paramount, which billed the 24-year-old in 1940 as "The Most Beautiful Man in the Movies," and "The Beautiful Blonde Viking God." The next year, he married his co-star, the vivacious, sexy Madeleine Carroll. But after only two films, Hayden left Hollywood in 1941, did some parachute training in Scotland at Donovan's suggestion, and in 1942, joined the Marine Corps. He went through boot camp and enrolled in Officers Candidate School in Quantico, Virginia. To complete his break with Hollywood and to minimize public recognition after the publicity of his enlistment ("Your face looks familiar," officers and others would say to him), he legally changed his name to John Hamilton. In the summer of 1943, the 26-year-old lieutenant, soon to be promoted to captain, transferred to the OSS. In his memoirs, Hayden wrote simply that "after a series of false starts in the direction of China, I was handed an enlisted man from the Navy who was fluent in Greek, telegraphy, and cipher, and we were dispatched to Cairo to harass the enemy."[127] Overseas they were separated. The enlisted man was sent behind German lines to help the guerrillas in Greece. Hayden went to Yugoslavia to aid Tito's Communist partisans against the Germans. The seafaring actor also operated a small fishing boat along the Dalmatian and Albanian coastlines, part of an OSS air rescue team and also ferrying supplies to OSS agents in Yugoslavia and southwestern Greece. For his exploits, Hayden was awarded the Silver Star Medal. Afterward, he returned to Hollywood, where his placid strength and taciturn, solid, weathered persona served him well in such films as *The Asphalt Jungle* (1950), *The Killing* (1956), *Dr. Strangelove* (1963), *The Godfather* (1971), and *The Long Goodbye* (1973).[128]

The enlisted man assigned to Hayden in the summer of 1943 was Seaman Second Class Spiro Cappony from Gary, Indiana, who had been recruited by OSS naval radio training school at Miami, Ohio, in July 1943. He and Hayden, who Cappony knew as Captain John Hamilton, were assigned as a combat team to be sent behind German lines in Greece. Cappony was sent first to the OSS Communications School at Area C in Prince William Forest Park. There, he learned OSS equipment, polished his telegraphy to 22 words a minute and practiced coding and decoding. He learned to tap the telegraph key with either hand, a skill that proved useful in Greece when he was wounded in the right arm. He credits the instruction at OSS Communications School with saving his life. "[In Greece,] the Germans would follow us. They had direction finders. I had my crystals and my radio in a suitcase. We learned how to use that, and we learned how to change our crystals to get the Germans off our backs. I learned how to set up my messages and the numbers and so forth. I had that all in my mind. I had to memorize all that stuff. I picked my [high] school song as my starting point. The guys in Cairo [the OSS radio base station], they knew my school song. So when they would get my messages, they would know it was me. When I got wounded [in my right arm], I had to transmit with my left hand, they wouldn't accept it. They didn't recognize my 'fist' [his telegraphic style]. I had a hard time convincing them it was me."[129]

Training at Area B in the use of the bayonet later served Cappony well in action in Greece in 1944. "I remember when I first went into Greece, we were there the first day, and I had my rifle. I heard a whistle. I look around, and I asked the captain [Captain James Kellis, head of the four-man SO team], I said 'Captain Jim, what's that sound?' He said, 'It's from the guerrillas. They've spotted a German. The Germans are coming at us.' When he said Germans and here it is my first day, and the Germans are coming at us, I got a tingle. I heard my captain say, 'Get your bayonet ready.' I tell you, when I got my bayonet ready, I started to think about my training that I had in Area [B]. I thought, 'Oh, my God, there'll be hell to pay up here.' But thank God, the Germans just came by. We were very, very quiet. There were about three or four hundred of them. They came up and looked around, but we had disappeared. It really scares you, because you are there for combat duty, but thank God for the combat training."[130]

Sterling Hayden [Captain John Hamilton] returned to the United States in November 1944, visiting Areas F and A, before being sent to France and Germany in 1945 hunting for anti-Nazi Germans to use against the Nazis. Spiro Cappony returned from Greece in late 1944, but did not have any money because he obviously had not received his pay while behind enemy lines. "Captain John Hamilton was my buddy. I met him in the Congressional Country Club. He says, 'Hey, Gus,' what are you doing here?' I said, 'I can't get out of here. I don't have any money.' He said, 'Get out and get drunk.' I said, I'd love to get drunk.' He said. 'Here's twenty bucks.' "[131]

Some Lighter Moments in a Tough Training Schedule

Despite the seriousness of the training, there were inevitably some lighter moments at the training camps. Some were funny, some ridiculous, some dangerous. Some resulted from carelessly by personnel inexperienced with firearms. One private on the cadre at Area B-5 in 1942 accidentally shot a hole in one of the walls of the arsenal where all the ammunition was stored. Two weeks later, another private there was caught using the bottom of the handle of a pistol like a hammer to tamp down a nail. He was caught and relieved of the weapon before it went off. For punishment, he was chewed out and confined to camp for a month.[132] A much less dangerous violation of regulations and procedures occurred at B-5, one afternoon in October 1942, when Private Albert Guay, the new assistant company clerk, who had only been in the Army for three weeks, was, as part of his duties, playing musical records over the camp's public address system after duty hours. "I was leafing through a bunch of records," he recalled. "We had a big stack of 78 rpm records." And I found 'The Star Spangled Banner.' I had no more than started to play it when the phone rang. It was Major Blogg, and he said: 'Who? Why? What for?' and 'Stop!' and 'Come over to my quarters!' So I went over to his quarters, and I told him that I had not had basic training yet, and I did not know that that you were not supposed to play the National Anthem except on special occasions. It brought the camp to a standstill."[133] At Area C in Prince William Forest Park in January 1945, Lt. Richard P. ("Scotty") Scott, who had just arrived, was "schooled" by Major Albert Jenkins, the camp commandant, because one Monday morning on the P.A. system, Scott played an inappropriate song for reveille. Others had played "Oh, What a Beautiful Morning," and such, but Major Jenkins did not approve of Scott's choice of "Drinking Rum and Coca Cola" for reveille.[134]

An extraordinary use of the public address system at Area C was made on June 6, 1944, D-Day, as the Allies landed on the coast of Normandy. On that historic occasion, the OSS Communications Branch relayed information on the landings directly from their base station in England. The men of Area C gathered in a recreation hall to follow the news. "We had another civilian named McEwen, an old Navy radio operator, who was one of us, an instructor in the code department," recalled Timothy Marsh. "He could copy [receive] up to 60 words per minute while talking at the same time. Like the old telegraph operators. Extraordinary! That day, D-Day, they tied in military circuits into one of the Rec halls, where we were, and he would listen and then tell us verbatim what was going on. He was telling us what was happening in the D-Day invasion as it was sent in Morse code from overseas. This was the only time that that kind of thing happened."[135]

Parachute training was hard work, but it sometime produced some humorous moments. In the summer of 1942, Area B instructors Captain Charles Parkin and Lieutenants Frank Gleason and Ed Stromholt and a couple of trainees went down to Area A and used it and the Marine Base at Quantico to make the five jumps necessary to qualify as certified paratroopers. "We got in one of those old B-10 bombers, and we took off," Gleason said. "When we were over the OSS camp, they would throttle it down to 90 m.p.h.; you could hear the 'beep, beep' of the warning indicator that it was about to stall, and we would jump. We did five jumps that day. The last one was from 500 feet. It was just: swing, swing, and bang, you're down. It was a forested area, but there was clearing, and we jumped into that clearing. There were two [OSS airborne] instructors with us. One was a Colonel [Lucius] Rucker."[136] The comings and goings of the OSS parachute trainees produced some confusion at the entrance to the Marine Base. As Parkin later recalled, "Each time, we would take off from a plane from Quantico, jump into clearings in Area A and walk back across the road to Quantico. Finally, the Marine sentry at the gate stopped me and said, 'Captain, I have seen you walk into the base three times today, but I've never seen you walk out. How are you doing this?'"[137]

For Parkin, the fifth jump did not quite go off as planned. His friend, Frank Gleason, remembered it. "The last jump, we were only 500 feet high. We jumped and we were practically on the ground. Charlie landed in a tree, he was upside down, hanging by his leg in that tree. It was quite a sight. I helped him get out of that tree. You've got to remember that those were crazy days. We were all young, and we were immortal and fearless, and we would do anything, anytime, anywhere."[138]

OSS high command liked to show off its paramilitary, commando-like training to certain dignitaries who wished to visit the camps. Sometimes such demonstrations went well. Norwegian OG member Arne I. Herstad from Tacoma, Washington, wrote to his fiancé of how more than a dozen of a "Norso" group, headed by Captain William F. Larsen, were called down from Area B-2 at Catoctin Mountain Park to Washington, as Herstead explained, to "put on a small program for the benefit of [Swedish-born, Norwegian] Crown Princess Märtha, Wild Bill Donovan, and a few more brass hats. I and 5 other fellows fired [sub]machine guns, 6 others fired pistols, and the rest of the boys were divided up for close order drill and a few other things we have had in our training. She talked with most of the boys and shook hands with all of us. Then we all had a real swell dinner with the princess of course. She's really swell, speaks excellent Norwegian. Very much more beautiful than her pictures."[139]

Sometimes the visits did not go well. In 1943, a stream of crisply uniformed staff officers—from colonels to generals—was escorted out to Areas A, B, and F to watch the SO and OG exercises or even to address the trainees. Such visits did not always sit well with the trainees, many of whom resented what they considered gratuitous pep talks by staff officers who would ride out the war in offices and clubs in Washington. Lieutenant Jack Singlaub recalled that at Area B his training group, which included many irreverent paratroopers and future Jedburghs, quickly became fed up with being forced to dress up and listen to such ceremonial visitors. On one particular occasion, they decided to deflate such pomposity. Halfway through the distinguished visitor's talk, the began to yell out a count in unison, concluding "forty-eight, forty-nine, fifty! Some Shit!" That ended the visits to that group in Area B.[140]

During one night field exercise at Area A in Prince William Forest Park, other trainees were simulated an assault including overcoming "German" guards and blowing up a dam. At a small concrete dam in the woods in Area A, a few enlisted members of the cadre pretended to be German guards. The attacking force of trainees would sneak through the woods, overcome the guards and set dummy charges to "blow up" the dam. Lieutenant William Dreux recalled one such exercise which went slightly awry. His cabin mate Lieutenant Robert Farley, who had experience fighting in the Spanish Civil War, led his attack force with particular gusto that night, using a flying tackle that left the "enemy" sentry sprawled on the ground and almost unconscious. The problem was that it was not an enemy sentry, but a staff colonel from Washington who had come down to observe the exercise. According to Dreux, the colonel never returned to observe any more such maneuvers.[141]

Practical jokes were not unusual. One night at Area C in the winter of 1943-1944, trainee Marvin Flisser of the Bronx decided to play a practical joke on his best buddy, with the help of some friends and the camp guards. "We had girlfriends [in Washington, D.C.]," Flisser said. "One day my friend [code named] Larry Lamont—now we didn't tell these girls where we were…. We made up a story—There are several different entrances to this little camp. I don't know, three or four different entrances. We made up a plan ourselves that we were going to play a trick on Larry, and we were going to tell him that his girlfriend, Helen, was looking for him. We told him, "Larry, Helen is at the gate, she's looking for you." He says, "What! I'm going to be court-martialed!" He was so frightened. So, he goes to one of the gates, and he says, 'I'm looking for Helen.' And they [the guards] say, 'Oh yeah, she was here, but she went to the next gate.' So he ran to the next gate, and said, 'I'm looking for Helen.' They say, 'Oh, she was here looking for you, but she went to the next gate.' We had him running around, all around the camp. We finally told him that this was all a ruse, a phony. (Laughs) Nobody knew who we were or where we were."[142]

One of the stranger elements of OSS training to many of the students was the role of the psychologists trying to assess them. Art Reinhardt remembered a very brief interview in a darkened tent at Area A in which the main question the psychologist with a pencil flashlight asked him was why he wore his sideburns so long. "I said I didn't know they were that long. That was the end of it."¹⁴³ Caesar Civitella remembers the first question that the psychologist asked him was: "'How did you feel just before you made your first parachute jump?' I said I had butterflies in my stomach. He said 'How did you know that? Did you ever eat a butterfly?' I said, 'Of course! With chocolate!' What kind of question was that."¹⁴⁴ Even more mysterious, civilian instructor Timothy Marsh remembered that sometimes at night, the psychologists used to project weird noises from their office in Area C. "We had this psychology tent, a cabin made up primary of psychologists, which was kind of remote from the rest of the camp. However, it had weird sounds coming from it at night, such as wolves chasing babies, children. You could hear them start a long way off and then getting closer and closer and closer. It was very unusual and unnerving. That was one of the things they liked to play late at night. There were other things that were unnerving, other sounds that were just unnerving....They were military [personnel]. We didn't know much about them, to tell the truth. But they seemed to be trying to unnerve people."¹⁴⁵

"Going into Town"

Despite the long hours for staff and trainees, there was some leisure time at the OSS training camps. Most of the recreation and leisure time activities, especially for the staff, came from visiting local towns on an evening pass. At Area B in Catoctin Mountain Park, the nearest town was Thurmont, Maryland, then home of about 6,000 people on the east side only a couple of miles from the entrance to the park. Initially, some of the cadre would occasionally go to Thurmont. Assistant Company Clerk Albert Guay, for example, attended the Lutheran Church there one Sunday in November 1942.¹⁴⁶ Some of the old time Thurmont residents remember seeing occasional soldiers, who were not local men home on leave, in the town during the war, but they did not know where these men came from.¹⁴⁷ Around Catoctin Mountain, rumors suggested that the sealed off area had become an officers training camp or later a facility for rest and rehabilitation for combat veterans.¹⁴⁸ A report from the OSS Security Branch in September 1943, indicated that the local residents were naturally aware that there was military activity in the area, "but there appears to be little concrete evidence of any special curiosity or interest exiting regarding the nature of that activity." "The entire Area [B] is ideally situated from the standpoint of physical security," the report concluded, "being far removed from any town or settlement of any size."¹⁴⁹

By 1943 in order to maintain secrecy about the training at Area B, personnel from the camp were no longer allowed to go to the nearby hamlets, like Thurmont. Instead, their evening passes permitted them to go to either "Hagerstown or Frederick. These were small but lively cities, each about a dozen miles away. Hagerstown's normal population of 40,000 was swollen by the influx of many people seeking jobs in the booming defense industry, particularly the Fairchild Aircraft Factory, which transports and other planes. Permanent staff members at Area B seem to have had regular brief excursions to Hagerstown on evenings during the week and on weekends. They could go to the movies, dance halls, roller skating rinks, bowling alleys, or local taverns, bars, cocktail lounges, and nightclubs, or, of course, the facilities provided by the United Service Organization (USO), which served the morale needs of U.S. military personnel.

There were nearly fifty enlisted men on the camp staff in Area B-5 in late 1942, and Al Guay, a recent college graduate serving temporarily as assistant company clerk, recalled that among them were "a rough bunch of guys up there." On his secondary day in camp, he noted in his diary that "last night a couple of guys got in a fight in Hagerstown & one got a fractured skull." As Guay told the author in October 2005, "I wasn't in town that night. But our recreation was to get into a truck and go to Hagerstown, Frederick, or Thurmont. We would go in town in the evenings, not just the weekends. I looked earlier this month to see if I could remember where we parked our truck. In Thurmont, it was in a square. We parked there and most of them went to the bars. I was not a barfly, so I didn't do that. I would walk around or go to a movie or a dance. There were a lot of USO dances around."[150] A number of the other enlisted men, Guay recalled, got mean when they got liquor in them.[151] "Sgt. got kinda loud & tough to a girl in a restaurant; finally got him outside to cool off," reads one diary entry. On a bus jammed with soldiers returning to camp from Saturday night leave, there was much trouble with drunks. One soldier "went after the major with a knife." Guay never saw that soldier again; he probably ended up in the stockade. Another time, two enlisted members of the cadre went absent without leave (AWOL). The soldier who slept in the bunk above Guay was, the diarist noted, "a real character. Has [in the past] been shot several times in brawls, stabbed, beat up, etc. A very nice fellow when sober: friendly, polite & retiring. But when drunk, he's loud, obscene & ornery." On 11 December 1942, the camp doctor "gave us a good sex lecture tonight, because too many men are getting venereal diseases."[152]

Since they were assigned to the secret OSS rather than a US Army unit, even the cadre did not have unit patches on their uniforms. Some simply let it go that way, others, like Guay went out and bought a shoulder patch. "It was just an Army shoulder patch with a star on it," he recalled. "We weren't supposed to let anybody know we were in the OSS. I bought the shoulder patch so I would just look like all the other soldiers. I was wearing a "Class A" [dress] uniform."[153] "When we would go downtown, people would ask us where we were stationed. We would tell them Camp Ritchie, Fort Detrick. But nobody from Camp Ritchie was ever around Hagerstown or Thurmont, and they would laugh and say, 'Oh, you mean up at the old CCC camp.'"[154]

Guay, who was from a suburb of Washington, D.C., made his first trip to Hagerstown on Thursday night, 29 October 1942, as his diary indicated: "Went to Hagerstown tonight. Not much of a place, though Shep Fields [a well known band leader] was there at a dance; cost me $1.65. Soldiers aren't treated very nicely. Gals gave civilians a better tumble [more attention]."[155] Two days later on Saturday night, he went back to town, saw a movie, and wandered around the streets alone for hours. "An old woman, whose 18-year-old son had just enlisted, stopped me on the street and asked me a lot of questions. I tried to make her feel better. She was pretty worried. I am homesick too. For the first time in my life, I am not able to come and go as I please, and I don't like it."¹⁵⁶ It took several visits to local towns before he began to meet young women. Accompanying different sergeants who had been at B-5 longer than he had, Guay began to meet some women, one at a bar, another at a USO dance. Finally, he did meet "a fairly attractive girl" named Dolly in Frederick and later a pretty young woman named Virginia at a dance in Hagerstown. He wrote to them both. He went dancing with Virginia and then to her house for ice cream and cake. On their second date, after bowling, he walked her home, and she let him kiss her goodnight. They went to a local USO talent show for their third date. He enjoyed it, but as he recorded in his diary, "She wouldn't let me kiss her tonight. She is too prudish that way. It does no harm to kiss a bit." A week later, he was back to see Virginia again, noting in his diary afterwards: "I'm getting to enjoy her company greatly. She hasn't let me kiss her anymore after the first time. If I'm around much longer, though, I will. I don't expect to be here much longer, I hope." He was trying to get into Officers' Candidate School. Seven days later, he went to Hagerstown again, "but when I called Virginia, she wasn't home. She had to go to a party given by her company. I sure wanted to see her tonight; I think of her a lot." The next day, however, he learned that he was being transferred, and the day after that the 19th of December 1942, he left, never to return.¹⁵⁷

Trainees, under pressure to conclude an intensive training curriculum in a specified number of weeks, had less leisure time than the staff. Trainee Spiro Cappony remembered that they were occasionally allowed into Hagerstown on weekend evenings "if we had behaved ourselves." "If we were good soldiers, and they were proud of us, they would reward us with a pass. They would take us in a bus [or more often an Army truck] and drop us off at a corner, forget about us for a few hours, and then, at 10:30 or 11:00 o'clock in the evening, they would come back and pick us up...."¹⁵⁸ Some of the men went to local bars and got drunk. Some would get into fights, sometimes with civilians but more often with other soldiers, because, as Cappony recalled, "they were jealous of us because we had money and also because we were kind of cocky. We were sure of ourselves. We knew we were strong, and we kind of poured it on a little bit." But most of the young trainees and staffers going to Hagerstown or occasionally Frederick, went to places to dance with young women: dance halls, USO halls, or even, if they had the money, to nightclubs. The single, young men were looking for female companionship, and young men in uniform were often very popular with young women. "We had a hell of a time" at the dance halls, Spiro Cappony recalled. He and others said that most of the young women did not ask what base they came from (and if they did, the men told them Camp Ritchie). "All they wanted was a good time," Cappony said, "and we wanted a good time."¹⁵⁹

Among the officers, urban socialites like Roger Hall, son of a Navy admiral, found the remote setting at Area B, terribly boring. Transferred in fall 1943 from Area F, the Congressional Country Club located only six miles from Washington, D.C., to the western Maryland mountain camp, Hall said the "B" in Area B apparently stood for "By God, it's a long way from nowhere."¹⁶⁰ "Everyone bitched more because of its remote locale," Hall recalled, "but it was no different in pattern" from other OSS training areas. There was still the detachment cadre, the instructional staff, and the student trainees. But because of the pressure to get the Operational Groups ready for the invasion of France the following spring, Hall said he and his fellow instructors, working under chief instructor Lieutenant Sam Robelards, worked five weeks straight without a day or night off.[161]

When they could, staff members from Area B obtained two-or three-day passes, which enabled them to go down to Washington, D.C., two hours away. An Army truck or bus would drop them off at the main bus station in Frederick, a city of more than 50,000. From there, they could take a commercial bus, or the enlisted men could hitchhike. Private First Class Albert Guay did both, and the handsome soldier in his snappy, Class-A uniform had some interesting experiences along the way, as he recorded in his diary. "Thursday, November 5th, 1942....Hitchhiked home; 4 women picked me up. There was a 16 yr. old blonde sitting next to me that was plenty O.K. They wanted me to visit 'em in D.C. tomorrow, but I'm not going to, I guess."[162] A month later, after being home on a three-day pass and heading back to camp, he recorded, "Tuesday, December 15th, 1942...Dad took me down to the bus. A girl, not bad either, sat down by me & was friendly. So I proceeded to make love to her. Mugged [kissed] her all the way to Rockville where she got off & wanted me to get off with her [but he did not]."[163]

Training Areas A and C and Neighboring Towns

At Training Areas A and C in Prince William Forest Park, the cadre and staff, as well as sometimes the students, would go into neighboring cities for relaxation. As at Area B by 1943, the closest villages—in this case Triangle, Dumfries, and Quantico—were "off limits" to the trainees, except for transit at the Greyhound Bus Terminal or the Richmond, Fredericksburg & Potomac Railroad Station. Officers of the cadre and instructional staff seem to have been exempt from this prohibition. Some of them lived there with their families. Bachelor officers were allowed to go into Triangle and talk with the local people. Lieutenant James Ranney, a telegraphy instructor at the Communications School in 1943-1944 explained, "The Commo guys didn't know a whole lot of information. We operated on a 'need to know' basis. Triangle was just a little crossroads. There wasn't a whole heck of a lot to do there."[164] The facilities of the U.S. Marine Base at Quantico could be used by permanently attached personnel but not by students or trainees.[165] Some of the permanent personnel and their families used the Marine base across the highway for its PX and movie theater. Captain Arden Dow and his wife attended the movies there, for example, often with Park Manager Ira Lykes and his wife. However, many of the cadre and instructional staff did not use the Marine Base. Army officer and communications instructor Jim Ranney recalled that he never had any interaction with that facility. "Marines," he said, "look down their noses at everybody else."[166]

When on a pass or leave, most of the OSS personnel went to nearby cities: Manassas, a dozen miles away; Fredericksburg, 20 miles; or Washington, D.C., some 35 miles. Communications trainee Art Reinhardt recalled that there would occasionally be a "liberty run" on Wednesday nights to Manassas or Fredericksburg. Trainee Marvin Flisser said "we used to go in the evenings, once or twice a week to

Manassas. We would go in an officer's car, but officers and enlisted men went together, and we would go to a place called the Social Circle, a nightclub or something like that. It was a big—like a barn, a big barn, and there was music there, a band.... There were a lot of girls and men there and a lot of booths and tables. The orchestra played only country music. All the people would get up and dance to this country music. We used to laugh. In between, the band would rest, and during that period, they would turn on regular music. Nobody danced during that period, only during the country music. And the girls would dance with girls, and they would walk around—they didn't do any square dancing particularly. They would just hold hands and walk." The men from Area C went out and danced with the women, "we did pretty well."[167]

Given weekend passes, men usually went to Washington, D.C. Sometimes they were driven to the bus stop at Triangle for the hour ride to the capital. More often, a canvas covered Army truck would drive them from the camp to the city. It would drop them off on Friday night or Saturday afternoon at Union Station and pick them up there on Sunday night. Some went in uniforms with the identification patches of their parent unit; others went in just plain khakis without insignia.[168] Sometimes members of the cadre and instructional staff from the training camps in Prince William Forest Park and Catoctin Mountain Park were taken over to Area F, the former Congressional Country Club, near Washington for additional lectures or instruction followed by some entertainment. Timothy Marsh, civilian instructor in telegraphy at Area C recalled that "We were there for some lectures, instruction on security, but we also had some socializing. We did sort of fraternize with the girls [from the OSS in Washington]."[169]

Some of the married members of the cadre and instructional staff at Areas A and C, those whose families accompanied them to the new assignment, lived in nearby cities, particularly Manassas, only about a dozen miles northwest. Timothy Marsh lived there at the Prince William Hotel with his wife and one-year-old daughter; so did Lieutenant Ellis Marshall, who was also married, and who was in charge of the radio repair department at Area C. Quite a few the staff lived in Manassas and commuted daily by car to Area C, although sometimes they had to stay overnight at the camp.[170]

Because of the secrecy surrounding Prince William Forest Park after the War Department took it over, and the fact that there were Army guards around the perimeter of the camp and gates as the entrances, there were various rumors in the local area about what was going on in the park. One rumor was that the park had been converted to a camp for conscientious objectors. One instructor there heard it so often in Manassas where he lived that he believed it was a cover story circulated intentionally[171] Many local residents believed that the camp housed German prisoners of war.[172] Reinforcing that rumor was the fact that some residents saw covered Army trucks driving along Joplin Road at night and they heard voices coming from inside in a foreign language, that they assumed was German.[173] There was some plausibility for this as there soon were German and Italian prisoner of war camps in the United States, a number of them in Virginia and Maryland, and at least one of those facilities located at nearby Fort Hunt, Virginia, was classified top secret.[174]

Life of a Wife of an OSS Instructor

One of the OSS families that lived in Triangle for a while was that of Arden and Dorothea Dow. In the spring of 1942, the newlyweds from Washington State had driven across the country to Training Area B, where Dow was assigned as instructor, and bride had joined four other young wives taking up housekeeping in a big old house a half hour away. That fall, Dow had been sent to Britain for training with SOE. When he returned in October, he was promoted to captain and assigned to be chief instructor at Area A-4 in Prince William Forest Park. He and his wife drove to Triangle and rented a house there. So did several of the other former instructors at Area B now assigned to Area A, including Joe and Rita Collart and Elmer ("Pinky") and Betty Harris, and Jim Goodwin, a graduate of B-2, now an instructor at Area A and his wife, Helene.[175] Art and Dodie Dow rented a tiny, furnished house with a living room, a kitchen, two bedrooms and a bath."I had an I.D. card, and I remember that at that time our address was Box 2601, Washington, D.C.," Mrs. Dow recalled.[176] "Here, I always had a hard time cashing a check, because they couldn't figure out why I was living here [in Triangle] but had a Box 2601 [in Washington, D.C.]....And I didn't know either. I couldn't explain it either. We [OSS wives] didn't really know what was going on at all."[177]

OSS staff couples frequently went to the big U.S. Marine Corps Base just down U.S. Route 1 at Quantico. They had privileges there and could shop there in the commissary for food and other supplies. They also went to see the latest movies at the motion picture theater on the Marine base. Frequently, the Dows went with the Park Manager Ira Lykes and his wife. "I remember the ranger [Lykes], who lived up in Chopawamsic [Recreational Demonstration Area] would come and go with us to the movies...his wife was there too," Mrs. Dow recalled. "I guess they were maybe in the thirties...very friendly. She was a small, petite woman with dark hair. He was a slender fellow with dark hair. He dressed in a [Park] Ranger's uniform, and that was amusing, because when we would go to the movies [at the Marine base], all the young officers there would be out on the steps or somewhere when we walked up. And they didn't' know whether to salute him or not. You could tell they were puzzled by that strange uniform [the forest green uniform and light tan, broad-brimmed hat]." They did not know whether he was a civilian, or a serviceman from another branch, or even another country's armed forces. They didn't know, but they would stand up."[178]

During the six months that they lived in Triangle, Captain Dow would leave his home each morning for Area A-4 and come home every evening. He did not tell his wife what he was doing there, but Mrs. Dow remembers that "he did tell me some things like—They didn't really know anybody's name. They [the trainees] were like a number or a nickname. And he did say it was amusing to see some of the servicemen [enlisted men] associate with the officers. They didn't know what rank they were [because all the trainees were dressed in the same fatigues]."[179] The Dows socialized with the Lykes and also with the other married officers on the staff at Area A, the Collarts, the Harrises, the Goodwins, as well as George White, the former Narcotics agent and instructor, and his wife, with whom the Dows went to the theater in New York City one weekend. These Special Operations Branch instructors did not socialize with the separate cadre or staff at the Communications Branch School at Area C.[180]

The Dows' first child, Sharon, was born on 15 March 1943, at the Marine base hospital. Three weeks later, Captain Dow, who had been given a new assignment in China, took his wife and infant daughter and Mrs. Dow's mother to the railroad station where they left for the cross-country trip to Washington state. He left for China a few days later. So for the next two years, the 25-year-old Mrs. Dow and their newborn daughter stayed with her parents in her hometown on the Washington coast west of Olympia. "It was very lonesome," Mrs. Dow recalled. "I would go weeks without word from him." Even then, she did not learn too much more than he was still alive. "The V-Mail, they had all been censored, always censored."[181] When her husband returned from China in May 1945, Mrs. Dow and the baby, now two years old, returned to Washington, D.C. After the war, Lieutenant Colonel Dow remained with the regular Army. He spent the next twenty years in the Army, in Korea, Taiwan, and the Panama Canal Zone, retiring in 1963 as a full colonel, by which time the Dows had two daughters. The family then moved to Athens, Georgia, where he worked for nearly twenty years as an administrator at the University of Georgia.[182]

OSS Relationship with the National Park Service

The National Park Service maintained a presence at Catoctin Mountain Park and Prince William Forest Park to keep an eye on the use of the property and to ensure that the military conformed to the terms of the use permit.

At Area B, the NPS official was Garland B. ("Mike") Williams. "I remember the Park Ranger, who would make frequent visits to our site," recalled Lieutenant Frank Gleason, an instructor there from the spring of 1942 to the spring of 1943. "He was just keeping a close look to make sure that we didn't tear up his forest too much. He was a bit upset to learn that we had caught a rabbit in the winter and cooked it for a rare treat.... Two soldiers had caught a rabbit running in the snow. We took the rabbit and skinned it and cooked it. He [Williams] reported me, because I was the officer in charge, and I was in on killing that rabbit. We killed a rabbit on his reservation. We're supposed to love every little creature, and we did too," he chuckled. "But soldiers are soldiers, and in the winter, it was something to do."[183]

Wildlife was abundant in the parks, but although the National Park Service had banned hunting, there were numerous instances when OSS cadre, staff or trainees got themselves some game. But sometimes the game got the staff, or at least their provisions. Instructor Frank Gleason remembered that "While living here, I had told [Sergeant] Joe Lazarsky that I loved Kielbasa [Polish sausage]. So we went up to Hazelton [his hometown in the coal mining region of northeastern Pennsylvania], and brought me back a big Kielbasa. I loved it. So I thought, I'll put it in the snow [to preserve it]. I went out and covered it up with snow. I came back about a day later. It was gone! Some bear or other animal had eaten it."[184]

Assistant company clerk at B-5, Albert Guay, an avid hunter like his father, planned to so some hunting. As he recorded in his diary, 23 November 1942, "talked to an old guy here who is going to show me where to go hunting. Said we might get a few crows, pheasants, rabbits, etc." Guay's father had sent him a shotgun. "The shotgun came tonight. As soon as I get a chance, I'm going out and do some shooting. First I have to find out where I can and can't go."[185] No one told him that hunting was not allowed anywhere in the park. Consequently, he did go out hunting. Sometimes he saw nothing, and other times they were just out of range. As he indicated on 29 November 1942, "Went hunting crows, but too cold & crows too smart. Saw lots, but out of range."[186]

The senior demolitions instructor at Area B, Captain Charles Parkin, recalled that he frequently saw Park Manager Mike William, who came by in his National Park Service uniform with his nameplate on it. "I would see him at least three of four times a week in our area," Parkin said. "He was just keeping his eye on us."[187] Gleason also remembered Williams, whom he called the park ranger. Gleason described him as 5'9" or 5'10," dark hair, nice-looking man, medium build, who looked like he was in his early 40s. He came up to the OSS B-2 camp regularly in his National Park Service uniform. "He would come up and see what was going on in the camp and how we were conducting ourselves." He mainly he talked with the camp commander, Major, later Lieutenant Colonel, Ainsworth Blogg, but he had a few brief conversations with Gleason, then a young lieutenant instructor. "See we cut down a number of trees to build this training facility that we had here [the "trainazium," a kind of extensive, wooden, jungle gym]," Gleason said. "Of course, I am sure that he was concerned about that. He counted every tree apparently![188]

Like many teenage boys, some of the young OSS men liked to whip around at high speeds in motor vehicles. Some of them did that in the park. Diarist Albert Guay noted an exciting afternoon that would surely have distressed Park Manager Williams. "Had my first ride in a jeep today and enjoyed it. Hit 60 m.p.h. down a dirt road." The next day, he had a second jeep ride, this time with a sergeant and a load of 125 pounds of dynamite they took to a magazine constructed in the park. "Had another ride in a jeep. Rode fast over a bumpy road, quite a bit of fun."[189] On at least one occasion, reckless driving nearly resulted in a catastrophe. On a frigid evening in mid-December 1942, one of Area B's canvas-covered Army trucks was bringing nearly a dozen staffers returning from leave back from the bus terminal in Frederick. On the high road that leads up the mountain sometimes running high above Big Hunting Creek, the truck skidded on a patch of ice, and spun around. Only a tiny ridge of gravel stopped it from going over the edge and plunging down a forty-foot bank into the creek below.[190]

At Training Areas A and C in Prince William Forest Park, Park Manager Ira Lykes also had his problems watching over the OSS, particularly after he was inducted into the Marines and had could only be physically present in the park on weekends. Civilian communications instructor Timothy Marsh remembered seeing Lykes, in his uniform and broad-brimmed hat, walking around the camp at Area C.[191] As he made his rounds, Lykes was usually accompanied by his big German shepherd dog, "Fritz."[192] As Dorothea Dow noted, Mr. and Mrs. Lykes did socialize with some of the married officers on the instructional staff at the OSS Training Area.

The park, of course, was sealed off from the public for the duration. The OSS took over the park north of Joplin Road (Route 619), and the Marine Corps occupied it south of Joplin Road and west of Route 626. Military security was pervasive, but the OSS was particularly secretive. Even Lykes, the park superintendent, had to be cleared by a sentry to get to and from his home and office.[193] The park headquarters was the old Smith family house, not far from Joplin Road on the north side about 1 ¼ miles from the old CCC work camp that became OSS Area A-4.[194] As with the other buildings in the park, the National Park Service gave it up for the duration. Instead, the park headquarters was shifted temporarily to a small, one-room building a bit off the south side of Joplin Road. Lykes and his secretary, Thelma Williams, continued to use the building regularly until January 1943. At that time, the Marine Corps made him a commissioned officer in charge of forestry and fire security at the Marine Base reservation.[195] From 1943 to 1945, Lykes worked at the Marine Base during the week. On the weekends, he would return to supervise the park.[196] During the week, the park office was staffed by Thelma Williams, the park's sole fulltime employee during the war.[197] Ms. Williams worked Mondays through Fridays, managing routine business and holding other matters or documents requiring Lykes' signature for his weekly visits. The small building included a woodburning stove for heat in winter. Serving in effect as acting park manager Mondays through Fridays, Ms. Williams shared the little building with a uniformed clerk that the OSS had assigned to the park office, who it was reported was "a secretive man given to drink."[198]

Lykes had one frightening moment during the war, when it appeared that the Marine Corps might take over his house. The quarters for the park superintendent were a big two-story house with a porch about a mile south of Joplin Road, a site where the NPS maintenance building on the south side is located today. Lykes continued to occupy it even though the National Park Service had officially given it up for the duration. In addition to the living room, dining room and kitchen on the first floor, Lykes had built a small, temporary office on the enclosed porch.[199] Since the National Park Service was no longer paying utilities for the park or its structures, the Marine Corps furnished the fuel and paid the electric and other bills for the house. Lykes did, however, experience at least one worrisome moment in regard to this house, during a visit by the commandant of the Marine Base at Quantico. "I'll never forget General Philip H. Torrey came in one day. I don't know why General Torrey liked me, but he seemed to. He came in one day for quarters' inspection, and he walked in and looked around, and the first thing he said was, what a damn good officers' club this would make. I didn't know how to respond to that, so I didn't respond. They never did turn it into an officers' club."[200]

As at Catoctin there certainly were episodes of cutting down trees and of shooting wildlife in Prince William Forest Park during the war. At Area C one day when, a group of nine or ten Communications Branch trainees were practicing at the rifle range near Quantico Creek, a flock of wild turkeys suddenly flew through the valley. According to one the trainees, "the command was given to fire at will and the entire group qualified [as marksmen], and they ate turkey for a week."[201]

A peaceful man who loved nature and had a sincere desire to protect the flora and fauna under his care, Ira Lykes had considerable philosophical differences with the Special Operations Branch of the OSS with its emphasis on secrecy, violence, gutter fighting, and winning at any cost. He certainly supported the war effort against the Axis, but the OSS often shocked his sensibilities—perhaps they sometimes did so purposely. At the very beginning of OSS's take over of the park, for example, in the spring of 1942, the first commanding officer of Area A, Captain William J. Hixson allegedly made Lykes an offer. As Lykes retold the story thirty years later, Hixson "came to me, and he said, we would like you to join the OSS and leave the Park Service. He said, we will offer you a full colonelcy. I said, what do I have to do for this? He said, well, we will drop you behind the lines in France and, he said, I might as well tell you, your chances of getting back are about 1 in 8 or 1 in 10. I said, thanks but no thanks, I don't think so. I don't think so. I think I have a career with the National Park Service when this thing is over. Of course, he used the old argument, what if it doesn't get over, you won't have a Park Service. But I think discretion is a better part of valor and here I was discrete enough to say so."[202] He declined Hixon's perhaps facetious offer.

Although Lykes helped the OSS with information about the park and the area and in the location and construction of new buildings and some of the other training facilities, he remained distressed years later about many of their methods. In 1973, long after his retirement, the 66-year-old Lykes was formally interviewed in an oral history project by the National Park Service. . The following is Lykes' remembered impressions of the OSS, which certainly convey a sense of his suspicion and criticism of it, even if some of it may have been misremembered or exaggerated :[203]

"Well, one day I had a call to come to Washington, and I think it was in, oh probably, the latter part of 1942, the middle of '42 [it was early spring 1942]. There the Director [of the National Park Service] and his staff advised me that the Office of Strategic Services, Colonel Donovan with his cloak and dagger boys, were looking for a place, and they thought Prince William Forest—Chopawamsic—was the ideal place to do it. So anyway, we would have to cancel all the leases on camp grounds, and they would move the OSS in there. I was to remain as the liaison between the Park Service and help them in any development they wanted done in the war effort because we were right in the beginning of World War II.

"I do recall that they—one of the processes was—I later found this out with my Marine Corps experience, that in recruiting men for OSS they would go into a military unit such as [a] Marine Corps platoon or Marine Corps battalion, and they would say, do you have any Italians or Germans in this organization or any of Oriental extraction in this, and when the commander would say yes, he [the OSS recruiter] would say, I would like to see them. Now this was done by the FBI in connection with the OSS officials and several other security agencies. They would get these individuals aside and they would say, do you have any relatives—for example, in Germany—Yes, I have a grandmother there. Well, how is she being treated? Well, the last we heard she was being poorly treated. Particularly if they were Jewish. Finally, they would get this boy, how would you like to get even with them?"[204]

"Then from that point on, they would test this man, and they would run a check on him and make absolutely certain that he was a loyal American. Then, they would take them to Washington from all around the United States and there at 2 o'clock in the morning, they would put them in a bus [or truck] with the windows all sealed so they couldn't see where they were going, and they would ride them down to Charlottesville and [from] Charlottesville to Richmond, and Richmond [to] somewhere else and finally wind up 35 miles away from Washington where they started. They [the recruits] had not the slightest idea of where they were. And here, they would be put into a camp with only three other men, plus one instructor. They were never allowed to mix generally within the camp. They had to stay in their group of four with one instructor present at all times, even sleeping in the barracks—eating and sleeping. Well, this is what made the place so ideal, because of the little tent building[s] we had there. They were winterized by the way, and they could put four in there with an instructor, and they didn't have to have any other services, except they would eat at the dining room, which they divided off into cubicles of four.[205]

"The [OSS Special Operations] officers there—they had one officer there who was a Britisher, a very stern looking individual with a ramrod back. He had been Chief of Police of Hong Kong [sic] for 20 years, and he knew everything there was to know about killing a man without a weapon. That was his instruction.[206] Another one [training course] was how to build a radio—or a transmitter out of an ordinary radio. Another one was how to demolish an automobile without being detected. Another one was how to blow up a bridge without being seen or how to make explosives out of ammonia and [word on audiotape unclear] acid and different things like this. This was taught to these men.

"Also, one of the training exercises was that you had to jump out of an airplane in the daytime. You [also] had to jump at nighttime over a designated place, and you had to jump at nighttime over anywhere. So they had an old C-23 [sic] down at Quantico [Marine Base], and I used to ride in the blister [plexiglass bubble] of that thing and take moving pictures of these men bailing out in the daytime. And then we would have a critique that afternoon of these men jumping out of an airplane. I think we killed two or three [sic; that is, two or three were killed in the jumps], but that was part of the game anyway.

"They had to take different lessons in how to bail out, and I built them a long cable with a pulley on it, and they would jump off a platform and come down and land and roll. This sort of thing. This was a tremendously involved business of which I want no more part, because I've seen men suffer as a result of this. For example, I remember we built a little place we called Little Tokyo in which we had one building, and we had to train the men how to get through the door of that thing and walk down a labyrinth in a blue light, a very, very dim light, while they were being fired at [and they had to fire back] in order to protect themselves.

"They taught them such delicate things as when they burst into a room of German staff officers, which one to shoot first. The one that stood up was the one that got shot first, and the one that had the wild look and stare on his face got shot last because he hadn't collected himself yet, and he didn't know what to do. This was the fine shadings in murder. Of course, we had to do it, because we had to fight fire with fire, and it was a really justified program."[207]

Twelve years later in 1985, at age 78, during another oral history interview, Lykes continued to be critical of the OSS Special Operations training program in the park. He told Susan Cary Strickland, an historian working for the National Park Service, that "Students were not permitted to gather in groups larger than four. An unexplained absence could result in imprisonment for the remainder of the war." In what one suspects was actually Captain Hixson pulling one over on a gullible Lykes as the OSS arrived, Lykes told Strickland that Hixson had called upon him to serve as a guide during a manhunt through the woods for a student who was AWOL (absent without leave). Hixson had given the park superintendent a .45 caliber pistol and told to "shoot first and ask questions later." Lykes told Strickland that he was very relieved that they had not found the AWOL student, at least while Lykes was with the search party.[208] During the 1985 interview, Lykes told Strickland that he was grateful to General Torrey, the commandant at the Marine Base, for getting him a position at base during most of the war, because it enabled the park manager, in Lykes' words, to get "away from the stomach-turning roughhouse of the OSS!"[209]

CHAPTER 8

OSS IN ACTION: THE MEDITERRANEAN AND EUROPEAN THEATERS

In war it is the results that count, and the saboteurs and guerrilla leaders in Special Operations and the Operational Groups, the spies in Secret Intelligence, and the radio operators in Communications did produce some impressive results. In this unconventional warfare, Donovan believed that "persuasion, penetration and intimidation... are the modern counterparts of sapping and mining in the siege warfare of former days." His innovative "combined arms" approach sought to integrate espionage, sabotage, guerrilla operations, and demoralizing propaganda to undermine enemy control and weaken the interior lines of communications and supply in enemy's rear before and during the assault at the front by conventional forces of the Allies.[1]

At the end of the war in Europe, General Dwight D. Eisenhower, the Supreme Commander of the Allied Expeditionary Force, credited the Special Operations of the American OSS and the British SOE with the very able manner in which the Resistance forces were organized, supplied and directed. "In no previous war," he added, "and in no other theater during this war, have Resistance forces been so closely harnessed to the main military effort....I consider that the disruption of enemy rail communications, the harassing of German road moves and the continual and increasing strain placed on the German war economy and internal security services throughout occupied Europe by the organized forces of Resistance, played a very considerable part in our complete and final victory."[2]

It has been estimated that during World War II, the total number of people who served in the OSS probably numbered fewer than 20,000 men and women altogether, less than the size of one of the nearly one hundred U.S. infantry divisions, a mere handful among the sixteen million Americans who served in uniform in World War II. Among the 20,000 OSSers, probably fewer than 7,500 served overseas.[3] The number of agents the OSS had behind enemy lines was far smaller. It remains undisclosed, but one indication of how many OSS agents may have been infiltrated as spies, saboteurs, guerrilla leaders or clandestine radio operators, is the number who took parachute training, the primary method of infiltration. In all, more than 2,500 men and dozens of women received OSS parachute training.[4] Yet, despite the comparatively small size of Donovan's organization and the even smaller contingent who risked, and sometimes lost, their lives in the shadow war, the OSS made significant contributions to victory in World War II.

The following two chapters aim not at being a full account of the OSS accomplishments overseas, which would be impossible in such a limited space.[5] Rather, within an overall context of the role of the OSS in foreign theaters of operation, the emphasis here is on the actions of OSSers whose preparation included training at Areas A, B, and C in Catoctin Mountain Park and Prince William Forest Park. Particularly important here are the achievements of the OSS and also how the spies, saboteurs, guerrilla leaders, and radio operators, who received at least part of their training at the camps in these National Park Service areas applied their training in their overseas missions and accomplishments.

The American Landings in North Africa, 1942

OSS's first opportunity to prove itself came in connection with the U.S. invasion of French North Africa in November 1942. As early as the late summer of 1941, Donovan's fledgling organization had begun placing a dozen agents, code named the "twelve apostles," in the collaborationist Vichy French colonies of Morocco and Algeria. A bevy of American businessmen and scholars with connections with France and its colonies, they were ostensibly given minor assignments with U.S. consulates, but these were covers for their clandestine missions. By January 1942 when President Franklin Roosevelt and Prime Minister Winston Churchill agreed on the invasion of North Africa (Operation Torch) in November, the agents were given the missions of obtaining intelligence and building "fifth column" resistance in Vichy French North Africa. They quickly established a clandestine radio network, gathered intelligence about defenses and the 100,000 Vichy French troops and their commanders, obtained maps of suitable air and sea landing sites, and sought through encouragement and financial inducements to gain support from resistance elements among the Riff tribesmen and other indigenous, Muslim, anti-French groups along the coast and in the mountains and the desert.[6]

In the United States, Donovan, with approval by the Joint Chiefs of Staff, sent a team of spies into the Vichy French embassy in Washington, D.C. in March 1942 to obtain code and cipher books. OSS operative Elizabeth ("Betty") Pack, code named "Cynthia," a beautiful aristocratic divorcee, and Charles Brousse, a press attaché at the embassy whom she had seduced, plus an unidentified safecracker, recruited by the OSS for his expertise in picking locks and opening safes, successfully photographed military and diplomatic codes and other secret documents from the safe in the Vichy French embassy.[7]

Summer 1942: As the time for the Allied invasion of North Africa grew near, OSS's Secret Intelligence agents joined the effort to try to persuade the Vichy French forces to support the landings. Special Operations agents sought to prepare sabotage units and recruit native resistance fighters. When 50,000 U.S. troops followed by 15,000 British soldiers landed at half a dozen locations along the North African coast beginning on November 8, 1942, OSS reception groups met the troops on many of the beaches and guided them ashore.8 Inland, OSS agents sabotaged military targets, cut off enemy communications lines, and were ready to guide American paratroopers at a designated safe drop zone using a top secret radio beacon. Although the paratroopers' planes never arrived because of false starts and high headwinds, other OSS efforts demonstrated their effectiveness in the field. Together with representatives from the U.S. Army and the State Department, OSS representatives helped convince much of the Vichy French officer corps in North Africa not to forcibly resist the American invasion.9 Despite some pockets of French resistance, the dangerous invasion, with troops convoyed thousands of miles to land on a hostile shore, was an overall success.10 The OSS received credit from Army Chief of Staff General George C. Marshall for its contribution to that victory through intelligence which was of high quality, abundant and accurate in its description of the terrain and the enemy's order of battle, that is, the identification and nature of the enemy Army, Navy, and air force units facing the Americans, the location of French headquarters and the names of officials upon whom the United States could rely for assistance in the administration of civil affairs. The OSS, particularly its SI branch, had proven itself to the U.S. Army's high command.11

With the successful Allied landings in French North Africa, the U.S. and British forces under overall command of General Dwight D. Eisenhower headed east toward German occupied Tunisia. OSS set up a regional headquarters in Algiers and worked with the British Special Operations Executive to aid the advance. In the process of gathering tactical intelligence and sabotaging enemy communication and transportation, OSS agent Carleton Coon, a Harvard anthropologist and authority on North Africa, led a group of some 50 American, French, and Arab guerrillas. Among other innovations, Coon is credited with inventing "detonating mule turds," plastic explosives specially shaped and colored like mule or camel dung and scattered along desert roads to disable German tanks and trucks.12 The Allied advance came to a temporary halt, however, when the German *Afrika Korps* launched a counteroffensive in February 1943, catching the American Army by surprise and driving them back through the Kasserine Pass. A desperate local commander ordered Coon and his guerrillas to try to stop German tanks with hand grenades and other weapons, but after planting a few mines, Coon declined to have his highly-trained specialists used as regular infantry against tanks, a decision later endorsed by the OSS.13

Jerry Sage, German POW camps, and "the Great Escape"

Misuse of OSS personnel in several incidents in North Africa also led to the wounding of several other OSS agents and the capture of at least two of them. Lieutenant Elmer ("Pinky") Harris, from Areas A and B, was wounded in action near Sabeitla, Algeria, but quickly recovered and was subsequently assigned to Allied and OSS headquarters in Algiers.14 Less fortunate were Jerry Sage and Milton Felsen, both alumni of Area B. In January 1943, Sage, by then promoted to major, had been sent to North Africa for SO work. But when the Germans in Tunisia counterattacked at

Kasserine Pass in February 1943, some local American commanders directed most of the OSS personnel there to the front. Carleton Coon's group had been one of these, but it had quickly withdrawn and none had been captured. Others were not so fortunate. One such group included twenty OSS agents that William J. Donovan and a 37-year-old assistant named Donald Downes, had assembled in the United States to conduct espionage and other clandestine activities in Generalissimo Franco's fascist but officially non-belligerent Spain. The possibility of a German occupation of Spain and a drive across the Straits of Gibraltar, with or without Franco's consent, was considered a major strategic danger to the Allies. The Americans in Downes' group were agents that he had trained at Area B at Catoctin Mountain Park. Among them were five former members of the Abraham Lincoln Brigade of American leftists who had fought against Franco in the Spanish Civil War in the 1930s. They knew Spain well and several of them were members of the Communist Party of the United States. This was an example of Donovan's willingness to use some communists as agents when they knew the area and had contacts with local Resistance leaders in Europe, many of whom were communists. The other part of Downes' group was composed of Spanish political refugees, members of the defeated Republican government, recruited by the OSS in New York and Mexico City.[15] Now, despite Downes' protest, most of his intelligence team was diverted from its planned mission to Spain to the front lines in French North Africa, where they joined Jerry Sage's Special Operations unit.

By happenstance in North Africa, Sage had enlisted one of the few African Americans to serve in the OSS. The United States military still kept blacks in racially segregated units in World War II, and the OSS did not officially recruit African Americans. But when Sage arrived in North Africa and sought a truck to transport his men and equipment, an ordnance officer responsible for vehicles would not let Sage take the truck without a driver from the motor pool. With the truck came a driver, an African-American corporal named Drake from an all-black transportation unit. Corporal Drake, whose Sage's memoirs identify only by his rank and surname, was from Detroit. He became part of Sage's OSS Special Operations team and quickly learned SO skills, including close combat, knife-fighting, and demolitions.[16] Sage and his unit recruited locals, trained them in the use of explosives, and planning missions to infiltrate enemy areas and destroy lines of communication and supply as well as ammunition depots.[17]

Attached to the U.S. Fifth Army, the SO team came temporarily under orders of a British infantry regiment. The English colonel ordered them to make a reconnaissance patrol. Sage was reluctant to do so because it was daylight and his team usually operated at night, but he accepted the order. They advanced stealthily in two sections. Sage moved into a wadi, a dry channel, with two sergeants, Milton Felsen and Irving Goff, both former Spanish Civil War veterans, who had received OSS training at Area B. Sage then motioned the other section forward. As soon as they arrived, the Germans, tipped off by Arabs, Sage later concluded, opened fire with artillery. Sage and Felsen were both wounded, Felsen more seriously.[18] Goff poured sulfa into Felsen's open wound and helped bandage both men. When they heard the clank of approaching German tanks,

Sage ordered Goff to escape with the other section. Goff looked back as he scurried away and the German soldiers approached: "The major, in an Abercrombie & Fitch brown jacket, was visible a mile away. He was silly, but great enough to stand up and divert them."[19] Sage's action allowed the other three OSS men, including Goff, to get away. When Sage and Felsen were about to be captured, both of them quickly buried all their OSS gear in the sand: pistols in shoulder holsters, Fairbairn daggers, a special belt and vest with hidden pockets, and their spy gadgets. Felsen, an enlisted man and seriously wounded was turned over to the Italians for medical treatment. "Get well," Sage told him, "because we're going back."[20]

As an officer, Sage was taken for questioning. He convinced his interrogators that he was a downed flier and consequently was taken to a prisoner of war camp for captured Allied aviators, first in Italy and then in Germany itself. Strong-willed, determined and imaginative, Sage escaped at least half a dozen times from such camps, but each time he was recaptured. Still, he earned the respect of his fellow prisoners and the nickname, "the Big X," prison slang for an escape or exit artist.[21]

During his two years in German prisoner of war camps, Sage later reflected that he had drawn upon his inner resources, which he said were his religious faith and "the superb training I had received under Donovan."[22] His paramilitary training was of great use, he said, when at one of the main POW camps late in the war, he was put in charge of turning a group of aviators into commandos to seize the camp in case the Nazis decided to liquidate all the prisoners. Teaching them the art of silent killing, the sentry-kill, and other lethal techniques from the OSS schools, Sage trained his thirty, hand-picked "kriegies," he said, into an effective "storm-trooper group."[23]

That had been while Sage was part of the planning for the large-scale breakout from *Stalag Luft III*, later the basis for a 1963 film, *The Great Escape* with Steve McQueen. But because of his escape record, Sage was removed to a more secure prison camp before the actual breakout occurred at *Stalag Luft III* in the spring of 1944. He was fortunate, because all but three of the 76 men who escaped, before the discovery of the tunnel stopped the remaining 174, were recaptured, and 50 the 73 who were recaptured were executed by the Gestapo.[24] Sage finally escaped successfully in January 1945, this time from a prison camp in Poland as the Red Army approached. From wireless radio transmitter in a hidden office of the Polish underground, he tapped out a message picked up by OSS base stations in Egypt, Italy, and England: "Jerry the Dagger is on the loose and coming home!"[25] He made his way home via the Ukraine, Turkey, and Egypt. Arriving at OSS headquarters in Washington in March 1945 a month before V-E Day, Sage received a warm, personal welcome from General Donovan, who declared happily, "I knew you'd get home early, Jerry![26]

Major Peter Ortiz, the most famous Marine in the OSS

Although the majority of the uniformed OSS SO and SI officers had received commissions in the Army or the Army Air Corps, there were a few naval officers and a couple of dozen Marines.[27] The most famous Marine in the OSS was Major Peter Julien Ortiz, who with two Navy Crosses and numerous other medals became one of the most decorated officers in the Marine Corps in World War II. Ortiz had been born in New York City in 1913 to an American mother and a French-Spanish father from a prominent French publishing family. Although he spent his childhood with his mother in California, his father insisted that he be educated in France. He studied in a lycée and then a French university, but in 1932, the rebellious 19-year-old youth joined the French Foreign Legion as a private. He was wounded in action against indigenous rebels in French North Africa, received many medals for heroism, and was promoted to sergeant and then acting lieutenant. In 1937, he returned to the United States and became an adviser on military affairs to film companies in Hollywood. Two years later, at the outbreak of World War II, he rejoined the Foreign Legion as a lieutenant, fought the Germans when they invaded France in 1940, was wounded and taken prisoner and held as a POW for fifteen months before escaping and returning to the United States.

In June 1942, Ortiz joined the U.S. Marine Corps as a private but after boot camp was awarded an officer's commission and soon became a captain. In December 1942, he was sent to Tangier, Morocco officially as assistant naval attaché, but that was a cover for his real assignment. He had already been assigned to the OSS to organize Muslim tribesmen and scout German forces to help prepare the assault on the German position in Tunisia. In March 1943, as the new U.S. Corps Commander, Major General George S. Patton, Jr., launched a major attack, Ortiz was wounded during an encounter with a German patrol behind enemy lines. A German bullet shattered his right hand, but Ortiz rolled on his other side and with his left hand tossed grenades which quickly silenced the enemy machine gun. His men dragged him to safety. Ortiz was brought back to the United States, and after surgery and recuperation, he was temporarily reassigned to OSS headquarter in Washington, D.C. That spring, he spent time training and helping instruct at the OSS's Special Operations training camps at Areas A, B, and F. In the summer of 1943, Ortiz was sent to England for Special Operations training with a multinational, "Jedburgh" team preparatory to being parachuted twice into German-occupied France in 1944, where eventually he too, like Jerry Sage, would become a prisoner of war, but at different German POW camps.[28]

OSS HQ, Training Camps, & Base Stations in MEDTO

During the winter of 1942-43, OSS established its main headquarters for the Mediterranean Theater of Operations (MEDTO) in Algiers. By February, OSS/Algiers had training officers for SI and SO operations, including a parachute school run by Colonel Lucius O. Rucker, a no-nonsense paratrooper from Mississippi who had run a similar school at Area A in 1942. His new parachute school just west of Algiers included weekly classes ranging from ten to seventy. They were mostly indigenous recruits, Spanish, Italian and French nationals, willing to become intelligence or special operations agents for the United States. Under an inter-Allied agreement,

British SOE continued to have overall responsibility for the special operations in the Mediterranean Theater, including those of the American OSS. The first unit of Italian SI recruits, three officers and nine enlisted men, arrived in March 1943, a second contingent in June. From the United States, representatives of MO, R&A, and X-2 arrived in mid-1943. The North African campaign ended in May 1943 when the German forces in Tunisia surrendered. In July, the first Operational Group arrived, it was composed of Italian Americans who had trained at Areas F, B, and A. During the winter of 1943-1944, French OGs began to arrive, so did a second Italian OG. In addition to the parachute school, there were also OSS schools run by the SO, SI, and Communications Branches for indigenous agents as well as for advanced training for American OSSers who had graduated from the training camps in the United States. Similar OSS training camps were established in Britain, India, and China. In addition to the extra training for foreign-speaking American SO and OG members, OSS instructors at the main OSS training camps in Algeria also trained Italian, Yugoslav, and French agents recruited for special operations work in their own countries.[29] Richard W. Breck, Jr., who after the war would play professional baseball for the Pawtuckett Slaters, as "Bobo Breck," and who had gone through SO instruction at Areas B and A, helped train foreign agents in demolition work at the Algiers camp. He later participated in the Allied campaigns in northern Italy.[30]

Among the OSS recruits at the training camp in Algeria was John ("Jack") Hemingway, son of the famous author. Young Jack had been only five when his parents divorced, and he had spent much of his youth in boarding schools. He later he dropped out of the University of Montana and Dartmouth College. After Pearl Harbor, the 19-year-old youth enlisted in the U.S. Army, and by the end of 1942, he was serving in the Military Police in North Africa. Like his father, young Hemingway wanted to see action and experience danger, so through friends of the family, he was able to leave the MPs and be reassigned to the OSS. First Lieutenant Jack Hemingway was welcomed into the OSS and assigned as an instructor and student at the organization's initial main training camp at Chréa, Algeria, in a cedar forest on a 6,000-foot mountaintop behind the provincial city of Blida. The location would later be moved to Koléa on the beach near Algiers, after local boys sneaked into the mountain camp and accidentally blew themselves up with an unexploded mortar round. Hemingway taught weapons usage to French and American agents for SI, SO, and OG. He remembered that "among the teaching staff was a number of very tough, young men. They had all been to the OS training schools in the States, and some of them had been through training in England as well. One of them impressed me especially. He was the only one of the younger men who had been in the field, and his toughness was not put on.... He had been commissioned in the field and, being several years older than I, he filled the spot one always has for a hero figure. His name was Jim Russell." Lieutenant James Russell had risen through the ranks in the Army. He would later take part in Special Operations missions in Sardinia, Corsica, and France.[31]

Stephen J. Capestro from Edison, New Jersey, had undergone SO training at Area A, but when he arrived at the SO training camp in Algeria, he found Jack Hemingway as one of his instructors. Capestro remembered the intensive parachute training and as well as field exercises day and night. The trainees were, for example, dropped off in the countryside in uniform with only a compass and a hand drawn map and told to find their way back. The indigenous people, Berbers and Arabs, were often hostile, Capestro said, seeing the United States as helping France control of its colony. They sometimes booed the Americans soldiers. Later Capestro reflected on differences he felt in training in the United States and in North Africa, particularly the night exercises. "The difference is, for example, taking training overseas behind enemy lines, or lines that aren't friendly lines [such as Arab North Africa] and doing training in the United States.... There's a hell of a difference. If you're on a highway in Virginia, and you hitch-hike home, there's no problem. In North Africa, I didn't know. If the people came out, they could have been enemies. They could have robbed me personally or attacked me out of hate for the Americans. Because on many nights on training sessions in North Africa, I saw them, their looks, and their snarling dogs. I could tell they were saying nasty, nasty things. On training missions if they found you alone, what would happen? They might have hijacked your wallet or worse. It concerned me. I just didn't know."[32]

Communications were vital, and following the invasion of North Africa, the OSS set up the first U.S. communications station in Algiers and transmitted all Army and State Department messages until an Army Signal Corps established a unit there. By March 1943, OSS had expanded its Algiers station into a major communication facility. It included a message center at SI headquarters in Algiers and a large Communications headquarters with a base radio station and a main receiving station nearby at Cape Matifou. Sarah ("Sally") Sabow, daughter of Hungarian immigrants from Bayonne, New Jersey, was a cipher clerk and one of six women sent to OSS MEDTO HQ in Algiers. Her boss was Major Peter Mero, one of the Commo Branch's main recruiters and entrepreneurs, and a frequent visitor to the CB School at Area C. From 1943 to 1945, Mero was in charge of all OSS communications units in the MEDTO. He and Sally Sabow married after the war.[33] Some distance away from these stations, a communications school was established to train indigenous agent-operators for clandestine work in their native countries. By May 1943, OSS Algiers was in direct contact with OSS clandestine stations throughout the region and was also linked to stations around the world.[34]

Another OSS radio base station was established at the eastern end of the Mediterranean in Cairo, Egypt. Lieutenant James Ranney, a former instructor at Area C, was stationed there and remembered the powerful transmitters, which relayed messages between OSS headquarters and agents in the field and also beamed a daily news broadcasts prepared by Morale Operations or other OSS units to Italy, the Balkans and other areas of eastern Europe. The station's equipment included half a dozen Hallicrafter HT-4 transmitters, each rated at 400 to 500 watts. There were four cage dipole antennas as well as two rhombic antennas and two Beverage "Wave" antennas, one of which was 3,000 feet long.[35]

The main OSS communications base in the Mediterranean remained at Algiers, until well into 1944, but after the invasion of Italy in September 1943, a station was established at the Italian Adriatic port of Bari and subsequently at Brindisi, when Bari became inadequate for certain operations due to mechanical and atmospheric difficulties.³⁶ In July 1944, the main headquarters for OSS communications in the MEDTO was shifted from Algiers to Naples, or more precisely, the 1,200-room Royal Palace of the Bourbon Kings, at Caserta, which in 1944-1945 served as the headquarters of the Supreme Command of the Allied Forces in the Mediterranean Theater. Naval Lieutenant Frank V. Huston, who trained at Area C, was in charge of communications training and the message center at Bari and Caserta and succeeded Peter Mero when the latter left for the United States in 1945.³⁷

Gail F. Donnalley, an OSS code clerk, who had spent a few weeks at Area C before being trained at OSS headquarters, served at CB base stations in Cairo, Egypt, and Bari and Caserta in southern Italy. He had a very small part in the negotiations for the surrender of the German troops in Italy. At Bolzano where *Waffen SS* Lieutenant General Karl Wolff, military commander in northern Italy, had his headquarters and was negotiating through OSS's Allen Dulles in Bern, Switzerland to surrender, the OSS had assigned a young, German-speaking Czech, code-named "Wally", as a radio operator in Wolff's headquarters to transmit the messages back and forth. They put him in the attic of Wolff's headquarters, and he would send messages from the German commander to the OSS base station in Caserta, where Donnalley, working at the OSS radio base station. "We would decipher it, then cipher it [with a more complex cipher] and send it on to Dulles in Switzerland," Donnalley recalled. "This went on for four to five months. On the day before the surrender [which occurred 2 May 1945], the S.S., who, of course, knew he was up in the attic, brought him down and beat him up and then sent him back up to radio the German surrender. They were frustrated and angry about having lost the war. After the surrender, 'Wally' was brought down to Caserta, where we in the OSS met him. He looked about 17 and was about 5 foot 5 and had curly hair. By that time, he did not look beat up. He did not speak English, or at least not to us. We shook hands and thanked him in English, and he mumbled something in his own language."³⁸

Donovan Joins the Invasion of Sicily

Although the OSS rejected use of its agents as combat infantrymen as a misguided waste of resources, Donovan, himself personally enjoyed being in combat at the front and frequently and needlessly exposed himself to its dangers. In the U.S. invasion of Sicily in July 1943, the OSS director and a few of his men accompanied the 1st Infantry Division, landing with them on the first day, and staying with them for a few days during their advance inland. Captain Paul Gale, a staff officer from the 1st Division whom Donovan later recruited for OSS, said Donovan kept pushing him to take the jeep farther forward. "General, we're getting where the Italian patrols are active," Gale warned. "Fine," Donovan replied. Soon enough, they ran into an Italian patrol. Donovan leaped up and fired the machine gun mounted on the jeep. "He was happy as a clam," Gale recalled. "We had a hell of a fire fight." But Major General Theodore Roosevelt, Jr., the division commander, subsequently chewed Gale out "for getting such an important man into such a bad position."³⁹ Donovan's delight in getting into dangerous situations became legendary. Years later, Irving

Goff, who had trained at Area B and had been with Jerry Sage's unit until Sage was captured, told oral historian Studs Terkel about the general's bravado in Sicily and Italy. "We moved from North Africa into Sicily. Donovan's on the boat with us. He's on the beach with us. He's in a foxhole with us. Hell, we hit Anzio on a PT boat together. German plane came down, Donovan's standin' there. He was a great guy, but he had foolish guts. I yelled at 'im, 'Get down, general!' He wouldn't get down, and bombs droppin' all around."[40]

In the invasion of Sicily, the U.S. Army command did not want to alert the Germans to the impending operation, and, therefore, refused to authorize a major OSS infiltration to lead guerrilla resistance. Nevertheless, in addition to Donovan's personal landing with the invasion force, some OSS teams were infiltrated behind enemy lines at the beginning of the invasion. Although some members of one team were captured,[41] other OSS teams did some effective work. Army Chief of Staff General George C. Marshall told a meeting of the Joint Chiefs of Staff at the completion of the Sicily campaign, that "the O.S.S. got in ahead of operations in Sicily and evidently accomplished things rather satisfactorily."[42]

The OSS subsequently played important roles in Sardinia, Corsica, and on the Italian mainland. With the invasion of Sicily, the Italian monarch Victor Emmanuel ordered the arrest of fascist leader Benito Mussolini and replaced him with Army Chief of Staff, General Pietro Badoglio, who immediately began secret but prolonged negotiations with the Allies over armistice and peace terms. Eventually, the Allies agreed to recognize the royal government of King Victor Emmanuel with Badoglio as prime minister. Meanwhile, the Germans quickly freed Mussolini, moved in 16 divisions and took control of most of Italy. The Allies were supported by the Italian government in exile and the disarmed Italian Army, but most effectively by the anti-fascist Italian Resistance groups behind German lines. The Allies invaded southern Italy in September 1943 after the Italian Government surrendered.

A Four-Man OSS Team Takes Control of Sardinia

While the main Allied planning in the Mediterranean in late summer 1943 concerned the invasion of the Italian peninsula, General Marshall saw a job for the OSS in the Italian island of Sardinia. There were almost 20,000 German troops still on the island in addition to an Italian garrison of 270,000 men. Marshall wanted to "give Donovan a chance to do his stuff without fear of compromising some operation in prospect. If he succeeds, fine, if not, nothing will be lost." Eisenhower, acknowledging that OSS was "a high level intelligence gathering agency," concurred.[43]

OSS's mission to capture Sardinia without the need for a full-scale Allied invasion was assigned to a four-man Special Operations team led by Lieutenant Colonel Serge Obolensky, who had trained at Areas B and A and then co-authored a training curriculum for Operational Groups and taught them at Area F.[44] By 13 September 1943 when the mission was parachuted into Sardinia, the Italian government had surrendered a few days earlier and the U.S. Army had landed at Salerno. The

Germans were taking over Italy and there were German units on Sardinia. Eisenhower wanted the OSS to obtain the surrender of the Italian forces on the island and if possible to get the Italians to harass the Germans who were departing for Corsica. The OSS team would carry letters from Eisenhower as well as the Italian King and Prime Minister to the garrison commander, General Basso, ordering him to surrender. Donovan apparently selected Obolensky because he believed the former Russia prince had both the social standing and the bravado to persuade the Italian general in charge to surrender to him as a representative of the Allied Force Headquarters.

The team consisted of Obolenksy, First Lieutenant Michael Formichelli from New York City, an original member of the Italian-American OG, whom Obolensky had known at Areas A, F and B, who would serve as an interpreter, plus two communications specialists, Second Lieutenant James Russell, SO, an instructor at the OSS training school in Algeria, and a British radio operator, Sergeant William Sherwood, SOE, who would relay information to the OSS base station in Algiers. Neither Formichelli nor Russell had ever jumped before, but both volunteered. Since, the OSS had no contacts on the island, this would be the most dangerous of infiltrations, a "blind jump," leaving them entirely on their own in enemy territory. On the night of 13-14 September 1943, the four men parachuted into the Sardinian countryside through an escape hole in the belly of a black-painted bomber. The drop went smoothly. Jumping into a combat zone at age fifty-two, Obolensky became the oldest combat paratrooper in the U.S. Army, but this was not a point he thought about at the time. Rather he was concerned that in the bright moonlight, the white parachutes might alert the island's defenders to their arrival. It did not, and the team members landed safely, buried their parachutes and assembled. While Russell and Sherwood were left in the valley to protect the radio equipment, Obolensky and Formichelli set off hiking through the night and early dawn for the city of Cagliari fifteen miles away. A friendly farmer told them that the nearest town was still occupied by the Germans, so they skirted it and using their map and a compass reached a railway station at the next town. In their U.S. Army uniforms and with submachine guns, the Americans emerged from a field into the railroad station to the astonishment of the passengers and waiting there and the local police officers at the station. Taking the initiative and acting boldly as if he had a regiment of troops waiting behind him, Colonel Obolensky demanded to see the officer in charge, and when he appeared declared brusquely: "I have a very important message from the King of Italy and General Badoglio to General Basso....Take me to him!"[45]

At the station, the watching crowd thought the American liberation of the island had begun and started to cheer. Uniformed members of the *carabinieri* politely led the two American officers to the local military commander, a colonel, who, cordially if cautiously because the Germans were still in the area, arranged to have them escorted to General Basso's headquarters. There the general anxiously consented to "follow the orders of my king." He agreed to surrender, but warily declined to attack the German troops which were leaving for Corsica and then northern Italy. Some of the Italian officers wanted to fight the Germans, but a fascist paratroop division mutinied when it heard of the surrender and shot the general's representative. Until the paratroopers' mutiny was quelled, the four Americans, now reunited, were kept in a safe house protected by armed *carabinieri*. Thirty-six hours after having left Algiers, the OSS team was able to radio headquarters that the mission had been accomplished: "except for the Germans retreating in the far north, Sardinia was ours."[46] A few days later, Brigadier General Theodore Roosevelt, Jr. arrived with a token occupation force and formally accepted the surrender of the 270,000 Italian troops on Sardinia.[47]

To Die in Corsica

A separate OSS mission to the nearby French island of Corsica proved more costly. Special Operations Branch had sent a four-man team—an American and three Corsicans—ashore from a submarine one night in December 1942. It was the first OSS secret agent team infiltrated into enemy-occupied Europe. As they arrived on the beach, they carried weapons, munitions, and a million French francs and Italian lire. Mussolini's Italian Army had occupied the Vichy French island the previous month. Disliking the occupying Italian fascist regime, French Corsicans supported the team, but the secret agents were eventually captured, tortured and executed in the summer of 1943.

After the Italian government in Rome surrendered on 8 September 1943, the French Resistance on Corsica rose up, 20,000 strong, seized many towns, and called upon the Allies to help them get rid of them of the occupying Italian and German forces. An expeditionary French force from North Africa landed along with a detachment OSS French and Italian speaking Operational Groups, consisting of two officers and thirty enlisted men that Eisenhower had requested to accompany the French as a token Allied force. The two officers were Major Carleton Coon and Lieutenant Elmer ("Pinky") Harris now recovered from the wound suffered in the North African campaign.[48]

Because of the earlier OSS mission, the local Corsican Resistance cooperated more with the OGs than with the French troops, whom they distrusted. The OSS proved both aggressive and heroic. On 25 September near Barchetta, a three-man OG team led by First Lieutenant Thomas L. Gordon of Brownsville, Pennsylvania, one of the four original OG officers trained at Areas A, F, and B, was on advanced patrol when it under heavy German mortar and artillery fire. The OSSers remained in position to cover the withdrawal of a French unit when enemy reinforcements arrived. A French captain observing it from a nearby hilltop, stated later that it was one of the bravest acts he had ever seen. Continuing to fire until the end, the three OSS men, Lieutenant Gordon, Sergeant Rocco T. Grasso of Babylon, New York, and Sergeant Sam Maselli, were killed by mortar fire. All three were posthumously awarded medals for bravery from the American and French governments.[49] Gordon was awarded the Distinguished Service Cross, the second highest military decoration for heroism of the U.S. Army (after the Medal of Honor); Grasso and Masselli the Bronze Star Medal.

During the twenty-five days that the Germans held their defensive perimeter while withdrawing through the post of Bastia, the OSS had the only Allied agent inside Bastia, sending information on enemy movements. Moving around the city to avoid capture, this indigenous agent was able to flash daily reports that enabled Allied planes to blast German armored units and ammunition caches as well as transport vessels in the harbor.[50] When the Germans completed their withdrawal, OSS established advanced OG and SI bases on Corsica from which they infiltrated agents and staged raids along the coast of German-occupied Italy.

OSS on the Italian Mainland

As the American Fifth Army and British Eighth Army began their agonizingly slow progress up the Italian peninsula from September 1943 through May 1945, they were aided by various branches of the OSS. The main Allied effort in the Italian campaign, a grinding war of attrition involving series of largely frontal attacks on successive German-fortified positions up the mountainous peninsula, was one of the least mobile and comparatively most costly of the war.[51] The Allies found themselves confronted by the skilful German Field Marshall Albert Kesselring and dogged German troops in formidable defensive positions. Under Lieutenant General Mark Wayne Clark, one of the war's most complex and controversial commanders, the Americans and their Allies fought ferociously lethal battles at Salerno, Anzio, the Rapido River, Monte Cassino, before finally arrivin in Rome in June 1944, just a few days before the invasion of Normandy. Thereafter, his resources drained for the campaigns in France, Clark put increasing reliance in 1944-1945 on coordination between his slowly advancing troops and the sabotage and intelligence from behind the German lines in northern Italy by Italian Resistance forces supplied and guided by agents of the OSS.

OSS agents went ashore soon after the U.S. Fifth Army's initial landing at Salerno, south of Naples, on 9 September 1943. Italian-speaking OSS agents from various operational branches, SI, SO, OG, X-2, MU, even R&A, served as interpreters, helped recruit Italians for supporting functions, and penetrated German lines and report on enemy units and deployments. From North Africa, OSSer Donald Downes attached his 90-man SI unit, including Irving Goff and other former leftists from the Spanish Civil War, who had been given paramilitary training at Area B, to Clark's Fifth Army. Mexican-born but Kansas raised, Sergeant Louis Joseph ("Luz") Gonzalez, 25, who trained at Area A in September 1943, worked in the Research and Analysis Branch office at OSS headquarters in Caserta, supervising a staff of 50 enlisted men and 100 Italian civilians engaged in translating Italian documents. He later described himself to his family as having been a "Rear Echelon Cloak and Dagger Kid."[52] The operational branches also recruited indigenous agents. At one point there were three different OSS units reporting back to OSS Mediterranean headquarters in Algiers but with practically no coordination. They had some great successes, such as getting four Italian agents into German-occupied Rome with a transmitter/receiver (Radio Vittoria), which produced important information, and a joint operation (code-named Simcol) between the Italian-speaking OGs and British SAS, which, with the help of the Italian Resistance movement, exfiltrated to safety by air, land, or sea at least 2,000 Allied POWs who had escaped or were released from Italian prison camps when Italy surrendered.[53]

It was in this period, that Joseph ("Jumping Joe") Savoldi, Jr., the Italian-born American football star and professional wrestler who trained at Area B, participated in the "McGregor" Mission, first seeking surrender but subsequently looking for information from Italian admirals and scientists about recent new weapons developed by the Axis. The McGregor team included Commander John M. Shaheen, former Republican party publicity director in Illinois, Ensign E. Michael Burke, sports promoter and later President and co-owner of the New York Yankees, Lieutenant Henry Ringling ("Bud") North, scion of the circus family, Savoldi and a few others.[54] Most of the team, including Savoldi, went ashore with the invading Army at Salerno in September 1943. Later they took high-speed PT (patrol-torpedo) boats to the island of Capri where they met with sympathetic Italian admirals and anti-fascist scientists and learned about the Germans' new radio-guided bomb as well as the Italians' deadly new magnetic-activated torpedo designed to explode underneath a ship, breaking it in two. This was an important coup for the OSS.[55]

Infiltrated Italian and American agents some ten to fifty miles behind German lines in southern Italy were particularly helpful in providing target information for Allied artillery and bombers.[56] One of the most effective of these was Peter Tompkins, 24, son of an affluent American family who had lived in Rome before the war. Infiltrated into German-occupied Rome, he made contact with the Italian Underground. Using the clandestine Radio Vittoria, he sent real time tactical intelligence about German plans and deployments, which proved more immediately useful than the Ultra decrypts which took two or three days to decipher and deliver. The intelligence provided by Tompkins and his Italian agents was credited by the U.S. Army with saving the Anzio beachhead with its 50,000 American and British troops southwest of Rome from being crushed by the Germans in early 1944.[57]

In a separate operation in 1943 and 1944, the OSS Secret Intelligence Branch, seeking strategic as contrasted to tactical intelligence, inserted and recruited agents in central and northern Italy. The overall effectiveness of that operation remains the subject of controversy.[58] A few of those indigenous operatives ultimately became double agents and were "turned" against the Allies by the Nazis, or were Nazi agents from the start. As the Allies advanced, the Nazis and Italian fascists also planted "stay-behind" agents for purposes of intelligence gathering and sabotage. The task of hunting them down belonged to OSS Counter-Intelligence, X-2, and particularly its chief in Italy, 26-year-old Lieutenant James Jesus Angleton, Jr., who had joined the OSS in September 1943, and gone through OSS training camps in Virginia and Maryland, including SO training at Area B. Dr. Bruno Uberti, a refugee from fascist Italy who trained at Area B with Angleton remembered Angleton as "extremely brilliant but a little strange....I would have liked to have been one of his friends," Uberti said, "but he never gave me the chance because he was so secretive."[59] The son of an overseas vice President for the National Cash Register Company, Angleton completed his training with a successful industrial espionage scheme in which he was able to infiltrate the office of the chairman of the Western Electric Company. Fluent in Italian because his father's office had been in Milan, Angleton was enrolled in Counter-Intelligence and soon dispatched to the Italian desk of that branch in London and subsequently in Rome, where by the end of the war, his special X-2 units were credited with capturing more than a thousand enemy intelligence agents. This was the beginning of a career that would make James Jesus Angleton, Jr., the most famous and controversial of the CIA's spy catchers.[60]

Operational Groups Join the Fight in Italy

The Operational Group for Italy was the first of the OSS OGs to be sent into combat. It consisted of nearly 150 officers and men, primarily second-generation, Italian Americans, who had been recruited from U.S. Army units, then given OG training at Areas F, B, and A before being dispatched in September 1943 to North Africa for additional training there and use in Italy.[61] A second Italian Operational Group was organized and trained at Areas F and B in October and November 1943. It then followed replaced the first group at the OSS training camps in Algeria. The OSS Italian OGs were under the overall command of Colonel Russell B. ("Russ") Livermore, a New York lawyer with whom Donovan had worked before the war. In the first group, the operations officer was Captain Albert R. ("Al") Materazzi, who trained at Areas A, F, and B, and been with the unit from the beginning. As the first OG unit to be trained, it had had its problems, Materazzi recalled. Area F at the Congressional Club had not been ready for them when they arrived and their training was rushed.[62] By the end of October 1943, the unit moved from Algiers to liberated Corsica with orders to harass Germans in northern Italy, while the main Allied forces were bogged down 200 miles to the south. Corsica is only 35 miles from the Italian mainland, and it became the headquarters for the first Italian Operational Group and the point of departure for its raids as well as some of the infiltration of SO and SI agents into central and northern Italy, usually by fast boats but by parachute drops if farther inland. The OGs took control of some of the smaller islands off the important port of Livorno (Leghorn), south of Pisa, and beat back raids by German commandos. Allied Forces Headquarters ordered the OGs to undertake coastal raids for intelligence, sabotage, and to draw German units away from the front line in the South. The Italian OGs staged a series of raids, probing behind enemy positions, capturing prisoners, destroying bridges, cutting rail lines, and exploding concrete shore gun emplacements before withdrawing.[63] Throughout the winter of 1943-1944, these OG teams gained the distinction of becoming the northernmost American troops fighting in the Mediterranean Theater. However, they paid a heavy price for their harassment of the Germans.[64]

The "Ginny" Mission and a German Atrocity

Most of the raids by the Italian-American OGs were successful, but the "Ginny" Mission in late March 1944 ended in disaster. Part of Operation "Strangle" to cut the main German supply line in Italy, the Genoa-La Spezia coastal railroad, the plan was to blow up a crucial railroad tunnel near *Stazione di Framura* some fifty miles south of Genoa. In the week to ten days it would take to repair the tunnel, Allied airpower would be able to destroy the long lines of backed up supply trains backed up far beyond the mountains.[65]

On a moonless night, a team composed of 15 Italian Americans from the New York-New Jersey metropolitan area, all of them trained at Areas F and B, set out in PT boats with 650 pounds of explosives to blow up the railroad tunnel. First Lieutenant Vincent ("Vinny") J. Russo of Montclair, New Jersey, one of the initial four OG officers to train at Areas A, F, and B, was in command; First Lieutenant Paul Traficante was his deputy. Among the 13 enlisted men was Technician 5th Grade Rosario Squatrito, (nicknamed "Saddo" and "Rosy Squat"), a quiet, young tool and die-maker, from Staten Island, New York, whose nephew would later write an account in his uncle's memory.[66] In the darkness, the fifteen men, hoisted their weapons, radio, explosives and other gear into three inflatable dinghies and paddled silently to shore. They headed inland toward the tunnel, and their radio operator soon reported back that they were "on the target." But suddenly enemy flares light exploded off the coast and shore searchlights began flashing across the water as German patrol boats raced toward the American PT boats. There was an exchange of machine gun fire, and the Americans headed back to Corsica. The PT boats returned the next several nights to the prearranged point, but there was no contact with the OG team. There had been no explosion at the tunnel, and the team had disappeared.

Not until a year later in April 1945 did OSS discover the fate of the Ginny Mission. Once ashore, the team had left the explosives in the rafts and sent scouts to locate the nearby tunnel. But it was almost dawn when the scouts returned, and Lieutenants Russo and Traficante decided to wait until the next night to approach the guarded tunnel and set the charges. Meanwhile, the team hid in an empty barn. They had hidden the dinghies well from the land side, but in the morning a fisherman saw them from the water and informed the local fascists. Alerted, the fascist militia and the German garrison sent out patrols and soon found the saboteurs in the barn about a mile from the tunnel. Although surrounded, the Americans opened fire from the barn. In the fire fight, a number of them were wounded, including their leader, Lieutenant Russo. Facing the inevitable, the group surrendered. They were interrogated, and on Sunday morning, March 26, only five days after they had landed, all fifteen, were taken in trucks to a open field along the coast called Punta di Bianca, southeast of La Spezia. There in two groups, their hands tied behind their backs, they were executed by a German firing squad and their bodies thrown into a large pit, their hastily dug, unmarked grave.[67]

Early in the war, Hitler had issued an infamous "Commando Order," directing that captured Allied commandoes or parachutists engaged in sabotage or guerrilla operations should be treated as spies instead of prisoners of war and summarily executed regardless of the fact that they were in military uniform and operating as Army units.[68] This was, of course, a violation of the Geneva Conventions about treatment of military prisoners of war. In regard to the Ginny Mission, a German communiqué at the end of March 1943 had stated that the commandoes had been "annihilated in combat," but the truth gradually emerged through subsequent message intercepts as well as local Italians. When he learned that the members of the Ginny Mission had been executed, Donovan ordered that the German general in command of the region, General Anton Dostler, was to be taken alive so that he could be tried as a war criminal. Dostler was captured at the end of the war, and although he pleaded that he had only followed Hitler's orders, this was not accepted as justification, and it was noted that he had continued to insist on the execution over the objections of several of his subordinate officers. In October 1945, Dostler was tried and convicted by a U.S. military commission in Rome, the first German general to be tried as a war criminal. Dostler was executed by an American Army firing squad near Naples, Italy at daybreak on 1 December 1945.[69]

Arming the Partisans against the Germans in Northern Italy

For nearly ten months, until the summer of 1944, U.S. Army commanders in Italy failed to understand how best to use the men of Donovan's organization, because many of the professional soldiers still viewed them with a mixture of jealousy and contempt as privileged but naïve amateurs. As Peter Tompkins, the highly successful SI agent in Tome, recalled bitterly, "Unfortunately, despite the fact that we had been at war over two years, the OSS had been granted little opportunity by field commanders to prove itself an effective weapon of espionage and sabotage behind the enemy lines; most of us were still being frowned on by the brass as a collection of madmen. I had therefore been obliged, by lack of facilities, or even of recognition, to operate more or less clandestinely not only from the Germans, but from our own side as well.[70] An official account by the OSS put it more diplomatically, stating that a summary of the operations of the Italian-speaking OGs from September 1943 to June 1944, showed that the type of operations they performed, "though highly successful and effective, was not strictly that for which they had been originally intended." "It was not until July of 1944 that AFHQ [Allied Force Headquarters, MEDTO] began to understand the manipulation of OGs as a weapon to weaken and disrupt enemy communications behind his lines."[71]

This realization by Allied theater commanders in the Mediterranean that they could and should use OGs effectively in connection with indigenous resistance movements, may have resulted from a combination of factors. One was their successes in Italy, another was their successes in France in the summer of 1944, and a third may have been because the Allied armies in Italy were reduced in strength in 1944 as veteran units were transferred to the new battle zones in France. Consequently, the American campaign in Italy was forced to place more reliance than in the past on anti-fascist partisans behind German lines in northern Italy. The Italian Resistance Movement was composed of a diverse spectrum of anti-fascist groups that had arisen spontaneously, but these parties joined together after the Allied landings in September 1944 under the National Liberation Committee (CLN) to cooperate with the Allies, which said it could put 90,000 partisans into the field if the Allies would supply them with arms and other material support.⁷² Dealing with Resistance movements in enemy-occupied countries was a new phenomenon for the Allies in World War II and one that was complex because it often interlinked political and military problems. The British secret service agencies sought to undermine communist, socialist and other leftist groups in the Italian Resistance. London's postwar political aim was for a conservative Italian government under a constitutional monarchy. In contrast, most of the American agents, unlike the British, were of Italian ethnicity, and even more importantly, unlike the British, the American OSS dealt equally with the partisans groups regardless of their political affiliation. The primary OSS strategy was to facilitate mutual relationships with all active partisan groups in order to maximize the military effectiveness of the Resistance against the German Army.⁷³

More than a hundred SO, SI, and OG teams, including both Americans and Italian nationals, were infiltrated behind German lines into northern Italy between 1943 and 1945. Among the Italian nationals trained and deployed by the OSS was Piero Boni. After the war, Boni, a member of the Socialist party, would emerge as the second highest ranking official in CGIL, Italian General Confederation of Labor, the largest confederation of labor unions in Italy. But in 1944, he was a 24-year-old labor lawyer from Rome, who had been earlier drafted into the Italian Army as a 2nd lieutenant, and after the Italian surrender in 1943, had returned to Rome, joined the Italian Resistance movement in the capital and met OSS's Peter Tompkins. In the summer of 1944, Boni was among those members of the Resistance selected and trained by OSS for intelligence and special operations missions in the North. Like many others in the Resistance, he was struck by the political impartiality of the OSS in dealing with the partisans from the left to the right.⁷⁴

Boni and the other members of his OSS training class underwent three weeks training by the Americans in July1944. The first two weeks were spent at an OSS school by the sea in Naples, where the instructors were Italian Americans, who had been trained at Areas F, B, or A, and the curriculum replicated that of those OSS training camps in Virginia and Maryland. "They gave us American cigarettes and my first chewing gum," Boni recalled."⁷⁵ Instruction was in Italian and the OSS recruits were trained in observation and reporting, transmission and reception using five-letter groups in code, and the effectiveness of various demolitions, particularly plastic explosives. They also practiced with American and British pistols, submachine guns, hand grenades and bazookas. "I particularly liked the bazooka," Boni said. "That was a new weapon, and it was good. Later [in northern Italy], we used the bazooka to shoot out a window at German headquarters!"⁷⁶ From Naples, the Italians were taken for a week of training and jumps at a parachute school staffed by British and American instructors at Brindisi. Captain Elmer ("Pinky") Harris had helped to establish the school in the winter of 1943-1944, but in April, Harris had become so ill with abdominal pains in what would eventually be diagnosed as intestinal ulcers that he was transferred back home to Bethesda Naval Hospital for recuperation.⁷⁷

After their training, Boni and five other Italian agents were parachuted into northern Italy in late July 1944 as the "Renata" Mission. They were soon performing their mission's goals: making contact with the local Resistance group, in this case near Parma, arranging for arms, food, and other supplies to be airdropped to the Resistance, and most importantly, obtaining information about the German forces in the area. By wireless radio or couriers, they information on enemy units, equipment, particularly armor and artillery, minefields, and movements of troops and military supplies, usually with dates and map coordinates.[78] The information sent back was valuable. One of their couriers took the German plans for reinforcing the Gothic Line, the main German defensive position in northern Italy. "We had an Italian engineer, who was working in the German Army headquarters, and he provided the plans to us, and we sent them to the Americans, so they knew exactly what the Germans were going to do, when and where."[79] Soon, however, the Germans learned about Boni and the others by capturing and torturing a team member until he gave them the information. A special brigade of SS troops was sent to find them, and on 17 October 1944, the team awoke to find the village surrounded. "We destroyed the radio," Boni said. "Eight men were killed in the fighting with the Germans, including the head of the Partisans... I jumped from a window... to the ground and then leapt into the river and thus escaped." Dodging Germans for a month, Boni and another survivor, code-named "Comandante Beretta," were finally able to get to the OSS command post in Siena in mid-November 1944.[80]

Despite incursions by the SS brigade, the partisan forces in the district of Parma had increased in size and momentum, and through Boni and Commandante Beretta, they requested a full-fledge OSS mission to operate between the partisan command in Parma and the advancing Allied armies. Allied Command quickly authorized the "Cayuga" Mission.[81] It was typical of the approximately thirty OSS Italian OG missions in northern Italy during 1944-1945. It was composed of Italian Americans and consisted of Captain Michael Formichelli, a veteran of Sardinia, and six enlisted men, all of them trained at Areas F and B. With the way cleared by Boni and Beretta, the initial elements of the "Cayuga" Mission parachuted into the Parma district on 23 December 1944. Back in the zone, Boni was reassigned to the "Rochester" mission, an SO mission, continuing the tasks of the earlier Renata Mission and including several Italians headed by another American, a Captain McClusky. The "Cayuga" and "Rochester" missions now operated on their own but kept in frequent contact with each other.[82] They provided much valuable information for the advancing American Army, such as enemy targets and where the retreating Germans were building defenses, erecting anti-tank ditches and emplacing mines.[83] OSS Florence praised "Coletti" (Boni's code name) for "the high level of intelligence you have been sending."[84] The praise came from Lieutenant Irving Goff, who years later recalled that "We had eighteen radio teams speaking German, French, English, Italian in northern Italy.... The intelligence we sent was called by Allied headquarters the best from any source. We had house-by-house.... We had an overlay map of all the German positions. The American Army knew where every German was."[85]

Formichelli's "Cayuga" mission established contact with the unified Resistance command in the Parma area, which directed thirteen partisan brigades, nearly 4,000 partisans. Despite having their radio broken and suffering a broken ankle on the drop into the mountains, Formichelli and his team of Italian-American OGs evaded or beat off several German patrols. The team was constantly on the move, traveling through the mountains at night, on foot or with horses or mules, stayed in private homes, churches, barns, even shepherds' huts. Formichelli held several meetings with partisan leaders, and after they accepted OSS leadership, he dispersed members of the mission to instruct the various brigades in guerrilla tactics, sabotage, and the proper use and maintenance of the weapons, explosives, radios and other materials. With his approval, 76 air drops were made bringing supplies to the partisans and the Cayuga Mission.[86] Preparatory to the final Allied offensive in April 1945, General Mark Wayne Clark called for a partisan uprising to immediately attack enemy transport columns, garrisons and encampments, blow up . enemy command posts and ammunition and gasoline depots, and cut enemy telegraph and telephone lines. At the same time, they should try to prevent retreating enemy from destroying bridges, power plants, and other facilities that would be needed by the advancing Allies.[87] The partisans, including those in the Parma area, did rise up and simultaneously staged night attacks on all German command posts in the area. They also attacked local garrisons and established roadblocks to hinder the German retreat. On 25 April partisans, according to plans approved by Formichelli, rose up in Parma, as in other northern cities, overcame the German guards and took over the city. In Parma, the 34th U.S. Infantry Division arrived that night. Formichelli arranged for the partisans to stage a victory parade through the city's streets and then surrender their weapons.[88]

During the five-month "Cayuga" mission, Formichelli subsequently reported that the partisans in his zone had been engaged in 182 actions, conducted 38 acts of sabotage, blowing up 6 railroad bridges and 7 highway bridges. They destroyed two trains loaded with arms and ammunitions, three locomotives, 41 trucks, and captured 57 trucks and numerous weapons and stocks of ammunition. They had attacked 43 enemy command posts and eliminated 26 of them, killed 612 enemy soldiers, wounded 750, and taken 1,520 prisoners, which did not include the enemy prisoners taken during the final stages of the partisan cooperation with the Allied Forces.[89]

Danger and Death in the Dolomites

What would become one of the most famous OG missions in northern Italy, the "Tacoma" Mission, began rather strangely. Both the commanding officer, Captain Howard W. Chappell, and his radio operator, Corporal Oliver Silsby, were members of a German-American Operational Group that OSS had recruited and trained in Maryland and Virginia for use in German-speaking countries. Why were these two members of the OSS German OG being deployed in Italy? The main reason was because of Chappell's persuasiveness and his ardent desire for action.

Howard W. Chappell was one of those extraordinary individuals that so typified the OSS. He came from Cleveland, Ohio, the first of four children of a post office worker and a nurse. Because of his mother's Prussian birth and ancestry, he was bilingual in English and German. A natural athlete, he played football in high school and college at Ohio State and Case Western Reserve and was also a Golden Gloves boxer. In June 1942, he joined the U.S. Army, earned an officer's commission, and served first in the Military Police and then as a parachute instructor at Fort Benning, Georgia. It was at Benning, that he gained his nickname, "Flash Gordon." This was due in part because of his movie-star looks—he was 6'2" tall, with broad shoulders, blond hair, and blue eyes. It was also because of his daring, even reckless, antics. In one episode, he allegedly perched on one of the horizontal bars of a jump tower, tossed out his parachute and leaped with it to the ground. In another, when some of his men were beaten up in a bar fight in notorious Phoenix City, Alabama, just across the state line from Fort Benning, he retaliated by driving a 2 ½ -ton Army truck through the front of the tavern. As a paratroop officer fluent in German, Chappell was recruited for the German OG being organized by the OSS. Like most other OGs, he trained at Area F and then Area B. In June 1944, he was sent to North Africa to await further orders.[90]

As the months dragged by, Chappell became tired of waiting to be deployed, and he asked Italian OG operational officer, Albert Materazzi for mission behind enemy lines.[91] He got his wish in December 1944, a month after his 24th birthday. It was an assignment to organize the partisans to block the German Army's vital supply and escape routes from northern Italy to Austria and Germany through the passes in the Alpine and Dolomite Mountains. This was the "Tacoma" Mission. The other two original volunteers were Corporal Oliver Silsby and Sergeant Salvadore Fabrega. Silsby, the radio operator, was from Detroit, Michigan. He had received his OSS communications training at Area C, then been sent to North Africa and subsequently made two jumps into Yugoslavia.[92] At 32, Fabrega was the oldest of the three; trained as part of the Italian Operational Groups, he was fluent in Italian and accompanied the mission as its interpreter. He had a complicated and somewhat unclear history. Born in Catalonia in 1913, he later left with his Spanish parents to spend four years in Germany before the family moved to Argentina. In his teens, he became a merchant seaman and traveled around the world. In addition to Spanish and German, he learned to speak Italian, French, and English. From 1936-1939, the former Catalan fought in the Loyalist Army against Franco's forces in the Spanish Civil War. Twice wounded, he left for France at the end of the war and joined the Foreign Legion. Later, he deserted and when France fell in 1940, he fled to England and then the United States, working in the merchant marine and becoming a U.S. citizen. Living in New York City, he joined the U.S. Army in 1942; the OSS recruited him into the Italian OGs and trained him at Areas F and B in 1943.[93]

The day after Christmas 1944, Chappell, Silsby and Fabrega leaped out of plane into a snow-covered Alpine clearing, 200 miles behind the Germans' front lines. They quickly began training the partisans in the raiding tactics, demolition work, and the use and maintenance of Allied weapons provided by air drops. Two raids successfully destroyed forty thousand liters of fuel and a locomotive with four cars of troops and material. Later, the team took in a couple of Austrian *Luftwaffe* personnel, who said they were deserters. But when it was discovered they were planning to rejoin their unit and inform on the team, their throats were cut. On 21 February 1945, three American sergeants of the "Tacoma" mission parachuted in to help with the partisans, Eugene Delaini, a weapons expert, Charles Ciccone, a demolitions man, and Eric Burchardt, a medical corpsman. When Italian Fascist militia moved into the local village and cut off the Americans' food supply, the team and its partisans sneaked down from the mountain at night and in a firefight, blasted their headquarters with machine gun fire and 18 rounds from the bazooka, a new weapon which the frightened fascists referred to as a "mysterious cannon."[94]

The raids and rumors about the presence of an American team caused the German headquarters in the area to dispatch initially more than a hundred, heavily armed Fascist troops to capture the "Tacoma" mission. The Americans and partisans moved through the snow in mule-drawn sleds, but the Americans were awakened the morning of 28 February with cries from the partisans that "Fascists are coming—hundreds of them!"[95] The Americans and partisans hit the Fascist troops with fire from rifles, submachine guns, light-machine guns, and mortars. There were now six American OSSers, plus nine rescued Allied aviators, and two dozen Italian partisans. During the day's action, 120 Fascists and two partisans were killed. That night, both sides withdrew; the next day, German troops arrived and blasted the mountaintop with artillery for six hours before overrunning it, only to find it the stone-hut vacant, except for a booby trap that the Americans left inside the door and which killed six Germans.[96] Field Marshal Kesselring, the German commander in Italy, concerned about the threat to the key German lines of supply and potential routes of withdrawal through the Alpine passes, now ordered that the partisans and their American leaders had to be captured at all costs. In the ensuing manhunt by 3,500 German troops, the combined "Tacoma" and "Aztec" missions found themselves surrounded on the morning of 6 March 1945; they had been betrayed. Everyone had to run for his life in different directions through the mountainous countryside. Most escaped, but half a dozen partisans and two Americans were captured. Radioman Oliver Silsby was the first to become exhausted and collapse. Captain Chappell stopped to help him, and both were captured, although Chappell was able to escape into the brush. Fabrega, who had dropped behind with two partisans, was also captured.[97]

Interrogated by the SS in Belluno, Fabrega stuck to the story that he was a downed American aviator and knew nothing of the OSS or the partisans. He, like the partisans captured with him was tortured, tied to chairs, beaten with clubs, and given electric shocks to various parts of the body. Fabrega did not disclose information. The next day, he and forty other prisoners were taken to the Belluno town square. The square was filled with townspeople, but all was quiet except the barked orders of the SS officers. A small German truck was backed under a tree; two youthful but beaten partisans stood in the rear. An officer brought Fabrega brought forward and asked the two youths if they knew him. Both said "no," despite the fact that they had been allied with the "Tacoma" mission for weeks and had shared the same stone hut with Fabrega only forty-eight hours earlier. Germans now beat them again and looped nooses over their necks, but neither said a word. Both looked up at the sky. The officer waved his arm, the driver pulled the truck forward, and the two bodies swung on the ropes. The silence was broken by a woman's scream.[98]

The rest of the Italian prisoners were then hanged, some by rope nooses but others savagely snagged by the throat on meat hooks. Fabrega kept to his cover story and continued to be starved and tortured. After eleven days, he was taken to an SS prison camp near Bolzano, 150 miles north. Startlingly while on the way, the Italian driver, "Sette," a chauffeur for the SS commander in Belluno, told Fabrega that he was also a spy for the OSS and tried to convince Fabrega to escape with him to the partisans. But the sergeant, either not believing him or realizing the importance of the kind of information the spy provided the Americans, declined. Instead Fabrega remained in the car all the way to the SS-run prison camp. There, he continued to undergo torture and interrogation.[99]

While Sergeant Fabrega and Corporal Silsby were prisoners, Captain Chappell had escaped capture, despite being wounded in the leg. He was caught again, but using the silent killing technique taught by William ("Dan") Fairbairn, he snapped the guard's neck and took his weapon. He later killed an SS lieutenant with a walking stick. He was able to regroup with the partisans and other American OSS team members, Sergeants Ciccone, Delaini and Burchhardt. When the Allied offensives began in April 1945, they and the partisans scattered roadways with four-pronged road spikes, which caused considerable damage to vehicular traffic, they tore down telephone poles and telephone and electric wire, and blew up bridges at Vas and Busche, and killed a couple of dozen enemy troops. On 24 April, seeking to get to a mountain pass to prevent retreating Germans, Chappell and two other Americans, aided by a blonde Italian countess, hid in boxes in the back of a truck, as their partisan driver narrowly got them though two German road blocks. Together with partisan groups in the area, they mopped up a number of small German garrisons and learning that a large convoy was headed north, they blew up a bridge just north of Caprile and set up a road block and a trap for the Germans in the narrow winding mountain road. When the convoy arrived, Chappell and the partisans opened fire from the high ground, killing some 130 Germans in fifteen minutes. The single-file convoy was trapped, and after the initial firefight, its leaders asked for terms of surrender. Chappell said unconditional surrender, and after a few minutes of threats and discussion, the 3,500 Germans surrendered. Among the prisoners were a number of SS men, including the notoriously cruel Major Schroeder, head of the SS in the region and responsible for the torture and executions in Belluno and elsewhere. When Schroeder surrendered, after first threatening to kill all the civilians he had as hostages, he was found to be carrying the weapon the Americans had given to a teenage partisan who had served as Chappell's assistant. The youth had been captured in March, and in torturing him in a vain attempt to get him to betray the Americans, Schroeder's SS men had cut off his hands and gouged out his eyes before executing him. The morning after the surrender of the German convoy, several officers of the 504th Panzer Division and other units told Chappell that they were very disturbed about being confined with SS and asked to be separated. Chappell granted their request. That night, he summoned Major Schroeder and his seven SS officers to his quarters, where they talked in German. Chappell's report tells what happened next:

We became quite friendly and even joked about how they had once captured me. We drank a little wine, and I learned the name of the spy who had disclosed my location prior to 6 March. *This man was later killed in an attempt to escape.*

We laughed about the fact that some of my equipment that had been captured was in his, and some of his officers, possession. He told me at this time that neither he nor any of his officers had ever committed any outrages and they regretted some of the brutalities that other Germans had committed.

Before he left he told me that he was glad that he had surrendered to me because all of his staff felt I would treat them as they would have treated me, if they had the chance. That was the way I felt about it. *All of them were killed that night trying to escape.*[100]

Over the next few days, as the Allies advanced north, Chappell, his men, and the partisans extended their roadblock farther south preventing the escape of more German troops. They tied up several German divisions and forced the surrender of 7,500 *Wehrmacht* troops. On 3 May, Chappell drove down to Feltre and welcomed the U.S. 85th Infantry Division and turned the German prisoners and his intelligence information over to the advancing American Army.[101] Both Fabrega and Silsby had survived the prison camp. In the final days of the war, Fabrega escaped in the confusion and went to Merano, where the top SS officials in the region were located. Brazenly, he walked into an SS barracks, announced he was a U.S. Army captain and told the Germans they were restricted to the barracks. His bluff worked, they stayed put, and when the U.S. 10th Mountain Division arrived the next day, Fabrega turned the city and the SS troops over to them, courtesy of a sergeant of the OSS. Thus ended the Tacoma Mission. All three initial members were awarded medals, Fabrega the Distinguished Service Cross, Chappell the Silver Star Medal, and Silsby the Bronze Star, for their heroism.[102]

Effectiveness of the Resistance in Italy

Although the most publicized, the Tacoma Mission's success was not unique.[103] Lieutenant George M. Hearn, a former football player from San Jose State College in California, who joined the OSS in December 1943, received SO and MU training in OSS training camps in Virginia and Maryland, and then ran boats along the Adriatic coast, bluffed a German commander in the city of Chioggia near Venice with the threat of an air attack, and accepted the surrender of a heavily armed garrison of 1,100 German troops on 24 April 1945.[104]

Sergeant Caesar J. Civitella, who had trained at Areas F and B with the OSS Italian OGs, was part of the "Spokane/Sewanee" Missions in the Tyrolean Alps in March 1945, which prevented German destruction of bridges and power facilities by seizing a hydro-electric facility supplying power to Milan, removed road mines, and forced the surrender of various German garrisons including the one guarding the Stelvio Pass in April 1945.[105] Across northern Italy, within days after the uprising had begun on 25 April, the patriot forces had taken over the cities of Turin, Genoa, and Milan, which formed Italy's key industrial triangle. Mussolini, his mistress, and some of his ministers sought to escape to Switzerland. Civitella and his unit were directed to go after Mussolini, capture him and hold him for trial.[106] But before the Americans could reach him, the Italian dictator and his mistress were captured by partisans near Lake Como and the Swiss border. They were summarily executed by partisans on 28 April and their bodies hanged upside down in downtown Milan. The German military governor of northern Italy, Waffen-SS Lieutenant General Karl Wolff, who had been negotiating via the OSS office of Allen Dulles in Bern since February, finally signed an armistice on 2 May 1945, six days before Germany surrendered. Those negotiations helped provide a crucial framework for the surrender of the German Army in Italy, and the fact that it was the OSS that Wolff approached indicates the importance that German High Command attached to Donovan's organization.[107]

This new form of warfare, the use of small teams of highly trained special operations combatants to supply, train, and direct indigenous insurgent groups, proved successful in Italy. In 1944, the Germans had used a hundred thousand Italian police and Fascist militia to contain the Italian Resistance. But by February 1945, the unified and Allied-directed Resistance movement had grown so strong in northern Italy that Field Marshal Kesselring ordered his commanders to suppress it, even if it meant bringing German combat units from the front.[108] Even so, entire regions of northern Italy had by the end of the winter been cleared of German forces, and the German commander-in-chief admitted that it was impossible to move troops or supplies through the area except in large, heavily-guarded convoys.[109] British General Harold Alexander, Supreme Allied Commander in the Mediterranean, estimated that Kesselring had detached as many as six of his nearly twenty divisions to control the partisans, and some historians agree with that judgment.[110] In addition to their main role in espionage and sabotage, the partisans, inflicted 2,700 German casualties and 3,800 casualties among Italian Fascist troops, according to official figures. But Kesselring argued that a more realistic estimate was some 5,000 German soldiers killed and 8,000 missing and presumed killed. His intelligence officers claimed the figures were even higher.[111]

The number of Americans in the OSS Operational Groups in Italy who were killed in action included three officers and nineteen enlisted men; several officers and enlisted men were wounded; one officer and seven enlisted men were captured. This was from an original contingent of seventeen officers and 126 enlisted men.[112] These figures did not include casualties in other OSS branches, for example, the mysterious death of Captain Roderick G. ("Steve") Hall, a 28-year-old Special Operations officer who had trained at Areas F and A in the winter of 1943-1944, parachuted into northern Italy, but was caught, tortured and executed by the OVRA, an Italian Fascist organization similar to the Gestapo, in January 1945.[113] Nor does it include the killing of Major William G. Holohan, deputy chief of SO in Italy, murdered in December 1943, allegedly by two American members of his own mission, the Chrysler Mission, as a result of personal animosities and differences over Holohan's policies towards the partisans in Northern Italy.[114] Many more Italians, of course, also paid with their lives in the shadow war against the Nazis. The Germans launched major anti-partisan offensives in the winter of 1944-1945 that included massacres of civilians, sometimes whole villages, for aiding the Resistance in its attacks on German soldiers. A total of some 35,000 Italians, including partisans, died, some 21,000 were wounded, and 9,000 were deported to slave labor camps in Germany, as a result of German reprisals and the anti-partisan campaign.[115]

At the end of the war in Europe, General Mark Wayne Clark commended the OSS Special Operations and Operational Groups for their roles in the "outstanding success of partisan operations in the areas where these men operated." The OSS operational unit with the U.S. 5th Army later received a Presidential Distinguished Unit Award.[116] In the long, frustrating Allied campaign in Italy, the OSS, like the British SOE, helped to compensate for the limited resources available to the Allied commanders in Italy by supplying and directing thousands of Italian partisan forces in the German's rear. The OSS did not alter the course of events in Italy, but they, particularly the OGs who called themselves "Donovan's Devils,"[117] did help, in ways far beyond their small numbers, diminish some of the difficulties confronted by the Allies in their advance up the Italian peninsula.

Directing the Resistance in France

To the U.S. high command, the Mediterranean theater was always peripheral to the main arena, the cross-channel assault on the German Army in France and the thrust into Germany itself. OSS, like British SOE and SIS, had an important role in that strategy. The role was firstly to establish contacts inside German-occupied France and provide intelligence about enemy strength and defenses and secondly, when the Allied invasions began, to lead French Resistance groups, the *maquis*,[118] and block or at least impede German reinforcements.

The Allies decided to invade northern France, with a smaller subsequent attack in the south, at the Quebec Conference in August 1943. The target date set for the early summer of 1944. As part of the planning for this major western drive against Hitler's Third Reich, the OSS created large bases in London and in Algiers from which it planned, trained, and directed the extensive operations with the French Resistance to aid the Allied invasions. Rivalry and distrust existed among the different political groups in the Resistance in France, as in Italy, and there was outright hostility between the Communist Partisans and the others. The Communists were also deeply suspicious of the OSS and the British SIS and SOE. The Allied agencies were often caught among these rivalries, but they were able in varying degrees to work with most of them with the frequent exception of the Communists.[119] Indeed, it was in part because of the internecine rivalries among the French underground, that General Eisenhower approved the use of British, American and French special operations teams in an attempt to direct the fragmented French underground to accomplish effective, coordinated attacks upon targets that would directly aid the Allied invasion and defeat of the German Army.[120]

Joint discussions in London in the summer of 1943 among OSS, SOE, SIS and the Free French forces of General Charles DeGaulle concluded that the main Resistance efforts *before* the D-Day should be directed toward obtaining intelligence as well as conducting sabotage of factories, power plants and fuel storage depots, in order to reduce the flow of war materials to the German forces. SO began infiltrating agents-- French-speaking Americans or French nationals dressed in civilian clothes--into occupied France by parachute or rubber raft in early summer 1943. Months before the Allied invasion, 85 OSS officers, enlisted men, and civilians worked behind enemy lines in France as part of the SO/SOE effort to build secret circuits among the *maquis* that would serve as nuclei for an eventual uprising of Resistance at the time of the invasion.[121] In the first six months of 1944, Resistance groups connected with the Allied agents sabotaged more than 100 factories producing war materials for the Germans, cut power lines, several railroad tracks, disabled more than a thousand locomotives and fomented strikes in coal mines.[122] Several agents were killed in skirmishes or captured and executed by the Germans, as were members of the French Resistance who had helped them.[123] Although SO initially suffered a considerable number of casualties, it became increasingly effective. By the time of the Normandy invasion, OSS SO had become as effective as British SOE as the two agencies organized a potent, armed French Resistance of some 300,000 men and women.[124]

Women did play an important part in espionage and sabotage in France. The *maquis* had long used women as well as men, and the British and Americans joined them. The risks were high. Thirteen SOE women agents were executed by the Nazis.[125] There were fewer American women agents and none was captured. The most famous, Virginia Hall, a Radcliffe graduate and former Baltimore socialite, was one of the most effective OSS espionage agents in France. This American socialite, who spoke fluent French, had been working as a code clerk in the U.S. embassy in London in 1940 when SOE recruited her and sent her, posed as a New York journalist, to Vichy France, where she organized and ran a very effective string of French undercover agents. She had a wooden leg, the result of a prewar hunting accident. It did not deter her, but it gave her a characteristic limp. "The woman who limps is one of the most dangerous Allied agents in France," the Gestapo declared. We must find and destroy her."[126] The Germans never did find her, but their intensive search forced her to leave France and return to England via Spain.

In late 1943, Virginia Hall transferred to the OSS SO Branch, and was code-named "Diane." Because of her wooden leg, she could not be parachuted back in, so OSS returned her to France by swift boat. She was landed on the Brittany coast at night, three months before the Normandy invasion. Dying her hair gray and hiding her limp under the full skirts and shuffle of her disguise as an elderly peasant woman, Hall moved around the countryside in the central region of France, living in different places to avoid the Germans who tried intensively to find her by triangulating her radio signals and offering rewards for her capture. She was co-organizer of the Heckler Mission, and working in the central France regions of Haute Loire and Le Puy in 1944, she financed, armed, and helped to direct a couple of thousand members of the *maquis*. In July 1944, three plane loads of arms, ammunition, and demolitions finally arrived, and these enabled the *maquisards* to destroy a number of bridges and tunnels and eventually to force the several thousand German troops out of Le Puy by sheer bluff. Her three battalions of Resistance fighters killed 150 German soldiers and captured 500 more, Finally in mid-August a three-man, multinational Jedburgh team arrived from OSS Algiers, and they organized a Resistance force at Le Puy of 1,500 men, a group, which Hall continued to supply with money and arms as she obtained them from Allied airdrops. For her heroism, she was awarded the Distinguished Service Cross.[127]

The D-Day invasion, June 6, 1944, was the largest amphibious invasion in history; 175,000 American, British, and Canadian troops landed in Normandy that day, followed by hundreds of thousands of others. An OSS Special Operations unit attached to the U.S. 1st Infantry Division, was scheduled to land at Omaha Beach at 4:30 a.m. two hours before the infantry assault began, in order to make contact with *maquis*, who knew the local area. Although their ship anchored within 300 yards of the shoreline in the predawn darkness, it proved impossible to land the OSS SO party. One of them, Major Francis ("Frank") A. Mills, a 29-year-old Oklahoman from the field artillery who had trained at Area A, looked on in horror at the slaughter on Omaha Beach. "Standing on the open deck, we could only watch as thousands of U.S. soldiers died during the day. We saw many small craft sink and many soldier's bodies floating past us with the outgoing tide.... God was with us that day and did not allow us to land two hours earlier as planned – or our OSS detachment would certainly have been destroyed." It was not until the next day that the OSS unit was allowed to go ashore with division headquarters. "We did land, moved across the beach the next day, and finally got ashore," Mills recalled '—making our way through bodies and body parts, and debris, and on and into the village of St. Mere Eglise... until contact was made with the French Resistance."[128]

Always seeking adventure, Donovan himself was also in the action. Against orders of the Secretary of the Navy, he persuaded an old friend to allow him in the forefront of the naval armada. From the cruiser *U.S.S. Tuscaloosa*, he and David Bruce, SI chief in London, watched the first waves of troops hit the beaches. In mid-afternoon, the two men went ashore at Utah Beach. That landing spot had been easily secured, but when Donovan and Bruce followed the infantry and OSS men into the hedgerow farm country, they suddenly came under enemy machine gun fire. Both dove to the ground, and, according to Bruce, Donovan turned and said, "David, we mustn't be captured, we know too much." Checking their pockets, they realized that neither had brought any of the OSS cyanide pills given to agents to avoid being tortured into revealing agency secrets, Donovan then whispered, "I must shoot first." "Yes, Sir," Bruce responded, "but can we do much against machine-guns with our pistols?" "Oh, you don't understand," Donovan said. "I mean if we are about to be captured I'll shoot you first. After all, I am your commanding officer."[129] Fortunately the Germans were pushed back by American units, and Donovan did not have to shoot either of them to avoid capture. Afterwards, Donovan returned to London and then to Washington, while Bruce set up an OSS field headquarters in a secured area of the Normandy beachhead.

The Jedburghs

The special operations plan to support the Normandy invasion was to use various Allied teams to lead the French Resistance in the disruption of enemy communications, attacks on troop movements and supply columns, and raids on enemy headquarters in order to impede the German opposition to the Allied advance. The most famous of these teams were the Jedburghs.[130] This was the code name for nearly one hundred three-man, multi-national teams, composed of two officers and an enlisted man as radio operator. The teams were drawn from Special Operations forces of the United States, Great Britain and France, and always included at least one native-speaking Frenchman. The "Jeds" went through advanced training together in Britain and served together, but each wore the military uniform of his country. Their mission was to direct, supply and coordinate the Resistance groups according to directives from the Allied high command. The American Jeds were OSS Special Operations officers or enlisted radio operators, who before advanced and team-training at Milton Hall and other locations in Britain had completed their OSS training in the United States, at Areas F, B or A for the officers and Area C for the radiomen. Between June and September 1944, 276 Jedburghs were parachuted into the war zone. Of these, 83 were Americans, 90 Britons and 103 Frenchmen. Most of them were dropped into France (with a few dropped into Belgium and Holland).[131]

In support of the Normandy Landings

In northern France, ten Jedburgh teams jumped into the countryside on D-Day, and some twenty more teams followed in subsequent weeks. Most went into Brittany, the large peninsula west of Normandy that was heavily garrisoned with German troops, which could pose a threat to the flank of the invasion force. Allied headquarters wanted the Jeds to direct thousands of French Resistance fighters and contain the Germans there by cutting off rail and roadways. They might also help conventional forces seize the big ports of Brest, Lorient, and Nazaire, which would be needed for supplying the rapidly expanding expeditionary force.[132]

Secretly airdropping OSS agents and supplies behind enemy lines in northern Europe from January 1944 through the end of the war, was the mission of the "Carpetbaggers," a couple of squadrons of special, black-painted, four-engine, B-24 "Liberator" bombers assigned to the U.S. 8th Air Force in Britain (OSS airdrops in the Mediterranean were conducted mainly from Algeria and later from Italy).[133] Flying alone at night, quickly and at low altitudes, usually not more than a mile high, the plane would drop propaganda leaflets as a cover for its real mission. "We'd be looking for a meadow, then flashes from a couple of flashlights would appear," recalled Lieutenant Eugene Polinksy, a navigator from Maywood, New Jersey. "We would drop down to 200 to 400 feet, open the bomb bay doors and send out supplies, munitions, etc. in parachute containers—or insert an agent."[134] The anonymous agents, whether men or women, were known simply as "Joes." He or she would slide back a cover from a round hole in the bottom of the fuselage, the "Joe hole," sit with legs dangling out and when given the signal would drop into the night. At 200 to 400 feet, the agent would be on a ground in a few seconds, barely enough time for the parachute to open.

One of the first, French-speaking Jedburgh teams dropped into Brittany right after D-Day was Team "Frederick" that included Sergeant Robert R. ("Bob") Kehoe, 21, from New Jersey, a graduate of Areas C and B, as the radio operator. It included a 26-year-old British major, Adrian Wise, who had been on two commando raids along the Norwegian coast, and a French lieutenant, Paul Bloch-Auroch, a 32-year-old reservist who had served in France and North Africa. Although unlike them Kehoe had never been in battle, he felt ready. "The experience at B-2 was a great morale builder," Kehoe recalled later, "and, when we departed in mid-December [1943 for Britain], we are in top physical condition."[135] In England, he spent several months training with his new teammates.

On the night of 9 June 1944, three days after D-Day, Jed Team "Frederick" parachuted into the Forêt de Duault, a forest in the north central part of the Brittany Peninsula. Their mission: connect with the French Resistance and prevent German troops or supplies from getting to Normandy by blowing up bridges and railroad lines and setting up barricades and ambushes on roadways. As they neared the drop zone, Kehoe remembered that after checking his equipment, he repeated the 23rd Psalm. The local Resistance set a triangle of bonfires to mark the drop site and when the three men reached the ground, gave them a warm cups of strong Breton cider. Kehoe quickly established radio contact with the OSS base station in London. Consequently, more than one hundred reinforcements from British Special Air Service (SAS), Frenchmen in this case, soon arrived. But the increased activity also led several hundred German troops to come looking for them. The Germans may have been tipped off by informers. They posted notices declaring that those found aiding parachutists would be shot, but informants would receive monetary rewards. The SAS dissipated into small groups and headed south, several were killed in skirmishes along the way. The Jed team decided to go it alone. With the Germans only a few hundred yards behind them, Kehoe buried the radio equipment so the team could move faster. With the help of courageous local French men and women, the team survived and even continued operations against the Germans. The first night on the run, as the team huddled in a concealed drainage ditch, their savior was a young woman, who like so many other female schoolteachers played a key role as guides, couriers, and coordinators in the *maquis*. Kehoe called them "the lifeblood of the Resistance."

For the next three weeks, despite German searches and massive sweeps, Jedburgh Team Frederick survived in the woods in their increasingly dirty uniforms amidst the coldest, dampest June in decades. At one point the Germans came within three or four feet from where the team members lay hidden in a briar patch, their guns drawn. "I was unable to prevent my hand [with the .45 pistol] from shaking constantly," Kehoe remembered. "Curiously, however, whenever a threat approached, the shaking stopped as my whole body became tense and alert. The body hormones apparently knew their job and did it well."[136] Recovering the radio and re-establishing contact with London, the team arranged for supply drops for the local organized Resistance and helped train the young *maquisards* in the use of the weapons and demolitions.

By mid-June, the team was working actively with an armed and organized local Resistance to prevent tens of thousands of German troops in Brittany from moving east to attack the Allied force in Normandy. German truck convoys were subject to Allied air attacks during the day, but were generally safe at night, until the Jeds organized, trained and supplied the French Resistance to attack them. As the team radio operator, Kehoe was the key link with headquarters. While the others slept during the night, he had to listen to one-way broadcast instructions from London scheduled from midnight to 3 a.m., because daylight airtime was reserved for quick two-way transmissions from the field operators. In rapid Morse code, the broadcast would first list those who had messages coming. If none for his team, Kehoe would sign off, otherwise, he listened until the message to him was sent and then later to the repeat to ensure accuracy. Accuracy was difficult but essential, especially in setting up airdrops. The message was, of course, enciphered. Although the base station's powerful transmitters sent a strong signal, it was often hampered by bad weather or enemy jamming. "The entire radio spectrum sometimes resembl[ed] a mass of screeching cow birds," Kehoe said.[137] At night, sometimes in the rain, the radio operator in the field would copy the text of the message by the light of a weak flashlight. He would then decipher enough to see if it were urgent, and if not would wait until daylight to decipher the rest using a sheet from the "one-time pad" which was burned after each use. The one-time paid was based on a memorized transposition system keyed to a signal at the beginning of the transmission of where to start the five-letter groups on the pad. The Germans launched more sweeps, including a massive one in which 4,000 troops were sent out against the Jeds and the Resistance. Several members of the local *maquis* were killed in these, but the Jeds had now learned to plan for such attacks and had arranged in advance for escape routes and clear responsibilities to avoid confusion and delay.

Jed Team "Giles," jumped into Brittany on the night of 8 July 1944. It included a British radio operator, a French captain, and an U.S. Army captain, Bernard M.W. ("Bernie") Knox. The 29-year old Knox had been born in Britain, graduated from Cambridge, fought on the side of the Republic in the Spanish Civil War, where he suffered a near fatal wound, then later moved to New York with his American fiancé whom he married in 1939. With the outbreak of the war, Knox joined the U.S. Army, became an officer and a U.S. citizen. Stationed back in England in 1943, he was recruited by the OSS, which trained the fluent French speaker for a Jedburgh Mission. When his combat experience in Spain became known, he was also made an instructor.[138] Dropped into Brittany, Knox's team was one of a dozen Jed teams, whose mission was to organize arms and finally help open the way for Allied armies to get to the port of Brest. When Patton's army broke through Avranches, Knox's team led others in a series of successful ambushes of elements of the hardcore, veteran German 2nd Parachute Division that tried to block the U.S. forces. They obtained a number of prisoners, many of whom had participated in atrocities against French civilians in the area. The French forces demanded they be turned over and quickly executed most of them.¹³⁹ His *croix de guerre* citation cited the Jeds arming and direction four thousand *maquis* to help liberate Brittany.

In mid-July, the Allies broke through the German lines at St. Lô and Lieutenant General George S. Patton's Third U.S. Army, led by its armored divisions, began its famous, fast-moving, long-ranging swing south and east to trap Germans in Normandy from the rear. Looking at Patton's plans for his giant right hook drive, his superior, General Omar Bradley allegedly asked "but what of your right flank?" To which Patton reportedly answered, "The French Resistants are my right flank."¹⁴⁰

Patton and Allied headquarters counted on the French Resistance, armed, trained, and directed by the OSS and the SOE to help protect Patton's right flank and his exposed lines of communication and supply and to help preserve the bridges and roads his armored columns needed for their rapid advance. On 2 August, London radioed new orders to Jed Teams in Brittany, Normandy and throughout northern France: "Allied advance will probably be rapid in your direction. Task is now preservation not destruction. Greatest importance…road [to] Morlaix, Sant Breuc, Lamballe. You will prevent enemy…demolition…on this and secondary roads…"¹⁴¹ Jed Team Frederick had by then 2,000 armed French *maquis* to assign to the highway to Morlaix and another 2,000 to watch and defend the roads, bridges and causeways on the secondary routes as well. While the main thrust of the Third Army swung southeast toward Paris, sizable armed columns also directed west into the Brittany peninsula toward Brest and the German submarine bases as St. Nazaire and Lorient. The armored columns counted on having the French Resistance keep the roads open for the American tanks, which they did. The Jeds met up with the lead tanks on the way to Morlaix, and the commander of the column quickly asked them for the *maquis* as guides in front and to protect his supply line behind. When the Germans in the Paimpol peninsula on the north coast began taking revenge on the local population and committing atrocities, Kehoe's team was able to convince the American general commanding the armored spearhead to divert some of his tanks and air support to protect the *maquisards* and other local residents against the heavily armed Germans. Thus the Allies aided the Resistance as the Resistance had been aiding the Allies. As a result of the Jeds, the French Resistance and Allied armor, the large German garrisons, with tens of thousands of troops in Brittany, were unable to provide reinforcements for the German effort to contain the Allied landings in Normandy.¹⁴² Years later, Major General John K. ("Jack") Singlaub, who had been a Jed on another team, called Team Frederick, with its British and French officers and Bob Kehoe, its American radioman, "one of the bravest, most effective Jed outfits in France."¹⁴³

A 23-year-old Vermonter of French-Canadian ethnicity named Paul Cyr was part of Jedburgh Team "George," which had considerable impact on the French ports in Brittany. The OSS had recruited him from the airborne school at Ft. Benning in 1943 and gave him SO training at Areas F and B. In Britain, he was assigned to a team with a French commander and a French radioman. They were dropped into southeastern Brittany three days after D-Day. Their first problem was that the overeager *maquis* immediately attacked the Germans before an adequate supply of arms and munitions arrived. Consequently, the Germans counterattacked, and the Jeds had to blow up five tons of explosives and escape with their French SAS men to avoid capture. The Germans tracked them with dogs and executed families who sheltered them. One or two of the Resistance leaders had been informants for the Gestapo. In a new district, where the Resistance was trustworthy, Captain Cyr and Team George led local *maquis* in numerous attacks on German troop convoys, railroads, canals and bridges. Through French construction engineers, Cyr's team obtained accurate plans for the German coastal defenses and submarine pens at St. Nazaire. Because of earlier reports of French traitors and a two-week gap in transmissions, London would not respond to their radio signals, believing the radio had been captured. So Cyr had to take the plans in person to Patton's Third Army headquarters. Allied air forces subsequently bombed the key parts of those U-boat facilities and significantly reduced the effectiveness of German submarine operations in the Atlantic. More immediately, Third Army headquarters drew upon the *maquis* directed by various Jedburgh teams, like Cyr's, to provide guides and to protect Patton's right flank and growing lines of communication and supply as the Third Army make its famous sweep eastward across France.[144]

Jumping into northwest Brittany in early July, just before the American breakout from Normandy was Captain William B. ("Bill") Dreux, a French American from New Orleans, who had trained at Areas F and A in the fall of 1943. In Brittany, he was accompanied by two Frenchmen in Jed Team "Gavin." The Germans who were all around, later identified and shot a butcher and veterinarian who sheltered Dreux. The Jed team sent back reports on German strength to Allied headquarters by radio or using young French girls on their bicycles as couriers. In one raid, Dreux led a group of *maquis* to try to punch a hole through the rear of German concrete bunker defenses on the Brittany coast. His group consisted of untrained French youths and some former French colonial soldiers from Senegal and Algeria. The Germans knew they were coming, drew them into a trap, and opened fire with machine guns. The colonial soldiers disappeared. The young Frenchmen stood fast but took a beating before Dreux could get them out. "I had made a tactical error," he recalled sadly, "for which others paid the price."[145] He learned from that experience, and subsequent patrols proved effective. Dreux led the liberation of more than one town, and his team's mission was hailed as a success when he returned to Milton Hall and ultimately to the United States.

Yet, after the war, as he grew older, became a successful lawyer and raised a family, Dreux began to raise questions about the impact of his actions. He could not forget the lives lost because of him: the Frenchmen who had been executed because they helped him, the teenagers shot down in an ambush into which he had inadvertently led them. The French villagers later reassured him that the defeat of the Nazis had been worth the price, but the deaths continued to haunt him.

Returning to Brittany in 1951, Dreux's thoughts went back to that airborne mission seven years earlier. "My mood was fatalistic on the plane [in 1944], but that was because there was nothing I could do about the flak, night fighters, or the pilot's navigation. Yet beginning with the moment I flung myself out of the bomber the rest was up to me—up to me and chance. It was not so easy to be fatalistic from then on...," he said in 1951 to a French priest, who had protected him in 1944. "I'd been thinking of those days behind the lines and that one of the qualities of that life was simplicity. All I wanted to know then, for example, was what lay around the next bend of the road, or whether that enemy machine gun in that clump of trees could cover the road to my left, or how soon it would be before I could stop and rest. Could I trust this Resistance leader? That barn we were hiding in, what was the best escape route if a German patrol came, and could we fight our way out?" "Most of the questions were like that," he reflected. "basic and uncomplicated. There was no need to consider a compromise between integrity and money, between moral courage and popularity. The answers often were not easy, but you got them fast enough."[146]

Jeds in Central and Southern France, July and August 1944

While some of the Jedburgh Teams parachuted into Brittany beginning in June, others jumped into central France in July to help the Allied advance eastward after the breakout from the Normandy lodgment. And still others parachuted into southern France in preparation for the 15 August Allied landings on the French Riviera and the drive north up the Rhone River Valley. Jed Team "James" was headed by Lieutenant John K. ("Jack") Singlaub, 24, a UCLA graduate and paratrooper, who had trained at Areas F and B in the fall of 1943. He was supported by a French executive officer and an American radioman, Sergeant Anthony J. ("Tony") Denneau from Green Bay, Wisconsin.[147] At 2 a.m. on 11 August Singlaub's team parachuted into the Correze region, the rugged, wooded hills of the Massif Central. The highway and railroad from Bordeaux northeast to Clermont-Ferrand and on to Lyons, was the main German escape route from southwestern France, and Allied headquarters wanted the tens of thousands of enemy troops in the German First Army Group to be trapped there.[148] Team James was met by leaders of units of some of the ten thousand members of the Resistance in the area. The French were anxious for revenge. The notorious German 2nd SS Panzer ("Das Reich") Division had been hampered in its movement through the area by the cutting off of the rail line and by constant harassment; consequently, the SS had committed extensive atrocities in Tulle and in Oradour-sur-Glane, where they killed hundreds of men, women, and children, machine gunning the men and burning the women and children locked inside the town church.[149] After meeting with the local Resistance commanders, the team headed for Egletons, a key town along the highway, where Gaullist and Communist forces, actively hostile to each other, were uncoordinatedly maintaining a stalemated siege of the German garrison, holed up inside a three-story concrete schoolhouse. Singlaub impressed the rival French forces with his skill and bravery. First, he helped shoot down a strafing German plane. Then, despite suffering bloody wounds to his ear and cheek from shell fragments, Singlaub grabbed a heavy Bren submachine gun, rushed forward, and from behind a shattered tree trunk, emptied two thirty-round magazines into the German gun crew sixty meters away, killing all of them. Team James forced the surrender of several small, isolated German garrisons along the main highway; they also convinced local French leaders to stage a series of small ambushes along the highway, particularly on the relief columns. The Germans' retreat was harassed for weeks. After seven weeks in the field, much of it with only minimal sleep, Team James, was flown back to England, their mission accomplished.[150]

Major William E. Colby, who would later become head of the CIA under Presidents Richard Nixon and Gerald Ford, was an "Army Brat," born in St. Paul, Minnesota, and raised at military facilities around the world, including China. On a scholarship, he went to Princeton University, spent a summer in France, and obtained a commission through Army ROTC. He had just finished his first year at Columbia Law School, when he was called to active duty in 1941. Bored with the field artillery, he joined the paratroops, from which OSS recruited him in September 1943. By December, the 23-year-old paratroop officer had completed SO training at Areas F and B and was bound for England as a Jedburgh.[151] The youthful, scholarly looking major was assigned to head Jed Team "Bruce" with two French soldiers. Their mission was to aid the French Resistance in the Yonne Valley some sixty miles southeast of Paris. Instead of being dropped into a rural area, the plane on a mid-August night, dropped them right into the center of a German-occupied town. The falling supply containers woke up the townspeople as well as the Germans, and the Jedburgh team had to dodge patrols for two days until they were able to reach the *maquis*. Although it was later discovered that the local Resistance had been run by a collaborator working with the Germans, by August 1944, Colby realized the Nazis would be defeated and he did not endanger the Jeds. Colby's team obtained airdrops of several thousand rifles as well as mortars, bazookas, and machine guns for the increasing number of Frenchmen and women joining the Resistance. They ambushed German patrols, attacked convoys, blew up supply depots, and destroyed several bridges across the Loire River to impede German units from the South from attacking the flanks of the Allied armies which were driving toward Germany.[152]

Years later an American Jedburgh named Aaron Bank would be celebrated as the "Founder of Special Forces," because of his role in creating the first of a continuing line of these special units in the U.S. Army.[153] Bank began his special operations career with OSS in World War II. A native New Yorker, Bank had worked as a youth as a lifeguard and swimming instructor in the Bahamas and southern France after high school. Traveling through Europe in the 1930s, he became fluent in French and German. He joined the U.S. Army as a private in 1939 and served in a transportation unit. But after completing OCS in 1943, he volunteered for OSS. The 40-year-old Bank took OG training at areas F and A, but was chosen to become a Jedburgh instead. He completed his Jed training in Britain. At the end of July 1944, Jed Team "Packard," with Captain Bank assisted by two Frenchmen, parachuted into southeastern France to aid the Resistance in impeding German troops and supplies. Bank's main task was to get the guerrilla leaders to make their ambushes short and furious—hit and run—and to watch their flanks while the action was underway. The Germans tried to track them down. Bank had some narrow escapes. But his mission was successful. Some 3,000 effective guerrillas were armed and directed, supplied by a dozen or more airdrops. They ambushed numerous convoys and imposed about a thousand or so casualties on the enemy. Like the other successful teams, they had helped a phantom Army arise, strike, and repeatedly melt away, helping to bleed the enemy economically and militarily. As the advance elements of the U.S. 7th Army overran Bank's territory, liberating the region, the head of the local Resistance invited Bank to address the cheering crowd from the balcony of the city hall in Nîmes. When they saw him is his U.S. Army uniform complete with a miniature American flag on it, the crowd went wild with shouts of *"Vive les Americains!"* It was, recalled Banks, who would soon be assigned missions in Germany and Indochina, "the zenith of our mission. It was the thrill of a lifetime!"[154]

Numerous other Americans served on Jedburgh Teams in 1944. Lieutenant Lucien E. ("Lou") Conein with Sergeant James J. Carpenter as radio operator, plus a French executive officer, deployed as Jed Team "Mark" from Algeria to southwestern France on the night of 16-17 August 1944. Born in Paris, Conein had been raised in Kansas City by an aunt. He served in the French Army, 1939-1940, then the U.S. Army, and transferred to the OSS in 1943, where he trained at Areas F and B, before being sent to Jed School overseas and then parachuted into France. There he worked with the *maquis*, including the Corsica Brotherhood, an underworld organization allied with the Resistance. Like Colby, Singlaub, and Bank, Conein would later become one of the most famous of the former Jedburghs, in his case emerging as one of the leading covert operators of the Cold War, first in Eastern Europe and then in Vietnam., where he worked for more than a decade and played an important role in the 1963 coup against South Vietnamese President Ngo Dinh Diem.[155] Other Jeds included Lieutenant Stewart J. O. Alsop, 30, in Jed Team Alexander in central France. (The Alsop brothers, scions of an old and wealthy Connecticut family and all Yale graduates, included Stewart, the Jed; John, 29, who served in France with an Operational Group; and journalist Joseph who did not serve in the OSS.[156]) Major Horace W. ("Hod") Fuller from Massachusetts, a graduate of Milton Academy and Harvard College, had joined the Marines in 1941 and been seriously wounded at Guadalcanal in 1942. Not content to ride a desk in Washington afterward, he joined the OSS in September 1943, trained at Areas F and B, and parachuted into southwestern France from Algeria in late June 1944, as the head of Jed Team Bugatti, which also included two French soldiers. Operating out of the Pyrenees Mountains, the team was successful in blowing up railroads, especially the main line from Spain which had been delivering iron ore to the Germans. When the Allies invaded along the Riviera, Fuller's team led an uprising of 5,000 *maquis*, ambushing numerous columns of Germans, delaying them or forcing their surrender. France. Fuller was awarded the Silver Star medal.[157]

Special Operations Teams in France

Much of the writing about the OSS in Europe has focused on the Jedburghs, but there were other OSS paramilitary teams in the ETO as well, SO and OGs. The most famous SO operative there was undoubtedly Marine Major Peter J. ("Pete") Ortiz, who having recovered from his wounds from North Africa and undergone SO training at Areas F and B, was parachuted twice into France in 1944. On the moonless night of 6 January 1944, Ortiz, a British officer and a French radio operator were dropped in Mission "Union" into the mountainous Haute Savoie region of southeastern France to help the 3,000 *maquisards* there on the Vercors plateau. Ortiz wore a Marine officer's uniform and the others their country's uniforms. They were the first Allied ground forces to arrive in France since 1940, and the French locals cheered them heartily. They arranged for the supply of weaponry and began training the Resistance in their use. Ortiz also liked to thumb his nose at the Germans, for example, by stealing Gestapo vehicles or making public appearances in villages in his Marine uniform. In the most famous episode, Ortiz entered a café where German officers were drinking and cursing the *maquis* and Allied operatives, flamboyantly threw back his cape revealing his Marine Corps uniform. He opened fire with a .45 Colt automatic in each hand, killing or wounding the Germans before disappearing into the night. For his activities in the Union" Mission, including leading many raids behind enemy lines between 8 January to 20 May 1944, Ortiz was awarded the Navy Cross.[158]

Promoted to major, Ortiz returned to the French Alps eight months later on 1 August 1944. This time, he parachuted in as commander of Union II Mission to lead the *maquis* in raids against the retreating Germans and prevent the enemy from sabotaging key installations. Ortiz's second in command was Captain Francis L. ("Frank") Coolidge, who had trained at Area E north of Baltimore and subsequently been an instructor at Area A-4 in Prince William Forest Park.[159] The other members of Union II included a French Army officer, and five U.S. Marines in the OSS. The Marine sergeants had been recruited by OSS in late 1943 from the 1st Marine Parachute Regiment. Sergeant Merritt Binns had been a parachute rigger and instructor at Quantico, Virginia. "They asked for volunteers, especially those who knew French," he recalled. A number volunteered for hazardous duty, and five of them after attending OSS SO courses at Area F and A, and then in England, accompanied Ortiz.[160] The team jumped at 400 feet, and one of the sergeants, Charles R. Perry from Massachusetts, was killed when his chute failed to open. Another, Robert La Salle, was injured so badly that he could not continue in the mission. The rest spent several days training the *maquisards* with the weapons that had been dropped from B-17s and then began attacking Germans. The Germans sent a force of nearly 4,000 to destroy what they believed was an entire Allied battalion, in the process, wiping out the entire village of Montgirod that had harbored the Americans and the *maquis*, burning the town and killing the villagers. The next day, 16 August 1944, the Americans were forced to move out in daylight, and while moving just below the ridgeline, they were confronted by an armored German patrol and were forced to take cover in the nearby village of Centron. The Germans started to surround the village and Ortiz and Sergeants Jack Risler and John Bodnar, who were the closest to the enemy, were pinned down by machine gun fire. On the other side of the village, Coolidge and the other two sergeants were able to escape by jumping into the Isere River. As the German infantry advanced, supported by the armored cars, Ortiz realized the situation was hopeless to him and also to the town, which might suffer the same fate at Montgirod. The French civilians urged him to surrender to avoid reprisals. Bodnar, a resident of Collegeville, Pennsylvania, later recalled that "Major Ortiz said that he was going to talk with the Germans and that we should try and sneak out while he did." Bodnar said, "Major, we are Marines. We work together, we stay together."[161]

Bullets splattered around his feet as Ortiz arose, then put his hands up and walked forward, requesting in fluent German to speak to the officer in command. When the German major appeared, Ortiz negotiated the surrender of his Marines in exchange for the safety of the town. He said they had only been passing through and that the villagers had not been harboring them. When Ortiz motioned the other two Marines forward, the German major could not believe that his battalion had been held off by only three men, but Ortiz, their automatic weapons, and a search of the town, convinced him. The Marines, and the French officer, who was captured later, were held in a naval prisoner of war camp near Bremen until it was liberated. Ortiz received a second Navy Cross for his actions. Coolidge and the other two sergeants were able to escape to American lines. After the war, Ortiz remained in the Marine Corps reserve, but as a civilian, he returned to Hollywood as both a technical adviser and a minor actor under his friend, director John Ford.[162]

Among the SO teams in the Loire region, Sergeant Herbert R. ("Herb") Brucker, was an OSS radio operator trained at Area C. Born in the United States, he had been raised since infancy in bilingual Alsace-Lorraine and became fluent in French and German. Returning to the United States in 1938, he joined the U.S. Army in 1940 and was trained as a radio operator by the Signal Corps before volunteering for the OSS in 1943. He had been sent briefly to Area B, but did most of his SO training at Area E, and presumably did his OSS Communications training at Area C.[163] He was then detailed to British SOE in January 1944 for additional training, was promoted to lieutenant, and was parachuted into central France on 27 May 1944 as the radio operator with and SOE team on the "Hermit" project of organizing a new circuit of French agents because the Germans had broken the last ring and executed several of them. He remained in France until September. The training, particularly the SOE training had been vital. "SOE training was far superior," he later conceded. "It made most of my OSS/SO stateside training seem amateurish."[164] Brucker later recalled one instance where he and his officer ran into trouble. Coming unexpectedly across a German roadblock on their tandem bicycle, they were inspected. When the Germans were about to find the pistols in their clothes and their radio in the suitcase, they shot their way out. They left the radio and their other supplies and ran as fast as they could. The remaining Germans continued to fire after them. Brucker knew he would be tortured and then executed if captured. That is what happened to the Hindu Indian woman SOE radio operator in the region before him. "I had now geared myself to immediate reaction if I was hit anywhere," he remembered. "As long as I'm hit anywhere but the feet I will continue running. If they hit the feet, I take the pistol and blow my brains out, so I kept on running and eventually found a safe house whose owner was part of the Resistance."[165] The team survived to continue its mission with great success. Brucker was promoted to lieutenant and awarded the Distinguished Service Cross. His next assignment was in China and after the war in 1952, he would become one of initial members and instructors of the U.S. Army Special Forces.[166]

Another group of SO men were assigned, like Brucker, to a joint SO/SOE mission in France. This one, Mission Freelance, was headed by an experience British captain and included a Canadian captain and four American officers, Major Edwin Lord, Lieutenant Richard Duval, Michel Block, and William Butts Macomber. The last came from an old upstate New York family including newspaper publishers, politicians, and judges. Young Macomber had graduated from Phillip's Andover Academy and Yale University, where he was President of the debating society and played varsity football and lacrosse and was captain of the wrestling team. After graduation, he joined the Marines, received a commission in December 1943, and transferred to the OSS, where he received SO training in Maryland and Virginia, and additional training in Britain. He was assigned as the weapons officer to the Freelance Mission which was parachuted into Montlucom in south central France. Macomber trained groups of *maquisards* in the use of automatic weapons, and they ambushed German convoys heading away from the advancing Allies.

The team heard rumors of surrender talks between German officers and Patton's 3rd U.S. Army. Lieutenants Macomber and Block with their *maquis* units met with German officers who made clear that they wanted to surrender to the Americans not the *maquis* and refused to have their units disarmed until they were escorted through masquis areas to the U.S. Army. The two and a few hundred *maquis* were in a town with 6,000 armed Germany troops. After several tense hours, they were eventually able to help connect representatives of Patton's Army with a German major general, named Elsar, who surrendered his 19,000 troops. It was an extraordinary capitulation, as Macomber wrote in his report. "To my mind it is one of the outstanding events in the overall story of the *Maquis* resistance in France. Of course Elsar's 19,000 were not militarily defeated by the *Maquis* which surrounded them. They were actually overcome by the joining[167] of the American Third and Seventh Armies, for this destroyed their escape route. Nevertheless... the nearest American regular troops that could be brought against them were those of the Third Army north of the Loire, and every bridge across the Loire was blown. If they had chosen to fight it out, it would have meant the diversion of sizeable forces and considerable cost in time, manpower, and material. Had there been no *Maquis* active, the Germans would almost certainly have followed this course. It is highly significant that the *maquis* so completely destroyed their nerves by continual sniping and ambush and by killing every prisoner which fell into their hands."

Jacques F. Snyder, the Saxophone playing, French-raised American soldier, who had joined the OSS and undergone SO training at Areas F and A in 1943, parachuted into France in 1944. His mission was for Secret Intelligence as well as Special Operations in the Grenoble area in the southeast.. He was accompanied by a Frenchmen, Jean Coppier, who had served in the *maquis*, but left France and enlisted with the OSS in Algeria. Synder was only a private, and Henry Hyde, a former international lawyer who ran SI in Algeria, decided to give him a captain's insignia to give him more status in dealing with the *maquis*. Synder and Coppier were parachuted into the mountainous region of southeastern France in May 1944. Since most of their mission was intelligence gathering about the German units in the area, they wore civilian clothing after the initial jump. Yet they also engaged in paramilitary operations, including attacking convoys and facilities with guns and plastic explosives disguised to look like potatoes and radioing for Allied bombing attacks. They brought in B-17 bombers which destroyed German ammunition and supply trains on the mail line between France and Italy. Rail traffic came to a standstill. Despite several close calls with the Germans and an ambush by communist partisans, Snyder and Coppier survived four months behind enemy lines until the 7th Army arrived in Grenoble in August 1944 and Private Snyder in his captain's uniform joined the French population in welcoming the liberators.[168]

Some other SO operatives were not as fortunate. One of them was Roberto Esquenazi-Mayo, a native Cuban and graduate of the University of Havana, who immigrated to the United States in 1941 to continue this studies, he enlisted in the U.S. Army in 1943, at age 22. He soon volunteered for the OSS, and after SO training in Maryland and Virginia, he was parachuted into southern France near Spain in August 1944. Unfortunately, he fractured his leg very badly in the landing and was unable to continue the mission. He was rescued by Spanish Republicans working with the French Resistance, who got him out of the country and to a military hospital in North Africa.[169] Another disappointed OSS SO officer was Lieutenant Roger Hall. A former instructor at Areas F and B, Hall had been sent to Britain and in late July 1944 was parachuted into Normandy near St. Lô to join a Jedburgh team, whose leader had been wounded. He arrived safely only to be told by the wounded officer that two hours earlier the advance elements of the U.S. Second Armored Division had raced through pursuing the Germans and that Hall had landed behind his own lines. His brief mission was over.[170]

Operational Groups in France, 1944

While the three-man, multinational teams called Jeburghs became famous, Donvaon's other combat units, the Operational Groups, were also contributing intelligence and sabotage to hinder the enemy. The OGs, usually composed of men from particular American ethnic groups, but sometimes including foreign nationals, were organized to fight in sections of a dozen men of more, and fought they did. The Jeds understood that. "The Operational Groups were not the 'glamour boys.' Those soldiers were doing the hard work and not getting much publicity," explained ex-Jed and former CIA Director William Colby in 1993.[171] The first 200 volunteers in the French Operational Group graduated from Area F in the fall of 1943. Major Alfred T. ("Al") Cox, who had succeeded Serge Obolensky as head of OG training at Area F, trained them hard. Cox was a strong and able athlete and an intelligent and commanding leader. Graduating from Lehigh University in Pennsylvania, with a civil engineering degree and membership in Phi Beta Kappa, he was captain of the football team, co-captain of the baseball team, and class President. From ROTC, he became an infantry officer and was instructing in guerrilla warfare at the Infantry School at Fort Benning, Georgia, when the OSS recruited him. He recruited other capable officers and instructors like Lieutenant Arthur ("Art") Frizzell from Fort Huron, Michigan, to help train and lead the French OGs.[172]

When the OSS French OGs had finished their training at Area F, the Congressional Country Club, Cox, as their commanding officer, took them to Area A at Prince William Forest Park, and then to Algeria. They arrived in Algiers in February 1944, and they went to the OSS parachute school there and continued their field exercises, preparatory to being dropped into France south of Lyons in August 1944 to aid the Allied landings of nearly 500,000 men along the French Riviera beginning 15 August. Additional French OG groups arrived subsequently, and they were sent to England, where under Lieutenant Colonel Serge Obolensky, they would join an OSS Norwegian OG unit and be deployed north of Lyons. More than twenty OG teams, most of them OSS French OGs but some in the southeast were OSS Italian OGs, were parachuted into occupied France in the late summer of 1944.[173]

Directing the French Resistance in sabotage and intelligence gathering, the OSS provided excellent tactical intelligence for Lieutenant General Alexander Patch and the 7th U.S. Army as part of the southern invasion force, "the best briefed invasion in history."[174] On Corsica, OSS radio operator, Edward E. Nicholas, III, trained at Area C, recalled the day the invasion of southern France began on 15 August 1944. "One morning I awoke to a mighty roar. Outside, the sky was full of planes from horizon to horizon and as far as the eye could see. The invasion of southern France was underway."[175] It proved a highly successful, if much less publicized invasion than that at Normandy. With the loss of 2,700 American 4,000 French military casualties, the Allied force captured 57,000 German soldiers, pursued the enemy up the Rhone River Valley, and quickly liberated the ports of Toulon and Marseilles. By October 1944, those two ports were handling over one-third of the more than one million tons of American supplies reaching Europe.[176]

Operational Groups Fight the Germans in France

While the Jeds and SO agents were in two-or three-man teams, the Operational Group sections ranged from ten to twenty uniformed and well-armed personnel. In all, 356 Americans in 21 OG teams parachuted into France in 1944. Most of them had trained at Areas F and B. The majority were flown from bases in Algeria, but a few departed from England.[177] OG Section "Percy Red" was composed mainly of Norwegian Americans, led by Captain William F. Larson and including Technician Fifth Class (T/5) Arne Herstad from Tacoma, who had been waiting in vain for a mission to Norway. Instead, beginning on the first of August 1944, the 18 OGs were parachuted into the Haute Vienne region of central France to connect with the *maquis* and impede the movement of German troops. Operating near Limoges, they blew up bridges, blasted highways, dug anti-tank ditches, and planted mines and booby traps. On 11 August the team blocked a German armored train by blowing up a bridge in front of it. But in the ensuring fire fight, Captain Larson was fatally wounded. Despite his loss, the Norwegian OGs continued their mission and eventually met up with Colonel Obolensky and OG "Patrick" before being flown back to London.[178]

Lieutenant Rafael D. Hirtz, who was born in Argentina but grew up in California, led OG Section "Donald" that was dropped into Brittany on 5 August. Their mission was to initiate guerrilla activity and to protect a large and particularly important bridge across the Pense River for Patton's armored columns. "We had about two weeks behind German lines before we knew that Patton's armor was coming through," Hirtz recalled. "I never took off my uniform, and we held the bridge so that when Patton's tanks did come through, they went right over it."[179]

At the beginning of September, Hirtz was one of nearly sixty members of OG Section "Christopher," combined with Jedburgh Team "Desmond" that were dropped by ten planes into the Poitiers area to harass and destroy German units fleeing southwest France toward the Belfort Gap into Germany. Section "Christopher's" mission was carried out although several of the men were wounded and 1st Lieutenant W. Larson was killed during an ambush on German troops (this was a different Larson than Captain William F. Larson, commander of the "Percy Red" Mission who was mortally wounded during an attack on a German armored train).[180] New York socialite and hotel magnate Serge Obolensky, former Czarist officer and Russian émigré after the Bolshevik Revolution, commanded OG Section "Patrick," two dozen members of a French OG, into central France to protect vital power stations from destruction by retreating Germans. Ironically this staunch anti-Bolshevik, was assigned to a Communist *maquis* group, to whom he had to give acknowledgment in the victory celebration following the liberation of the town of Chateauroux.[181]

Private Emmett F. McNamara, a 22-year-old radio operator from the Roxbury district of Boston, was part of Section Lindsay that parachuted into south central France on 17 August 1944. The OSS had recruited him in 1943 after he graduated from the Army's Signal Corps school at Fort Monmouth, New Jersey and claimed that he could speak French. He spent a couple of weeks at Area B learning OSS codes and radio equipment, then more than a month at Area F doing more practicing and also going through the close combat, weapons, and field craft training with the French Operational Group. He earned his paratrooper's wings in England, where in the plane on his way to his first jump he recalls sitting across from Lieutenant Colonel Serge Obolensky, who was in his fifties. "I looked across at him and said to myself, 'If this old man can do this, so can I.' He did, he jumped right out, and so did I, right behind him." In mid-August, McNamara jumped into France. The unit landed in rough terrain and several men were injured, including the commander of the "Lindsey" Section, 1st Lieutenant P. Earle, who broke his leg in three places. McNamara and fellow radio operator, Robert ("Bob") Vernon from Idaho, established contact with London, taking turns with one of them turning the hand cranks to generate power for the transmitter and the other tapping out the Morse Code on the telegraph key. The 20-man section, accompanied by the *maquis*, quickly accomplished its primary mission, obtaining the surrender of a German of 120 troops around the hydroelectric plant at La Tuyere. Some of the Americans and members of the French Resistance then set up an ambush for a German truck convoy near St. Fleur, but the *maquis* had underestimated the number of German troops in the convey which actually numbered more than 500 soldiers. "They outnumbered us and started firing," McNamara recalled. "Most of the French ran away. Emile [Private Emile G. Roy] got up and started to run, but he got hit in the leg. I carried him over to the truck, we had hidden behind a hill, and we got away."[182] Later the American ambushed a smaller column and also captured a German unit before the American armies arrived.

Most of the OSS French OGs were sent into southern and central France as Company B from OSS Algiers, under the overall command of Lieutenant Colonel Alfred T. Cox.[183] Between June and September, 14 OG teams were infiltrated into southern France to support the 7th Army's invasion. Ten were French-speaking teams, three Italian-speaking and one mixed. Lieutenant Erasmus Kloman, a Princeton graduate who had trained at Areas F, B, and A among others, was given responsibility for logistics for the OG teams from Algeria and the selection of sites for their infiltration.[184] In their after-action reports, the OGs indicated that the training they had received prior to their actual operations had been more than adequate. Many remarked that what they faced in the field provide easier than the problems continually worked out during their practice period. However, they wished that there had been more stress on French military nomenclature, maintenance and repair of radios for radio operators, and the operation and maintenance of all types of foreign vehicles and weaponry. The OG teams found the American bazooka and the British Gammon grenade invaluable in ambushing enemy tanks and other vehicles. But the Marlin submachine gun was held to be unreliable, and the OGs believed that each section should be supplied with Browning Automatic Rifles as well as light machine guns.[185]

Major Cox personally led OG Section "Lehigh," whose mission in August was to coordinate all the Operational Group campaigns along the west bank of the Rhone River. He found that as the Allied success became more certain, the *maquis* overconfident and despite OG recommendations wanted to attack German Army units directly, in which case they were usually defeated, instead of operating in true guerrilla fashion—hit and run ambushes and raids and slashing at the enemy flanks. Or they wanted the glory of liberating French towns and cities rather than doing the necessary work of harassing and impeding German columns.[186] One of those it helped coordinate was Section "Louise" headed by Lieutenants Roy K. Rickerson and W. H. McKenzie III, who had been trained along with their dozen men at Areas F and B. Five days after being parachuted in on 18 July, Rickerson's section used nearly 200 pounds of plastic explosives and destroyed an important suspension bridge which collapsed into the Rhone River. That same day, McKenzie and his men blew up a railroad viaduct that collapsed onto a highway below, severing both lines; in another mission, they derailed a train carrying 16 tanks and 5 box cars filled with enemy supplies. On 29 July, with the *maquis*, the section ambushed a column of 400 Germans, killing or wounding near 100 and destroying a tank and six trucks. Later 37mm anti-tank guns were airdropped, and Rickerson and his men rolled them into the mountains and ambushed retreating German units below. On 25 August, after an attempted ambush of a convoy in a valley near Chomerac turned into a trap sprung by the Germans, the *maquis* fled, leaving the Americans to fight their way out. They did, but Rickerson received some superficial if still bloody wounds. At the end of August as the Allies moved northward, Rickerson, in an extraordinary bluff, told a German colonel in command at Chambonte that he and his men were surrounded by advance elements of the American and French armies. The bluff worked, and 3,800 troops surrendered to the handful of Americans. It was one of the largest single captures by an Operational Group in France.[187]

Section "Lafayette" was one of the OSS Italian OGs sent into southern France in August 1944. Lieutenants O.J. Fontaine and L.L. Rinadi commanded a dozen men, including Technician Fourth Class (T/4) Caesar J. Civitella, a young Philadelphian and paratrooper, who had joined the second group of OSS Italian OGs and trained at Areas F and B before being sent to Algeria for further training. Section Lafayette parachuted into the western Rhone River Valley on 29 August 1944 to harass the enemy and protect installations valuable to the advancing Allied armies. They quickly linked up with Lieutenants McKenzie and Rickerson of the Louise Section and were there when Rickerson bluffed the German commander into surrendering 3,800 men to the Americans. At the beginning of September, the combined units moved up the river, following the overall commander, Major Cox, arriving at Lyon to take part with the *maquis* in the celebration of the liberation of that city. Afterwards, the Italian OGs were transported by ship from Marseilles to Naples and then by truck to Tuscany, where they would participate with the Italian partisans in the Resistance in northern Italy over the winter of 1944-1945.[188]

OSS and the Allied Drive across France

Coordination with the *maquis* behind enemy lines had been quite successful. During 1944, some 523 OSS Special Operations agents and OG troops were infiltrated into occupied France. Of these 85 were SO agents and radio operators, 83 were Jedburghs, and 355 were members of Operational Groups. The OSS worked with the *maquis* in the largest uprising of resistance forces in history.[189] As the Resistance rose up against the Germans following the Allied invasion, the OSS/SOE operation dramatically increased the delivery of side arms, machine guns, bazookas, mortars, ammunition and explosives, medical supplies, food rations, shoes and uniforms. Before the invasion, these were supplied clandestinely at night. But between June and September 1944, responding to calls from radio operators with teams of Jeds, SO agents and OG sections, OSS sent in mass flights of hundreds of B-17 and B-24 bombers to air drop 5,000 tons of such supplies to the French Resistance.[190]

The result was of great assistance to the success of the Western Allies. The uprising of this civilian Army aided and directed by the OSS and the SOE impeded the Germans through controlled sabotage, ambush and irregular combat. By harassing lines of communication and supply, the Resistance diverted whole German divisions from the front. Within a week after D-Day, the Resistance had accomplished the destruction of 800 strategic targets ordered by General Eisenhower's headquarters. They cut all the field telephone lines and forced the Germans to rely mailing on messages sent by radio, which the Allies, could intercept and decode using "Ultra." The OSS and the *maquis* provided regular intelligence on German units, enabling, for example, the Allied to stem a projected German attack at Baccarat and to hamper the German retreat at Dienze. At least 5,000 *Wehrmacht* soldiers were redirected to try to stop major attacks by OSS-led Resistance units in Correze alone. In a major multiplier effect, with simply a handful of OSS and SOE agents, the Resistance liberated all of France south of the Loire and west of the Rhone, forcing the surrender of 20,000 German troops. Through the destruction of bridges, waterways, railroads, isolated garrisons and sometimes the actual engagement of troops, large sections of German military manpower, artillery, armored units and materiel were delayed or diverted from opposing the Allies armies. At Normandy, this gave the Allies added days and sometimes weeks before German reinforcements arrived. In one case, Resistance groups, directed and supplied by Jedburgh Team "Amonnia," under American Captain Benton McDonald ("Mac") Austin, a French executive officer, and American radioman Sergeant Jacob B. ("Jack") Berlin, were able to slow down, through blasted bridges and railroad lines, a crack German armored division, the 2nd SS Panzer ("Das Reich") Division, as it struggled to reach the invasion beachhead from southwestern France. Six weeks after D-Day the 2nd SS Panzer Division had advanced only one hundred miles and was still two hundred miles from Normandy. With the help of OSS or SOE directed *maquis* forces, several other German divisions were delayed for nearly a month.[191]

As the Allied armies drove across France, OSS Special Operations personnel were attached to the General Staff Section of the headquarters of each American Army and Army Group to coordinate with the OSS agents and the *maquis* on the flanks and in front of the advancing armies.[192] Effectiveness often depended upon the attitude of the regular Army Military Intelligence (G-2) officer—unsympathetic in the 1st Army, which eventually banned the OSS, supportive in Gen. George S. Patton's 3rd Army, and most supportive in the south of France from Lt. Gen. Alexander Patch's 7th Army. OSS provided 79 percent of the pre-invasion intelligence on the German Order of Battle for the 7th Army, including location and condition of German defenses, even camouflaged ones, with what General Patch praised afterwards as "extraordinary accuracy."[193] Such close-in work took its toll, and between August and October 1944, of the American and French agents operating in front of the U.S. 7th Army to provide tactical intelligence, 10 were killed, 15 wounded, and 39 captured, some of whom were tortured and later died.[194]

Among those captured was Lieutenant Jack Hemingway, son of the famous author. He had joined the OSS in North Africa and trained there under Lieutenant James Russell, Special Operations, who had himself trained at SO camps in Maryland and Virginia and been part of Serge Obolensky's four-man mission to Sardinia in 1943. In his first combat role, Hemingway, together with Russell, and two French radio operators, was parachuted into the south of France in mid-August 1944 assist local *maquis* and report on enemy defenses. Jack Hemingway was an avid fisherman like his father, and he took with him on the parachute drop not simply his pistol but a rod and reel and a fly fishing box.[195] Much of the mission proved disastrous—the two Frenchmen were seriously injured on the jump, the radio was broken, and later, a dozen, teenage members of the resistance were captured with the OSS explosives and tortured and killed by the Nazis. Hemingway narrowly avoided capture when a German patrol saw him fishing in a stream, but luckily mistook him for a solitary French peasant and passed him by.[196]

On a second mission, this time with an OSS Secret Intelligence team attached to the 3rd Division of the U.S. 7th Army, Hemingway found himself amidst the fluid battle lines in eastern France. One night in November 1944, he was helping to infiltrate an indigenous agent through and behind German lines. As one of his superiors, Peter M.F. Sichel, a member of the famous wine grower and merchant family, explained later, "Jack did not follow instructions and stayed with his agent much too long, thereby ending behind the German lines.... Jack was wounded in the leg and captured. We did not know if he had his little [OSS] notebook on him, and had to change all kinds of future missions, and agonized over him, not knowing if the Germans knew of his activities. Fortunately he had no compromising material on him and the Germans took him as a junior officer of the 7th Army."[197] Lieutenant Hemingway spent the rest of the European war in German POW camps. Afterwards, helped by a courageous war record, a *croix de guerre*, and what Jack called "a nice set of scars," the often stormy relationship between the son and the irascible novelist improved, especially as Jack remained in the Army after the war.[198]

Piercing the Third Reich

As the American Armies moved through eastern France to the German border, OSS Secret Intelligence (SI) sought to obtain effective tactical and strategic information useful to the western Allies. Within their own country, the Germans could rely upon secure telegraph and telephone lines instead of wireless radio transmissions that the Allies had intercepted and successfully decoded through "Ultra." OSS SI units with some of the American armies were able to obtain valuable tactical information by infiltrating German agents through the German Army lines. One OSS SI section that was particularly effective was that in Lieutenant General Alexander M. Patch's U.S. 7th Army, which had advanced north up the Rhone River Valley from its landings on the French Riviera in August 1944 and arriving at the Moselle River, on the south end of the Allied line, some fifty miles from the Rhine in December. Patch had benefited from use of French-speaking OSS SI and SO agents since the Riviera landings, but now as his army approached Germany, he needed German-speaking agents to learn what was going on behind the *Wehrmacht's* lines. So joining Henry Hyde, the head of SI with the 7th Army, and Peter Sichel, his chief assistant, were two ethnic Germans, Lieutenants Charles A. ("Carl") Muecke and Peter Viertel, recruited from the Marine Corps in the Pacific. The bright and talented son of working-class German parents who had immigrated to Queens in New York City, Muecke had graduated with honors from the College of William and Mary and jointed the Marines. Viertel had been born in Dresden but moved with his family in the 1930s to Hollywood, where his father was a movie director and Peter wrote screenplays, including Alfred Hitchcock's *Saboteur* in 1942. After the U.S. entered the war, he joined the Marines.[199]

For native-speaking German agents to send back through the German Army lines, they concluded that the best source would be recent German POWs. The use of POWs for such a purpose was a violation of the Geneva Convention as well as of regulations from General Eisenhower's office; nevertheless, General Patch after a meeting with Donovan, quietly side-stepped these prohibitions, and the four-man OSS SI section began recruiting anti-Nazi German POWs, and even a few German-speaking women, as agents. If they passed all interrogations, they were then taken to a secret training area for indoctrination, training, and the memorization of a cover story.[200] When needed, the agents were lead up to the front lines by SI members from the 7th Army or one of its divisions and then sent forward through the German lines at night. Normally, this went relatively smoothly, although as the Hemingway's case, there sometimes were problems either with the Americans or the Germans. So they also added short-range parachute drops in which the German agents were parachuted behind enemy lines at night disguised as German soldiers with passes showing they were on leave from their units and on their way to or from home. The 7th Army SI unit handled three dozen infiltrations into Germany in the winter of 1944-1945. Two-thirds returned with valuable intelligence within a week or two to a prearranged place.[201]

These German agents provided important tactical information. For example, Peter Sichel said, they determined where the German air force was launching its jet fighters, since Allied aircraft could not fight any nearby airbase for them. A German POW agent learned that the jets were hidden under trees near the autobahn, the superhighways and Munich, and they took off and landed on the autobahn near Augsburg and Munich. The Allies then bombed them under the trees and also destroyed the nearby autobahn. In another instance, a German POW agent located the long-range cannon, the "Big Bertha" that the Germans were using to shell Allied lines, but which aerial reconnaissance could not find. It was hidden on a railway car in a tunnel and only rolled out to fire. The Allies then bombed the tunnel shut. The agents also provided warnings of the German Army's build up for its surprise attack through the Ardennes Forest in December 1944. "We also obtained information about the buildup which led to the 'Battle of the Bulge,'" Sichel recalled. "We knew about this last build up of the Germans. We had sent people behind the lines to get information of what units were there, the [German Army's] order of battle."[202]

Deep penetration of the German police state proved, at least until the chaos of 1945, more difficult than tactical infiltrations. British and American services both viewed it as exceedingly difficult. The first American espionage unit to penetrate deep into Germany and Austria, a team that infiltrated through Yugoslavia in July 1944 had been wiped out within six weeks. In fact, all the twelve OSS teams sent into the Third Reich from the Mediterranean Theater had been captured or killed.[203] In the beginning of September 1944, the first OSS agent to be parachuted into Germany from England was dropped into the German homeland. Jupp Kappius was one of the German Jewish refugees who had fled to the United States from Hitler's genocidal regime. The OSS recruited him and in 1944 trained him and his wife, Anne, in OSS training camps in the United States and England to be infiltrated back into their homeland to create an underground resistance movement and commit acts of sabotage. In the industrial cities of the Ruhr area of Germany, Kappius began organizing among disaffected workers. To avoid using a radio to communicate with Jupp Kappius, his wife, disguised as a German Red Cross nurse, served as a courier, twice traveling between the Ruhr and neutral Switzerland with vital information. Another German woman refugee, Hilde Meisel, who completed her OSS training with Anne Kappius, and was code-named "Crocus," was sent to Vienna, where she set up an intelligence network. However, on her trip back through the mountains to Switzerland to report on her success, Meisel as spotted by an SS patrol as she approached the Austrian-Swiss border. She was almost to the border when a sharpshooter hit her, shattering her legs. She feel to the ground, and seeing the SS patrol running towards her, Hilda Meisel put a lethal, cyanide pill into her mouth, bit down on its shell, and died instantly.[204]

OSS casualties in France

Success by the OSS, SOE and the Resistance in France in 1944 did not come cheaply. It was paid for in blood, by hundreds of *maquisards* and by dozens of members of the OSS and the SOE and those who worked for them. The OSS casualty rate, particularly among the small teams was higher than normal combat statistics for frontline infantry units. OSS had 523 special operations personnel fighting behind enemy lines in France in 1944, all of whom had received at least preliminary training in the OSS camps in Maryland and Virginia. Of these 83 were Jedburgh officers or radiomen, 85 were SO officers and their radio operators, 355 were members of OG sections. Of the 523, a total of 86 were casualties. Of these 18 were killed; 17 were missing, or captured, although very few became prisoners of war; and 51 were wounded. This meant an overall casualty rate of about 17 percent, almost one out of every five or six in combat.[205]

Of the 83 American Jedburghs, 5 were killed in action, 6 percent, almost double the 3½ percent overall death rate; 3 were missing or captured; and 6 were wounded.[206] The injuries of some were quite severe. Captain Cecil F. ("Skip") Mynatt, Jr., the commander of Team Arthur parachuted into eastern France in August, fractured his spine in the landing. Captain Douglas ("Doug") DeWitt Bazata, head of Team Cedric in eastern France was badly wounded in action. Major Cyrus ("Cy") E. Manierre, Jr., headed team Dodge, which was dropped into southeastern France in late June. Manierre was later captured and brutally tortured, and although liberated in May 1945, he never fully recovered from those beatings by the Gestapo.[207]

Some of the American Jeds died in their parachute or weapons accidents, some were slain in combat, and some were executed. The parachute of Sergeant Lewis ("Lew") F. Goddard, a radio operator part of Team "Ivor" that dropped into central France, failed to open and he was killed instantly when he hit the ground. Lieutenant Lawrence E. ("Larry") Swank, a West Pointer, had joined OSS served as a demolitions instructor at Area B before going over to Britain to become a Jed. As the commander of Team "Ephredrine," Swank and two Frenchmen parachuted successfully into the Savoie department of southeastern France on the night of 12-13 August 1944. Subsequently in an accidental shooting by inexperienced *maquis*, Larry Swank was killed.[208] Sergeant Lucien J. Bourgoin, a radio operator, was killed in action with Team 'Ian" in central France. All three members of Jedburgh Team "Augustus" were captured and executed in August 1944. The commander of Team Augustus was Major John H. Bonsall, a Princeton University graduate from Morristown, New Jersey. He had become an artillery officer in 1941 and an OSS Special Operations officer in 1943, undergoing training at Areas F and B that year before being sent to England to become a Jed. In Britain, the 25-year-old Bonsall was teamed up with Sergeant Roger Côté, 21, a radio operator from Manchester, New Hampshire who had graduated from Area C, and Captain Jean Delviche of the French Army.[209] On 15 August, they were dropped about one hundred miles northeast of Paris, near Soissons in Picardy, behind the lines of the rapidly retreating German Army. A few members of the Resistance greeted them, but Bonsall soon radioed London that it was "impossible to form *maquis* now due to one, too many Boche [Germans]; two, lack of good hiding areas; three, very few arms."[210] Indeed, movement by these three Jeds was extremely risky because the entire area was still contested, with German units in rapid retreat, pursued by advance elements of Patton's Third Armored Division less then six miles away. The fast-moving American armored forces had already overrun the area originally assigned to Bonsall's team, and the Jeds found themselves behind American lines. Instead of considering their mission over, however, Team Augustus decided to re-cross the fluid lines, get back behind the fleeing German forces and organize the *maquis* there to attack German convoys.

On the stormy, rainy night of 30 August 1944, dressed as French civilians and with false identification papers, the three Jeds borrowed a horse-drawn peasant cart from a friendly farmer and with their weapons and radio under a load of hay, they started off on the back roads. They had traveled more than a dozen miles when, late that night, as they arrived at the village of Barenton-sure-Serre, three miles from their destination, the storm had grown so fierce and the night so dark, that the Jeds probably did not see the three German tanks posted at a check point at the rural intersection until they were at them. Around 11 p.m., some residents of the nearby village heard a dozen shots. Half an hour later, the tanks, elements of either the 9th SS Panzer Division or the 116th Panzer Division moved out, continuing the German retreat eastward. In the morning, the villagers found three bodies at the intersection, all with bullet holes in the back of the head. Bonsall and Deviche lay side by side; Côté, who had apparently made a break for it, lay face down, his arms outspread, a dozen yards away. The Germans had stopped the wagon, probably found the weapons and radio in the back, concluded the men were spies and decided to execute them summarily. In the morning, the villagers buried the three in unmarked graves beside the local church and put a French flag over them; the *maquis* arrived and held a military ceremony. Later that day, the U.S. 3rd Armored Division swept through the area on the heels of the retreating German forces.[211]

Among the 355 members of OSS Operational Groups in France, 10 members were killed, 4 missing or captured, and 40 were wounded.[212] One of those killed was First Lieutenant Paul A. Swank from Houston, Texas, who despite the same last name was no relation to Larry Swank who was killed in a shooting accident. Paul Swank was second in command of OG Section "Betsy." His unit had parachuted into the Department of the Aude between Toulouse and the Pyrenees Mountains on 11 August. Their mission was to block German reinforcements from interfering with the Allied invasion of the Riviera scheduled to begin four days later. Working with the *maquis*, they destroyed a number of bridges and railroads. On 17 August, the section attacked garrison of Germans guarding a huge warehouse of supplies for the German Army. When it was learned that the Germans were rushing up reinforcements from Couiza, Paul Swank and four enlisted men, along with eighteen *maquisards* sought to head them off by blowing up a bridge across the Aude River near Quillan. Approaching the bridge, Swank discovered that a column of 250 motorized German infantry was too close to the bridge to get the explosives attached to its columns. Instead, Swank quickly set them into a cliff alongside the road. As the German trucks approached, he set off the explosives, spilling down rocks and debris and forming an improvised road block substantial enough to slow but not prevent enemy passage. Leaping out of the trucks and deploying for action, the German infantry opened fire. Swank ordered his men to escape to the nearby hills, while he and a sergeant remained to cover their retreat. In the ensuing firefight, the sergeant was hit in the hand and foot, but managed to withdraw under covering fire from the other OGs and the *maquis* who had taken position behind rocks on the hill. Swank continued to hold off the Germans. Although hit eight times in the arms and chest, he kept firing as long as he could hold up his weapon. He was still conscious when the Germans reached him and an officer shot him in the head. The senior American sergeant, a regular Army NCO, said afterwards of Lieutenant Swank, "I've never served under a better, more considerate man. He had more guts than the rest of us put together. His loss to those that knew him is irreparable."[213]

Paul Swank's commanding officer wrote that "The German officers later remarked to inhabitants of a neighboring village that they had never seen a man fight as bravely or as long until killed."[214] Inspired by Swank's courage and self-sacrifice, the twenty remaining men, led by Technician Fifth Grade Nolan J. Frickey of New Orleans, Louisiana, laid down such a heavy fire, which killed or wounded nearly fifty Germans, that the enemy retreated, concluding that they faced a much larger force. This enabled the rest of OG Section Betsy and the *maquis* to capture the German warehouse, which contained enough food rations for one million troops for ten days. Technician Fifthy Class (T/5) Nolan Frickey received the Silver Star Medal, and Lieutenant Paul Swank was posthumously awarded the Distinguished Service Cross.[215]

OG Captain William F. Larson and at least three OG enlisted men were killed in action in central and southern France in August 1944. Captain Larson, a member of the "Percy Red" mission, was, as mentioned earlier, mortally wounded in action against a German armored train near Limoges. Technician 5th Grade Raymond Bisson of Rochester, New Hampshire, and Sergeant Camille A. Barnabe from Woonsocket, Rhode Island, had both participated in several successful bridge demolition and other actions behind enemy lines since their infiltration on 26 July. On the night of 3-4 August, Bisson, crawled 60 yards under heavy machine gun fire and destroyed a locomotive with his bazooka in the enemy-held Annonay Station near Ardeche. Five nights later, he helped placed demolitions that destroyed a main railroad bridge near St. Etienne, southwest of Lyons. His friend, Sergeant Barnabe was killed on 10 August, when the Germans bombed the village of Vanosc, a *maquis* stronghold and the headquarters for the OG section. Disregarding the rain of bombs, Bisson administered medical aid to many of the wounded. A little less than three weeks later, on 28 August, Bisson was in a car driving to rejoin the section preparatory to attacking an enemy column near St. Julien Moline-Molette, when the automobile was strafed by a fighter-plane, he was killed instantly. Sergeant Bernard F. Gautier from Union City, New Jersey, and Technician 5th Grade Robert D. Spaur from Georgetown, Kentucky, members of an OG section, parachuted into southern France on 7 August. Their section conducted successful raids, but five days after their arrival, when the section ambushed a German motorcycle patrol leading a troop convoy near Rialet, d. Spaur was killed instantly in the firefight. After the initial firing, Gautier ran out onto the road to prevent the Germans from setting up a machine gun. He was shot in the back and killed by a wounded German. The surviving Americans and the *maquisards* subsequently killed all the Germans. Both T/5 Raymond Bisson and Sergeant Bernard F. Gautier were posthumously awarded the Silver Star Medal for bravery in action.[216]

A Hero and a Tragedy: Chris Rumburg's Death

"Show me a hero, and I'll show you a tragedy," F. Scott Fitzgerald declared, and the heroic death of former Washington State University football star and OSS Area A instructor, Ira Christopher ("Chris") Rumburg, certainly confirmed that epigram. Rumburg was literally larger than life. When he arrived at Washington State University in 1934 from Spokane, he was 6-feet, 3 inches tall and weighed 190 pounds. A natural athlete and leader, he worked his way up to be in his senior year, captain of the football team, champion heavyweight wrestler, and President of the student body. In football, he not only showed his leadership but his toughness. Throughout the Cougars' season, he was plagued with injuries, including a bruised back, but he kept playing until finally a battered leg aggravated an old shinbone injury put him out of action for the final two games of the season against Stanford and Oregon. Rumburg was also cadet commander of university's ROTC battalion, and he received a lieutenant's commission in the Army reserves upon graduation. He served first at Hunter Liggett Military Reservation in California, and he served as an infantry instructor.[217]

Called to active duty in World War II, Chris Rumburg, like several other former WSU athletes and reserve officers, Art Dow, Joe Collart, "Pinky" Harris and Jerry Sage, was invited to become an OSS Special Operations instructor by Ainsworth Blogg, the first commanding officer at Area B. In the spring of 1942, Rumburg had driven across country from Seattle to Washington, D.C. with newlyweds Art and Dodie Dow. He had served as an instructor at Area B and then Area A in 1942-1943.[218] By 1944, he was in England and a lieutenant colonel assigned to the headquarters of the 264th Regiment of the 66th Infantry Division, possibly as an OSS adviser. In mid-December 1944, the Germans in a surprise counterattack broke though the American lines in the Ardennes Forest in Belgium and France. Reinforcements were rushed in. The 66th ("Black Panther") Division was dispatched to relieve the 94th

Division and contain large numbers of enemy troops in in the Brittany-Loire area. On Christmas Eve, Rumburg and 2,200 men from the division boarded a transport ship that was to take them from Southampton, England to Cherbourg, France for the trip to the front. The transport was the *SS Leopoldville*.[219]

An old Belgian passenger liner refitted as a transport, the *Leopoldville* was jammed with American troops that wintry evening as she chugged across the channel accompanied by another troopship and four escort destroyers in a diamond formation. Many of the men on board, most of them young soldiers, 18 to 21, were singing Christmas songs. They were within six miles of Cherbourg, when shortly before 6:00 p.m., in the early dark, there was an enormous explosion and the entire ship shook. A German submarine had snuck through the escort screen and slammed a torpedo into the ship's hull. An estimated one hundred of the troops were killed instantly. But that was just the beginning of the tragedy. Steel beams snapped, tables and equipment had been thrown into the air and crashed down upon helpless men. Water gushed into the hull and soldiers struggled to get out of the rapidly filling lower compartments. Many of the interior metal stairways from them had been twisted into a mass of steel and splintered wood. Only two steel ladders remained for evacuating survivors from the large troop compartments below. A destroyer stayed to help, but the rest of the convoy steamed to Cherbourg to escape the submarine. The destroyer evacuated most of the stretcher cases. A British cruiser came alongside and urged the men packed along the railings to jump across. But in the choppy sea, some fell between and were crushed. The cruiser pulled away. The soldiers waited for aid since tugboats were due from Cherbourg, but around 7:30 p.m., the *Leopoldville* suddenly lurched heavily to starboard and swiftly sank, stern first. Then the soldiers began leaping into choppy waters of the English Channel, where the swells were up to twelve feet high. The ship went down quickly. Tugs, Coast Guard cutters and PT boats began to arrive from Cherbourg, searching for survivors in the dark and frigid waters. About 500 soldiers are believed to have gone down with the ship, another 250 are believed to have died from injuries, drowning, or hypothermia. The ship's captain and four of his crew went down with the ship, but most of the Belgian crew had fled, taking to the lifeboats around 7:30 p.m. without indicating that the ship was sinking or lowering the remaining floater nets, rafts, and lifeboats for the American soldiers. The ship then went down quickly. In all 763 American soldiers perished, including three sets of brothers. It was the worst transport sinking experienced by an American infantry division.[220]

Lieutenant Colonel Rumburg had been in the officers' quarters on the upper decks when the torpedo had exploded below the water line. Rushing down to assess the situation, he helped lead dazed and wounded men through the debris to the main deck. Several times with his enormous strength, he carried men on each of his broad shoulders. Later, as he and other officers peered down into a jagged, gaping hole in the metal floor down into the "E" deck, they heard a voice calling for help. They flashed down lights into the swirling water four to eight feet below, but could not see the source of the calls. Rumburg shed his coat, slid into the water below and swam to the sound of the voice. He found the soldier trapped in debris, and while Rumburg was pulling it aside, a timber fell and crush his hand, severing two fingers.

Nevertheless, he was finally able to drag the debris aside and pull the sputtering soldier to a position below the hole. Before the men above could get ropes down to them to pull them up, a sudden surge of incoming water swept them upwards and smashed their heads against a bulkhead. Rumburg was dazed, but the soldier was knocked unconscious and out of the colonel's arms. Rumburg tried again to find him, but after his earlier exertions, more than 15 minutes in the freezing water and the loss of blood from his wounds, the colonel was too numb and exhausted to sustain himself let alone resume the hunt for the missing soldier. They got a rope around his waist and after half a dozen attempts finally got the 200-pound colonel back up through the hole. He was carried up to the infirmary. Later, when it became evident that the ship was going down, he climbed out on deck.[221]

A letter from a friend who joined the 66th Division shortly afterwards and heard about his heroism from men who had survived described Rumburg's final moments: "One of his men was about to enter the water without his 'Mae West' [life preserver]. So Chris took his off and gave it to him. Chris then jumped in the water and swam around getting his men to the rafts and seeing that they stayed calm. He found one fellow that was having trouble getting to the raft. So he helped him to the raft and helped shove him on. Then after using up all the great strength that God gave him, his hand slipped from the side of the raft and sank from sight. I have heard about a lot of acts of courage during this war... but none greater than this."[222] Chris Rumburg is credited with saving the lives of more than one hundred of his fellow soldiers. His body, like those of 500 others, was never found, and in the end, he was one of the nearly 800 American soldiers who perished on the Leopoldville, a tragedy the details of which were kept from the public for more than fifty years.[223] Chris Rumburg was posthumously awarded the Bronze Star and Purple Heart, and celebrated by Washington State University with a memorial fund created in 1974 and in 1997 at Fort Benning, Georgia, by veterans of the 66th Infantry Division in their Leopoldville Disaster Monument. In death, his heroism was commended by the commander of the 66th Division, Major General Herman F. Kramer, who wrote that "the memory of his deeds will remain long in the minds of scores of men he succeeded in saving from a similar fate."[224]

Blowing Up Bridges in Greece

Greece was occupied by German troops in the spring of 1941, the government, including the king fled. Almost immediately a Resistance movement emerged to fight the occupiers. The country was plunged into abject misery, including a shortage of food as Germany reaped the foodstuffs, and the Resistance grew despite severe reprisals. Even though the majority of the 500,000 to 2,000,000 members of the Resistance were leftists rather than communists, the Communist party dominated the National Liberation Front (EAM/ELAS) and the Andarte as the partisans were called. A smaller Resistance group, the National Republican Greek League (EDES), was a more conservative, royalist organization. By 1943 sporadic civil war broke out between the communists and the royalists, as they fought each other as well as the

Germans. In November 1942, a British SOE team parachuted into central Greece and with the help of the Resistance blew up a viaduct which carried the main north-south railway line between the key cities of Athens and Salonika. This was the most spectacularly successful sabotage operation in German-occupied Europe since the beginning of the war, and it demonstrated the potential for future special operations missions in Greece and elsewhere. Although SOE had senior responsibility in the Balkans, OSS began its operations there in 1943. The main OSS objectives of working with the Resistance in Greece were to hamper the shipment of vital supplies to Germany, to impede the effectiveness of the Germans in Greece, and to keep the numerous German divisions of occupation troops there from being transferred to fight against Allied armies in Italy or in France.[225]

OSS was, from the beginning, willing to work with the both factions of the Resistance, whichever one had control in their particular area. After a few Secret Intelligence operations, the first major OSS mission into Greece was the "Chicago" or "Evros" Mission under Special Operations officer Captain James G.L. ("Jim") Kellis, an extraordinary man in an organization of unique individuals. Kellis had long had a vision of a role in that area of the world. A Greek American from Yorkville, Ohio, he had left his hometown after graduating from high school, traveled widely in the eastern Mediterranean and studied at the renowned St. Athanasse International College in Egypt. With the outbreak of World War II in 1939, Kellis joined the Army Air Corps and was commissioned a second lieutenant. He was working on routine assignments at an air base in Florida, when the United States entered the war in December 1941. Kellis, then 26, became obsessed with the idea of Americans helping to liberate occupied Greece. Soon, he completed a detailed plan for American special operations there and submitted it to Army Intelligence in the War Department. In the fall, he was recruited by the Special Operations Branch of OSS. He spent the winter of 1942-1943 at SO training camps at Areas F, B, and A. By May 1943, he was at the OSS eastern Mediterranean headquarters in Cairo, and not long afterwards he completed a detailed proposal for blocking railroad shipments of large amounts of chrome ore that neutral Turkey was shipping to Nazi Germany. His plan was to blow up two key railroad bridges in northeastern Greece. It would not be an easy task as there had been no previous SOE or OSS infiltration in that area and the region was heavily guarded by troops from the nine German and eleven Bulgarian divisions in Greece.[226]

Because of the lack of knowledge about partisan groups in the area, OSS ruled out infiltration by air or sea. Instead, an advance team of Kellis and two other men would try to penetrate the German-guarded, Greek-Turkish border on foot. Kellis was accompanied by two other Greek Americans, already in the SO Branch, and both from the Navy. Seaman Spiro "Gus" Cappony, 20, from Gary, Indiana, was the team's radio operator. Cappony had trained at OSS Area C and then with Marine Captain John Hamilton, (actor Sterling Hayden) at Area B. But after that training, their proposed joint mission was cancelled; Cappony had been sent to Cairo and Hayden was sent to Italy where he skippered gun-running ships to partisans in Yugoslavia, and Cappony was sent to Cairo. After assignment to Kellis's mission, he took parachute training. The other member of Kellis's team was Petty Officer

Michael Angelos of Chicago, whose specialty was demolitions. Angelos had received his OSS demolitions training at Areas A and B. Typical of OSS informality, Cappony and Angelos referred to their commanding officer as "Captain Jim."[227]

Moving stealthily at night and dodging German patrols, the three men—Kellis, Cappony, and Angelos—sneaked across the border into Greece in early 1944. They were dressed in civilian clothes and carried bags containing their American uniforms, a radio, some weapons and money in local currency as well as gold coins. All three could speak Greek fluently, and a friendly fisherman put them in contact with the local leader of the *Andarte* partisans. He was suspicious of these newcomers, but when he realized that the Americans could supply his men with arms and ammunitions, he agreed to help Kellis in his mission. Cappony radioed the news to the OSS office in Cairo and also made contact with the OSS base station in Bari, Italy. But the team had already been betrayed. An informer had notified the Germans and patrols approached the village. When the partisans identified the informer, Cappony later recalled, they "dragged him out naked, made him march up the hill, dig his own grave, and then shot him in the head."[228] The American team withdrew from the village and moved to the top of the highest mountain in the area. There they lived in caves or old Greek Army tents and subsisted on bean soup, black bread, and an occasional piece of cheese. "I went from 180 to 118 pounds," Cappony said. "The menu wasn't too good."[229] Relying on women and girls, whom the Germans permitted more freedom of movement than men, Kellis established an intelligence and courier network. Since he found the partisans largely unarmed and untrained, he recruited former officers and noncoms from the Greek Army and set up a guerrilla training school.

Arriving at local villages wearing their American uniforms with a U.S. flag patch on the shoulder, Kellis, Cappony, and Angelos would receive an enthusiastic welcome. The villagers were excited to see the Americans, especially Greek Americans who could speak the native language fluently. But in their sojourns as they moved closer to their targets, the Americans and the accompanying partisans often had to fight they way out of surprise attacks by German patrols. Once, they escaped only by leaping off a cliff into the trees and brush below.[230] Having established a working relationship with the partisans, Kellis summoned the remaining section of the mission. They came ashore at night in March 1944, delivered by the OSS Maritime Unit in a *caïque*, a small, schooner-like vessel the Greeks used for trading all across the Aegean Sea. The new team members were Navy Lieutenant John ("Johnny") Athens from Tulsa, Oklahoma; Gunnery Sergeant Thomas L. Curtis, a tough old Marine from Boston; and Chief Petty Officer George Psoinos from Lowell, Massachusetts. Unlike most of the OSS recruits, Curtis was a career military man. He had joined the Marines in 1935 and by 1942, the tough, brawny sergeant was an instructor at the Marine base at Quantico teaching men who would become part of the Marines' 1st Raider Battalion. Later that year, he was one of the first Marines transferred to the OSS. His first assignment for OSS Special Operations Branch was to train its recruits at Area A in paramilitary combat and at Area D in amphibious warfare. In September 1943, "Gunny" Curtis was sent to Cairo to be part of the reinforcements for Kellis's SO mission the following year.[231] The reinforcements brought ten tons of weapons, ammunition, demolitions, food, clothing, medicine, another supplies. These were hauled from the seacoast in horse-drawn wagons to the team's mountaintop headquarters. With such equipment, the partisan force grew to more than 300 men, and Kellis began a rigorous training program geared to his sabotage mission. One group of partisans was trained by Gunny Curtis and others in the use of weapons. The other was trained by Angelos and others in the use of the new plastic explosives. The two groups were kept apart from each other, and neither was told of their objective.

Their objective was the destruction of the two main railroad bridges over the Evros River, one crossing into Bulgaria, the other heading west across Greece toward Yugoslavia. To distract the Germans, Kellis also decided on a diversion attack on a small bridge that was near the German garrisons but fifty miles from the real targets. In addition, deceptive rumors were circulated to the local Gestapo, who knew that there was an American team in the area, but could neither find them nor ascertain their objective. On 28 May 1944, the diversionary team, led by Psoinos and Cappony, and the two teams attacking the main bridges arrived at their target areas and saw the bridges for the first time. "We had no difficulty avoiding the German guards," Kellis reported, "but my first sight of the [Svilengrad, Bulgaria] bridge was a shock. It looked too big and too substantial for us to destroy, even with twelve hundred pounds of plastic [explosives]. The structure was built of steel and reinforced concrete. It had four massive concrete piers, each of which was eight by ten feet. The overall length was 240 feet, and a single railroad track rested on a heavy steel frame which was solidly set in concrete arches." Kellis and Athens realized that the explosives would have to be "very carefully placed and thoroughly tamped to utilize the full force of the explosion."[232] Around midnight on 29 May, Kellis's men eliminated the sentries and carefully and quietly planted the explosives. They were making the connections with the prima cord on the center of the bridge, when something awakened the Germans in their barracks. "They fired a flare and opened up with a machine gun and sub-machine guns in the general direction of the bridge," Kellis wrote.[233] But the Germans were too late, the saboteurs set the fuses, ran from the bridge, and a few moments later the entire structure was shattered in a series of enormous explosions.

Meanwhile, the other main mission, headed by Gunny Curtis and Angelos, had launched their nighttime assault on the Alexandroupolis railroad bridge in Greece, a 100-foot long bridge supported by a single, eighty-foot high pier. They evaded a roving German patrol and when they arrived at the bridge itself, they found that the thirty guards were Greek policemen. Gunny Curtis held a submachine gun to the stomach of one of the Greek policeman, announced in fractured Greek: "I'm going to blow this bridge and I'll do the same to you if necessary."[234] He then gave the frightened policeman the option of helping him. The man agreed and so did some of the others. As the Americans and some of the Greeks began to put 500 pounds of explosives in place, a German patrol arrived and started shooting. The partisans accompanying the Americans were able to hold them off, while the explosives were placed. Then Curtis shortened the time delay from nine to three minutes and sprinted away as fast as he could. Three minutes later a blast rocked the ravine, sent the bridge framework into the air, and shot a column of flame and smoke that could be seen for twenty miles. Fifty miles away, Cappony and Psoinos and their group of partisans had created a diversion at a bridge not far from the German garrison. With their missions accomplished, all three teams fled into the night and successfully evaded capture by the Germans.[235]

Kellis's mission was a complete success. The destruction of the two main bridges severely restricted the flow of chrome ore so important to the German war effort. The team had completely destroyed the bridge at Alexandroupolis. It took the Germans nearly five weeks to replace it. The Svilengrad Bridge was replaced with a temporary structure in three and a half weeks, but that was later washed out by the river's spring floods and never again carried any appreciable amount of freight shipments, and the supply of Turkish chrome to Germany was dramatically reduced.[236]

For the rest of 1944, the main OSS activity in Greece was the joint SOE/OSS Operation "Smashem" aimed at hampering the withdrawal of 80,000 German troops to be used against the major thrust of the western Allies following the Normandy and Riviera landings in France.[237] The main OSS effort in conjunction with the partisans was the work of eight Greek-American Operations Groups, less than 200 men in all. The Greek government in exile had requested them. The OG teams were recruited from the U.S. Army's 122nd Infantry Battalion, itself a special unit known as the "Greek Battalion," half of whose members were Greek Americans and half were Greek nationals in the United States who wanted to fight the Germans.[238] Like the other OGs, the Greek Americans and Greek Nationals trained at Areas F, B, and A before going overseas for additional training in North Africa. When they left the United States in December 1943, they were in high spirits, dressed smartly in the new trim, "Eisenhower jackets" and paratrooper jump boots, and singing in both English and Greek. Communications officer Theodore Russell said "We looked good, acted good, and the biggest thing, we felt good. Officers from other outfits would ask me, 'Who are you guys?' Security told us to say that we [were] truck drivers; they knew that wasn't the case."[239]

Beginning in April 1944, they infiltrated into Greece by parachute drop or fishing boat, and connected with partisan groups at strategic points. Led by officers such as Captain George Verghis and Lieutenants Nicholas G. Pappas and John Giannaris, they severed rail lines, blew up bridges, planted mines and fired bazookas and mortars at trains and truck convoys.[240] Two enlisted men received Silver Star Medals in the guerrilla campaign. Despite being hit by a rifle bullet in the leg and then by grenade fragments in the head, Corporal Spero Psarakis from Brooklyn, New York, had continued to attack a German Army billet guarding the Athens-Salonika Railroad. With his submachine gun, he killed all eight Germans in the command center, which enabled the rest of his unit to destroy seven bridges along the line.[241] The other was T/5 Gus L. Palans, 23, from Burlington, Vermont, who like the others trained at Areas F and B in 1943; despite heavy machine gun fire, he helped prepare and blow a charge that destroyed a vital railroad line.[242] Between September and December 1944, the OGs, supported by local guerrillas and supplied by air from Cairo, mined roads, ambushed more than a dozen trains and dozens of truck convoys, destroyed 15 bridges, blew up six miles of railroad track, and killed nearly 400 enemy soldiers.[243]

The Germans sought to capture, torture and kill partisans and the Allied teams, and also took reprisals against civilians. Among the 150 men in the Greek-American OGs, one was killed, and two officers and nine enlisted men were wounded.[244] As Greek OG Group II, attacked a pillbox and outpost guarding the Salonika-Athens railroad line north of Lamia, Sergeant Michaelis Tsirmulas was mortally wounded by German machine gun fire. His commander, Lieutenant Giannaris, ran to help him and was nearly killed when he stepped on a mine. Subsequently, Giannaris spent the next two years in Army hospitals. Almost sixty years later, the Defense Department awarded all the members of Operations Group II Bronze Star Medals.[245] SO units also continued active during 1944. After blowing up the Evros River bridges, James Kellis and his SO team attacked German transportation facilities throughout northeastern Greece. Kellis was wounded and so was Seaman Spiro Cappony, his radio operator, Cappony had been shot in the arm, and back at camp, the wound had begun to fester. Since the team at that time had no antiseptics and to prevent infection and gangrene, a pot of boiling olive oil was poured over the wound to cauterize it. Cappony fainted from the pain and still carries the scars. He recovered, but for some time he had to tap out his telegraph messages with his left hand. Kellis was decorated and Cappony was awarded a Bronze Star.[246] By December 1944, the bulk of the German troops had surrendered or left, and the OSS OG and SO teams were withdrawn, having provided much secret intelligence, effectively supplied guerrillas, and performed a major contribution in delaying the removal of German troops to the crucial theaters of the war.

Amidst Warring Factions in Yugoslavia

In Yugoslavia, the goals of the OSS were to help the resistance forces to sabotage railroad lines carrying supplies into Germany, to tie down tens of thousands of German occupation troops and prevent them from being used on the front lines against the Allied armies. In addition, the OSS in Yugoslavia rescued aviators who had been shot down on the bombing runs from Italy to the Ploesti oilfields in Rumania, and it also sought to deceive Hitler into thinking that the American invasion might occur in the Balkans instead of in Normandy. The Germans had invaded Yugoslavia in the spring of 1941 and occupied the most populated areas of the country, but a substantial Resistance movement quickly emerged and took refuge in the sparsely settled mountains. The movement split between the Partisans under Communist Party leader Josip Broz ("Tito"), a Croatian, with their core area in the mountains and forests of northern Bosnia, and the Royalist Chetniks under Yugoslavian General Draža Mihailović, a Serb, who operated out of their base in the wooded mountains of Montenegro in the south.[247] Bitter enemies, they fought each other as much as the Germans. Among the undercover operations by the Western Allies, Britain initially had a monopoly in the Balkans, but in the late summer of 1943, Donovan was able, despite SOE objections, to begin sending OSS operatives into Yugoslavia to establish connections and sources of information independent of the British. In the third week in August, two OSS Special Operations officers were airdropped, one into the headquarters of each of the two hostile factions. Army Captain Melvin O. ("Benny") Benson was infiltrated into Tito's headquarters; and Marine Captain Walter R. Mansfield was air dropped into Mihailović's camp.

Parachuting alone into the area near Mihailović's base just before midnight on 19 August 1944, Mansfield was a highly regarded SO officer. A Boston native, Harvard graduate and a former member of Donovan's law firm, the 32-year-old Mansfield had joined the OSS as a civilian. But he had attended Marine Reserve Officer class and also learned demolitions and guerrilla warfare at OSS SO training areas in Maryland, Virginia, and England.[248] Accompanying him were 15 canisters filled with small arms, radios, and three tons of ammunition. A few minutes after he landed amidst the bonfires of the drop zone, he was surrounded by a small group of ragged-looking men with black beards. "I told...their leader that I was an American," Mansfield recalled, "whereupon they all began to shoop, holler, and kiss me (black beards and all) shouting 'Zdravo, Purvi Americanec' (Greetings, first American). I mustered up my Serbian to reply, 'Zdravo Chetnici'—the first American had landed."[249] Mansfield was later joined by Lieutenant Colonel Albert B. Seitz and Captain George Musulin, an American of Serbian ancestry. All three were much impressed by the Chetniks.[250]

Allied action in Yugoslavia remains controversial. Leftists among the British SOE mission attached to Tito emphasized the superiority of his forces, overstating the communist partisans' numbers and accomplishments, while denigrating the Chetniks.[251] Although London cut off supplies to Mihailović, the OSS argued that both Yugoslavian factions were effective and should be aided in their separate areas of control—Tito in the north and west, Mihailović in the east and south. Captain Mansfield wrote strong endorsements of the Chetnik leader. Yugoslavs loyal to the monarchy and Mihailović had been among the foreign groups trained at OSS Area B. But at the Tehran conference in November 1943, Stalin and Churchill backed Tito and insisted that Roosevelt cut off all support for Mihailović.[252] Despite Donovan's protests, the American OSS mission to Mihailović were forced to leave the Chetniks in the early months of 1944.[253]

Tito and his Partisans had their admirers in the OSS. Captain Benson was the first, followed by Lieutenant Colonel Richard ("Bob") Weil, Jr., 27, a former President of Bamberger department stores, a division of R.H. Macy and Company, who accompanied one of the OSS mission's to Tito.[254] In November 1943, Lieutenant George Wuchinich, a second-generation American from Pittsburgh whose parents had been Orthodox Serbs from Slovenia, and who had trained in 1942 at Areas B, A, C, D and RTU-11, led the "Alum" Team that was parachuted into Partisan-held territory in the mountains near Ljubljana in Slovenia in November 1943, the first OSS team to arrive in northern Yugoslavia (Tito's headquarters was farther south).[255] Wuchinich was accompanied by a Greek-American radio operator, Sergeant Sfikes, and four other enlisted men. He found the Partisans suspicious of both the British and the Americans. But when Wuchinich was finally allowed to meet the local general and accompany the Partisans into battle against the Germans, he became glowing in his reports. Indeed, he compared them to the dedicated, long-suffering Continentals in the American Revolutionary War.[256] Finally, in June 1944, Wuchinich gained enough trust to be allowed to pursue his assigned mission—to secure daily reports to OSS on the main Balkan railroad system which ran through Maribor at the Slovenian-Austrian border before dividing into separate main lines to Italy and Greece. Trekking through the mountains, they established an observation post overlooking Maribor and then returned to camp. From 30 June through 4 August, the observation post sent as much detailed information about troops and supplies going through the throat of the southeastern European rail network as the Allies could desire. The Germans finally located it, killed the radio operator and seized his equipment, but the Allies had gotten the information during period immediately following the Normandy invasion, which is when it was most needed.[257] Wuchinich's team also gained valuable information from a deserter about the development and proving ground for the new "flying bomb," the V-1 "buzz bomb," rocket the Germans began to launch against England in mid-1944; and they helped rescue more than a hundred downed Allied aviators. Wuchinich's reinforced team did suffer casualties, however; at least two of the Americans were killed.[258]

Activity by the OSS increased dramatically in Yugoslavia in 1944, especially support for Tito and his Partisans. The number of OSSers attached to the Partisans grew from six in late 1943 to 40 men in 15 different missions in 1944. Major Frank Lindsay's SO team destroyed a stone viaduct carrying the main railroad line between Germany, Austria and Italy, impeding German reinforcements and supplies. From January to August 1994, Donovan's organization sent detachments of Yugoslavian-American Operational Groups, together with some Greek-American and other OG sections, all of them trained at Areas F, B, and A, to accompany British commandos on a series of raids on German garrisons along the Dalmatian coast of Croatia. There was a dual purpose in this campaign. One was to draw off German troops who were being used in a major offensive designed to crush Tito's Partisans. The other purpose was to deceive Hitler into thinking that the main invasion by the Western Allies might come in the Balkans instead of France.[259] Corporal Otto N. Feher, from Cleveland, the son of Hungarian immigrants, was a member of the Operational Group team that helped raid and defeat the German garrisons on the sizable islands of Solta and Brac between Dubrovnik and Split. "They told us from the start, there's no prisoners. You get caught, you're dead," Fehr said. He also reported that nearly one quarter of his 109-member contingent (perhaps the contingent he originally trained with) were casualties during the war.[260] The raids, together with the aerial attacks on German forces by Allied aircraft, assisted Tito in narrowly escaping capture. The OSS also kept supply lines open from Bari by which to sustain the Yugoslav Resistance.

Sterling Hayden in Tito's Partisans in Yugoslavia

OSS delivered agents and supplies to Yugoslavia in two ways—by air or by sea. By air, it was initially by parachute drops, although increasingly rough airstrips were built, first for small planes and then for two-engine transports. The majority of supplies came in by boat, despite the Germans' control of the coastline and patrol of the seacoast. OSS Maritime Unit was in charge of the seaborne ferrying, but it was actually performed by many men from Special Operations. The trip was 150 due north from Bari to the island of Vis, the OSS/SOE base nearly 50 miles from the Yugoslavian coast and was made in fishing boats and small schooners in an overnight run from Bari. One of the main skippers was 27-year-old actor and seaman Sterling Hayden (Captain John Hamilton, USMCR) of OSS Special Operations. After joining the Marines as a private in 1942, Hayden went to OCS and received an officer's commission (he also legally changed his name to John Hamilton to avoid publicity), was transferred to OSS Special Operations and underwent paramilitary training at Area B in the summer of 1943.[261] The initial plan was for Hayden and radioman Spiro Cappony, who underwent SO training with him at Area B, to go to Greece, but although Cappony wound up in Greece, Hayden was sent to Italy to skipper boats to Yugoslavia. Hayden was an experienced seaman who had spent nearly a dozen years sailing out of New England on various vessels from schooners, to fishing boats, to freighters. Now Hayden, along with Captain Melvin ("Benny") Benson, Lieutenant Robert Thompson, and Danish-born Captain Hans Tofte, co-instructor with William ("Dan") Fairbairn at Areas F, B and A, ran Operation "Audrey," whose mission was to ferry supplies to anti-fascist partisans and retrieve rescued Allied airman from Yugoslavia as well as western Greece, and Albania.[262] With his clandestine cargo, Hayden sailed his old 50-foot sailboat with a diesel engine at night and though storm-tossed waters, evading high-speed German E-boats, Stuka dive bombers, and other hazards, and making ten successful trips back and forth across the Adriatic, taking supplies and often returning with downed airmen who had been rescued by the partisans and the OSS.[263]

Intrigued by Tito's partisans, Hayden accompanied a band of them in the summer of 1944 deep into Croatia and in one episode within 20 miles of the border with Hungary joined them in the summer of 1944 and decided in the summer of 1944 to see some action inland and himself had been infiltrated deep into Croatia in the summer of 1944 to see them in action and in one mission, as far as twenty miles from the Hungarian border, nearly 200 miles inland, to rescue downed Allied aviators. He was impressed by their dedication to their mission for Yugoslavia and their support for the American OSS men, who they contrasted with British "imperialists." "It is impossible," Hayden later told a group of OSS trainees, "to work with the Partisans and not be completely moved by their determination and sacrifice."[264]

Some people in OSS believed that Hayen, Benson, Thompson and Tofte, the Operation Audrey supply team, had become too close to Tito's communist partisans and lost their objectivity. Most of the team was removed from the assignment in early 1944, and Tofte himself was sent back to the United States, officially for insubordination. Donovan personally liked Tofte and thought his operation had been quite successful. So in August 1944, the OSS director had Tofte transferred from SO to SI and sent to London where he became second in command of intelligence procurement.[265] Sterling Hayden returned from the interior of Yugoslavia in July 1944, and resumed the sailing trips to the Dalmatian islands, in one of which he was ambushed on Korcula and almost captured. In the fall, he came down with jaundice and nervous exhaustion and was shipped back to the states in November 1944. After a month's rest, was sent to the OSS unit with the U.S. 1st Army as it pushed forward from eastern France into Germany. For his gallantry in the face of the enemy, Hayden was awarded the Silver Star Medal.[266]

The OSS effort in Yugoslavia was a success to the extent that its support of the Resistance did help keep many German divisions there and not at the main Allied fronts and it also helped rescue thousands of downed Allied aviators and aircrews. But given the political decisions made by the Big Three, Churchill, Stalin, and Roosevelt, the proportion of support went increasingly and overwhelmingly to Tito's Partisans instead of Michailović's Chetniks. In the summer of 1944, Tito's Partisans were again on the attack—against the weakened Chetniks as well as against the Germans. OSS re-established its contact with and support of Mihailović that summer, primarily through a new unit created to help rescue downed airmen. By the end of the war, some 2,000 downed airmen had been rescued and evacuated via Chetnik or Partisan controlled areas of Yugoslavia.[267] The majority of these airmen were Americans shot down during U.S. 15th Air Force's bombing raids from Italy against the heavily defended Axis oilfields and refineries in Ploesti, Romania. Most were crews of B-24 "Liberator" bombers, but some were pilots of their fighter escorts. OG member Otto Feher remembered the Resistance bringing in a Tuskegee Airman, the first black pilot he had ever seen, who had eluded capture by the Germans for several weeks.[268] Another 1,000 airmen had been rescued by OSS SO in the rest of the Mediterranean Theater, a total of 3,000 skilled Allied airmen rescued to fly again.[269] Allied support of the wartime guerrilla operations, first of the Chetniks and then of the Partisans, had included the equipping of tens of thousands of guerrillas. They had held down 35 Axis divisions, including 15 German Army divisions that might otherwise have been deployed in Italy, France, or the Eastern Front.[270] But Allied favoritism towards the Partisans and especially the Red Army's direct assistance in the fall of 1944 helped Tito create a communist state in postwar Yugoslavia.[271] Similarly, although a small OSS mission worked with the rival communist and non-communist resistance movements in tiny, neighboring Albania, primarily to rescue survivors from downed American planes, it was the communists who came to dominant the country in the postwar era.[272]

Disaster and Death in Czechoslovakia

Unfulfilled hopes, unforeseen setbacks and faulty planning led to the worst disaster in the history of the OSS. The place was central Slovakia. The time was the fall and winter of 1944-1945. The impetus was an uprising of several thousand partisans and the revolt of two divisions of the Slovak Home Army against the Nazi collaborationist regime as the Anglo-American armies pushed eastward toward Germany and the Red Army pushed the Wehrmacht westward towards the Czechoslovakian border. The Czech government in exile in London under President Edvard Beneš flew in a regular Army general to take charge of the 1st Czechoslovak

Army in Slovakia and appealed to the Allies to aid the revolt which rapidly grew to more than 60,000 soldiers and partisans.[273] To assist the Slovakian insurgency and rescue downed pilots, as well as to establish an intelligence network in Czechoslovakia and neighboring Austria and Hungary, the OSS's Special Operations headquarters for the Mediterranean Theater assembled two OSS missions of more than two dozen OSS personnel.[274]

Chosen to command the SO's "Dawes" Mission to aid the Uprising and help rescue downed airmen was Navy Lieutenant J. Holt Green, scion of an old and prominent family from Charleston, South Carolina. Holt Green was a graduate of the Harvard Business School and had managed the family's textile mills in North Carolina before joining the Special Operations Branch of the OSS in early 1943. He and most of the other SO members of the team had trained at Areas F, B or A; so had some of the SI members in addition to the SI schools; and the radio operators had, of course, also trained at Area C. Green was not inexperienced. Overseas, he had participated in several missions to Yugoslavia. Most of the enlisted men in the Dawes Mission were the kind of ethnic Americans with roots in the occupied countries that Donovan had seen as potential "shadow warriors" conducting espionage and sabotage behind enemy lines. Master Sergeant Jaroslav ("Jerry") G. Mican, a native of Prague, had emigrated to Chicago in the 1920s, become a U.S. citizen, earned bachelor's and master's degrees and taught foreign languages in a Chicago high school. Politically active in Chicago's Czech and Slovak communities, Mican knew influential political figures in Illinois and in the Czech government in exile, including Vojta Beneš, brother of the former Czech President. Although 42, Mican had enlisted in the Army as a private and then joined OSS/SO. It was Mican who selected some of the other Czechoslovakians for the mission: Sergeant Joseph Horvath, 24, who had immigrated from Slovakia to Cleveland, Ohio, with his parents in 1928; and Czech-born Sergeant John Schwartz (code named Jan Krizan), of SI, who had escaped from the Nazis in 1940, fled to New York, and joined the U.S. Army before being recruited by the OSS. The two radio operators were both Chicagoans: Army Private Robert Brown would accompany Holt Green, who he had previously served on a mission in Yugoslavia, Navy Specialist First Class Charles O. Heller, also a Czech speaker, would be the radioman for Schwartz. This was the first team sent in as the "Dawes" Mission, and of the six, only Schwartz would survive.[275]

On an old landing strip near Banská Bystrica the center of the uprising in western Slovakia, two, four-engine B-17G "Flying Fortresses" of the 15th U.S. Air Force landed on 17 September 1944, with the six Americans and five tons of arms and ammunition. The crowd applauded, but downed American airmen waiting to be rescued, took one look at Green in his naval officers hat and asked "What the hell is the Navy doing here?"[276] The two planes returned to Italy with 15 Allied airmen rescued by the partisans. Green and his team set up a headquarters and established communications with Bari. Three weeks later, a second and much larger OSS contingent of arrived on 7 October. Six B-17s with 32 P-51 fighter escorts filled the landing field. They brought in more than a dozen additional OSS personnel, plus twenty tons of supplies: submachine guns, bazookas, ammunitions, explosives, communications equipment, medical supplies, and food and clothing. They took back 28 downed American airmen. The second contingent included SO, SI, Commo, and Medical Corps personnel, because the OSS regional headquarters in Italy now sent more SI personnel to spread an intelligence network.

A former Austrian, Lieutenant Francis Perry, an SI officer from Brooklyn, was assigned to return to his native Vienna, 120 miles to the southwest, and recruit an espionage network. Captain Edward V. Baranski, SI, an ethnic Slovak from Illinois, who had been urging SI to infiltrate agent teams into Czechoslovakia for some time, was in charge of OSS SI's "Day" Group, which was ordered to establish a ring of local spies in German-occupied Slovakia. His three SI team members included indigenous Slovak civilian agents, Anton ("Thomas") Novak and Emil Tomes and civilian radio operator Daniel Paletich, a Croatian who spoke Slovak. Another group of SI agents, whose mission was to build a circuit of agents from Budapest, Hungary, 100 miles to the south, included two ethnic Hungarians from New York City, Lieutenant Tabor Keszthelyi and Sergeant Steve J. Catlos, plus Private Kenneth V. Dunlevy, an SI radio operator and cryptographer. Special Operations sent along Lieutenant James Harwey Gaul, son of a prosperous Pittsburgh family, who had done archaeological excavations in Slovakia while a graduate student at Harvard. He was to assist Holt Green. To instruct partisans in the use of American submachine guns, bazookas and other weapons as well as plastic explosives, SO weapons and demolition experts Captain William A. McGregor, former head of the lacrosse team at the University of Maryland, and Lieutenant Kenneth Lain, who had been an athlete at the University of Illinois, were included. Air Corps Lieutenant Lane B. Miller from California had been a B-24 "Liberator" pilot, who had been shot down and rescued by partisans in Yugoslavia, was to be in charge of the airmen rescue mission. Naval photographer Nelson B. Paris came along to record the historic mission with still and motion picture cameras. Learning about the mission, Associated Press war correspondent Joseph Morton from St. Joseph, Missouri, gained OSS permission at the last minute to accompany the group, and he climbed on board carrying his portable typewriter.[277] Of these 13 men, only five, McGregor, Lain, Novak, Catlos and Dunlevy, would emerge from the mission alive.

Neither the Slovaks nor the OSS regional headquarters in Bari, Italy had anticipated just how quickly, forcefully and successfully the Germans would act, although Holt Green by radio had advised against sending in the second team and subsequently asked for an evacuation. Hitler recognized the danger the revolt posed to the supply lines to the Wehrmacht trying to stop the advancing Red Army, already in Poland, from getting to Germany itself, and he dispatched five veteran divisions, with artillery and armor, to crush it. When they quickly smashed the rebellious units of the Slovak Home Army, the partisans scattered, and SS, Gestapo, and special anti-partisan units hunted down partisans and those who had aided them. The Slovaks and the Dawes Mission had hoped that the Red Army, 200 miles away would break through the German defenses in the Carpathian Mountains, but the Soviets did not get through and liberate Slovakia until March 1945.[278]

OSS regional headquarters did eventually try to rescue the Americans, but bad weather prevented the flights, and then on 26 October, the German Army took Banská Bykstrica and the airfield. Moving out ahead of the Germans, Holt Green decided to split the group of Americans, which had grown to 37 including the OSS teams and downed U.S. aviators, into four sections, hoping to reduce casualties and chances for capture. Like the partisans, the Americans headed for the Tatra Mountains to the north to await rescue by the Russians.[279] The winter of 1944-1945 was cold and cruel, one of the worst in Europe in decades. Rain, mud, and then ice storms battered those seeking refuge in the mountains. Food was scarce. During a march of more than eighty miles along the mountain ridges in the direction of the Russian Army, the American OSS men and aviators lost members a few at a time. Then in mid-November, exhausted and freezing, all of the airmen along with two OSS officers chose to go down to a village and surrender. More were captured later as they tried to obtain food. As Christmas drew near, the remnants of the OSS mission, plus some British SOE and SIS agents, found shelter in a mountain hotel near Velny Bok, just north of Polomka, Sergeant Joe Horvath's birthplace. Holt Green organized a Christmas Eve party. On Christmas day they set flares for an expected airdrop of food and other supplies, but it did not arrive. On the next morning, 250 German troops of a special anti-partisan SS unit stormed up the mountain, overcame the partisan guards and captured Green and most of the Americans and British agents in his group after a firefight in which Green and James Gaul received gunshot wounds in their arms. The group had been betrayed by one of their partisan guards. Five members of the group, who had been quartered in a hut higher up the mountain, were able to avoid capture. The escapees included two Americans, Sergeant Steve Catlos and Private Kenneth Dunlevy; two British agents; and 24-year-old Maria Gulovich, a Slovakian schoolteacher and partisan, who had been hired by Holt Green as an interpreter and guide. She would help lead that small group to safety.[280]

Most of the members of the American and British missions, whether captured at Velny Bok or earlier, were taken 200 miles west to Mauthausen, near Linz, Austria, one of the infamous Nazi concentration and death camps. (The downed airmen, except for Lane Miller of the OSS mission, were taken to regular POW camps, and liberated at the end of the war.) The group from Velny Bok arrived on 7 January 1945, and Berlin sent special SS and Gestapo officers to interrogate them. Under the personal supervision of the camp commandant, SS Colonel Franz Ziereis, most of the British and American captives were tortured while being interrogated. The commanding officers were apparently tortured first. Captain Edward V. Baranski had his hands tied behind his back, his wrists attached to a chain hanging from a beam above, then he was hoisted upwards so that his whole weight pulled on his backward bent arms. He writhed in pain while being interrogated. Lieutenant Holt Green was put in a crouching position, his hands bound beneath his thighs, behind his knees. An interrogator struck him with a heavy whip across the face and back until they were bloody. The English major was tortured with what was called the "Tibetan prayer mill," three or four wooden rings, which when strongly pressed together, crushed the victim's fingers. The torture for these and other captives went on for two weeks. Berlin ordered them executed as spies, despite the fact that the military members had remained in uniform during the entire mission.

Beginning on the morning of January 24, 1945, the American and British prisoners—all of them that day or over the next three months, accounts differ—were taken one at a time to a windowless, underground bunker and shot in the back with a pistol by the camp commandant himself. The dead included British Major John Sehmer and several members of his SOE mission, among them a 30-year-old Slovak-American woman, Margita Kocková, a teacher who had returned to Slovakia and been assigned by the headquarters of the 1st Czechoslovak Army to be an interpreter for the British SOE team. The members of the American mission who were executed at Mauthausen included Captain Baranski, Lieutenants Green, Gaul, Keszthelyi, Miller, Perry, Sergeants Horvath and Mican; radio operators Brown and Heller; Navy photographer Paris; and AP correspondent Joe Morton, who had joined to report the story of the Dawes Mission. Two indigenous civilian members of the mission, Slovak Emile Tomes and Croatian Daniel Pavletich, were captured and killed in Slovakia. Their deaths brought to 14 the total number of members of the Dawes Mission who were killed.[281]

The group that had escaped capture at Velky Bok because they were farther up the mountain continued their wintry trek eastward through the mountains led by guide Maria Gulovich. Two weeks and fifty miles later, she and American Sergeants Steve Catlos and Kenneth Dunlevy, together with two members of the British mission, finally met the advancing Red Army. But instead of being rescued, they were interrogated by the Soviet secret police, who considered them possible spies and prevented them from contacting their own forces. Held anonymously in Soviet custody, the group was taken to Romania on the way to the Soviet Union. But at Bucharest, were there was an Allied mission, they were able clandestinely to contact an American general, and a group of GIs in jeeps came and whisked them to safety. The team brought with them their Slovak guide, Maria Gulovich, who later received a Bronze Star Medal for bravery and eventually became a U.S. citizen. Holt Green and James Gaul, the two commanders of the Dawes Mission, were posthumously awarded the Distinguished Service Cross.[282]

SS Colonel Franz Ziereis, the commandant at Mauthausen, was captured and mortally wounded by an American patrol on 23 May 1945, some sixty miles south of the camp from which he had fled a the approach of Patton's 3rd U.S. Army. Deputy Commandant, SS Lieutenant Colonel Georg Bachmayer, had shot his wife, children, and himself, the day Germany surrendered, May 7th. Torturer Walter Habecker was located by the British in 1947, arrested, and incarcerated in a military prison, where he later hanged himself. Some of the other interrogators and torturers were tried and hanged. Of the two top Nazis under Hitler, responsible for the executions at Mauthausen, as well as genocide against Jews and others and numerous other war crimes, SS chief Heinrich Himmler was captured and committed suicide with poison in May 1945. Ernst Kaltenbrunner, Himmler's chief subordinate, who oversaw the SS, Gestapo, and the methods of liquidation of those in the Nazi camps, was tried, convicted, and sentenced to death at Nuremberg, with Donovan and several OSS men and others in the audience at the sentencing. Kaltenbrunner was subsequently hanged.[283]

Trapping German Troops in Norway

The Norwegian government in exile had asked the United States as well as Great Britain for assistance, and part of the help given by President Roosevelt was OSS support for the Norwegian Resistance. A number of Norwegian nationals received training from the Special Operations Branch at Areas B and A. Crown Princess Märtha, accompanying President Roosevelt, had visited Area B in 1942 while Norwegian Army Lieutenant Edward Stromholt was instructing them. Back in Norway, cooperating with the local Resistance, some of these saboteurs blew up German supply depots and sank a German steamship with an underwater mine. In addition, SO cooperated with SOE in organizing and supplying the Resistance with 450 tons of weapons, explosives, and other supplies by parachute drop from Britain, fast boat from Scotland, or overland from neutral Sweden.[284]

In addition, a Norwegian Operational Group was formed in 1943, when OSS recruited nearly 80 Norwegian Americans from among the U.S. Army's 99th Mountain Battalion at Camp Hale, Colorado.[285] The OSS Norwegian OG received its OSS training at Areas F and B that fall, and as Corporal Arne Herstad from Tacoma, Washington, wrote to his fiancé in October, a group of them were called down to Washington, D.C. to stage an exhibition for Princess Märtha.[286] That winter they underwent more training in Scotland, but they waited in vain for a mission to Norway. Instead, a number of them were parachuted into southern France in support of the Allied invasion of the Riviera in August 1944, among them, their new OG commander, Major William E. Colby, a future Director of Central Intelligence, who had gone into the Yonne Valley southeast of Paris as head of Jedburgh Team Bruce.[287]

Finally, they were assigned a mission in Norway in early 1945. The goal was to trap some 150,000 German troops concentrated in the Narvik-Tromsc area of northern Norway and prevent them from being transported back to defend Germany.[288] Since the Allies now controlled the sea, and the roads were blocked with snow, the only escape route was the single track Norland Railway running down to Trondheim. The OG mission was to intermittently destroy bridges and lines of railroad track to keep the thousands of German soldiers from getting to the war zone. Major Colby divided the men assigned to him into two groups. He would lead the 33-man main party, called Norso I. Lieutenant Roger W. Hall, who had been an instructor for various OGs at Areas F and B and who had been parachuted briefly into Brittany in 1944, would bring in a second group, Norso II, consisting of 19 men, a month later at the next light-moon. Norso II.[289]

To evade the Germans, air drops had to be made in moonless nights, but severe storms forced back attempts in January and February. Despite heavy winds, on 25 March 1945, eight planes left Britain headed toward the Arctic Circle, carrying the white-clad skiers of Norso I, their weapons, demolitions, and supplies. The groups sat tight-lipped as the planes then headed east to northern Norway. Only four of the planes got through to the drop zone, twenty men and half the equipment. Two of the planes and men that had been forced back, returned over Norway the next day, but both crashed in the stormy weather, killing all aboard. Without them, the party of twenty, led by Major Colby and Lieutenant Tom Sather, a native Norwegian who had moved to Brooklyn and become a U.S. citizen, joined by several members of the Resistance set out for their target, the railroad bridge north of Tangen, thirty miles over the mountains. They trekked on skis, dragging a massive sled with 180 pounds of explosives, through snow and sleet storms in temperatures that dropped to minus 20 degrees. Finally reaching the bridge several days later, they scaling down ice-covered cliffs several hundred feet high and bent to their task. Lieutenant Glenn J. Farnsworth, a demolitions expert who, aside from Colby, was the only other non-ethnic Norwegian in the unit, guided the men in setting the charges.[290] Colby recalled the moment vividly in his report a written only few months later:

> Quickly Farnsworth, Sergeant Myrland, Cpl. Kai Johansen, and Sgt. Odd Andersen set the charges under the long, I-girdered bridge. They planted all we had, enough for four bridges that size.
>
> It is difficult to blow up steel—most often it simply bends out of shape. But the second Farnsworth touched the wires and the TNT went off, the structure vanished. The noise was awful, rocking back and forth between the hills. Even the softening lake seemed to jump, and it did crack with a boom like distant thunder. The happy men stood around with smiles on their grimy, weary faces. At last they had done something, and the Nordland railway was stopped.[291]

After escaping pursuing Germans, the OSS team went farther down the line and simultaneously blew up more than a dozen sections of track over more than a mile with 30 pounds of plastic explosives. "Then came the Germans like violated bees," Colby wrote. The commandoes fled in the dark, but the Germans pursued them night and day for several weeks. Out of rations in the sparsely settled area, the men lived on a grut, an unappetizing Norwegian concoction of flour and water, obtained from scattered farmhouses. There were several firefights before Colby and his team finally made it to safety, their mission a success. The disruption of the crucial Nordland rail line interrupted the southward flow of enemy troops, reducing it to a trickle: from one battalion a day to only one a month. By the end of the war, there were still 100,000 German troops, 12 divisions, trapped in Norway.[292] The OSS Norso I team was the first Allied fighting unit to operate in Norway since 1940. When the war ended in May, Lieutenant Hall and his Norso II unit, which had been delayed, arrived to help with the surrender of German units.[293]

Afterwards, the combined units, two dozen men in their U.S. Army uniforms and the American flag, marched along smartly in several victory parades to the great acclaim by the Norwegian people, who cheered the "fabulous Norsos." The War Department considered the difficult mission worthwhile. Colby and Sather were awarded Silver Star Medals and several of the other men received Bronze Star Medals. "Eleven of our men and fifteen Air Corps men had paid with their lives for our mission," Colby wrote later. "We all hoped that our efforts had made these sacrifices worthwhile and helped to end the war by even a few minutes."[294]

Effectiveness of OSS and the Resistance in Nazi-Occupied Europe

Like any new organization, the OSS had its successes and failures, but it emerged from World War II with considerable renown for its daring deeds and achievements. The European Theater received the greatest public attention, although the OSS had first proven itself in North Africa. The multinational Jedburgh teams garnered the most publicity. But the SO teams and OG sections were clearly also important, and they, like the American Jedburghs and some of the spies of SI and the counterspies of X-2, had generally received at least part of their OSS training at Area B in Catoctin Mountain Park or Areas A or C in Prince William Forest Park.

Major Alfred T. Cox, head of the OSS French OGs in 1944 conducted a post-operation critique, and all the responding OG members expressed the opinion that their extensive training had been effective. Some of the men indicated that the field training problems, half of which had been conducted at night, were more difficult than actual operations. Generally, the men of the French OGs believed that more emphasis on operating and maintaining foreign weapons and vehicles would have been helpful. So would more familiarity with methods of instruction to use with indigenous guerrilla fighters. The OGs found most of their equipment and weapons satisfactory, but the SSTR-1, radio transmitter/receiver, while compact, lightweight and having the necessary range was not durable enough for the rough handling it received and the OGs recommended more training for the radio operators on its maintenance and repair.[295]

Nevertheless, the men who operated the wireless telegraphy (W/T) radios that were the lifelines between the teams in the field and their regional headquarters and supply bases, were often able to work miracles with their sets. Sergeant Caesar Civetella of the OSS Italian OG had the greatest praise for his radioman, Technician Fifth Class (T/5) Joseph P. Seliquini, radio operator for the section on the Nancy Mission in southern France in 1944 and the Spokane/Sewanee Missions in northern Italy in 1945. Seliquini was from Philadelphia, Civetella recalled, and he was good in music. "They said if you were good in music and math, then CW [Continuous Wave; i.e. Morse Code] was easy for you. It was for Joe...He was a T/5 and one of the few enlisted men who received the Legion of Merit, which is normally reserved for officers. He got it because on his missions, he did not miss one message either to or from base. He had a perfect record."[296]

Virtually all the OSS agents—Jedburghs, SO teams, and OG sections—agreed that the units had been inserted much too late to achieve maximum effectiveness with the *maquis*. Every day spent training the members of the French Resistance would have made them more combat effective. The situation in France, was rife with political disagreements among partisan groups, from the left to the right of the political spectrum. In his final report on debriefing of the Jedburgh teams in southern France, Lieutenant Colonel Kenneth H. Baker also stressed the need for members of OSS teams to be fully briefed on the political situation in the country and among the Resistance groups and if possible the military operations should be strictly separated from political considerations.[297] William Colby, who headed a

Jedburgh team in France and an OG mission in Norway, and much later became CIA Director, said in an interview in the 1990s, that "in the Jedburghs, they taught us how to sneak around and shoot and use knives. But there was absolutely no training in the politics of the problem: how to get along with people. So I read Lawrence of Arabia's book, and he had all sorts of things in here about how you get along with a strange culture, how to relate to them and handle yourself, how to defer and suggest. You don't take command, you don't boss people, you just have to work your way through it. It was good training in the basic principles of how you get along."[298]

Inserting agent teams or military units far behind the battlefront to engage in unconventional warfare was an audacious experiment. OSS was learning as it went along. Evidence of its improving learning curve is provided by the declining casualty rate among OSS units by the final year, even as more OSS personnel were being put into the field.[299] The wartime experience of the "shadow warriors" of the OSS showed the importance of individuals and small groups in leading and coordinating considerable numbers of armed civilians in the Resistance. Those young partisans could not replace conventional forces. But properly used, they could provide valuable intelligence information, harass the enemy and weaken his morale, temporarily interdict enemy lines of communication and supply and force the enemy to withdraw thousands of troops that were needed elsewhere to deploy at least temporarily to try to crush or at least contain the Resistance. Thus the OSS demonstrated the importance of the "war in the shadows."

"The success and speed of the Allied Armies in the Battle of France are now recognized as due in large measure to the activity of the French Resistance," Donovan wrote to the Joint Chiefs of Staff in February 1945 as the war in Europe neared its end. "The Resistance impeded the movement of German reinforcements towards the original Normandy beaches and guarded the flanks of the American armies driving to the Seine in the north and the Vosges from the south. It diverted whole German divisions from the front and harassed the enemy behind his lines. It supplied continuous strategic and tactical intelligence on the enemy situation, prevented German demolition of vital installations and assisted in isolating and mopping up enemy units bypassed by the Allied advance.... OSS dispatched 187 secret agents into France." "When the landings in Normandy became imminent," Donovan continued, "steps were taken to assure maximum possible coordination of Resistance activity with the actual plans and needs of Allied armies... A similar cooperative arrangement was established with the [landings in the south of France] ... Throughout July, August, and September [1944] interferences with German movement by rail and road throughout France and severance of vital power and communication lines were widespread and continuous, and as a direct result, the Germans were greatly hindered both in moving troops and in bringing supplies."[300]

A detailed report by U.S. Army (G-3) Operations about the work of OSS and British SOE with various Resistance movements in the Mediterranean and European Theaters of Operations, while also laudatory, was more specific in its assessment. The critique was written by an experienced professional soldier, Brigadier General Benjamin F. Caffey, Jr., an accomplished staff officer and commander (he led the 39th Regimental Combat Team in the capture of Algiers) who was a longtime, personal friend of Eisenhower and whom Eisenhower considered "a most able, even brilliant officer."[301] "Resistance groups, alone, cannot win a campaign or a battle," Caffey concluded, "but they are capable of rendering important assistance to regular forces."[302] His report noted that Resistance groups could conduct sabotage and engage in guerrilla activities, but they could not engage the enemy successfully in conventional offensive or defensive operations. "Every time they have attempted the latter [engage the enemy in conventional operations], whether in France, Italy, Yugoslavia, or Greece, they have been soundly defeated." Even in 1945, Caffey made a point that was vitally important, although it was often forgotten in subsequent years. It was essential, he said, that the local civilian population be friendly. If not, it was virtually impossible for special operations detachments to accomplish their missions. That had been demonstrated in the failed attempts to infiltrate missions into Austria, Hungary and Bulgaria. In those cases, the local population not only would not protect the agents, but invariably reported their presence to the enemy. In its overall assessment, while voicing the regular officers' familiar complaints about the paucity of professional officers in the OSS and British SOE and the consequent unfamiliarity with discipline and properly coordinated staff work, the Army's G-3 report declared that "while OSS and SOE are hampered by poor staff work, their personnel in the field have done remarkably well. They deserve credit and appreciation for their fine work.... This method of warfare has a vast potential in obtaining military strategic and tactical objectives." General Caffey warned his colleagues that "No commander should ignore this potential."[303]

The significance of this unconventional warfare as a contribution to Allied victory was recognized by major commanders in both the Mediterranean and European Theaters of Operations. It was acknowledged by American generals Mark Wayne Clark in Italy, Omar Bradley and George Patton in northern France, and Alexander Patch in southern France.[304] German commanders, from Field Marshal Albert Kesselring in Italy to Field Marshals Gerd von Rundstedt and Walther Model in France, certainly recognized the seriousness of the problems they faced from increasingly assertive and effective Resistance forces organized, armed, and led in their rear by the Special Operations forces of the British SOE and American OSS.[305] "The Resistance surpassed all our expectations," General George C. Marshall, chief of staff of the U.S. Army, and the man responsible for American military strategy, is reported to have stated in 1946, "and it was they who, in delaying the arrival of German reinforcements and in preventing the regrouping of enemy divisions in the interior, assured the success of our landings."[306]

General Dwight D. Eisenhower, Supreme Commander of the Allied Expeditionary Force, in his contemporary praise for the combined Special Operations effort that coordinated the Resistance, declared: "In no previous war, and in no other theatre during this war, have Resistance forces been so closely harnessed to the main military effort." He applauded the way the Resistance forces had been, in his words, "so ably organized, supplied and directed," and, in addition, he gave special credit to those radio operators and others responsible for communications with occupied territory. He also congratulated OSS and SOE for the "excellent work carried out in training, documenting, briefing and dispatching agents."[307] Looking back in his memoir, *Crusade in Europe,* Eisenhower went even further, asserting that "the Resistance had been of inestimable value to the campaign.... Without their great assistance, the liberation of France and the defeat of the enemy in western Europe would have consumed a much longer time and meant greater losses to ourselves."[308]

CHAPTER 9

OSS IN ACTION: THE PACIFIC AND THE FAR EAST

Although the most publicized achievements of the OSS occurred in Europe and North Africa, Donovan's organization also contributed to the war against Japan in the Far East. That contribution was mainly in the China-Burma-India Theater (CBI), which American veterans of the CBI often call the "forgotten war" of World War II. But it was an important war and one in which the OSS made significant achievements. In the beginning, after the Japanese had pushed through most of Southeast Asia, they were finally stopped at the border of India. With the Burma Road severed, the Americans turned to an airlift and astonishingly supplied the Chinese by making thousands of flights "over the Hump," across the Himalaya Mountains. OSS-led guerrillas, 10,000 Kachin tribesmen, helped undermine Japanese control in Burma, and the OSS established contact with other resistance movements in Thailand and Indochina. Most importantly, the fighting in China itself tied down the bulk of the Japanese Army throughout the war. Most of the members of OSS Special Operations, Operational Groups, and Communications, and many in Secret Intelligence, who served in the Far East had obtained at least part of their training at Training Camps A, B, or C in Catoctin Mountain Park and Prince William Forest Park.

OSS Director William J. Donovan had a long and strong interest in the Far East, dating back to his prize-winning senior thesis at Columbia on Japan's emergence as a world power. In the interwar years, he made several trips to Asia, and a month after the Japanese attack on Pearl Harbor, Donovan established an office in Honolulu for liaison with the Army and Navy in the Pacific.[1] Within three months, he dispatched a representative to China to "improvise an underground apparatus."[2] Donovan hoped his organization would play an important role in the war against the Japanese Empire.

General MacArthur Snubs the OSS

OSS tried but failed to gain significant access to the island-fighting war in the Pacific. General Douglas MacArthur, commander in the Southwestern Pacific, would have nothing to do with the OSS. He sneered at Donovan's offers of assistance, insisting on exclusive control of all forces under his command and holding

Donovan's collection of amateurs in disdain. Whether MacArthur's 1942 decision was made for practical or personal reasons, or both, it effectively excluded the OSS for most of the war.[3] Donovan was not even able to outflank MacArthur, at least initially. In April 1943, the OSS chief sent an agent to try to convince Vice-Admiral William F. ("Bull") Halsey, whose naval forces assisted MacArthur, to allow OSS into the Southwest Pacific Area. But Halsey was not persuaded and finally told the man to "Get the hell out of here!"[4]

In the winter of 1944-1945, some OSS personnel, most of them trained in the National Parks in Maryland and Virginia, were sent to the Philippines as MacArthur's forces landed first on Leyte and subsequently on the main island of Luzon. Some of these may have been with MacArthur's authorization, others perhaps not. Delivered at night by submarines, they were deployed under the authority of Admiral Chester Nimitz, commander of the U.S. Navy in the Pacific. Army First Lieutenant Donald V. Jamison, a Native American, was put ashore on Luna, La Unión, in the Philippines in the fall of 1944. An OSS Special Operations officer, undoubtedly trained at the SO camps at Areas B and A, the 22-year-old Jamison was directed to engage in reconnaissance and demolition work behind Japanese lines. Later recalling the fierce battles in the Philippines, he said that he had first learned guerrilla skills as a boy on the Rincon Indian Reservation near San Diego. His father was a Seneca-Cayuga Indian from upstate New York, but his mother was a member of the Luiseno Band of Mission Indians in California. There he learned marksmanship, riding, and hunting in the wild. Recruited and trained in World War II first by the Army and then by the OSS, he was landed by submarine in the Philippines. He worked with Filipino resistance groups to hinder Japanese lines of communication and supply and impede the enemy's opposition to the landings and advance of the U.S. Army. Afterwards, Jamison received several medals from the Philippine government and began a lifelong friendship with Ferdinand Marcos, a wartime guerrilla leader, who later became President of the Philippines.[5]

Admiral Nimitz Welcomes OSS Frogmen

In the Pacific Ocean Area, Admiral Chester W. Nimitz, the theater commander and commander-in-chief of the Pacific Fleet, was less rigid than General MacArthur, although he too was reluctant, at least initially, to include the OSS. In 1943, Donovan offered OSS personnel for espionage, sabotage and "black propaganda" against Japan and its outposts in the Pacific, but neither Nimitz nor the Joint Chiefs of Staff (JCS) was interested.[6] The next year in April 1944, when Donovan met with Nimitz at Pearl Harbor and showed him the list of the OSS's specially trained units, the only group that interested the admiral was one from the Maritime Unit. Nimitz told Donovan, "I can use your swimmers."[7]

The OSS operational swimmers, or "frogmen," were part of the Maritime Unit that Donovan had established in 1942 with its training facilities first at Area A and then Area D on the Potomac River, and finally in the Bahamas and off California. Navy veteran and deep sea diver, John P. Spence from Tennessee, had been recruited by the OSS in 1942 for training in small boat handling and underwater demolition. He underwent SO paramilitary training at Area B and then combat swimming and demolition training at Area D. He remained at D as an OSS/MU instructor through the end of 1943. Spence was subsequently sent to the Bahamas where he trained frogmen for deployment in the Pacific. He is recognized by the OSS and by the Navy as "one of the first combat swimmers in the United States," and in 2001, the Naval Academy celebrated him as "the last surviving member of the original five OSS combat swimmers."[8]

When OSS established its West Coast schools in California in 1944 to prepare members from the various branches, SI, SO, MO, and MU, for service in the Far East, Marine Lieutenant Elmer ("Pinky") Harris, the Washington State alumnus from Ketchikan, Alaska who was one of the original SO instructors at Areas B and A, was assigned temporarily as an instructor at the Underwater Swimming School at Catalina Island near Los Angeles. An able instructor, regardless of the subject, Harris had taught paramilitary techniques at Area B, then parachute skills at Area A before being sent in SO units to North Africa and Corsica. Afterwards, in early 1944, he was transferred to Brindisi, Italy to establish a parachute training school. Following medical treatment in the United States for severe abdominal pains, Harris was sent as an underwater demolitions instructor to the OSS school at Catalina Island in the summer of 1944.[9]

The OSS Maritime Unit had developed splayed, rubber swim fins for its swimmers and adopted a self-contained underwater breathing apparatus, SCUBA, invented by Dr. Christian Lambertsen, who subsequently joined OSS. Donovan lent OSS Maritime Unit Group A to the Navy's Underwater Demolitions Team Number Ten (UDT-10), and the two teams served together in the Pacific with Nimitz's approval in 1944 and 1945. The OSS frogmen of MU-Group A, trained by John Spence, Elmer Harris, and others, participated jointly with Navy UDT-10 in pre-landing inspections and obstacle destruction in more than half a dozen Japanese held islands.[10]

Operational swimming was a dangerous business. Three members of a five-man OSS/Navy frogman team were lost in one of their first Pacific missions, the exploration of the Japanese fortified island of Yap in the western Carolines in August 1944. In the middle of the night, the submarine *U.S.S. Burrfish* launched the swimmers in two rafts from about two miles out, but paddling in, the men found a reef a quarter mile from shore. While two men held the rafts at the reef, the other three swam in to reconnoiter. The three had not returned when dawn approached and their comrades, believing them captured, returned to the submarine which submerged and left. The Japanese had captured them and transferred them by ship to the Philippines, but they were never located by the Allies. All five members of the team were awarded the Silver Star Medal, the military's third highest decoration for bravery, three of them posthumously.[11]

Before going to Yap, the ill-fated swimming team had already explored the Japanese defenses on Peleliu, and the Navy chose that island as the target for the Marines invasion in mid-September 1944. Nineteen-year-old Marine Sergeant Patrick Finelli from Newton, Massachusetts, had already been trained in demolitions and booby traps when the OSS obtained him as an operational swimmer in the summer of 1944. Recruited in California, he may have been trained first by Lieutenant Elmer Harris at Catalina Island. Beginning on 12 September, three days before the scheduled invasion of Peleliu, dressed only in swim trunks, sneakers and leather gloves, young Finelli and other OSS and Navy swimmers spent several days setting off more than a thousand demolition charges to clear entryways through the coral reef. "It was hot, thirsty, itchy, and terrifying work," Finelli recalled. "The Japanese had their own swimmers hiding explosives in the coral reefs." While in shallow water close to the beach, the men were shot at from shore. Most of the work was done during the day, but the Americans made one dive on a moonlit night. "That's about the most frightening thing I've ever done," Finelli said. "Your every movement creates phosphorescence, and every time you rub up against something you think it may be a big fish that wants to eat you or a Jap swimmer who wants to kill you." On the day of the invasion, 15th September 1944, the swimmers used demolitions on the reef and also on shore. Several Japanese, armed with knives and bayonets, made a suicidal banzai charge at them. In the knife fight, Finelli suffered a number of cuts but survived and was treated for his wounds a naval hospital in Hawaii. After the war, Finelli worked for Polaroid until his retirement in 1993, and he continued to swim regularly at 83 years of age.[12]

Over the next several months, the OSS swimmers, working jointly with the Navy's UDT-10, participated several other important campaigns, including the seizure of Ulithi, the invasion of the Philippines at Leyte and Luzon. In all sixteen of the combat swimmers in the joint OSS/Navy team were killed in action.[13] The OSS "frogmen," plus the Navy's Underwater Demolitions Teams along with two or three other naval special operations units of World War II, are officially recognized as the forerunners of today's Navy SEALS.[14]

Reginald Spear: OSS Agent Extraordinaire

One of the most extraordinary OSS Officers to serve in the Pacific, as an operational swimmer as well as a secret agent, was Reginald G. ("Reg") Spear, a precocious young inventor and bold adventurer from California. After OSS training, Lieutenant Spear eagerly went on missions including frogman operations against Japanese fortified islands, penetrating a prisoner of war camp in the Philippines, and searching behind enemy lines in China for a missing espionage agent. Some of his missions were ordered personally by President Franklin D. Roosevelt.

Although born and raised in California, Spear was descended from a prominent family in England and Canada.[15] At 18, Spear enlisted as a private in the U.S. Army. His high test scores and technical ability led to his being sent to Officers' Candidate School at the Army Ordnance facility at Aberdeen, Maryland. In 1943, as a second lieutenant in the Ordnance Corps, he was recruited by the OSS, and then underwent training at Area F, the former Congressional Country Club, and RTU-11 ("the Farm"); he concluded his OSS training at Area B at Catoctin Mountain Park in January and February 1944.[16] His British and Canadian connections—Winston Churchill knew of his family—and his own talent and abilities led young Spear to be summoned several times to meet with President Franklin Roosevelt, sometimes at "Shangri-La," the Presidential retreat at Catoctin Mountain Park. Roosevelt sent him on secret missions to work with the U.S. Navy and the British in the Pacific and in Asia.[17] "I had a reputation," Spear recalled. "They sent me for intelligence in the Pacific to work between the Americans and the British."[18] After his training concluded at Area B, the 20-year-old Spear and a fellow OSS Army officer arrived at Nimitz's headquarters in Hawaii. The Admiral thought they should be in Navy uniforms if they were going to work for him. So he made them naval lieutenants. But according to Spear, Edward Layton, Nimitz's intelligence chief, said, "If this young man gets caught by the Japanese in his activities, they will cut his head off. However, if he has an extremely high rank for someone so young, then they will believe that he must be someone special and treat him with more consideration." So they gave Spear a naval captain's uniform, equal in rank to an Army colonel, with eagles on his lapels.[19]

President Roosevelt swore Spear to secrecy about the missions he gave him, and Spear has maintained his silence on those to the present day.[20] But Spear was willing to discuss operations he undertook for Nimitz. In late summer 1944, following training in operational swimming off Maui, Spear made a clandestine inspection of Japanese occupied island of Peleliu, long before the planned invasion. A submarine put him ashore one night, and from native residents and from his own observations, he learned that the Japanese had heavily fortified the island with tunnels, caves, and concealed weapons bunkers. Spear returned with this information and its implication that it would be a very difficult assault.[21] But the invasion of Peleliu proceeded on schedule, nevertheless, and the Americans suffered more than 7,000 casualties, making Pelilu one of the costliest invasions in the Pacific, foreshadowing the dug-in defenses and high casualties the following year on Iwo Jima and Okinawa.[22]

Spear also took part in a nighttime reconnaissance of Yap. The first mission to Yap in August, which had cost the lives of three swimmers, failed to produce useful intelligence. This time, the submarine commander hove to only a mile out, and Spear and his entire team paddled all the way to shore. They were soon discovered by a couple of Japanese sentries. But when one of the sentries used his rifle for a stranglehold and broke the trachea of a captive swimmer, the other Americans quickly killed the sentries with barehanded techniques Dan Fairbairn had taught them. Stacking the Japanese uniforms neatly in a pile and shoving the corpses into the outgoing tide so it would look like suicide, the OSS team moved out and surveyed the small island's defenses. Well before dawn, they returned to their injured comrade and paddled back to the submarine. The pharmacist mate could not save the wounded frogman, who subsequently died and was buried at sea.[23] Nimitz canceled the invasion of Yap as unnecessary, and the bypassed island remained a harmless, isolated Japanese outpost until the end of the war.

Nighttime reconnaissance on Japanese-fortified islands was dangerous enough but walking into a Japanese internment camp for Allied prisoners of war in broad daylight, as Spear would do in December 1944, called for even steadier nerves. Only a few weeks before MacArthur's forces were scheduled to land on Luzon, the Allies feared that the Japanese might massacre POWs and other internees rather than let them be liberated and possibly testify about war crimes. In mid-December, Japanese guards had indeed murdered nearly 150 Americans at a POW camp on Palawan, crowding them into wood-covered, air-raid ditches, pouring and igniting gasoline, and killing most of the prisoners.[24] Spear's assignment on Luzon was to determine the possibilities of an impending massacre of the civilian and military prisoners at the Santo Tomas facility in Manila and whether the prisoners were fit enough to assist an attempt to liberate the camp by American paratroopers and Filipino guerrillas.[25]

On the night of 4 December 1944, shortly before MacArthur's forces began to land on Luzon, Spear paddled ashore from an American submarine to learn about the conditions of the prisoners at Santo Tomas.[26] The spy's cover was to pose as a junior assistant mining engineer from Canada working for a British-run, Filipino company in Luzon that was mining gold the Japanese wanted. There was a problem in the mine and the British manager was one of the internees at Santo Tomas. The American labels in Spear's civilian seersucker suit had been replaced by ones from Victoria, Canada. He carried falsified identification cards and he wore a red armband that indicated that he was a friendly civilian authorized to visit the camp. When Spear arrived at the entrance, he found the guards at the gate busy with a large number of Filipino women. Spear joined the line and when he reached the gate, a harried Japanese guard looked quickly at his identification, glanced at his red armband, and then peered into Spear's bag, which contained a packet of rice and twelve packs of cigarettes. "The guard reached in and took them all," Spear said. "I argued with him. I got two packs back and went into the camp."[27] There were three thousand American and other internees and prisoners of war in the camp. Because of his red armband, Spear was allowed to walk unhampered over to their area. For forty-five minutes, he talked with members of the POW executive, committee, including the mining company manager, and was able to get answers to the questions he had been given. Now he had to get that information back to the OSS.

The nearest clandestine radio station was deep in the rugged mountains north of Manila in a guerrilla hideout called "Victory Hill" operated by Filipinos and some American servicemen who had escaped when the U.S. forces in the Philippines surrendered in May 1942. Spear took a train to a town at the base of the mountains. From there, two guerrillas accompanied him on the long hike up the mountain, first along a narrow trail, then sloshing two miles up a shallow river, and finally up another trail until, behind some large boulders, they reached the guerrillas' camp. Arriving around midnight, Spear was so exhausted that he laid down and fell asleep. In the morning, he wrote out his report on the POW camp. He went through the list of questions, including the key ones. Was it believed that the Japanese would murder the prisoners? Answer: No. Was the condition of the prisoners critical? Answer: Yes. The radio operator sent them out immediately.[28]

His mission accomplished, Spear and his guide walked back down the mountain and from town, he was taken by truck to the coast where the submarine was supposed to wait for him. Spear was a little late. The submarine was not there. Its skipper had been unwilling to remain any longer. "We radioed the sub, and it came back, but farther up the coast," Spear said. "I had to run three or four miles up the beach to reach it."[29] Two months later, the prisoners at Santo Tomas POW camp were successfully rescued on 16 February 1945 by the U.S. Army and Filipino guerrillas.[30]

Later in 1945, Reginald Spear was sent on a secret mission behind Japanese lines in northern China. A Chinese businessman in New York City, Dr. Konrad Hsu, an authority and an entrepreneur in radio technology, had been selling equipment to the British clandestine services in China and India, and the British wanted to recruit him and tap the influential Hsu clan in northern China as an agent network. The British Secret Intelligence Service initially planned to run the operation using one of the Canadian government's communications networks rather than those of the British or Americans, perhaps to deceive the Chinese spy network. In the end, however, SIS needed the financing that only Donovan's organization could provide. But although OSS funded the "Oyster" project (code-named after Konrad Hsu's fondness for shellfish), SIS alone coordinated the operation and maintained direct contact with Hsu in regard to it.[31] Reg Spear was called when one of SIS's Chinese contacts lost contact with his wife, who was serving as a top British agent. "They sent me into China to find her," he explained.[32] And he did find her.[33] Once again, mission accomplished.

For his extraordinary service during World War II, Reginald Spear was awarded the Navy Cross, Distinguished Service Cross, Silver Star, and the Legion of Merit.[34]

China-Burma-India Theater (CBI): The "Forgotten" War

In the Far East, World War II began in 1937 with the Japanese invasion of China. Over the new few years, Tokyo expanded its control over most of the urban-industrial areas of the coast and along the main rivers. The occupied areas of China were forced to supply Japan with foodstuffs, raw materials, and industrial goods. Beginning in December 1941, when Japan attacked America, British, and Dutch territories, Tokyo rapidly extended its empire south and west in the Pacific and through Southeast Asia to the India-Burma border. Despite these widespread conquests and their subsequent defense against counterassaults launched by Anglo-American forces, the bulk of the Japanese Army remained engaged in China throughout the war.

The Western Allies gave priority to defeating Nazi Germany. The campaign against Japan received fewer resources. There Roosevelt's strategy emphasized island-hopping Army and Navy offensives through the Pacific toward the Japanese home islands. He viewed China's main role as tying down and grinding up as many Japanese Army divisions as possible, as the Soviet Union did to the German Amy. That would result in fewer Japanese soldiers to fight the Americans in the Pacific.[35] The United States would supply China with loans, weapons, material, and advisers, but not many American troops. There was considerable fighting in China, but it usually saw the Chinese on the defensive.[36] For most of the war, the Chinese were reluctant to suffer the heavy casualties generated by offensive operations. Primarily through conscription of young peasants, the Chinese maintained armies totaling between three and four million men. Inadequately supplied, trained, and led, they still, by their very presence, tied down about 1.2 million Japanese troops.[37]

Even in areas still under Chinese control, power remained fragmented. Generalissimo Chiang Kai-shek's Nationalist Government had to contend with various personal factions, regional warlords, and Mao Tse-tung (Zedong) and his band of communists.[38] The communist forces had sought refuge in the mountains of the North, but like Chiang's government had spies everywhere. Jockeying for position while the Americans defeated Japan in the Pacific, both Chiang and Mao sought primarily to strengthen their forces for their inevitable postwar struggle for control of China.

With the China-Burma-India Theater (CBI) designated as of secondary importance and thus not receiving comparatively few Americans troops, Donovan believed it was ripe for unconventional warfare. With the Japanese empire overextended, the OSS would seek to harness the latent opposition to the conquerors. Much of the areas to be contested in the CBI, particularly in Southeast Asia, were either sparsely populated or actual jungle, and this, together with thinly spread occupation forces, made the situation conducive for guerrilla warfare by indigenous groups, organized, armed, and directed by the OSS.[39] Donovan found some grudging acceptance from Lieutenant General Joseph Stilwell, the top American commander in the China-Burma-India Theater. A crusty old soldier, "Vinegar Joe" Stilwell had the unenviable task seeking to get Chiang Kai-shek to use the support the United States was providing for offensive action against the Japanese instead of allowing it to be dispersed through corruption or rivalries or stockpiled for postwar use.[40]

Detachment 101: OSS Success in Burma

In the dark days of early 1942, as the Japanese drove back the Allies everywhere, Donovan sought out Stilwell about a role for the OSS in the Far East. The result was OSS Detachment 101, the first composite SO/SI group, which ultimately proved to be one of the greatest successes of Donovan's organization. Its activities in Japanese-occupied Burma between 1942 and 1945, resembled, perhaps more than those of any other OSS detachment, the mission and capability of the modern Special Forces of the U.S. Army.[41] It is often cited as the first unit in U.S. military history created specifically for conducting unconventional warfare operations behind enemy lines.[42]

Detachment 101 began in April 1942 with Major (later lieutenant colonel) Carl Eifler and two dozen men. A barrel-chested, no-nonsense, law enforcement and reserve Army officer, Eifler was the son of an oil field worker in Los Angeles. A high school dropout, he served in the Army as a private, then later worked for the Los Angeles Police Department and subsequently the U.S. Customs Service, doing undercover work to catch smugglers on the Mexican border. In 1940, he was appointed chief Customs Inspector in Hawaii. He had become a reserve Army officer, and was called to active duty in 1941. He first commanded a company in the 35th Infantry Regiment in Honolulu and after the Pearl Harbor attack, a military police unit guarding enemy alien detainees in Hawaii. Six-feet, two-inches tall and weighing 250 pounds, Eifler was an imposing figure in his early forties. Strong as a bull, he had a bellowing voice, gruff demeanor, and fierce temper. His energy and enthusiasm had impressed Stilwell who responded to Donovan's request by recommending Eifler to head the unit being sent to him in the China-Burma-India Theater.[43] The rest of the cadre was quickly chosen, mainly through personal acquaintance. From the 35th Regiment, Eifler picked his First Sergeant, Vincent Curl, a tall Midwesterner, and his executive officer, Captain John Coughlin, a West Pointer. Coughlin chose several others he knew and trusted, including Lieutenant William R. ("Ray") Peers, an infantry officer and a UCLA and ROTC graduate. At Fort Meade, Maryland, outside Washington, D.C., Eifler and Coughlin were guests of General M.B. Halsey and his wife. Mrs. Halsey, later called the "Mother of 101," took an interest in the fledgling detachment and recommended several promising recruits, including Lieutenant Floyd R. Frazee, who had been a jeweler and was worked with small tools; and Jack C. ("Jack") Pamplin, a civilian attorney, who subsequently volunteered for the Army was made a sergeant and was assigned "detached duty" to Eifler's Detachment 101. While in Washington Eifler also recruited four other lieutenants, Bill Wilkinson, Frank Devlin, Harry Little, and Phillip Huston, plus Sergeant Allan Richter and some other sergeants, and a Chinese American named Chun Ming.[44]

The initial group of Detachment 101 was split into two sections for training. Eight men went to SOE's Camp X outside Toronto. That section included Eifler and most of the other officers, Coughlin, Devlin, Frazee, plus Sergeant Curl, Chun Ming, and a man from Donovan's headquarters identified only as "Ben." They trained for two weeks under British instructors at Camp X. Meanwhile, at OSS Area B at Catoctin Mountain Park, nearly a dozen other members of Detachment 101 went through two weeks training under American instructors headed by Charles Parkin. The Catoctin trainees included a couple of Detachment 101's officers, including Lieutenant William ("Ray") Peers, plus nine sergeants. Among the sergeants, many of whom would later become officers, were Vincent Curl, John C. ("Jack") Pamplin, Irby E. Moree, George T. Hemming, and Donald Eng. Allen Richter was off buying radio equipment in New York City, but augmenting the trainees was an officer who was not a member of Detachment 101, Lieutenant Nichol Smith, who would be sent to France and subsequently to Thailand, This was apparently the first group to be trained at Area B.[45]

Reunited after their separation at the two camps, Eifler's two dozen men departed from Norfolk, Virginia in May and arrived in India in July 1942. The original objective of the group had been to conduct intelligence and paramilitary operations in China, but when Stilwell flew down from Chiang Kai-shek's wartime capital in Chungking (Chongqing), he directed Eifler to set up a base in northern India, learn how to operate in the jungle, and penetrate Japanese occupied Burma. Stilwell had only minimal resources, and he needed help in preparing a campaign to retake the Burma Road, the main overland supply route across the mountains to China. Meanwhile, supplies were being airlifted from Assam, India to Kunming, China by hundreds of transport planes flying through the mountain passes of the Himalayas (over "The Hump" as it was called).[46] The Japanese were attacking the transport planes from their airbase at Myitkyina (pronounced "MITCH-in-aw") in northern Burma. Go in behind enemy lines and blow up the road and railroad bridges that enable the Japanese to supply the Myitkyina airbase, Stilwell ordered. A man of few words and short temper, "Vinegar Joe" allegedly dismissed the Detachment 101 commander by stating, "Eifler I don't want to see you again until I hear a boom from Burma."[47]

Detachment 101 would ultimately give him those "booms." After a shaky start, it established itself as "the most effective tactical combat force in the OSS."[48] But first it faced formidable obstacles in the Burmese tropical jungles and the victorious Japanese Army. Northern Burma, an area larger than New England, contained rugged hills, mountains and thick, largely unexplored jungle. Movement on the ground was torturously slow, trails had to be cut through the thick vines and underbrush. For much of the year, the weather was a major impediment. In the spring temperatures soared above 100 degrees with high humidity. The summer brought torrential rains of the Monsoon, producing rot and rust. Diseases—malaria, dysentery, cholera—were rampant.[49] Ubiquitous mosquitoes, blood-sucking leeches, and deadly snakes were constant hazards. Nicol Smith, an Area B graduate who would stop at Detachment 101 Headquarters on his way to China with a contingent of Free Thais, recalled being awakened at night by the deep roar of a tiger and the high-pitched shrieks of gibbons and of finding a coiled and angry King Cobra under the table.[50]

Without combat experience and with only their prior training, the initial group set up a secret base in and old tea plantation near Nazira in northern Assam, India just across the border from Burma. They recruited and trained some Anglo-Burmese and native Burmese to serve as intelligence agents, radio operators, and saboteurs. At the Detachment 101 training school at the secret base, American and native instructors offered at minimum a basic three-month course before the agents went into the field. As in the United States, longer training was required for more specialized skills such as radio operation.

Detachment 101: Communications in the Jungle

With agents being infiltrated over thousands of square miles of mountainous jungles, Detachment 101 faced the problem of establishing a communications network. In the signal unit, Donald Eng and Allen R. Richter solved the problem. Eng had been at Area B, but Richter, alone among the two dozen, had been at neither Area B nor Camp X. Instead, while the paramilitary training was going on, Richter, who had been in the communications industry, had been assigned to purchase radio supplies directly from stores in New York City. By the time the others had graduated from the training schools, Richter had assembled the materials—tubes, wires, another components—that they would need.[51]

From the OSS base at Nazira, the target area, Myitkyina airfield was 150 air miles away by air or 400 miles by land. Richter and Eng set up a base station at Nazira, but what was needed were relatively lightweight portable wireless radios that the agents could carry and that could transmit and receive messages over the mountains. The two of them designed a prototype by December 1942; it weighed 50 pounds including the battery and carrying case. The transmitter was a straightforward crystal oscillator and amplifier, usually a pair of 6V6s; radio tubes depended on what they had on hand; the receiver was a three-tube regenerative design using assorted materials. They built the containers for the prototype radios out of materials they scrounged from local airports. "We used the aluminum belly skins from crashed planes," Richter recalled. "It still needed to be put into something strong, yet lightweight, so it could be carried. The manager of the tea plantation came to the rescue by supplying us with boards used to make wooden apple crates. It worked out fine."[52] "We called it the Burma Radio," Richter said proudly, "and it could transmit 1,500 miles!"[53]

The forerunner of the OSS SSTR-1, "suitcase radio," it made Detachment 101 self-sufficient. Natives and Americans were trained how to use it, but one of the complaints that instructor Jack Pamplin had when he returned to Washington was that although most of the "commo" operators being sent from Area C were fast enough, they were not adequately trained in how to repair and maintain the equipment in the harsh jungle environment. "Stress [the] fact that a fast operator is not necessarily a good one," Pamplin advised OSS "commo" instructors, "for the Far East, a resourceful operator is the ideal."[54] At the same time, the chief of the OSS Communications Branch, Colonel Lawrence ("Larry") Lowman, praised the work of the "commo" men in Burma for their innovation and for the delivery of intelligence and other information even in the most adverse conditions. "Our work has, we believe, played a unique, successful and important part in the realization of the overall OSS objectives in this war."[55]

Detachment 101: Taking on the Japanese in Burma

Beginning early in 1943, Detachment 101 made a series of long-range penetrations deep into the jungle in Japanese occupied North Burma by airdrop, the first airdrops made by the OSS. One of these teams blew up the railway leading to Myitkyina in eighteen places, but one member of the team was killed by a Japanese patrol, another captured, and in a premature explosion, a Burmese saboteur blew up himself as well as a bridge.[56] Soon the detachment became more proficient. In August 1943, the "Knothead" mission, commanded by now Captain Vincent Curl, assisted by now Lieutenant Jack Pamplin, dropped into the upper Hukawng Valley, less than a hundred miles from the Japanese airfield. By the end of 1943, OSS had six such permanent, if mobile, bases in the North Burma area, one of them positioned an agent atop a hill only ten miles from the Myitkyina and overlooking and reporting on activity at the airfield. Each base was staffed by a combat nucleus of eight or ten Americans. These Detachment 101 teams learned the region and recruited indigenous people as guides, spies, and guerrillas.

The OSS teams also served the 10th Air Force, rescuing downed aviators and providing detailed locations of targets that the Japanese had carefully hidden in the jungle: key bridges built just under the river surface, munitions and petroleum depots covered by camouflage netting or the jungle canopy, and underground bunker and aircraft hangers.[57] By the end of 1943, 80 percent of 10th Air Force's targets in the area resulted from OSS information.[58] "The target designation by our men has been most accurate, and the Air Force [pilots] are finding these targets without ever seeing them," Peers reported. "We receive a message that four furlongs from X road junction along the Kamaing Road, 60 yards in, there is a group of 300 Japs and 15 supply bashas, these all in the jungle. This is given to the Air Force who plot it and designate it by aerial photo. Their flights go over and thoroughly bomb and strafe this area, and as a result, huge clouds of black billowy smoke issue forth, showing the presence of petrol and various other stores. The Japs know that these cannot be seen from the air and know they must be designated by somebody on the ground, and as a consequence, our people are very much sought after by the armed forces of the Mikado. However, not only do we designate the targets, but we also give them [the Air Force] their results. The best one we have had thus far is one target designated southeast of Kamaing in which 30 cart loads of dead Japs were hauled away."[59]

Crucial to the success of Detachment 101's missions was the recruitment, organization, arming and direction of indigenous agents and guerrillas. In Northern Burma, the OSS recruited primarily Kachins ("Kah-CHINs"), fiercely proud and able mountain people. They despised the Japanese invaders, and the Americans drew on that hated. Some OSS officers, like Jack Pamplin explained that the Kachins' loyalty to the OSS reflected the fact that the Americans treated them with respect, unlike their previous overlords, the British.[60] Others, like Carl Eifler believed that the OSS purchased their loyalty by supplying with what they wanted: food, weapons, medicine, silver coins, and opium.[61] Whatever the reasons, the OSS was able to mobilize eleven thousand guerrillas, whom they called the "Kachin Rangers."[62]

Small, wiry tribesmen who were natural hunters, Kachins served as guides, spies, and warriors. They could follow invisible tracks through the jungle or across towering mountains. In keeping with tradition and for hacking through the jungle and other purposes, each warrior carried a long curved sword called a dah. But they now also learned to use the weapons the Americans supplied: rifles, submachine guns, and grenades. They had their own aggressive way of fighting the Japanese. When Stilwell expressed skepticism to one Kachin tribal leader about how many Japanese he had killed. The Kachin emptied a bamboo tube he carried and out spilled a pile of human ears. "Count them and divide by two," he told the startled general.[63]

The Kachins showed the Americans how to survive in the jungle and how to surprise and kill the enemy there. They constructed home-made booby-traps with trip wires and crossbows. OSS in Washington devised a diabolic anti-personnel device, a small hollow spike topped with a .30-caliber rifle cartridge and a pressure detonator. The device was buried below the surface on a trail used by the enemy; when stepped upon, it fired the bullet straight up through the foot and possibly the rest of the body.[64] The Kachins also used dagger-sharp, pointed bamboo sticks several feet long, called panji, which they implanted at an angle in the jungle undergrowth on either side of a trail where they planed an ambush. When the Kachins, using submachine guns supplied by the Americans, attacked the front and rear of a column, the Japanese soldiers in the middle would dive for cover, impaling themselves on the deadly, spear-like panji.

Among the OSSers who witnessed such an ambush was Lieutenant John C. Hooker, Jr., from Atlanta, Georgia. Hooker had trained at Areas A and F in 1944, and taken part in Maritime Unit raids along the Arakan coast of southern Burma, before being assigned briefly to Detachment 101 in early 1945. For several weeks, Hooker participated in airdrop supply missions, and then in March1945, he jumped near Lashio and spent nearly three weeks with a field team and their native guerrillas. As the Kachins prepared an ambush, Hooker watched them cut and plant "panji" stakes along side. Then, Kachin teams with British Bren submachine guns hid themselves at each end of a 200 yard stretch of trail. "The action was quick," Hooker recalled. "The Japanese platoon-size force of about fifty men entered the site, and in five minutes all were dead. The advanced guard of the element was picked off by Kachin snipers a half mile south of the ambush. When the smoke cleared I had emptied both twenty round magazines of my gun. The [Kachin] Rangers went among the dead clipping off ears.…Each man had a bamboo tube on a cord slung around his neck where he stored his trophies."[65]

Donovan Lands behind Enemy Lines

Both Donovan and Eifler were strong-willed, self-assured, competitive men, who enjoyed the thrill of danger. Eifler did not take criticism easily, so when on a visit to Detachment 101 Headquarters in December 1943, the OSS chief chastised him about ambiguities in his operational reports, Eifler bristled and replied with a challenge: "Would the General like to go behind the lines and see for himself?" Donovan paused, smiled tightly and snapped, "When do we leave?" "First thing in the morning, sir," Eifler replied.[66]

The trip in a small unarmed, unescorted plane would carry the two men 150 air miles behind Japanese lines. Normally senior officers were not sent behind enemy lines for fear of their being captured and tortured to provide high-level information. In accepting Eifler's challenge and indulging his own sense of honor and adventure, Donovan, who knew many of the highest operational plans and secret intelligence sources of the Allies including the breaking of the Japanese and German codes, was taking an enormous and unjustified risk. His capture by the Japanese would have been a disaster for the Allies.

The night before the flight, Donovan shared quarters with Lieutenant Colonel Nicol Smith, a former author who had trained with Ray Peers at Area B, served in France, and was currently escorting a group of Free Thais to China. Smith wondered why the OSS director was risking so much. "General, aren't you risking your life?" Smith asked at last. "Everything is a risk," Donovan replied. "My boys are risking their lives every day." Years later, one of his biographers, Richard Dunlop asked Donovan the same question, and the general told him that he had been carrying an L pill, one of OSS's lethal cyanide tablets.[67] In the morning, Donovan asked Smith to hold his wallet and identification papers until he returned. "If anything goes wrong, it'll be just as well if I'm incognito," Donovan explained. "That's an understatement, General," Smith replied.[68] After breakfast, Donovan at first refused the parachute Eifler offered. "I'll ride the plane down if we crash," he said. "I can't afford to be captured." To which the boastful Eifler retorted, "General, if we land within fifteen feet of the enemy, I will bring you back. Please put on your chute."[69]

With both wearing parachutes, Eifler flew the little two-seat plane over the dense jungle, past Japanese outposts, and in about two hours landed at the short camouflaged airstrip of OSS camp Knothead, on the opposite side of a mountain range from Myitkyina airfield. On the ground, Donovan spent several hours talking with Captain Vincent Curl and his men and, through an interpreter, some of the Kachin. The visit over, Donovan and Eifler left in a hair-raising takeoff, as the little plane, overloaded by a combination of new fuel and the burly Eifler and Donovan, each weighing over 200 pounds, barely cleared a gap in the trees at the end of the airstrip. Elated, Donovan returned to Nazira in a jubilant mood, but Lieutenant Colonel John Coughlin, executive officer of Detachment 101, who had been away, pulled him aside and demanded: "General, what were you thinking about to go in there with Carl?" Donovan replied simply, "I had to." "You should have considered more things than your damned honor," Coughlin snapped. "If I'd been there, I would have reminded you of every one of them."[70] Subsequently, citing medical reasons, Donovan sent Eifler back to the United States and replaced him with Lieutenant Colonel Ray Peers.[71] The burly Eifler returned to Washington in 1944 a legend as a successful jungle commander and the "deadliest colonel." He gave lectures at OSS training camps in the United States and then was given a series of important assignments by Donovan for missions, including training teams of Korean and American saboteurs to be sent into Korea and Japan itself, but the war ended before Eifler's teams, trained in the United States, could be deployed to the Far East.[72]

Detachment 101: Driving Back the Japanese

Beginning in 1944, Detachment 101 went beyond intelligence gathering, sabotage, and harassment of the enemy to provide direct assistance to a major Allied offensive to capture Myitkyina airfield and drive the Japanese out of Burma. The main out of the plane when he announced "I'm going behind Jap lines." Unlike Detachment 101 Executive Officer, John Coughlin, Smith said that with that trip into danger, Donovan "earned the lifelong loyalty of everyone at 101." Anglo-American offensive included Merrill's American "Marauders" and the Wingate's British "Chindits" but their received valuable assistance from the OSS and their Kachin tribesmen. The few hundred Americans and their indigenous guerrillas ambushed enemy troops, severed their lines of communication and supply, undermined Japanese resources and morale, and provided scouts to guide the spearheads of the attack,"[73] Under Peers command, Lieutenant Vincent Curl's team in the jungle and other field units began to organize attack groups of Kachins to coordinate with the conventional forces. Kachins also were assigned to provide intelligence about Japanese deployments and to guide and assist the advancing columns of Brigadier General Frank D. Merrill. By April 1944, Peers reported to Donovan that the collection of intelligence had been surpassed by the "sharp increase in the actual combat functions of our patrols."[74]

Among those leading such combat-oriented Kachin patrols was Lieutenant Joseph E. Lazarsky from Wilkes-Barre, Pennsylvania. The former sergeant and demolitions instructor at Area B had graduated from OCS and been sent to China in 1943, but after six months there, he was summoned by Peers to Burma. Lazarsky speaks modestly and in short and somewhat elliptical sentences when he describes his role. "A plane got shot down. Ray Peers got left with no one to take over the drops into the jungle....I taught some Americans how to run a drop into the jungle. We dropped throughout Burma, 45 agents—Anglo-Burmese, Kachin....Once in the Burma jungles, we tied up with the Kachins. I had demolitions. I recruited Kachins and was made commander of the 1st Kachin Battalion. I had seven Americans and five Britishers with me. My sergeant major was a top Kachin. He spoke English. He had been in the Kachin Rifles in the British Army. But earlier in the war, before I got there, he had been captured by the Japanese, who tortured him to try to make him talk. They gave him the hot water treatment. They poured scalding hot water down his throat. He lost his voice almost completely. Only a whisper. We were going to lay an ambush, and he and I discussed it. We ambushed the Japanese at Lashio and many other places in Burma."[75]

OSS and its indigenous guerrillas went behind hit and run tactics to stand and fight engagements against regular troops in the final Allied campaigns against the Japanese in Burma in 1944 and 1945. Under Peers, Detachment 101 expanded the recruitment and training of the Kachins, and it organized them into virtually a small Army, nearly 10,000 tribesmen in ten battalions directed by officers like Lazarsky, Pamplin, and Curl. Through the use of mobilized indigenous guerrilla forces and the provision of combat assistance as well as intelligence information to the spearheads of advancing conventional forces Detachment 101 was instrumental in the first major Allied military success in North Burma, the defeat of an elite Japanese division and the capture of Myitkyina airfield in August 1944. The role continued as additional Allied conventional troops pressed forward to capture Bhamo and Lashio.[76]

In the attack on Lashio, the key to the Burma Road, Lazarsky, of the 1st Kachin Battalion, led off, attacking Japanese infantry and motorized columns on the Burma Road itself. The Japanese chased him with infantry, artillery and tanks. Although withdrawing, he continued to ambush his pursuers. When Lazarsky reached his resupply airfield, his unit dug in and beat the Japanese back in a three day battle.[77] The heaviest prolonged fighting by Detachment 101, and some of the heaviest fighting in all of Burma, was done by the 3rd Kachin Battalion. Among junior officers was Lieutenant Roger Hilsman, a West Pointer, who had arrived with Merrill's Marauders, been wounded, and subsequently joined Detachment 101.[78] At Lawksawk, facing a thousand Japanese in a fortified position, the 3rd Kachin Battalion of roughly equal size surrounded the field fortifications. The Kachins first attacked directly amidst withering enemy fire and hurt but did not overcome the enemy. The OSS called in air support, but although the fighter bombers damaged the fortifications and inflicted many casualties, the Japanese remained entrenched if still surrounded. Finally in desperation, the Japanese defenders counter-attacked in a banzai charge against one segment of the Kachin line, 700 Japanese soldiers against 400 Kachin Rangers and Hilsman and the other American officers. But showing extraordinary discipline and courage, the Katchin tribesmen from their positions in the jungle withstood repeated charges over several hours by the Japanese soldiers, and the Kachins eventually gained a costly but important victory in the siege at Lawkswak.[79] The campaign to reopen the Burma Road was complete with the capture of Lashio in March 1945. Thereafter, the OSS unit and its guerrilla battalions attacked scattered enemy forces in eastern Burma and sought to block the flight of the Japanese into Thailand. The Japanese Army in Burma surrendered in Rangoon on 28 August 1945.

Detachment 101 amply demonstrated the possibilities of unconventional warfare that Donovan advocated and the multiplier effect of innovative, energetic, well-trained special operations leaders. It had begun with only two dozen Americans in 1942 and part of 1943 before the dramatic expansion of 1944 and 1945. Even at its peak strength, Detachment 101 had only 131 officers and 558 enlisted men, with 120 Americans serving out in the jungles at any given time. A dramatic multiplier, they mobilized, armed, supplied, and directed an indigenous guerrilla force of 10,000 men. They played an important role in defeating the Japanese in Burma.[80] In addition to their role as guides, rescuers, intelligent agents, and guerrillas, the men of Detachment 101 killed 5,500 Japanese soldiers, killed or seriously wounded and estimated 10,000 other Japanese soldiers; blew up 51 bridges, derailed 9 trains, destroyed or captured 277 trucks or other vehicles, and demolished 2,000 tons of ammunition, gasoline or other Japanese supplies.[81] The cost was 27 Americans, 338 indigenous guerrillas, and 40 native espionage agents killed.[82]

Early in the final campaign, Carl Eifler told instructors at Areas A, E, and F that the most important part of OSS training was to inspire students with the organization's mission and the need for flexibility and innovation to achieve it. He stressed the importance of "aggressiveness and a driving energy to get the OSS's job done." Given the current urgent demand for men in the field, Eifler explained in July 1944, "the selection of the right men for the jobs, aggressive men with drive and determination, is more important than the training we can give them." Peers, who had succeeded him as Detachment 101 commander, disagreed with the latter part. While Eifler did not consider providing trainees with background knowledge of a country, its society, politics, and culture, to be very important, Peer saw it as essential and recommended increased background, a kind of area studies, be provided by to American trainees in OSS schools in the United States.[83]

For its heroic and effective action in clearing northern Burma, Detachment 101 received a Presidential Distinguished Unit Citation. The citation, issued by General Dwight D. Eisenhower, as Army Chief of Staff in 1946, represented particularly high praise from the head of the Regular Army: "The courage and fighting spirit displayed by the officers and men of Service Unit Detachment No. 101, Office of Strategic Services, in this successful offensive action against overwhelming enemy strength, reflect the highest traditions of the armed forces of the United States."[84]

OSS Detachment 404 Raids the South Burma Coast

While OSS Detachment 101 fought in the mountainous jungles of northern Burma, OSS's Arakan Field Unit of Detachment 404 sent its nearly 200 OG, SI, and MU personnel on more than three dozen missions from 1944 to 1945 on raids along the mangrove filled Arakan coast of southern Burma.[85] Detachment 404 was headquartered in Kandy, Ceylon (now Sri Lanka) as was the Allies' Southeast Asia Command under British Admiral Lord Louis Mountbatten, with which it coordinated its actions. With eventually 595 personnel in Ceylon Detachment 404 was responsible for OSS operations the southern Burma coast, Thailand, southern French Indochina (Cambodia and southern Vietnam), Malaya and the Dutch East Indies (Indonesia). Headed first by Col. Richard P. Heppner, Detachment 404 included not simply the Arakan Field Unit, but personnel from most of the OSS branches. Its Research and Analysis Unit, responsible for finding information about industrial targets for bombing in Japanese occupied countries in its region, was headed by Cora Du Bois, a specialist in Southeast Asia, who had bean an instructor at Sarah Lawrence College with degrees in anthropology from Columbia and the University of California when the OSS recruited her.[86]

The Operational Group section of the Arakan Field Unit of Detachment 404 had trained at Areas F and A. Headed by Major Lloyd E. Peddicord from Dotham, Alabama, with Captain George H. Bright as its operations officer, the OG included 19-year-old Lieutenant John C. Hooker, Jr., from Atlanta, Georgia, and Lieutenant Louis A. O'Jibway, a full-blooded American Indian from Sault Ste. Marie, Michigan. The entire team had trained at Areas F and A in the early summer of 1944.[87] Subsequently, from December 1944 to February 1945, first from Ceylon and later from Akyab, Burma, they were transported in Maritime Unit fastboats. These were similar to Navy PT (Patrol Torpedo) boats but shorter and without torpedo tubes. From these fastboats, the OSS teams were put ashore in their rubber boats and searched the shoreline and nearby villages in advance combat scouting groups before the main units arrived by landing craft. On some of their nighttime reconnaissance trips, their mission was to determine if Japanese troops were in force on or near the beach. Sometimes they were. At Ramtree Island, on the night of 19 January 1945, Bright's team killed two Japanese sentries on the beach, and enemy mortar shells soon started raining down on them. Bright, Hooker, and O'Jibway and their teams quickly paddled out to the waiting fast boats and sped away. The OSS men returned the next night in a 110-foot, heavily armed, British motor launch to a nearby river. Bright was in charge. Hooker's team, armed with Browning Automatic Rifles and M-3 submachine guns, was scattered around the forward deck. With muffled engines, the launch moved up the narrowing river through the jungle. About 3 a.m. they reached an area so narrow the tree branches scrapped the sides of the boat. "Suddenly all hell broke loose. Unseen enemy were firing on the boat from both sides of the stream," Hooker recalled. "Small cannon, possibly 37 millimeters, fired into the bow and twice more into mid ship. All aboard the boat opened fire, while the British guns [six Lewis machine guns] swept the sides of the river, and we concentrated on the areas where there were flashes. Putting the boat in reverse, we backed down the river. The stream was too narrow for us to turn around until we were about a mile down stream and out of range of the ambush."⁸⁸ One of Hooker's men was mortally wounded when a canon shell exploded in his face. He was the only married man in Hooker's team. The first sergeant's ear drums were burst by the blast. Captain Bright was hit in the chest by a rifle bullet. Several British gunners were hit and two other Americans suffered surface wounds from bullets and shrapnel. It took them five hours to get back to a hospital ship that was part of the British invasion force. Hooker, O'Jibway, and several of the others on the teams were later sent to fight first with Ray Peers in the Burma jungles and then to train and lead OSS OG Chinese commandoes against the Japanese in China.

Penetrating the Tangled Situation in Thailand

Situated between the British colonies of Burma and Malaya and the French colony of Indochina in 1941, Thailand (previously known as Siam), was the only independent nation in Southeast Asia at the outbreak of the war. The Bangkok government under Premier and Field Marshal Phibun Songkhram officially allied with Japan and declared war on Britain and the United States. The Japanese Army used the country as a springboard to invade the British colonies of Burma to the west and Malaya to the south. Because of Bangkok's alliance, Tokyo kept only a small number of troops there and allowed Thailand to retain nominal independence.⁸⁹ While Great Britain reciprocated by declaring war on Thailand, the United States did not. Instead, the State Department chose to view Phibun's as a puppet regime and Thailand as an occupied rather than a belligerent nation.⁹⁰

In Washington, officials at the Thai Legation rejected the Phibun government's capitulation and collaboration with Japan. Members of the legation declared themselves to be "free Thais," committed to continuing the struggle against the Japanese. Secretary of State Cordell Hull supported their position and referred them to Donovan's organization.⁹¹ On 12 March 1942, the legation submitted a proposal to have a group of young Thai nationals in the United States trained and infiltrated into their homeland for subversive operations. Lieutenant Colonel Kharb Kunjara, the air attaché at the legation, met with Lieutenant Colonel Garland H. Williams of Special Operations, to arrange for the young Thai student volunteers to be trained, equipped and deployed by Donovan's organization. The first group of thirteen Thais began OSS training on 12 June 1942, first SO training at Areas B and A, then radio training at Area C, followed by Parachute training at Fort Benning, and Maritime training at Areas A or D.⁹² The Free Thai legation, angry at the collaborations policies of the Phibun government, declared that the larger goal of these trainees was to penetrate "into Thailand proper to organize subversive works and to pave the way for the final push of the United Nations Armed Forces, to drive the Japanese back to their own little islands."⁹³

Although Thailand had little if any strategic military importance to the U.S. War effort in Asia, and there were no major battles fought there, the OSS paid considerable attention to it and developed extensive operations there during the final year of the war. In some respects this was due to the OSS's need for a significant role against Japan, but more influential was the U.S. political effort to avoid British imperial expansion into the country in the postwar era and to achieve an independent pro-American Thailand. The OSS mission in Thailand, taken in coordination with the State Department, was primarily a political mission of U.S. foreign policy.

Heading that mission was Captain Nicol Smith, formerly a successful author of adventure travelogues and who had pre-war experience living in the Far East as well as other areas around the world. Smith joined the OSS in early 1942 and trained with Ray Peers and the Detachment 101 group at Area B in April 1942.[94] After further SI training, Smith had been sent on an espionage mission to Vichy France. Returning in December 1942, he was assigned to equip and lead the first and second groups of OSS trained Thai nationals, 21 in all, to China for eventual clandestine deployment in Thailand.[95] The young Thais, who had been graduate students at Harvard, M.I.T. and other leading American universities, had completed their training by mid-January1943, as plans for the mission had evolved, and were in a holding area in or near Prince William Forest Park.[96] In March 1943, Smith and two OSS instructors from Area B, Frank Gleason and Joseph Lazarsky, took the new Thai agents to India and over the "Hump" to China.[97]

From China, these agents were to be infiltrated to help what appeared by spring 1943 to be a growing anti-Japanese underground in Thailand, some of it encouraged by dissident members of the government in Bangkok. The primary Allied contact was Phibun's main rival, the pre-war Finance Minister, Pridi Phanomyong. He, like most Thais, had opposed collaboration with the Japanese. Obstacles resulting from the frustratingly complex political situation in China precluded Smith from dispatching agents into Thailand until June 1944, but even then, two of the infiltrated agents were killed and six others captured and imprisoned.[98]

Advances in Thailand

The next month, July 1944, with the Japanese being driven back in Burma and the Pacific, Pridi was finally able to topple the collaborationist Phibun government through political maneuvering. OSS parachuted two new Thai agents into their country. Although one was captured, the other, who had close family ties with Pridi, met the new head of the government. Shocked, the agent learned that not only was Pridi now head of the government, but he had long been and still was in fact the head of the underground and clandestinely leading the "Free Thai" movement within Thailand.[99] He had kept that subversive role secret in order to avoid repressive measures by the small, but well armed Japanese force in Thailand. Donovan met a Thai delegation in Ceylon in January 1945, and with Pridi's approval, an OSS mission of two American officers, one from SI and the other from SO, soon arrived in Bangkok by seaplane and fast boat.[100] To counter British postwar claims based on the country's alliance with the Japanese, Pridi suggested a Thai uprising supported by an invasion by two U.S. Army divisions. The Joint Chiefs of Staff did not want to divert such large numbers of troops, numbering up to 30,000, but under OSS prodding, they agreed to have Donovan's organization arm the Thai underground and use them for intelligence gathering. Such efforts would enhance the U.S. foreign policy goal of an independent, pro-American Thailand and help prevent the British from absorbing Thailand as a colony. In June 1945, when JCS approved the plan, several American OSS officers and enlisted men were parachuted into Thailand. One was killed by the Japanese and some of the others were captured by Thai police. But the remainder established bases in the northern and southern parts of the country to run intelligence networks and to arm and train Thai guerrillas. The OSS goal was not to prepare them to fight the Japanese as in Burma, but to "retain and increase the good will of Thailand toward America in this politically critical area."[101]

Delays in JCS approval meant the expanded program was never fully implemented, but OSS made a start. By mid-August 1945, OSS had only seven fully effective SO guerrilla bases. Marine Lieutenant William Butts Macomber, Yale graduate, who received his SO training in Maryland and Virginia in early 1944, before being sent to England and then parachuting into France as part of an SO/SOE to reinforce the "Freelance" circuit, may have headed one of these. He had reported to OSS Detachment 404 headquarters in Kandy, Ceylon, in 1945 and subsequently participated in combat operations along the Burma-Thailand border.[102]

Near Prae, at one of the OSS SO camps established in the summer of 1945 to train Thai guerrillas, OSS Sergeant Steve Sysko, 20, a Polish American from Springfield, Vermont, who had been trained at Areas A, B, C and F, noted in his diary that his SO team had trained and outfitted 300 guerrillas in three weeks, but they did run into some unusual problems. On the way to one of the drop zones for supplies, Sysko noted, "We ran into some wild elephants. In order to escape from them, we ran down the hill. Elephants cannot go down hill very fast because they have to feel with their front feet. We escaped and returned to camp. We did not get the drop." Later the four-man team and its Thai hosts rode elephants carrying them and their equipment to new campsites, one step ahead of the Japanese.[103]

Many of the young Thais trained at the various OSS camps in Maryland and Virginia in 1942, were parachuted into Thailand in 1944 and 1945 for espionage work, and they organized networks and sent back useful intelligence. They also helped rescue downed aviators. One of the most dramatic rescues was of one of the P-40 pilots on Major General Claire Chennault's "Flying Tigers." Lieutenant William ("Black Mac") McGarry, an ace with 10 victories, had been shot down during a daring raid on a Japanese airfield in northern Thailand in March 1942. His squadron mates had seen him bail out and wave to them from a clearing it the jungle, but that was the last seen or heard of him.[104] In late 1944, as Nicol Smith began to move his agents into Thailand, Chennault asked the OSS to find out if McGarry were still alive. Within four days, Smith's agents learned that he being held in a small POW compound, and Smith launched an operation to rescue him. Put in charge was Wimon Wirayawit, a Thai national and M.I.T. graduate student, who had enlisted in the Free Thai movement in the United States in March 1942, coincidentally the very month that McGarry was shot down. Now, on the night of 9 September 1944, the former M.I.T. graduate students parachuted near the POW camp, becoming the first OSS-trained Free Thai to be successfully infiltrated into Thailand. Working with local Thais and the camp commander, who willingly told the Japanese that McGarry had died, Wimon Wirayawit rescued the American aviator and escorted him to a safe house near Bangkok. By arrangement, two OSS PBY "Catalina" seaplanes flew at night from Ceylon to the Gulf of Siam for the retrieval. Wimon obtained a Thai Customs Patrol boat to take McGarry, two OSS officers, and five Free Thai agents, down river and out to meet them. One of the OSS men was feverish and frequently shouted deliriously in English as the boat moved down river that night. To prevent the Japanese from hearing him, whenever an enemy patrol boat would come near, the Free Thais would go on deck and exuberantly sing and dance traditional Thai songs. Moving out into the gulf, Wimon's boat delivered McGarry and the others to the PBYs, which flew them back to Ceylon. The rescue mission was a complete success, and McGarry was soon back in China flying with Chennault's airmen again against the Japanese.[105]

Like many other OSS officers who served in Southeast Asia, Nicol Smith emphasized the need to deal fairly and with respect with indigenous people recruited and trained to work for the OSS. He attributed the success of the overall OSS mission in Thailand the ability of OSS personnel to operate diplomatically with an increasingly wider circle of influential people and to exhibit patience, tact, and respect in leading the Free Thais. "Never, in the months of isolated comradely association with this small group," Smith said, could he "afford for one minute to treat them as anything other than brothers-in-arms and full equals." He slept in the same room with them, ate the same food. Any hint or small suggestion evident in his words or behavior that he considered himself superior, would have meant failure to his mission.[106] A summary of OSS Achievements prepared by OSS Schools and Training Branch in 1945 concluded that "operations in Thailand have been brilliantly successful."[107]

The contributions of Pridi and the Free Thai movement, as well as U.S. pressure, led to the recognition by the Allies of wartime Thailand as having been an occupied nation rather than a Japanese ally. At the end of the war, OSS officers advised Pridi to delay signing the Anglo-Thai peace treaty as unduly harsh (Britain had abandoned demands for annexation of Thailand's Kra Peninsula leading to Malaya, but insisted on the right to have British bases there). Pridi did reopen negotiations, and with the support of the State Department, his successor, the former Free Thai envoy, obtained a treaty in 1946 that maintained complete Thai sovereignty. The OSS effort on behalf of an independent Thailand had been a success.[108]

OSS, the Japanese, the Viet Minh and the French in Indochina

During World War II, French Indochina—today Vietnam, Laos and Cambodia—was still governed as a colony by collaborationist Vichy France while occupied by the Japanese Army. Discontent arising from the occupation and the starvation caused by exportation of most of Vietnam's rice to Japan, provided an opportunity for Vietnamese anti-colonial nationalists to mobilize against the French as well as the

Japanese. President Roosevelt's wartime policy favored decolonization and independence for former colonies even if in phases, such as U.N. trusteeships. But Britain and France wanted to re-establish control over their colonies in the Far East. By early 1945, with victory assured and attention focusing more clearly on the postwar order, the Roosevelt administration's fervent opposition to restoring Europe's colonies began to wane.[109] When Roosevelt died in April 1945, just as the European war was ending, the Allies were experiencing increased friction with the Soviet Union. Roosevelt's successor, Harry Truman, and the State Department were more interested in Europe and China than Southeast Asia. At the founding of the United Nations in San Francisco in May 1945, the Secretary of State reassured the French that Washington had never officially questioned Paris's sovereignty over Indochina.[110]

In the Far East, much of the OSS was largely unaware of these gradual and unofficial shifts in U.S. policy. Official U.S. policy toward Indochina, which had been governed by a Vichy French regime that collaborated with the Nazis and the Japanese, was for independence. While the war continued, OSS operatives in the field focused on fighting the Japanese. But at the same time, they were suspicious of the French, both the colonial administrators, police, and soldiers of Vichy and the Free French representatives of de Gaulle's new government being sent out from Paris to re-establish French control throughout Indochina as Japan was defeated. But Japan still needed to be defeated, and American civilian and military authorities in China, including the OSS, had difficulties working with the French, because de Gaulle's representatives argued over command and because the Vietnamese would not cooperate with the former colonialists. That left primarily the Viet Minh, a nationalist, anti-colonialist organization, that since 1944 had been actively leading resistance to the Japanese in Vietnam and helping to rescue downed American aviators.[111] The Viet Minh was headed by Ho Chi Minh.

Ho Chi Minh and the OSS

The OSS connection with Ho Chi Minh, longtime communist and Vietnamese nationalist leader, has remained controversial. Ho wanted American support against the French as well as the Japanese. His motivation in seeking U.S. support in 1944-1945 has been much debated as has the OSS motivation in working with him.[112] The OSS had worked with communist partisans in Italy, France, Yugoslavia and other countries. Donovan took a pragmatic view about working with the communists, accepting a broad range of allies in support of the OSS goal, which, he said, was "the earliest possible defeat of the Axis." In Vietnam, many of the OSS officers saw Ho as the main leader of a resistance movement that included both communist and non-communist nationalists. It was an insurgency against both the French colonialists as well as against the Japanese occupiers, and seemed supportive of the official U.S. policy of decolonization. Indeed, Ho and the Viet Minh were far more valuable in creating agent intelligence networks, rescuing downed aviators, and fighting the Japanese than were the French, who suspected that the United States was trying to replace them in Indochina, and whose primary interest was less in defeating the Japanese than in getting in French troops to restore Paris's control of Indochina.[113]

From its regional headquarters in Kunming, China, OSS began to penetrate northern Indochina. This was deemed necessary after the Japanese took control from the Vichy French in the spring of 1945 and previous sources of intelligence from Hanoi were cut off. SI officer Marine Lieutenant Charles Fenn, a British-born, American journalist, who had joined the OSS in 1943 and been trained in Maryland and Virginia, met in March 1945 with Ho Chi Minh, who agreed to allow OSS agents and radio operators in Vietnam in exchange for arms and medicines for the Viet Minh.[114] In April, a new and ultimately highly controversial OSS officer arrived in Kunming to head SI operations in Indochina. Army Air Corps Captain Archimedes L. A. Patti, a 31-year-old Italian American, had attended training camps in Maryland and Virginia, and in January 1944 had landed at Anzio when the Allies thought the landing would lead directly to Rome. Returning to Washington, he was put in charge of SI's Indochina Desk. Now in Kunming, the Army and the Air Corps wanted him to get information about Japanese military units and installations as well as to conduct sabotage missions against key railroad lines. Patti met with Ho Chi Minh, who with his Viet Minh agents and guerrillas already well established and in control in several provinces inside the country, seemed to offer the fastest way for OSS to set up operations there. Like many others, Patti's sympathy for the Vietnamese and disdain for the French would intensify as the Vietnamese courted the Americans while the French berated them.[115]

General Albert Wedemeyer, Stilwell's successor in China, approved the dispatch of two OSS/SO missions into Indochina in July 1945 to sabotage railroads, gather enemy target and deployment information and work with and each train some 50 Vietnamese guerrillas. Although composed mainly of Americans, they would include a few European Frenchmen as well. The teams were code-named Deer and Cat. The main SO mission, the Deer Team, parachuted into a jungle village about 75 miles northwest of Hanoi in mid-July. The commander was Major Allison K. Thomas. A native of Lansing, Michigan, who had lived in France for a while before the war and knew French well, had become a lawyer in Michigan. He joined the Army, became an officer, and because of his knowledge of France had been recruited from the Army by OSS. He served with distinction in X-2 Counterintelligence with Patton's Third Army in France and Germany, then transferred to China in the spring of 1945. He was then about 30 years old. His second in command was Lieutenant René Défourneaux, an American who grew up in France, The enlisted men included Sergeant Lawrence Vogt, the weapons instructor; Sergeant William Zielski, the radio operator; Sergeant Alan Squires, photographer; Private Paul ("Hoagy") Hoagland from Romulous, New York, the medic; and Private Henry ("Hank") Prunier, the interpreter.[116] He 21-year-old Prunier, son of a Worcester, Massachusetts family of Franco-American descent, who spoke fluent French and had a working knowledge of Vietnamese, or Annamese as it was called in French, because he had studied it at Berkeley for a year. Most of these men had trained at SOE camps in Britain and then at the West Coast OSS camps at Catalina Island and Newport Beach, California.[117] Hoagland had received medical training at other facilities and Zielski, who had been

OSS "commo," probably had received some training at Area C. In July of the previous year, Zielski had been part of an OSS team that parachuted into northern France. His team had sneaked into the occupied city of Brest by hiding in empty wine barrels alongside full casks being delivered to the German garrison. Despite the dangers, Zielski radioed important intelligence information to London while moving around behind enemy lines during the following two months. For his heroism, Zielski had been awarded the Distinguished Service Cross by General Eisenhower.[118]

In Vietnam, the Deer Team parachuted into the jungle camp of Ho Chi Minh, Vo Nguyen Giap, his military chief, and other Viet Minh leaders, who received them with open arms and a banner in English, "Welcome to our American Friends."[119] Ho, who was not at the reception, was ill, possibly suffering from malaria, dengue fever, and/or dysentery. Hoagland, the American medic, gave him quinine, sulfa drugs, and other medicines, and he soon recovered.[120] The Viet Minh turned over some POWs they had liberated in an attack on a Japanese garrison and they provided enemy target and deployment information that Zielski radioed back to Kunming. Thomas, like Patti,, concluded that the Viet Minh were the most effective force aiding the Americans in obtaining intelligence and in fighting the Japanese. So at the end of July and beginning of August, the Deer Team brought in American weapons, built a small training facility with barracks, assembly hall, radio room, and firing range, and began training some 40 guerrillas, that the Viet Minh labeled the "Vietnamese-American Force."[121] They showed them how to use grenades, mortars, bazookas, and primarily how to shoot the M-1 rifle and the American carbine.[122] Although Thomas eagerly pursued the training and all the Americans participated in it, some of the team members were discontented. Many of them agreed with Sergeant Vogt who complained that he "had volunteered to kill Japs, not to be a drill sergeant." Lieutenant Défourneaux distrusted the communists and disagreed with Thomas's leadership. Nevertheless, they continued to train the Viet Minh until 15 August 1945 when Tokyo accepted the Allied terms of surrender.[123]

Closely linked to them was the Cat Team consisting of SO personnel, Captain Charles M. ("Mike") Holland and two sergeants, John Burrowes and John L. Stoyka, the radio operator. They had parachuted to Ho Chi Minh's camp on 29 July several days after the initial Deer Team had arrived. There they met Ho, Giap and the other Viet Minh leaders. Moving out into the jungle and setting up their own camp to pursue their own mission, they were captured in mid-August by the Japanese. Stoyka, who had served with a Jedburgh mission in France and then with an SO team in North China, managed to escape. Viet Minh villagers guided him to the Deer Team. Possibly because Stoyka had escaped with news that Holland and Burrowes were prisoners, the Japanese released the two Americans unharmed in Hanoi on 31 August, and Ho had welcomed them warmly as representatives of the United States.[124]

After the Tokyo Surrender

Tokyo's acceptance of the Allied terms of surrender on 15 August 1945, led to the dispatch of OSS missions to rescue POWs and interned civilians from Japanese internment camps. Immediately after the news from Tokyo, OSS in Kunming had sent a nine-man SO/SI rescue team to the border area of northern Indochina under Major Aaron Bank, graduate of OG training at Areas F and A, who had served as a Jedburgh in France in 1944. He had arrived in China in June 1945, and been scheduled to lead a team into Laos in August to disrupt Japanese lines of communications, but it was cancelled at the last minute.[125] Bank's new assignment was to search for hidden POW camps in the jungle, investigate possible war crimes, and report on general conditions in the area.[126]

Tokyo's decision to surrender had left a power vacuum in Indochina. It also provided an unprecedented opportunity for Ho Chi Minh and the Viet Minh. Ho announced the formation of the National Liberation Committee of Vietnam and on 20 August, the Viet Minh took over Hanoi. On 22 August 1945, Patti and the Quail Mission were sent to Hanoi to rescue prisoners and assess the situation there. Bank was waiting at Gai Lam airport outside Hanoi when Archimedes Patti and the Quail Mission arrived. Patti had immediately freed the POWs in an adjacent prison camp, and then linking up with Bank's team, they drove into Hanoi. The Americans, the first Allied "force" to arrive in Hanoi, were welcomed, but the Viet Minh and the unruly crowd denounced the French. Patti was a controversial figure, later accused of overreaching his authority and he was eventually recalled. "He was very volatile, a glory seeker," recalled Henry Prunier a member of the "Deer" Team under Allison Thomas that had parachuted into Ho Chi Minh's jungle camp several months earlier. "We were not glory seekers. In his book, he wrote that the 'Deer' Team was just an experiment. I could have shot the son-of-a-bitch. He was experimenting with our lives!"[127]

Ho Chi Minh, the charismatic, fatherly-looking, nationalist and communist leader of the Viet Minh sought to use the presence of American OSS officers, like Allison Thomas, Archimedes Patti, Aaron Bank, and Mike Holland, to indicate U.S. support for the Viet Minh' anti-colonialist movement. Thomas had his unit accompany Viet Minh guerrillas parading through jungle villages behind a Viet Minh flag.[128] Holland had allowed Ho to portray him as a representative of the United States in Hanoi, and in September, Holland and Bank, eventually finding themselves in Hanoi with no transportation back to their camp, accepted Ho's offer of his car and accompanied him as he drove south, stopping at towns and cities, speaking to cheering crowds, each time gesturing to the two OSS officers in their American military uniforms as evidence of U.S. support for the goal of Vietnamese independence. He left them after a rally at Hué, the old imperial capital but provided them with a car and driver to take them back to their camp in Laos.[129]

Patti, believing that he was still following Donovan's order in April not to help the French restore control, favored the Viet Minh and sought to mediate between them and the French. But times had changed. Not all OSS officers agreed with Patti or favored the Viet Minh over the French. Paris-born, American Jedburgh, Lieutenant Lucien E. Conein, who had OSS training at Areas F and B, had spent much of the summer of 1945 providing similar guerrilla training to French troops along the Indochinese border and arrived in Hanoi shortly after Patti, but was entirely supportive of the French.[130] More importantly, OSS headquarters in

Kunming by the end of August had reprimanded Patti for not offering more help to the French members of the team and their efforts to re-establish French control. The next day, 31 August, the French met with the Viet Minh and began to inform them of France's demands. Ho responded on 2 September 1945, the day Japan formally surrendered aboard the *U.S.S. Missouri* in Tokyo Bay, by quoting the U.S. Declaration of Independence and proclaiming the establishment of an independent, Democratic Republic of Vietnam. The Communists also started outlawing their rivals and eliminating, by imprisonment or execution, many of their opponents. The U.S. Government did not recognize Vietnamese independence, in fact it followed a neutral and then supportive stance toward French sovereignty. Ho and the Viet Minh had sought to use the OSS support in the fight against the Japanese in 1945 to solidify their position in Vietnam as well as to obtain recognition from the United States. But the key to the August 1945 Revolution, the seizure of Hanoi, Saigon and many provinces, as well as Ho's declaration of independence was not the OSS but the fact that the Viet Minh seized the opportunity in the power vacuum when the Japanese suddenly surrendered.[131] By the end September, the Chinese government had recognized French sovereignty in Indochina and on 28 September 1945, the Japanese commander formally surrendered to the Chinese, the Americans—and to the French.

"Cochinchina is Burning"

Cochinchina, the area of Vietnam south of the 19th Parallel, was, under the Potsdam Agreement of July 1945, to be occupied temporarily by the Anglo-American South East Asia Command, headed by the renowned Lord Louis Mountbatten. It was to accept the surrender of the Japanese and maintain order there until the future of Indochina was determined. Using Indian troops, the British were in charge of the occupation, but their first priority seemed to be the facilitating the deployment of French forces to reassert Paris's control over its former colony, as the British were doing in Burma and Malaya. In the midst of this, a small OSS/SI team began arriving in Saigon on the first week in September 1945 to rescue POWs and report on political developments. "Operation Embankment," was headed by Major A. Peter Dewey, 28, scion of a prominent Republican family. Dewey's father was a congressman, and Thomas E. Dewey, New York governor and 1944 Presidential candidate, was a relative. As a youth, Dewey had been educated in Switzerland and studied French history at Yale. He served as a secretary to the U.S. ambassador in Berlin, and then became a Paris correspondent for the *Chicago Daily News*. During the German invasion of France in 1940, Dewey worked with an ambulance unit, then supported the Free French under Charles deGaulle.[132] He returned to the United States, and according to Capt. Charles Parkin, Jr., an instructor at OSS Area B in Catoctin Mountain Park, Dewey served as a civilian instructor there for a while in 1942.[133] In the fall, Dewey enlisted as a lieutenant in the U.S. Army and subsequently served on intelligence missions in Africa and the Middle East. Dewey joined the OSS in Algiers in 1943. In August 1944, he parachuted into southern France, heading a team that with the local *maquis* gathering intelligence and then during the invasion, captured 400 Germans and destroyed three tanks. Returning to Washington in October, he worked in OSS headquarters for several months, until he was chosen for deployment to Saigon after the Japanese surrender.[134]

Dewey and his team arrived in Saigon the first week in September, but by the middle of the month, they had run into severe difficulties with the British and the French. Dewey had been meet with the Viet Minh, and he vigorously objected to the British commander's policy of evicting the Viet Minh from the government offices, police stations and military barracks they had seized from the Japanese and Vichy French. Most of the American team was disgusted when the British allowed the French, whom they had rearmed, to take over Saigon and inflict brutal reprisals on the Vietnamese. In retaliation, the Vietnamese across the political spectrum joined in a general uprising and open guerrilla warfare throughout the city.

Under pressure from the British and French, OSS headquarters ordered Dewy back from Saigon. In the midst of the bloody uprising against the French, local guerrillas ambushed and wounded OSS Captain Joseph Coolidge of the OSS Research and Analysis Branch. The next day, 26 September, the day Dewey was scheduled to leave, he was killed when guerrillas with machine guns opened fire on his jeep. The Vietnamese guerrillas had apparently mistaken him for a French officer. In Hanoi, when Ho Chi Minh learned about Dewey's death, he rushed to U.S commander, General Philip Gallagher, and expressed his deep sorrow. He sent letters of condolence to Dewey's parents and to President Harry Truman.[135] Although there were no more OSS casualties in Indochina, by January 1946, when the British Army handed over full authority in southern Vietnam to the French and departed, the brutal guerrilla war around Saigon, involving assassinations on both sides, had caused more a thousand casualties, Vietnamese and French. In Dewey's final report to OSS, written the day before he was killed, he had concluded: "Cochinchina is burning, the French and British are finished here, and we ought to clear out of Southeast Asia."[136]

OSS and the Tangled Web in China

From the time the U.S. entered the war in December 1941, Donovan's headquarters in Washington had high hopes for a major role in China. But the situation in China was a tangled web of competing interests, among both the Chinese and the Americans. Viewing OSS as a threat, rivals managed to limit its operations for most of the war. Not until the final year of the war did OSS achieve a breakthrough in China and begin to contribute significantly to the defeat of Japan.

Despite numerous entreaties and emissaries, Donovan was unable to obtain an effective foothold in China for his organization until 1943. CBI Theater Commander General Joseph Stilwell allowed Detachment 101 in Burma, but declined to permit Donovan's autonomous, centralized intelligence and special operations agency to operate in China unless it was approved by Generalissimo Chiang Kai-shek. Chiang would not approve it without the recommendation of his intelligence chief, General

Dai Li, who headed a combined intelligence and secret police organization with perhaps 300,000 agents, and who jealously sought to prevent any independent intelligence operations in China.[137] Since Chiang and his generals were using American military and economic assistance for domestic purposes or stockpiling it for the postwar civil war with Mao Tse-tung and the Communists rather than using those weapons and supplies to help defeat the Japanese as the Americans wanted, Chiang Kai-shek did not look favorably upon an independent American intelligence agency reporting directly to Washington about his government and Army.[138]

When Army intelligence had refused in late 1941 to allow Donovan's organization or the Office of Naval Intelligence to be part of its mission to China, the Navy had made its own arrangement, directly with Chinese spymaster, Dai Li. The head of the Navy's mission was Captain Milton ("Mary") Miles, a veteran of the Yangtze River patrol in the 1920s.[139] His assignment was to gather intelligence about Japanese air and naval bases, coastal shipping, and the weather moving eastward across China toward the ocean, all of which would aid the U.S. Navy in the Pacific. Miles arrived in April 1942, established a working relationship with General Dai Li and, since both Donovan and the Army's Intelligence Mission had been rebuffed, Miles' "U.S. Naval Group, China," became the only American intelligence organization initially allowed into China by Chiang Kai-shek.[140]

Donovan originally tried to work with Miles and Dai Li, by designating Miles as head of the OSS in China, but he had to go further to win over Dai Li. It was the millions of dollars that OSS had available, much of it in "unvouchered funds" for clandestine operations, which did not require accounting, that made Donovan's organization so attractive to Dai Li.[141] In a friendly gesture in December 1942, Donovan got Miles and Dai Li to agree to his plan for sending OSS Special Operations (SO) teams to China to arm and instruct Chinese guerrillas. Unknown to Dai Li and Miles, Donovan and Alghan R. ("Al") Lusey, an American businessman with much experience in China who later became head of the Secret Intelligence Branch (SI) in China, secretly planned to use the SO men eventually as SI spies, connected by a clandestine radio network what would also be set up under the rubric of Special Operations. Even the American SO operatives themselves did not know of Donovan's and Lusey's ulterior motives. The idea was to give the Chinese as many SO men and materials as possible, in exchange for, what was to be kept secret: a foothold in China for future intelligence operations.[142]

OSS's new position in China was formalized in April 1943 in a secret treaty establishing the Sino-American Special Cooperative Organization or SACO (pronounced "SACKO"), among the U.S. Navy, Army, OSS and the Nationalist Chinese Government. The United States agreed to provide weaponry for 85 Chinese Special operations units and establish 13 SO Schools to train the Chinese recruits. In return, the Americans would be allowed to establish several weather monitoring stations, some W/T broadcasting units, and four intelligence stations along the southeastern coast of China. But to get his foothold in China, Donovan had to agree that Dai Li was head of SACO and Miles was his deputy; Dai Li retained control over the activities of the Chinese SO units that were to be trained, armed, and deployed by the Americans.[143] However, despite Dai Li's goal of closely regulating OSS operations in China, the SACO agreement in the long run actually enabled Donovan to maneuver among the Navy, Amy and the Chinese Government and ensure that OSS would survive and even expand in China.[144]

OSS and the SACO Training Camps in China

Under the SACO agreement, more than a dozen training camps were established in 1943 to prepare Chinese recruits for special operations with indigenous guerrillas behind Japanese lines. Both the Americans and the Chinese provided instructors, but the bulk of the instruction was done by the U.S. military personnel. Not many of the Americans spoke Chinese, or at least more than a few words or phrases, so interpreters, usually local Chinese but sometimes Chinese Americans, translated their instructions. The United States delivered all the weapons, munitions, and other materials. Miles later claimed that the SACO force totaled 97,000 Chinese and 3,000 Americans and that it killed 71,000 Japanese troops. This was a dubious assertion. The official number of graduates was 26,800, and the number of Japanese casualties they inflicted may well have been fewer than their own numbers. Later some of them, particularly those recruited from Dai Li's intelligence and secret police forces were used for internal control. But recruits from the occupied areas of eastern China had seen the suffering the Japanese had caused in their locales, and they were eager to strike back at the invaders.[145] Most of the 2,500 Americans who worked with SACO were from the Navy because of Miles, but a few were from the OSS, because of Donovan.

At least three of the SACO special operations training camps included former instructors from OSS stateside schools, especially Area B in Catoctin Mountain Park and Area A in Prince William Forest Park. SACO training camp known as Unit 1, the Xiongcun Ban in Chinese, was established in Xiongcun, Xiu xian, Anhui, in the mountains five miles south of Huizhou (Shexian), about 200 miles west of Shanghai. Because of Captain Miles, the commanding officer and most of the dozen American instructors were naval personnel, sailors or Marines.[146] The instructor in demolitions and subsequently apparently chief instructor and possibly eventually camp commander, was Major Charles M. Parkin, Jr., 28, U.S. Army Corps of Engineers and OSS.[147] Parkin had been with the OSS since early spring 1942 when he had been assigned as an instructor at Area B in Catoctin Mountain Park.[148] He arrived in China in June 1943 and served at SACO camp Unit 1 through February 1945.[149] The American instructors, through interpreters, taught weaponry, close-combat, demolitions, intelligence gathering and cryptography, sabotage and guerrilla warfare. The Chinese instructors were assigned for political indoctrination. The headquarters, classrooms and accommodations were located in a small rural temple and pagoda, but because this was the first SACO training camp, it took six months to get it set up and start classes.[150] The training cycle lasted three months for each class. The graduates did conduct several successful sabotage operations. One unit raided the Shanghai airport, destroying five fighter planes and a number of fuel tanks. Another unit, consisting of nearly 100 men, sent out in September 1943, reported that they had wrecked a train, assassinated a "puppet" collaborationist Chinese provincial governor and chief of secret police.[151] Parkin later reported that the camp had trained 7,000 Chinese guerrillas and that in five operations, its graduates were responsible for providing valuable intelligence, severing the Hangchow-Kinhwa Railroad in 250 places, rolling up 600 kilometers of telephone/telegraph wire, wrecking 3 trains and 5 pillboxes, and killing some 100 Japanese soldiers.[152]

The SACO camp known as Unit Two, Hongjian Ban in Chinese, was established in June 1943, near Changsa in Hunan Province, south central China. Naval officers and chief petty officers made up the majority of the American staff, but later that year two OSS Army officers arrived. Recently commissioned following Officers' Candidate School, they had formerly been sergeants and instructors at Area B, now Lieutenants Leopold J. ("Leo") Karwaski and Joseph E. ("Joe") Lazarsky, both from the coal mining region around Wilkes-Barre, Pennsylvania. They were assisted by 14 enlisted men. The aim, as in the SO camps in Maryland and Virginia, was to recruits in American and other weaponry and demolitions, field craft, sabotage, and guerrilla operations. Unit Two handled 300 Chinese recruits at a time, and, even though Karwaski and Lazarsky later left for more hazardous duty, it had established a reputation as the most productive of the SACO training centers. It produced 2,200 SO graduates between 1943 and 1945.[153]

"I was there for six months training Chinese soldiers in central China," Lazarsky recalled. "Leo and I saw that these Chinese from the Chinese Army divisions we were training didn't know anything. They didn't have much of a training program [in the Chinese Army]. We started from scratch. Anything we taught them was new to them. At first they got rather obsolete weapons, the Springfield '03 rifle, then later the M1, Garand. After I left [for Burma], Leo was teaching the Chinese about plastic explosives—looked like flour. [It could be stored in flour sacks to deceive the Japanese] He baked a cake with it. You could eat a touch of it, but not much. Four or five of the Chinese ate the cake and died."[154]

OSS officers, Major Arden ("Art") Dow, from Washington State, who had been an instructor at Area B and chief instructor at Area A-4, and Captain Frank Gleason from Pennsylvania, who had instructed in demolitions at Area B, were in charge at SACO Unit Three or Linru Ban. It was located just south of Loyang in Honan Province in east central China. Dow, 27, was the camp commander and Gleason, 25, his executive officer. They had come highly recommended. In February 1943, a key officer attached to the staff of Admiral Ernest O. King, chief of naval operations, wrote enthusiastically to Miles about the OSS men coming to China and described Gleason, an Army Engineer, instructor at Area B, and a graduate of SOE's sabotage schools in England, as "the best grounded man in the U.S. on industrial espionage [and sabotage]."[155] From Assam, India, Dow and Gleason had flown over the Hump, arriving at Kunming, in Yunan Province, the OSS China headquarters in July 1943. From there, they had flown to the wartime capital the Chinese Nationalist Government in Chungking (Chongqing), a city of three million on the Yangtze River in Sichuan Province in west central China. Chungking was also the site of Stilwell's headquarters, OSS's liaison, Miles's Navy Group, and the U.S. Embassy.

In late August, Dow and Gleason had left Chungking by Army truck convoy and six weeks and nearly two thousand miles of dirt road later, they arrived at the site of the training camp in early October. It was in an area around an old, abandoned Buddhist temple in the mountains of Fengxue si. Although several Chinese officers were assigned, primarily to observe the Americans and to provide ideological indoctrination for the Chinese trainees, the Americans ran the instruction. "It was just we two Army officers and eighteen Navy petty officers," Gleason recalled.[156] The Navy NCOs were there because Captain Miles. The school's mission was to train "armed special services units" to destroy Japanese lines of communication and supply in the area between Pinghan, Longhai, and Jinpu. The men were trained to be infiltrated and blow up railroads, bridges, and airfields. Given the rough and ready training, the camp commander gained the sobriquet "Rowdy" Dow, and he called the SACO camp, "Pact Rowdy."[157]

At the Training Camp of "Rowdy Dow"

"When we arrived on the scene in early October [1943]," Gleason later reported, "we found that we had to convert our training camp from a site which had been one for the monks, to one which could be considered an American type military establishment." It took them about a month to construct two rifle ranges, each with 30 targets, plus two grenade ranges, four demolition ranges, a combat firing range, a pistol range, and an arsenal. As Gleason put it in his report when he had returned to the United States nearly two years later, "Surprisingly, this work was interesting to us newcomers to China, for we were already learning the necessity to improvise when the required material could not be procured locally or in China. Our supply lines were over two thousand miles long and practically inaccessible. When we needed gasoline and oil for our generators, we found it much easier to buy the material *from the Japanese* [who sold it on the black market]. In fact, it would have taken months by any other manner."[158]

The first class at the camp consisted of some 400 Chinese trainees. They were recruits supplied by various Nationalist Army generals in what was called, the "First War Area." Daily and weekly training schedules were modeled after the training at Areas A and B in Maryland and Virginia. In the six weeks course, students would receive a total of 35 hours practice with a submachine gun, 29 hours with a carbine, 14 hours with a .38 caliber revolver, 16 hours practice throwing hand grenades, 18 hours of unarmed, close combat instruction, and 24 hours of demolitions and sabotage, plus lesser amounts for aircraft identification, observation, and reporting. In addition there were 42 hours of field problems in daylight and 15 hours of night problems, a total of 267 hours of instruction and practice.[159] Each class lasted between six and twelve weeks. In all, five classes graduated from the training school, Unit Three, between 1943 and 1945. In the early summer of 1944, the school had to move westward because of a major Japanese offensive. It re-located in the next province, Shaanxi, first in Hu county and later in a place called Ox Winter (Niudong). It was at that time in the summer of 1944 that the first class of Chinese guerrillas graduated from the camp went into action against the Japanese.[160]

OSS instructors had mixed reactions to the Chinese recruits at their training camps, as Dow indicated in March 1944. Although many of the recruits were not in good physical condition when they arrived at the camp—large numbers suffered from scabies, conjunctivitis, and ulcerations—the recruits had strong legs and great physical endurance. They could march for more than thirty miles a day and even climb mountain trails without becoming exhausted. They were tough and ferocious fighters when they wanted to be, Dow wrote. They were in their element at night. While many were eager to learn and eagerly engaged in field exercises such as ambushes, they took a cavalier attitude toward classroom instruction.[161] Many of the Chinese recruits, Dow complained, "seemed to think it was all a big joke." The students skipped classes at will and when they did attend, frequently slept through the lectures. Many openly cheated on written exams. They carelessly disassembled their weapons without permission or supervision. In the process, they often lost springs, firing pins, and other small but essential parts, with the result that the weapon became inoperable. Language was a problem because of the shortage of interpreters and the numerous local dialects. But the Americans did try to teach the Chinese trainees the technical skills to fight the Japanese—close combat, weapons usage, demolitions, radio communication. The instructors drew upon their own training and courses. "Our OSS training was very effective," Gleason said later. "We blew those bridges. We did it with what we learned in our training here [in the United States] and in England [at SOE school]. In China, we had classes where I taught Chinese how to destroy mechanical equipment. I felt fully prepared."[162]

Chinese recruits were proud of the new American weapons, which gave them great status, and they learned to shoot them well. However, the Americans sometimes became disillusioned and discouraged. After a visit by Dai Li to the Unit Three SACO training camp in February 1944, Dow was convinced that "General Dai and the men of his organization are interested in just one thing—getting all the arms and equipment they possibly can. I believe the training by we Americans is merely a cover to get more equipment."[163] Dai Li had assigned Chinese agents from his organization at Units Two and Three and other camps to provide compulsory political indoctrination for the guerrilla trainees to keep them loyal to him and Chiang Kai-shek. The indoctrination was originally scheduled for a full month of the three-month training cycle, but under protest from the Americans, it was reduced to certain afternoons. Maintaining their distance between the two types of training—the practical and the ideological—the Chinese political instructors were forbidden by Dai Li to fraternize with the Americans. At Unit Three, the Linru Ban training camp, the Chinese instructors lived in the Buddhist temple, while the Americans lived in a nearby twelve-room, Western-style house they had constructed.[164]

The Americans planned to launch the Chinese trainees on some simple guerrilla operations behind enemy lines after their graduation, but Dai Li's organization never provided proper clearance while the Americans were there. Nevertheless, with Dai Li's authority, Dow and Gleason did accomplish two missions from the camp against the Japanese in late 1943 and early 1944. But because of Dai Li's orders, they had to special Chinese groups supplied from around Loyang rather than graduates from the camp. In the first mission behind Japanese lines, they successfully destroyed a steam turbine power plant of the Chiao Tzoa Mines located some 75 miles north of Loyang. In the second mission, in January 1944, a raiding party of nearly 300 Chinese guerrillas and saboteurs, blew up a long railroad bridge across the Yellow River between Kaifeng and Sinsiang in Honan Province. Frank Gleason later reported to Washington, the raid "was a definite indication that the Chinese could operate successfully, if the word would be given from Chungking."[165] Gleason's was a familiar American complaint; Chiang's government simply did not want to risk of major offensives, when they could sit and wait until the Americans had defeated Japan in the Pacific. Gleason later told OSS officials back in Washington, D.C. that at least in late 1943 when the Japanese were preparing for their big offensive in 1944, the Chinese forces were "not doing much outside of a few 'Rice Campaigns.' These brief expeditions were more economic than punitive, and were organized at harvest time to penetrate Jap lines, collect as much rice as possible, and retreat at once."[166]

Expanding with the help of Chennault's Air Force

Although the some of the SACO Chinese guerrillas made successful raids, restrictions by Dai Li and other Chinese generals precluded it from becoming a vehicle for OSS expansion, which, even while it frustrated Donovan and his operatives, did not upset Chungking. Through Donovan and Lusey's secret plan, the OSS did begin to set up a communications network in China and some of the SO operatives were eventually converted into SI agents. But SACO was a failure for the OSS in terms of establishing an independent intelligence network and the kind of SO and OG teams and missions that had proved so effective in Europe and the Mediterranean. For a larger OSS role in China, authorized by President Roosevelt at the Cairo Conference in November 1943, Donovan, after an explosive meeting with Dai Li in Chungking in December, turned next to Air Force General Claire Chennault.[167] The 14th Air Force was the main instrument of U.S. offensive action in China. Like OSS, it had the dual role of enhancing the defensive and offensive power of the Chinese Army and of participating in the blockade of Japan—by attacks on enemy shipping in Chinese waters. Donovan had also heard rumors that Stilwell might be recalled because of his disputes with Chiang Kai-shek, and that provided a further reason for offering the OSS's intelligence gathering potential to Chennault, an old rival of Stilwell's. Chennault saw Donovan's connection with the White House as a way of further undermining Stilwell.[168] So while in China in December 1943, Donovan met with Chennault and began a plan for a special, joint OSS/Air Force unit to gather intelligence about tactical targets of opportunity for the 14th Air Force.

By secret agreement in April 1944, Donovan and Chennault created the 5329th Air and Ground Forces Resources and Technical Staff, AGFRTS (or, as it was informally known to its members, "Agfarts"). The group was composed of both Air Force and OSS personnel with the latter including large numbers of SI agents, some of them derived from Chennault's own intelligence officers. Operating outside of the restrictions imposed by SACO, the AGFRTS agents infiltrated behind Japanese lines. Through networks of coastal shipping watchers as well as through interrogation of enemy prisoners of war, they produced immediate and valuable target information which they supplied to OSS, to the 14th Air Force, and to the U.S. Navy in the Pacific.[169]

Major Arden ("Art") Dow played an important role in the expansion of the OSS activities under the cooperative agreement between Donovan and Chennault. After six months training Chinese guerrillas, Dow left SACO Unit Three, "Pact Rowdy," in April 1944. Accompanied by Sergeant Anthony ("Tony") Remineh, an OSS radio operator and code clerk trained at Area C, Dow went on a special mission to the forward echelon headquarters of the 14th Air Force. Up in the war zone, Dow had been instructed to explain to the airmen how OSS could work behind enemy lines generating sabotage and guerrilla activities as well as obtaining intelligence about weather and target selection. Upon Dow's return to Kunming, he was expected to draw up a plan for implementing such joint operations between OSS and the 14th Air Force. The plan was to include OSS personnel from Special Operations, Secret Intelligence, Morale Operations, Research and Analysis, and Communications, and include the establishment of "a powerful radio station to be used in relaying such intelligence and weather reports as they would turn over to us."[170] Upon completion of the mission, Dow did drew up such a plan, Chennault accepted it, and from 1944 to 1945, OSS provided widespread tactical intelligence to pinpoint enemy targets for the American airmen. By infiltrating behind Japanese lines and by studying all sources of information, they helped American bombers and fighter-bombers locate and strike Japanese troop concentrations, supply depots, and rail and road traffic, river and coastal shipping. More than two-thirds of the tactical intelligence received by Chennault's headquarters came from OSS. As for the OSS in China, Donovan's strategy was, in his words, "using Chennault air raids as cover for our operations."[171]

Eventually, AGFRTS became the center of OSS China operations not simply for Secret Intelligence and Special Operations, but for Morale Operations and Research and Analysis as well.[172] As it expanded beyond the operational branches, it included women as well as men. Most of the American OSS women working in China were preparing "black" propaganda in Morale Operations or working as clerks or secretaries in administrative sections. Most of them had worked first in Kandy, Ceylon. The number of American OSS women in China probably did not exceed twenty, most of them at OSS headquarters in Kunming or at the OSS office in the China Theater headquarters in Chungking.[173] Among the numerous MO members in Kunming was Elizabeth ("Betty") Peet MacDonald (later McIntosh), former war correspondent in the Pacific, who had joined the OSS in 1943 and received training at Area F, the former Congressional Country Club. Now she produced forged orders, wrote fictitious Japanese newspaper reports and other documents to undermine enemy morale. (Male artist William Smith, also in MO there, prepared anti-Japanese cartoons and rumors and printed leaflets.) Tens of millions of pieces of MO disinformation were airdropped by Chennault's planes in occupied areas. At Kandy, Ceylon, and later Kunming, China, Julia McWilliams, a Smith College graduate, was in charge of the registry where intelligence and other reports were filed. She later married Paul Child, chief of the OSS's small, graphics and presentation branch in the China Theater, and after their postwar transfer to Paris, became famous as Julia Child for teaching Americans about French cooking.[74]

OSS and the 1944 Japanese Ichi-Gō Offensive in China

In the debate between Chennault and Stilwell over the proper U.S. strategy in China—whether the priority should be for an air or a ground campaign—Stilwell had argued that as soon as American airpower began to have serious impact, the Japanese Army would simply march in and occupy its forward airbases—unless there was a strong American supplied and trained Chinese Army to oppose them. Stilwell's pessimistic prediction came true in the middle of 1944. That spring Tokyo launched the last and largest Japanese offensive in the war in China. Involving nearly half a million Japanese troops and lasting from April to December 1944, Operation *Ichi-Gō* cut a broad swath southwards across central and southern China. The Japanese quickly captured the southeastern airbases from which American airmen had been harassing Japanese forces, sinking Japanese ships, and bombing the Japanese homeland. By pushing the Chinese Army back, the Japanese also opened a land route directly to French Indochina. The Emperor's Army then pressed westward toward the inland airbases, Kunming and the Nationalist capital of Chungking.[75]

The collapse of the Nationalist Army and its failure to impede the Japanese advance shattered Washington's hopes that China would play an active role in the defeat of Japan. Instead, China's wartime importance was now seen as limited to its ability to tie down a million Japanese troops. If Nationalist China survived and the American forces from the Pacific were in 1945 or 1946 able to achieve beachheads on its coast, China might also serve as a staging ground for U.S. bomber attacks and an amphibious invasion against the Japanese home islands. Meanwhile, U.S. policy was to ensure that China did survive the *Ichi-Gō* offensive.

Chiang Kai-shek pressured Roosevelt into replacing Stilwell in October 1944. A fighting general not a diplomat, "Vinegar Joe" had become a bitter foe of the Chinese Nationalist President and commander-in-chief. Stilwell's successor was Lieutenant General Albert C. Wedemeyer, a military planner and a more tactful officer. His command was reduced to China and northern Indochina, with Burma, Thailand, and southern Indochina assigned to a separate command. Wedemeyer pulled the OSS under his command as an independent agency. This removed a major part of it from Chinese control and Dai Li's domination through SACO, even if it did not end Dai Li's spying upon OSS and its agents. Consequently, there were three OSS agencies opening in the China Theater: Detachment 202, based in Chungking, to coordinate OSS operations in China and northern Indochina; SACO; and AGFRTS, but increasingly all OSS personnel were brought under Wedemeyer's command, and Detachment 202 was empowered to create independent new operations behind enemy lines, as it would do extensively by 1945.[76]

OSS, like other American agencies, did all that it could to help block the Japanese *Ichi-Gō* offensives. Despite all efforts, by November 1944, the Japanese had rammed through Hunan and Kwangsi Provinces in southeast China an headed inland toward

Kweilin and Liuchow. Chiang's government expected the Japanese to be deterred by the mountain ranges. But the emperor's troops loaded up mules and surged from the low rice paddies of Kwangsi through the Natan Pass onto the high plateau of Kweichow Province south of Chungking. The Japanese had pushed hundreds of miles westward, and they now pointed directly at the provincial capital at Kueiyang in Kweichow Province and the heart of Chiang's government's communications system. General Wedemeyer had arrived expecting to hold and strengthen eastern China, but he quickly learned that Kweilin and Liuchow were being abandoned, and panic was sweeping Chungking several hundred miles away. The U.S. embassy was advised to evacuate dependents. Wedemeyer insisted that the OSS be given the opportunity to use its demolition and sabotage skills to try to block, or at least slow down, the Japanese advance.

Consequently, OSS-trained teams aided Chinese efforts to defend the Nationalists' capital. They constructed obstacles to help block the enemy advance and set booby-traps in buildings abandoned by the Chinese. One OSS team severed telephone and lines in 156 places, dynamited seven bridges, and halted rail traffic by tearing up 524 sections of track. Two OSS enlisted men in the field radioed in a message, first in code and then more urgently in clear text, that a large enemy force was crossing the river near Yiyang, and 14th Air Force fighter-bombers hit the Japanese soldiers in the open, causing hundreds of casualties. A few days later, the same two OSS sergeants called in the airmen to destroy two dozen sampans carrying Japanese troops across the Siang River.[177]

Before Wedemeyer arrived, the OSS had for some time been tied up in what Major Frank A. Gleason called "enough red tape to throttle a cow."[178] Every OSS action was required to be approved up the Chinese chain of command, but approval was often not forthcoming. But OSS increasingly gained leeway. Because of tensions between the OSS and Dai Li and Milton Miles, the OSS instructors had been recalled from the SACO training schools in the spring of 1944, and transferred to OSS SO and SI operations out of Chungking. In early May, plans for SO operations under the SACO-OSS Agreement were drawn up by Dow, Gleason and the others for training small groups of Chinese agents, no greater than twenty five, for operations in large cities in occupied China.

Gleason's OSS Advanced Base 21 was established at Kweilin in Kwangsi Province in southern China and twelve Chinese agents trained in special operations were deployed in the area of neighboring Canton. Three targets for sabotage were selected, and in mid-August 1944, Major Gleason, commander of the base, sent Captain Leo Karwaski and two other Americans went down to Shuihing to set up a forward operating base. A week later, the Chinese agents were sent to Shuihing and then dispatched in three groups to their targets. Nothing was heard from them until late December 1944, when they radioed back that they had been successful in their sabotage.[179]

Toward the end of August 1944, the Japanese pushed forward their Corridor Campaign, driving toward Luichow in Kwangsi in several different columns. Like the Chinese Armies and the 14th Air Force, the various OSS teams, including the one led by Gleason and Karwaski were forced to retreat. Working through AGRFTS,

Gleason's unit, Advanced Base 21 of Detachment 202, was given the opportunity to show what a small, skilled and dedicated OSS team could do. With the blanket approval of General Wedemeyer and with the cooperation of the 68th Air Wing of Chennault's 14th Air Force, Gleason obtained permission from the Chinese Commander in the Fourth War Area, Marshal Chiang Fa Kwei, to destroy all the railroad networks emanating from the key cities of Kweilin and Luichow in an attempt to halt or at least slow down the rapid Japanese advance.[180] Gleason's new superior at AGFRTS, a Colonel Bowman, was an old Regular Army officer who was direct and frank and without sugar coating, gave Gleason "the first direct and uncomplicated picture" of the situation, which was disaster. Equally important, Bowman gave Gleason his assignment and then "from that time on we were left strictly alone."[181]

With little more than a dozen Americans, Gleason's OSS unit carried out extensive operations over an area of some 250 square miles between 20 September 1944 and December 1944, when the Japanese curtailed their offensive Sometimes operating only a few miles away from the advancing enemy lines, the little SO team never lost a man, although, as Gleason recalled, in blowing one of the bridges "we didn't get far enough away. The rocks came pouring down on us. We quickly scurried underneath the jeep, or we would have been killed."[182] During four months in the field, the unit blew up more than 150 bridges, erected more than 50 road blocks, and destroyed three dozen river ferries, blew up one tunnel, and smashed an assortment of locomotives, railroad cars, trucks, Army barracks, and machine shops. They also eliminated stockpiles of munitions and gasoline supplies to prevent them from falling into enemy hands. One such episode produced perhaps the biggest explosions in south China.

Setting off the Biggest "Booms" in South China

Amidst a frigid cold snap at the end of November 1944, Time magazine reporter Theodore White drove south from Chungking to get a first hand look at the panic in front of the Japanese advance. As he drove 500 miles south, a never-ending stream of refugees trudged past his car fleeing north. Arriving at Kueiyang amidst the disorganized, routed Chinese troops, White found what he called the only military unit that "had any real coherence and purpose." It was the OSS group of 15 American officers and enlisted men commanded by Frank Gleason, a young, red-headed major from Wilkes-Barre, Pennsylvania.[183] With Gleason now in this OSS demolition team, which Gleason unilaterally designated "Detachment 21," were Lieutenant Leopold ("Leo") Karwaski also from the Wilkes-Barre area, Captain Stanley A. Staiger of Klamath Falls, Oregon, Sergeant Graham Johnston, an ex-jockey from New Canaan, Connecticut and Sergeant Paul Todd from Kalamazoo, Michigan. The 16 members of the team soon learned that the Chinese were impressed by rank and title, and they often elevated themselves to "colonels" when they needed help from Chinese soldiers or civilians.[184]

Gleason, Karwaski and the rest of the OSS team had stopped teaching explosives and were now exploding them, doing what they could to hamper the Japanese. OSS officials later described the 24-year-old Gleason as "a boyish, aggressive, courageous and intelligent young officer... with enthusiastic impulsiveness."[185] Using his ingenuity since he had been given no funds, Gleason managed to expand his convoy from three to nine trucks. He used local Chinese interpreters, including a streetwise, 12-year-old orphan boy who had originally joined the team as a kind of mascot. With their help, Gleason recruited Chinese peasants to assist him as needed. Gleason's mission was one of scorched earth, to do whatever possible to impede the Japanese advance. In addition to destroyed bridges, roads, telephone lines and the like, Gleason sought to deny the enemy food and equipment, by authorizing peasants to pick up the weapons and other equipment abandoned by the fleeing Nationalist soldiers, and by encouraging them to forage for food and other supplies in the towns soon to be occupied by the Japanese.[186]

In early December 1944, only 20 miles from the spearhead of the advancing Japanese Army, Gleason heard rumors when he arrived at Tushan, 140 miles south of Kueiyang, of a large supply of weapons and ammunition in the nearby hills. A preliminary reconnaissance revealed three enormous ammunition depots, each one made up of 20 to 30 giant warehouses, with each warehouse some 60 yards long, nearly two-thirds the length of a football field. Every warehouse was stacked to the rafters with weapons and munitions, much of it American made. The hoard included pistols, rifles, machine guns, mortars, fifty artillery pieces, plus millions of rounds of small arms ammunitions, thousands of mortar shells and artillery rounds, and 20 tons of explosives. The regional authorities had been storing these for domestic use in the postwar civil war. With bureaucratic inefficiency, the Chinese staff had continued to horde the weapons and munitions even as the local Nationalist troops ran out of ammunition and left the area. Now those staffs, like the Chinese Army, were in full retreat, and the enemy was only 20 miles away. The Japanese would soon capture what Gleason estimated to be 50,000 tons of valuable military supplies.[187]

Rather than let this treasure fall into the hands of the enemy, Gleason decided to blow it up. It was a painful decision. Gleason came to it reluctantly. He realized that "this was our material, painstakingly flown over the 'hump,' laboriously hauled over poor roads by thousands of coolies." As he said later, "The tragedy was that this was our material, but it would have been a greater tragedy had it fallen into the hands of the Japanese."[188] Gleason sought authorization from Chungking to destroy the giant cache if necessary, but by the time the authorities authorized him to prepare to destroy it, the enemy was less than a day away. Gleason and his men spent most of that day placing charges and fuses all around the three depots. About 4:00 p.m., with the Japanese advance patrols only a couple of hours away, Gleason finally received permission to blow up the warehouses.

Lighting the fuses, Gleason and his men set off explosions in the warehouses, one right after another. Each blast rocked the countryside. Jammed with munitions and explosives, the buildings quickly erupted in fiery red and yellow. Inside, the crackling heat "cooked" the ammunition, popping off thousands of bullets and sending artillery shells whistling through the air. Flames and smoke surged into the sky. As the Americans left in their trucks, Gleason and the others turned back for a final look. "When the last dump when off, there was a single column of black smoke that went straight up, with the most terrific sound you ever heard... a black column 100 yards thick holding up the overcast like a pillar."[189] It was, Gleason said, "like Dante's 'Inferno.'"[190]

On the way out of town, the Americans stopped to blow the bridge across the Fung River, furthering impeding the advance of the enemy column. The Japanese a few hours after Gleason left. The warehouses continued to burn and explode for the next three days. The enemy could merely watch as the precious munitions, weapons, and other supplies were lost to them.

"My mission was clear," Gleason recalled, "'To block the Japanese advance for 90 days,' this came from Wedemeyer himself. We accomplished the mission and stopped the Japanese advance into the interior of China."[191] Indeed at Tushan, the Japanese Army, for a variety of reasons, did come to a halt. They held the city for a week, then turned around and picked their way back across the scorched area that Gleason and his team had helped deny them. Contracting their lines, the Japanese dug in for the winter at Hochin, half halfway between Kweiyang and Liuchow.[192] Gleason was recommended for the Legion of Merit and the rest of his team for the Bronze Star medal, the fourth highest military decoration of the U.S. Army.[193] Theodore White continued to praise Gleason and his team, including an account of Gleason's mission in his 1946 non-fiction account of the war, Thunder Out of China, and used it for the basis of a 1958, novel, which Hollywood made into a film, The Mountain Road, in 1960, starring James Stewart. Although Gleason was one of the advisers on the film, he says that most of the movie is "pure fiction."[194]

Building an Expanded Communications Network in China

Crucial to expanded, independent OSS operations in China was the further development of its communications networks. The distances were vast compared to countries in Europe, and although the Army and Navy trunk lines were used when possible, most of the coverage, especially with the proliferating field teams and sub-stations, were maintained by the OSS itself. The Communications Branch had to maintain the lines of communication, staff the base and sub-base stations, furnish radio operators for field missions, and supply and maintain communications equipment.[195] "Commo" personnel came in from Area C in the United States, flown in along with the equipment, "over the Hump," but to train indigenous radio operators, the Communications Branch established a radio operator and code training school in Kunming. In late 1944, with more independence, OSS in China was able to abandon the time-consuming double-transposition coding system that the Army had required and replace it with its own more efficient system of "one-time" pads. Demonstrating the importance of the communications network, communications gear made up the majority of equipment sent out by air from headquarters in Kunming to OSS units in the field: 50 percent of shipments as contrasted to 30 percent in the form of demolitions and 20 percent in rations, weapons, and field equipment.[196]

From Washington, Colonel Lawrence Lowman, chief of the Communications Branch, reported real progress: "In the Far East....OSS communications networks are operating directly with the 14th Air Force in China.... Operating out of Kweilin, OSS communications teams were continually responsible for furnishing vital target information and weather data and for activating timely air missions against enemy ground troops and river and coastal traffic. Established many times deep behind enemy lines in China, these OSS communications teams have been directly responsible for the destruction of thousands of Japanese troops and substantial quantities of enemy material and support....The clandestine communications networks of the Office of Strategic Services have been set up without duplicating established Army, Navy or commercial facilities. We have concentrated on developing the specialized equipment and techniques of operation required of clandestine radio activity. Our work has, we believe, played a unique, successful and important part in the realization of the overall OSS objectives in this war."[197]

OSS communications also played a role in the brief extension of U.S. military assistance to the Chinese Communists in North China. From the south, Chiang Kai-shek had forbidden U.S. contact with Mao Tse-tung and his Communists in the mountains of Yenan in Shenshi Province. But while the Nationalist Armies were disintegrating in front of the Japanese advance, Mao's Communist guerrilla armies established effective control in various rural areas of the country. In late 1944, the new U.S. ambassador to China, Patrick Hurley, a Republican stalwart and former diplomat, endorsed Roosevelt's effort to get the Communists and Nationalists to work together against the Japanese.[198] A U.S. military mission, the "Dixie" Mission of 18 men, headed by Colonel David D. Barrett of Army Intelligence (G-2), arrived at Yenan in July 1944.[199] The OSS was in charge of communications, and Captain Paul Domke and Sergeants Tony Remineh and Walter Gress, both trained at Area C, quickly established radio contact with Chungking, a thousand miles to the south. Initially transmission was in Morse code, but later they were given a sophisticated, lightweight, radio transmitter-receiver developed by OSS, and they could utilize voice radio.[200] The American Yenan radio station, codenamed "Yensig," also received data about weather flow from Siberia to the Pacific from U.S. Navy personnel stationed in the Gobi Desert. The radio men at Yenan also tried to determine whether there was radio traffic between the Moscow and the Communists in Yenan, but given the Dixie Mission's relatively poor equipment, they failed in that regard. However, Remineh did determine that the Communists had a second radio station some ten miles away from the headquarters. When Americans went out behind Japanese lines, Remineh created special machine codes for them. He forwarded their encoded messages from the field to the main relay station in Chungking.[201] As the Yenan radio transmissions could be widely intercepted, the most sensitive material was coded and sent by messenger.

While providing the Americans with access to their intelligence and guerrilla networks behind enemy lines in other parts of China, the Chinese Communists also made major requests, particularly for communications equipment from the OSS. Some lightweight radio sets were delivered, but the major requests went unfulfilled and the entire communications project collapsed after OSS ran into major problems with Mao and the Communists.[202] Mao's guerrillas in the field had seized and detained members of a demolitions team that had parachuted into the East Shansi area and arrested the OSS "Spaniel" Team, a five-man SO unit, airdropped near Fuping in Hopei (Hebei) Province. The Spaniel Team was headed by Major Frank L. Coolidge, who had been with Peter Ortiz, when Ortiz was captured in France, and then had taught for a while at Area A. With Coolidge in China was an American captain and two enlisted men, plus a Chinese or Chinese-American agent. The Spaniel Team had parachuted into the Fuping area with the purpose of using the local Communists' contacts with collaborationist Chinese "puppet" generals to provide intelligence and help conduct sabotage against the Japanese. Angered, Mao decreed that all future OSS missions into Communist areas would also be detained until they obtained prior clearance from Communist headquarters in Yenan. Despite U.S. protests, the Spaniel Team was held by the Communists for four months and only released in at the end of the war. Unauthorized OSS missions could prove highly embarrassing to Mao, because they could begin to expose the Communists' myth that they were more aggressively fighting the Japanese than the Nationalists.[203]

OSS "Commo" Men in the Field

The experiences of the Communications, or "Commo," men in China varied as much as the topography of that vast country. It depended on where you were, what you were doing, and when you were doing it. John W. Brunner, 21, from Philadelphia, Pennsylvania, had been trained in cryptography in Washington and Area C, in the summer of 1944. He arrived in Kunming in February 1945, was assigned to the OSS headquarters as a cipher clerk, and remained there until the end of the war. He worked with other cryptographers, all enlisted men, in a locked code room on the second floor of the communications center. "I hardly ever saw an officer," he said later. "There was a captain downstairs, who was in charge. I saw him only a couple of times. I worked upstairs, coding and decoding. Only a couple of people were allowed in there. We were told that even if General Donovan arrived, he would have to have permission of the officer in charge to enter. We should not let him in without the captain's permission."[204]

Coding and decoding messages between Kunming and Washington, Brunner saw many very important documents. "I personally saw two top secret 'eyes only' messages from Truman to MacArthur," he said. "MacArthur had been sending advance teams into China. But he was not allowed to send his people here before the invasion [the Army's projected amphibious landings and capture of Canton and other ports on the south China Coast planned for the winter of 1945-1946]. We were arresting them. Truman's messages cautioned him against sending his teams into China before the invasion. Truman sent a copy to the OSS, which is what we got... In the cables I saw, Truman sent MacArthur a cable saying that he was not authorized to operate on the mainland of China. MacArthur disregarded this and sent another team into China. We caught them and arrested them. We sent them to Washington. Truman sent another cable to MacArthur stating, 'I am your commanding officer, and you will obey orders.' This was in April or May 1945. So the Truman-MacArthur conflict began long before the Korean War. This, however, is never mentioned anywhere in any of the books. But I was in charge of the message center and had access to the 'eyes only' correspondence, and I saw these."[205]

A new major OSS communications base station was established in April 1945 at a 14th Air Force base at Chihkiang between Kunming and Chungking AGFTS was absorbed by the OSS, and this OSS Chihkiang Field Unit was given responsibility for communications with all the old AGFTS installations south of the Yangtze River. OSS also created a new Field Unit at Hsian (Xian), and that new base station assumed former AGFRTS responsibilities north of the Yangtze. Among the half dozen OSS "Commo" personnel sent to staff the new Chihkiang, base station in 1945, were Lieutenant James F. Ranney from Akron, Ohio, and Private First Class David A. Kenney from Encampment, Wyoming. After being recruited by the OSS in 1943, Ranney had spent a week at Area B, then served for several months as an instructor at Area C before being sent to Egypt and then Italy in 1944.[206] Kennedy, who had shared a cabin at Area A with a Sioux/Lakota Indian named Iron Moccasin, and a Greek American named Jerry Codekas, had been trained at Area C and then sent to Base Station Victory in London in the summer of 1944. In 1945, both volunteered for service in China. Ranney and Kenney worked at the Chihkiang base station until the end of the war, Ranney as second in command and Kennedy as an operator.

"I did a lot of radio operating," Kennedy recalled. "We had very heavy radio traffic between agents in the field. We would contact people in forward areas behind Japanese lines. We would print it out and then sent it back to headquarters in Kunming."[207] As at most OSS radio installations, the transmitter and receiver stations were located some distance from each other to avoid interference. The receiver station and some of the officers' billets were located in a former civilian building on the outskirts of Chihkiang near the long runways for the big cargo planes of the Military Air Transport Service. The transmitter station, which contained a powerful BC-160 and four smaller transmitters, was in a small frame building near a shorter airstrip used by P-51 "Mustang" fighter planes of the 14th Air Force. In order to repair and maintain their communications equipment, the "Commo" men, being thousands of miles away from their main sources of supply, often scavenged for wire, aluminum, and other parts from the remains of aircraft that crashed on landing.[208] At Hsian (Xian), Robert L. ("Bob") Scriven, who had been an instructor at Area C, and who then helped construct Base Station Victor in London, headed a "Commo" group which established the new Hsian base station in spring 1945. When the Japanese surrendered and the war ended, Scriven and his team was flown to Peking (Beijing) and directed to install a new base station within three days. They did, and it was on the air in by the end of the third day.[209]

Many "Commo" men worked out in the field, close to or even behind enemy lines. Nineteen-year-old Private Arthur ("Art") Reinhardt from near Buffalo, New York, had such a position. Having trained as a radio operator at Area C in the summer of 1944, he arrived in Kunming in October 1944. At the Kunming communications center, he worked briefly doing decrypting with a Hagelin M209 machine, a Swiss designed portable rotor cryptographic machine system. Soon, however, he was sent into the field to establish a sub-base radio station at Suich'uan airbase, Kiangsi Province, southeastern China. In the wake of the Japanese *Ichi-Gō* offensive, this had become the most forward airbase of the 14th Air Force. At the sub-base station, Reinhardt and the other operators would receive coded information sent in from behind enemy lines by OSS SO and SI teams. Three times a day, these field team operators with their SSTR-1 "suitcase" radios, would send coded messages providing information on local weather conditions, on Japanese forces, and on targets for the planes of the 14th Air Force in China or of the Navy in the Western Pacific. At the sub-base, the communications men would decrypt the messages using the M209 machine and send them in enciphered Morse code to OSS China headquarters in Kunming. From there, the most important information would be forwarded to OSS headquarters in Washington, D.C. As the Japanese *Ichi-Gō* offensive continued, however, the Suich'uan airbase came under regular bombardment by enemy planes from Canton. "They used to bomb us two or three times a week," Reinhardt recalled. "They would bomb the airfield, dropping 500 pounders and also cluster 'banana bombs,' which were antipersonnel....In December of 1944, Suich'uan was bombed thirty times, thirty straight nights."[210]

When the Japanese Army came down and overran the Suich'uan and other U.S. airbases in the area in January 1945, the OSS contingent moved east and established a new sub-base station at Ch'angt'ing in Fukien Province. From there, Reinhardt's new assignment was with a three-man field team that included Robert Bell from Secret Intelligence Branch as the commander, Reinhardt as radio operator/cryptographer, and Maurice Mao, as interpreter. They operated near a village called Ningtu not far from the Japanese lines, and they reported their findings three times a day to the sub-base station at Ch'angt'ing. Reinhardt lived in the field doing this for five months, but by June he had become so seriously ill and plagued by high fevers, that he was sent back to Ch'angt'ing for a few weeks to recuperate. Afterwards, in late June 1945, Reinhardt was dispatched to another sub-base station, considerably northeast near Shang-joa, south of Shanghai. From that sub-base station, located in an old abandoned Chinese temple in a very isolated and remote locale, Reinhardt and his comrades received radioed reports from a network of coast watchers, Chinese agents, recruited, trained in radio operation and a special code, and equipped with OSS agent radios. The Chinese agents sent information on the assembly of Japanese convoys in massive Hangchou Bay south of Shanghai. Their communiqués were recoded and radioed to Kunming, which, in turn, forwarded summaries to the U.S. Navy, whose submarines, ships, and planes would attack the convoys. Admiral Chester Nimitz, naval commander-in-chief in the Pacific, issued a commendation for their work, including the sinking of every ship in a twenty-six ship Japanese convoy. But for Reinhardt, as for numbers of others who served in rural China, there was a price in ill health. For five years, Reinhardt suffered from recurring high fevers and diseases caught in China until these were finally successfully diagnosed, and he was cured in 1950.[211]

All the OSS teams that went out into the field had their radio operators, whose job, like Reinhardt's, was to maintain communication, no matter what the difficulties—technical, atmospheric, or tactical—with their base stations. The head of Special Operations in northern China in 1945, Lieutenant Colonel Frank B. Mills, a career Army officer and leader of OSS SO teams in Europe and the Far East, later wrote of them: "The radio operators provided the essential communications link between the operational teams and the supporting base, and they were not only superb radio operators, but were some of the best combat soldiers we had in France and China."[212]

Expansion of OSS missions in China in 1945

When General Wedemeyer replaced General Stilwell as Theater Commander in China in the wake of the failure of the Chinese Armies to stop the Japanese *Ichi-Gō* offensives, he took a much more favorable view toward the OSS than his predecessor. Wedemeyer also sought to improve the Chinese Armies through increased supplies, retraining, and a greater American role in training and command. Chiang Kai-shek agreed to this, and he similarly consented to give OSS at least some authority to conduct comparatively independent intelligence and combat operations behind enemy lines. Consequently, Wedemeyer authorized an expanded role for the OSS. He included OSS/China chief Colonel Richard P. Heppner, 37, a Princeton man and one of Donovan's law partners, in theater policy meetings, and he agreed that OSS could prepare for major efforts to undermine the Japanese Army. With the war in Europe nearing its conclusion in the winter of 1944-1945, many veteran Jedburghs, SO teams, OG detachments, SI agents and other OSS personnel there were given the choice of returning to regular military service or volunteering for OSS missions in China. As a result, OSS/China received not just newly-trained operatives from the United States but large numbers of experienced OSS veterans.[213] OSS/China would grow dramatically in 1945, from a mere 106 agents in late October 1944 to a peak of 1,891 in China in July 1945.[214] Along with the rapid growth, Heppner and Wedemeyer in early 1945 reorganized OSS/China into a solid branch structure.[215] The Secret Intelligence (SI), Special Operations (SO) Branches as well as the Operational Groups (OGs) assigned to Detachment 202 began immediately to train and deploy and ultimately to demonstrate their capabilities under the new authority and additional resources.

SI Expansion and a Dangerous Mission on the China Coast

The main strategic intelligence mission of OSS did not get under way until General Wedemeyer established OSS as an independent command with full authority over its own personnel in China at the beginning of 1945. This gave the SI Branch freedom of action. New forward bases were established as well as the development of an intelligence network in north-central China rivaling the former SACO-AGFRTS-OSS chains in the south. An OSS headquarters for North China was established at Hsian (Xian) in Shensi Province, which was also the site of a forward airbase for the 14th Air Force. Lieutenant Colonel Gustav Krause, SI, former administrative assistant to Colonel Heppner, was the commanding officer, responsible for OSS as it moved for the first time into North China. Representatives of all OSS operational branches were there in Hsian, Secret Intelligence, Special Operations, Morale Operations, Counter-intelligence, Services and Supplies, two Communications Officers, Captains Allen Wooten and Benjamin Adams, and a representative of Schools and Training, Captain Eldron Nehring, who had been a chief instructor at Area A at Prince William Forest Park, was in Hsian to advice SO and SI teams on how to train saboteurs, guerrillas, and spies.[216]

Combination SO/SI teams were to be sent out from Hsian into half a dozen northern provinces to recruit, arm, train and direct Chinese guerrillas behind enemy lines. But there was also a even more secret part to this operation, devised by OSS Headquarters in Washington, it was for Chinese or Korean agents to be included with SO teams parachuted behind enemy lines in northern China. But from there, these agents would leave the teams and gradually work their way up into the Japanese Empire's "Inner Zone," providing intelligence from Manchuria, Korea, and eventually Japan itself. Supported by the local Nationalist commanders and other anti-Japanese Chinese, the OSS unit in Hsian soon sent the first OSS field teams to operate in north China since the beginning of the war.[217] In addition to its cooperation with SO, SI had its own missions, sending its agents directly as SI teams to obtain tactical information about Japanese Army deployments, collaborators, prison camps, and such. The "Phoenix" and other operations were so successful in North China, and even Manchuria, that they returned with bags of military files from various Japanese headquarters in those areas.[218] The "Chili" Mission provided target information for American airmen. Half a dozen OSS SI teams went out into Shansi and Anhui provinces, one of them led by Captain George S. Wuchinich, a Pittsburgh native of Serbian ancestry who had trained at every OSS camp in Maryland and Virginia in 1942, and then been sent as a liaison officer with Tito's partisans in Yugoslavia. Along with Captain William Drummond, Wuchinich obtained considerable intelligence in Anhui (Shaanxi) Province. After the Japanese surrender in mid-August 1945, Wuchinich, an old left-winger, decided on his own, without orders from headquarters, to make contact with the Chinese Communists. Wuchinich believed he had an ideological affinity with them, having fought with the Abraham Lincoln Brigade in the Spanish Civil War and later alongside Tito and his Communist partisans. But when he and his three teammates arrived and camped at a Buddhist temple, they found themselves in a disputed area, amidst a battle among Communists, Nationalists, and former Japanese puppet forces. The Communists won, seized Wuchinich's team and their radio. Their signal to OSS headquarters went dead. Held captive by the Communists, it would be weeks before they would be heard from again and even longer before they were released unharmed.[219]

In the South China, after MacArthur's troops had taken Manila in February 1945, General Wedemeyer ordered his intelligence assets, including the OSS, to make an intensive survey of the South China coast in preparation for possible future capture of Canton (Guangzhou) and the other ports of Guangdong Province in South China and then a drive north toward Shanghai. The plan, code-named "Operation

Carbonado," was for American amphibious landings there from the Philippines. Such an invasion by the Americans combined with guerrilla operations and an eastward thrust of rejuvenated Chinese Nationalist Armies, it was hoped, would divert Japanese troops from being taken back to the home islands and would also provide closer airbases for B-29s to bomb Japan.[220]

OSS-trained Chinese commandos, the plan envisioned, would parachute behind Japanese lines along the coast and pave the way for MacArthur's amphibious invasion. Preliminary to these operations a detailed survey of Japanese emplacements in the coastal and immediate inland areas was needed. Chosen to head this SI survey of the south Chinese coast from Hong Kong to Hainan Island, a mission, code-named "Akron," was now Lieutenant Colonel Charles M. Parkin, an Army Engineer, OSS/SO officer who had been an instructor at Area B and most recently at SACO Training Camp Unit 1. Parkin had earned his American parachutist wings at Quantico and Area A and his British parachutist insignia at an SOE jump school in England.[221] The Navy had earlier sent in several coastal reconnaissance teams from the sea, but all of them had been killed or captured. The OSS decided to parachute in behind Japanese lines. It was the first such airborne operation in China. Parkin divided the area into three zones. Each would be reconnoitered by a four-man team composed of an OSS officer, a radio operator, a photographer to take photographs of the beaches, obstacles, fortifications and other relevant infrastructure, and a Chinese interpreter.[222] In the spring of 1945, Parkin and his team parachuted into the area south of Macao. In twenty days behind enemy lines, the OSS teams covered 400 miles of coast and the immediate hinterland.[223] Picked up by Navy PBY seaplanes at prearranged points along the coast, the three teams returned safely with their valuable information, maps and photographs. Parkin prepared the "Akron" Mission report and delivered it personally to General Wedemeyer in Chungking. The OSS won high praise from the theater commander for the speed, accuracy, and thoroughness with which this task had been carried out.[224] Wedemeyer then ordered Parkin to take the report directly to the Joint Chiefs of Staff in Washington. Parkin flew from Chungking to Washington with the top secret report in a locked briefcase, chained to a handcuff on his wrist. The planned invasion of the South China coast never occurred, of course, Japan surrendered in August. But Parkin's courage and success in the "Akron" mission led to his being awarded the Bronze Star and Legion of Merit when he had returned to Washington.[225]

Special Operations Missions behind Japanese Lines

The Special Operations Branch in China obtained, trained, armed and led hundreds of Nationalist Chinese guerrillas. Small teams of SO officers and enlisted men led guerrillas deep behind enemy lines in attacks on Japanese garrisons and in sabotage operations against bridges and other vital points in the major rail and road systems used to transport enemy troops and supplies. During the final year of the war, OSS deployed 33 SO teams, a few of them almost 500 miles behind Japanese lines in China. With daring and ingenuity, these American leaders, most of whom were graduates of OSS training camps in Maryland and Virginia and veterans of guerrilla operations in the European Theater, did important damage far beyond the number of OSS men involved to impede Japanese in 1945.[226]

Commanding SO Operations in northern China from Hsian in 1945 was Major Francis Byron ("Frank") Mills, an Oklahoman, field artillery officer, graduate of OSS Area A, who had participated in the D-Day invasion, and then, as head of the SO section with the U.S. 1st Army, coordinated SO agents with the French *maquis* as the 1st Army pushed across France to the German border.[227] In late 1944, he returned to the United States and at the OSS West Coast training camp on Catalina Island near Los Angeles, he received more SO training geared to the Far East.[228] He flew over the "hump" and arrived in China early in 1945. Arriving at the U.S. airbase and OSS headquarters in Kunming, Mills was confronted by a sign American personnel had installed. It read "China is No Place for the Timid."[229]

With the *Ichi-Gō* offensive still in progress, Mills was immediately put in charge of an SO combat unit of nearly twenty men, and told to take them and their supplies by truck convoy more than 1,000 miles north and destroy the U.S. airbase at Laohokou before it could be overrun and utilized by the Japanese. Most of his team were also OSS veterans, mostly from Europe but a few from Burma. They were given pistols, carbines, submachine guns, grenades, mortars, and enough explosives to blow up the airfield. They also had to carry their own gasoline, since their route, the main north-south highway in western China, was a two-lane dirt and gravel road with no gas stations along it. Most of the traffic consisted of Chinese peasants traveling on foot or by cart and a few Chinese Army trucks. Mills recalled that the Americans, as they drove slowly through the twisting rural road over mountains and valleys, were often the first "white-colored people" that many of the peasants had ever seen ("white devils" they were sometimes called).[230] Finally after ten days, having crossed twelve mountain ranges, Mill's convoy arrived at Chengtu in Szechwan Province, a forward base of the 14th Air Force. There, they learned that the Japanese had shortly earlier overrun and captured the U.S. airbase at Laohokou.

Mills proceeded to the new OSS regional headquarters in Hsian (Xian) in Shensi Province, site now of a U.S. fighter-bomber base. Mills would oversee some nearly 30 SO teams being created to conduct guerrilla and sabotage operations in the Japanese-occupied area north of the Yangtze River, an area that included Peking (Beijing) and was as large as Europe. The OSS compound at Hsian was located in a former Seventh Day Adventist mission about a mile outside the city. Several hundred yards square, it was surrounded by a stone and clay wall, about eight feet high. Inside were several rather simple one-story structures. OSS had brought gasoline generators to provide electricity for lights and radios. There was no running water, it had to be brought from wells and boiled and sterilized. There were only outdoor toilets and showers. The men slept in sleeping bags on canvas Army cots in large, squad-sized Army tents. Mosquito nets helped protect against insects, lizards, and scorpions. At first they ate canned Army food, but later they hired Chinese cooks to combine their rations and local foodstuffs. Outside the compound there was a path several yards wide with rows of life-size stone statues of ancient figures presumably guarding an important tomb. Years later, the tomb of the First Emperor was discovered with 6,000 life-size terra cotta warriors buried under a huge mound of earth not far away from the OSS compound in Hsian (Xian), which was the old imperial capital and the end of the famous "silk road."[231]

Among the SO or SI teams were many men who had trained at Maryland and Virginia and who had served in Europe. Major James G. ("Jim") Kellis, had led the "Chicago" mission that destroyed major railroad bridges in Greece and Bulgaria and impeded shipments of chrome from Turkey to Germany. Major Paul Cyr, a member of Jedburgh Team "George," had parachuted into Brittany just before D-Day and spent two months with the French Resistance attacking railways, bridges, and German troop columns. Captain George Wuchinich had led team "Alum" into Tito's camp in Yugoslavia.

Like Mills, many of these SO officers brought their radio men with them from Europe to the China. These OSS radio operators had trained at Area C in Prince William Forest Park. Among them were William H. Adams, who had served in Jedburgh Team "Graham" in the Basse Alps; Arthur Gruen, with Jed Team "Miles" in Gers Province; and John S. Stoyka, with Jed Team "Basil" in Doubs Province in eastern France. In the Far East, Sergeant Stoyka would served first in North China and then be sent to Indochina, as a radioman for the "Cat" Team. After they were captured by the Japanese in Vietnam, Stoyka had escaped, found another American team and reported the location of the other prisoners, an act that saved their lives. Mills praised these veteran communications men. "The radio operators," he later emphasized, "provided the essential communications link between the operational teams and the supporting base, and they were not only superb radio operators, but were some of the best combat soldiers we had in France and China."[232]

Herbert R. Brucker, one of the OSS radiomen in France, had served with such distinction there with a British SOE team on the "Hermit" project of organizing a new circuit of local agents in central France from May to September 1944 that he had been promoted to lieutenant and received the Distinguished Service Cross.[233] Volunteering for a new SO assignment in the Far East, Brucker was sent briefly to Peer's Detachment 101 and then to Detachment 202 in China. With his Alsatian accent, his beret and his new goatee and mustache, he quickly became known as "Frenchy." In China, he was assigned to Captain Leon Demers and Team Ibex. Working in Colonel Mills' region, they trained fifty Chinese warrant officers as guerrillas and as the 4th Marauders, they led them behind Japanese lines. Their ultimate objective in August 1945 was to obtain Japanese records in Kaifeng, Henan Province, but the city was guarded by two enemy divisions and tanks, and it was not until after Tokyo's decision to surrender on 15th August that the mission was accomplished.[234]

China was a very foreign country and culture for most of the OSS, and indeed for most of the members of the U.S. military who served there in World War II.[235] Except for a few OSS men who been missionaries or the sons of missionaries in China, almost none of the Americans could speak any of the different Chinese dialects, which varied widely from Cantonese in the South to Mandarin in the North. Aside from a few words or commands, they had to rely upon hired interpreters, usually local Chinese but sometimes Chinese Americans, to communicate with the Chinese. Nor, of course, did they know Japanese. The OSS recruited Japanese Americans for the Secret Intelligence Branch to work as translators of Japanese radio messages or for captured documents or prisoners. Captain Chiyoki Ikeda, for example, was with the SI Branch at Hsian. China was, as Mills said, a strange new country to them, but what was exciting was the many unknowns they faced, learned from, and coped with.[236]

Because the areas in China were so much larger than the European countries, the SO teams there were expanded to generally at least four officers, SO and SI, and a handful of enlisted men who served as weapons and demolitions experts, a radio man, and a medic. Added to these were one or two Chinese interpreters and the Chinese guerrillas, the team would recruit, train, arm, and lead. In China, the SO teams were code-named after creatures, like "Elephant," "Dormouse," and "Spaniel," and they were given a great deal of freedom in their training of indigenous guerrillas and in their operations behind enemy lines. The commander of each team was given an area of operations, intelligence on Japanese or Communist forces in the area, and most importantly, contact information about local commanders of the Chinese Nationalist Army who had been performing some guerrilla operations. It was hoped that those commanders would provide guerrillas, who the OSS would arm, train, and direct in more aggressive and effective special operations against Japanese targets.[237]

Like many others, Mills eventually became rather cynical about the Chinese guerrillas. They knew that the combat experienced Americans were good fighters and effective in sabotage and guerrilla warfare. But "what they really wanted," Mills wrote, were "weapons, supplies, equipment and money—preferably gold—for their own purposes. Patriotic? No, not by our standards. What the Chinese guerrilla leaders were after was power—weapons, ammunitions, and food, in that order... .This was the only politics in China. The U.S. had these instruments of power, and the Chinese naturally wanted all they could get. As we learned, it was the only way the guerrillas could be persuaded to do anything at all. Human lives meant little to them. There was an overabundance of people in China who had none of these power resources, and who could be exchanged for U.S. support."[238] This support and other resources were largely stored for later use in the civil war between the Nationalists and the Communists for control in China, a war which both sides seemed to see as inevitable after the Japanese were defeated.

Destroying the Main Yellow River Bridge

For the top priority target in North China, the main railroad bridge across the Yellow River, Major Mills selected a six-man team, headed by Major Paul Cyr, A 24-year-old Vermonter, who had trained at OSS Areas F and B. Mills had known him in England and then in France, where Cyr had become the first of the OSS Jedburghs to have been awarded the Distinguished Service Cross, the second highest military decoration of the U.S. Army, after the Medal of Honor.[239] Now the mission of Cyr's

Team "Jackal" was to blow up the key bridge on the main north-south railroad line from Peking to Hankow, a line that transported Japanese soldiers and supplies from Manchuria to South China and even Southeast Asia. It was a railroad line the Japanese had constructed, and its most vulnerable point was the Yellow River bridge they had built at Kaifeng. If that bridge were destroyed, Japanese troop trains would have to detour hundreds of miles and many days to the east to get across the river. The Japanese had built a sturdy new bridge of steel and concrete, and they guarded it well, positioning antiaircraft batteries at several points along the bridge and at each end and garrisoning 10,000 troops in the area. The 14th Air Force had tried had tried many times to take out the bridge in low level daylight attacks, but despite losing many planes, they had been unable to destroy it. Even the bombs that hit the sturdy span created relatively minor damage that Japanese engineers had repaired within a few days. As Mills later put it "I decided to put the first OSS [SO] team into this area to knock out the bridge for a longer period of time. It was a high priority target and we were competing in a sense with the 311th Fighter/Bomber Group to see who could get there first and do a good job of blowing the bridge up."[240]

The initial drop of Team "Jackal" included Major Cyr; his second in command, Lieutenant Albert Robinchaud; radio operator Sergeant Berent E. Friele, who had been the radioman on Jedburgh Team "Gerald" in Brittany; photographer Navy Specialist Jerry Welo, who would provide photographs of the mission and of other targets for the air force, and Boris Chu, the Chinese interpreter. This would be the most dangerous kind of infiltration, because little was known about the area or the Chinese reception group, who would either protect or betray them. "The best thing we had going for us," Mills wrote, "was the sympathetic and supportive attitude of the Chinese people, mostly the peasant farmers and villagers. In any guerrilla war these are the people who must be convinced that your cause is just, and they must also believe that if they support you their lives will somehow be better....the Chinese people...hated the Japanese invaders intensely for the barbaric treatment of the people.... rape, torture, mutilation and decapitation were common, and the people knew that the U.S. fight against this enemy was just."[241]

After more than a month of making radio and messenger contact through Chinese channels, the team was flown 500 miles east along the Yellow River on the moonlit night of 22 May 1945, and when the six fires in a "T" formation were seen, the men jumped into the darkness. Photographer Jerry Welo was the only one injured; he broke his ankle, was treated and later returned to Hsian. The reception group was friendly and a base camp was established in the countryside near Changti. Hidden in a closed horse cart, Cyr was driven into Hsinhsiang, the center of a Japanese force of 10,000, to meet his main contact, a Chinese puppet regime general, who was playing a double game, pretending to be a Japanese puppet but really responding to orders from the Nationalist government. A second group of four OSS men including Captain Edward B. Zarembo, second in command, another photographer and another interpreter were parachuted in a week later, together with more supplies to arm the Chinese guerrillas.[242] Lieutenant Robichaud set up a firing range and a training schedule and began the instruction of several hundred Chinese peasant soldiers and officers supplied to them by the puppet general. For some of the

Chinese, it was the first time they had used firearms, and they flinched when firing the carbines and rifles. Cyr was ready to begin operations in late July, but the local Chinese Nationalist commander did not want any offensive operations in his area. Nevertheless, the Chinese guerrillas trained by Team "Jackal" made hundreds of cuts in local railroad lines.[243]

Despite the reluctance of local Chinese leaders to make a major attack, Major Cyr insisted on attacking their main target, the Yellow River Bridge at Kaifeng. Cyr, Zarembo, Friele, and Chu set out on 24 July with a group of seventeen Chinese guerrillas. "These seventeen picked men I had were river thieves," Cyr told Mills later, "as tough a breed as I ever hope to see."[244] The team arrived and set up a temporary camp twenty miles upriver from the bridge. Down around the bridge were some 8,000 Japanese troops garrisoned around the southern end of the bridge, 1,500 at the north end and 200 on a sand island in midstream. The bridge itself was defended by heavy machine guns and antiaircraft guns. Cyr's Chinese guerrillas had lived in the area so they knew the guard routine on the bridge. The plan was to build a wooden boat which would float the saboteurs and their demolitions downstream at night. The Americans, and the Chinese guerrilla leader, Major Tien, surveyed the bridge and the area, sketching routes to approach and escape. On 6 August, as they prepared for the attack, Major Tien insisted that the Americans stay behind, as once the bridge was blown and the Japanese searched for the perpetrators, the westerners would stand out in the population and make it easy for the Japanese to capture everyone involved.

It was rainy and windy on the night of 9 August, providing perfect cover. Major Tien and the guerrillas shoved the heavily laden boat off into the dark and drifted downstream. When they arrived at the bridge, the rain and wind muffled any noise as the men scrambled up the bridge supports. Each man took his 72 pounds of explosives to one of the six columns—steel tubes filled with concrete—that made up each pier of the long bridge. Silently, they inserted the fuses, 50 feet in length, which would burn for 25 minutes, even underwater. When all were in place, Major Tien ordered his men to swim ashore as he personally inspected and lit each fuse. Half an hour later, a Japanese troop train arrived. The locomotive was almost across when the first charges exploded. A section of the bridge collapsed, six cars went down with it and the locomotive slid backwards into the river with a giant hiss of steam. As Japanese soldiers began to clamber out of the remaining cars, the second charge exploded, followed by a series of charges, one after another. Entire sections of the bridge collapsed into the river, and the whole troop train of twenty cars containing some 2,000 enemy soldiers plunged into the swirling black waters. On shore, the troops of the Japanese garrison ran around in confusion, shouting and screaming, and in the chaos the saboteurs escaped. Team "Jackal" and its guerrillas had accomplished its mission in a spectacular fashion, destroying the most important bridge on the Yellow River and a Japanese troop train as well.[245]

OSS Special Operations in South China

As spectacular as the destruction of the Yellow River bridge was, the majority of the SO operations in North China had not been as effective as the SO Teams in South China. The Chinese guerrillas and even the Chinese high commands in the Nationalist Army War Zones in North China obstructed the SO teams, seeking American training, supplies, funds and prestige, but impeding through petty evasions and delays most major American-led operations. The bulk of SO accomplishments in China in 1945 were in the South China, along the Changsha-Hengyang-Kukong Corridor where they received the overall support of the regional commanding officer General Hsieh Yueh. Still, most of the commanders of SO teams in the South as in the North felt frustrated despite their achievements.

One of the first SO missions in South China in the last year of the war was that of the "Muskrat" team headed by Marine Captain Walter R. Mansfield, 32, a Boston native, Harvard graduate and former member of Donovan's law firm, Mansfield had been trained by the Marines but also took OSS training in Maryland, Virginia and England.[246] In Europe, Mansfield was air dropped into Mihailović's Chetnik camp in Yugoslavia. Now Mansfield directed the "Muskrat" operations of four SO sub-teams against Japanese line of communication and supply in Hunan Province in south central China from February to May 1945. Although they trained nearly 800 Chinese guerrillas, Mansfield was quite unhappy with the results. He found them poor marksmen and unwilling to fight. As he reported about the third Chinese colonel sent to him, "Colonel Tan proved to be a pleasant, mild mannered leader but like all guerrilla leaders I have met, he thought only of capturing Jap goods and kept postponing operations until he could find a spot where he could isolate a small group of Japs and destroy them."[247] In three months in the area, Mansfield reported that the major accomplishment had been in training Chinese guerrillas. In operations, most of the fighting had been done by the Americans. The "Muskrat" team had staged 11 ambushes or clashes with Japanese troops, caused 277 casualties to the enemy, sent 27 prisoners back for interrogation, destroyed 2 bridges, 2 sampans, 3 artillery pieces, 22 trucks, and 2 Japanese warehouses and captured a large amount of miscellaneous Japanese equipment.[248]

Beginning in May 1945, however, with the war in Asia and the Pacific moving towards its conclusion, the Chinese Armies, both Nationalist and Communist, began in many areas to go on the offensive. Nationalist support for OSS operations increased during the summer, and by July 1945, nearly two dozen SO units were conducting guerrilla warfare in the south of the country. They cut railroad lines, destroyed bridges, harassed enemy troop movements, and guided aerial attacks by the 10th U.S. Air Force fighter-bombers on trains and truck convoys. This guerrilla activity forced the Japanese to divert significant numbers of troops from their front lines in China or from being sent to fight the Americans in the Pacific.[249]

The "Elephant" Team was the most effective SO field unit in South China. Its commander was Captain Walter C. ("Clark") Hanna, Jr., a former artillery officer who had been recruited by the OSS in 1943, and trained as an SO agent in Maryland and Virginia, and as a Jedburgh in England. He had been parachuted into the French

Alps region with Jed team "Sceptre" in August 1944.[250] Team Elephant, included Captain Hanna, three lieutenants, two radio operators, Sergeants R.E. Baird and H.V. Palmer, and two American interpreters, who could speak Chinese.[251] They spent a month trekking by foot though the mountains and by early June reached their destination, a mountaintop headquarters of the regional National guerrillas, five miles northeast of Kiyang in central China. They were welcomed by the guerrillas with the freshly severed heads of two Japanese soldiers. Torture for information and then beheading were all too common for captives obtained by either the Chinese or the Japanese.[252] Hanna and his men established a base there, a site to receive supplies by airdrop, and a training camp. They needed food, medical supplies, weapons and demolitions, but for several weeks they received no airdrops and then the first drops in late June included no weapons, ammunition, money or radios. In his final report on the mission, written in September 1945, Hanna expressed frustration and bitterness toward the OSS headquarters hierarchy in Kunming, which had repeatedly failed to re-supply his team during its months of active guerrilla operations in the field.[253]

Nevertheless, Hanna's Team "Elephant" and the Chinese guerrillas it trained engaged the Japanese and indeed were in continuous aggressive contact against the Japanese troops in their area throughout the summer. They were credited with killing 764 Japanese soldiers, destroying 501 rails, three locomotives, 22 trucks, five bridges, 70 barrels of gasoline, numerous phone and telegraph lines cut or destroyed, six warehouses burned, 200 bombs destroyed and one airfield put out of commission. Its intelligence led to 250 air strikes on various Japanese emplacements. Some of the Americans were wounded but none killed. A doctor parachuted in to perform an appendectomy on one of the Americans. Among the 756 Chinese guerrillas trained by the Americans, several dozen were killed or wounded in the engagements with the Japanese.[254] Hanna was promoted to major and awarded the Army's Legion of Merit; the other American members of Team "Elephant" received the Bronze Star Medal.[255]

Silver Stars for Team Dormouse's "Mice"

For extraordinary gallantry and effectiveness in blowing up a key Japanese railroad in South China, three officers in SO Team "Dormouse," received Silver Star Medals. They were Captain Raymond E. ("Ray") Moore, Lieutenant Jack Matthai, and Lieutenant James Fine from Monongahela, Pennsylvania. As SO officers, all three had been trained at the SO schools in Maryland and Virginia. Captain Moore, 26, from Plainfield, New Jersey, had left Georgetown University enlisted in the Army, became an officer and a paratrooper, was recruited by OSS and trained at Areas F and B. After further training in England, he went on two Jedburgh missions n France in 1944. Returning to the United States in November, he married his hometown sweetheart, and in February 1945 arrived in China.[256] After sitting around in OSS headquarters in Kunming for a month, Moore and several other SO officers complained about the delay in getting into action. "Do you fellows want to wait a few weeks and parachute in, or would you rather walk three hundred miles to your target area?" the colonel asked. "We'll try going overland, sir," they answered. "That seems to be the quickest way to get there."[257]

Team "Dormouse" was commanded by acting Major Benton McDonald ("Mac") Austin of Savannah, Georgia, who had commanded Jedburgh Team "Ammonia" in southwest France. Captain Moore was his second in command. The other Americans on the initial team included Captain Everett T. ("Ev") Allen of New York City, a former Jed in southern France, and as radioman, Sergeant Vincent M. Rocca from New York City, the radioman. Rocca had served with Moore on Jedburgh missions in France. The original Chinese interpreter for the team was "Casey" Wong.[258]

It took the "mice" of the "Dormouse" mission five weeks by primitive railroad, truck, and on foot to get to their combat zone in Hunan Province. Captain Allen and Sergeant Rocca both came down with malaria and had to be sent to the nearest U.S. Army field hospital. The team continued with a Burmese/Chinese radioman, nicknamed "Chicago," and a Mr. Chu, 51, a highly educated Chinese man who had studied at Columbia University and traveled widely. There were Japanese all around the area and one day, Moore himself was operating the new "Joan-Eleanor" VHF radio system for ground-to-plane voice communication when an American voice suddenly came on the same frequency from nearby to ask who and where they were. Moore told him only that they were waiting for a drop. He said he was too and asked if he could be of assistance. As Moore later recalled, "I replied 'No, thanks.' To which he said, "Roger, good luck. I'll be seeing you, kid.' It wasn't until later that I realized I had been talking to an American-speaking Jap who was trying to find out where we were."[259] The drop went off successfully on 9 July and brought three new SO team members: Lieutenant Jack Matthai a civil engineer from Baltimore, who had spent several months with just his interpreter, the two of them alone surveying the area; Lieutenant James Fine from Monongahela, Pennsylvania, who had done SO work in Burma and Thailand; and Sergeant. Tom Tracey from Jersey City, "a crack radioman" to replace their Burmese/Chinese operator.[260]

With a civil war going on between two neighboring Chinese warlords, Generals Chiang and Wang both of whom appeared more interested in eliminating each other than fighting the Japanese, the rebuilt Team "Dormouse" gained support from them because the Americans had the ability to produce supply drops of new weapons and ammunition. On 28 July 1945, the Americans and their Chinese guerillas launched their first attack on their main target, the key Japanese railroad between Hengyang and Kweilin. While one section attacked a Japanese guard garrison, two other sections blew up 900 feet of track around the curve, but Japanese engineers rebuilt it within two days. Knowing Americans were in the vicinity, the Japanese placed a price of $500,000 in Chinese Nationalist currency (equal to U.S. $1,000) on the head of every OSS man brought in dead or alive, sent out special teams after them, and reinforced their protective garrisons along the railroad. The average Japanese soldier was in deadly fear of the Americans' submachine guns and bazookas and was afraid to leave their protected posts and pillboxes. The Americans and their Chinese guerrillas continued a series of smaller raids, while planning a major attack on the railroad. One of the new arrivals, Lieutenant Jack Matthai, who had been a civil engineer in civilian life, planned a major explosion that would shower down an estimated 300 tons of dirt and rock onto the tracks where they cut through a hill. He calculated that it would take two weeks for the debris to be taken out of the narrow passageway. Because of heavy Japanese patrols at night, it would have to be a daylight attack. Matthai and Fine, another new arrival, led the demolition party of ten men in planting 400 pounds of explosives along sides of the hill, with two rifle squads to cover them. Major Austin, the C.O., and Captain Moore headed separate groups with rifles, light machine guns and bazookas to pin down the approximately 400 Japanese troops dug in at both entrances to the cut. When the firing started, all but one of Matthai's ten native demolition carriers ran away, so Matthai, Fine and the one of the remaining native laborers, personally hauled the 400 pounds of explosives into final positions. Meanwhile, Moore's section was pinned down by heavy mortar and machine gun fire while advancing across broken country within two hundred yards of the Japanese positions.[261]

Moore's Silver Star citation reported what happened next. "Captain Moore tried to maneuver his men into firing positions, but failed due to accurate enemy fire. Seeing his men were helpless, he crawled back to his machine guns, through enemy fire; while bullets kicked around him, and led them to the enemy's right flank. In order to place effective fire on the enemy, it was necessary for him to set his section of machine guns in plain view, so that the pressure on the rest of the team would be released. A mortar shell knocked out one machine gun and Captain Moore put it back into position and effectively manned the gun while mortar and small arms fire searched for him. His accurate fire forced the enemy to keep low and only return a weak and inaccurate fire. His personal bravery, tactical skill and devotion to duty were responsible for the withdrawal of his section with few casualties and were immediately responsible for the success of the team's mission."[262]

Then as Moore later described it, "the earth was shaken by the most terrific blast I'd ever heard. I looked over toward the cut, and it seemed as though the earth were heading skyward. All I could see was a great filthy black umbrella-shaped cloud that must have been a hundred feet high. When the dirt and rocks fell, a great swirl of dust hung over the area.... We beat it back to the village rendezvous, and there I was delighted to see Mac, Jim and Jack—the latter two covered from head to foot with dirt from the blast."[263] The blast was even more successful than they had hoped. Although almost a dozen of the Chinese guerrillas had been killed in the attack, the blast had buried the tracks under more than 450 tons of debris. It would and did take more than two weeks to cart it away and reopen the railroad line. The regular Chinese Nationalist Army was drawing near and the Japanese would be denied the vital transportation line when they needed it most.

It was not until the next day, 12 August 1945, that radioman Tom Tracey learned from OSS Kunming that the Japanese government was offering to surrender, which it did on 14 August 1945. The next message was to be ready for evacuation. "It's hard to describe the emotions of our small group of Americans hundreds of miles from anywhere when we received this news that it was all over," Moore said later. "More than anything else, it meant going home to love and peace and all the things we had been longing for and dreaming about night and day. Surprisingly enough, we did very little celebrating. From that minute on, all of us were mentally at home, and we began to hate every minute that kept us from being there."²⁶⁴ Returning back to OSS headquarters in Washington, D.C., Moore, Matthai and Fine were each awarded a Silver Star Medal; Radioman Tracey the Bronze Star; and Captain "Mac" Austin the Legion of Merit for the success of the "mice" of Team "Dormouse."²⁶⁵

OSS Chinese OG Commandos

An important OSS attempt to create elite fighting units among the Chinese Nationalist Forces, particularly after the failure of the Chinese Armies in the face of the Japanese *Ichi-Gō* offensive, was Donovan's plan to create units of OSS Chinese Operational Groups (OGs) of American trained commandos. The belief was not simply that properly trained and armed and with American veterans as leaders and advisers, these commandos could wreck havoc with Japanese lines of communication and supply as the European OGs had done behind German lines. There was also a hope that such units would encourage the Chinese Nationalist Army as a whole and fight more effectively. Although approved in January 1945, the program did not get started until the first Chinese OG commando training camps were established in April. The goal was 20 company-size commando units within the OSS Chinese OG, a total of 3,000 Chinese paratroopers, led by about 400 Americans. But actual cooperation by the Nationalist Chinese Army was slow and strained and only six of the approximately 180-man, Chinese commando units, totaling about 800 Chinese soldiers, had been fully trained and begun to conduct combat operations by the end of the summer when the war suddenly ended.²⁶⁶

Most of the 160 American officers and 230 enlisted men assigned to the OSS Chinese Operational Groups were veterans who had parachuted into German-occupied France in the summer of 1944.²⁶⁷ A few had staged amphibious raids from Ceylon against the coast of Japanese-occupied Burma. In addition, officers and enlisted men were recruited from OSS replacement centers in the United States. Almost all of the American personnel in the OSS Chinese OGs had received their stateside training at Area F and then either Area B or Area A in Maryland and Virginia. Many of them received additional training on Catalina Island on the West Coast. They served as leaders, advisers, radio operators, medics and other specializations with the OSS OG Chinese commando units. Since very few spoke Chinese, there were half a dozen interpreters assigned to each 180-man unit. In overall command of the Chinese OGs was Lieutenant Colonel Alfred T. ("Al") Cox, who had been one of the original OSS OG leaders in 1943. He and then fellow Lieutenant Joseph Alderdyce had helped Serge Obolensky write the OG curriculum and field manual, basing it on translations of European manuals as well as their own observations of SO and OG training at the OSS camps in Maryland and Virginia.²⁶⁸ In 1944, Cox, working from OSS Mediterranean Headquarters in Algiers, and then with Team "Lehigh," Cox had been responsible for the successes of the more than a dozen OSS French and Italian OG teams parachuted into southern and central France in the summer of 1944.²⁶⁹

After the success of the clandestine units working with the Resistance in France, most of the American Jedburghs, SO and OG teams, were given the choice in late 1944 of remaining with the OSS and being sent on similar missions in China or leaving the OSS and being assigned back to regular duty in the service, Army, Navy, Air Corps or Marines, from which they came. "I don't know how many opted not to say in OSS," said Captain Arthur P. ("Art") Frizzell from Fort Huron, Michigan, who had been commander of the first OG section in France in June 1944 and who became commander of a battalion of three Chinese OGs, the "Blackberry" Mission. "My own personal point of view is that it was the greatest group of guys ever assembled."[270] Private Emmett F. McNamara from Boston, who had been a member of French OG Section Lindsey in France and then served with Chinese OG 1st Commando, recalled "All volunteered to go to China. No one would look at the other guy and say, 'I'm not going. Are you going? Yeah, I'm going.' So we all got over to China."[271]

Under Colonel Cox, the American veterans set up the OSS Chinese OG training camp at Iliang, some thirty miles east of Kunming. The eight-week training course was modeled after the OG courses in Maryland and Virginia, Areas F, B, and A, tailored to conditions in China. Cox got instructors from the OSS camps in the states, particularly those who had previously received Army training or taught at the Infantry School at Fort Benning. The course was completed with parachute training at the separate OSS Airborne School at Kunming airfield under the command of Marine Captain Elmer ("Pinky") Harris, the Ketchikan, Alaskan native, who had by January 1945, arrived in China after OSS service at Areas B and A, the Mediterranean Theater, and MU instruction in California. His chronic abdominal pains had finally been identified as the result of intestinal ulcers, and after an operation in California, he was deemed fit for full service and dispatched to Kunming as C.O. of the new OSS Air Operations Unit and Parachute School. For his service in the Mediterranean Theater and the China-Burma-India Theater, Harris was later awarded the Legion of Merit.[272] Chief instructor at the parachute school in China was airborne veteran, Lieutenant Colonel Lucius O. ("Ruck") Rucker from Mississippi, who taught OSS paratroopers at Area A and at Fort Bragg in 1942 at Algiers, in 1943 and 1944, and now at Kunming in 1945. Rucker trained the first Chinese paratroop units in that country's history.[273]

The first commando recruits send by the Chinese Nationalist Army to the OSS presented a problem. Only about one-quarter of the ill-nourished, peasants conscripted into the Nationalist Army and sent to the OSS school at Iliang were physically qualified or otherwise prepared for the demanding paratrooper and commando training. For recruits they accepted, OSS replaced their ragged uniforms with American combat uniforms and boots, and built up their physical condition with extra rations, good food, and physical exercise. Ellsworth ("Al") Johnson, medic with the 2nd Chinese Commando Unit, who had served in the Patrick Unit under Serge Obolensky in the French OG, recalled how instead of one lean meal of rice a day that they received in the Chinese Army, they were fed three good meals a day by the Americans with plenty of meat and vitamins.[274]

When the Chinese recruits reached good physical condition, they began their paramilitary training with American weapons and equipment. Americans accompanied them and had Chinese interpreters, but most of the orders were conveyed and issued to the Chinese soldiers by their own Chinese NCOs and commissioned officers. Most of the recruits had never had basic training in the Chinese Army. Setting up a rifle range, the OSS instructors trained the recruits first in the use of Springfield '03 bolt-action marksmen's rifles. This was difficult for those Chinese whose arms were too short to hold the weapon properly, and the kick of the fired weapon was so strong that some Chinese would be knocked down. After about four weeks, the Chinese soldiers were introduced to the infiltration course in which they had to crawl along the ground while machine guns fired live ammunition over their heads. Lieutenants John C. Hooker, Jr., along with Larry A. O'Jibway and four other member of the OSS amphibious unit that had been trained at Areas F and A and served along the Burma coast, had been assigned to the 10th Chinese Commando in April 1945. After four weeks of training the Chinese, Hooker, was surprised when it came to the infiltration course. "There we found that Chinese soldiers really do not want to die, and, therefore, we Americans had to crawl with them and jab them in the butt with rifles to get them through the course."[275]

Most of the peasant conscripts in the Chinese Army could neither read nor write, so compass and map-reading was quite difficult. Al Johnson reported that when a group of the trainees went out on such a field exercise, "invariably a scouting party had to be sent to find them."[276] Parachute training came at the end. By then they were toughened up through running, push-ups, and calisthenics. The instructors built a jump platform, and the men learned to jump and roll when they hit the ground. Finally, they were taken to an airfield outside of Kunming and made their four jumps to earn their Chinese jump wings. It was the first time they had been in an airplane. Eight weeks of OG training produced Chinese commandos, who seemed proud and self-confidant in their ability to fight.

Some of the Americans had their doubts. The situation in China was completely different from that in Europe. In France, Italy, Greece, Yugoslavia, or Norway, the OG sections or SO teams parachuted into an existing and already functioning underground Resistance network. OSS simply helped it operate more effectively and coordinate with the Allied armies. But in China, OSS had to start from scratch and recruit, train, and operate underground guerrilla and espionage groups behind enemy lines. Sergeant Ellsworth ("Al") Johnson was one of the skeptics. Son of an Army sergeant of Swedish descent who married a young woman from Michigan, young Johnson had grown up near Grand Rapids. At 19, he was drafted into the Army in early 1943, went through basic training and then medical and surgical training in Texas. OSS recruited him as a medic for the French Operational Group. He trained with them at Area F and then Area B. The group received parachute training in North Africa, then additional training in England.[277]

In August 1944, Johnson was part of Serge Obolensky's OG Section Patrick that operated successfully with the *maquis* in central France for a month. Afterwards, most of the group, except the over-age Obolensky, volunteered for OG duty in China. They arrived in Kunming in early 1945, and in combination with some other former French OGs were designated the 2nd Commando. In mid-summer, they flew deep behind enemy lines, as part of the "Blueberry" Mission to interdict Japanese lines of communications and supply. Looking around at the men in his C-47, "I could not help but wonder what I had gotten into again," Johnson wrote later. "I kept telling myself we had trained these Chinese as best we could. We had provided them with arms and food, and put all we could into making them a viable force. Nevertheless, as I surveyed the 30 some men around me, I could not help but feel that any fighting we might get into would be done in large part by the Americans."[278]

In July 1945, half a dozen OSS OG Chinese Commando units were parachuted behind Japanese lines to impede the enemy's lines of communication and supply, including cutting off river-born rice shipments to Japan, and provide intelligence from various provinces in South China.[279] The missions were code named after a fruit. The first airborne mission in the history of the Chinese Army was the "Apple" Mission of the OSS OG 1st Commando Unit, commanded by Captain Vernon G. ("Vern" and "Hop") Hoppers from Spartansburg, South Carolina, who had been head of OG Section Justine in France, the second French OG to deploy. He and Captain Arthur P. ("Art") Frizzell from Fort Huron, Michigan, had drawn lots in June 1944 to determine which would lead the first OG to jump into France. Frizzell won, so in China, it was Hoppers' turn to lead the first mission.[280] The Apple Mission parachuted from 14 C-47 transports into an area near Kai Ping, south of the West River (Xi Jiang) to interdict road and river traffic. One Chinese paratrooper broke his arm when it became tangled in the static line of the man in front of him, another landed in a pond and drowned. Learning from informants of the arrival of the 180-man unit, the Japanese immediately dispatched 500 troops and pursued them relentlessly for four days until the unit reached a more defensible position in Loting. Hoppers organized an intelligence network and started destroying cargo sampans along the West River. On 5 August, the 1st Commando launched an attack on Japanese fortified positions at the junction of the West and Namkong Rivers. The after action report indicated that in this, the commandos' first engagement, they displayed "courage and fearlessness" driving out the Japanese and inflicting 25 casualties upon the enemy while suffering only 10 casualties of their own.[281] Private Emmett F. McNamara from Boston, radio man who had been on the "Lindsey" Team in France, recalled the battle years later. "When we attacked, most of the Chinamen held back, but a few went up with us. One of our Chinamen got hit. The others left him there, but Cahill [Private John A. Cahill of Victor, New York] got up and ran down the hill toward the Japanese, picked him up, threw him over his shoulder and carried him back to safety. He got a Silver Star [medal] for that."[282]

The 8th, 9th, and 10th Commandos, had been formed into a provisional battalion on the "Blackberry" Mission. Captain Frizzell was in overall command. The C.O. of the 10th Commando ("Banana") was 6'6"-tall, 200-pound, Captain George Gunderman, an approximately 35-year-old, OG officer, recruited from paratroop school at Fort Benning and then trained at Area F and A before being sent to China.[283] Six of the 17 Americans, who accompanied the 154 Chinese soldiers, were from an OG amphibious unit that had trained at Areas F and A and that had served in Detachment 404's crash boats along the south Burma coast. In the 10th Commando, Lieutenant Hooker was assigned to the 60 mm mortars; Lieutenant O'Jibway to a machine gun unit and then to a rifle section.

The Blueberry Mission was flown into an old U.S. airfield at Liuchow, nearly 400 miles east of Kunming. The members of the mission subsequently sailed down the West River (Xi Jiang) for nearly 200 miles in sampans and junks. Their initial assignment was to assist the Chinese Nationalist Offensive. Specifically, they were to support an attack by the Chinese Nationalist Army's 265th Regiment of the 89th Division to capture Tanchuk airfield from the Japanese. Thereafter, they were to move eastward along the West River ahead of the advancing Chinese Nationalist armies. The commandos planned a coordinated attack with the commander of the Chinese regiment. On the morning of 3 August, the 10th Commando together with the 8th Commando, headed by Capt. William H. McKenzie, who had helped lead OG Section Louise in France, attacked and captured the high ground overlooking the airfield. The 9th Commando was held in reserve. But when the commandos then fired a flare signal for the regular Chinese regiment to advance to support them, the Chinese soldiers and their officers ignored the agreed upon signal and remained in their secure positions in the rear. Hooker recalled that Chinese officers in the commando units also disappeared as soon as Japanese mortar shells began to fall.

In the 10th Commandos' sector, Hooker's mortars battered the enemy from his hilltop position, while O'Jibway's rifle teams stormed a further hill. The Americans had taken the high ground. In mid-afternoon, a Japanese officer with a Samurai sword led two dozen soldiers in a suicidal "Banzai" charge toward Hooker's mortars. Hooker's Chinese commandos killed or wounded all but the leader. "The officer was charging directly at my position," Hooker remembered, "and I watched a Chinese soldier shoot him three times with an American carbine. He never slowed down. I threw my '03 Springfield to my shoulder and hit him in the center of his chest when he was twenty yards from me. The bullet slammed him back down the hill. The Chinese soldier looked at me and should 'Ding How, Ding How,' which translated means 'Very Good, Very Good.'"[284] The Americans continued to fire on the Japanese from the two hills until they began to run out of ammunition.

After holding their own position for several hours under heavy enemy rifle, machine gun, mortar and artillery fire, awaiting the support that never came, the commandos had to withdraw. The losses were high: 22 killed in action and 31 wounded, some of them mortally. Ultimately the death toll reached 38. Many of the Chinese commandos had performed honorably. The Japanese, who had lost 164 killed and an unknown number wounded, withdrew during the night. The Chinese regular regiment occupied the town and the airfield the next day. As a result of this action, it was decided, as it had been in North Africa, that OSS units should not be used for spearhead frontal attacks to encourage regular infantry units, but rather should be utilized as intended for the kind of clandestine interdiction and harassment behind enemy lines for which they had been trained.[285]

Meanwhile, the "Blueberry" Mission by the 2nd Chinese Commando was assigned to interdict road and river traffic along in the Paoching-Hangyang-Chansha area of Hunan Province. It launched its offensive before the decision at headquarters not to have OSS spearhead attacks. Its 160 Chinese and 20 Americans were commanded by 40-year-old Captain John E. Cook, who had been Obolensky's second in command in Section Patrick in France. In China, most of the Americans on this mission had been OSS French OGs, who had trained at Areas F and B in 1943. Half of them had been with Cook in Section Patrick. Now in late July 1945, they parachuted near Chakiang, deep behind Japanese lines. There were a few injuries but no fatalities, and the group was met by their advance party, headed by Captain Roy K. Rickerson from Bossier City, Louisiana, formerly with OG Section Louise, which had captured 3,800 Germans in France. A few days before the 2nd Chinese Commando parachuted in, Rickerson and his small team had run into a small patrol of Japanese soldiers led by an officer. "We killed the entire squad except the officer," Rickerson explained. "I was about to take a bead on him, when my M-1 [rifle] jammed. The Jap raised his Samurai sword to try and sever my head from the rest of me, so I raised my rifle to ward off the blow. This did the trick, but the tip of his sword came down and caught me in the back. The fortunate part was that the blow unjammed my gun, and I was able to finish him off."[286]

As Rickerson had arranged, the 2nd Chinese Commando established their camp in an old Buddhist Temple in a village some 50 miles east of Hengyeng. Sergeant James E. Gardner, from Ogdenburg in upstate New York, was the radioman. He had been trained in the Army Signal Corps, been recruited by OSS and taken Communications Branch training at Area C before joining the French OGs for training at Areas F and B. [287] Gardner had accompanied Section Patrick in France. Now as one of the two American radio men of the 2nd Chinese Commando, Thomas F. McGuire from Lakewood, Ohio, was the other, he established contact with Kunming. Two Chinese members of the unit worked the hand cranks to generate electrical power for his transmitter/receiver. Since they had to live off the land, the unit carried a million yen with them to purchase, food, lodging, information and intelligence. Many of the streams were polluted with human and animal waste used as fertilizer, so the team medic, Sergeant Ellsworth ("Al") Johnson, purified the drinking water with chlorine tablets. Meats in the outdoor markets were often fly covered, so the Americans relied upon chickens and pigs kept penned in their own compound. Villagers often stared at the Americans in disbelief, having never seen an American soldier, whom they called "Megwo-Ping."

Captain Cook met with General Wong the commander of local Chinese guerrilla force and General Chiang, head of the 10th Chinese Army, and they chose as the first target a garrison reported to contain some 300 Japanese soldiers. But the Japanese commander had learned all about the Americans from informants and obtained a thousand reinforcements. The outpost's defenses included snipers, machine guns, mortars, and artillery, distributed in pillboxes, trenches, and caves. The Chinese and American plan was for the 2nd Commando to assault the heaviest defenses in the front, while 200 of General Wong's guerrillas attacked the flanks. General Chaing's soldiers operated on the flanks to provide assistance where needed and to repel any Japanese counterattacks or reinforcement from a larger Japanese garrison four hours away.

At dawn on 5 August 1945, the 2nd Commando began the attack, with the advance elements led by Lieutenants Larry A. Drew from Los Angeles and Burke E. Whitney from Tenafly, New Jersey. Whitney took the Chinese mortar and demolitions men, and several American sergeants, off from the main body and established their position on the reverse slope of a rise overlooking the garrison. A Japanese sniper in a tree sent a bullet into the lungs of Sergeant Hasbrouck B. ("Hob") Miller of Gloversville, New York, who was to target the mortars. Lieutenant Drew rushed up, grabbed Miller by the legs, and dragged him to the safe side of the hill. Staff Sergeant David G. Boak and his .50 caliber machine gun teams, set up on another hill. The 2nd Commando was supposed to hit hard and fast, but when the Japanese began firing mortars, the Chinese troops left the line sought huddled there for protection. The whistling shells hit a number of them. Medic "Al" Johnson was hurrying there when a shall exploded 100 feet in front of him; one of the Chinese commando was hit and bounced 10 feet into the air.

Because of heavy enemy fire, Captain Cook could not get his troops in the main frontal position to make their attack. So he sent Lieutenant Drew, and Sergeants Roy Gallant and James Gardner, the radioman, along with several Chinese commandos around to take out the machine gun nest. That meant going over a dune-like hill, across a small valley and attacking the concrete pillbox on the next rise. Despite the loss of at least one Chinese commando, they got to a small shack about 150 feet from the pillbox. Lieutenant Drew told Gallant "That pillbox has to go. It's causing us too much trouble." Roy Gallant was a lumberjack of French-Canadian ancestry from Athens, Maine, a quiet, gentle person, but a man of great courage and enormous physical strength, his chest was as big as a barrel and his upper arms as wide as tree trunks.[288] He had been in combat with the Patrick Mission in France. Now hunched in the shack in China, he reflected for a few moments, looked at the pillbox, and said quietly, "if Jim [Gardner] goes with me, and about four other Chinese commandos, I figure I could get the job done." What happened next was later recounted by Gallant, Gardner and Drew to "Al" Johnson who wrote it:

"Jamming a full clip of rounds into his Thompson [submachine gun], and looking over at Jim, Roy left Larry standing at the shack and headed out to complete the task. Before Roy tried to take the pillbox, he circled around behind it to see how many Japs were in the area. To his great surprise, he saw a goodly number had dug foxholes and many were beginning to advance toward us. After determining the strength of the enemy, Roy and Jim took their Chinese [commandos] and crawled up as close to the pillbox as they could get without being seen. Because the slits were so small, Roy was afraid he could not slip a grenade inside. Just as Roy was about to make his first attempt, a Jap face appeared staring at Jim through the slit. Capt. Cook had been watching the whole action of Roy and Jim from back at the dune where I was attending the wounded. When he noticed the face in the slit, he placed an accurate shot, killing the Jap and saving Jim's life. All of this confusion allowed Roy to sneak close enough and drop the grenade inside the pillbox. After a few seconds, he heard a muffled thud. Just to make sure, he slid another one in. This took a few seconds too long, and Roy received a shot from a sniper's gun. Those [snipers] that we had killed before were replaced by others. Roy was hit in the shoulder, but the sniper was killed by Jim Gardner, who was near Roy at the time. Lt. Raf Hirtz [an Argentine-born American who had served in France and joined the 2nd Chinese Commando] had come upon the scene, and between him and Jim, they were able to bring Roy back to where I was standing."[289]

While Johnson treated the wounded, giving them morphine, patching up their injuries, and having them carried to the rear, Drew returned to the mortar position, finished setting them up, and started arching rounds over the hill into the Japanese emplacements. The shells landed with loud, "whumps," causing much havoc and destruction. But the Japanese mortar rounds were getting closer, and the outnumbered Americans and their Chinese commandos, without any support from the Chinese regulars or guerrillas, decided to withdraw. Captain Cook, using his "walkie-talkie," a portable two-way radio, gave the order, and the groups slogged the 25 miles back to their base at the Buddhist Temple.

They had received no help from their Chinese allies. General Wong's guerrillas started out on their flanking maneuver when the 2nd Commando began its attack, but when the first enemy mortar rounds began to explode, the guerrillas stopped, retreated and simply watched the battle from a hill about a mile away. General Chiang and his conscript troops from the Chinese 10th Army never even moved from their secure positions far out of range. Even some of the American-trained Chinese commandos refused to advance. "I witnessed Capt. Cook holding a .45 Cal[iber]. pistol to the head of several Chinese commandos," "Al" Johnson said, "forcing them to got into battle." Cook yelled at them: "Get out and fight, or you'll get shot."[290]

The attack had been a failure, and Cook was furious with the Chinese. He later reported that the maps they provided had been completely inaccurate and that in the predawn darkness, their local Chinese guide had moved them to point 3,000 yards from the Japanese lines when they wanted to be only 800 yards away. Worst was the refusal of many of the Chinese to fight. "Chinese officers will not lead or command during enemy fire," Cook reported angrily. "Americans try to command, lead and browbeat soldiers to get up close for attack without much success. This position could have been taken easily with a forceful assault. –Col[onel] Chiang–completely scared when nearing the enemy." This last was a reference to the Chinese colonel of one Commando unit, who had refused from the beginning of the battle to order his troops forward.[291] Unsupported by the Chinese guerrillas or the regular Chinese Army, the 2nd Commando's attack had been repulsed. Four Chinese commandos had been killed in action and nine wounded, several of whom later died; two Americans, Miller and Gallant, had been wounded but survived. More than 30 Japanese bodies were observed, before the 2nd Commando withdrew, and it was later estimated that nearly 100 of the enemy had been killed. Nevertheless, the assault was a bitter failure.

Next day, August 6, as the OG made plans to evacuate the wounded. But while radioman James Gardner was contacting Kunming, the news came in that an atomic bomb had been dropped on Japan. "The war is over—the war is over," he kept saying, and the team was ready to celebrate until they learned that they were instructed to remain in position for another 30 days, because it was feared that the Japanese in the interior of China were not ready to surrender. "Wouldn't you know it?" "Al" Johnson recorded. "Our war is to last an additional 30 days longer than everyone else. We could be killed during a time when peace was established. None of us were extremely happy."[292]

Later, after news arrived of Tokyo's offer to surrender, the 2nd Commando was ordered to proceed to Hengyang, where the Japanese troops from the area were assembling for a formal surrender. On the way, the OG was harassed by Chinese guerrillas, many of them like General Wong, turned out to be Communists, and by bandits, all of whom wanted the Americans' weapons and equipment. At Hengyang, they learned about promotions for some of them as well as awards. Sergeant Ellsworth ("Al") Johnson, the medic, received the Bronze Star, and the Chinese Memorial Badge, to add to the Liberty Medal he had earned in France and eligibility for the French Legion of Honor.[293]

Quite a few of the OSSers who had trained Chinese Commando units never got to take them into action. Arne I. Herstad and the other OSS Norwegian OGs, who had served in France and then trained the 7th Chinese OGs, had the 7th Commando ready for its first assigned mission, when the war ended. As Herstad wrote from China to his sweetheart, Andi Kindem, whom he had married when he returned from France the previous year, "…the end of the war stopped everything. I was alerted for 43 days and never got to go anyplace. I was scheduled to go to Shanghai Disappointing to say the least."[294]

The OSS Chinese OGs were now disbanded. The Chinese soldiers were sent to the Chinese Commando Command established in the Nationalist Army. The American OGs were to make sure that their own weapons, munitions, radio, medical and other equipment, went to the Chinese Nationalist Army and were not taken by the Chinese Communists. This they did and the Americans were flown back to Kunming in late August. After nearly a year in China, for some even after six months, many of the men suffered from dysentery and yellow jaundice and other ailments; malaria was common, but kept largely under control through the use of Atabrine tablets. John Hooker recalled that Captain Gunderman was so afflicted that he could no longer walk. The 21-year-old Hooker was offered a promotion to captain, but was too debilitated to accept it. He was 40 pounds underweight and debilitated by malaria and dysentery. "I was quite yellow, my teeth were all loose, and I felt like hell," he recalled."[295] He and O'Jibway were treated in Kunming and Calcutta and traveled all the way back to the Congressional Country Club together. In China, not long after the exiting Americans' arrival in Kunming, their weapons were confiscated and the men were confined to the OSS compound outside the city. The civil war between the Nationalists and the Communists had resumed. There was much confusion and killing, and at night, they could hear gunfire and artillery fire in the distance. "We were at their mercy and none of us felt at ease," "Al" Johnson recalled. "When the news came that we were to leave, we were all happy to go. We had had enough of war and all the stupid things that go along with it."[296]

OSS and Korea

In August 1945, OSS also began to expedite a long held goal of establishing intelligence agents in Korea. As a result of OSS expansion in China in 1945, Colonel Heppner reorganized the organization's regional structure on 1 August 1945, creating three new commands. OSS Southern Command would be headed by Lieutenant Colonel William R. ("Ray") Peers, a graduate of OSS Training Area B, C.O. of Detachment 101 in Burma since late 1943. The Hsian (Xian) Field Unit in northern China, under Lieutenant Colonel Gustav Krause became the OSS Central Command.[297] Entirely new was the OSS Northeastern Command, based in Tuchao, and headed by Captain Clyde B. Sergeant. Its sole mission, code named "Eagle," was to train and dispatch OSS/SI agents into Korea. Those agents were 100 Koreans living in China, who had been selected by Kim Ku, head of the so-called Korean Provisional Government, based in Chungking and recognized and largely controlled by Chiang Kai-shek.[298] Korea had been conquered and maintained as part of the Japanese Empire since 1895. The Japanese forced Koreans to work as laborers and other menial positions in Japan and their empire. The leaders of the Korean nationalists in exile were Kim in China and Syngman Rhee in the United States.

As early as January 1942, Donovan had written to Roosevelt of his hope to use Koreans to operate against the Japanese in Manchuria, Korea, and Japan itself. Most influential with the OSS was Kim's rival, Syngman Rhee, an ardent nationalist, formerly imprisoned by the Japanese, who had studied at Harvard and earned a doctorate at Princeton, converted to Christianity and would eventually become President of the Republic of Korea.[299] Lobbying in Washington for U.S. support for an independent Korea, Rhee, had received little support from the State Department or the Joint Chiefs of Staff. Consequently, in 1942, he had gone to the OSS, establishing what would become a lifelong friendship there with one of Donovan's deputies, M. Preston Goodfellow, a former newspaper publisher. It was Goodfellow, in charge of Special Operations, who had authorized OSS training camps at Catoctin Mountain Park and Prince William Forest Park. The 66-year-old Rhee allegedly began his interview by suggesting that he be given parachute training and be dropped into Korea to raise a resistance Army against the Japanese. Goodfellow demurred but asked Rhee to help recruit young Koreans living in the United States and abroad who could be trained by OSS for resistance in Manchuria and Korea.[300] Rhee agreed and Colonel Carl Eifler, organizing Detachment 101 which he believed in March 1942 would go to China, was flooded with applications from Korean Americans. He had to turn them down when his mission was changed to Burma.[301] Furthermore, Chiang and Dai Li, supported by the US Army and State Department, opposed the idea. Goodfellow apparently ignored JCS and State in early 1942 and sent a group of Korean Americans up to Toronto to be trained at SOE's Camp X.[302] A year later, Rhee had renewed his plea, arguing not only that the Koreans living in China and the United States could be trained and infiltrated as spies and guerrilla leaders, but that they could help rescue downed American aviators as the aerial bombardment of Japan escalated from airbases in China and the Pacific.[303]

OSS training camps at Catoctin Mountain Park and Prince William Forest Park trained some groups of ethnic and native Koreans. In May 1943, there were a few Korean-American trainees at Area B. OSS recruit Peter Sichel remembered playing poker with them in the evenings. They were excellent poker players, he said, and always won.[304] The U.S. Army and Marines captured a number of Koreans in the Pacific or the Far East who had been conscripted into the Japanese Army mainly to work as laborers. Many of these Korean prisoners of war were brought to the United States and held in a POW facility at Camp McCoy in Wisconsin. In 1944-1945, the OSS, using a Korean American, recruited a number of volunteers from among the Korean POWs interned at Camp McCoy for infiltration and subversion of the Japanese in Korea or Japan itself. Men with sufficient patriotism, intelligence, and hatred of the Japanese to become good agents were, if they were accepted, sent to the OSS secret, secluded training facilities on Catalina Island, which was declared off limits for tourists for the duration.[305] Captain Robert E. Carter, 26, from Alexandria, Virginia, who had been with the OSS German OG at Areas F and B in late 1943 but then was reassigned for advanced training in Britain, was put in charge of two of the Catalina training camps, Howland's Landing and 4th of July Cove, in 1945 (the main training camp was at Tonyon Cove, 3 miles from the town of Avalon, but it and its Korean trainees were kept strictly separated from the other two). Under the supervision of Major Vincent Curl, Colonel Carl Eifler's representative, Carter and his staff spent seven months providing intensive training and field exercises for nearly two dozen Koreans, who Carter described as "dedicated, serious students." "They were to go back into Korea for sabotage, espionage, communication, and opposition to the Japanese."[306] A different group, non-English speaking Koreans, but probably Korean exiles approved by Syngman Rhee, rather than the POWs, underwent paramilitary training at Area C-1 in Prince William Forest Park in the summer of 1945. Many of these men later went on to become high officials in Rhee's postwar government of South Korea.[307]

Meanwhile in China, the OSS had to accept Chiang Kai-shek's backing of Rhee's rival, Kim Ku, and began training Koreans selected by Kim Ku in the early spring of 1945 for the "Korean Provisional Army" under General Li Bum Suk, and, as part of the Eagle Project, to send Koreans as guerrillas and intelligence agents into Manchuria, Korea, and even Japan.[308] The training camp was a small village near the Yellow River by Hsian (Xian) in Shensi Province. The force was primarily composed of Koreans who had escaped or been captured from the Japanese armies and labor battalions. Early in the project, OSS SI chief Colonel Paul L. Helliwell, 30, an attorney and banker, had sent a new OSS recruit up to report on General Li and the project. Helliwell's representative, this brand new arrival from the United States, was a 27-year-old lieutenant from upstate New York, a Brown University graduate, who had trained at Area E and Catalina Island. When after his inspection, the young lieutenant reported to Helliwell that Li was neither receiving intelligence from Korea nor had any operational plans to infiltrate his supporters there, Helliwell outraged with the young lieutenant sent him away. The young lieutenant soon came down with dysentery and may or may not have subsequently engaged in some combat missions, as his memoir written years later asserts. By the time he wrote that memoir in 1974, E. Howard Hunt had completely lost his credibility as well as his reputation. Two years earlier, in 1972, ex-CIA officer Howard Hunt, the author of several works of fiction as well as actual perpetrator of several "dirty tricks," helped organize the Watergate break-in that led to the downfall of President Richard Nixon.[309]

At least one of the American members of the OSS unit training Koreans near Hsian (Xian) for the Eagle Project, that Hunt had denigrated in 1945, had gone on to a distinguished career. Chester L. Cooper, a Bostonian, who left graduate school at Columbia to enlist in the Army as a private, had joined OSS in India out of boredom with his Army service as a sergeant there. At the Eagle project training base, OSS gave Sergeant Cooper a lieutenant's uniform to impress the Koreans. On a visit to the training camp at the beginning of August 1945, Donovan who had seen Cooper as a sergeant two weeks earlier in Kunming, recognized him but wondered at the lieutenant's uniform. "We did this for the Koreans, Sir," Cooper answered. "Well, you've made a lot of progress over the past two weeks," Donovan replied with a wink. "Keep up the good work!" So began Chester Cooper's long and distinguished career in intelligence and diplomacy—with the OSS, the CIA, State Department, and the White House.[310]

Donovan was enthusiastic about the Eagle Project for OSS penetration of Manchuria, Korea, and Japan. He also had other plans for infiltration of Japan. One was the Napko Project, a clandestine operation to train Korean Americans and Korean POWs to be deployed into Japan for intelligence and sabotage in advance of the planned U.S. invasion of the Japanese home islands at the end of 1945. Eifler later said that Donovan had in the summer of 1944 assigned him the Napko Project, to train at Catalina, ten groups of Koreans, drawn from POWs from Camp McCoy, to have them return to Korea led by himself and thirty American officers, foment rebellion there, and finally penetrate Japan, where millions of Koreans worked in an impressed labor force. Eifler's representatives, like Vincent Curl and Robert E. Carter, continued to train such Korean teams through the summer of 1945. The majority of these Koreans were trained at Catalina Island in 1945, although it is possible that some may have been among the Koreans trained at Area C that year.[311] One controversial and unsupported account contends that Donovan had already infiltrated agents into Japan, a six-man OSS team, presumably Koreans, Japanese or Korean or Japanese Americans, into Honshu, the main island in July 1943.[312] Colonel Carl Eifler claimed that his ten units of OSS trained Korean POWs turned SOs, were ready to leave their Catalina Island training camps, led by himself and thirty American OSS officers in August, and head for Japan itself. However, the mission was cancelled when the Japanese surrendered.[313] OSS Army Captain Howard Chappell, the German-American paratrooper who headed the SO Tacoma Mission in the Italian Alps with Sergeant Fabrega and Corporal Silsby, in 1944 and returned to the United States with a Silver Star Medal, claimed that he had a similar mission, but with Japanese Americans. According to Chappell, Donovan appointed him commander of a secret unit that was training a group of Japanese Americans on Catalina Island to parachute into Japan in the fall, establish a base, transmit intelligence and then wreck havoc behind enemy lines when U.S. Army and Marines landed on the beaches of the southern home island of Kyushu, an invasion that had been scheduled for 1 November 1945. Chappell said after the war that given the rivalry between MacArthur and Donovan, he often fantasized how great it would have been after sabotaging behind Japanese lines, for him to meet MacArthur as he stepped off the landing craft, "we could have said, 'General, we'd like to deliver Japan to you courtesy of Wild Bill Donovan.'"[314]

The Japanese, of course, surrendered before the planned invasion took place. Donovan was at Hsian (Xian), China, on 7 August, the day after the atomic bomb was dropped on Hiroshima. The next day, the Soviet Union declared war on Japan and invaded Manchuria. As the Red Army surged south, Donovan told Wedemeyer that the Eagle Project must move forward quickly. "If we are not in Korea and Manchuria when the Russians get there, we will never get in."[315] The Air Force dropped an atomic bomb on Nagasaki the following day. On the 10th, the Japanese cabinet offered to surrender if the emperor was retained. Donovan was on his way to Washington when Tokyo accepted unconditional surrender on 15 August in the Far East (14 August in the United States, V-J Day). Gen. Douglas MacArthur and U.S. forces began occupying Japan on 27 August, and the surrender ceremony was held on the Battleship *U.S.S. Missouri* in Tokyo Bay on 2 September 1945.

Rescuing POWs

The sudden Japanese capitulation and the quick end of the war in mid-August 1945 caught the OSS by surprise. Donovan's organization had made significant contributions in the jungle warfare in Burma. But in China, the OSS was just beginning to stage a number of coordinated major projects—in Manchuria, Korea, and South China—that promised important results. But they had not yet been launched, and now with the surrender, they would not have any significant impact on the defeat of Japan. Still the OSS hoped to obtain some recognition of its important work in China. With the news of the sudden Japanese decision to surrender, OSS/China chief, Richard Heppner cabled Donovan "Although we have been caught with our pants down, we will do out best to pull them up in time."[316]

Heppner immediately asked General Wedemeyer to airlift OSS commando and intelligence teams into key areas throughout Japanese-occupied China, Manchuria, and Korea. The OSS teams would raid various Japanese headquarters and seize vital documents and individuals, Japanese as well as "puppet" collaborators, some of whom would be charged as war criminals. They would safeguard American and National Chinese Government interests in China. They would land in strategic spots in Manchuria "in order that we may be on ground before arrival of Russians," and in Korea "in order that our interests may be protected before the Russian occupation."[317] On 12 August, Heppner ordered OSS teams into strategic spots in Mukden (Shenyang) and Harbin in Manchuria and Weixian (Weihsein) in China's Shandong Peninsula. They included OSS personnel from SO, SI, Medical Branch, and the Communications Branch.[318]

Major James G.L. Kellis from Illinois, who had spent the winter of 1942-1943 at SO training camps at Areas F, B, and A, and in 1944 had led the SO "Chicago" mission blowing up bridges in Greece, had joined SO China in 1945 along with two of his Greek-American radio operators, George N. Psoinos and Michael T. Angelos (Spiro Cappony had returned to the United States). Kellis had been training SO Team "Greyhound" in the mountains around Hsian in June and July waiting to be sent to the area around Peking (Beijing). Now with Tokyo's surrender imminent, Kellis, the two radio operators, and two Chinese interpreters were dispatched to Peking to arrange for the surrender of the Japanese forces there and the entry of the Allies. They parachuted into an area about fifty miles east of the city on the night of 12 August 1945. Dodging Japanese and puppet troops for several days, they entered the heavily guarded city on 16 August clad as Chinese puppet soldiers. Given refuge by a Chinese puppet general, the team negotiated through him with the head of the Japanese garrison for its pending surrender. Kellis also negotiated with the leader of the Soviet Army then marching on the city. His radio operators relayed vital intelligence information as well as the proceedings of the negotiations to the OSS base in Hsian (Xian). For their actions, Kellis received a second Legion of Merit, the Silver Star and Bronze Star Medals; Psoinos and Angelos received Bronze Star Medals.[319]

On 15 August 1945, the day Tokyo announced it would surrender, Wedemeyer ordered all agencies in his theater to give top priority, to locating and rescuing Allied Prisoners of War and civilian detainees held by the Japanese in camps.[320] Rescuing POWs posed more of a problem in Asia then in Europe as not only were the distances so much greater but rescuers would be flying into relatively unknown territory far behind enemy lines and might be confronting enemy forces that were prepared to kill the prisoners and their would be rescuers. There were intelligence reports in the days after 15 August that kamikaze planes were still attacking the U.S. Navy, that the emperor was unable to enforce his own cease-fire order, and that some fanatical Japanese officers had vowed to continue the war. It was uncertain, to say the least, whether rescue teams would meet resistance from prison camp commanders and guards when the teams parachuted or flew in to demand their surrender and the release of their prisoners. In revenge against the U.S. bombing of Japanese cities, the emperor's troops had been executing Allied prisoners for several months during the summer. Intelligence reports reported a recent increase in atrocities against Allied POWs, including public executions in Malaya, Thailand, and Taiwan. The Allies feared that the Japanese might try to massacre prisoners so they could not testify against earlier atrocities.[321] Although other agencies would assist, there was really only one U.S. military organization in Asia that was trained, ready, and equipped to carry out such missions: the OSS.[322] OSS carried the brunt of the operation with the Air Force transporting the rescued people out after the OSS had found them, gave them food and immediate medical treatment, and secured their release. OSS teams were well suited for parachuting into camps behind enemy lines and taking control before reinforcements arrived. In addition, these missions provided a cover for OSS intelligence and other operations in areas OSS was already seeking to penetrate before their takeover by the Russian or Chinese Communists.[323]

OSS organized nearly a dozen of these "Mercy Missions." Each was code-named after a bird. The initial missions were in North China. The "Magpie" Mission, headed by Major Ray Nichols flew into Peking (Beijing) on 17 August under arrangements made by Major Kellis and his team. They found 624 Allied POWs, among them the Navy commander who had headed the American forces on Wake Island captured by the Japanese on 8 December 1941, and four airmen from the Doolittle raid on Tokyo, who had been kept in solitary confinement almost continuously since their capture in April 1942. All four airmen were in serious condition, and one of them was almost dead with beriberi.[324]

Unlike the mission to Peking, most of the rescue missions did not have a prearranged reception committee and flew in not knowing what to expect. The "Duck" Team flew to the Shantung Peninsula on 17 August, arriving at a landing strip near a prison camp not far from Tsingtao, where 1,500 civilian Allied internees, men, women, and children, including 200 Americans, were held in a former Presbyterian mission surrounded by barbed wire. Heading the "Duck" Mission was Major Stanley A. Staiger, 24, an infantry officer from Klamath Falls, Oregon, who had undergone OSS SO training at Areas B and A and then served with Frank Gleason's "Detachment 21" in its successful mission helping to stop the Japanese advance by blowing up all the ammunitions and weapons warehouses in front of it in January 1945.[325] Now at the camp near Tsingtao, the sullen Japanese guards let the Americans into the camp and, as the Office of War Information press release described the scene, a "surging mass of prisoners," flocked around Staiger and his team, "wringing their hands, embracing them, pounding their shoulders, kissing them."[326]

Rescuing Allied POWs on massive Hainan Island off the south China coast was the goal of the "Pigeon" Mission. The Japanese there were still firing on American reconnaissance planes. The mission was led by Major John ("Jack") Singlaub, a paratrooper from Los Angeles, who had trained at Area F and B, been a Jedburgh in France and then volunteered for duty in China. He spent several months preparing an SO Chinese guerrilla team to impede a possible advance into China by a Japanese Army division in Vietnam. Now he was sent to Hainan Island. With only two day's notice, he put together a nine-man team, including his regular radioman, 1st Sgt. Anthony J. ("Tony") Denneau from Green Bay, Wisconsin, who had been with him since France, and an interpreter, Lt. Ralph Yempuku, a young Japanese American soldier from Honolulu, who had served with OSS Detachment 101 in Burma.[327] On 27 August 1945, a C-47 from the 14th Air Force flew them across the South China Sea, and at midday, they jumped from 500 feet and landed less than a mile from one of the military compounds. Confronted by a Japanese lieutenant and two truckloads of soldiers, Singlaub took a haughty, conquering attitude and through his interpreter, Lieutenant Yempuku, commanded the Japanese officer to halt his men and to take Singlaub to the commander of the Hashio POW camp. "We are here to help the Allied prisoners," Singlaub declared summarily. "The war is over."[328]

The Americans spent the night in quarters surrounded by armed Japanese guards. In the morning, when the Japanese colonel arrived and said he had just received word of the pending surrender, Singlaub treated him brusquely and demanded to see the senior officers among the POWs. When they arrived, "we had trouble containing our emotions," Singlaub recalled. The Australian colonel and the two Dutch officers were "little more than skeletons." Flesh and bone, no muscle tone at all. Deeply sunken eyes, milky and unfocused, stared from skulls that looked like death's heads. The Dutch colonel's neck and arms were scarred from repeated beatings. "I became aware of a faint, sweet-sour odor, something like fermentation, which I soon realized as the stench of starvation."[329]

The OSS rescue team found hundreds of Allied POWs on Hainan Island, many of them in terrible physical condition. There were 500 Australian and Dutch POWs in the Hashio prison camp. The Americans learned that nearly 200 had died of malnutrition, malaria, dysentery, beriberi and beatings since they had arrived in 1942. The team found that at least eight American airmen had been captured at Hainan after being shot down. At a Japanese naval facility at Sanya, three American airmen had been held since March 1945. They had been badly injured when their planes were shot down, and the Japanese had refused to treat their wounds and burns. Their guards had beaten them systematically in revenge for the bombings of Japan. They were put on starvation rations, and two of the American airmen had died as a result. Five American airmen had been publicly executed after being paraded through streets of island towns that had been bombed.[330]

The Pigeon Team's radio had broken in the parachute drop, and radioman Tony Denneau had to jury-rig a setup using the OSS crystal in a Japanese set to contact OSS headquarters in Kunming on an emergency frequency. Because the Japanese transmitter was so powerful, Kunming initially thought it was hoax, and only after a lengthy procedure of prescribed challenges and responses did they accept the team's coded message. After OSS and Army reinforcements arrived, Singlaub recovered two dozen additional Allied prisoners, including one American airman, who had been holed up in a mountain camp of Chinese Nationalist guerrillas and another American aviator with Chinese Communist guerrillas on the island. A few days later, a British destroyer arrived to begin evacuating the first contingent of rescued POWs to safety.

Manchuria: Rescuing General Wainwright; Clashing with the Soviets

With the surrender of Japan, one of OSS's first assignments was to get teams into Manchuria ahead of the Russians.[331] Initially this was done through Mercy/Rescue Missions flown in to rescue Allied POWs. The "Flamingo" Mission was to be dropped into the city of Harbin, deep in Manchuria, but the Soviets arrived before the scheduled drop, refused clearance, and the mission was cancelled. However, the "Cardinal" Mission under Maj. James T. Hennessy was parachuted near Mukden, the industrial center of Manchuria on 16 August 1945. An angry and suspicious Japanese patrol captured them when they parachuted into a field outside the POW camp, disarmed and beat them, most savagely the Japanese-American interpreter, Sergeant Fumio Kido.[332] But the beatings stopped when an officer arrived, and two days later, the Japanese commander surrendered and the Americans liberated 1,321 Americans, 239 Britons, and a few Australian, Canadian, and Dutch. On 27 August Hennessy and his men finally located a small camp at Sian, a hundred miles north of Mukden and rescued ten captured Allied generals and other high officials, including British Lieutenant General Sir Arthur E. Percival, former commander at Singapore and U.S. Lieutenant General Jonathan M. Wainwright, who had surrendered American forces in the Philippines in 1942. A week later, the gaunt 62-

year-old Wainwright sent a grateful tape recorded message to the OSS commander, which began: "This is General Wainwright speaking. Greetings, General Donovan. I am speaking from Hsian, China, where I have just eaten my first good American breakfast since the war."333 With the "Cardinal" Mission was Captain Roger F. Hilsman from OSS Detachment 101 in Burma. His father, a career Army officer, had surrendered the U.S. command on Negros Island in the southern Philippines two months after Wainwright had given up on Corregidor. Young Hilsman had joined the "Cardinal" Mission hoping to find his father. He did locate Colonel Hilsman at Hoten prison camp in Manchuria, where he had shrunken to 100 pounds, and his hair had turned completely white.334

The OSS rescue missions were a tremendous success. As historian Ronald Spector has so aptly express it: "A small number of determined young Americans, launched into the vastness of Asia, had performed notable feats of courage and improvisation that would be long remembered by those whose lives had been saved or imprisonments ended by the appearance of a lone B-24 bomber and a handful of parachutes."335

OSS casualties in China

Remarkably, there had been few American OSS combat casualties in China. Of course, there were not many OSS personnel in China until the final months of the war. The OSS contingent there had grown slowly, from 144 in December 1943 to 300 in February 1945. After Wedemeyer authorized expansion and independent action, OSS grew rapidly from 800 in April 1945 to nearly 2,000 four months later at the end of the war.336 According to some OSS China veterans, probably no more than half of these, some 1,000, ever came near the Japanese.337 Many of OSSers in China came down with various diseases, some bad enough to be sent home. Some died in accidents, and at least two were killed in Canton after the end of the war in an altercation with a fellow officer.338 But the combat fatalities among the perhaps 1,000 American OSS agents engaged with the enemy numbered only five, although probably several hundred of the Chinese Nationalist guerrillas trained by the OSS were killed in action, and unknown numbers of Chinese spies employed by OSS, who simply stopped reporting, had either fled or been found out and executed.339

Among the five OSS men killed in action, four died fighting the Japanese. Two of them Captain Thomas C. Blackwell and Lieutenant John Allen—died as a result of a bizarre accident. As Special Operations officers and paratroopers, both had been trained at OSS schools in Maryland and Virginia. A salesman before the war, "Blackie" Blackwell had joined the OSS in 1943. His perseverance had gotten him through OSS training; his engaging personality and good humor won him many friends; and he had been served in Special Operations in France. Arriving in China in April 1945, he was sent in a convoy from Kunming to the OSS station in Chihkiang, an airport city and one of the routes to Chungking. There he waited while his unit, Team "Ermine," was being formed. On 26 June 1945, Blackwell and another OSS officer, Lieutenant John Allen, the latter who had originally been scheduled for the Dormouse Mission, volunteered to accompany a Morale Operations air drop of thousands of propaganda leaflets over Japanese held area. Flying over the city of Changsha, the C-47 suddenly veered to evade anti-aircraft fire, and the two men who had been sitting in the open doorway tossing out leaflets suddenly plunged out into the air. They parachuted to the ground, but the Japanese surrounded them. It was later learned that Blackwell had pulled out his pistol and been killed in a shootout with the Japanese. Allen was captured, dragged through the streets and tortured. Later, still badly battered, having been put on a Japanese troop train headed for regional headquarters; he was killed along with many of the Japanese soldiers, when the train was attacked by American fighter-bombers.[340]

Most well-known among the OSS fatalities in China was Captain John M. Birch. In the closing days of the war, the most successful, yet also the most tragic, OSS mission operating out of Hsian was the R2S mission led by this 27-year-old intelligence officer from the 14th Air Force, who had been assigned to OSS Secret Intelligence beginning with the AGFRTS group in Kunming. Birch spoke Chinese, knew China well, and had earned many friends there. He was also an ardently evangelical young man, a fervent Baptist, with a hatred of the Japanese and what they were doing to the Chinese. Since the summer of 1945, working out of Hsian (Xian), Birch had established and obtain excellent information from a dozen intelligence nets, reporting in by radio from Peking [Beijing] the Yellow River basin and the Shandong Peninsula. The Shandong Peninsula, which stretches out from northern China like a finger pointing at Manchuria and Korea, was geographically and strategically important for transportation and communication among China, Manchuria and Korea, and the Chinese Communists wanted to deny it to the Chinese Nationalists or their American ally.[341] After the Japanese surrender, when the OSS rescue and intelligence missions began, OSS/Hsian ordered Birch to go to Shandong Province to seize enemy documents and to obtain information on airfields from which American POWs could be flown. Birch and his team, which consisted of four Americans, seven Chinese, and two Koreans, set out on 20 August for the Shandong Peninsula.[342]

The Communist 8th Route Army was ordered to intercept Birch's team and its patrols stopped them several times. Finally on 25 August, at Huang-kou station in the Xuzhou [Suchow] area, a group of Communist soldiers surrounded Captain Birch and Lieutenant Dong Qinsheng (Tung Chin-sheng), who had been separated from the others, and ordered them to give up their weapons. Birch became increasingly irritated, emphasized that the war was over, denounced the Communists as "bandits," and claimed that he was on a peaceful mission simply to survey airfields for POW rescue, not to interfere in any struggle between the Kuomintang and the Communist Party. The soldiers opened fire and both Birch and Dong were hit and fell to the ground. Despite the pain, Dong lay there, was left, and survived, although he lost a leg and an eye, but Birch continued to resist. Birch's hands were tried behind his back. He was dragged away, murdered and mutilated almost beyond recognition.[343] General Wedemeyer protested directly to Mao Tse-tung and Chou En-lai, demanding a response and also the release of all the other OSS personnel the Chinese Communists had detained. At the beginning of September 1945, the

Communists released the remaining members of Birch's team, who had been captives for two weeks and four OSS officers of the Spaniel team, who had been held for four months. The fate of Captain George Wuchinich and his team from the Chili Mission was unknown, and it was not until late September 1945 that Communists finally released them.[344] Birch's murder and the detention of the other OSS men was powerful evidence of the increasingly aggressive policy of the Chinese Communists.[345]

In the Far East, OSS produced some of its most impressive results of the war. Most importantly, in Burma, Detachment 101 played a crucial role in harassing and sabotaging enemy supply lines when the Imperial Japanese Army controlled that country, and later, it helped guide Merrill's Marauders and the Allied advance into Burma. Some 500 Americans in OSS Detachment 101 mobilized 9,000 Kachin and other native peoples in Burma. They were credited with directly inflicting 15,000 casualties on the enemy, including the known deaths of 5,447 Japanese soldiers and another 10,000 enemy soldiers believed to have been killed or seriously wounded. In addition, another 12,000 enemy casualties resulted from air attacks instigated by Detachment 101.[346] In other areas of Southeast Asia, OSS played smaller if still important roles. It was influential in linking Thai patriots from the United States with anti-Japanese resistance forces within Thailand. In French Indo-China, OSS officers continued to follow the original global aims of supporting the most active resistance movements against the enemy, even if they included communist partisans, and not encouraging re-colonization by former imperial powers. The result of such actions by officers in the field in the last year of the war was that the OSS was initially brought into open support for Ho Chi Minh and the Viet Minh in their efforts against the Japanese and subsequently, if temporarily, Indochinese resistance against French reassertion of control of the former colony. Since the French did re-establish control for another decade (ultimately the Viet Minh triumphed in Vietnam), the OSS legacy in Indochina was a mixed one, disdained by the French, for example, but offering the possibility of a pro-American relationship with the Independence Movements and ultimately independent nations. In Vietnam, as in India and Burma where many members of the OSS missions were sympathetic to the movements for independence from British rule (as well as Thailand where the United States opposed British designs on the country), the OSS established a very positive reputation among the ultimately triumphant nationalists.

OSS's role was limited for much of the war in China by politics within the Nationalist Chinese government as well as service rivalries within the American armed forces there. But in the final year of the war, OSS experienced the beginning of a dramatic growth and impact in that theater. With the end of the war in Europe in the spring of 1945, American military efforts shifted to the Far East. As the OSS reported in June 1945, "the activities of OSS China are expanding rapidly as the strategic picture in that theater unfolds. Because of the fluid 'lines' throughout China there are almost limitless opportunities for effectively organizing and supplying resistance behind the Japanese."[347] All of the OSS branches were at work: Special Operations teams conducting raids and sabotage on enemy supplies and supply lines and providing target information for the U.S. Army Air Forces, Chinese Operational

Groups being trained and equipped, Secret Intelligence Branch extending intelligence chains; Research and Analysis Branch also supplying intelligence information, Communications Branch building extensive communications networks, and Morale Operations Branch expanding operations to undermine enemy morale. All of these conditions in China, OSS headquarters in Washington concluded "present OSS with perhaps its best opportunity for large-scale operations, particularly since the American Army is so far not engaged there in any important numbers. With the lessons learned in Europe, and at Detachment 101 [in Burma], OSS in China will be able to coordinate all its weapons against the enemy as never before and will operate through the air, on water, and on land... the scope of the OSS effort in China is already impressive."[348]

That buildup was cut short by the sudden emperor's decision in mid-August 1945 to surrender. But in the Allied victory, the OSS received due credit from the American military commander in China, Lieutenant General Albert Wedemeyer. In General Orders issued from his headquarters in Chungking, Wedemeyer praised OSS personnel for the "outstanding performance of duty in their vital missions" and said their achievements constituted "a record of extraordinary heroism, resourcefulness, initiative and effective operations against a ruthless enemy in the Orient." There were fewer than 2,000 OSS people in China in 1945, but this small force was officially credited with direct responsibility for killing more than 12,000 Japanese troops.[349]

OSS in China proved important in numerous other ways. It helped rescue downed American aviators there as in Southeast Asia. OSS intelligence information directed American Air Forces to mobile tactical targets in China and provided the U.S. Navy with enemy shipping information that enabled American submarines and carrier planes to sink thousands of tons of shipping bound for Japan. The OSS was also credited with a successful sabotage campaign behind enemy lines against bridges, railroad lines and other transportation facilities, telephone and telegraph lines, which hindered enemy reinforcements and supplies and which required redeployment of significant numbers of Japanese forces from the frontline to rear areas in attempts to eliminate the OSS-led guerrilla activities. OSS-trained Chinese paratroop commando units had had begun actively to engage the Japanese. At the end of the war, OSS SO parachute teams located, airdropped in, and protected Allied POWs from possible massacre in Japanese prison camps. As General Wedemeyer concluded "the record of the achievements of the officers and enlisted men of the Office of Strategic Services will constitute a chapter beyond parallel in the history of the accomplishments of the United States Armed Forces in their successful prosecution of the war against the Japanese on the Asiatic Continent."[350]

The sentiment of many OSSers was perhaps best summarized by Sergeant "Al" Johnson, a medic who had served with OG teams in France and China. Looking back at their experience in World War II, he concluded: "We had been given a job and had done it well... not just an ordinary assignment, but one that brought a small group of men together to penetrate the enemy lines many hundreds of miles. We learned to live under the noses of the enemy to create an organization that could work underground, to live off the land, and to cause large-scale damage, not only to material things but also to psychologically impair the enemy so that he was unable to function effectively. We were ordinary men with ordinary backgrounds doing an extraordinary job under very difficult circumstances."[351]

CHAPTER 10

POSTWAR PERIOD: END OF THE OSS AND RETURN TO THE PARK SERVICE

OSS may have won its battles in the field, but it lost its final campaign—in Washington. It was better prepared to fight armed enemies overseas than bureaucratic enemies in the nation's capital. The OSS was terminated within a month after the end of World War II.

To a great extent, this was due to Donovan himself. Always the romantic adventurer, he spent much of his time away from Washington in the war zones. Unlike desk-bound, bureaucratic spymasters, Donovan, a World War I hero, enjoyed being on or near the front lines. His restless energy, dynamism, and personal attention inspired the men and women of the OSS. An inspiring and visionary leader, he conceived of America's first centralized intelligence and special operations agency, forerunner of the Central Intelligence Agency and Army Special Forces.

Yet, Donovan was inattentive as an administrator and uncompromising as a bureaucrat.[1] Fascinated with strategic visions and actual field operations, he was bored by organizational detail. He personally disdained office work and left the daily running of his agency to others, as he had done at his law firm. Consequently, the administration of his rapidly burgeoning, worldwide organization was often chaotic. Because the OSS recruited imaginative, free-wheeling, assertive individuals, the organization was probably inherently difficult to manage, but Donovan's absences and inattention made it more so. While that management style contributed to individual and local initiative, it also left the organization vulnerable to its critics. Indeed, significant discontent existed even within the top echelon of the OSS itself.[2]

One of Donovan's dictums was that because he was dealing with a new form of warfare, he would rather see OSS people think imaginatively and fail than be constrained to narrow, traditional, routine responses out of fear of failure. "I'd rather have a young lieutenant with guts enough to disobey an order than a colonel too regimented to think and act for himself," Donovan often said.[3] But while this endeared him to many OSSers and produced results, it also produced some mistakes, which helped to make the organization vulnerable to its enemies in Washington. And Donovan certainly had his enemies in the nation's capital. In the ongoing bureaucratic turf battles, the OSS threatened a number of powerful old line, congressionally mandated agencies, including the War and Navy Departments, the

State Department, and the Federal Bureau of Investigation. In contrast, the OSS was a temporary, wartime agency, created under the President's executive authority as commander-in-chief. Its main assets, the support of President Franklin D. Roosevelt and the unrestricted funds he unilaterally authorized, were a distinct advantage to the new and temporary agency and a stimulant to the jealousy and opposition of its powerful competitors.

Battle in Washington over the OSS

Donovan believed the OSS would serve as a model for a permanent central intelligence agency in the postwar period. But his organization, composed primarily of wartime civilian volunteers rather than professional soldiers or spies, lost the immediate battle for permanence in the bureaucratic arena in Washington, which in the immediate aftermath of the war saw little need for such a permanent, centralized clandestine agency as the OSS. The predominant mood in America was for demobilization and the enjoyment of peace. Donovan had made too many enemies and in the end proved incapable of achieving his goal of maintaining the OSS as a permanent central intelligence agency with himself as its head.

By the fall of 1944, it was evident that the Allies would be victorious and the war was reaching its final phases. The White House's Bureau of the Budget began making plans for liquidation of the temporary wartime agencies and for reduction of the federal government to peacetime conditions. The Army and Navy began serious planning for the postwar missions, size and structure, including military intelligence. The State Department sought responsibility for foreign political and diplomatic intelligence, and the FBI claimed authority in the area of domestic intelligence and counterintelligence operations. In such a contentious climate, Donovan made a major bid for a permanent central intelligence agency.[4]

"When our enemies are defeated," Donovan wrote to President Roosevelt on 18 November 1944, "the demand will be equally pressing for information that will aid us in solving the problems of peace."[5] Fissures were already emerging in the wartime alliance of the United States, Great Britain, and the Soviet Union, and Donovan was beginning to envision intelligence networks to deal with the rivalries and differing power relationships and goals of the former Allies in the postwar world. Donovan's proposal was for a "Central Intelligence Service." While departmental intelligence bureaus would continue to collect and analyze the tactical intelligence directly related to the duties of their agencies, the new organization would focus on strategic intelligence, coordinating, collecting, and producing intelligence and reporting directly to the President. It would provide the Chief Executive with the foreign intelligence necessary to plan and execute national policy and strategy. From the beginning, Donovan believed not only that the new agency should produce intelligence from various means, including espionage, counterespionage, and the kind of investigations conducted by the OSS's highly successful Research and Analysis Branch, but that it should, like the Special Operations and Morale Operations branches, have the authority to conduct "subversive operations abroad" and perform "such other functions and duties reliant to intelligence" as the President might direct.[6]

This new, permanent, and independent Central Intelligence Service would have its own budget and authority for "procurement, training, and supervision of its intelligence personnel." In the OSS, Donovan noted, there existed personnel already trained and experienced in the missions of the new organization. Warning that these assets should not be lost, Donovan suggested that the President not wait for the end of the war to establish the new agency. "There are common sense reasons," he wrote to Roosevelt in November 1944, "why you may desire to lay the keel of the ship at once."[7]

The fight against the launching of a new central intelligence service began immediately. When Roosevelt circulated Donovan's proposal to established government agencies with intelligence functions, they sought to block it through memos, lobbying and delay, but some went further and started a smear campaign in the press. On the morning of 9 February 1945, Donovan opened the front door of his Georgetown home, picked up the morning newspapers and was appalled to read the front-page headline on the Washington *Times-Herald:* "Donovan Proposes Superspy System for Postwar New Deal: Would Take over FBI, Secret Service, ONI and G-2 to Watch Home, Abroad." Washington correspondent Walter Trohan's story ran on the front page of this and other newspapers of the isolationist, anti-Roosevelt, McCormick-Patterson chain, including the *Chicago Tribune*, and *New York Daily News*. Announcing that the President was considering a plan from Donovan for a central intelligence service, the newspaper account then launched into a sensationalist diatribe against the OSS, Donovan, and his proposal. Trohan inaccurately claimed that the proposed agency would "supersede all existing Federal police and intelligence units" and he predicted that it would create "an all-powerful intelligence service to spy on the postwar world and to pry into the lives of citizens at home." Likening the proposed agency to the Nazis' dreaded secret police, Trohan warned that it could become an "American Gestapo."[8]

Donovan's plan for a postwar, worldwide intelligence service received balanced coverage from much of the press and even favorable editorial comment from internationalist newspapers such as the *Washington Post*, *New York Herald-Tribune*, and *New York Times*.[9] But the Trohan story and its accusations caused a sensation and, as intended by the FBI and military intelligence agencies that leaked it to the press, prevented any chance of the President immediately supporting Donovan's proposal.[10]

Roosevelt had made no promises to Donovan about a postwar agency. He had accepted the need, at least in wartime, for effective and centralized intelligence, although he had limited some of Donovan's wartime proposals in response to appeals from competing agencies. In regard to Donovan's proposal for a permanent OSS type organization, the President was willing in November 1944 to circulate the idea, but he certainly kept his options open. On 5 April 1945, two months after the furor caused by the original Trohan article, Roosevelt resumed consideration of the issue of postwar intelligence. Just before leaving for a vacation at Warm Springs, Georgia, the tired and ailing President asked Donovan to assemble the heads of the ten dozen federal units concerned with foreign intelligence and internal security to "contribute their suggestions to the proposed centralized intelligence service....so that a consensus of opinion can be secured."[11] Roosevelt's managerial style was to solicit various opinions, even allowing people with opposing views to think they had his support, all the while keeping his own counsel and deferring his own decision, and certainly not disclosing it, until he had to do so. Clearly he was listening to the other agencies as well as to Donovan on the subject. It is unclear what Roosevelt would have done about the proposal if he had lived. But while at Warm Springs, Roosevelt suddenly died from a cerebral hemorrhage on 12 April 1945.

President Harry S Truman and the OSS

With his death, Donovan and the OSS lost their most important patron. The new President, Harry S Truman,[12] was no friend of Donovan or his organization. Indeed, Truman felt no obligation to sustain Donovan and the OSS beyond the end of the war. A pro-New Deal, but fiscally conservative Democrat from Missouri, Truman was suspicious of Donovan and his free-wheeling, expensive wartime agency from the beginning. That skepticism was reinforced by Bureau of the Budget Director Harold B. Smith, whom Truman considered an efficient and honest public servant opposed to empire building and increased government spending. In several meetings in late April 1945, Smith cautioned Truman about taking hasty action approving any of the plans for a postwar intelligence system and suggested that the President allow the Budget Bureau to assess the situation and recommend a sound and well-organized strategic intelligence system. Truman agreed.[13]

The new President was thus already suspicious of Donovan when he first met the OSS director on 14 May 1945, and increasingly, he came to dislike him.[14] In a brief entry in his appointment book, Truman noted snidely that Donovan had come in to "tell how important the Secret Service [sic] is and how much he would do to run the government on an even basis."[15] Donovan's own recollection of that meeting was that Truman told him that "the OSS has been a credit to America. You and all your men are to be congratulated on doing a remarkable job for our country, but the OSS belongs to a nation at war. It can have no place in an America at peace." When Donovan argued that the OSS would be even more important in the troubled peace, Truman declared, in words similar to the Walter Trohan newspaper story, "I am completely opposed to international spying on the part of the United States. It is un-American. I cannot be certain in my mind that a formidable and clandestine organization such as the OSS designed to spy abroad will not in time spy upon the American people themselves. The OSS represents a threat to the liberties of the American people."[16]

Truman's hostility toward Donovan and his organization may have come from a variety of sources. Some of them may have been matters of policy, such as the concerns of Truman and Smith about the costs and dangers of a permanent central intelligence agency. Donovan later blamed Truman and the Army and Navy for blocking his plan.[17]

Other reasons for the new President's hostility may have been more personal, including Truman's attitudes toward Donovan himself.[18] From Truman's alleged remarks to Donovan at their first meeting in May 1945, it appears that he may also have been influenced by the articles in the McCormick-Patterson newspaper chain and a key, anti-OSS source Trohan had used. Shortly after taking office, President Truman had been informed of a secret, scathing internal report on the OSS authored by Colonel Richard Park, Jr. of Army Intelligence (G-2). Park, who had served briefly as a military aide to President Roosevelt, may have mentioned his report when he met with Truman on 13 April 1945, the day after Roosevelt's death. Certainly, he soon sent the new President a copy of his hostile report.[19]

Classified 'Top Secret," the Park Report on the OSS ran 56 doubled-spaced pages.[20] According to Park, President Roosevelt had asked him to make an informal investigation of the OSS and report on his findings and conclusions. His report, Park said, was gathered "in an informal manner and from personal impressions gained on a tour of the Italian and Western Fronts." It was almost entirely a compilation of denunciations of the OSS, most of the accusations being based on second-hand information, suspicions or rumors, and most were from unnamed sources. The report's overriding theme was that the OSS was a body of amateurs in contrast to established intelligence professionals and consequently had made major mistakes in recruitment, training, and security procedures as well as in the handling of its funds. All of these alleged errors, Park concluded, had embarrassed the United States, especially its professional agencies, wasted enormous sums of money, and resulted in "badly conceived, overlapping, and unauthorized activities," including the capture and execution of its agents. Park cited accusations from unnamed sources in Army and Navy Intelligence, the State Department, and the FBI, and he declared as further evidence of OSS's unworthiness that General Douglas MacArthur and Admiral Chester Nimitz had banned it from their theaters. Finally, Park concluded, drawing on accusations from a number of unnamed critics that "many improper persons have penetrated into the O.S.S." As he put it, "the Communist element in the O.S.S. is believed to be of dangerously large proportions," and at the same time, "the O.S.S. is hopelessly compromised to foreign governments, particularly the British, rendering it useless as a prospective independent, postwar espionage agency."[21]

Park's accusations obviously reflected the views of OSS's critics, particularly its rivals in the old line intelligence services. Donovan apparently never saw Park's report, although he knew many of its assertions through the Trohan article, which had drawn upon it. While the OSS certainly did make mistakes, Park's "report" was hardly an independent assessment; instead, it was filled with rumor, half-truths, innuendo, and other distortions. For example, Donovan had recruited a few communists or communist sympathizers because they had been in particular regions, such as Spain, or because they could organize and work with anti-Nazi communist partisans in occupied countries, but they were never more than a handful. Duncan Lee, a Yale graduate, formerly of Donovan's law firm who was Donovan's executive assistant at OSS and later head of the China section of SI, was the only communist in or near the leadership. Communists certainly were not in the organization in "dangerously large proportions" as Park put it, and congressional hearings following the Trohan articles left the OSS relatively unblemished. But conservatives' suspicions remained.[22] In fact, the OSS leadership was in fact dominated by business-oriented lawyers and businessmen. As for the British influence, while it had been substantial in the founding and initial development of Donovan's organization, the OSS and British SOE and SIS soon became rivals, diverging in many of their goals and methods. B the winter of 1944-1945, OSS had clearly established an aggressive independence along with a skeptical attitude and sometimes hostile relationship toward the British Ally and its postwar goals.[23]

One example of Park's use of innuendo is his suggestion of massive waste by the OSS. As he put it, "If the O.S.S. is investigated after the war it may easily prove to have been relatively the most expensive and wasteful agency of the government. With a $57,000,000 budget, $37,000,000 of which may be expended without provision of law governing use of public funds for material and personnel, the possibilities of waste are apparent. There are indications that some official investigation of O.S.S. may be forced after the war. It is believed the organization would have a difficult time justifying the expenditure of extremely large sums of money by results accomplished."[24] In fact, no such massive irregularities were ever discovered by the media or the government. Although OSS, like other wartime agencies, surely had made errors and wasted funds, most of Park's indictment of overall incompetence, mismanagement, and waste, is simply incorrect.[25]

Within the 56-page diatribe against Donovan's organization, Park devoted three pages to kind words about two OSS branches. Special Operations, he said, had "performed some excellent sabotage and rescue work." Research and Analysis Branch "has done an outstanding job." Both had been "the subject of commendatory letters from theater commanders and others."[26] SO and R&A, Park said, had elements and personnel that "can and should be salvaged" by the War Department and the Department of State. This brief exception did little to offset the overall thrust of Park's harangue, which also included the danger of a "new secret, world-wide intelligence agency which would control all other U.S. intelligence agencies." Park said Donovan's proposal had "all the earmarks of a Gestapo system." Indeed, he warned that both Congress and the press "may be interested in exposing what many claim to be an American Gestapo."[27]

The "Gestapo" quote and other hostile sections of Park's report had been lifted almost verbatim by Walter Trohan in his denunciatory newspaper article in the McCormick-Patterson press.[28] Not surprisingly, Park's recommendations matched those of the traditional intelligence agencies. The OSS should be terminated, Donovan replaced, and those components and personnel determined to be worthy should be transferred to appropriate agencies such as the War Department and State Department. The postwar intelligence/ counter-intelligence operation should be a cooperative arrangement modeled on the collaborative structure established by the Army G-2, ONI, the State Department, and the FBI for use in the Western Hemisphere during the war.[29]

Donovan's Dilemmas

As Donovan recognized, the long-term future of OSS was in jeopardy by the summer of 1945 even as the role of the OSS expanded dramatically in China. Informed by the military chiefs, Truman now knew that the primary intelligence successes of the war had been through cryptology - the deciphering of enemy codes through MAGIC and ULTRA. This had been the result of signals intelligence breakthroughs achieved by the intelligence sections of the U.S. Army and Navy in the case of MAGIC, the reading of the Japanese codes, and the Poles and the British in the case of the German codes.[30] It had not been due to espionage or research and analysis which were the focus of OSS's secret intelligence efforts. Furthermore, unlike its rivals, the OSS was, after all, a temporary war agency created by executive order with no underlying foundation in congressional statute. Unlike the departments and bureaus that opposed it, OSS had no long traditions, no influential alumni, no established personal and economic connections with Congress. Because it had been a secret organization, the American public knew little about it. The first burst of publicity, the Trohan series about the threats to Americans posed by Donovan's plan for a postwar super-spy agency, had cast it in a bad light at a time when the emphasis was on forthcoming demobilization and the reduction of the expanded and costly wartime governmental bureaucracy.

Economizers in government led by Senator Harry Byrd, a powerful Virginia Democrat, pressed for reduction of agencies and personnel and cut backs in expenditures and taxes. The OSS's rivals and the McCormick-Patterson press resumed their attacks on Donovan's organization. Its proposed budget for the next fiscal year was cut almost in half.[31] OSS was in an extremely precarious position. As Thomas Troy, a CIA historian, has written, in mid-1945, Donovan "had been stymied by the big four [State, War, Navy and the FBI], ignored by the President, and was unsupported in the Congress. He had been smeared by the press—he [allegedly] harbored Communists, was controlled by the British, was rebuffed by heroes MacArthur and Nimitz, traveled with self-seeking bankers, financiers, industrialists and socialites, had squandered money, and was marked for sensational exposé."[32]

Once again Donovan hurt his own cause by refusing to accept a compromise. Given the formidable opposition, his deputy, Brigadier General John Magruder, had advised Donovan in May to concede temporarily on both the independence of the director and the timing of the creation of a peacetime central intelligence agency. But Donovan refused.[33] He took a hard line. In a letter several months later to Samuel Rosenman, one of Truman's key advisers on postwar reorganization, Donovan declared bluntly: "I understand that there has been talk of attempting to allocate different segments of the [OSS] organization to different departments. This would be an absurd and unsatisfactory thing to do. The organization was set up as an entity, every function supporting and supplementing the other. It's time for us to grow up, Sam, and realize that the new responsibilities we have assumed require an adequate intelligence system."[34]

Instead of negotiating within the bureaucracy, Donovan launched a publicity campaign to build political support by informing the public of the OSS's contributions to winning the war. This presented somewhat of a paradox: publicity for a super secret organization. When recruited, OSS personnel had been sworn to secrecy, taking an oath sometimes reinforced by threats of imprisonment for violation on the grounds of national security. They could not even tell their families what they had done during the war. Most of the former OSS men and women maintained that secrecy for decades—until the declassification of the records of the OSS by the CIA beginning in the 1980s. Some remain unwilling to talk fully about their wartime secret service even to the present day. But beginning in August 1945, Donovan and a group of writers and publicity persons on his staff began a media blitz, declassifying reports and providing interviews with OSS heroes to obtain credit and public recognition for the difficult and sometimes highly dangerous work of an organization about which the public knew virtually nothing.[35] By early September, sympathetic magazines and newspapers, including the *Washington Post*, *Chicago Daily News*, *New York Times*, *Life* and the *Saturday Evening Post*, supported the OSS with news stories of its heroic exploits, including the recent rescues from Japanese prison camps of General Jonathan Wainwright, aviators from the Doolittle Raid, and thousands of other American POWs, and editorials supporting its "Daring Exploits" and what one of them called "Our Priceless Spy System."[36]

It was already too late. Initially, the Budget Bureau had assumed that the liquidation of OSS would take place over several months so that its most valuable assets could be saved and distributed to other agencies. Donovan too had assumed even in mid-August that OSS had at least several more months to continue achievements and publicize its part in the Allied victory. But in the third week of August 1945, the White House, having already ordered the immediate end of the Office of War Information (OWI), the government's unpopular wartime propaganda agency, suddenly directed that OSS too be terminated as quickly as possible.[37] Based on previous discussions, the Budget Bureau agreed that OSS's Research and Analysis Branch would go to the State Department. The newly-imposed time constraint led the Bureau to recommend that the rest of the OSS be turned over to the War Department "for salvage and liquidation." On 4 September 1945, Budget Director Smith submitted a general intelligence reorganization plan to President Truman. Predictably, Donovan fumed when he heard about it, and he protested vigorously. But Truman ignored him, and on 13 September, the President authorized the Budget Bureau to draft an Executive Order terminating the OSS. According to Smith's diary entry that day, "The President again commented that he has in mind a broad intelligence service attached to the President's office. He stated that we would recommend the dissolution of Donovan's outfit even if Donovan did not like it."[38]

Although Truman may have been thinking of a broad agency attached to the Executive Office, the Bureau of the Budget and the old line departments were thinking of departmental units coordinated by an advisory group. There was deliberately no place for Donovan. Indeed, since some of the OSS assets would continue by being transferred to other agencies, and since Truman did want some kind of postwar strategic intelligence coordination in order, as he said many times, to prevent another Pearl Harbor surprise attack, the decision to dissolve the OSS may well have been based at least in part to force out Donovan.

Sudden End of the OSS

On 20 September 1945, Truman signed Executive Order 9621, dissolving the OSS effective 1 October 1945. The functions, properties and personnel of the Research and Analysis Branch and Presentation Branch were transferred to the State Department. All of the other functions and properties and most of the nearly ten thousand persons in the OSS were transferred to the War Department, which was interested mainly only in those with military experience in Secret Intelligence (SI) and Counter-Intelligence (X-2). Most of the OSS personnel would be discharged, including Donovan himself. The functions of the OSS Director were transferred to the Secretary of War, who "shall whenever he deems it compatible with the national interest, discontinue any [OSS] activity transferred by this paragraph and wind up all affairs relating thereto."[39]

In a separate and rather chilly letter of appreciation to "My dear General Donovan," written for the President by the Budget Bureau staff, Truman acknowledged Donovan's "capable leadership" in "a vital wartime activity." The letter assured Donovan that "the peacetime intelligence services of the Government are being erected on the foundation of the facilities and resources mobilized through the [Office of Strategic Services] during the war." Rubbing salt in the wound, Truman's letter stated that that transfer of the OSS Research and Analysis Branch to the State Department marked "the beginning of the development of a coordinated system of foreign intelligence within the permanent framework of the Government."[40]

The historic episode that occurred in the Oval Office at 3 o'clock, 20 September 1945, was recorded by Budget Director Harold Smith in his diary: "When I gave the President the Order on OSS for his signature, I told him that this was the best disposition we could make of the matter and that General Donovan . . . would not like it. I showed the President our communication with [Secretary of State] Byrnes and indicated that the State Department was willing to accept certain of the OSS functions while the rest would go to the War Department. The President glanced over the documents and signed the Order. He commented, as he has done before, that he has in mind a different kind of intelligence service from what this country has had in the past."[41]

It was an abrupt end to the OSS. As a result of an oversight in the wording of the Executive Order, the agency had to be dismantled within only ten days.[42] Donovan and his key assistants spent those few days microfilming the files of the director's office from the previous four years in order to preserve that record.[43] On 28 September 1945, Donovan bade farewell to the OSS staff in Washington. The sad ceremony was held in large building at the foot of the hill below the OSS buildings, in the Riverside Skating Rink on Rock Creek Drive. Closing out the wartime experience of America's first national intelligence agency, the 62-year-old Donovan reassured the faithful crowd of about 2,000 men and women:

> We have come to the end of an unusual experiment. This experiment was to determine whether a group of Americans constituting a cross-section of racial [ethnic] origins, of abilities, temperaments and talents, could risk an encounter with the long-established and well-trained enemy organizations.

How well that experiment has succeeded is measured by your accomplishments and by the recognition of your achievements. You should feel deeply gratified by President Truman's expression of the purpose of basing a coordinated intelligence service upon the techniques and resources that you have initiated and developed...

When I speak of your achievements, that does not mean we did not make mistakes. We were not afraid to make mistakes because we were not afraid to try things that had not been tried before. All of us would like to think that we could have done a better job; but all of you must know that, whatever the errors or failures, the job you did was honest and self-respecting.....

Within a few days each one of us will be going to new tasks, whether in civilian life or in government service. You can go with the assurance that you have made a beginning in showing the people of America that only by decisions of national policy based upon accurate information can we have the chance of a peace that will endure.[44]

Two days later on October 1, 1945 the OSS ceased to exist. Some of the civilian staffers had already left and returned to their prewar occupations or taken new positions, but for most of the OSS personnel, the change had been so sudden and unexpected that the majority simply waited to see what would happen to them as the State and War Departments made the crucial decisions. Meanwhile, most of the OSS military personnel returning from overseas were put up temporarily in holding areas, primarily at Area F, but some also at Area C, until they could be mustered out or accepted new assignments in the military.

OSS Veterans Return to the Training Camps from Overseas

Some of the OSS training camps had been used as holding areas for returnees from overseas since late 1944 when significant numbers of OSS veterans came back from Western Europe and North Africa. Although some of the returnees were housed at Areas A and C in Prince William Forest Park (Area B at Catoctin Mountain Park was being relinquished by OSS at that time), the majority were sent to Area F at the former Congressional Country Club in Bethesda, Maryland, because of its hotel-like facilities and its closeness to Washington.[45]

Hundreds of OSS veterans flocked back from Europe in 1944-1945 and spent time at the holding area at Area F or overflow areas at A and C before being reassigned to the Far East, or mustered out at the end of the war. In December 1944, Donovan ordered that Areas F, A, and C, be made available for the relaxation and recreation of OSS personnel returning from overseas.[46] While military training was provided at the camps for those being sent to the Far East, other returning veterans enjoyed a period of rest and relaxation, particularly in the second half of 1945. Dances and other forms of entertainment were held at all three areas. Returning from England to Area C in late 1944, radioman Roger Belanger from Portland, Maine, found a much different atmosphere then when he had left. "They had a couple of dances in the mess hall," he recalled. "WAVES from Arlington Farm—a Navy communications station, where they also handled secret work—[came over] in buses blacked out so they could not see where they were going. A bus load of them came in late 1944, came in their uniforms. We had a band. I think they were musicians from the group.... We also went to a place in Manassas, the Social Circle, a dance hall, and they served local beers. They took us down in a bus on Friday or Saturday night. We had a Class A pass and could leave the base anytime we were not on duty. They would also take us in a truck to Washington, D.C. and pick us up Sunday night in front of Union Station there. We had a good time."[47] By the beginning of summer 1945, however, OSS headquarters became concerned, and the Communications Branch was asked "to discontinue dances at Areas 'A' and 'C' because of the serious embarrassment which might arise if [there was] an official investigation of such recreation at 'Secret' areas."[48] However, dances, movies, lectures, and sightseeing trips continued on a regular basis at Area F, now the official center for OSS returnees from the combat zones.[49]

An OSS softball league was organized that included games between teams from various OSS branches or between training areas. On 27 June 1945, a hotly-contested match between the undefeated teams of Area F and Area C was played at Prince William Forest Park. Pitching for Area F was Captain Joseph Lazarsky, former instructor at Area B and now recently returned from combat duty with Detachment 101 in Burma. As the sports reporter for the Area F newspaper reported, "The lead exchanged hands every inning until the final, when 'Captain Joe' pitching his first game in three years and facing the heaviest hitter of Area 'C' proved himself a real 'money' player and put out the side in one, two, three order. Area F beat Area C, 14 to 12. A week later the "Holdees" at Area F won the league championship for the first half of the season. But two months later, as the second half of the league season drew to a close, Area F had plummeted. On 14 August 1945, the team was playing the Reproduction Branch softball team in Potomac Park in Washington, D.C., when in the sixth inning, the city learned of the Japanese decision to surrender and "pandemonium broke loose—whistles, shrieks, horns blowing, etc., heralding the end of the war." As the Area F newspaper reported, "the boys [were] fearful the game would run into extra innings and postpone their taking part in the general revelry and victory celebration." "Captain Joe" Lazarsky kept "pitching steady ball," but Reproduction Branch scored two home runs in the seventh, one on an error, and won the game. Area F fell to fifth place, and it was Area C, from Prince William Forest Park, with the only undefeated 5-0 record that won the championship.[50] With the sudden end of the war, all military training at OSS camps was terminated and all except Area F were closed. At the former Congressional County Club, the focus shifted to informational programs to help with the transition from service to civilian life—speakers were brought in to describe job and educational opportunities, the G.I. Bill, and wartime insurance and pensions—as more and more OSS veterans returned from overseas and were mustered out.[51]

OSS tried to take care of its veterans returning from the war zones. Private First Class David A. Kenney, a radio base station operator who had trained at Area A and C and served in Europe and the Far East was in Chichiang, China, on 15 August 1945, the day Tokyo announced it would surrender. He was sent with an OSS team to

Nanchang to investigate reports of Japanese atrocities for war crimes trials. In the middle of the investigation, he was suddenly summoned to Shanghai, where the OSS commanding officer handed him a telegram and told him that his mother had died a few days earlier. "He made me a courier, and they let me fly back to the States by plane instead of by boat," Kennedy recalled, noting that air travel was highly restricted. "OSS had respect and compassion for its people." Kennedy was mustered out of the service two days before Christmas in 1945.[52]

Navy Seaman First Class Spiro ("Gus") Cappony, a radio operator who trained at Areas C and B with actor Sterling Hayden and then was sent on a bridge-blowing OSS mission to German-occupied Greece, returned to Washington, D.C. Cappony, who had been wounded, was the first member of the mission to return to the States, and he said that Secretary of State Cordell Hull immediately summoned him to his office and questioned the 20-year-old Greek American about the political situation between contending partisan groups in northern Greece. After the interview, Cappony went to Area F. "I had not been paid. I didn't have any money," Cappony recalled. "Captain John Hamilton [Sterling Hayden], he was my buddy. I met him in the Congressional Country Club. He says, 'Hey, Gus, What are you doing here?' I said, 'I can't get out of here. I don't have any money.' He said, 'Go out and get drunk.' I said, 'I'd love to get drunk.' He said, 'Here's twenty bucks.'" He subsequently got paid and was granted a 30-day leave.[53]

"Commo" Branch radio operator Arthur ("Art") Reinhardt came back to the United States from China at the end of 1945 with two dozen other OSS veterans and more than a thousand GIs jammed aboard the troopship, *Marine Angel*. Crossing the Pacific, he slept on deck in an Abercrombie and Fitch sleeping bag supplied by the OSS. He never forgot the scene when the vessel arrived back in the United States on 16 December 1945.

"As we were coming into port, into Tacoma, a big USO [United Service Organization] boat comes up with all the [USO] girls and the flashing lights and flags and everything, and the loud speaker says: 'There are twenty-four OSS members aboard ship, and they will disembark immediately! We are going to pull the launch alongside, and you will get off!' So, that's what happened. They said, 'We have a train waiting for you.' Some of the GIs [on deck] remarked, 'Who are these bastards?' And we casually climbed down onto this launch, or this fairly large boat, which took us in to Fort Lewis, I guess. We had a big steak dinner. We were taken out to the train. It turns out that twenty-four of us had two Pullman cars [deluxe railroad cars], which were attached to a troop train....By the time we arrived in Washington [D.C.], there was probably only six of us left, because the other guys would peel off along the way, saying, 'Well, you know, I'm only 200 miles from home.' We were coming across the country, and we had the time of our life. I can remember in Billings, Montana, 10 degrees below Zero, and we were running down the street in our shirt sleeves to buy some milk. We were happy troopers to get back." After spending a couple of days at Area F, Reinhardt was granted leave to go home to the family farm outside of Buffalo and was mustered out a month later in January 1946. The former farm boy turned communications specialist, who had spent more than fourteen months in war-torn China, much of it behind enemy lines, was only 20 years old.[54]

In the wake of the Allied victory in Europe in May 1945, Donovan, in addition to heading the OSS and battling for a permanent central intelligence agency, served from June to November 1945 as Deputy U.S. Prosecutor for the International War Crimes Tribunal at Nuremberg. The Chief U.S. Prosecutor Robert H. Jackson, an Associate Justice of the U.S. Supreme Court, asked him to serve because of all the work that the OSS Research and Analysis Branch was doing after the German surrender in gathering written and visual evidence to document the Nazi war crimes.[55] In the pre-trial period, Donovan, who spoke German, personally questioned Hermann Göring, the second-highest ranking Nazi. But during the summer, Donovan and Jackson became deadlocked over best method of prosecuting leading Nazis. Jackson decided to rely solely on contemporary documentary evidence, which provided overwhelmingly convincing proof of guilt. Taking a larger view, Donovan contended that basing the case only on written documents, while legally convincing, would not have the impact that seeing and listening to witnesses of the war crimes—perpetrators, observers, surviving victims—would have upon public opinion both in Germany and in the world. As Donovan explained, he "wanted to make it clear to the German people that the trials were not a retribution but were because of the Nazis' unprecedented outrage against humanity."[56] Donovan would not change his position, so just before the trials began in November 1945, Jackson dismissed him from the prosecutorial team. The OSS had been terminated the previous month, but Donovan remained technically on active duty as a Major General until January 1946.[57] After watching the trials begin in Nuremberg, Donovan returned to the United States, to his horse farm in Berryville, Virginia and his apartment at 2 Sutton Place in Manhattan, and resumed his law practice on Wall Street.

President Truman invited Donovan to the White House in January 1946, to receive an Oak Leaf Cluster to the Distinguished Service Medal, he had won in the First World War. The President credited the former OSS chief with great service in "secret intelligence, research and analysis, and the conduct of unorthodox methods of warfare in support of military operations."[58]

Transitions

With the termination of OSS in October 1945, the remains of Donovan's former agency were divided organizationally between the War and State Departments. The minority, 1,362 of the OSS civilian men and women in the Research and Analysis and Presentation Branches, were transferred to the State Department in a new unit called the Interim Research and Intelligence Service. The rest—9,028 other OSS civilian and military personnel—were transferred to the War Department. There a reluctant Brigadier General John Magruder, a former OSS deputy director, was put in charge of a remnant entitled the Strategic Services Unit (SSU). With the support of Secretary of War Robert Patterson and Assistant Secretary of War John J. McCloy, who was a friend of Donovan's and a firm believer in central intelligence, SSU had been created to keep the OSS personnel, records, budget, and assets separate from the Military Intelligence Division (G-2) of the U.S. Army.[59] McCloy ordered Magruder to disband the paramilitary units, Special Operations, Operational Groups, Detachment 101, and the Jedburghs, as well as Morale Operations and the rest, but to preserve the OSS's intelligence assets, including their field stations and foreign agent networks concerned with fascism and communism in Europe, the Middle East, and the Far East, over to SSU.[60] In anticipation of escalating tensions with the Soviet Union, Donovan's organization had during 1945 been identifying or planting networks of foreign agents in areas of Eastern Europe and China and Korea occupied by the Red Army as well as in countries around the globe deemed to be of strategic interest to the United States.

Over the following months, the size of SSU would shrink rapidly, as thousands of former OSSers were let go from civilian service or discharged from the armed forces. Magruder's deputy and successor, Colonel William W. ("Buffalo Bill") Quinn, was a career soldier, but as head of G-2 for the Seventh Army in southern France he had come to appreciate the value of the OSS. Years later, recalling the immediate post-war period, he wrote that "My initial business was primarily liquidation. The main problem was the discharge of literally thousands of people. Consequently, the intelligence effort more or less came to a standstill."[61] Five out of every six OSS men and women left, and numbers of Donovan's old organization plummeted in three months from some 10,000 down to just under 2,000 by the end of 1945.[62] The Strategic Services Unit was reduced to perhaps one-twentieth of the dimensions of the OSS, and the sources of U.S. foreign intelligence contracted accordingly.

It was that situation—the immobilization of the OSS assets, the threat that Congress would reduce the intelligence budget and that the Army might absorb the SSU into the G-2 bureaucracy, as well as disagreement between State, War, and Navy on one hand and the Bureau of the Budget and the FBI on the other—that led to a new coordinated if not yet centralized intelligence agency, the Central Intelligence Group (CIG).[63] The Joint Chiefs of Staff (JCS) submitted a plan that would save SSU and its assets and its espionage and liaison with friendly foreign intelligence services by putting SSU under a proposed National Intelligence Authority, a collaborative body drawing upon the intelligence units in several governmental departments. In December 1945, Truman had chosen the JCS plan over State's attempt to dominate national intelligence, and on 22 January 1946, the President created the Central Intelligence Group.[64] Two days later, he appointed as the first Director of Central Intelligence, Rear Admiral Sidney Souers, a former Missouri businessman and a reservist serving as Assistant Chief of Naval Intelligence. Donovan publicly criticized it as lacking centralization, independence, and civilian control.[65] In June 1946, with the appointment of Lieutenant General Hoyt S. Vandenberg, a career officer in the Army Air Forces, the small CIG, previously an empty shell, began to absorb the much larger SSU and its clandestine foreign intelligence activities. For more than a year, the Central Intelligence Group served, under several different directors, as a coordinating intelligence body, until the establishment in 1947 of the Central Intelligence Agency (CIA).[66]

OSS's Legacy and the Central Intelligence Agency

The emergence of a "Cold War" between the United States and the Soviet Union and the reorganization of the U.S. military and national security establishment led through the National Security Act of 26 July1947 to the creation of the truly independent, centralized, and congressionally mandated Central Intelligence Agency (CIA). The Director of Central Intelligence reported directly to the President. Donovan, who had come out of the war with enhanced public prestige and as an authority on foreign intelligence, had, along with former OSSers like, attorney Lawrence R. Houston and ex-chief of the London station, David K.E. Bruce, helped shape the final bill, which was largely a reprise of his own proposal at the end of the war. Since it was largely a return to Donovan's proposal in 1944-1945, the former head of the OSS was delighted with the intelligence sections of the National Security Act.[67] In the Congressional debate, considerable concern was expressed about whether the new agency would become a "Gestapo," spying on and arresting American citizens, and as a result Congress denied the CIA law enforcement or subpoena powers and prohibited it from any internal security role in the United States.[68] All agreed that the primary mission of the CIA was to collect and analyze intelligence from other countries. The OSS headquarters in the old National Institute of Health buildings at 24th and E Streets NW in Washington, D.C., became the first headquarters of the CIA in 1947 until the new CIA campus complex was built in Langley, Virginia, in the 1950s. (The old OSS temporary structures, the "Q" and "M" Buildings, down the hill were later torn down as eyesores, and ultimately replaced by the Kennedy Center.) With many former OSSers employed by the new central intelligence agency, often via the transitional units of SSU and CIG, it used many of the OSS's training methods and operational techniques regarding intelligence gathering and analysis. Peter Sichel, an SI agent with the U.S. Seventh Army in France and Germany in World War II, continued doing secret intelligence there for the next seven years. "I was in Berlin from 1945 to May 1952," he recalled. "I did the same job, just different employers, OSS, SU, CIA."[69]

Little attention was paid in the public discussion or congressional debate in 1947 to a clause in the National Security Act authorizing the CIA to perform "other functions and duties related to intelligence affecting the national security." But almost from the beginning that provision was used as the legal basis for "covert operations," which soon became a major, if later highly controversial, part of the activities of the CIA.[70] When in the summer of 1948, the National Security Council authorized the CIA to conduct "special operations" that agency quickly began to develop covert operations capability under Frank G. Wisner, head of its new Office of Policy Coordination. A former OSS Secret Intelligence officer in Istanbul and Bucharest, Wisner had been at the State Department since the OSS was terminated.[71] The CIA's operational branch employed then or soon afterwards a number of former OSS SO or OG personnel either directly, as in the case of William Colby, Al Cox, James Kellis, Joseph Lazarsky, Hans Tofte, and Lucien Conein, or indirectly on detached service from the Army such as William R. ("Ray") Peers, Leopold ("Leo") Karwaski, and others. "Basically, the Agency hired people who had already been trained. They hired from the OSS and from the military," recalled Caesar J. Civitella, a former member of the OSS Italian OGs, who served with the Army's Special Forces from 1952 until his retirement in 1964, and then worked for the CIA from 1964 to 1983. "When I was in the military, I did attend an 'E and E' course at 'the Farm.' The instructor was an old OSS guy. They did use former OSS training at 'the Farm.' "[72] Civitella was referring to the CIA's large, wooded, swampy training reservation in southeastern Virginia. Located outside of Williamsburg, it is much larger and quite distant from the country house training site, RTU-11, that OSS referred to informally as "the Farm."

Although the Central Intelligence Agency was institutionally the direct descendant of Donovan's wartime organization, veterans of the Office of Strategic Services, like other Americans, divided over many of the policies pursued by the CIA since its creation in 1947. The greatest divisions came over the subversive role of political and military covert operations. Donovan, unlike his British counterparts, had united intelligence gathering and analysis and covert action in a single agency, and the CIA maintained his organizational model. As the Cold War mentality came to dominant Washington in response to the threat seen as posed by the Soviet empire and the challenges of communist movements around the globe, the clandestine operations branch of the CIA became an important instrument in U.S. foreign and national security policy.

It was not just the agency itself, but many in the foreign policy and national security establishment who supported that concept that Donovan had celebrated as effective in World War II. In addition to the institutional and practical grounds for employing clandestine political, psychological, and guerrilla operations beyond its missions of intelligence collection and analysis, the CIA claimed also to inherit the OSS's mantle of moral justification for such actions. During the global struggle against the evils of fascism in World War II there had been no questioning of the ethicality of the OSS's foreign intervention and subversion in countries occupied by the ruthless forces of the Nazi and Japanese regimes. Many former OSSers in leadership positions in the CIA, including most prominently Allen Dulles, Frank Wisner, Richard Helms, William Colby, James Angleton, Thomas Braden, and much later, William Casey, took the Donovan model and applied it aggressively in the "Cold War" era.[73] But while the model and even some of the personnel remained the same, the world in which they operated had changed dramatically.

In World War II, OSS had actively encouraged and supported resistance movements against the foreign armies of occupation and Nazi or Japanese regimes that had seized and controlled their countries. But in subsequent decades, the CIA routinely encouraged and supported resistance movements to undermine national governments, some of them popularly elected, deemed too sympathetic to the Soviet Union and hostile to the United States. The justification was based on a "Cold War" mentality which stressed that any gain in influence by the totalitarian regimes in the Soviet Union and Communist China was a loss for the United States. To numbers of former OSSers, in or outside of the CIA, as to many other Americans, the ends seemed to justify the means, and they generally supported the actions of the agency.[74] The CIA and its intelligence gathering and covert operations continue to enjoy endorsement by many surviving OSS veterans.[75]

There has also been, however, considerable dissent by OSS veterans against what they see as misuse of the CIA and a betrayal of the legacy of the OSS. As strident anti-communism and vengeful McCarthyism in the 1950s pushed the agency to exercise its covert operations in support of regimes of political repression and right-wing dictatorships and to engage in heavy-handed intervention in the politics of underdeveloped nations, dissent was increasingly suppressed within the agency. But dissent continued to exist among some OSS and CIA veterans.[76] "The CIA is an entirely different group than the OSS, although the early CIA was built on the bones of the OSS, so to speak," explained Rafael D. Hirtz, a former OSS officer who had served with Operational Groups in France and China. "The unfortunate thing, what happened to the CIA, at least in my belief, is that it became too political," Hirtz said in an interview in the 1980s."[77] Peter Sichel, who served with the 7th Army OSS unit in France, and later worked for the CIA in Berlin, Washington, and Hong Kong, resigned from the CIA in 1959. "I resigned because they were doing things I did not like," he said. "They were sending people into nonexistent [resistance] groups in China. I did not like it when they did it in the Ukraine, but I accepted it then. Not the second time. They were sending them to their deaths or lifetime imprisonment. This was Cold War madness."[78]

Divisions among OSS veterans over polices of the CIA have continued to the present day.[79] Former intelligence officer R. Harris Smith had decried the fact that the legacy of liberalism and dissent in the CIA became a forgotten part of its history, and he argued that it was important to understand that history in order to restore legitimacy to an institution under attack for its failures and for yielding to political pressures.[80] This plea was somewhat echoed in 2007 by *New York Times* reporter Tim Weiner in his book, *Legacy of Ashes: The History of the CIA*. This latest attack argues that from the beginning of the CIA in 1947, and despite efforts from Allen Dulles to William Casey to link its spirit to the old OSS, the CIA was essentially unable to fulfill its role as America's central intelligence service. There were, according to Weiner, only a few halcyon moments: When Richard Helms, an OSS veteran and head of the CIA, confronted President Lyndon Johnson and then President Richard Nixon with the facts about the Vietnam War that they did not want to hear and when Robert Gates headed the CIA and "kept calm and carried on as the Soviet Union crumbled." But as Weiner concluded, in the world the United States faced in 2007 and possibly for years to come, the country needs an effective espionage service against its enemies. "The CIA someday may serve as its founders intended," he hoped, for in the long run, "we must depend upon it."[81]

The Special Forces: Another OSS Legacy

Like the CIA, the U.S. Army's Special Forces are also a legacy of the OSS. Because Donovan's wartime agency was technically a civilian rather than a military organization, the U.S. Army lineage organization does not officially acknowledge a direct heritage between the OSS and the Army's Special Forces. However, the heritage owed by today's Special Forces to Donovan's Special Operations, Operational Groups, and Maritime Unit is widely recognized, even by the Special Forces themselves.[82] OSS SO teams, and particularly the Operational Groups of the OSS, were designed, like today's Special Forces, to have the ability to conduct lengthy, long-range penetration deep into enemy area in order to organize, train, equip, and direct indigenous guerrilla forces. Unhampered by official criteria, the U.S. Army's main historical study of its Special Forces concludes that the OSS, while not a strictly military organization, left a legacy of knowledge, which together with the experience of a few officers in guerrilla warfare "was instrumental in the creation of the Special Forces in 1952. This gave the Army a formal, continuing capability for unconventional warfare for the first time in its history."[83]

With the demise of the OSS in October 1945, although some of the intelligence and communications assets had been retained by the War Department and State Department, the Jedburghs, SO, and OGs, were demobilized. With the onset of the Cold War, the Army briefly considered creating special operations units in 1947; however, much of the Army leadership, still suspicious of the OSS's achievements and focused primarily on big unit conventional warfare, quickly dismissed the idea.[84] But by 1950 with the war in Korea and the escalation of the Cold War around the globe, the United States Government sought to develop means for unconventional warfare—psychological as well as guerrilla operations—as well as dramatically increased conventional and nuclear war-making capabilities to confront the Soviet Empire and Communist China.

The Army had initially supported and cooperated with the CIA's office of covert operations, offering it assistance in the field of guerrilla warfare by setting up a training course at the Infantry School at Fort Benning, Georgia. When the Korean War broke out in 1950, the Army lent some of its personnel to the CIA engaged in guerrilla activities behind enemy lines in Korea and in mainland China. But by the spring of 1951, Army leaders began to express concerns about the secrecy and relative autonomy with which the CIA was operating and whether its covert operations were actually effective. The Army's suspicion and hostility was reciprocated by the CIA. Both were also jockeying for primary responsibility for conducting the expanding unconventional warfare.[85]

Brigadier General Robert A. McClure, chief of the Army's Psychological Warfare Office, and especially his assistant, Colonel Aaron Bank, a former OSS Jedburgh, worked to convince senior managers of the Army and the Joint Chiefs of Staff that the military should explore the development of Special Forces capabilities for behind the lines guerrilla operations. On 14 April 1952, under control of the commanding general of the U.S. Third Army, McClure was able to establish the U.S. Army's Psychological Warfare (PSYWAR) and Special Forces Center at Fort Bragg, North Carolina. This marked the unofficial beginning of the U.S. Army's Special Forces.[86]

It was a small if significant beginning and it stemmed directly from the OSS. Aaron Bank, who later be celebrated as the "Founder of Special Forces," had trained at OSS Areas F and B as well as SOE camps in Britain and served as a Jedburgh in France and an SO/SI rescue team leader in French Indochina. Remaining in the Army, he had been in the Counter-Intelligence Corps in Europe, and when the

Korean War broke out, he had been sent to Korean with an Airborne Regimental Combat Team. In 1951, he had been brought back to Washington, as an assistant to General McClure as the Army resumed it study of unconventional warfare. Bank helped persuade the Army to include guerrilla warfare, and in the spring of 1952, he was appointed the head of the Special Forces Department in the new PSYWAR Center at Fort Bragg and commanding officer of the new 10th Special Forces Group, the first Special Forces unit, which was targeted for Europe.[87]

Training, doctrine, and mission of Special Forces were based specifically on OSS concepts, especially the Operational Groups, rather than on those of the Airborne, the Rangers, or the First Special Service Group.[88] To recruit a staff for the Special Forces Department at the PSYWAR Center at Fort Bragg and to help him lead the first Special Forces unit, Bank began with two other OSS veterans in the Army. Lieutenant Colonel Jack T. Shannon, originally from Iowa, had joined OSS, trained in SO, and jumped into France on a three-man, Inter-Allied Team "Bergamotte," in July 1944, and subsequently fought behind Japanese lines with Detachment 101 in Burma in 1945.[89] He was brought to Bragg as Bank's deputy. Lieutenant Caesar J. Civitella from Philadelphia had trained at Areas F and B as an enlisted man in the OSS Italian OG and fought in southern France and northern Italy in World War II. Civitella had rejoined the Army in 1947 and served in the intelligence section of the 82th Airborne Division before being sent to Officers Candidate School in 1951 and receiving his commission. Civitella was assigned as a recruiter and as an instructor teaching courses in guerrilla warfare and in air operations and support.[90] The three-man cadre quickly recruited a staff. Among the first of their volunteers were three OSSers. Lieutenant Colonel James Goodwin, from upstate New York, had been a student and then instructor at Area B in 1942, then served with OSS in Europe. Captain Herbert R. Brucker, Alsatian born U.S. citizen and OSS trained radio operator, had served on a three-man SOE mission in France in 1944 and then an SO team in China in 1945. Brucker had been assigned to the Army's Counter Intelligence Corps in occupied Germany after the war, then to the 82nd Airborne Division. Recruited for Special Forces, Brucker was made head of security. Subsequently, he taught clandestine operations and trade craft (dead letter drops, eliminations, etc.) in the 10th Special Forces Group He provided training not only for that group, but also developed the first "escape and evasion" training course for U.S. Navy aviators in case they were downed behind enemy lines.[91] The staff also included other OSSers, like Lief Bangsboll and Jack Hemingway, plus Army veterans of World War II and the fighting in Korea.[92]

The initial assignment was to develop an eight-week course on guerrilla warfare, escape and evasion, and subversion of a hostile state. The only existent Army Field Manual related to the topic, FM 31-21, *Organization and Conduct of Guerrilla Warfare*, had been prepared largely on the basis of the personal experiences of Army officers in the Philippines, like Lieutenant Colonels Wendell Fertig and Russell W. Volkmann, who had refused the order to surrender and, although untrained in unconventional warfare, had led Filipino resistance groups on Mindanao and Luzon throughout the Japanese occupation. Volckmann had worked with Bank to convince the Army about the value of special forces and was part of staff. The manual was used in the Special Forces curriculum, but, as Civitella recalled, "most of the stuff we used was from the OSS." For constructing the curriculum, the cadre drew heavily upon OSS assessment techniques, lesson plans, and after action reports and interviews. Plus in their lectures, they drew on their own personal experiences. Bank had personally brought many classified OSS documents down from Washington in the trunk of his car. Civitella had put them in the safe from which they were used in building lesson plans, sometimes verbatim, as in the case of the description of the OG Command.[93]

Similar to the OG sections, the basic Special Forces unit, an A Team, consisted of approximately a dozen men. Recruits had to volunteer for parachuting as well as hazardous duty. Many of them came from the Rangers, then being deactivated. The new unit was super secret, even within the Army. As in OSS, the early training of the Special Forces emphasized self-confidence and élan as well as individual skills: operations and intelligence, light and heavy weapons, demolitions, radio communications, and medical aid. Each enlisted man specialized in one area but also learned the rudiments of the others. There were also courses in organization of resistance movements and operation of their networks, agent training to include espionage and sabotage, guerrilla warfare, codes and radio communication, survival, the Fairbairn method of hand-to-hand combat, and Rex Applegate's instinctive, point-and-shoot, pistol firing technique.[94] Aaron Bank was fanatical about training, according to former Jedburgh, Major General John K. Singlaub, "He believed that Soldiers must have expert knowledge of their weapons systems—so much knowledge and firing the equipment should be 'second nature.' Then the Soldier could focus solely on the mission."[95]

To provide adequate room for field exercises, Bank obtained the use of one of Fort Bragg's satellite training areas, Camp Mackall.[96] There, as at Areas A and B in World War II, the trainees became adept at demolition work, guerrilla tactics, and ambushes. The swampy areas with abundance of snakes, frogs, and rodents proved useful in practicing survival strategies, living off of nature. But for a more practical exercise in clandestine activity, the training staff decided to take the more than 600 members of new 10th Special Forces Group outside an Army base, and into a rural, forested civilian area which would include residents, barns and safe houses, places to meet, gather and train indigenous guerrillas, as well as targets for simulated sabotage such as bridges, railroad tracks, telephone and electrical lines and power stations. For the realistic operations for the Group's A Teams, they chose Chattahoochee National Forest in the Appalachian Mountains of northern Georgia. Special Forces used the mountainous timberlands of the U.S. Forest Service in late 1952 the way OSS had used the woods of the National Park Service in World War II.[97] In the summer of 1953, Donovan visited Fort Bragg to inspect the troops of the new 10th Special Forces Group. Impressed after a day of briefing and demonstrations, Donovan recognized that they had adopted the traditions, heritage, and legacy of OSS Special Operations. "You have revived precious memories," he said as he left. "You are the offspring of the OSS.[98]

The emphasis on the training of the Army's Special Forces in the early 1950s was on supporting guerrilla operations and that was because the mission that was envisioned was helping anti-communist, anti-Soviet resistance groups operate in non-Russian territories of Eastern and Central Europe occupied by an invading Red Army, just as OSS groups had aided anti-fascist resistance groups behind German and Japanese lines in World War II. Indeed, after rioting broke out in cities in East Germany in the summer of 1953, Colonel Bank and the 10th Special Forces Group was sent to Bad Tölz, in the Bavarian Alps near Munich (a new unit, the 77th -later 7th--Special Forces Group was created under former OSS officer Jack Shannon, based in Fort Bragg; it was later sent to Korea).

There was little or no attention given in these early years to counterinsurgency operations *against* guerrilla forces. That would come in the late 1950s and early 1960s and would initiate a doctrinal battle within the U.S. Army about the proper role of Special Forces.[99]

From Guerrilla to Counter-Guerrilla Warfare and the "Green Berets"

"A lot of people think Special Forces came into existence in 1952, but we were not officially recognized [then]. We were a small, secret group. It was not until 1961 when [President John F.] Kennedy recognized us that we came officially into existence," said Caesar J. Civitella, a former OSSer and Special Forces member. "We existed in fact, and some of our people had been killed, but officially we didn't exist before 1961."[100] As the Cold War shifted to the Third World and Communists sought to control independence movements and anti-colonial guerrilla wars through "wars of national liberation," President Kennedy saw the Special Forces as potentially effective agents of counter-insurgency. Kennedy provided the Special Forces with increased resources and status and allowed them to become the first American soldiers to wear berets, for which they became popularly known as "the Green Berets." In the 1960s, Special Forces groups, now emphasizing counter-insurgency tactics, and the multiplier effect of having 12-man Special Forces teams that could assist or lead indigenous forces of up to 300 or 500 men, advised or actually fought against Communist-inspired guerrillas in Latin America and Southeast Asia. Although the Army's Special Forces were cut back after the Vietnam War, they regained strength in the 1980s. The following decade, they were part of an all-service, quick-response command, the U.S. Special Operations Command created in 1997 that included an expanded force of Army Rangers as well as Navy Seals, and special operations units in the Air Force and Marines.[101] During the first decade of the twenty-first century there was a major new emphasis placed on the use of Special Forces.

Colonel Aaron Bank retired from the Army in 1958 but was named Honorary Colonel of the 1st Special Forces Regiment in 1986 and remained so until his death at the age of 101 in 2004.[102] Like Shannon, Civitella, Brucker and others, Bank was a personal link between the OSS and the Special Forces, and he credited the OSS with being the predecessor of the Special Forces. "Although it occurred indirectly, there is no doubt that OSS gave birth not only to CIA but also the Green Berets in 1952,"

Bank wrote. "All the Green Beret training programs, maneuvers, concepts, and conduct of operations were based on those of OSS. I feel that Donovan should have bestowed upon him the honor and credit for this continuity of the heritage and traditions established through his genius and foresight."[103]

Return of the Parks: Prince William Forest Park

With the end of Donovan's organization on 1 October 1945, its successor, the War Department's Strategic Services Unit (SSU) began to liquidate the OSS's facilities. The Schools and Training Branch under Colonel Henson L. Robinson had been shutting down training facilities for several months and quickened the pace since August.[104] Area A, the Advanced Special Operations training camp, had been terminated as a training facility in early 1945. Sub-camps A-2, A-3, and A-4 were completely closed, all personnel had left, and in October the contents of the sub-camps were being inventoried for return to the National Park Service. At A-4, the headquarters for Area A, no trainees or instructors remained but there was a small staff of officers and enlisted men required to complete the closing of Area A. On 4 October, Colonel Henson L. Robins, chief of the Schools and Training Branch, estimated that the job would be completed by the middle of the month.[105]

The process of shutting down Area A and evaluating and preparing the site for return to the National Park Service had been going on since January. After the end of training there, the staff had destroyed a number of instructional items directly related to the OSS's sabotage and espionage operations, which would be of no use to the National Park Service. Among these were scaling ladders, firing mechanisms, door sets for lock pickers, and model girder bridges, railroad tracks, and radio towers, which had been used for teaching trainees where to place explosive charges.[106]

In February 1945, under the supervision of Captain James E. Rodgers, the last commanding office of Detachment A, Technical Sergeant John T. Gum, a longtime employee of the Civilian Conservation Corps (CCC) at the site and subsequently liaison between NPS and the OSS at Area A, had prepared a relevant inventory.[107] Having been at the park since the late 1930s, Gum, in his multiple wartime roles as a representative of the CCC, the NPS, the Army, and the OSS, had supervised much of the OSS construction in Area A. When Area A finally shut down completely, Sergeant Gum was transferred to Area C, the OSS Communications Branch facility, in the eastern sector of the park. In addition, the NPS Park Manager, Captain Ira B. Lykes, U.S. Marine Corps, Reserves, who performed the dual role of forestry officer at the Quantico Marine Base during the weekdays and park manager on weekends, was directed by NPS headquarters to provide a final inspection of the sub-camps, when the OSS finally left. Lykes' inspection would confirm whether the War Department had performed its responsibilities under the leasing agreement, which were to restore the grounds and facilities to their prewar condition except for changes that were acceptable to the National Park Service. In addition to those conditions, NPS wanted to make sure that the approximately 1,000 to 1,500 more acres purchased by the War Department for Area were included when the park was returned to the Department of the Interior.[108]

Preparing the inventories in 1945, the Army distinguished in Area A between properties of the National Park Service under the Department of the Interior and of the Civilian Conservation Corps. Until the early 1940s when it was terminated, the CCC had operated under the War Department or temporary New Deal agencies. OSS sub-camp A-4 was an old CCC work camp, and the Schools and Training Branch believed that National Park Service had in its words, "nothing to do with the buildings and equipment" there, so they were not covered by the leasing agreement. The OSS inventory listed 24 buildings in the CCC camp that had been taken over by OSS and remained there. The inventory listed contrasting uses made of them by the CCC and the OSS. Often these were similar: the CCC headquarters building, for example, became the OSS headquarters; the CCC barracks, each holding 40 men, became the OSS barracks; and the CCC oil house became the OSS Motor Pool Office. But sometimes they were put to new uses: The CCC infirmary became the OSS Guard House; a CCC structure listed only as "Building," became the OSS Post Exchange and Recreation Hall. The mess hall had been retained, but it was remodeled and its kitchen augmented with new equipment, including three gas ranges, two portable electric refrigerators, two steam tables, a deep fat fryer, an electric peeler, and a dishwashing machine.

OSS had erected nearly a dozen new buildings in Area A-4, and they remained there at least at the time of the inventory in early 1945. Among them were a one-hundred man, standard Army latrine, actually a latrine/washroom that included toilets, washstands, showers and hot water heaters and that replaced the CCC's two portable pit latrines. There was also an auto washing building, carpenter shop, storage house, commissary, small code room/storage building, a sentry box, and most expensive, a ten-room Bachelor Officers' Quarters (BOQ) accommodation. At A-4, OSS had also constructed a rifle practice range with ten standard targets.[109] A report on A-4 by Sergeant Gum in October 1945 indicated that the National Park Service would not ask that these new or remodeled buildings be returned to their previous condition, because "prior to our occupancy, the National Park Service had no definite plans for using this area [this former CCC facility] as a summer camp."[110]

Unlike the former CCC facility, the other OSS sub-camps at Area A had previously been built and used by the National Park Service as summer camps. The OSS had not removed or destroyed any of the NPS buildings, but it had made changes, which it considered necessary improvements. Most importantly, most of the buildings in all the sub-camps as well as the CCC facility had been winterized. Insulated wallboard had been installed inside and the outside creosoted. Wood or coal-burning heating stoves had been added. OSS installed 29 hot water tanks in the washroom/latrines, mess halls, and infirmaries, as well as the headquarters, guard house, instructors' quarters, students' quarters, and Bachelor Officers' Quarters. Other improvements included a new sewer field; porch enclosures for a few buildings, and upgrading of the kitchens and mess halls in each sub-camp.[111]

A number of new structures had been built in Area A. At A-2, these included several wooden buildings: four 25-man latrine/washrooms, an enclosed 20 by 40 foot Pistol House; and an 8 by 8 foot guard house. At A-3, a latrine/washroom was the only new building, but a demolition area was constructed there for practice with explosives. At A-5, OSS had constructed a 12 by 16 foot Boat House on the little lake on the South Fork of Quantico Creek for practice in water crossings and boat landings.[112]

The OSS understood that it would have to "obliterate" the facilities it had constructed for use of weaponry and explosives. The Special Use Permit granted to the War Department stipulated that "structures of a purely military technical nature" would be removed by the War Department and "the site restored as nearly as possible to its condition at the time of issuance of this permit." The dangerous facilities to be "obliterated" and restored to prior condition included the Demolition Area, where trainees had practiced with explosives, the two brick or cinderblock magazines for storing ammunition, and outdoor pistol, rifle, and hand grenade ranges and a mortar range.[113]

Other military structures in Area A to be demolished were a "Mystery House" or "House of Horrors," and three "Problem Houses." The Problem Houses may have been used to pose particular challenges for the trainees to solve or they may have represented a village in enemy territory. The Mystery House contained a number of pop-up targets, resembling armed enemy soldiers, to test the trainee's nerves and skill in instinctive pistol firing in frightening conditions.[114]

Area C in Prince William Forest Park

At the two sub-camps, C-1 and C-4, in Area C, the OSS Communications Training School, located in the eastern sector of Prince William Forest Park, formal training of any sort stopped prior to 1 October 1945, and the staff was ordered to prepare the camps for closing. Colonel Robinson expected that would be completed by 1 November, and the personnel involved then reassigned or transferred.[115] Sergeant Joseph J. Tully had been part of the C-1 cadre since his return from the London base station in November 1944. Years later, he remembered that in September 1945, he and a couple of the other enlisted men at Area C were kept busy transferring the remaining ammunition and weaponry from the magazine and arsenal at the training camp to the Corps of Engineers post at nearby Fort Belvoir, Virginia. "We made a trip every other day. A couple of guys would load up a truck—a two and a half ton—and take it over. We took over all the excess ammunition—small arms stuff and grenades—and weapons too, mostly 'Tommy guns,' rifles, and .45s."[116]

Donovan's organization had maintained and even improved existing National Park Service buildings at the two sub-camps in Area C. The Army winterized the buildings, made some changes to various lodgehalls, and dramatically upgraded the equipment and facilities at the mess halls and kitchens. In addition, the OSS erected several new wooden structures. At C-1, these included most prominently a radio repair shop, but also a radio transmitter building, a 25-man latrine/washroom, a guard house, two 8-by-8 foot sentry boxes, and an undesignated, portable wooden structure, 16 by 16 feet, labeled simply, "Plywood Building," which may have been a classroom. C-4 was the primary training area for the Communications Branch. There, the main new construction, much more extensive than at C-1, included a

Recreation Hall, originally called a "Multi-Purpose Building," which cost $24,000, and a Bachelor Officers' Quarters, which cost $12,000. Other new structures included a 100-man latrine/washroom, three 16-by-16 foot "Plywood Buildings," possibly serving as classrooms, and an addition to the mess hall. The Army had also brought in a prefabricated, metal Quonset Hut. To house even more trainees, the OSS provided 25 combination wood and canvas tents, each holding four men, creating a so-called "tent city" capable of housing 100 men, for a total cost of $5,000. The "tent city" was dismantled, but most of the new wooden buildings were retained by the National Park Service. For weapons practice in Area C, the OSS had constructed a outdoor pistol range, a rifle range, and a hand grenade range. Like the magazine and arsenal, these weapons ranges were to be totally obliterated before the facility was returned to the National Park Service.[117]

Wartime Hazards and Mysteries in the Park

To return Prince William Forest Park to the National Park Service, official procedure required that the Army first certify that Training Areas A and C were "surplus" property, that is, no longer needed by the Army. That process began on 16 October, when Lieutenant Colonel Ainsworth Blogg, former commander of Area B who had spent most of the war as Executive Officer of the Schools and Training Branch, informed the SSU Counsel's Office that Areas A and C were being closed by the Schools and Training Branch and would be considered surplus to the branch's needs as of 1 November 1945. Referring to the terms of the lease, Blogg reported that the areas' "present condition is equal to or better than the condition when received by OSS. Certain changes in landscaping have been made which may not be desirable from the standpoint of the National Park Service and certain buildings have been constructed, and certain changes made in existing buildings, but by and large the property is in excellent condition."[118]

Although the War Department subsequently declared the property surplus as of 1 November 1945, a hitch occurred in returning the area to the National Park Service because of unexploded shells and other hazards that remained from the OSS use of the park.[119] Colonel Blogg had declared that portions of the area could be used without limitation by the National Park Service "except for one area which was used for a mortar range and which contains unexploded shells." Schools and Training Branch had requested the Corps of Engineers at Fort Belvoir for assist in a "decontamination of this area."[120] There were actually three abandoned mortar target ranges in Area A, and the process of getting Army Engineers with metal detectors to locate and remove unexploded shells, "duds," or other hazardous items from those and other areas in the park had begun a year earlier in October 1944. Sergeant John Gum tried throughout 1945 to get the Post Engineer's Office at Fort Belvoir to check Area A for "unexploded booby traps, anti-personnel mines, and dud mortar shells."[121] Colonel Robinson pressed the Engineers again in January1946.[122] But was not until August 1946, NPS was able to obtain written assurance from the Army Engineers' Bomb and Shell Disposal unit at Fort Belvoir that Areas A and C were "long since cleared" of unexploded mortar shells, grenades, booby-traps and other explosives.[123]

The area was returned to the National Park Service in 1946. However, occasionally over the following years, a "dud" shell, fragment, or other remnant from the OSS training in World War II, would be found in the park. Long after the war, NPS rangers part of a rifle-launched grenade in a building roof, the tip of a bazooka round in a mound of each, and most mysteriously, a giant stone with an iron pipe that looked like a gun barrel, protruding from it.[124] This last may have been a mock tank for trainees to practice using destroying with bazookas or improvised devices such as "Molotov cocktails," bottles filled with gasoline and a flaming cloth wick.

Perhaps the strangest remnant that remained, at least temporarily, from the War Department's occupation of Prince William Forest Park was an odd barrack-like building in the southeastern part of the park. Robert A. Noile was six years old when the OSS left in 1945. He grew up across Old Joplin Road, only 300 yards away from the park. Together with other neighborhood youngsters, including the children of Ira Lykes and John Gum, he frequently played in the park immediately after the war and for several years afterwards. The youngsters simply crossed the street and entered the park to play in the woods and the abandoned buildings. On hot days, they would swim in a big, lidless wooden water tower or in the lake. But what still puzzles Noile, now a ranger in one of the nearby county parks, is a weird, barrack-like building that they discovered. What made this structure unusual was that the floor and the walls up to the windows were covered inside with sheet metal. Similarly intriguing, there were carvings in German on the wooden frameworks around the windows. "We played in those barracks," Noile remembered, "and we had no idea what that [metal sheathing] meant. We kept seeing carvings on the wood, names. We didn't know German. I understand now that it was German, but we didn't know [then] what it was."[125]

No written record of such a structure has appeared, but a couple of different explanations of the use of such a building are plausible. One is that it was a simulated Nazi interrogation cell used in training spies and saboteurs to resist or escape or to test their abilities to withstand interrogation by the Gestapo. Another explanation is that after the OSS abandoned Area A in January 1945, German POWs might have been brought and housed there temporarily. Recent disclosures have revealed that the Army and Navy ran a super secret POW holding and interrogation facility (known only by as "P.O. Box 1142") for special prisoners—submarine crews or high ranking officers or scientists—not far away at Fort Hunt in Fairfax County, Virginia, in what is today part of the George Washington Memorial Parkway.[126] Until more evidence becomes available, the function of the metal-sheathed room, long since demolished, at Prince William Forest Park remains a mystery.

Return of the Park to the National Park Service in 1946

"After I was mustered out [of the Marines] in January 1946, I immediately went back to the job," former Park Manager Ira Lykes recalled. "Of course, OSS had closed up and left a lot of fine development in some of the campgrounds, [but] they left an awful mess in some other places."[127] One day shortly after he returned, he was walking over a meadow, and he found "one whole field had signs all around it, 'Do Not Cross—Mine Area.' Well, I, of course, wondered immediately what to do. OSS had been disbanded....." Lykes drove to Fort Belvoir to see a colonel named Hogg, who had worked on the Alcan Highway to Alaska in World War II. The Park Manager wanted Hogg to remove them if they were still there, and he also asked whether the Army Engineers could help restore and improve the property, specifically whether they could build a network of internal roads to connect the cabin camps, because at the time it was a nine mile ride over external state roads to get, for example, from cabin camps that were actually not far apart in the northeastern and southeastern parts of the part. Lykes had hoped the OSS would do this, but they had not. Now, however, Hogg agreed that the Army Engineers could complete that construction if Lykes could provide the materials. As part of the postwar combat training exercises, units of combat engineers from Fort Belvoir were being sent to nearby Fort A.P. Hill to build roads and bridges and other facilties. But afterwards, they would demolish them, so a new unit could do the same thing. This was wasteful and bad for morale, Hogg said, but if Lykes could rename the projects he wanted and link them to military missions, the Army engineers could build them. So Lykes ordered military roads and bridges and asked for a landing strip that would become a play field. A dam for a swimming pool was called an impoundment, a riding stable was termed an advance post, a watch tower was as an observation post.

Lykes obtained $35,000 for materials the first year from the National Park Service, and the Army Engineers arrived in September 1946 and immediately went to work building roads, bridges, and dams and other facilities. As they worked, Lykes was surprised to see small planes fly overhead and drop bags of flour simulating bombs and shells and combat conditions. The Army Engineers often worked as fast as Lykes could mark out the projects. Many of these engineer troops were African-American soldiers and Lykes said that when they learned that the cabin camps in the northeast end of the park were for underprivileged black children from the capital, "believe me they really put their heart and soul in it, because they felt they were helping their own people and, most importantly, they knew that what they were doing was not going to be torn up by the next troop that came down, that it was going to stay. It was something lasting and something of value."[128]

This innovative partnership continued from 1946 until the outbreak of the Korean War in 1950. By February 1948, the Army Engineers had completed 47 miles of internal roads, plus trails, two bridges, an earth filled dam and swimming area, an administration building and traffic circle, a central service area and two ranger stations. They had also demolished unusable portable and temporary buildings left by the OSS near Cabin Camp 4 and relocated the OSS-constructed Boat House from Cabin Camp 5 to Cabin Camp 2.[129] More work was done over the next two years, including a new entrance road, this one about a quarter of a mile from U.S. Route One. Lykes estimated that over the nearly four-year period, the labor and equipment used by the Corps of Engineers in this "practice work" would otherwise have cost the National Park Service well over one million dollars.[130]

With the return of the park to civilian usage in 1946, Lykes had begun immediately to restore it, resume camping, and to give it a new name. Chopawamsic Recreational Demonstration Area now seemed archaic, cumbersome, and inappropriate as it was no longer an RDA but a permanent part of the National Park System. National Park Service officials concurred with the idea of a new name, but instead of Lykes' suggestion of "Old Dominion" to link the park to the entire State of Virginia, they recommended naming it in honor of the county. Local authorities concurred, and on 20 August 1948, the facility was renamed Prince William Forest Park.[131]

Chopawamsic Recreational Demonstration Area and its successor, Prince William Forest Park, were originally designed only for organized and extended group cabin camping, not for casual day usage. Until 1951, when the Pine Grove Picnic Area was constructed, there were no day use facilities in the park. During World War II, the park had been, of course, closed to the public while the OSS was there, but even after it reopened, there was little if any day usage. The few roads within the park were unpaved and made of practical if uninviting, rough gravel. The memory among local residents of the forbidden access during the war and the armed sentries, also surely deterred prospective visitors. If these were not enough, signs posted around the perimeter of the park after the war warned: "Federal Reservation: Closed except to persons holding camping permits."[132]

The relationship of the park with the military and national security did not end in 1946 when the NPS resumed control. The Army Engineers had helped with improvements from 1946 to 1950. In the early 1950s, when the Central Intelligence Agency expanded its covert operations and abandoned its use of Army training facilities, it briefly considered using the old OSS training area. In the opening months of 1951, during the massive military build-up that accompanied the Korean War, the CIA informed the National Park Service that it intended to take over the entire park as a training area. Opposition led by local civic leaders including the editor of the *Washington Star* and the district's influential member of Congress, Representative Howard W. Smith, turned back the CIA's bid to resume the training of spies, saboteurs, and other covert operators in the park.[133] Instead, the CIA built its own training facility for clandestine services on ten thousand acres at Camp Peary, near Williamsburg, Virginia.

Relations with the Marine Corps

Given its proximity to the U.S. Marine Corps Base in Quantico, Virginia, Prince William Forest Park has had a continuing relationship with the Marine Corps and its parent organization, the Department of the Navy. In World War I, Congress had appropriated funds for the Navy to purchase a site for a Marine Barracks not far from Washington, D.C. The site chosen was in the Virginia woodlands at Quantico, some 35 miles south of the capital. The barracks were built and thousands of Marines were trained there in the First World War. By 1920, the facility was expanded to included a centralized officer training program, at the new Marine Corps Schools, which eventually developed into today's Marine Corps University.[134]

The National Park Service arrived at what would become Prince William Forest Park in 1935, and from the beginning consulted with officials at the Marine Corps base in regard to plans for the recreational area, because the park and the base shared a common boundary for a considerable distance. NPS and the USMC also worked together to find what Marine officials had called "mutually advantageous" ways to utilize these "contiguous areas under Federal control."[135] In 1938, for example, the Department of the Interior granted the Department of the Navy permission to build a dam and reservoir on the Chopawamsic Creek in the park land to store water for the Marine base. For the next few years, until it became unnecessary after the United States entered the war in December 1941, the Park Service provided permits for military training exercises by Marines on park land.[136]

In World War II, the Marine Corps expanded dramatically, growing from 25,000 officers and men in 1939 to 143,000 by mid-1942 and nearly 500,000 by the end of the war in 1945.[137] As early as 1938, the Marines began conducting military maneuvers in the adjacent, NPS wood lands.[138] After the declaration of war in December 1941, and the rapid growth of the Marine Corps, field maneuvers on the park lands became so frequent in May 1942, Park Manager Ira Lykes complained that the Marines "have assumed the right to enter upon the area without advising or even consulting this office."[139] At the same time, Lykes continued to maintain a cooperative relationship with the personnel of the Marine base. He emphasized the park's recreational value to military personnel, including fishing in its well stocked streams and ponds.[140] In the hot, dry summer of 1942, Lykes voluntarily taught Marines the latest techniques in fighting forest fires so they could more effectively control fires when they began in the woodlands of their reservation.[141]

During most of the war, Lykes actually served in the Marine Corps while continuing his now limited responsibilities at the park, which, since it was occupied by the OSS, was closed to the public for the duration. Maj. Gen. Philip H. Torrey, commanding officer at Quantico, obtained a commission for Lykes as a first lieutenant in the Marines and an appointment to serve at the Marine Base at Quantico as a provost marshal or security officer, particularly responsible for the forestry program, for the prevention or control of fire on some 94,000 wooded acres of the Marine base. That included approximately 5,000 acres of the park land south of Joplin Road that the Marines used as an extension of their base during the war and declined after the war to relinquish it to the park service. That was an area, Lykes later recalled, that the Marines called the "Guadalcanal Area" after the fierce jungle battles with the Japanese on the Pacific Island of Guadalcanal in the winter of 1942-43. The Marines had 13 firing ranges, mockup villages, and other installations for field exercises.[142] The little village of Joplin, a hamlet with a general store and a schoolhouse, near the intersection of old Joplin Road and Breckinridge Road, was torn down by the Marine Corps. All that remained is a cement slab, the foundation for the old schoolhouse, and some of the big oak trees.[143]

Assigned to active duty on 3 April 1943, Lykes reported directly to the Commanding General of the Marine Barracks, Quantico, for what was called Engineering duty. The National Park Service officially put Lykes on furlough due to military service.¹⁴⁴ But because the base and the park were adjacent, Lykes, as he later recalled, could "serve two masters."¹⁴⁵

Lykes remained at his new duties with the Marines on weekdays during the war, but he would resume his responsibilities as manager of the park on weekends. As indicated in Chapter Seven, Lykes remained in the park manager's house just south of Joplin Road for the duration of the war, and he maintained social and professional relations with the commanding officers of the OSS training camps, such as Capt. Arden Dow at Area A and Maj. Albert Jenkins at area C, as well as the Marine Corps, Maj. Gen. Phillip Torrey. Lykes kept an eye on the park property, the area north of Joplin Road being occupied by the OSS and south of that road being used by the Marine Corps.

During weekdays, the sole administrative employee of the park was Ms. Thelma Williams, Lykes' secretary. She worked in a one-room temporary headquarters for the park off Joplin Road. Sharing the office with her was a clerical employee of the OSS, a secretive man, who Lykes reported was given to drink. Thelma Williams would fill Lykes in on the week's developments when he returned to the office at the end of each week.¹⁴⁶

At the end of the war, the Marine Corps relieved Captain Lykes from active duty and put him on terminal leave on 3 December 1945, and he returned to his former position as park manager with NPS on that date, more than a month before his terminal leave from the Marines ended and he received a Certificate of Honorable Discharge from the Marine Corps on 21 January 1946. In fact, by that time, he had already been back at work at the park fulltime since the beginning of December 1945.¹⁴⁷

Expansion of the Marine Base into the Park

The Marine Corps' wartime expansion of a force of 500,000 was accompanied by major enlargement of the Marine base at Quantico. What had started out in 1917 as a base covering five miles of forest bordering the Potomac River, had grown during the next several decades to a vast military reservation and training area covering nearly 100 square miles and a community of more than 12,000 military and civilian personnel, including families.¹⁴⁸ That expansion led to a rather longstanding controversy about control over the southern area of the park. At issue were 4,862 acres south of Route 619, the old Joplin Road. As the Marine base expanded during World War II and acquired some 50,420 acres west of U.S. Route 1, it also sought utilization for training purposes of the 4,862-acre tract of adjoining park land south of Route 619. That part of the park had never been developed for recreation and had preserved as a wilderness by NPS with only a one-room temporary office and a small maintenance area in it.¹⁴⁹ Talks between Marine Corps' generals and National Park Service officials over the tract began as early as 1941, and by 1943, when the park north of Route 619 was already being used by the OSS, the Marines were actively utilizing the park land south of Route 619. Park Manager Lykes had conversations with some Marine generals in which the possibility of the Marine Corps purchasing and adding to the park, approximately 1,900 acres of Quantico Creek watershed lands west of the park in exchange for a like amount in the wilderness tract on the Chopawamsic Creek watershed in the south part of the park being added to the Marine Corps Reservation.[150] The National Park Service worked on a land transfer agreement. Due to wartime needs, and in anticipation of such an exchange and permanent agreement, the Marine Corps was issued a temporary permit in 1943 to use the entire 4,862 acre-tract of park land south of Route 619 for the duration of the war plus six months.[151]

During the war, the Marines used the Park Service tract south of Route 619 for military field exercises, lumbering, and other operations, but although the 1943 agreement was technically terminated in 1946, the entire tract of nearly 5,000 acres appeared to be a fixture of the Marine Corps Schools. Negotiations began at the level of the Secretaries of the Interior and the Navy by their respective legal and real estate divisions, but the talks soon broke down.[152] They resumed, however, and reached an agreement. Based on that agreement, Congress, on 22 June 1948, Congress adopted Public Law 736. The 1948 legislation authorized the transfer of nearly 5,000 acres of park land to the Navy to be used by the Marines; it also authorized up to $10,000 for the purchase of 1,500 acres of privately owned land in and around the park; and it provided for the transfer to the park of 1,139 acres that the War Department had purchased during the war.[153] Although the War Department's land was transferred to the park, the $10,000 for the purchase of additional park lands was never appropriated, and consequently, the transfer agreement between the park and the Marine base was not carried through. Instead of resolving the issue, the law itself became the subject of continuing dispute between the Navy and Interior departments, which disagreed over questions of jurisdiction, funding, and the special use permit that had been granted to the Marine Corps during the war.[154]

The dispute would continue long after Park Superintendent Ira Lykes left Prince William Forest Park in 1951. Lykes left to become superintendent of Shiloh National Military Park, at the site of the Civil War Battle of Shiloh, at Pittsburgh Landing, Tennessee. There in addition to managing the national park and cemetery he raised funds for and produced and directed an award winning documentary film, Shiloh— Portrait of a Battle. In 1956, he was assigned to NPS headquarters, where he served in the park planning office as chief of the Park Practice Program, established to expand aspects of the park and recreation program. He retired in 1963, but was called back to continue that work. When his wife, Betty, died in 1968, Lykes sold their home in West Springfield, Virginia, and moved to Lake City, Florida, where he taught classes at the Forest Ranger School there and Lake City Junior College.[155]

Resolving a Half Century Jurisdictional Dispute

The dispute between the Marine Corps Base and Prince William Forest Park over the southern part of the park south of old Joplin Road continued for half a century. The Marine Corps continued to use the nearly 5,000 acre special use permit tract. But the National Park Service, claiming that the Navy Department had not fulfilled an obligation to purchase and transfer some 1,500 to 1,900 acres surrounding the park in exchange for the southern part it had occupied, declined to provide the Marine Corps with permanent legal jurisdiction over the contested tract of 4,862 acres south of old Joplin Road. There were several attempts at solutions, one a personal attempt at a compromise by NPS Director George B. Hartzog, Jr., but advocates of all or nothing on both sides defeated such attempts at a compromise solution.[156] By the 1980s, after the new Interstate 95 had opened up the area to commercial development, it was clear that soaring property values had made prohibitively expensive any acquisition and donation of land by the Navy to the National Park Service.

Robert ("Bob") Hickman became superintendent at Prince William Forest Park in October 1994. He had been a member of the Park Service since 1973, and had been site manager of NPS section of the Baltimore-Washington Parkways from 1984 to 1992. In 1987, he has served as acting Chief Ranger and Assistant Superintendent at Catoctin Mountain Park.[157] At Prince William Forest Park, Hickman found that the contested jurisdiction issue took up more of his time than anything else. "We began to explore each option again," he said. "We went to the Marine Corps to see if they would be interested in dividing the property in the spirit of the 1948 law—1,500 to 1,700 acres for the park, and the Marine Corps to get the rest," Hickman recalled." The Corps was interested. Negotiations started with Brigadier General Edwin C. Kelley, Jr. and concluded with his successor, Brigadier General Frances C. Wilson. On 10 March 1998 in Lejeune Hall on the Marine Corps Base, she and Hickman signed a breakthrough Memorandum of Understanding.[158]

Emphasizing mutual goals of both the park and the base and providing a scenario of future actions to achieve these, the Memorandum of Understanding of 1998 sought to settle the land issue by fulfilling the 1948 legislation without any cost to the government and to resolve the longstanding dispute over boundary and jurisdictional issues. Shifting to cooperation, representatives of the park and the base agreed to develop the watersheds jointly, keep the clearing of forests to a minimum, and provide joint recreational use of Breckinridge Reservoir and along mutual border, including development of a "Green Corridor," a strip extending 300 feet on either side of Route 619, which runs between the park and the base.[159] In addition to jointly addressing common concerns, the parties understood that they would resolve the jurisdictional dispute by splitting the contested acreage. Park Service would regain complete public use of a sizable block of property south of old Joplin Road (Route 619), and the Marine Corps would receive permanent jurisdiction over lands to the east, south and west of that property.[160]

With the two federal entities now in agreement, Congress authorized the transfer of jurisdiction over the land, and President George W. Bush signed the authorization into law in December 2002.[161] Under the actual jurisdictional transfer agreement signed between the two federal entities in August 2003 and made effective 22 September 2003, the jurisdictional confusion perpetuated by the half-century old dispute was finally resolved. In effect, the park gained or regained full jurisdiction over about 1,700 acres south of old Joplin Road (Virginia Route 1619) that it had lost to the Marine base as a result of World War II. The Marines kept nearly 3,400 acres.[162]

"The Marines under the 1948 legislation were to receive 5,000 acres," Bob Hickman said in reflecting on the agreement. "Instead the park retained lands and gained their full use for the public, and gained some Marine Corps' land, and the Marines received 3,398 acres that they could use exclusively for their purposes. And a confusing patchwork of jurisdictions was simplified to a point which each agency could pursue its mission more effectively."[163]

Prince William Forest Park Today

Prince William Forest Park continued to evolve in the decades after the war. Especially since the 1960s, its mission had grown from preservation and group camping to include day use. Organized camping resumed immediately after the war, but the subsequent transformation into an increasingly suburban area and the large number of daily visitors, led to a new and equal emphasis on day-use facilities. Ira Lykes's successors as park superintendent shifted the emphasis away from long term organized camping in permanent structures for a few character building organizations toward facilities for tent camping, hiking, motoring, picnicking, fishing, canoeing, and swimming for larger numbers of users. The result was the creation of an internal scenic driving looping through the park, picnic grounds, tent campgrounds, a nature center, and a Visitor Center.[164]

In 2007, nearly 250,000 visitors a year came to Prince William Forest Park, which under Park Superintendent Robert Hickman continued to offer day usage as well as overnight camping in the cabin camps and in a concession operated trailer park. In addition, the park has preserved one of the largest representative samples of Piedmont Forest in the National Park Service. It also acts as a sanctuary for many different kinds of plants, animals, birds and other forms of wildlife, and it protects most of the Quantico Creek watershed.

More than 150 of the 250 structures in the park are listed on the National Register of Historic Places, some of them constructed by the Civilian Conservation Corps in the 1930s and some by the Office of Strategic Services in World War II. The Visitor Center includes information about the Prince William Forest Park for the tens of thousands of visitors who stop by each year. There is information about camping and recreational usage of the park, material about the resources of the park, its physical characteristics, its flora and fauna. There is also material about the history of the area from prehistoric times when Native Americans used the natural resources in the park to the present day. The visitor's center also contains materials and artifacts concerning the history of the park itself. It begins with its founding as the Chopawamsic Recreational Demonstration Area in the 1930s. The World War II role of the park is visually represented through a number of artifacts—radios, codebooks photographs, and some mortar shell fragments—as well as a plaque provided by the OSS Communications Branch Veterans to commemorate the OSS communications school at Area C. All of this helps to inform visitors of the park's role as a site for training of OSS special operations personnel, secret intelligence agents, and radio operators during the war. For as this study has indicated, the wartime use of Prince

William Forest Park in Virginia—by the OSS and by the Marine Corps—had important impacts on the park itself as well as on the American victory in World War II.

Returning Catoctin Mountain Park to the NPS

OSS Training Area B, which had provided Basic Special Operations courses in 1942 and 1943 and advanced Operational Group courses in 1943-1944, had been abandoned as a training camp by the OSS on 20 February 1944.[165] But other wartime users replaced the OSS at Catoctin Mountain Park. A new Special Use Permit was issued by Acting Secretary of the Interior Abe Fortas on 31 May 1944 authorizing the use "for war purposes" of the two cabin camps (Camps Nos.1 and 2) and the old CCC work camp, as well as 1,800 acres in the undeveloped northwest portion of the park. The undeveloped area would be used by Army trainees at Military Intelligence Training Center at Camp Ritchie as well as the Marines who remained at Cabin Camp No. 1 (Misty Mount) to provide security for the Presidential Retreat. Camp Ritchie was overcrowded, and the initial idea was for some of its trainees to be accommodated at Camp No. 2 (Greentop), but that never occurred as Ritchie's housing problem had ended.[166] Neither Cabin Camp No. 2 nor the CCC Camp buildings (Round Meadow) were occupied by the Army after 31 May 1944 except by two or three soldiers assigned as a fire guar. Occasionally in 1944 and 1945, small continents of intelligence trainees from Ritchie conducted day or night maneuvers in the undeveloped northwest section of the park, but that stopped in the summer of 1945. Catoctin Park Manager Garland B. ("Mike") Williams reported that in August 1945, that the Army said it had no further use for the park and intended to declare it surplus property and return it to the National Park Service.[167]

With the end of the war and the end of the OSS in October 1952, Mike Williams moved quickly to have the park restored to its normal condition under the lease agreement. Unlike Ira Lykes or John Gum at Prince William Forest Park, Mike Williams had remained a civilian in the National Park Service throughout the war. At 55 in 1945, Williams had served as fulltime custodian of the park throughout the war.[168] At the end of the war, he resumed management of the property. That meant deciding what wartime changes to keep and which to eliminate, and to get the Army and the Marines to fix up or otherwise restore the facilities they had used. Even as Williams began this task, however, he suffered a setback. His residence, the Park Manager's House originally built in 1939, burned to the ground on 21 October 1945, the result of a chimney fire.[169] Williams, his wife, Grace, and their three sons moved temporarily to a house on east Main Street in nearby Thurmont.[170] The Park Manager's Residence, now called Quarters 1, was rebuilt and reoccupied by the Williams family in 1947. It still exists today behind the Visitor Center.[171]

Wartime occupation of the park by the OSS had an important impact on Catoctin Mountain Park. For security purposes, the Army had added obtained 288 additional acres to the park's original 9,832 acres.[172] Mike Williams reported favorably in October 1945 on a number of improvements the OSS had made. It had winterized

Cabin Camp No. 2 (Greentop), the buildings being sealed overhead and along the side walls with plaster board. It had installed a large forced-air heater for the central wash house. But, there were also changes that Williams said were of no use, or even dangerous, in regard to the civilian use of the park. The Army Engineers would have to eliminate these and restore the sites to their prewar conditions. A demolition area with bulldozed surface and its protective embankment and observation pits was one of these, so were the obstacle course, and the outdoor pistol and rifle ranges. A number of military-related wooden structures would have to be removed, including the "trainazium" resembling a giant "jungle gym" and "Dan" Fairbairn's "House of Horrors." As a result of OSS demonstrations of heavy explosives in one section of the park, a vacant, old house had collapsed, and wreckage needed to be taken away.[173]

Condition of the park roads, mostly gritty dirt roads, had long concerned Mike Williams. They were susceptible to serious erosion, even washouts from sustained heavy rains, and they suffered even more under the heavy vehicular traffic of the military, both the OSS and the Marines. Williams emphasized the need for improvements, particularly on the narrow, twisting graveled road between Thurmont and Hagerstown. When he could not get help from the military, he worked with local, state, and federal authorities to convert it into a modern highway. Through a federal-state program, Williams was able to get construction started in the spring of 1944 on what would become State Route 77.[174]

Army Engineers had the responsibility for restoring the area to the condition at the time the War Department had it taken over in 1942, or to the satisfaction of the National Park Service. The first thing that Williams wanted when the war ended was to make sure that park was safe. That meant ensuring that all remaining booby-traps, mines and unexploded munitions, most of them in the undeveloped northwestern quadrant, were eliminated. By the end of September 1945, the Post Engineer at Camp Ritchie, certified the area as "decontaminated."[175] To ensure safety of the civilians who would use the park, Williams insisted on personally inspecting the area, which he did in February 1946, accompanied by a Bomb and Shell Disposal Team. No unexploded munitions were found, and the area was deemed safe for public recreational use.[176] In 2000, however, a mortar round was found in the undeveloped northwest area, and as a result, the Corps of Engineers began a new series of field studies.[177]

At the end of World War II, the Army classified Cactoctin Mountain Park as "surplus to the War Department needs," effective 31 October 1945. Withdrawing the handful of fire guards, the Army officially abandoned Catoctin Mountain Park. Mike Williams accompanied Army Engineers as they covered the area and inspected all buildings and other structures used by the OSS. The Engineers prepared estimates for removing the military structures that Williams deemed were unsuitable and for clearing and grading the demolition areas, magazine, pistol and rifle range 'trainazium," "House of Horrors," taking away the wreckage of the destroyed vacant house, and for general cleanup of the area. The cost was estimated at nearly $41,000, almost as much as their original construction.[178] Williams asked the Corps of Engineers what the Army intended to do with the buildings and equipment in the old CCC work camp, and he was delighted to learn that the Army Engineers considered it to be the property of the National Park Service. Although it was not suitable for a group camp, the CCC camp could benefit the NPS management of the park in a number of ways.[179]

A Restoration Survey of the "Catoctin Training Center" by the Army Engineers in October 1945 listed the changes that had OSS had made at the park and what the Engineers, in consultation with Park Manager Mike Williams, estimated needed to be done to restore the area while keeping the changes that the NPS would accept.[180] Williams accepted a number of the new buildings and improvements at Camp No. 2 (Greentop) as well as the CCC work camp, particularly the winterizing, and improvements in the mess hall and kitchen facilities. He also accepted the replacement of chemical toilets with flush toilets in several barrack as well as the construction of a 25,000-gallon concrete water reservoir with wooden roof, a pumping station and water lines, additions to the waste disposal facilities, including an 8' by 18' by 10' concrete septic tank, a pump house and 7,500 feet of sewer lines.[181] The Army Engineers noted that in regard to eliminations Williams wanted made to Camp No. 2 (Greentop), since the Army's use of that facility had been terminated in October 1945 so that it could be reassigned, at least temporarily to the Marines, it was not clear whether the Army or the Marines should be required to make that restoration.[182]

Marines Move Up to Greentop

With NPS's permission, the Marine Corps' security detail for the Presidential Retreat had moved up the hill in January 1946 from Cabin Camp No. 1 (Misty Mount) to Cabin Camp No. 2 (Greentop), which the War Department had declared surplus. The move to Greentop was a result of the Corps' need for places of rest and rehabilitation for survivors of the bloody fighting on Iwo Jima and Okinawa. The Marine Corps said the "cool, quiet, and healthy environment" of Catoctin Mountain Park seemed ideal.[183]

Williams heartily approved the idea, particularly because since the OSS had stopped using Camp No. 2 in early 1944, the condition of the facilities without any maintenance or rehabilitation by the Army was "rapidly deteriorating." Since they moved into the park in June 1942, the Marines had kept Cabin Camp No. 1 (Misty Mount) in excellent condition and made many improvements there, winterizing the camp, constructing a repair garage and a combination recreation hall and movie theater, and keeping all the buildings maintained and in excellent repair.[184]

Both the Secret Service and the White House were agreeable to the Marines occupying Camp No. 2 (Greentop). Consequently, in mid-September 1945, representatives of NPS and the Marine Corps met and agreed on terms of a new use permit. Signed on 4 October 1945 by Acting Secretary of the Interior Abe Fortas, the agreement revoked the May 1944 "war purposes" permit and granted a Special Use Permit authorizing the Marine Corps to use Cabin Camp No. 2 (Greentop) for "rehabilitation and security purposes." The new permit for the Marines extended through 1 May 1947, subsequently revocable by the Secretary of the Interior.[185]

On a cold January 5th in 1946, the Marines vacated Camp No. 1 (Misty Mount) and moved up the hill to Camp No. 2 (Greentop). During the previous six weeks, Marines had come up to Camp No. 2, re-stained the buildings, repaired windows and doors, replaced worn-out plumbing, installed a new power line, and cleared the camp area and vicinity of dead and downed timber resulting from two severe sleet storms the previous winter. The Marines planned to spend $25,000 to improve Camp No. 2, and over subsequent months, they erected a service garage and a portable building which served as a combination movie house, recreation hall and post exchange.[186] With NPS approval, the Marines kept many of the improvements made by the OSS at Camp No. 2 (Greentop), such as the winterization of the buildings, the heating apparatus in the central washhouse, and the improved mess hall and kitchen.

The Marines concurred with NPS that the purely military remnants of the OSS should be removed. They did not need the ones at Greentop: the "trainazium," the "House of Horrors" pistol house, and the rifle range. In other areas of the park, the military remnants included a demolition observation area, grenade range, munitions magazine, obstacle course, and guard houses. Mike Williams and the Associate Director of NPS, Arthur Demaray, insisted that the Army remove all of these, as well as the wreckage of an old house destroyed during OSS demolitions exercises, and restore all those sites to their original condition.[187]

Initially, however, the Corps of Engineers, contended that the Army no longer had responsibility, at least for the area of Camp. No. 2 (Greentop). That responsibility, the Engineers asserted had ended when the War Department's permit occupancy was terminated in September 1945, the area declared surplus and the Marine Corps subsequently occupied it. By March 1946, the OSS field exercises areas of the park in its northwest quadrant had been declared free of unexploded shells, but Williams and the National Park Service were still trying to get the Army Engineers to raze the purely military buildings, rifle range, obstacle course, and demolition observation area in the park.[188] On another subject, however, the question of ownership of the old CCC work camp was finally resolved when the War Department formally transferred it to the Department of the Interior, as part of the Army's official return of the park to the National Park Service on 14 May 1946.[189]

Moving the Marine Corps Out of Catoctin

The Marine Corps remained in Camp No. 2 until their permit expired in May 1947. Although the Marines made improvements at Greentop, as they had at Misty Mount, and kept it in good condition, Williams had his share of difficulties with the Marines, combat veterans and the security detail, which moved into Camp No. 2 in 1946. The Park Manager held off making formal complaints for some time due to the work the Corps had done in restoring the camp. But violations of the use permit continued to mount and confronted the Park Service's responsibility for preserving the natural and historical resources of the park.

In the Use Permit issued to the Marines, the Secretary of the Interior had outlined methods of proper disposal of garbage and other materials, but the Marines, despite warnings, had not burned their rubbish nor buried their garbage since a garbage disposal field was established for them in October 1943. Instead, they had simply dumped their garbage and trash in the open, and the growing amounts attracted buzzards, crows, vermin and stray dogs and disrupted the area. Worse was an episode in September 1946. While, Mike Williams was absent for several weeks on an NPS assignment, the Marines, without Williams knowledge or approval, built a parking lot and two roads through the park to their camp at Camp No. 2 (Greentop). The Use Permit provided that trees in the park would be protected, but in this project, the Marines cut down nearly two dozen trees, including half a dozen large red oaks. The venerable stone fences in the park, some of them more than a century old, were also part of the area's protected heritage. Williams had previously warned Marine officers against their men cutting trees and destroying stone fences, particularly historic ones and those marking boundaries. But while he was away, the Marines not only cut down the trees to make way for the new roads, but to provide gravel for the roads and parking areas, they appropriated stone from NPS stockpiles, and when that was not sufficient, they tore down several historic stone walls and ground them into gravel. Greatly irritated by what he found when he returned, Williams filed a formal complaint with the Marine major in command, holding him responsible for seeing that there be no more such violations.[190]

Expiration of the Marines' Use Permit was coming up in May 1947, so Mike Williams began negotiations with the Marine Corps in January for an early return of Cabin Camp No. 2 (Greentop) to the National Park Service. In March, they came to an agreement that the Marine Corps would accept the cancellation of the use permit and turn over some its trucks, a tractor, and several water heaters and gas ranges, and NPS would accept the condition of Cabin Camp No. 2 as it was. That meant that NPS would be responsible for eliminating the OSS rifle range and the Marines' recreation hall/movie theater and rehabilitating Greentop. By letter of 28 March 1947, the Department of the Interior accepted Camp No. 2 (Greentop) in its existing condition from the Marines and cancelled the Corps' Special Use Permit.[191]

Consequently, the Marines left Catoctin Mountain Park at the end of March 1947. After that the Marine contingent guarding the President was to be transported from the Marine Barracks in Washington D.C. to Catoctin as needed.[192] Upon the Marines' departure, the newly created Department of Defense removed from Camp No. 2 the 300-foot rifle range and play field that the OSS had established and the combination recreation hall, movie theater, and post exchange building that the Marines had constructed.[193] Thus in the spring of 1947, the final occupation and use of Catoctin Mountain Park by the military as a result of World War II came to an end.

Returning Catoctin Mountain Park to Civilian Use

Following the end of the war, the National Park Service sought to expedite the reopening of Catoctin Mountain Park to the public. In January 1946, when the

Marines moved up the mountain to Greentop, Mike Williams hoped to have Misty Mount ready for youth from the Salvation Army to camp there that summer.[194] But there would be no camping at Misty Mount until the summer of 1947. In regard to Greentop, the Maryland League for Crippled Children, unhappy with their temporary wartime summer camps in Pennsylvania, started pressing in early 1946 to be allowed to return to the facilities that had been customized for their disabled youths. When the Maryland League met resistance, its officials enlisted Secretary of Agriculture Clinton Anderson, who wrote an endorsement to the White House, adding there was a rumor that the Marines were not planning to leave but were instead augmenting Greentop as a regular summer retreat for the Marine Corps Commandant. Secretary of the Interior Julius A. Krug rejected the rumor but acknowledged that accommodations had been improved for the commandant's visits.[195]

Like the National Park Service, the Truman White House also wanted Greentop reopened for the crippled children.[196] The public learned about the Presidential Retreat at Catoctin as soon as wartime censorship was removed.[197] President Harry Truman, however, seldom used Shangri-La, preferring the Presidential Yacht *Williamsburg* or the warmer weather and deep-sea fishing off the U.S. Navy base on Key West, Florida. It was not until President Dwight D. Eisenhower that a Chief Executive resumed regular use of the Presidential Retreat, which Eisenhower renamed "Camp David," after his grandson.[198] In the summer of 1947, after the Marines had left, the Maryland League for Crippled Children and the Salvation Army resumed their summer camp programs at Greentop and Misty Mount for the first time since OSS and the Marine Corps had occupied the cabin camps in 1942.

In response to an invitation from the Maryland League for Crippled Children, President Truman visited the handicapped young campers in August 1947 on their first return to Greentop.[199] Some residents in nearby Thurmont remember him during such occasional visits to Catoctin slipping away from the Secret Service and driving around behind the wheel of a convertible with the top down on a hot summer day.[200]

Postwar Division of Catoctin between the Federal and State Governments

The federal government had originally built the Recreational Demonstration Areas, including Catoctin and Chopawamic RDAs, with the expressed intention of eventually turning them over to the states for state or local parks. In 1943, the Roosevelt Administration announced that this would be done in all except half a dozen RDAs, which would be included in the National Park System. Both Catoctin RDA (later Catoctin Mountain Park) Chopawamsic RDA (later Prince William Forest Park) were among the half dozen included in the National Park System. It was not their OSS usage, however, but rather accessibility from the nation's capital as well as other neighboring cities that may have been the determining factor in their case.

Plus at Catoctin, the establishment of the Presidential Retreat there mandated federal jurisdiction.[201]

When Maryland hunters, fishers, and other area sportsmen learned that the federal government intended to retain Catoctin Mountain as a permanent part of the National Park System, they began to mobilize to try to reverse that decision and persuade Washington to transfer the land to Maryland as a state park. Some were opposed to the no-hunting policy on the federal RDAs, and local gamesmen had long considered Mike Williams unduly committed to preservation over expanded public usage. Emphasizing the need for "healthful outdoor enjoyment," the League of Maryland Sportsmen adopted a resolution in July 1944 calling for the Park Service to transfer Catoctin RDA to the Maryland Department of Forests and Parks. In June 1945, the Maryland State Forester, supported by U.S. Senator Millard E. Tydings (Dem. Md.), began pressing NPS to transfer the land to the state at the end of the war.[202]

President Truman announced in December 1945, however, that the park would be permanently retained by the federal government and continued as part of the National Park System because, as he put it, of "the historical events of national and international interest now associated with the Catoctin Recreational Area," presumably the wartime meetings there between Roosevelt and Churchill. Truman said this was in accord with "the position expressed by the late President Roosevelt before his death."[203]

The legislative process proved extremely slow in the formal designation of Catoctin's new status, in large part because of continued opposition by many sportsmen and Maryland officials, including Joseph F. Kaylor, Director of the Maryland Department of Parks and Forestry. Blocked in 1947 and 1948, the legislation finally passed in 1949 with the support of social service agencies in Washington, Baltimore, and around the region, and with major emphasis on providing general public access to expanded recreational facilities in the park, fishing, swimming, boating, hiking, picnicking, as well as camping.[204] Nevertheless, pressure from Marylander hunting and fishing enthusiasts and others continued, and the National Park Service sought to way to accommodate the needs and desires of the regional public with the Park Service's legislative mandate for natural preservation of the area as well as of the Presidential Retreat.[205]

To resolve the continuing tensions with Maryland, NPS Director Conrad Wirth proposed having the southern half of the park—the area south of State Route 77—transferred to the state of Maryland. On 11 June 1954, the National Park Service, turned that 5,000-acre area, containing Cunningham Falls, the historic Catoctin Iron Furnace, and a considerable part of Big Hunting Creek, over to the State of Maryland. With that division, the northern part of Catoctin Recreational Demonstration Area was designated Catoctin Mountain Park on 12 July 1954, and the area south of Route 77, now managed by the Maryland Department of Natural Resources was opened on 4 July 1954 as Cunningham Falls State Park.[206]

Garland B. ("Mike") Williams, who had been part of the federal development of the park at Catoctin since 1935 and served as custodian of the property while it was taken over by the military during World War II, remained as manager of Catoctin

Mountain Park when it became part of the National Park System. He continued in that position in charge of Catoctin Mountain Park until his retirement at the age of 63 in 1957.[207]

Catoctin Mountain Park Today

After World War II, Catoctin Mountain Park continued to follow its conservation and recreation mandates. Day use recreation emerged with the establishment of the Chestnut and Owens Creek Picnic Areas in 1966. Emphasis has continued to shift towards day use with the increasing suburbanization of both the Washington and Baltimore metropolitan areas. In 2007, more than 500,000 recreation visitors went to Catoctin Mountain Park, less than 8 percent staying overnight in tents or cabins. The former Blue Blazes Visitor Center Contact Station, now the Catoctin Mountain Park Visitor Center, also served as Park Headquarters in the 1950s, until Headquarters moved in the 1960s to its current location as Camp Peniel, the former Church site, which was under lease to the Army during World War II. The Visitor Center still provides information as well as housing artifacts from all periods of the park's existence.[208]

Most of the original structures built at Greentop by the Works Progress Administration in the 1930s for the National Park Service, and which were also part of the OSS's Area B-2, remain in use today. However, the Dining Hall burned and was rebuilt with a Recreation Hall added to it. The Maryland League for Crippled Children, renamed the League for the Disabled of Baltimore, continues to use the camp each summer. The Frederick County Public School System hosted a residential Outdoor School at Greentop from 1957 to 1996. Budget cuts reduced the weeklong residential environmental program to a day use activity for fifth and sixth graders.

Misty Mount, where the Marines had stayed from 1942 to 1946, was used for years after the war by the Girl Scouts. The Washington County Public Schools ran an environmental education camp at Misty Mount until moving into their own camp at Clear Spring, Maryland, in 1979. After being closed for four years, Misty Mount reopened in 1983 with the aid of a friends group called CAMPER. The cabin camp is now used by a variety of small to medium-sized camping groups.

After serving as a Civilian Conservation Corps work camp in the late 1930s and then as OSS Area B-1, Camp Round Meadow hosted the first Job Corps Camp in the United States from 1965 to 1968 and a Youth Conservation Corps residential camp from 1971 to 1978, as Catoctin Mountain Park continued its tradition of youth conservation programs. In the 1960s, The District of Columbia Public Schools began to use Round Meadow for their Summer Nature Camp, and that program has continued to the present.

Mel Poole became superintendent of Catoctin Mountain Park in 1997. A native of Norfolk, Virginia, he joined the National Park Service in 1978 and served in numerous assignments around the country, from fighting forest fires in California to battling oil spills in Alaska. Eventually, he was posted at the National Capital Region as a Natural Resources specialist. From 1990 to 1997, he was Park Manager for the Presidents' Parks, responsible for the parks immediately adjacent to the White House and to the NPS's duties in the Executive Mansion itself. Since 1997, he has been responsible for Catoctin Mountain Park and the NPS's relations with the Presidential Retreat there.[209] Camp David, of course, continues as the Presidential Retreat with increasing visits from foreign heads of state and media presence for each such visit. It remains a place of rest and relaxation for the President of the United States.[210]

Return of Area F to the Congressional Country Club

Use of the Congressional Country Club by OSS and its successor ended in February 1946. Negotiations for return of the club to its civilian owners had begun when the war and the OSS ended. Despite the damages caused by the OSS's paramilitary training on the club's golf courses and other grounds and facilities, the club's board of governors had learned in November 1945 that the War Department had paid the club rent amounting to $120,400 during the war. With that income, the club, which had been hurting at the end of the Great Depression, had paid all its bills, including interest on the mortgage, and still had $46,000 remaining. The President of the club was a tough negotiator and got the government to agree to provide and additional $187,000 to restore the clubhouse and the golf course, both of which had been badly damaged. During the OSS occupation, trees had been downed, the fairways were criss-crossed with barbed wire obstacles and roads, the refreshment stand on the 13th tee had been blasted, and more than fifty cabin tents and a few Quonset huts dotted the grounds and tennis courts. The club sold the cabin tents and Quonset huts and launched into a major repair operation.

The clubhouse needed much work, including repair of the front marble steps where an OSS Santa Claus had driven his jeep right up into the main ballroom on Christmas 1945. The clubhouse reopened with a grand ball in April and the golf course the following month. Membership in May 1946 had already grown past 500. The Congressional Country Club emerged quite successfully from its wartime occupation by the OSS.[211]

Postwar Careers of OSS Veterans

"You know it was an amazing organization," recalled John W. Brunner, a China Communications veteran, about the OSS. "The people I served with later became lawyers, judges, Ph.D.s, all very prestigious."[212] Indeed, the OSS did include extraordinarily talented, self-motivated, and accomplished men and women. Not surprisingly, many of them went on to distinguished postwar careers. After earning a doctorate at Columbia University, Brunner himself served as a professor and chair of the German Department at Muhlenberg College for more than thirty years in addition to becoming a published authority on OSS weaponry.[213] Captain Bernard M.W. ("Bernie") Knox, who led Jedburgh Team "Giles" and French *maquis* in attacks on German paratrooper positions in France and then led Italian partisans in northern Italy, earned at Ph.D. in classics at Yale under the GI Bill after the war. He taught at Yale and subsequently became Director of Harvard's Center for Hellenic Studies. A renowned scholar and translator of classical Greek drama and literature, Knox wrote or edited nearly two dozen books and provided the introduction and notes to Robert Fagles' prizewinning translations of Homer's *The Illiad* and *The Odyssey* in the 1990s.[214]

Lieutenant Henry Deane McIntosh, SO, who jumped into France as a Jedburgh, received a Silver Star Medal, and then served behind Japanese lines in China, returned to resume studying medicine at the University of Pennsylvania. Becoming a distinguished cardiologist, he taught at the medical schools of Duke, Baylor, and the University of Florida, was President of the American College of Cardiology, founded an organization to make cardiac pacemakers available to needy patients around the world. For his achievements, he received the Presidential Citation at the White House in 1986.[215]

Nearly two dozen men who served in Donovan's organization later capped their postwar careers as U.S. ambassadors, including David K.E. Bruce, wealth and socially prominent head of OSS Secret Intelligence in Europe, later U.S. Ambassador to England, France, and Germany; William B. Macomber, Jedburgh in France and Special Operations (SO) officer in Burma and China, ambassador to Jordan and Turkey; and Richard Helms, Secret Intelligence (SI) agent in Europe, ambassador to Iran. Donovan himself was appointed ambassador to Thailand in the early 1950s. Arthur Goldberg, first chief of the OSS Labor Branch in Europe subsequently served in the 1960s as Secretary of Labor, Supreme Court Justice, and U.S. Ambassador to the United Nations.[216] A number of OSS veterans later served as Presidential advisers, among them Roger Hilsman, Arthur Schlesinger, Jr., Walt Rostow, Carl Kaysen, Douglass Cater, Clark McGregor, Arthur Goldberg, C. Douglas Dillon, and a host of others.

There were many lawyers or future lawyers in the OSS, a number of whom later pursued successful careers as attorneys or members of the judiciary. Walter R. Mansfield, SO, who parachuted into Yugoslavia and later "Muskrat" Team behind Japanese lines in south China, became a federal appeals court judge in New York.[217] Edgar Pritchard, SO, became a prominent attorney in Arlington, Virginia; Jedburgh William Dreux resumed his law practice in New Orleans and litigated cases before the U.S. Supreme Court. Howard Manning, the last commanding officer of Area C, returned to North Carolina to resume his law practice, first in Chapel Hill and then Raleigh.[218] Turner McBaine, SO, who worked out of Cairo, later became counsel to Standard Oil Company of California.[219] J. Evelle Younger, a former FBI section chief recruited by OSS's counter-intelligence branch (X-2) for the Far East later became District Attorney for Los Angeles in the 1960s and Attorney General of California in the 1970s.

Of the two most famous OSS close combat instructors, William E. ("Dan") Fairbairn and Rex Applegate, the former, already a legendary figure among close combat enthusiasts, returned to Great Britain after the war. He retired from the British military shortly thereafter and reportedly started a martial arts school with

George de Rewelisko, another SOE instructor in Britain and Canada in unarmed combat.[220] In retirement, Fairbairn frequently gave demonstrations and received numerous honors. He died in 1960 at the age of 75. Rex Applegate stayed with the U.S. Army in the military police, eventually retiring as a lieutenant colonel. In 1976, Applegate updated his 1943 book, Kill or Get Killed, for use as a close combat manual for the Marine Corps as well as for popular sales.[221] Applegate became a founding member of the International Close Combat Instructors Association, marketed books and films on the subject and also designed a line of knives. He spent much of his time promoting his pistol shooting method to police departments and pistol enthusiasts. He died in 1988 at 74 years of age.[222]

A number of OSS Special Operations officers had careers with the Federal Bureau of Narcotics. Lieutenant Colonel Garland H. Williams, SO, who had helped Donovan and Preston Goodfellow establish the Special Operations training camps in Maryland and Virginia, returned in 1945 to the Federal Narcotics Bureau, where he headed its Intelligence Division until his retirement in 1954. He subsequently advised intelligence operations in the Army and the State Department.[223] George H. White, another undercover federal narcotics agent, who had served as an SO instructor at Areas B and A in 1942 before being transferred to the OSS Counter-Intelligence Branch (X-2), returned to the Federal Bureau of Narcotics in 1945. He served as district supervisor in several of the major cities in the United States. In addition, beginning in 1952, he worked part-time for the CIA in its secret experimental drug program, using the synthetic hallucinogen, LSD, in a search for a "truth serum," a program that White and the CIA continued, without success through the mid-1960s. White retired from the Federal Bureau of Narcotics in 1965 after a thirty-year career.[224] Both Williams and White had been with the Federal Narcotics Bureau before the war; after the war, they were joined by a young recruit, Major Howard W. Chappell, SO, 27, who after training at Areas F and B, had won acclaim for the "Tacoma" Mission behind German lines in the Italian Alps. After returning home and running his own business for a year, he joined the Federal Narcotics Bureau in 1947, and was given undercover assignments infiltrating organized crime, drug syndicates in New York City, Washington, D.C, and Chicago. His successes led to his appointment as agent in charge in Los Angeles. Chappell thrived on danger, and in addition to his administrative duties, he continued to do undercover work, this time in Mexico. Chappell left the bureau in 1961 and spent the next decade as executive officer and President of the Board of Public Works for the City of Los Angeles.[225]

Colonel Carl Eifler, SO, the boisterous, barrel-chested, first commander of Detachment 101 in Burma, had come to Special Operations from the Army Reserve and a law enforcement career with the U.S. Customs Service. Eifler remained in the Army until 1947. The new CIA tried to recruit him, but its doctors turned him down, as his 1943 head injury led him to suffer massive headaches and some erratic behavior, and he would be vulnerable to seizures and strokes for the rest of his life. Nevertheless, after retiring from the Army, he returned to the U.S. Customs Service in Hawaii, this time as Deputy Collector of Customs. In 1952, he went back to college and earned degrees in psychology and divinity from Jackson College in Hawaii. Retiring from the Customs Service and returning to the mainland, he went on to obtain a doctorate in psychology from Illinois Institute of Technology in 1963. Settling with his family in Monterey County, California, he opened a practice in clinical psychology there.[226] He was inducted into the Military Intelligence Hall of Fame in 1988, and in 1997 a new sports complex was named after him at the Military Intelligence School at Fort Huachua, Arizona. Eifler died in 2002 at the age of 95.[227]

Among Hollywood celebrities in the OSS, John Ford, Field Film Unit, and Sterling Hayden, SO, went back to the studios in Los Angeles. So did Peter Ortiz, SO. Ford soon turned to directing a series of heroic cowboy and cavalry pictures, starring John Wayne or Henry Fonda, which came to personify the legendary western genre.[228]

Sterling Hayden (aka Marine Captain John Hamilton), even more solid, weathered, and fatalistic than before, wandered from studio to studio in a series of dark, film noir movies in the 1950s.[229] He was forced to testify before the House Un-American Activities Committee, where he revealed his sympathies with Tito's communist partisans in Yugoslavia and his brief postwar affiliation with some American communists. Abandoning those friends and publicly naming them to the committee, intensified Hayden's cynicism and self-contempt, and except for his role as Gen. Jack D. Ripper in Stanley Kubrick's *Dr. Strangelove* (1963), Hayden spent most of the 1960s sailing his schooner, fighting a child custody battle in court, and writing his autobiography, *Wanderer*. He staged a come back in the 1970s, and died at the age of 70 in 1986. Peter J. Ortiz, whose medals for Special Operations exploits behind German lines in North Africa and France made him one of the most decorated Marine officers of World War II, returned to Los Angeles after the war and became involved in the film industry, advising on films about the OSS and playing small parts in more than a dozen films between 1949 and 1957. Ortiz died in 1988 at the age of 75 and was buried in Arlington National Cemetery.[230]

Some of the other celebrities who had worked for the OSS faded into relatively obscurity not too long after the war. Joseph ("Jumping Joe") Savoldi, SO, the Notre Dame football star and professional wrestler, who was part of the "MacGregor" Mission in Italy, returned to the wrestling circuit. But arthritis soon ended his career in the ring, and he became primarily a manager and promoter. While a promoter in Chicago, Savoldi discovered, trained, and gave a start in professional wresting to an able, young American black man, Houston ("Bobo Brazil") Harris, who would later become the first successful African-American professional wrestler, winning the World Heavyweight Wrestling Championship in 1962, integrating the sport and becoming known as "the Jackie Robinson of professional wrestling."[231] From 1952-1961, Savoldi refereed a few wrestling matches, made a few guest appearances on TV programs, had his own sports radio show for awhile, and worked in the insurance business. In 1962, at 54, he returned to college, earned his teaching credentials and then mentored troubled boys and worked as a high school science teacher in Henderson, Kentucky until his death there in 1974 at the age of 65.[232]

Moe Berg, the professional baseball player and linguistic genius, who as a civilian spy helped OSS ferret out information on the status of German development of an atomic bomb, returned to the United States and resigned from the OSS. In 1945, he was only 43, but he apparently declined any regular employment for the rest of his life. President Truman awarded him the Presidential Medal of Freedom, but for an unexplained reason, Berg declined to accept it. A lifelong bachelor, he subsequently lived with different family members and seems to have become increasingly idiosyncratic. He continually implied that he was still a spy, doing work too secret to reveal. After his brother forced him to leave, Berg moved in with his sister in Belleville, New Jersey, where he remained until his death in 1972 at the age of 70, still living on the legendary accomplishments of the first forty years of his extraordinary life.[233]

Back to Business

Many of the former OSSers went into private enterprise. Lieutenant Colonel Ainsworth Blogg, SO, first commanding officer of Training Area B and, for most of the war, deputy director of the OSS Schools and Training Branch, returned to the insurance business in Seattle. Rafael D. Hirtz, SO, born in Argentina, raised in Europe and California, who had served behind the lines in France and China, moved to Washington, D.C. and became a grain exporter.[234] OSS's key spy in Rome, Peter Tompkins, SI, remained an expatriate in Europe, writing magazine articles, screenplays, and controversial books, not only his spy memoirs but most notoriously, his assertion that his mother had had a passionate affair with George Bernard Shaw.[235] Reginald ("Reg") Spear, SO/SI, a young California inventor with a British and Canadian family background, who had trained at Area B and then gone on special missions in the Pacific and Far East for President Roosevelt, returned to civilian life and California after the war. He continued his schooling and also his inventing. His first invention was when he was 12 years old and created a retractable dog leash. Since World War II, he has been credited with more than 40 inventions. Some involving naval ordnance such as 7.2-inch HVAR multiple rocket launchers and the Sidewinder Missile were invented jointly with William B. McLean. But most were invented independently or with a variety of others. Those inventions included the solar cell, the Spear 360 movie camera used by high-flying spy planes, the U-2 and the SR-71, a dock-loading gantry for loading and unloading seagoing tankers, and a processing for liquefying natural gas for ease in transporting and storing it. In his 80s, Spear continued to do consulting and to work on a memoir of his experiences with the OSS in World War II.[236]

Serge Obolensky, SO, former Russian nobleman and New York socialite who had married and divorced Mary Astor daughter of real estate tycoon, John Jacob Astor IV, and who had led Special Operations teams into Sardinia and France, returned to high society in Manhattan. Hotelier Conrad Hilton offered the 55-year-old socialite a position directing public relations and promotion for the Plaza Hotel. It was similar to Obolensky's prewar position at the St. Regis Hotel across the street, and he took it. Obolensky remained there, a fixture of high society and socializing with glamorous women including Ginger Rogers, Marilyn Monroe, and Jacqueline Kennedy Onassis. He died at 87 in 1978.[237]

M. Preston Goodfellow, the former New York newspaper publisher, who had established the SO training camps and served as an OSS Deputy Director of

Operations for the rest of the war, subsequently engaged in politics, diplomacy and business. Through his wartime friendship with Korean nationalist Syngman Rhee, Goodfellow after the war helped play a political kingmaker role in South Korea and in the process built a lasting personal and commercial relationship.[238] During the American-Soviet military occupation of Korea immediately after the war, Goodfellow was an economic and political advisor to the U.S. military governor in 1946, supporting a separate South Korea and Rhee, who became its first President.[239] After leaving the War Department, Goodfellow, retained an interest in newspapers as publisher of the *Pocatello* [Idaho] *Tribune*, and he helped for over forty years with the Boy's Clubs of America, but his main activity was as President of Overseas Reconstruction, Inc., a Washington based firm offering personal contacts and providing advice for American investment overseas and foreign investment in the United States.[240] A successful entrepreneur, Goodfellow died in 1973 at the age of 81.[241]

A number of other SO/OG members went into business. Elmer ("Pinky") Harris, Washington State University alumni who, taught at Areas B and A, and then served in Italy and China, returned to his native Alaska after the war and helped pioneer tourism to the Alaskan wilderness as a part of a group of business people involved local aircraft companies.[242] Ellsworth ("Al") Johnson, medic with OG units in France and China, returned to western Michigan, went to a community college on the GI Bill, married, raised a family of a boy and a girl, and went to work for a company handling wholesale beauty supplies and rose from salesman to general manager. After retirement, he made and sold doll houses and their furnishings, and at 85 was working as a starter on a golf course.[243] Arne Herstad, a member of the OSS Norwegian OG, who parachuted into France and helped train OSS OG Chinese commandos in the Far East, returned home to Tacoma, Washington. He had married, Andi Kindem, his sweetheart whom he had met in Michigan, and they raised a family. He worked as head sawyer with a lumber company for a decade before teaming up with his brothers to form a construction company that built quality homes on what is now called the Herstad Addition in Tacoma. He died in 1981.[244]

Albert ("Al") Materazzi, OG, one of the leaders of the Italian Operational Groups both in their training in Areas A, B, and F, and as operations officer for their amphibious raids from Corsica, returned to the United States in 1945 and was immediately transferred back to the U.S. Army Map Service and put in charge of reproduction research. In August 1945, he testified at the war crimes trial of German General Anton Dostler, who had ordered the summary execution of fifteen members of the OSS "Ginny" mission in Italy. Dostler was convicted and executed. After leaving the service, Materazzi returned to his prewar research in graphic arts, spending twenty years with a firm that manufactured chemicals for the lithographic industry and later managing quality control at the U.S. Government Printing Office. Retiring at 65, he continued to work until he was 75 as a consultant on quality control on the lithographic and photographic reproduction processes. In a long and distinguished career in that field, he was founding member and President of four of the industry's technical associations. Even in his 90s, he continued to write about the OSS in Italy and the German atrocities committed.[245] Richard W. Breck, Jr., SO, who had left Harvard, trained at Area B, and gone on three missions behind enemy lines in Italy, spent 1946, as a baseball pitcher with the Pawtucket, Rhode Island, Slaters, a farm team for the Boston Braves. "Bobo Breck," pitched until 1947 when he broke his leg. Later he took night courses at MIT and worked for a series of defense contractors including the Little Company with its contracts for the hydrogen bomb, and Raytheon on new radar systems and the Hawk and Sparrow missiles.[246]

Many members of the OSS Communications Branch (CB) went back to or into the communications industry after the war. Lawrence Lowman, head of the Communications Branch, returned to his prewar position as a vice President of the Columbia Broadcasting System in 1945. Owner William S. Paley, put him in charge of preparing CBS Television.[247] Peter G.S. Mero, one of the leading recruiters of radio operators and other "commo" people, returned from heading OSS communications in Italy, to head the Communications Branch in the new Strategic Services Unit.[248]

Subsequently, Mero went back to Chicago, married Sarah ("Sally") Sabow, a former OSS cipher clerk who had served in Algeria and Italy. He became an executive in the Pioneer Electric Corporation, invented a successful teleprinter, and later worked with the American Red Cross and with Radio Free Europe.[249] According to CB veteran Allen C. Richter, "He had many hats in many closets."[250] Richter, a key member of the communications team with Detachment 101 in Burma, ended the war in Washington as assistant chief of CB's Plant and Engineering Division. After the war, Richter worked for a communications equipment manufacturing firm as foreign sales manager, selling transmitters and receivers to the Chinese. At one point soon after the war, he formed a consulting company. One of its aspects was to help independent radio stations prepare for television. "I ran into Larry [Lowman] in Grand Central Station one day," Richter recalled. "….Larry asked me if I wanted to come to work for him at CBS-TV, as anything I wanted. But I said no. T.V. was kind of shaky then. In hindsight, it was not a good decision."[251] Instead of CBS, Richter joined the American Red Cross in Washington and helped run their telecommunications system for fifteen years. After that he ran the Washington offices first of Robert Shaw Controls, which tested instruments, and then the Singer Sewing Machine Company. He retired in 1971.[252]

Some of the communications instructors at Area C returned to the burgeoning telecommunications industry. Captain James F. Ranney, had taught International Morse Code at Area C before serving in the Mediterranean and China theaters. After the war, he went back to Ohio, his home state, and worked as an engineer at various radio stations, ending up as chief engineer at a radio station in Cincinnati. Subsequently highly active in the Communications Branch's alumni group, Ranney edited, together with his son, a collection of recollections by OSS "Commo" veterans, the *OSS CommVets Papers*.[253]

Timothy Marsh, had been a civilian instructor at the Communications School. By the end of 1944, when the number of trainees declined, most of the civilian instructors left. Some enlisted as civilian radio operators in the Merchant Marine, and several of them lost their lives on torpedoed vessels. But Marsh had a wife and infant daughter, and instead of the Merchant Marine, he went to Chicago and found jobs with a radio manufacturer and then a radio station. But after a few years, however, the condition of his young daughter's health, led Marsh to return to his native Tennessee. Initially, chief engineer at a radio station in Jackson, he later was employed by the Motorola field office for thirteen counties, subsequently taking ownership of the dealership until his retirement in 1979. During much of the next two decades, in Shelbyville, Timothy Marsh and his wife, Helen, served as the historians/archivists for Bedford County, Tennessee.[254]

Countless OSS radio operators who underwent training at Area C returned to the private sector. Roger L. Belanger, who had returned from CB's London base station to become part of the cadre at Area C, went home to Maine and worked first as a salesman and then designing marine instruments and finally for an engineering company building ships. He then turned a hobby into a business by building scale model ships. He has constructed more than 150 hand-made wooden ship models for private collectors and museums.[255] Joseph J. Tully, who remained a lifelong friend of Roger Belanger, used the benefits of the GI Bill to learn tool making. Joining a company in Palmyra, New Jersey, that did rotogravure printing—high speed rotary press printing used for long print runs of colored material such as stamps and magazines—he advanced over the years from toolmaker to tool machine designer, plant engineer, and wound up as sales manager before his retirement in 1988.[256] David Kenney, a radio operator who had trained at Areas A and C and served in England and China, returned home to a ranch in Wyoming owned by his two brothers. "I learned to fly under the G.I. Bill," he said. "Later I got my degree in aeronautical engineering from Northrop's university in Los Angeles County." He worked as an aircraft engineer for Northrop and later Rockwell aerospace companies in Los Angeles until his retirement in 1983.[257]

After the war, Spiro Cappony, the naval seaman and OSS radio operator, who trained with actor Sterling Hayden and then helped blow up bridges in Greece in the "Chicago" Mission, went to college at night under the GI Bill, while working at the Pentagon as a naval reservist called to active duty. The new CIA offered him a position in 1947, but he declined. Instead, he returned to Gary, Indiana, his hometown, married, and, like his family before him, ran a restaurant. He also spent much of his postwar career in the construction industry working for Pangere Construction Company, which he continued to advise even after his retirement.[258]

Marvin S. ("Mark") Flisser, the Brooklyn College graduate, who as a trainee at Area C was soon transmitting 40 words a minute, double the normal rate for operators, had not continued in the Communications Branch. Instead, the assessment staff at Station S identified him as having leadership qualities and an IQ of 140, he was sent to Officer's Candidate School. Returning a second lieutenant, OSS headquarters assigned him to help research Nazi war crimes. Mustered out of the Army at the age of 24, he returned to New York City, joined a stock brokerage firm as a trainee, and ultimately wound up as a partner in a firm that included H. Ross Perot. After thirty years on Wall Street, he retired, went to California and earned a doctorate in nutritional science. Then, he founded, and for fifteen years managed, a nutrition company based in Stamford, Connecticut. He retired once again, first to Scarsdale and then at age 78 to a retirement community in Monroe, New Jersey.

From there, he audited courses at nearby Rutgers and Princeton Universities.[259]

A young soldier named Jack S. Kilby, from Jefferson City, Missoui, who served in radio repair shops with the OSS Communications Branch for Detachment 101 in northeast India and then Detachment 202 in western China, returned home and with the help of the GI Bill graduated from the University of Illinois and then with a Masters Degree in electrical engineering from the University of Wisconsin. He went to work for Globe Union and then in 1958 for Texas Instruments as a research engineer. Continuing the OSS philosophy of thinking outside the box, he invented that year a working integrated circuit, which formed the basis for the microchip industry. For that discovery, he was eventually awarded the Nobel Prize in Physics.[260]

Continuing to Serve in Uniform

A lot of the men in Special Operations and Operational Groups remained with the Army after the war. At least two of them became generals. William R. ("Ray") Peers, who had trained at Area B and spent the war with Detachment 101 in Burma, made a career in the Army, although he was several times assigned on detached service to the CIA, after its founding, helping to establish its paramilitary training program and directing covert operations in southern China from secret bases in Burma and Taiwan during the Korean War. Returning to the Army, he held several intelligence and other staff positions. When the Vietnam War began, he advised the Joint Chiefs of Staff on counterinsurgency special operations. In 1967, he commanded the 4th Infantry Division, and in 1968, as a lieutenant general, he was in charge of 50,000 U.S. troops in the corps-level I Field Force plus South Vietnamese and South Korean troops. In 1969, Peers was selected to investigate the My Lai Massacre. His conclusion, the "Peers Report," was condemnatory of the action and highly critical of the cover up, and he recommended the court-martial of two enlisted men and thirty officers from lieutenant to major general. Peers died in 1984 at age 69.[261] John K. Singlaub, who trained at Areas F and B, was a Jedburgh in France and the leader of an SO rescue mission in China, also made the Army his career as a combat infantry and special operations officer. During the Korean War, he served first on detached duty with the CIA and then as an infantry battalion commander with the 3rd Infantry Division. With his emphasis on special operations, Singlaub commanded the Joint Unconventional Warfare Task Group (MACSOG) in Vietnam. He was instrumental in establishing the Army Rangers training center at Fort Benning. While chief of staff of the U.N. Command and the 8th U.S. Army in South Korea, Singlaub publicly challenged President Jimmy Carter's 1977 decision to withdraw U.S. troops from Korea, and Carter relieved him of command. Singlaub retired that year as a major general.[262]

Instructors Joseph H. Collart, Charles M. Parkin, Jr. Leopold ("Leo") Karwaski, and Frank A. Gleason, Jr., who served abroad after being instructors at Areas B or A, all remained in the Army and spent the next two decades with the Corps of Engineers. Joseph H. Collart, who had been one of the instructors at Area B in 1942, had left the OSS in early 1943 and joined the Airborne Engineers, serving in Europe.

By the end of the war, Collart was a lieutenant colonel and commanded a battalion.²⁶³ Twenty years later, Collart was a full colonel and from his headquarters in Hawaii in charge of Army Engineers in the Pacific. In 1965 with the announcement of the buildup of U.S. ground forces in South Vietnam, Collart and his staff were called upon to go to Vietnam and then prepare estimates for the financial costs of constructing the bases for the combat units being scheduled to arrive. Because of quick and detailed estimates and budget requests, the Army's 18th Engineer Brigade was able to get the facilities built in time.²⁶⁴ After returning from China, Charles Parkin did graduate work at Cornell and held a variety of positions with the Army Engineers. He retired as a lieutenant colonel and then spent in various positions in the business sector as a production control engineer in Chicago, a manager of copra and citrus plantations in the Caribbean, and running his own carpet business in Florida. Following his third retirement, Parkin turned to writing books, one of rockets and another on an 18th century nautical explorer, and founding a maritime museum in Oregon.²⁶⁵ Leo Karwaski, one of the "three ski's" instructors at Area B, had served in China. After the war, his primary career with the Army Corps of Engineers, but he reportedly worked periodically on detached duty for the CIA inThailand in the early 1950s and in Vietnam in the late 1960s. Karwaski retired a colonel and is buried in Arlington National Cemetery.²⁶⁶

Frank Gleason earned a Master's Degree at Harvard after the war and served in various civil and military positions with the Engineers, concluding his career as chief engineer at the big U.S. Army base at Cam Ranh Bay in South Vietnam. After retiring as a full colonel in 1971, Gleason spent fifteen years at Georgia Tech University in various business and financial administrative positions. Following his second retirement in 1985, he worked buying and selling companies and doing insurance work, and after his final retirement in his mid-80s, he lectured on history to senior citizens and tutored children in elementary school. Still lively and engaging, he declared gleefully in December 2007, "I'm 87 years old. I've been all over the world and had a lifetime of experiences. I tell people that 'I've seen the monkey and heard the owl!' "²⁶⁷

Arden W. ("Art") Dow, an instructor at Areas B and A, who then served in China, moved, after the war, from the Army infantry to the airborne forces, and was stationed in Korea, Taiwan, and Panama. His wife, Dorothea ("Dodie") Dow, gave birth to their second daughter in 1952, while they lived in Panama. Dow retired as a full colonel in 1963, and he became business administrator for the College of Business of the University of Georgia. He died in 1982; Mrs. Dow continues to live in Georgia; she has ten grandchildren.²⁶⁸ John Hooker, who trained at Areas F and A, before serving in Maritime Unit raiding parties along the Burmese coast and then with the OSS Chinese OG commandoes in China, spent a postwar career in the Army, with Military Intelligence in Japan and Germany and commanding an infantry battalion in Germany. Retiring as a full colonel in 1963, he worked for the National Security Agency in Vietnam for five years. Subsequently, he managed commercial radio stations, owned and outdoor advertising company, and worked in a variety of capacities for a Florida police department. He also taught Civil War history.²⁶⁹

A number of OSS SO and OG personnel stayed with the Army and eventually worked with the Special Forces. Several were pioneers in the Special Forces when it was secretly created in 1952, among them Aaron Bank, Jack Shannon, Caesar Civitella, Herbert Brucker, and James Goodwin. Jerry Sage, Washington State University alumnus, instructor at Areas B and D, who had been captured in North Africa and spent the next two years trying to escape from German POW camps, finally successfully in early 1945. After the war, he served on the headquarters staff of the U.S. European Command in Germany and became an expert on the detainment, repatriation, and immigration of Displaced Persons, testifying before congressional committees on the subject. He was an instructor at West Point for three years and, except for the dissertation, completed a doctorate at Columbia University. After a year as a battalion commander during the Korean War, he taught at the Army's advanced schools, and in the early 1960s, he was assigned to the Special Forces, first in Vietnam and then in Europe. Working out of the Bavarian Alps, his unit was prepared to parachute behind Soviet lines and lead resistance forces in case of a Russian invasion of Western Europe. After retiring as a full colonel in 1972, Sage did graduate work in education and taught civics instruction at high schools in Columbia, South Carolina, while enrolled in a doctoral program (his second), this time in education, being voted the outstanding teacher for the entire state. In 1980, he moved to Alabama, where he raised money to help the needy and handicapped through schools and training centers, worked for the Special Olympics, and continued to give "enrichment lectures" in schools, civic, youth, and church groups. Arguing against peer pressure and gangs and what he called the "sheep complex" that herded young people into bad habits. "I try to stress the importance of having the courage to choose the harder right over the easier wrong," Sage said, "and to lift others up rather than tear them down."[270]

Paul Cyr, as a young Army captain from Vermont, had been one of the stars of the OSS. Trained in Special Operations, he had won the Distinguished Service Cross on Jedburgh Team George which harassed the Nazis behind their lines in France. He then went to China where his Team Jackal blew up the mile-long bridge across the Yellow River. Settling in Indiana, this war hero ran for Congress in 1950, but lost. Later he went to Washington as an assistant to Senator Homer E. Capehart, a Republican from Indiana. In 1963, he became an important civilian employee of the new Army Matériel Command, acting as a liaison with Congress for one of the Pentagon's largest supply agencies. He later worked for the Federal Energy Administration. He retired in the 1980s and died in 1994.[271]

Working for the CIA

Four subsequent Directors of Central Intelligence had been with the OSS. Allen W. Dulles, SI in Switzerland, was appointed by President Eisenhower and served as DCI from 1953 to1961. Richard M. Helms, SI in Europe, worked his way up in CIA and was appointed DCI by President Lyndon Johnson, serving from 1966 to 1973. William E. Colby, who had trained at Areas F and B, had been a Jedburgh in France and then SO in Norway. After a few years as a lawyer, Colby joined the CIA in 1950 and was appointed director by President Richard Nixon, serving as DCI from 1973 to 1976. He died in a canoeing mishap at 76 in 1996. William J. Casey, had broken his jaw in a training accident at Area B, but went on to a top position in SI London. After the war, he became a leading New York tax attorney, a wealthy investor, and a major figure in the Republican Party. President Ronald Reagan appointed him DCI, and he served from 1981 to 1987.

The CIA naturally recruited a number of former OSS personnel to serve careers with the new agency. Helms and Colby were two of them. Peter M.F. Sichel, SI with the 7th Army, worked for the CIA for a dozen years, serving as base chief in Berlin, chief of European Operations in Washington, D.C., and then station chief in Hong Kong. He resigned in 1959, in a disagreement over policy, and joined the family wine business, in the process building "Blue Nun," into a major international brand. Joseph Lazarsky, SO, former Area B instructor who served overseas in Burma and China, finished college under the GI bill after the war and then spent twenty-five years in the Far East with the CIA, beginning with duty in Taiwan, India, and Indonesia. After being Deputy Station Chief in Saigon in the early 1970s, Lazarsky served as Station Chief in Seoul, Jakarta, and Manila, retiring in 1978.[272] Oliver M. Silsby, Jr., from Detroit, who had original trained as a radio operator but then taken SO training and parachuted into Yugoslavia and with Howard Chappell into northern Italy, where he was eventually captured by the Nazis, returned home after the war and graduated from the University of Michigan under the GI Bill. He also attended the Institute for Political Studies in Paris. He spent nearly thirty years with the CIA's Directorate of Operations, serving in France, Belgium, and Luxembourg, before being sent to Laos from 1968 to 1970. He helped establish an guide the CIA's terrorism bureau in the 1970s, before his retirement in 1979.[273] Caesar Civitella, who trained at Areas F and B with the OSS Italian OGs and then served in France and Italy, was a pioneer with Army Special Forces from 1952 until his retirement as a major in 1964. He then worked for the CIA until 1983. In 2008, he was honored with a special award from the U.S. Special Operations Command.[274] Lucien Conein, who trained at Areas F and B, was a Jedburgh in France and also was infiltrated into Vietnam, technically remained in the Army and rose to the rank of lieutenant colonel, but he was on detached service with the CIA almost continually from 1947 to 1968. His assignments carried him from Eastern Europe to Iran to South Vietnam, where he wound up as liaison in the coup that overthrew President Ngo Dinh Diem. In 1972, he rejected an offer from E. Howard Hunt, another ex-OSSer and retired CIA officer, to join the "plumbers," a secret team that bungled the Watergate burglary. From 1973 to 1984, Conein ran secret operations for the Drug Enforcement Administration. When he died in 1998, the New York Times called Conein, "the last of the great wartime spies."[275]

A lot of OSS people continued to work in the intelligence field after the war, even if not always with the CIA. Erasmus ("Ras') Kloman, SO, who had trained at Areas A, B, C, D, and F, and served in the Mediterranean Theater, pursued careers in government, academia, and private business after the war. He earned an M.A. from Harvard and a Ph.D. from the University of Pennsylvania, served sequentially with the CIA, the Department of State, and the Foreign Policy Research Institute at the University of Pennsylvania, and subsequently with several corporations as a public affairs officer and executive. After retiring in the mid-1980s, he painted and wrote travel books as well as a memoir of his service in the OSS, Assignment Algiers.[276] Like scores of others, who had been with the Research and Analysis Branch, Louis J. ("Luz") Gonzalez, who as an Army sergeant trained briefly at Area A, before being sent to Italy, wound up after the war with the State Department's Bureau of Intelligence and Research. The Mexican-born Gonzalez, who knew both Spanish and Italian and became a U.S. citizen when he joined the Army in 1942, worked at the State Department until his retirement.[277] Jacques F. Snyder, the French-speaking, Saxophone-playing American GI who was recruited by OSS, given OG training at Areas F and A, and later parachuted back to France where he successfully carried out a combination special operations and intelligence mission, remained in the Army with Military Intelligence after the war. He served in France and Cambodia, part of the former French Indochina, until a hard parachute landing and some compressed vertebrae ended his active military career in the 1960s. Subsequently as a civilian employed by the Defense Intelligence Agency, he worked in Vietnam and then mainly in Europe. He retired from the Defense Intelligence Agency in 1991. His son recalled that "All the time [I was] growing up, I never knew what my father did for employment. When we would ask what he did for a living (for school paperwork, applications, etc.), he would say 'government employee.' "[278]

Women of the OSS

More than a few of the OSS women later worked for the CIA. Virginia Hall, the "limping lady" the spy and resistance leader, who so infuriated and successfully evaded the Gestapo in France, came back to the United States after the war and was awarded the Distinguished Service Cross. She was the only civilian American woman to receive that award in World War II. Later, she joined the CIA in 1951 and was assigned to the fledgling covert action office as an intelligence analyst. The following year, Hall became one of the first female operations officers. She interviewed refugees and exiles from communist countries, prepared political-action projects, and planned resistance and sabotage activities to be carried out behind Red Army lines in case of war with the Soviet Union. Several of her assignments as an operations officer were overseas. In 1966, having reached 60, she retired, spending the rest of her life at her Maryland home until her death at 76 in 1982.

When Elizabeth ("Betty") Peet MacDonald left the Morale Operations unit at OSS headquarters in Kunming, China in 1945, her husband, Commander Alexander MacDonald, decided to remain in the Far East and establish an English-language newspaper in Thailand. She returned to the United States, and they later divorced amicably. She resumed her career in journalism, writing for Glamour magazine, and producing a memoir of her service in the OSS.[279] Subsequently, she married Richard P. Heppner, former head of OSS in China and a partner in Donovan's law firm. She did some work for Voice of America and the State Department. Then in 1958, Dick Heppner died of a heart attack at 49 after completing a world tour as Deputy Assistant Secretary of Defense for International Security. A year later, Allen Dulles hired Betty Heppner for the CIA. In 1959, most of the women at CIA, she reported, were assigned to office work. Only a few did research and analysis, and even fewer, like Virginia Hall, were assigned to operations and then mainly at lower levels. Betty Heppner saw the agency gradually transformed as times and the culture changed and bright, college-trained, computer-smart women arrived and moved into positions formerly held only by men. Betty Heppner moved up too and eventually she also ran agents. She cited William E. Colby, former OSSer and DCI from 1973-1976, as an outspoken champion of equal opportunity for women. By 1994, the professional employees at CIA were chose to equal in number, 59 per cent male, 41 per cent female. Betty Heppner (Betty MacIntosh after she married Frederick MacIntosh a World War II fighter pilot) retired from the CIA in 1973, and subsequently wrote an account of women in the OSS and CIA, A Sisterhood of Spies.[280]

Aline Griffith, the beautiful, young Hattie Carnegie model from Pearl River, New York, who had worked in OSS Counter-Intelligence (X-2) in Madrid gathering information from social circles and running her own net of spies, stayed in Spain after the war. In 1948, she married a Spanish nobleman and became Aline, Countess of Romanones.[281] From her homes in Spain and with her social contacts and sharp insights, she continued to work for American intelligence, the CIA, for nearly the next twenty years, and then wrote two memoirs about it.[282]

Several other OSS women worked for the CIA, according to Betty MacIntosh, but among the most successful was Eloise Randolph Page, who had been an executive secretary to Donovan. Toward the end of the war, OSS headquarters sent her to Brussels to head the counter-intelligence (X-2) office there. Later, she had a career with the CIA, eventually becoming a top executive and the first woman chief of station in Athens. As Page told MacIntosh, "historically, I suppose you could say that the women of the OSS prepared the groundwork for their sisters who came after them in CIA."[283]

Aside from the CIA, numerous other OSS women had various postwar careers. Hélène Deschamps Adams, who as a French teenager had spied for the French Resistance and for the OSS, married an American Army officer and came to the United States as a war bride in 1946. They had one daughter. Forest E. Adams died of a heart attack in 1951. Mrs. Adams never remarried, but taught French and wrote two memoirs of her wartime experiences. She died in New York in 2006 at 85.[284] Barbara Lauwers Podoski, a native Czech who had joined the U.S. Women's Army Corps, and, as a German linguist, served with the OSS's Morale Operations Branch distributing effective propaganda to German soldiers in Italy, returned to the United States. After the war, she sold hats, worked as a dental assistant, did broadcasts for Voice of America, and finally found a permanent job she liked with the Library of Congress.[285] Maria Gulovich, the young Czech schoolteacher, partisan, and guide, who led two American survivors of the ill-fated OSS mission to Slovakia to safety, in 1945 was brought to the United States at Donovan's request and enrolled at Vassar

College, from which she graduated in 1948. She later moved to southern California, married an American businessman, worked had two children, and worked in a real estate business. She was awarded the Bronze Star for her heroism in World War II.[286]

One female OSS member, unknown to the public during the war, would become a celebrity chef. Her name was Julia Child, and through her books and then her television program, she became the best known chef in America. Julia McWilliams, a big, bright, happy, young woman from California, met her future husband, Paul Child, while both were working at OSS China headquarters in Chungking. They married in Washington right after the war. He took her to Paris where he was assigned by the State Department, and there she spent a decade studying French cuisine. Beginning in 1961 with her book, *Mastering the Art of French Cooking*, and subsequently with her own television show and later on *Good Morning America*, Julia Child demystified French Cooking for the American public. She went further than that. Child taught Americans not just how to cook but how to think deeply about food, a quality missing from many of the glitzy TV chefs who have followed her.[287]

OSS "Commo" people in the CIA

Quite a few of the people from OSS Communications Branch later worked for the CIA. Bob Kehoe from New Jersey, the radioman on Jeburgh Team Frederick in France, who won the Distinguished Service Cross, returned to Rutgers University to finish his degree under the GI Bill and then earned a Masters Degree in International Relations at Columbia. He spent thirty-five years at the CIA in the training and in management. He earned a Ph.D. in Far Eastern Studies from American University in 1970.[288] Gail F. Donnalley, an OSS code clerk, trained in Area C as well as downtown Washington, served in Egypt and Italy, and had a small part in the German surrender in Italy. After being mustered out, he returned to Ohio Wesleyan University, earned his degree, and worked for the U.S. Bureau of the Budget for ten years. Then in 1955, he was recruited by the CIA where he worked in the agency's communications section until his retirement in 1980. Subsequently, he worked for Apple Computer Company until his second retirement. Even after that, he kept working, serving in the ombudsman program for the State of Virginia.[289]

Art Reinhardt, who trained at Area C and spent fourteen months as a radio operator in China, became a HAM radio operator and continued communications work after the war. For five years, he also struggled with recurring high fevers and diseases he had caught in China, until they were finally successfully diagnosed and cured in 1950. Reinhardt continued his role in developing communications for America's secret intelligence organizations. He pioneered in establishing the Diplomatic Telecommunications Service for the Department of State, and he subsequently worked for communications in the Central Intelligence Agency, retiring from the Agency in 1976. "I'm very grateful to the OSS," he said later, "because it turned out that it gave me a career. That was totally unexpected, and I wouldn't trade it for anything. And it made me forget I was ever in regular military service!"[290]

Donovan Preserves the Legend

William J. Donovan, the dynamic chief of the OSS, had been only 62 in 1945 when the war and the OSS ended. He lived another 14 years. During the postwar period, in addition to resuming work at his law firm on Wall Street, he remained keenly interested in the gathering and analysis of foreign intelligence.[291] He kept up his own sources of information through trips abroad and contacts within the American intelligence community. A vigorous Cold Warrior, he warned repeatedly of the dangers the Soviet Union posed to the interests and values of the United States.[292] Recognizing the importance of global economic development for the United States and other nations both for material purposes and to counter communist subversion, Donovan helped form a corporation to encourage economic regeneration of countries devastated by war and vulnerable to Soviet influence. The World Commerce Corporation had representatives in nearly 50 countries and became heavily invested in manufacturing, exporting and importing, and may also have been involved in gathering intelligence.[293] Donovan vigorously supported the Marshall Plan to stimulate economic recovery in Western Europe, and he championed European unification to block Soviet influence.[294]

Donovan still had political ambitions. He launched a campaign for U.S. Senator from New York in 1946, but it was blocked by Republican Governor Thomas E. Dewey.[295] There was speculation in 1947 when the CIA was created that Donovan might be its first director, but there was little chance of that. Truman had never liked Donovan, and the feeling was mutual.[296] The President appointed the current head of the Central Intelligence Group, Rear Adm. Roscoe H. Hillenkoetter, who served until 1950. In the 1952 election, Donovan vigorously supported Eisenhower, and the former OSS chief was disappointed at not being appointed Secretary of Defense or Director of Central Intelligence (DCI). Named DCI was former OSS bureau chief in Switzerland, Allen Dulles, the brother of Secretary of State John Foster Dulles.[297]

Instead, Eisenhower appointed Donovan as U.S. ambassador to Thailand.[298] The OSS had been influential in Thailand during the war, and Donovan had been celebrated by that country, as well as many others.[299] At 70 years of age, the former head of the OSS served as Ambassador to Thailand from 1953 to 1954. "They're scraping the bottom of the barrel to send out an old guy like me," he told one former member of Detachment 101 when he was appointed. But he was just joking. "It's good to have one more challenge to meet," he had added.[300] In fact, with the communists in control in China and with the French capitulating to the communist Viet Minh in Vietnam, Donovan's main role was to help prevent communist infiltration into Thailand and to integrate Thailand into a regional defense system, the Southeast Asia Treaty Organization (SEATO), led by the United States.[301] Donovan had gone beyond Foster Dulles's diplomatic arrangement and recommended that SEATO should have a unified, military-political command structure, like NATO, for which Donovan hoped he would be the first theater commander. But his idea was rejected, and at 71, frustrated, tired, ailing, and with public service straining his income, Donovan resigned. Back home, he continued to tout his idea for SEATO, but he also took a larger view toward the struggle against Ho Chi Minh and the communist regime in North Vietnam. "It is not essentially a military matter," he told a meeting of former OSS alumni in 1954. "It is a political struggle which must be won in the stomachs of the hungry and in the minds of the people. In Washington, they think that American military might is the solution to the problem, but any intelligence man knows this is not true."[302]

Two years later, at 73 and suffering from increasing arteriosclerotic impairment of blood supply to and atrophy of the brain, Donovan had a stroke. In 1957, he suffered a massive stroke and lost control of his mental faculties. Learning of Donovan's rapidly deteriorating health, President Eisenhower awarded him the National Security Medal, the highest civilian award, to go with his military Medal of Honor, Distinguished Service Cross with Oak Leaf Cluster, and Distinguished Service Medal. Donovan had yet another stroke, and, at the President's insistence, he was transferred from New York to the Pershing Suite at Walter Reed Hospital in Washington, D.C. It was there that Donovan died on Sunday, 8 February 1959, at the age of 76.[303] Hearing of Donovan's death, Eisenhower lamented: "What a man! We have lost the last hero."[304]

CHAPTER 11

SUMMARY AND CONCLUSION

The 65th anniversary of the opening of OSS training camps for spies, saboteurs, guerrilla leaders, and clandestine radio-operators in the National Parks—in particular Catoctin Mountain Park and Prince William Forest Park—occurred in 2007. Although the training camps were closed and the OSS terminated in 1945, the valuable contributions to the Allied victory made by those facilities and by Donovan's organization itself are an important part of the history of World War II. William J. ("Wild Bill") Donovan believed that intelligence, deception, subversion, and psychological and irregular warfare could spearhead the Allied liberation of Europe and the Far East, and he crafted a novel instrument to serve that purpose. Like the secret agency itself, much of its history was cloaked in silence and mystery. The American public remained only partially aware of the OSS, its members, their training, their missions and their accomplishments until the 1980s when the CIA began to declassify the records of Donovan's organization. Subsequently, OSS veterans, sworn to silence, began to feel free at last to talk about their experiences in training and serving in America's first centralized intelligence and clandestine operations agency. Most of the remaining OSS files, including personnel files, were not declassified until 2008, more than half a century after the end of World War II.[1]

Particularly during the Cold War, with its extensive intelligence and counterintelligence operations and clandestine actions on both sides, the public became fascinated with the shadowy world of spies and secret agents. Before the cynicism of recent years, secret agents were seen as glamorous. Popular novels and films reflected that view. Sometimes they noted the institutional dichotomy between the civilian spies and the rowdy, covert action agents, whom the less combat-oriented members of the OSS sometimes referred to as the "Bang-Bang Boys."[2] But more often, particularly the sensational ones produced for the mass market, merged espionage, counterespionage and covert operations in a mélange of action, most famously in Ian Fleming's debonair James Bond-007 series, but also in the tense, suspenseful *Mission Impossible* episodes originated by Bruce Geller, and the action-filled technothriller films starring Tom Clancy, Tom Cruise, or Matt Damon. Aside from the three postwar films, *O.S.S.*, *Cloak and Dagger*, and *13 Rue Madeleine*, which celebrated the OSS and Robert DeNiro's recent film, *The Good Shepherd*, which attacked both it and the CIA, the OSS itself has seldom provided the basis for Hollywood films. Because until relatively recently the full extent of the operations of Donovan's organization had not been made public, the OSS has been portrayed mainly through historical, biographical, or autobiographical works rather than through the movies.

While the most popular topics concerning the OSS for the public and scholars alike have been the cloak and dagger work of the spies and counterspies, and the behind enemy lines operations of OSS guerrilla leaders and saboteurs, the least explored area of the OSS has been its training schools. The present study, commissioned by the National Park Service to help understand the role of the National Parks in the OSS's activities in World War II, provides considerable new light on that aspect of the OSS—and indeed on the CIA and the Special Forces which inherited some of its personnel and adopted much of the training techniques of Donovan's organization.

Training Spies, Saboteurs, and Agent Operatives in the Parks

With its cardinal principle of secrecy, the OSS established its training camps in secluded yet accessible areas, most of them rural Maryland and Virginia within two hours drive from the organization's headquarters in Washington, D.C. Here as in many other matters, OSS initially drew upon the experience of the British secret services. Donovan's Special Operations (SO) Branch replicated the British Special Operations Executive's (SOE) penchant for rugged, isolated terrain for toughening up its covert operators for paramilitary missions behind enemy lines. It set up Training Areas A, B, C, and D in secluded woodlands. The only deviation was Area F, which was established on the grounds of the former Congressional Country Club for the Operational Groups. OSS's Secret Intelligence (SI) Branch replicated British Secret Information Service's (SIS) use of country estates as schools for introducing recruits into the murky world of espionage. Thus, it established Training Areas E and RTU-11 ("the Farm") in spacious manor houses with surrounding horse farms. Yet some members of each of the two American branches trained at the other's facilities. This was particularly true in the teaching of rugged survival and close-combat techniques at the Special Operations training camps at the two National Parks, where men preparing to be spies or other operatives sometimes joined the military recruits who were being trained physically and psychologically for clandestine raids from forest or mountain hideouts upon enemy outposts, command centers, or vital communication or transportation facilities.

The appeal of Catoctin Mountain Park and Prince William Forest Park, then known as Catoctin and Chopawamsic Recreational Demonstration Areas respectively, was precisely because of their location not far from Washington, their comparative isolation in rural areas, their existing camp facilities, and the fact that they were already federal property. That meant they could be obtained quickly and easily in the spring of 1942. With war declared, the War Department simply demanded that the Department of the Interior lease those lands of the National Park Service to it for military purposes for the duration of the war. The two parks had cabins for accommodation, woods in which to practice hit and run attacks on enemy targets, and open meadows for firing ranges, demolition work, and other field exercises. With nearly 10,000 acres each, the two parks were sizable enough to cloak the secret training that would be provided there, yet they were only one or two hours away from OSS headquarters.

The first three OSS training camps were established in the two parks in April and May 1942. Training Area B for the basic paramilitary course was created in Catoctin Mountain Park in northwestern Maryland, 70 miles north of Washington. Training Areas A and C were established thirty-five miles south of Washington in Prince William Forest Park. Area A for the advanced courses in special operations was located in the cabin camps in the western part of Prince William Forest Park. Training Area C, a school for preparing clandestine radio operators, was established in the cabin camps in the northeastern sector of Prince William Forest Park. At the end of the war, Schools and Training (S&T) Branch's only complaint about the facilities for Areas A and C at Prince William Forest Park was that OSS had to make a considerable number of changes to winterize them for its year around training, since they had originally been built as summer cabin camps.[3] Although S&T found the mountainous terrain of Catoctin Mountain Park useful for paramilitary training exercises at Area B, it concluded that the location a full two hours north of Washington was somewhat too far for efficient coordination, and that Franklin Roosevelt's use of his Presidential Retreat there during the summer considerably curtailed the paramilitary training exercises when he was in residence.[4]

Although additional OSS training schools for other operational branches of the OSS were subsequently established, Areas A, B, and C in the two National Parks served as the primary training sites for the Special Operations and Communications branches. Areas B and A also served as subsidiary training sites for the commando-like units of the OSS Operational Groups (OGs) after their initial training at Area F, the former Congressional Country Club in Bethesda, Maryland, acquired by OSS in 1943. The lakes in Area A served as the training site for waterborne infiltration practice before the acquisition of Area D on the eastern bank of the Potomac River and the establishment of the OSS Maritime Unit. The fields of Area A were used for parachute practice and low altitude jumps before OSS parachute training was relocated to the Army's main parachute school at Fort Benning, Georgia.

In the summer of 1942, the Secret Intelligence Branch acquired a country estate in Maryland 20 miles south of Washington as a training school called RTU-11, or "the Farm." The following year, the newly established Schools and Training Branch established Area E, ultimately consisting of two country estates north of Baltimore, which served as training sites for a general introductory course for OSS recruits of various operational branches (as would sub-area A-3 in Prince William Forest Park). Area E eventually served mainly the Secret Intelligence, Counter-Intelligence, and Morale Operations branches.

The majority of the 13,000, or more, men and women in OSS, however, did not go to the training schools of the so-called operational branches. The clerks, typists, office workers and other administrative and support personnel, as well as the scientists and engineers of the Research and Development Branch and the scholars and other analysts of the Research and Analysis Branch, most all of these civilian employees, had been employed because they already had the skills required.[5]

In the winter of 1944-45, as the war in Europe neared its end, and the U.S. Army began plans to transfer many troops to the Far East, most of the OSS operational branch training sites in Maryland and Virginia became holding areas for returning veterans awaiting reassignment to Asia or other purposes. Most of the OSS's Far Eastern training programs had shifted to the agency's new training schools located on Catalina Island and Camp Pendleton Marine Corps base in southern California. These West Coast schools were modeled after those in Maryland and Virginia. With the Japanese surrender and the rapid termination of the OSS in October 1945, all of the OSS training sites were returned to their former owners. They were given back without the firing ranges, demolition areas, "houses of horrors," and other facilities that the OSS had built for the rough and tough training of the Special Operations teams (SO) and Operational Groups (OGs).

Aims and Methods

"Set Europe ablaze!" was the goal enunciated by Prime Minister Winston Churchill when he authorized the creation of the British commandos and Special Operations Executive (SOE) forces, and it became part of Donovan's grand vision of the OSS as well, not just a centralized intelligence agency but also one that acted to subvert the enemy. It was widely believed at that time that the Germans' success in conquering much of Europe so quickly was not simply due the capability of their armies but also to the effectiveness of their spies, saboteurs and sympathizers ("fifth columnists" in the term of the day), who undermined the ability of the targeted nations to resist Hitler's forces. Churchill and Donovan sought to turn that technique against the Axis. They would use spies, propagandists, saboteurs, commando raiders, and guerrilla leaders to inspire, supply, and direct resistance movements to conduct subversive activity and raids behind enemy lines in the Axis-occupied countries. What Churchill meant by his famous phrase was to set German-occupied Europe ablaze with the fire of subversion by indigenous resistance movements supplied and directed by the Allies. The German Army's lines of communication and supply would be hampered by subversive efforts by these Allied-led local partisans. Eventually, when the Allied conventional armies were raised and assaulted Hitler's Empire from the front, the Allied agents played a crucial role in sabotaging the German Army's supply lines with explosives they set as well as by bombs dropped by Allied aircraft they directed to the supply depots, assembly points, troop trains and convoys and other tactical targets.

Such unconventional warfare was made possible largely by two technological developments: the airplane and the radio. Airplanes facilitated the delivery of spies, saboteurs, guerrilla leaders and other personnel as well as weapons and supplies into enemy-held territory. Agents and supplies were generally parachuted in at night from low flying, black painted bombers. Radio, or more precisely the wireless transmission (W/T) of telegraphic messages by short-wave radio signals, provided a means of communication between regional headquarters and the spies and agents behind enemy lines. The idea was to obtain strategic and tactical intelligence and to engage in sabotage and other subversive activities behind the enemy lines. The regular military was suspicious, even hostile, to Donovan's group of civilians and former civilians. They disdained the absence of military discipline and protocol in the OSS and the inattention to the precision of dress that the regular military required. But the professional soldiers made a mistake in so easily dismissing Donovan's neophyte crew, since these were glorious amateurs, who were talented, eager, daring, and innovative, and most importantly, were in the forefront of new approaches to intelligence operations and unconventional warfare.

Donovan's vision of unconventional warfare, encouraged by the British, was broad and bold. He wanted to carry the war to the enemy right away and behind their lines in weak spots in occupied territory. Initially, he planned a combined centralized intelligence and subversive operations agency that would include more than gathering and coordinating intelligence and staging guerrilla and commando operations behind enemy lines. It would also use information and technology, especially radio, as weapons. Foreign radio broadcasts would be beamed at Allied, neutral, and enemy-occupied countries with news of the positive efforts and achievements of the Allies and negative, disinformation ("black propaganda") to undermine the morale of the enemy forces and civilian population. Donovan lost the positive propaganda entity in a bureaucratic battle to the Office of War Information, but he kept the black propaganda aspect, which became the domain of OSS Morale Operations Branch (MO). The centralized gathering and analysis came from the spies of the Secret Intelligence Branch (SI) and the rings of local agents they would recruit and run, and from one of Donovan's primary innovations, the Research and Analysis Branch (R&A), the scholars and others who used the foreign language newspapers, economic and political reports, and other published material in the Library of Congress as well as material obtained from agents overseas to provide comprehensive assessments of key industrial, political, and military targets for Allied bombers, commandos, or saboteurs.

The concept of deploying commandos, saboteurs, and guerrilla leaders behind enemy lines assumed organizational form in the Special Operations Branch and the Operational Groups. Despite considerable support from President Roosevelt and a number of influential friends among economic, political, and social elites, Donovan had his enemies. The Wall Street lawyer and his organization of amateur soldiers, spies, and intelligence analysts, raised hackles among professionals in established and competing agencies, including especially the Military Intelligence Service, the Office of Naval Intelligence, the FBI and the State Department. Donovan had originally envisioned the agency providing primarily centralized strategic intelligence to various clients from the President himself to the Joint Chiefs of Staff (JCS), to particular military and civilian departments. He also hoped to have saboteurs and guerrilla leaders, and British type military commando units which he daringly hoped he would sometimes be able to lead personally on raids. But while the President and occasionally the JCS valued the intelligence that SI and especially R&A provided during the war, it became clear by 1943 that what some military theater commanders wanted more from OSS was tactical intelligence about the enemy forces deployed against them that could be used immediately. That involved running rings of agents near the battle zone. The U.S military developed its own commando-like units—Army Ranger units, Navy Underwater Demolition Teams, and Marine Raider battalions, primarily for short-range penetrations, spearheading advances. The armed services limited Donovan's Special Operations and Operational Groups mainly to deep penetration, working with partisan resistance groups far behind enemy lines. Thus, the missions Donovan's organization had originally conceived of and trained for were altered somewhat during the course of the war.

OSS training also evolved, but much more slowly. Training methods for these paramilitary forces came originally from the British Special Operations Executive (SOE) forces, which provided instructors, manuals, equipment and the aura of having already conducted operations behind enemy lines. The first American special operations instructors were trained at British SOE's secret Camp X at Oshawa, near Toronto, which one of them referred to as the "Oshawa School of Mayhem and Murder."[6] They, like most of Donovan's uniformed personnel, were citizen soldiers at that time rather than career soldiers, often they were reserve officers. Some were military police officers, some civilian law enforcement officers, some, particularly in the case of demolitions instructors, were engineering officers. The influence of the law enforcement officers/instructors quickly waned as it became clear that their orientation had been towards apprehending law breakers, while the OSS/SO curriculum was designed to teach trainees how to create damage and avoid being caught by local police or military forces. The British emphasis, carried over to OSS, on extreme secrecy and the "cloak and dagger" aspects of training, also seem to have become less important as time went on, and although not abandoned, they were de-emphasized in contrast to the increasing importance on practical techniques of accomplishing the mission whether espionage, sabotage, commando operations or guerrilla leadership.

Charismatic and visionary, William J. Donovan, more than anyone else, was responsible for creating America's first central intelligence agency, and through his Special Operations teams and Operational Groups, he was a major progenitor of the Special Forces. Yet, he was an abysmal administrator. Uninterested and perhaps unable to manage a growing organization that had so many different missions and branches, Donovan frequently fled to the war zones and left the daily management to others. He built the organization by recruiting intelligent, able, and innovative people and then largely letting them find places for themselves. The branches essentially operated autonomously. "I ended up disliking Donovan," recalled H. Stuart Hughes, Harvard trained historian and grandson of 1916 Presidential candidate Charles Evans Hughes, who worked in Research and Analysis Branch. "He was, I think, responsible for a certain wild style of administration and the sense that everything was chaotic. I remember that Sherman Kent [Yale historian and head of European Division of R&A] at some point had been reading Shakespeare. He found the passage, 'Confusion now has made his masterpiece.' He laughed and said, 'That's us!' "[7]

It was in response to such a haphazard arrangement and the problems of building a training program at the same time that OSS itself was evolving that the Schools and Training (S&T) Branch was formed in the winter of 1942-1943. The S&T Branch spent the rest of the war seeking to coordinate and to the best of its ability to standardize at least some of the training policies among the schools and training camps of the various operational branches—especially the often competing Secret Intelligence and Special Operations branches. S&T never did completely control them, and the branches remained the dominant influences on their trainees throughout the war. Indeed, they remained more attuned to evolving developments in the war due to their own agents in the combat zones. Although Schools and Training Branch had official authority over the instructional program, including the training schools, the curriculum, written and visual teaching materials, and the staff and instructors at the training camps, most of the staff and instructors came from the operational branches. Their loyalty remained with their branches, and most of them sought to be assigned overseas. It was not until near the end of the war that Schools and Training Branch obtained authority over the training camps that the various operational branches had established overseas, and even there, S&T had difficulty imposing its will. In practice, Schools and Training Branch served more as a managerial agency—overseeing and allocating among the training camps—than the key instructional agency. As instruction became less general and more specialized, it derived largely from the operational branches themselves. Overseas, the training camps were dominated by their regional detachments.

The trend in instruction over the course of the war moved from more generalized training in the early days, when it was unclear how individual recruits would ultimately be used, toward more specific training aimed at particular types or locations of missions. Because of the pressure to produce agents, the basic courses in both SI and SO were three to four weeks of intensive training. Graduates then went on to advanced and more specialized courses.[8] Yet attempts by SI and SO to tailor training of individual students to their future missions, were generally fruitless, in part because the area "desks" at OSS headquarters often did not know the missions of particular individuals in advance. So there always remained general aspects to the training. They deliberately included the kind of physical and intellectual demands designed to test the individuals and weed out those unsuited either physically or emotionally for the demands of operations behind enemy lines. These physical and mental demands were also designed to create in those who graduated as members of the OSS operational branches, a sense of self-confidence, élan, and belief in themselves, their ability, and the mission of their elite organization. The OSS paramilitary training, as in other elite military organizations—rangers, paratroopers, Marines—was in part designed to impart the proud, can-do spirit of an extraordinary organization.

Aggressive physical toughening had greater emphasis in the paramilitary training camps than in the more subtle training schools of Secret Intelligence, Counter-Intelligence, and Morale Operations. The men and women of the latter three branches were often older and civilians, in contrast to the trainees in SO, OG, and CB who were required to have be in the armed services. All of the branches learned some basic aspects of the others' skills, but the training that occurred in the two National Parks, was primarily geared to SO, OG, and "Commo" work. (The Communications Branch was a technical service, and its training course for its own personnel, required a mastery of OSS's specialized equipment, codes, and high-speed wireless transmission. It course for its personnel generally lasted three months.[9]) In addition to the physical toughening, the training courses at the two National Parks included a mastery of weapons. Most of the military recruits had already received basic training in the armed services. OSS trainees had to achieve a level of proficiency far beyond the standard Army training. They had to learn to operate and maintain not only a variety of standard American weapons but also various weapons from Allied or enemy countries. They learned to use specialized OSS weaponry—knives, grenades, pistols, rifle and submachine guns, some with silencers. To bolster their confidence, overcome combat fear, and simply give them skills to survive in the war's killing zones, they learned quick and effective means of pistol shooting (the "instinctive" method of firing off pairs of shots from the hip) as well as a hundred ways of disabling an enemy in unarmed combat using jiu-jitsu, kick-boxing, karate, and other forms of martial arts.

The OSS schools taught other skills as well. For sabotage, the students learned about various forms of explosives, including the new malleable but stable and highly explosive "plastic" compounds. They studied how to use such demolitions to destroy, railroad tracks, trains, bridges, tunnels, supply depots, industrial plants. For intelligence gathering, they gained knowledge about how to identify enemy units by their particular insignia, what to look for in military or industrial facilities. They were taught how to obtain and direct rings of indigenous agents. For guerrilla leadership, they learned how to recruit and work with local guerrilla resistance groups, how to train, lead, and supply them. SO and OG trainees practiced raids against simulated enemy outposts, power plants or bridges. The students were taught how to create miniature cameras out of matchboxes, how to sketch particular facilities, how to operate one of the wireless, radio/telegraph sets carried in what looked like a regular suitcase. They learned learn how to maintain cover even if captured, how to resist interrogation, and, if necessary, how to break the coated cyanide pill (the "L" for lethal pill they carried) in their mouth before revealing the names and locations of other agents or other vital information. "They gave us three [kinds of] pills," said George Maddock, a member of an OSS team that jumped into southern France in 1933, "one to give us energy, one to wake us up, and another one to kill us in case we were captured."[10]

Organizationally, the OSS personnel who ran the training camps were divided into two staffs: one for administration and maintenance of the camp and the other for instruction. A commanding officer was put in overall charge of the camp, but he dealt almost as an equal with the chief instructor. As with the vast majority of uniformed personnel in Donovan's hastily built organization, most of the men who staffed the paramilitary camps, as well as the OSS recruits who trained there, had previously been civilians. Donovan and his chief subordinates were successful business and professional people, and they recruited men and women who showed initiative, imagination, intelligence and adaptability, people who could think imaginatively, "outside the box." They also wanted people who were reliable, and so they frequently counted upon personal connections and background for recruiting, particularly those who would become commissioned officers. This personal network contributed to the OSS's reputation for being filled with socialites, of being "Oh-So-Social." Although there was some truth to this as, the presence of Vanderbilts, Morgans, Whitneys, Mellons, and the like in the upper ranks attested, the vast majority of men and women who worked for the OSS came from the college-educated middle class. Some of the rank and file, especially those recruited from among the draftees and volunteers in the enlisted ranks of the military, came from the high-school educated, working class. What most of them had in common was that they scored high on intelligence tests and had already showed considerable ability and initiative. Many of them were adept in at least one foreign language. Those in the Communications Branch generally had some prior radio or telegraphy experience, a good number were short-wave radio hobbyists, known as "Hams." With a few exceptions, most of the members of the OSS were not career military people. Even those in uniform in Special Operations, Operational Groups, and the Maritime Unit had generally been civilians who had became part of the armed forces only because of the war. On the whole, the regular military establishment was leery of Donovan and what it considered his free-wheeling, improvised group of amateurs. With its quasi-civilian status and its notorious lack of attention in its military branches to standard Army protocol and discipline, the OSS was indeed a most unmilitary military.

Although the training camps at Areas A, B, and C, at Catoctin Mountain Park and Prince William Forest Park were organized as military detachments and were filled with uniformed personnel, both staff and trainees, they were most unmilitary in their decorum. There was no saluting or marching and few distinctions between officers and enlisted men. An atmosphere of informality and individual self-responsibility rather than ceremony and formal discipline pervaded the OSS and the training camps as well. The uniforms, weaponry, munitions, and tactical problems may have been military, but the emphasis was not on following orders but on individual skill, initiative, and imagination to achieve success in the mission.

Critiques of Training

Initially, Donovan's organization received advice, teaching aids, equipment, and even some instructors, from the United Kingdom, but it had its differences with its British counterparts. These differences involved both the OSS's organization, which included both intelligence and special operations, and in its goal, which was only to defeat the Axis, not to restore the British Empire. OSS had its own American missions and style. It was geared toward Americans not Englishmen, as the informality and lack of military discipline illustrated. Starting with the British model, the Americans gradually developed their own training system, evolving both by plan and by trial and error, primarily learning by doing. It was a new organization feeling its way along. In general, its training was effective in one of its major goals:

preparing agents psychologically, physically and to respond rapidly and appropriately to unpredictable situations. Nevertheless, there were issues that needed to be resolved.

As the OSS expanded during the rapid American mobilization of 1942-43, it faced the fact that some of the recruits who volunteered for overseas operations proved to be unfit for the physical and emotional demands. An elite organization, emphasizing heroism and hazardous duty attracted volunteers who craved the excitement and glory. But some of such volunteers lacked the emotional stability or the physical stamina for dangerous service behind enemy lines. Instructors tried to identify and weed out such characters and many trainees were dismissed and sent back to their armed forces. But some of the unstable got through training and were dispatched overseas before their unsuitability was discovered. Consequently, Donovan's office in 1944 initiated a major new psychological program to assess candidates for overseas duty even before they began their training.

The psychological assessment program, as it ultimately evolved, proved remarkably effective. In 1942 and 1943, many OSS recruits had found the interviews with psychologists perplexing and even a waste of time. As one student reported in 1942, he and the other students at Area B were "somewhat bewildered and made uncomfortable by our interviews with the psychological staff. The questionnaires given out by these men seemed pointless and naïve to us all."[11] Two years later, a radio-operator recruit had the same kind of senselessness after being interviewed for less than a minute in the psychologist's darkened tent in Area A. The psychologist waived a little pencil flashlight around, "asked a few things: where you were born, what you're interested in, and various others things. One question he asked me: 'Why do you wear your sideburns so long?' I said, 'I didn't know they were that long.' That was the end of it....It was strange, a little disorienting."[12]

By 1944, the OSS had expanded and perfected its assessment techniques. It established an Assessment Center, Station S, in a country estate in Fairfax County, Virginia. There recruits were held and observed through a series of written and verbal tests and practical field exercises. Over a three day period, the potential agents for dangerous overseas missions were observed as they worked, played, talked and went through three dozen lifelike situation tests. In the last twenty months of the war, OSS teams of leading psychologists and psychiatrists, using radical methods and working in secrecy, developed a novel and successful method for assessing personalities and predicting an individual's performance on the kind of unpredictable situations prospective agents would face in the field. They employed simulations and situational exercises to identify and evaluate knowledge, behavioral traits, skills, competencies and weaknesses. According to an OSS report, the assessment program succeeded in "screening out the 15-20 percent who were obviously unfit."[13] The evaluation teams learned that beyond the specific skills and training, what makes an effective saboteur in France, an able spy in Germany, a good commando in Burma, a reliable undercover radio operator in China was a secure, capable, intelligent and creative person who can deal effectively with uncertainty and considerable stress. The effectiveness of the OSS's predictability with reasonable accuracy based on their assessment performance charts contributed to the success of the OSS. It also contributed to the postwar publication of the technique and its adoption by other government agencies as well as a number of corporations. It is still being used today.[14]

There were other gaps and difficulties along the way, some of which were quickly addressed and some of which were not so readily resolved. Francis ("Frank") Mills, a major in the field artillery, arrived at the OSS training camps outside the nation's capital in 1943 and could mainly recall the self-defense and silent killing instruction by the famous British expert, Colonel Fairbairn. All OSS trainees who saw him remembered the extraordinary skills of that otherwise unassuming, bespectacled, older Englishman. Mills said his group did not receive any training in intelligence gathering activity. "We were in special operations, fighting with the sympathetic forces behind enemy lines. We knew that," he said. "[But] we were given very marginal, almost no real training in how guerrillas were supposed to operate. So, we were given what little training the Army or OSS had to offer."[15] Erasmus ("Ras") Kloman, who entered OSS as a Princeton graduate and a young lieutenant, recalled a number of problems in Special Operations training in the winter of 1943-44. Most importantly, it was never clear what his mission would be, and thus the training could not be matched to it. At first based on his knowledge of French, he was assigned to training as a SO agent who would be parachuted into occupied France. But that assignment was changed to SO in Yugoslavia. When he actually arrived overseas, he was sent neither to Yugoslavia or France but was given a series of administrative assignments in Egypt, Algeria, and Italy. A lot of his training in particular skills, for example a couple of days of Morse code, half a day of lock-picking instruction, he considered too brief to be adequate. He considered it "a little bit of this and a little bit of that in case it might come in handy someday."[16]

In fact, a major complaint by many trainees and indeed by officials in the Schools and Training Branch, was that neither the students nor S&T knew, particularly in 1942-43, what kind of mission the operational branches and their regional desks had planned for particular students. Thus it was not clear to the students how any given topic related to their future mission, if at all, and the instructors at did not know either. (This was not true, of course, for the foreign trainees, the Yugoslavs, Norwegians, Thais, and Koreans, for example, who knew they would be infiltrated back into their home countries.) One response by Schools and Training was to establish a more generic form of training for the American trainees. Although Kloman was rescheduled to be sent to Yugoslavia, he had never been given instruction in Serbo-Croatian languages, nor had he been briefed on the political and military situations in that country. His superiors said that everything would become clear when he reached the Yugoslav desk in Cairo. "I supposed," he wrote later, "it was assumed I would pick this up once I went abroad."[17] Very much concerned about this problem, Schools and Training Branch did seek to link the regional desks in particular OSS operational branches—especially Special Operations, Secret Intelligence, Operational Groups—with particular individuals and groups of students, the better to gear their instruction toward their ultimate missions for OSS. Another problem, albeit one that conflicted with the desire for secrecy, was that especially in the early years, students were ignorant about the overall organization of the OSS and its various and sometimes competing branches.[18] Schools and Training Branch did later add an introductory course to give students a sense of where they fit in the larger organization and how the different branches could compliment and work in support of each other.

OSS tried to make the training as realistic as possible, despite the fact that the exact situations agents would face in the field could not always be foreseen, and in any event, many of those situations could not be adequately duplicated in the camps. Firing at a cardboard target was not the same as shooting at an enemy who was trying to kill you. Instructors tried to increase the realism by using live ammunition and explosives. They designed a rigorous obstacle course with small explosives set off by trip wires. They forced students to crawl under barbed wire with machine gun bullets zipping over their heads. Fairbairn built a mystery, pistol house, or "house of horrors" as it was called by the students, at Areas A and B. Students would be awakened in the middle of the night, given a pistol and ordered to kick in the door to the mystery house and rush though its darkened rooms and corridors, responding instantly and accurately with their Colt .45 to suddenly illuminated enemy mannequins and pop-up silhouettes of German soldiers. In addition, Fairbairn, who had mastered jujitsu, judo, knife-fighting, taught awed trainees what one of them recalled were "100 Ways to Kill a Person without Firing a Shot."[19]

The most frequent complaint of the students was of being "held" too long after completing their training. When they graduated, they were at their peak of enthusiasm and self-confidence and ready for their mission overseas, but OSS then confronted the problems of obtaining space on ships and planes going abroad. Priorities were lost among a welter of inter-service rivalries, bureaucratic confusion and the overall demands of logistics upon an already overburdened global transportation system supplying America's armed forces. The new graduates were frustrated by the endless delays, and their enthusiasm and readiness eroded the longer they remained unassigned after graduation. Schools and Training officials tried to remedy this by sending them to additional courses, if there was space for them, or letting them go on leave, but sometimes they were kept in camps that were not at full capacity at the time.

Making Training Realistic

Few of the OSS instructors in the Stateside training camps had any actual combat experience, at least until late in the war, and this was worrisome. As an espionage or morale operations student at Area A complained after graduation in 1942, "with the exception of Capt. White [from the Federal Bureau of Narcotics], no single instructor had any major experience with undercover work. Consequently, the lectures seemed rather lifeless. As a graduate of this course, I still have no idea of how to deal with 'black market' operations, false entry, financial operations, or any of the present day operational problems."[20] To ameliorate this inexperience, several instructors were sent to Great Britain in the early fall of 1942 to gain firsthand experience at the British schools staffed by instructors, some of whom had worked behind enemy lines. Lieutenant Frank Gleason, a demolitions instructor at Area B in Catoctin, attended an industrial sabotage school in England for two months and learned how to blow up steam turbines, power plants, and factories. "When I left, I was a trained terrorist," he recalled in 2005, "but in a worthy cause!"[21]

Initial American instructors subsequently gained combat experience in the field, but they seldom returned to the United States as instructors. Some who had been overseas at British schools, like Frank Gleason and Charles Parkin in late 1942, or combat veterans, like Carl Eifler and Allen Richter from Detachment 101 in Burma in 1944, did give guest lectures upon their return.[22] It should be noted that most of the U.S. Army's officer instructors in the first years of the war lacked combat experience, although some of the old-time NCOs had seen combat in World War I.

In place of the general lack of experience by instructors, OSS sought, as the war went on, to incorporate lessons its agents derived from experience in the field and apply them to the curriculum in the training schools. But generally the field agents were too busy to write reports on recommendations for further training back home. Operational branch officers declined S&T's requests for copies of reports from overseas units as breaches of security and useless extra work.[23] This was a slow process for S&T, and apparently the lessons could be implemented into the curriculum more rapidly by the operational branches themselves (agents' field experiences relayed directly via branch headquarters to instructors) than by the more pedagogically oriented and centralized Schools and Training Branch. The training at the stateside camps was seen as a form of basic and mid-level advanced OSS training. For more advanced, specialized training, including instruction from veterans of the combat theaters, OSS first relied upon British SOE schools, and subsequently on overseas training schools established by the OSS operational branches themselves. SO and SI, for example, set up schools in North Africa, Italy, England, and China. In 1944, Schools and Training Branch was given official authority over these overseas OSS schools, and it then sought to coordinate OSS training at home and abroad. Some members of OSS in England and in China argued by late 1944 that S&T's role in the United States be limited to assessment screening and providing general indoctrination and basic military training. They argued that recruits would benefit from then being sent to specialized finishing schools overseas where they would be immersed in conditions in that theater of operations and brought into direct contact with operatives from the field.[24] This was not adopted, although Schools and Training did assume organizational responsibility for the overseas training schools that the operational branches had established.

OSS, like the Regular Army, was developing new curricula and training manuals to meet the new forms of warfare and to use the new weapons, munitions, and equipment, such as plastic explosives, a variety of weapons with silencers, bazookas, suitcase-size wireless transmitters and receivers. The OSS syllabuses and manuals were clear about the initial aims and methods of training in Special Operations and Communications in the training areas in Prince William Forest Park and Catoctin Mountain Park. The initial part of the training was to provide both physical conditioning as well as a sense of self-confidence and spirit in the organization and its purpose. It was also designed to weed out the unfit. For those who remained, it was to provide them with elementary skills in most of the areas they would need, plus advanced skills in their specialization. OSS did modify and adapt the curriculum in light of what its operatives learned in the field in the combat zones.

"The overall layout of training by OSS was really good," concluded Allen R. Richter, who was part of the initial Detachment 101 communications contingent. "When we got overseas to Assam [India], we followed the same ideas. We would get our recruits and keep them together, but separated from the others, which meant they would sleep, eat and train there in their own little compounds. The advantage of that would be that everyone was doing their own thing, and not mixing demolition with radio and other activities--specialization. We copied Area C overseas." Richter recalled that at least from 1942 to late 1944 when he returned to the United States, the specialized training program of Detachment 101had been influenced by the Communications Branch back home and its training school at Area C, not by Schools and Training Branch, the umbrella training organization of OSS. "We had nothing to do with Schools and Training [Branch]," in India, Richter concluded.[25]

Problems of Schools and Training Branch

Back home, the Schools and Training Branch suffered its own problems. "Someone recently likened Schools and Training to an island of ignorance with darkness on both sides of it," bemoaned the new chief of S&T in October 1943, Lieutenant Colonel Henson L. Robinson. "We are trying to run a group of schools without knowing anything about the number of students we must train, the type of missions our students will have, or what happens to them after they get to their eventual destinations."[26] In a lengthy report, Colonel Robinson included some examples of what led to S&T's frustration:

> We are suddenly informed by one of the [Operational] Branches that next Monday there will be 80 students to be trained for a very special mission; who must be kept segregated in a separate area; who will have to have special training in demolitions along with some other subjects that have not yet been decided upon; and a request that we rig up some models of various power plants, etc., for these students to play with.
>
> Or, we are told a large group of Japanese, Thailanders, or Balkans [Yugoslavs] may be expected week after next and must be put in a separate area. A group now in process of formation is a good illustration. After various meetings, in none of which was any representative of Schools and Training included, a plan was evolved. Somebody was to recruit a hundred officers and fifty wireless operators. Operational Groups agreed to furnish some of their officers to give the group a short course in demolitions and small arms.
>
> Communications agreed to furnish some [telegraph] key sets and a few instructors to train the wireless operators. Quite by accident, later, we were told that we might expect to have 150 people suddenly dumped on our hands and it was up to us to find some place to put them. We tentatively agreed that,

if and when the plan matured, we would put the group in Area F. Without further warning or advance notice, about 120 officers and men arrived at Area F, bag and baggage.... The camp commander suddenly was confronted with the necessity of feeding and housing 120 people for whom he had drawn no rations or prepared any accommodations. He complained, justly, and we complained vociferously.... So far we have received nothing.[27]

Schools and Training wanted to be involved from the inception of plans that could involve its training camps. Its leadership also desired reports on the successes or failures of the former trainees in actual operations abroad or lectures by returning field veterans to instructors in the training camps in the United States so that training could be adjusted and improved to reflect actual conditions in the field.[28] Since it could not obtain such branch reports, S&T sought similar information on its own, conducting a series of interviews with OSS operatives returning from overseas in 1944-1945.[29] These individuals had numerous suggestions for S&T's instructors. In 1949, the CIA summarized a number of them as it built its own training program, modeled largely on that of the OSS. Many of the returning OSS veterans in 1944-1945, had contended that OSS as a whole and the training schools in particular put too much emphasis on what Major Peter Dewey, returning from France before his assignment to Indochina, called "too much 'cloak and dagger' creepiness in the training." Dewey advised that the training "approach should be more matter-of-fact."[30] A number of the field veterans complained that there was too little training in observing and reporting compared to cover and security. "Discipline, power of observation, military perspicacity, and common sense are the sine qua non of life behind the lines," reported an SI agent from Greece."[31] Different agents sometimes offered opposite views of the same issue. One SO instructor in Ceylon declared "natives being trained as [special operations] operatives must be treated with friendliness and respect. There is no other way." But an SI agent from a neutral European country stated flatly *"Never* trust a man the first time," and the chief organizer of a sabotage team warned that "friendly elements in the police can supply information of great value, but in nine out of ten cases the friendly policeman is a dangerous agent provocateur."[32] Despite S&T's efforts, the operational branches continued throughout the war to view the Schools and Training Branch merely as a support unit to provide instruction facilities for them as needed. Although S&T was given some additional authority, the operational branches remained predominant in operations and in the training of their agents throughout the war.

Value of OSS Training

Many of the American agents overseas attributed their success at least in part to the value of what they had learned at the OSS training camps in the United States. They credited their achievements to the physical training, specific skills and techniques, and the self-confidence and faith in themselves and the organization, and the value of their mission. Not surprisingly, those who stayed in the armed forces or later joined the CIA continued to draw upon and replicate techniques from the OSS training camps.

Major General John K. ("Jack") Singlaub who as a young Jedburgh had trained at Areas B and F and then SOE schools in Britain, served with distinction in France and China, and after the war wound up his career by commanding all U.S. Army troops in South Korea, reflected in 1996 on what he had learned in his OSS training. "These were individual skills that are perhaps useful but are most important for training the state of mind or attitude, developing an aggressiveness and confidence in one's ability to use weapons," he said. "One of the most important aspects of the training was that it gave you complete confidence." By the time he and his colleagues jumped into France in 1944, Singlaub said, "we had complete confidence that we could survive if we had a weapon. We were good. I mean, we hit targets in very dimly lighted places.... We were taught this 'instinctive fire'.... [All of] that gave you an ability to concentrate on your mission, and not worry about your personal safety. That's really a great psychological advantage. I used that later in training my units when I was a battalion commander and later, a Battle Group commander."[33]

After the war, Robert R. Kehoe was employed by the CIA's Office of Training and Education. The young New Jersey native had been a Jedburgh team radio operator in France after completing Commo training at Area C and SO training at Area B, and SOE instruction in Britain. "The experience at Area B-2 was a great morale builder," he said later, "and when we departed in mid-December [1943], we were in top physical condition."[34] He incorporated much of OSS training for the CIA.

Relating his personal experiences in a postwar memoir, Lieutenant Jerry Sage, who had spent more than two years in German POW camps after being captured in North Africa in February 1943, emphasized the importance of what he had learned in the OSS training camps, particularly Area B, where he learned while also instructing. Using techniques he learned at the OSS training school, he had escaped half a dozen times, but each time was recaptured in Germany. One of Sage's most stressful moments and one in which, he said, his OSS training came to his rescue occurred in spring 1944, when he was brought back to *Stalag Luft III* after having been caught and beaten by the Gestapo. He was soon confronted by the irate camp commandant, a rather old colonel, under pressure from his Air Force superiors and the Gestapo to prevent any more escapes. As Sage recalled, *Kommandant* von Lindeiner went into a rage and pulled his pistol out of its holster, his hands shaking. "I'd learned from Dan Fairbairn that nobody is dangerous who just tells you to put your hands up and holds his pistol firmly on you," Sage wrote later. "You can finally trick him and get close enough to disarm him in a number of ways. I knew how to do that.... I stood up slowly, fixed him with my eyes, and walked very gently toward him—with no threat and no bombast. Very quietly and calmly I said, 'Be reasonable.' This was in my poor German but he understood me. 'You would never forgive yourself, if you killed an unarmed man like this.'" It worked, the tension of the moment was broken, and the commandant went back to his office.[35]

At *Stalag Luft III*, where he was part of the plan for what became known as the "Great Escape," rumors spread in 1944 that the Nazis might kill the POWs if the Allies reached Germany. The senior American officer asked Sage to train a hand-picked group of men to try to seize the camp if the Germans started such an operation, or at least to avoid being killed without a fight. Sage drew upon Fairbairn's instructions on "silent killing," the dispatching of sentries with knives, other instruments, or bare hands, to train a selected group of his fellow prisoners to take over the POW camp in case the Germans started "liquidation proceedings." That did not happen, but Sage did escape successfully in January 1945 from a German POW camp in Poland, returning home via the Ukraine and Egypt.[36]

OSS training was equally effective in the Far East according to many veterans who served there. After a tour of duty as a demolitions instructor at Area B from 1942 to early 1943, Lieutenant Frank Gleason, SO, was sent to China. There he helped instruct Chinese commandos and he personally helped impede a Japanese advance by blowing up bridges and several warehouses of stored weapons and munitions to keep them from falling into enemy hands. After a successful postwar career in the Army Corps of Engineers, he retired as a full colonel. Asked how effective OSS training had been in China, Gleason asserted, "It was very effective. We blew those bridges. We did it with what we learned in our training here and in England. In China, we had classes where I taught Chinese how to destroy mechanical equipment. Joe Lazarsky used it against Japanese in China and with [Ray] Peers in the jungles of Burma. I felt fully prepared....Most of the students who graduated from the OSS training camps in Maryland and Virginia thought highly of their preparation there."[37]

Lazarsky, who later spent a career with the CIA, concurred in regard to OSS training. "The training in weaponry and demolitions was effective. So was building self-confidence and the ability to get things done." Lazarsky had also used such training to prepare indigenous agents in the Far East. "It was very effective [training]," he said. "If you debrief a Thai agent, they would tell you that. Even after the war, they would say thank you. [One of them said] 'You know what you and Leo [Karwaski] taught me about demolitions—we could *not* have gotten that anywhere else.'"[38]

"Training is not spectacular work," Schools and Training Branch acknowledged in its typewritten history. "It means doing a sound teaching job, adjusting sights to fit circumstances, and keeping right on doing it." Certainly there were some brilliant instructors who spiced the programs with their personalities and operating experiences, "but the bulk of the work was done by hundreds of lesser known instructors and administrators who stuck to the grind, class after class."[39] Operating like the OSS itself which was created in haste and without American precedent and which was impelled with a tremendous drive for speed, production, and results, the Schools and Training Branch sometimes appeared confused and indecisive, as S&T acknowledged. Yet, training areas and programs were indeed developed almost overnight to fit the evolving needs of Donovan's organization and other wartime developments. To meet suddenly increased quotas, the capacity of training areas was from time to time doubled in size, sometimes by putting new sub-camps into operation, sometimes with the creation of "tent cities" to accommodate additional students. Yet, Schools and Training also admitted that "only toward the end of World War II was OSS beginning to approach the kind of training that was really adequate for the complex and hazardous operations carried out by OSS personnel."[40]

Size of the OSS and the Task of Training

At its peak in December 1944, OSS included 12,974 uniformed and civilian personnel worldwide. This included nearly 8,500 men and 4,500 women; approximately 7,500 of these (including 900 women) served overseas.[41] Intelligence branches composed 26.8 per cent (3,484 persons) of the total. Operations, including the OGs, made up 23.7 per cent of the total. Miscellaneous units comprised 22.8 per cent. For some reason, the Communications Branch was listed within the Miscellaneous Category. It was the largest segment of that category. Communications Branch personnel on December 31, 1944, numbered 1,728 persons and represented 13.2 per cent of total OSS personnel. Administrative Services [support services: including Research and Development, to Security, Special Funds, Medical Services, Procurement, as well as other branches, including Schools and Training] comprised 16.5 per cent of the OSS.[42] Of the nearly 13,000 members of the OSS, approximately 4,000 were civilians and some 9,000 were uniformed personnel.[43]

In summarizing Schools and Training's achievements, the branch's postwar history emphasized the numbers that the organization had handled in the last two years of the war. Between January 1944 and the end of the war, the Assessment Stations screened and evaluated 5,300 candidates, the Basic Espionage Schools graduated more than 1,800 operatives, and the Advanced School at RTU-11 ("the Farm") graduated 800 men and women; the Special Operations Schools trained 1,027 men.[44] These figures did not include the trainees in 1942-1943, nor did they incorporate the numbers of trainees in the specialized groups over which Schools and Training had divided or little control, such as the training of military recruits for the Maritime Unit, the Operational Groups and the Communications Branch. S&T's official historians concluded that "like the other branches of OSS, though falling far short of perfection, Schools and Training on the balance somehow accomplished a creditable task. Men were trained and sent against the enemy. Men did accomplish results that substantially contributed to the war effort."[45]

CIA adopts OSS Training

Effectiveness of the OSS training was confirmed by the fact that its successors, the CIA and Army Special Forces, adopted much of it. "The [Central Intelligence] Agency picked it up almost 100 per cent," explained Lazarsky, who subsequently spent twenty-five years with the CIA. "They took the manuals, instructional materials, and that right into the Agency. You know, the COI [Office of the Coordinator of Information] and the OSS started it from scratch. The Agency would have been foolish not to have adopted their training."[46] Indeed, William R. ("Ray") Peers, who as a young lieutenant had trained at Area B in spring 1942 before leaving for the jungles of Burma as one of the early leaders of Detachment 101 in Burma, later served in Taiwan as chief of a CIA program for training Chinese agents to be infiltrated into mainland China, 1949-1951.[47]

Although former OSSer Frank Wisner's covert operations office was the driving force within CIA for its first few years as well as one of the main recruiters of former OSS SO personnel and OSS training methods, when former Army General Walter Bedell ("Beetle") Smith became Director of Central Intelligence in 1950, he began to emphasize intelligence gathering and analysis. Smith established a relationship of confidence and trust with Truman similar to that of Donovan and Roosevelt. He quickly recruited former OSSers, some from Secret Intelligence but mostly from Research and Analysis to prepare the basis for what became the national intelligence estimate that the DCI would present to the President. Called back William Langer, former OSS chief of R&A, from Harvard, and Langer recruited a number of former OSS staffers to assist him. Ray Cline, had been a young Harvard graduate when OSS enlisted him in 1943 for R&A's Current Intelligence office. In 1950, he became the CIA's the first chief of the new Estimates Staff. The National Estimates Board members included several former OSSers: Langer; Calvin Hoover a Duke University expert on the Soviet Economy and a former member of OSS Secret Intelligence; Sherman Kent of Yale, who been with Langer at R&A; and a number of non-OSS veterans. Kent was reluctant to leave Yale to join the National Estimates Board, and Cline later recalled talking to him in 1950 at the temporary CIA headquarters in an old OSS building, across a scarred old wooden desk inherited from OSS. "I told him that so few people in the new CIA knew what intelligence analysis was all about and such threatening situations existed in the world that he was needed." "I do not know whether this influenced him," Cline said, "but he came, stayed, and [as Langer's successor] built the National Intelligence Estimates into a significant element in decision-making."[48] General Smith also brought in as deputy director of CIA Allen Dulles, former OSS Secret Intelligence chief in Switzerland, who had resumed a law practice but also maintained his Washington connections. Dulles would replace Smith in 1953 and serve as Director of Central Intelligence until 1961.

For the indoctrination and initial training of field agents, CIA has continued to rely in part upon OSS paramilitary style training to evaluate recruits and build self-confidence and élan as much as imparting usable skills. While the agency first relied upon Army bases, by the early 1950s it established its own top-secret, 10,000-acre paramilitary training facility at Camp Peary in the woods near Williamsburg, Virginia. It continues the rough and tumble type of OSS special operations training there to the present day. But unlike the exclusively male trainees at the rugged OSS Special Operations training camps in the National Parks in World War II, there are now women as well as men engaged in military-style training and simulated Special Operations exercises at the CIA's "boot camp" that is known in agency variously as "The Farm," "Isolation," and "Camp Swampy."[49]

Valerie Plame Wilson, a CIA covert operations officer, became famous when officials in the administration of President George W. Bush blew her cover after her diplomat husband challenged a key rationale they had put forward for the 2003 U.S. invasion of Iraq. She opened her best-selling 2007 memoir with a description of the paramilitary exercises she had participated in as a 22-year-old trainee for the CIA in 1985. In the climactic field exercise, her team of three male and two female trainees, each carrying an eighty-pound backpack containing survival materials and ammunition and each toting an M-16 automatic rifle, spent a clammy late fall night practicing what she called "escape and evasion from an ostensible hostile force—our instructors." At dawn, they linked up with another group of trainees at the designated landing zone, but soon found themselves under simulated attack by hostile forces. Magnesium flares exploded around them amidst the sound of machine gun fire and the noise of exploding artillery shells. Adrenaline flowing, M-16s blazing, they rushed to the helicopter, which whisked them off to safety.[50] Earlier, Plame Wilson's CIA training had included personality tests and stress tests, many of them derived from the OSS, an introductory course providing an overview of the organization, more tests and courses, and most appealing to the CIA students as to their OSS predecessors, talks by case officers about their direct experiences in the field. Without wartime pressures, CIA provided a much longer training period than OSS. After three months of introductory training, the future intelligence analysts and operational case officers were assigned as "interims" in various departments, after which they were sent to a three-month, military-style course at the "Farm." It was tough and demanding and, according to Plame Wilson, although "the Agency clearly understood that we were rarely, if ever, going to be called upon to use these skills" the managers maintained the paramilitary course because it "forged an esprit de corps that would last throughout one's career" and it provided yet another chance for the Agency to assess "a new employee's strength of character, ability to work in a team, and dedication—all skills critical to success in the Agency, no matter what your career path."[51]

From the beginning, the CIA had also adopted the OSS's communication system. "The agency kept on the OSS radio training and equipment," Joseph Lazarsky stated firmly.[52] But it was even more than that. Looking back on the antecedents of the Agency, Ray Cline declared in 1976 that "one of Donovan's lasting achievements for central intelligence was securing the right of independent encrypted radio and cable communication with all of his field units." This achievement of a separate and effective network, Cline concluded, was "essential for clandestine intelligence collection operations, and an indispensable precedent for building up the magnificent professional staff of communications operators, which later gave CIA the advantage of prompt, secure links to the field with regular staff communications or clandestine radio nets that neither the State Department nor the military agencies could rival."[53] Cline knew whereof he spoke, for in the course of his long career, he worked not simply for the OSS and CIA but also for the Pentagon and the State Department.

OSS's paramilitary operations behind enemy lines had impressed a number of influential U.S. military commanders, and their support, Cline surmised, was one of the key reasons why the OSS was able to maintain its separate communications network.[54] One of those commanders was General Dwight D. Eisenhower, the Supreme Commander of the Allied Expeditionary Force, who in May 1945, with the defeat of Hitler's regime, declared that the value of the OSS "has been so great that there should be no thought of its elimination."[55]

Special Forces: Successor to the OSS

Although the OSS was eliminated in October 1945, its legacy included the Army's Special Forces as well as the CIA, and those Special Forces, know from the 1960s through the 1990s as the "Green Berets," also adopted many OSS training procedures. When the U.S. Army established in first Special Forces unit in 1952, it followed the training and traditions of the OSS Special Operations and Operational Groups. The commander of the first Special Forces unit, Colonel Aaron Bank, later celebrated as the "father of Special Forces,"[56] had received his initial OSS training at Areas F and B, before serving in France, Germany, and Indochina. In 1952, much of Bank's initial cadre was composed of former OSSers, including Jack Shannon, Caesar Civetella, and Herbert Brucker, and they prepared the training curriculum for the first Special Forces Group, which was established at Fort Bragg, North Carolina.[57]

Like OSS paramilitary training, initial training of the Army's Special Forces emphasized self-confidence and élan as well as individual skills with weapons, demolitions, field craft5, and at least rudimentary ability with communication equipment and medical treatment. There were also courses in organization of resistance movements and operation of their networks, agent training to include espionage and sabotage, guerrilla warfare, codes and radio communication, survival, instinctive pistol firing, and the Fairbairn method of hand-to-hand combat and silent killing. Although many of the initial recruits came from the Rangers which were being deactivated, more than fifty came from OSS veterans. Most of the training was done at Fort Bragg and its satellite, Camp Mackall, with its woods and swamps. But the final extensive field exercise simulating clandestine operations behind enemy lines was held in Chattahoochee National Forest in the Appalachian Mountains of northern Georgia. Banks and the other former OSS officers used the mountainous timberlands of the U.S. Forest Service just as the OSS had used the forests of the National Park Service in World War II.[58]

Drawing on the legacy of elite Army units, including the Rangers, Paratroopers, and various Army Raider units, the U.S. Army's Special Forces today also embrace the aura of the OSS's combat teams. Their tough, hard-boiled, daredevil self-image was augmented by ultra-demanding physical training, thriving on danger, and achievements in the field. Through the daredevils of Donovan's Special Operations teams and Operational Groups, OSS is widely recognized as a forerunner and an ancestor of today's Special Forces, indeed, some of the OSS emblems are incorporated into insignia worn by troops in today's Special Operations Command.[59]

Achievements of the OSS

The reputation of the OSS rested in part on its accomplishments and in part on the aura of "Wild Bill" Donovan himself, who President Eisenhower eulogized as "the last hero."[60] But in part the organization's reputation derived from the legend it created after the war. It was a romantic legend emphasizing individualism, innovation, heroism, and glamour. In keeping with traditional American images, the tale focused on amateur adventurers bent on excitement, glory, and victory in a crusade for law and order, justice and democracy. Although it may have irked many professionals in the armed forces and the old line government intelligence bureaus, the legend of the OSS helped establish a cult of romanticism about secret agents that contributed to popular support for dark arts of espionage and special operations for decades afterwards. Both the OSS and later the CIA helped to foster that image for their own purposes. But that meant that controversies over the CIA's clandestine activities would sometimes lead to disputes over the nature of the OSS and its relationship to the CIA.[61]

The deliberately crafted image of the OSS, like that of the dominant narrative of the American war effort itself, emphasized heroism, self-sacrifice and significant contributions to Allied victory. Understanding that the legend accentuated the achievements and minimized the problems in the organization, one can still appreciate the value and the historic role of the OSS. Although the military intelligence agencies and some other have remained skeptical of the glamorous history of the OSS, and while it is true that the Allies would have won the war without it, there is considerable evidence, as this and other studies have shown, that Allied victory was expedited and many Allied lives saved by the extraordinary efforts of the men and women of Donovan's comparatively small but highly dynamic organization. Despite its brief existence, the OSS did have a lasting impact.

Although the public has been fascinated by the spies and saboteurs, the real world probably has few "James Bond" characters. Instead, one of the most important an contributions of OSS was the unglamorous work of the men and women, studying and writing in the Research and Analysis Branch (R&A). They were little known by the public and unheralded by the media. It was a Donovan innovation, a group of civilians expert in particular areas, not working for any particular department, but rather gathering data on specific topics from as many sources as possible, analyzing this material, and generating strategic intelligence reports. They collected disparate scraps of information and tried to assemble them into a meaningful mosaic. Working primarily in Washington, D.C., the more than 900 scholars in this path-breaking unit, included many persons destined for future fame. Among them were Crane Brinton, Ralph Bunche, August Hecksher, H. Stuart Hughes, Charles Kindelberger, Herbert Marcuse, Walt Rostow, and Arthur M. Schlesinger, Jr. They produced reports on everything from the state of enemy morale and weapons production to the most effective targets for Allied bombing attacks, such as the Nazis' synthetic oil plants. The detailed reports R&A made of economic, geographic, strategic and political aspects in various countries not only proved valuable during the war but were still being used by intelligence officers of the U.S. Army,[62] and undoubtedly the CIA as well, for years afterwards. Donovan's R&A demonstrated that much valuable intelligence could be obtained from seemingly mundane published sources and how civilian scholars, working with libraries and other resources, could play an important role in obtaining, summarizing, and evaluating intelligence data. Despite the problems achieving inter-and even intra-agency cooperation and access to information, R&A's Current Intelligence Office began the process of what would under CIA become the preparation of the centralized, summarized, regularly submitted National Intelligence Estimate.

Spies and Intelligence

OSS was denied direct access to the most important intelligence breakthroughs of the war—the American MAGIC and the British ULTRA decrypts of enemy coded wireless messages—and this limited OSS to less vital information. British Secret Intelligence Services (MI-6) dominated Anglo-American human espionage in Europe until 1944, when OSS's Secret Intelligence Branch began to achieve independent results from its own spy handlers and the rings of indigenous agents. An exception was the OSS success in 1942 in Vichy French North Africa where because of French distrust of the British, it was the OSS which was able to establish an extensive network of agents there who not only provided vital information for the U.S. invasion in November but negotiated with the Vichy French to limit resistance to the American landings.[63] By the last year of the war in Europe, SI officers and their agents, were able to provide accurate and useful Battle Zone intelligence. An Army G-2 staff member of the Joint Intelligence Committee of the U.S. Joint Chiefs of Staff, recalled that at the meetings of that bickering inter-service intelligence committee, "the Army and OSS both claimed a universal competence... .The Army had no hesitation about contradicting an OSS political or economic estimate. OSS delighted to expose deficiencies in the Army's order of battle [Army intelligence's identification of the enemy units in the battle zone]."[64]

There were numerous instances where Army commanders were able to utilize effective OSS intelligence to supplement their own G-2 staff reports. OSS's chief agent in occupied Rome, Peter Tompkins, provided information about an impending German counterattack on the Anzio beachhead that enabled Allied commanders to sustain their position against what was supposed to be a surprise attack. Most exemplary among the Allied commanders using and coordinating with the OSS was Lieutenant General Alexander M. Patch, head of the Seventh U.S. Army. Information from OSS agents helped convince him that he could risk initial landings in southern France with only three U.S. divisions and a small Allied airborne force. In Patch's subsequent drive through the upper Rhone River Valley, intelligence from OSS revealed a hole in the German defenses that enabled his forces to race 150 miles around the enemy's left flank. By pinpointing the location of the German commander's only remaining armored division, OSS agents led to its destruction by Allied airpower and subsequently helped the Seventh Army push forward, later eliminate the Colmar pocket, and finally drive into Germany. In March 1945, an OSS agent in a German uniform provided key tactical intelligence, the location of a German Panzer division, that allowed the Ninth U.S. Army to cross the Rhine River at a location where there was little chance of a counterattack by German armored forces.[65]

OSS's Secret Intelligence and Counter-Intelligence Branches proved effective in both Europe and Asia. In China, although hampered by Chiang Kai-shek's own spymaster, Dai Li, and the Chinese intelligence and surveillance system, OSS's Secret Intelligence Branch produced significant results by the last year of the war there. It was responsible for identifying a high percentage of the targets attacked by the bombers and fighter-bombers of General Claire Chennault's Army Air Forces, and relaying information from its coast watchers to Admiral William Halsey's fleet that led to the destruction of significant amounts of Japanese shipping.

In Europe, Secret Intelligence and Counter-Intelligence was also effective. OSS's X-2 Branch that handled counter-intelligence and counter-espionage seems to have been more effective than the Army's Counter-Intelligence Corps in ferreting out enemy agents planted behind advancing American armies. The most spectacular achievement of OSS Secret Intelligence, however, was certainly the accomplishments of Allen Dulles in Switzerland. Dulles obtained some of the best human intelligence coups of the war through his top level contacts within the German foreign ministry, general staff, and military intelligence agencies in Berlin, most importantly career foreign service officer and anti-Nazi, Fritz Kolbe who had been rebuffed by the British. Dulles's contacts with disaffected Germans provided much valuable economic, political, and strategic information. The topics included the location where the V-1 and V-2 rockets were being developed, the spying of the Albanian valet to the British ambassador to Turkey, who as "Cicero" was selling secrets to the Nazis, foreknowledge of the German generals' conspiracy against Hitler in 1944, and solicitations to Dulles that eventually led to a negotiated German Army surrender in Italy a week before the Nazi regime capitulated in Berlin. In regard to Dulles's main German agent, Fritz Kolbe, Richard Helms, retired Director of Central Intelligence, wrote in a memoir published in 2003 that "Kolbe's information is now recognized as the very best produced by any Allied agent in World War II."[66]

Although Donovan championed centralized intelligence and authorized the scholars and spies to make it work, his own combative nature led him to take special interest in the paramilitary teams fighting behind enemy lines. Driving by his sometimes misguided sense of the demands of personal honor and perhaps also by a thrill of danger, Donovan went ashore in American landings on Sicily, Anzio, and Normandy, and recklessly flew in an inspection tour deep inside Japanese-occupied Burma. He envied his paramilitary forces, hailed their accomplishments, and sought to make sure that they were well trained, equipped, and supplied. Donovan took a personal interest in the development of special weapons, explosives, and espionage devices and materials developed by various support offices to service different branches of OSS. The most noted was the Research and Development Branch under chemist Stanley P. Lovell, who Donovan liked to call his "Professor Moriarity." Some of the projects were ludicrous—the idea of bats carrying small incendiary bombs over Tokyo, for example. But others were so effective they continued to be used, in different forms, to the present day: magnetic limpet mines, self-contained underwater breathing devices, waterproof watches, swim fins, small mines shaped like insignificant camel, donkey or horse droppings (later in the Vietnam War, the CIA adopted the idea and used simulated tiger droppings to conceal small, fist-sized sensitizer/transmitters to signal enemy movements along jungle trails). The "Liberator" pistol, a cheap one-shot .45 caliber pistol, designed for killing a sentry or other solitary individual and obtaining his weapon, was distributed by the OSS behind Japanese lines in China. Later during the Cold War, they were distributed by CIA in the Congo, and a 9mm version went to anti-communist tribesmen in the mountains of Laos and Vietnam.[67]

Other development offices had their successes as well as problems: producing forged passports and identity papers and paraphernalia for spies, some of which passed inspection and some of which did not; as well as matchbox cameras and various mechanisms for hiding secret messages. In those days when electronics was based on the vacuum tube and home radios were sizable pieces of furniture, OSS Communications Branch developed some extraordinary pieces of equipment. Among these were the famous "suitcase radio," the SSTR-1, a portable transmitter-receiver and power supply that could be packed into a suitcase or three small packages, which became the standard equipment for OSS field agents behind enemy lines around the globe. A small, shorter-range wireless set, the SSTR-3, could be carried in a briefcase. These wireless telegraphy transmitter/receivers proved highly effective, when they were not damaged in the aerial drop, as too often happened. OSS also developed and deployed in the last year of the war, a small, hand-held radio communicator, which enabled an agent on the ground to communicate by voice with a plane circulating high over the area in a very high frequency system, codenamed "Joan-Eleanor," which could not be detected by enemy direction finding equipment.

For the protection of its agents who frequently worked in stealth, OSS created effective silent, flashless pistols and even submachine guns, so the agents could fire without betraying their position. Seeking to impress President Roosevelt with the OSS's latest invention, Donovan once sneaked one of the new silenced .22 caliber pistols into the Oval Office in a shoulder holster while carrying a small bag of sand. While the Chief Executive was dictating to his secretary and looking away, Donovan pulled out the weapon and fired an entire, ten-round clip into the bag of sand in the corner without the President hearing a sound. With his handkerchief around the still hot barrel, Donovan handed the pistol to the President and explained that he had just fired ten bullets into the bag of sand. Shocked, the wide-eyed President quickly composed himself, then inspected the weapon, thanked Donovan for the new gun and offered his congratulations to its developers. Then regaining his sense of humor, he joked, "Bill, you're the only black Republican I'll ever allow in my office with a weapon like this!"[68]

OSS and the Multiplier Effect

The OSS itself particularly hailed the work of the daring, action-oriented paramilitary teams organizing, training, supplying, and directing indigenous resistance groups behind enemy lines. These were the Special Operations teams or two or three agents and the Operational Group sections, usually of ten to twenty men each, sent in when more substantial, self-sustaining units were needed. The Operational Group sections, somewhat like commando units, were generally ethnic, foreign-speaking Americans drawn from the ranks of the wartime Army of citizen-soldiers. It was one of Donovan's great insights that from America's multiethnic population, he could recruit commando-like units familiar with the language and cultural of countries occupied by the Nazis. While the SO teams were more oriented toward liaison with local Resistance, the OG detachments were combat units themselves and were more oriented toward direct combat engagements in guerrilla warfare. Both SO and OG, however, engaged in sabotage and subversion usually in coordination with indigenous resistance groups. In total, about 1,500 members of Special Operations teams and Operational Group detachments were infiltrated behind enemy lines.[69] Most of the SO and OG personnel were trained at Area F and also either Area A or B. Their missions would not have been possible without the clandestine radio-operators of the OSS Communications ("Commo") Branch, who kept their teams in contact with their base stations, sometimes under the most adverse conditions. Most of these combat radio operators, like Robert Kehoe, Spiro Cappony, and Art Reinhardt, had been trained at Area C. As Major Frank Mills later wrote of the radio operators who had accompanied the Jedburgh teams from Europe to the Far East in 1945, "These radio operators provided the essential communications link between the operational teams and the supporting base, and they were not only superb radio operators, but were some of the best combat soldiers we had in France and China."[70]

Paramilitary teams demonstrated what would later be called a "multiplier effect." OSS had dispatched less than 200 agents in France, and according to Donovan, they armed and organized more than 20,000 men and women in the local Resistance.[71] Other small groups of agents in Italy, Greece, and the Balkans played similar roles. A few Americans, two or three in a Special Operations team or a dozen or so in an Operational Group detachment, were inserted behind enemy lines, and then trained, supplied and directed local resistance groups numbering hundreds, even thousands. In the Mediterranean and in Europe, these paramilitary OSS teams infiltrated and fought with distinction in North Africa, Sicily, Sardinia, Corsica, Italy, Albania, Greece, Yugoslavia, Hungary, France, the Low Countries, and Norway. There they organized, supplied, and directed local partisan bands in hit and run raids and in destroying key bridges, railroad lines, and tunnels to impede German efforts.[72] In 1944 following the Allied landings in France, they were particularly active in seeking to block or delay hundreds of thousands of German reinforcements from trying to drive back the Allied liberators. In addition to seriously interfering with the sending of German reinforcements, these OSS teams also rescued more than a thousand downed Allied fliers, who were then able to continue in the air war against the Axis.

The accomplishments of the Special Operations teams and Operational Groups in the European and Mediterranean theaters ranged from Serge Obolensky's inducing the surrender of a 300,000-man Italian garrison on Sardinia, to the multinational Jedburgh and SO teams that helped impede German reinforcements to the Normandy invasion area and then directed the French *maquis* in protecting the exposed right flank of Gen. George Patton's Third Army as it rushed across northern France. The chief of Army Intelligence (G-2) in the European Theater of Operations estimated such actions may have saved the lives of as many as twelve thousand Allied soldiers, and reported to Supreme Allied Headquarters, "You can be satisfied that the OSS has already paid for its budget in this theater."[73] Throughout the European and Mediterranean Theater, the OSS paramilitary operations included both SO teams and Operational Groups, and their effectiveness was certainly disproportionate to their small size. Small teams totaling 200 Greek-Americans, led by officers like Jim Kellis and Johnny Athens, inflicted 1,400 casualties on German units while suffering only 25 casualties themselves. They destroyed key bridges, halting Turkish chrome shipments to Germany, cut railroad lines, severed communications links, and tied down large numbers of German units in Greece for a year and a half. In southern France, a French-speaking American OG team led by Roy Rickerson blew up bridges and railroad viaducts and blocked the Rhone River canal and with his dozen troops and a hundred Resistance fighters, forced the surrender of a contingent of 3,800 German soldiers. Aaron Bank and his team armed and directed some 3,000 French *maquis* against the Germans supply lines. William Colby parachuted with white-clad OG team into the snow-covered mountains to destroy a key railroad line in Norway. Actor turned Marine and SO officer Sterling Hayden skippered supply vessels past German patrol boats to deliver much needed supplies to the partisans in Yugoslavia. Teams of Italian Americans, many on missions organized by Albert Materazzi, helped the Italian Resistance cause enough problems in northern Italy that the Germans had to dispatch several divisions from the frontlines to try to suppress them.

Because of their location behind enemy lines, the SOs and OGs also became involved in sending intelligence information, particularly tactical information, especially concerning German troop and supply movements and vulnerable transportation targets (although unpredictable flying conditions, coordination, and the Army Air Corps' own priorities, often made it difficult to get fighter-bombers to the target in time—or to obtain supplies on schedule[74]). A number of women agents were used by the OSS, as well as by local Resistance movements, particularly as spies and liaison personnel. The most celebrated American woman SO agent was Virginia Hall, the "limping lady" with the artificial limb, feared and hunted by the Gestapo, who provided valuable information for the Allied invasion and also organized and trained three battalions, several thousand resistance fighters, in the *maquis* for guerrilla warfare in support of the Allies. In 1945, she became the only civilian woman in the war to be awarded the Distinguished Service Cross, America's highest medal for bravery after the Medal of Honor.

Assessing the OSS

Although many in the Regular Army were skeptical, General Eisenhower and his top assistants came to understand the value of the OSS. A report by one of Eisenhower's trusted subordinates, Brigadier General Benjamin F. Caffey, in the U.S. Army's Operations Division in early 1945 concluded after a study of coordination of Resistance movements throughout Europe by the OSS and SOE that "Resistance Groups, alone, cannot win a campaign or a battle, but they are capable of rendering important assistance to regular forces." This, Caffey said, included forcing the enemy to deploy large numbers of troops to protect lines of communications and supply, disrupting the flow of vital supplies and information to the enemy by destroying his lines of communication and supply, and alerting advancing forces to hidden defenses such as gun emplacements and minefields, and also freeing advancing conventional troops from the need to clear up pockets of resistance. General Caffey, like other regular officers, criticized the lack of experienced, career officers among the unconventional units, but his overall assessment was positive. "While OSS and SOE are hampered by poor staff work, their personnel in the field have done remarkably well," His conclusion was that "They deserve credit and appreciation for their fine work... This method of warfare is a vast potential in obtaining military strategical and tactical objectives. No commander should ignore this potential."[75]

In occupied France and elsewhere, the sight of armed and uniformed American soldiers deep in enemy occupied territory lifted the spirits of the villagers and swelled the ranks of the Resistance, especially when parachuted loads of weapons and other supplies began to follow. Ralph Ingersoll, a member of the staff of General Omar Bradley, commander of U.S. forces in the Normandy invasion, concluded that the German commanders had to assign at least half a dozen divisions to counter the *maquis* during the invasion. "The [OSS led] French Resistance was worth at least a score of divisions to us, maybe more."[76] German generals in France and Italy were quite concerned with the seriousness of having to face not only the Allied armies in their front but increasingly assertive and effective Resistance forces in their rear.[77] General George C. Marshall, U.S. Army Chief of Staff, declared flatly that "The Resistance surpassed all our expectations, and it was they who, in delaying the arrival of German reinforcements and in preventing the regrouping of enemy divisions in the interior, assured the success of our landings."[78] "Without their great assistance," General Eisenhower concluded, "the liberation of France and the defeat of the enemy in western Europe would have consumed a much longer period of time and meant greater losses to ourselves."[79]

In Southeast Asia, the most widely heralded OSS operations were conducted by its famous OSS Detachment 101 in Burma. Led by men trained at Areas B and X, it conducted some of the organization's most difficult and most successful operations. By the end of the war, 700 Americans, who served in Detachment 101, headed first by Carl Eifler and later William Ray Peers, had mobilized, trained, supplied, and directed more than 10,000 Kachin tribesmen. The number of Americans in Detachment 101 who actually parachuted into the Burmese mountain jungles to lead those 10,000 Kachins numbered less than 200. They harassed the Japanese Army's lines of communication and supply, thus helping to protect the American construction of the new Burma Road.[80] They severed enemy supply lines, targeted camouflaged supply depots for American bombers, rescued downed fliers[81] and forced the Japanese to maintain large numbers of troops to try to protect their lines of communication and supply. Ultimately, as in Europe, the OSS-led resistance groups provided major assistance to advancing Allied armies, this time American, British, and Chinese armies pushing the Japanese out of Burma.[82]

Concerning the relationship of OSS Detachment 101 and the U.S. Army, Toni L. Hiley, curator of the CIA Museum, likes to tell the story of what she calls "the two handshakes." When the first battalion of the Army's raider unit known as "Merrill's

Marauders," penetrated into Burma in March 1944 and started working their way through the jungle, an OSS agent from Detachment 101 suddenly stepped out of the jungle onto their path and welcomed them. The two units then worked together as the OSS teams and their Kachin guerrillas guided the Army units through the jungle, while simultaneously protecting their flanks against surprise attacks from the Japanese. Hiley contends that the OSS's successor, the CIA, and the Army are still working together. The CIA had its agents into Afghanistan seventeen days after the terrorist attacks on the United States on 9/11/2001. A few days later, those agents greeted the first of the Army's Special Operations teams when the soldiers arrived.[83]

The success of one part of Detachment 101, 300 Americans and 3,200 native guerrillas, in the spring of 1945 against two Japanese divisions, killing 1,300 Japanese soldiers and routing 10,000 others while losing only 47 men, who earned a Presidential Distinguished Unit Citation for Detachment 101 in January 1946 from General Eisenhower, who was then Chief of Staff of the U.S. Army.[84]

Only gradually did OSS expand its role in the Far East, but when it eventually obtained considerable success. There OSS agents also worked with indigenous groups to subvert Japan and to rescue hundreds of downed fliers behind Japanese lines in Burma, Thailand, Indochina and China itself.[85] OSS missions into Thailand contributed to the overturning of its initially pro-Japanese government. Donovan's organization got into China belatedly in the spring of 1943. The OSS was often defensive about its effectiveness in China, declaring, correctly, that its role there was delayed and restricted by internal squabbles within the multi-branch American mission there and most importantly by restraints imposed by Chiang Kai-shek's chief of intelligence and secret police, Dai Li. Nevertheless, OSS Detachment 202 with headquarters in Kunming made increasingly contributions to Allied victory by helping the Chinese forces tie down hundreds of thousands of Japanese troops, many of whom would otherwise have been sent to fight the Americans in the Pacific. OSS ran rings of agents who provided useful targeting information from behind Japanese lines for "Flying Tiger" General Claire Chennault's 14th U.S. Air Force as well as Admiral Chester Nimitiz's Pacific Fleet. Special Operations teams sabotaged railroads, bridges, tunnels and munitions warehouses. Frank Gleason's team blew up 50,000 tons of weapons, munitions, and explosives to prevent them from falling into Japanese hands. Paul Cyr's Chinese guerrillas destroyed a heavily guarded, vital railroad bridge across the Yangtze River that American bombers had been unable to take out.

"There were fewer than 2,000 OSS people in China in 1945, and at least half of them were desk people rather than combat people and never got near the Japanese," recalled John W. Brunner, a veteran of the Communications Branch in China in 1944-1945. "But with less than a thousand combat personnel, we were officially credited with the killing of more than 12,000 Japanese." The official report supports Brunner's contention.[86] The OSS transferred many of the successful SO and OG teams from Europe to China in 1945, and Arden Dow and Frank Gleason and others were training the first Chinese commandos, paratrooper units that had just begun combat missions when the war ended. With news of the emperor's decision to surrender, SO teams flew or parachuted into Japanese-run POW camps in China, took control of them, and rescued thousands of military and civilian captives, including surviving fliers of the "Doolittle Raid" on Toyko, and General Jonathan Wainwright, who had been a prisoner of the Japanese since the capture of the Philippines in 1942. At the end of the war, OSS had been preparing espionage and paramilitary teams of Koreans and Japanese Americans to be infiltrated into Japan to assist the planned U.S. invasion of the home islands in 1946. In his official commendation of Donovan's organization, the U.S. commander in China, Lieutenant General Albert Wedemeyer, praised the OSS personnel for the "outstanding performance of duty in their vital missions" and concluded that their achievements constituted "a record of extraordinary heroism, resourcefulness, initiative and effective operations against a ruthless enemy."[87]

The much touted successes of the OSS with indigenous resistance groups became a major part of its legacy in the postwar world. Under former OSSer Frank Wisner and others, the CIA tried without success in the late 1940s and 1950s to apply the same techniques to building resistance groups in the Soviet Union, Communist China, and North Korea. Later in the 1960s, the methods were applied to wholly different strategic situations in other parts of the world, sometimes with disastrous results. Even some former OSSers asserted that the CIA's emphasis on clandestine paramilitary action in the postwar era as a part of the OSS legacy mismatched what in fact had been a unique experience of World War II, when the OSS like the western Allies received widespread, enthusiastic reception from groups seeking to overthrow the hated Axis conquerors.[88]

Costs, Failures, and Accomplishments

It remains unknown how many people were infiltrated or otherwise worked in enemy territory for the OSS in World War II. The Special Operations Branch reported that it infiltrated some 1,600 operatives. But comparable figures about the precise number of uniformed personnel in Operational Groups sent behind enemy lines or the number of men and women working as spies for the Secret Intelligence Branch have not been disclosed. Working behind enemy lines is hazardous work, and it took its toll, even though those total figures, if they include indigenous agents as well as Americans, may never be known.[89]

In all, Americans in the OSS suffered some 450 casualties. Of these, 143 were killed in the line of duty. More than 300 were wounded, including a handful who were captured and were not executed. These figures do not include a large and unknown number, probably in the hundreds, of foreign agents working in various capacities for the OSS who were killed or wounded.[90]

Despite numerous successes, the OSS paramilitary campaigns in the various theaters in World War II were not without their failures. Attempts to sustain infiltrated paramilitary missions in areas that were not overwhelmingly supportive of the Allies and where the Axis were in complete control such as Austria, Hungary, Bulgaria, and southern Serbia failed because the local population would not protect them, indeed invariably reported them to the enemy.[91] When the Germans quickly crushed an uprising in Slovakia, Holt Green's OSS team there was aided by many local people but was ultimately betrayed by others. The Gestapo captured, tortured and executed seventeen members of Green's "Dawes" Mission at Mauthausen concentration camp.[92]

Mistakes, treachery, reprisals, bad luck can be lethal. Guerrilla warfare is neither a romantic pastime nor an inexpensive form of combat. Even in countries where the population overwhelmingly supported the Allies and overall the missions were successful, there were casualties. In France, 21 of 276 Jedburghs were killed in the summer of 1944. At least three teams were ambushed shortly after landing and most of the members killed. Jedburgh Team "Augustus" became trapped between fluid front lines in eastern France and when they ran into a German roadblock guarded by a Tiger tank, all three were killed, two Americans and a Frenchmen. In Italy, an entire detachment of Italian American OGs, 13 men of the "Ginny" Mission, landed from PT boats to blow up an important railroad tunnel were betrayed by a fascist sympathizer and were lined up, executed, and buried in a mass grave. In China, Captain John Birch, an Air Corps intelligence officer, working for OSS was brutally murdered by Chinese Communists, and in Indochina, Lieutenant Colonel Peter Dewey, head of the OSS mission in Saigon, was killed in an ambush by Communist guerrillas, who apparently mistook him for a French officer.

OSS achieved a great deal with a comparatively few people. Its successes were disproportionate to its numbers. Its secret intelligence and research and analysis did provide useful information, strategically and tactically, and revealed the limitations of traditionally departmentalized intelligence. It underscored that intelligence gathering and analysis were more than simply the domain of the military and that centralized analysis was a vital process in national security. In addition, its paramilitary agents in the field, possibly never more than two thousand Americans working deep behind enemy lines, also proved of considerable assistance to the Allied cause, particularly in providing target information, impeding lines of communication and supply and distracting and diverting enemy resources to try to find and destroy them. They did so far beyond their numbers because of the multiplier effect in generating, arming, and directing ultimately tens of thousands of indigenous forces. At its peak in December 1944, Donovan's organization consisted of only 13,000 Americans. This was slightly less than one full-sized U.S. Infantry Division of 14,000 men, and the U.S. Army deployed 90 such infantry divisions, plus 16 armored divisions. The 13,000 in the OSS was but a handful compared to the 16 million Americans in the armed forces in World War II.

The fledgling organization won most of its battles overseas, but it lost its most important bureaucratic battles in Washington. Yet, although short-lived and comparatively small in size, the OSS did, in fact, leave a substantial legacy in many areas of American life. It had important impacts on the National Parks in which it trained. It was instrumental in the development of the U.S. Army's Special Forces. And it was the forerunner of the Central Intelligence Agency. In that latter respect, OSS was the only one of the more than one hundred temporary "war agencies" that ultimately survived the postwar demobilization, albeit after a two-year hiatus, and reached independent status.

Relationship of the OSS and the National Park Service

The OSS not only contributed to the eventual emergence of the CIA and Special Forces, its main institutional legacies, but it had an important impact upon the two National Parks, which provided a home for some of its most important training camps. The superintendents of the two parks—Ira B. Lykes at Prince William Forest Park and Garland B. ("Mike") Williams at Catoctin Mountain Park—attested to the effects as they sought to preserve the parks while also benefiting in the long run from the military's wartime occupation. During the war, the superintendents and their superiors worked assiduously with OSS training camp commanders to ensure that the military abided by the terms of the special permit that allowed them exclusive use of the property during the war. That meant ensuring respect for the National Park Service's mission of preservation of the land and the resources and facilities that Congress had designated as worthy of maintaining in the public interest. But this mission had to be carried out in the extraordinary circumstances resulting from the U.S. declaration of war and America's full scale mobilization in World War II. The War Department took over the two parks for OSS training purposes, brought firearms and explosives into the woods and declared the area to be off limits to the public for the duration of the war.

It took considerable insight, judgment and tact as well as a keen sense of purpose for the civilian National Park Service to make the best of the dramatically changed situation when the users of the park were the U.S. Army and Marine Corps instead of charitable civilian organizations such as the Maryland League for Crippled Children, the Salvation Army, and the Boy and Girl Scouts of America. Yet, Lykes and Williams performed admirably in supervising the military use of the properties and seeking to ensure the preservation of their natural and historical resources. The two park managers built relationships of respect with the military camp commanders. They encouraged a positive and cooperative relationship between these government agencies, each recognizing that the other had a job to do in difficult times. The park managers filed complaints when violations of the permit became too frequent or egregious, such as the shooting of wildlife and the destruction of trees and historic stone fences. Yet, they could also be helpful to the military in dealing with the local authorities over issues about road use and the control of forest fires. Most importantly, unlike the military which was there only temporarily, the NPS took the long-term perspective on developments in the area that the National Park Service had been federally-mandated to preserve for the benefits of the American public.

Legacy of the OSS in the Parks

Despite all the firing of weapons, blasting of explosives, and digging of trenches and other emplacements that disrupted the area, the military also contributed to a number of lasting improvements in the two parks. As required by the NPS, the purely military structures were dismantled at the end of the war—the rifle ranges, pistol houses, mortar and grenade ranges, demolitions areas, and munitions storage bunkers. A few old houses had been destroyed, some trees cut down, and at Catoctin some old stone fences had been ground up for gravel. But this had to be measured against the added value to the parks as a result of their wartime occupation. The 1,139 acres acquired at Prince William Forest Park by the War Department and the 288 acres acquired at Catoctin Mountain Park during the war helped to round off the two parks. At the end of the war, when the National Park Service resumed full control of the majority of acreage in the two parks, it also acquired the old CCC work camps, which had been the prewar property of the War Department, and which were improved during the war and deeded to the NPS afterwards by the War Department. Similarly, the OSS had winterized the cabins, dining halls, recreational halls and other facilities at the NPS's group camps. It had erected some new buildings and other facilities that the NPS retained. Utility systems—electrical, water supply, sewerage and waste treatment—had been modernized and expanded, and during OSS occupancy, there had been some rearrangement and improvement of the interior road system.

The two National Parks experienced some other significant changes as a result of the wartime experience. Catoctin Mountain Park became internationally famous as the site of the Presidential Retreat, President Roosevelt's Shangri-La, later renamed Camp David by President Eisenhower, and as the site of wartime meetings between Roosevelt and Churchill and subsequently many important international meetings, including the site for the Camp David Peace Accords between Egypt and Israel in 1978. Although the majority of federal Recreational Demonstration Areas from the 1930s were turned over to the states after the war, Catoctin was retained by the U.S. Government in large part because of the Presidential Retreat created there. Because of its proximity to the nation's capital, Prince William Forest Park was also retained by the federal government. However, as a result of the wartime occupation of the southern part of the park some 5,000 acres appeared to become a fixture of the U.S. Marine Base at Quantico, for half a century. Only in 2003 did a mutually acceptable resolution divide the sector, providing nearly 3,400 acres for the Marines and 1,700 acres for the park. Meanwhile, in the postwar years, Prince William Forest Park had maintained a positive relationship with the U.S. Army. Superintendent Ira Lykes developing a cooperative arrangement with the Corps of Engineers at Fort Belvoir in which the combat engineering troops moved a number of their practice exercises to Prince William Forest Park and in the process constructed roads, dams, and bridges in the park at minimal expense to the National Park Service.

OSS Veterans

The majority of OSSers, like other American veterans, came home to resume their life in postwar America, to complete their education, to get married, to get ahead, and to make up for the years they had given for their country. But they did not forget their wartime service, sense of common purpose, patriotic duty and achievement, and above all, the comradeship through good times and bad. Within a few years, they, like many other veterans, sought to resume those friendships and commemorate their achievements. There was a feisty, independent character to the OSS, and its veterans liked to reminisce about those days of service in the unorthodox, individualistic, innovative and generally anti-bureaucratic organization that Donovan had created. After the war, many of them became highly successful.

What impressed the present author about the OSS veterans he interviewed for this study beyond the fact that they were highly intelligent, articulate, and accomplished was that they remained intellectually and often physically active, many of them through their 80s and early 90s—some were auditing classes at nearby colleges, others doing volunteer work, teaching young people or aiding others. From their postwar resumés, it is clear that they remained a highly able, self-motivated, achievement-oriented, and very special group. They had become highly successful in business, law, academe, diplomacy, government, the military or intelligence work. Some became Presidential advisers, a number became U.S. ambassadors. Some remained in the Army, generally in engineering, infantry, military intelligence or Special Forces. Many of the communications people went into radio or television work in the private sector; others continued in government. A number of OSSers worked for the CIA in intelligence, special operations, communications, or the training.

While Donovan and other former leaders of the organization sought to build the public memory of the OSS through a series of books, articles, and Hollywood films in the postwar era, other ex-OSSers held reunions and formed their own veterans' societies. Their variety illustrated the continued influence of the different specialized units of the OSS. Veterans of the Burma campaign formed the Detachment 101 Society. Those who had served in the Communications Branch established the CommVets Association. Many members of the Secret Intelligence Branch joined the larger Association of Former Intelligence Officers. Those in Special Operations or Operational Groups could become members of the U.S. Army Special Forces Association. Members from any branch of OSS could also join the more broadly based OSS Society.[93] Most of these groups held their annual meetings and other reunions in Washington, D.C., occasionally at the Congressional Country Club (the former Training Area F), which maintains an exhibit of photographs from its wartime occupation by the OSS. The CommVets have maintained a special interest in the site of the Communications School, Training Area C, in Prince William Forest Park, organizing a number of reunions and group visits to the old training camp and supporting the creation of an OSS exhibit in the Visitor Center at Prince William Forest Park. A wayside exhibit on the OSS is also planned at the site of former Training Camp B at Catoctin Mountain Park.

Commemorating the OSS

Spies, saboteurs and other agent operatives work behind enemy lines, but there has always been a duality in clandestine activities between the murky world of espionage and the openly perilous environment of the special forces. That duality is generally reflected in different types of museums: there are those that exhibit the history of spies and those that feature the rowdy daredevils of various types of special operations. The aura of the OSS has proven attractive and useful to other institutions, not just its direct heirs, the CIA and the Special Forces. While the most visited exhibit featuring the OSS is in the privately-operated International Spy Museum, which opened in 2002 in Washington, D.C., the official memory of the OSS is maintained by a number of institutional museums. The Airborne and Special Operations Museum completed in 2000 and the U.S. Army's JFK Special Warfare Museum at Fort Bragg, both in Fayetteville, North Carolina, celebrate the OSS Special Operations teams and Operational Groups, as well as one of the predecessors of today's Special Forces.[94] The National Navy UDT-Seal Museum on North Hutchinson Island at Fort Pierce, Florida, commemorates the operational swimmers of the OSS Maritime Unit as a direct ancestor of today's SEALS.[95] The new National Museum of the Marine Corps that opened in 2006 on the Marine Base at Quantico, Virginia, includes a highly favorable reference to Marines like Peter Ortiz who served in the OSS: "Audacious Marines with icy nerves parachuted behind enemy lines in Europe with the Office of Strategic Services."[96]

The CIA hails the OSS as its predecessor in its official publications, its website, and in its private museum at CIA headquarters. The museum is not open to the public, but a tour of the OSS exhibit and other aspects of the Agency's history is part of the indoctrination of every new CIA recruit. In the white marble entrance hall to a main building at the CIA headquarters, one wall contains a bas-relief of Allen Dulles, OSS veteran and the Director of Central Intelligence who was responsible for the construction of the new CIA campus-like headquarters at Langley, Virginia. A few feet away from the bas-relief of Dulles is a display case with a book containing the names of CIA officers who lost their lives in the service of the nation. Etched into the white marble on the opposite wall are the names of 116 members of the OSS who were killed in the line of duty in World War II. Overlooking the names of the American dead in the entrance hall to the CIA headquarters, is a single statue, a life size, bronze figure of Major General William J. ("Wild Bill") Donovan.[97]

Although the Central Intelligence Agency continued to draw upon the aura of Donovan and his glorious and successful citizen-spies, analysts, saboteurs and guerrilla leaders after World War II, OSS veterans, like many other Americans, differed over many of the CIA's actions, for example, the use of paramilitary covert action to overthrow, popularly-elected, anti-American governments during the Cold War era. There had been considerable popular trust in the government and the military in the era of World War II, and much of it continued until the challenges of the late 1960s and early 1970s. The OSS belief was that secret intelligence and covert operations could be combined for worthy purposes. In the era of prolonged U.S.-Soviet tensions termed the Cold War, when these were combined to overthrow or attempt to overthrow governments, some of them popularly-elected, some OSS veterans supported their use as a necessary aspect of the anti-communist containment policy. But other former OSSers contended that such actions ultimately hurt American interests and represented a distortion of the OSS legacy. It was, the latter argued, an entirely different strategic situation from World War II. The Second

World War was a declared war of full national mobilization, and the Allies and OSS had been welcomed by the populations against hated Axis occupying forces and received great cooperation from the pubic and organized resistance groups.[98] In more recent years, influenced by failed endeavors, fumbled coordination, bumbling spies, rogue agents, extra-legal actions, and even betrayals, many Americans have become more skeptical, even cynical. While the present study was being completed between 2005 and 2008, the discussion in the OSS Society's electronic bulletin board and chat room as well as the author's interviews with many OSS veterans demonstrated the continuation of an ongoing division among OSS veterans over the foreign, national security, and intelligence policies of the United States.[99]

Recently, there have also been some suggestions that the "war on terrorism" would be better conducted not by direct action, man-hunting teams of the American military's Special Operations Command but in the long run by less glamorous missions which would establish security, encourage economic and political development and spread persuasive messages to win over the local population. By 2006, the Army was considering doing so by incorporating Special Forces with civil affairs, and psychological operations. But Max Boot, a senior analyst at the Council on Foreign Relations, and some others, have gone further to suggest removing unconventional warfare from Special Operations Command and assigning it to a resurrected Office of Strategic Services. This new OSS could include Army Special Forces, civil affairs and psychological operations, plus the CIA's Special Activities Division, consolidating, as Boot told a Congressional Committee, "all the key skill sets needed to wage the softer side of the war on terror" as well as having the ability to employ indigenous personnel from particular regions on a much larger scale than currently.[100] Thus, the debate over the OSS and its legacy continues.

What remained in the National Parks when the last members of the OSS left the after World War II? There were no echoes of marching orders or other military commands. The noise of gunfire and demolitions explosions had long faded away. The softer sounds—the grunts accompanying jujitsu training and the quiet tapping of the telegraph key—had also vanished. Memories remained, of course, of the strenuous training exercises but also of the comradeship and the leisure time at the recreation hall or in neighboring towns and cities, and perhaps also a fond memory of outdoors and the stillness and beauty of the woods in the National Parks. The ever present emphasis on secrecy had prohibited trainees from revealing their true identity even to each other. They went by fictitious names, and they told local residents—and even their own friends and relatives—that they were just regular members of the armed services. The local people generally accepted that explanation which seemed plausible given the extensive number of military facilities in Virginia, Maryland, and the District of Columbia, and the millions of young men in uniform in wartime America.

Most dramatically, of course, the OSS occupation of those two National Parks during the war meant the closing off of their recreational lands and interior roads to the public for the duration. The press reported only that the two parks had been taken over for the "war effort" and that the public was barred from entry. The nature of their contribution to the war effort was not disclosed. Those who drove to or by the park entrances would see them guarded by armed sentries and, of course, Army vehicles, especially closed Army trucks, were observed on the roads. It was rumored that the facilities were interning prisoners of war, which seemed plausible as there were German POW camps established in a number of rural areas. Although local residents around Catoctin National Park were not aware that the park was being used as an OSS training camp, preparing spies and saboteurs, they did observe that President Franklin Roosevelt was made a number of visits to the park. They assumed that he had established a secure, wartime retreat there, an assumption verified within weeks after the end of the war, when the press revealed the Presidential Retreat to the entire world.

"Wild Bill" Donovan, visionary, charismatic leader, and incurable romantic, and the freewheeling OSS that he molded in his image—bold, innovative, "dashing, slightly madcap, and highly effective"—will undoubtedly always remain fascinating.101 It has only been in recent years, however, with the declassification of the OSS records beginning in the 1980s, the spate of books on the OSS, and in the 1990s, the fiftieth anniversary of the war effort and the celebration of the veterans as the "Greatest Generation," that the people in and around Catoctin Mountain Park and Prince William Forest Park have begun to learn the full story of the OSS in the nearby National Parks.

Now it is hoped that they can also learn more specifically about the contributions of the two National Parks to the war effort. It is a fascinating story of spies, saboteurs, guerrilla leaders, and clandestine radio operators who began their rugged training in these formerly cheery campgrounds and who then went forth, sometimes deep into enemy territory, to a secret, shadow war to help defeat the Axis powers in World War II. It is a story well worth remembering and commemorating.

ENDNOTES

INTRODUCTION

1. Michael Warner, *The Office of Strategic Services: America's First Intelligence Agency* (Washington, D.C.: Central Intelligence Agency, 2000), 1; the CIA website (www.cia.gov); and Thomas F. Troy, *Donovan and the CIA: A History of the Establishment of the Central Intelligence Agency* (Frederick, Md.: University Publications of America, 1981), vii, 409-410.

2. Alfred H. Paddock, Jr., *US Army Special Warfare: Its Origins: Psychological and Unconventional Warfare, 1941-1952* (Washington, D.C.: National Defense University Press, 1982; rev. ed., University Press of Kansas, 2005), 23-25; and Aaron Bank, *From OSS to Green Berets: The Birth of the Special Forces* (Novato, Calif.: Presidio Press, 1986), 205-206. Because OSS was a civilian organization with military personnel merely assigned to it, the official lineage of the Army's Special Forces, specified by the U.S. Army's Center of Military History (www.army.mil/cmh/lineage/branch/sf), does not credit the OSS as a predecessor. But Special Forces veterans see OSS's combat units in Special Operation, Maritime Unit, and Operational Group Branches as their predecessors and so do many Special Forces organizations. See, for example, the websites of the U.S. Army's John F. Kennedy Special Warfare Museum at Fort Bragg, at www.soc.mil/swcs/museum; and Airborne and Special Operations Museum at www.asomf.org/museum, both in Fayetteville, N.C.; and Troy J. Sacquety, "The OSS," *Veritas: Journal of Army Special Operations History*, 3.4 (2007): 34-51. The Pentagon allows veterans of those OSS combat units to wear the distinctive Special Forces shoulder tab; www.tioh.hquda.pentagon.mil/Tab/SpecialForcesTab.htm.

3. Warner, *Office of Strategic Services*, 9.

4. On the peak figure of 13,000 OSS personnel in December 1944 (with 5,500 in the USA and 7,500 overseas), see *War Report of the OSS (Office of Strategic Services)*, 2 vols. (New York: Walker and Co., 1976), I, 116; hereinafter, *War Report of the OSS*. This is a declassified published version of the original institutional history of the OSS, "War Report, Office of Strategic Services," that was completed on 5 September 1947, by former OSS personnel working for its successor, the Strategic Services Unit. The executive officer of the History Project was Serge Peter Karlow; the chief of the History Project was Kermit Roosevelt, a former OSS officer and a grandson of President Theodore Roosevelt. An OSS document from May 1945 signed by General Donovan gives the following breakdown: "OSS has a total personnel of 12,816, of which 6,939 are members of the Armed Forces. Of a total of 2,593 officers and 6,346 enlisted service personnel, 2,192 are Army officers and 5,817 Army enlisted men." Maj. Gen. William J. Donovan to the Adjutant General, 15 May 1945, subject: Recommendation for Promotion [of Col. M. Preston Goodfellow], p. 3, in Millard Preston Goodfellow Papers, Box 2, Biographical Material Folder, Hoover Institution, Stanford, Calif.

5. Brett J. Blackledge and Randy Herschaft, Associated Press, "Newly Release Files Detail Early US Spy Network," 14 August 2008, http://www.washingtonpost.com/wp-dyn/content/article/2008/08/14/August; Spy Files Include a Justice, a Baker, and a Filmmaker," Newark (N.J.) *Star-Ledger*, 15 August 2008, A4.

6. Warner, *Office of Strategic Services*, 9; the figure of perhaps one third of the 13,000 being trained as agents is from Erasmus H. Kloman, *Assignment Algiers: With the OSS in the Mediterranean Theater* (Annapolis, Md.: Naval Institute Press, 2005), 10.

7. Other colloquialisms, such as "Oh-So-Secret" and "Oh-Shush-Shush," parodied the pervasive wartime secrecy of the OSS. Junius Morgan dispensed clandestine funds at the London office; Paul Mellon was an administrative officer in Special Operations Branch; his brother-in-law, David Bruce, also a millionaire, was chief of the London Office. At OSS headquarters in Washington, D.C., Henry Morgan (Junius's brother and also a son of J.P. Morgan) headed the Censorship and Documents Branch; William Vanderbilt served as an executive officer in Special Operations; and Alfred DuPont headed a section for espionage projects in France. Smith, *OSS*, 15-16.

8. John Chamberlain, "OSS," *Life*, 19 November 1945, 119-24; Corey Ford and Alistair McBain, *Cloak and Dagger* (New York: Random House, 1945); Arthur Goldberg, "Top Secret," *Nation*, 23 March 1946, 348-50; Paul Cyr, "We Blew the Yellow River Bridge," *Saturday Evening Post*, 23 March 1946; Stewart Alsop and Thomas Braden, *Sub Rosa: The O.S.S. and American Espionage* (New York: Reynal & Hitchcock, 1946); Elizabeth MacDonald [later Elizabeth MacDonald McIntosh], *Undercover Girl* (New York: Macmillan, 1947). Among the Hollywood feature films about the OSS, the most notable were *Cloak and Dagger* (Republic, 1946); *O.S.S.* (Paramount, 1946); and *13 rue Madeleine* (20th Century Fox, 1947). See also James I. Deutsch, "Representations of O.S.S. Espionage in Hollywood Films of 1946," paper presented at the annual meeting of the Society for Historians of American Foreign Relations, Georgetown University, Washington, D.C., June 22, 1997.

9. Corey Ford, *Donovan of OSS* (Boston: Little, Brown, 1970); Smith, *OSS*; Edward Hymoff, *The OSS in WWII* (New York: Ballantine, 1972); Joseph E. Persico, *Piercing the Reich: The Penetration of Nazi Germany by American Secret Agents During World War II* (New York: Viking, 1979); Anthony Cave Brown, *The Last Hero: Wild Bill Donovan* (New York: Times Books, 1982); Richard Dunlop, *Donovan: America's Master Spy* (New York: Rand McNally, 1982); Smith, *Shadow Warriors*; Thomas F. Troy, *Wild Bill and Intrepid: Donovan, Stevenson and the Origin of the CIA* (New Haven, Conn.: Yale University Press, 1996).

10. The Central Intelligence Agency began declassifying the OSS records and transferring them to the National Archives in the 1980s. By 1991, some 5,000 cubic feet of OSS documents from the CIA and the U.S. State Department's intelligence office had been opened for scholarly investigation. See Lawrence H. McDonald, "The OSS and Its Records," *The Secrets War: The Office of Strategic Services in World War II*, George C. Chalou, ed., (Washington, DC: National Archives and Records Administration, 1992), 78. By 1997, declassification by the CIA had brought to collection to some 7,000 cubic feet, with another 300 cubic feet of records received from CIA in 2000. Interviews by the author in 2005 with John E. Taylor and Lawrence H. McDonald, archivists, Modern Military Records Branch, National Archives II, College Park, Md. As indicated above, in August 2008, CIA declassification resulted in the release of another 750,000 documents.

11. Tom Brokaw, *The Greatest Generation* (New York: Random House, 1998).

12. A sampling of such prominent autobiographies would include, General William R. Peers and Dean Brelis, *Behind the Burma Road: The Story of America's Most Successful Guerrilla Forces* (Boston: Little, Brown, 1963); Allen W. Dulles, *The Secret Surrender* (New York: Harper and Row, 1966); William Colby and Peter Forbath, *Honorable Men: My Life in the CIA* (New York: Simon & Schuster, 1978); William J. Casey, *The Secret War Against Hitler* (Washington, DC: Regnery Gateway, 1988); Major General John K. Singlaub with Malcom McConnell, *Hazardous Duty: An American Soldier in the Twentieth Century* (New York: Summit Books, 1991); and Richard Helms with William Hood, *Look Over My Shoulder: A Life in the CIA* (New York: Random House, 2003).

13. For example, Franklin A. Lindsay, *Beacons in the Night: With the OSS and Tito's Partisans in Wartime Yugoslavia* (Stanford. CA: Stanford University Press, 1995); Dan C. Pinck, *Journey to Peking: A Secret Agent in Wartime China* (Annapolis, Md.: Naval Institute Press, 2003); Charles Fenn, *At the Dragon's Gate: With the OSS in the Far East* (Annapolis, Md.: Naval Institute Press, 2004); Richard W. Cutler, *Counterspy: Memoirs of a Counterintelligence Office in World War II and the Cold War* (Washington, D.C.: Brassey's, 2004); Erasmus H. Kloman, *Assignment Algiers: With the OSS in the Mediterranean Theater* (Annapolis, Md.: Naval Institute Press, 2005).

14. Peter Grosse, *Gentleman Spy: The Life of Allen Dulles* (Boston: Houghton Mifflin, 1994); John W. Brunner, *OSS Weapons*, 2nd ed. (Williamstown, N.J.: Phillips Publications, 2005); Elizabeth P. McIntosh, *Sisterhood of Spies: The Women of the OSS* (Annapolis, Md.: Naval Institute Press, 1998); Robin Winks, *Cloak and Gown: Scholars in the Secret War, 1939-1961* (New York: William Morrow, 1987); Barry M. Katz, *Foreign Intelligence: Research and Analysis in the Office of Strategic Services, 1940-1945* (New York: St. Martin's, 1999); Maochun Yu, *OSS in China: Prelude to the Cold War* (New Haven, Conn.: Yale University Press, 1996); Christof Mauch, *The Shadow War against Hitler: The Covert Operations of America's Wartime Secret Intelligence Service* (New York: Columbia University Press, 2003); Will Irwin, *The Jedburghs: The Secret History of the Allied Special Forces, France 1944*

(New York: Public Affairs/Perseus, 2005). For general overall accounts of the OSS in World War II, R. Harris Smith, *OSS: the Secret History of America's First Central Intelligence Agency* (Berkeley: University of California Press, 1972) and the meticulously researched, Thomas F. Troy, *Donovan and the CIA: A History of the Establishment of the Central Intelligence Agency* (Frederick, Md.: University Publications of America, 1981) are still the most detailed scholarly studies. But the most readable popular treatment of the OSS (at least in the European and Mediterranean Theaters) at the present writing is Patrick K. O'Donnell, *Operatives, Spies, and Saboteurs: The Unknown Men and Women of World War II's OSS* (New York: Free Press/ Simon & Schuster, 2004).

15. "OSS: 60th Anniversary Issue," *Studies in Intelligence: Journal of the American Intelligence Professional* (June 2002): 1-196, published by the CIA's Center for the Study of Intelligence; CIA Museum, Langley, Va.; U. S. Special Operations Command (USSOCOM) Award to Caesar J. Civitella, OSS Italian Operational Group, May 19, 2008, USSOCOM HQ, MacDill AFB, Tampa, Florida. The USSOCOM also adopted the OSS emblem, the spear point as its own emblem (www.ussocom.mil).

16. The sole study available is a history of the OSS's Schools and Training Branch (S&T) prepared by that branch itself in August 1945, probably by Major Keith P. Miller. Nearly forty years later the typescript was declassified under the Freedom of Information Act and published as William L. Cassidy, editor, *History of the Schools and Training Branch of the OSS* (San Francisco: Kingfisher Press, 1983). Although helpful as a guide, Cassidy's volume, based on a single source, is of limited value as a history of training in the Office of Strategic Services.

CHAPTER 1

1. Gerhard L. Weinberg, *A World at Arms: A Global History of World War II* (New York: Cambridge University Press, 1994), 48-150; Allan R. Millett and Williamson Murray, *A War to Be Won: Fighting the Second World War* (Cambridge, Mass.: Harvard University Press, 2000), 44-90.

2. The term "fifth columnists" originated in the Spanish Civil War of 1936-39, when rightist General Emilio Mola, leading four columns of troops against Madrid, the capital of the leftist Spanish Republic, declared that he had a "fifth column" inside the city. The belief that "fifth columnists" were a major factor in the fall of the western democracies from Norway to France, although widely believed in World War II was, in reality, vastly exaggerated, and as revealed afterwards as a myth. There were Nazi sympathizers and spies, but their role was minimal in the German military victories. Louis de Jong, *The German Fifth Column in the Second World War* (Chicago: University of Chicago Press, 1956).

3. Christof Mauch, *The Shadow War Against Hitler: The Covert Operations of America's Wartime Secret Intelligence Service* (New York: Columbia University Press, 2003), 20-21. There is no consensus on the origin of the nickname "Wild Bill," with suggestions ranging from the football field to the battlefield. There are references to "Wild Bill" Donovan in the American press during World War I, but no record of that nickname in the Columbia College yearbooks, student newspaper, or sports journals. The 1905 Columbia yearbook, called him "quiet or always making a fuss," and he won prestigious speaking award and was voted second best looking man in his class. Donovan's official graduation from Columbia Law School was 1908, but he subsequently insisted that it was 1907 and after he became a Trustee (1921-27), Columbia accepted his version. See seacjs@aol.com [Kyle Bradford Smith], "'Wild Bill' Donovan Response," *OSS Society Digest*, Number 1022, 6 May 2005, osssociety@yahoogroups.com, accessed 7 May 2005.

4. On Donovan's background, see various biographies, including Corey Ford, *Donovan of the OSS* (Boston: Little Brown, 1970); Anthony Cave Brown, *The Last Hero: Wild Bill Donovan* (New York: Times Books, 1982); and Richard Dunlop, *Donovan: America's Master Spy* (Chicago: Rand McNally, 1982); plus other works such as Thomas F. Troy, *Donovan and the CIA: A History of the Establishment of the Central Intelligence Agency* (Frederick, Md.: University Publications of America, 1981); *Wild Bill and Intrepid: Donovan, Stephenson and the Origin of the CIA* (New Haven, Conn.: Yale University Press, 1996). On Donovan's being a non-smoker and virtual teetotaler, see "Note on his terrific health and energy as assets," 3 September 1945, typescript notes by his assistant, Wallace R. Deuel, a journalist who prepared a series of postwar articles on Donovan, located in Wallace R. Deuel Papers, Box 61, Folder 6, Library of Congress.

5. In 1940, amidst U.S. defense mobilization, Hollywood included Donovan's role in World War I in a feature film, *The Fighting 69th* (Warner Bros., 1940), a vehicle for James Cagney and Pat O'Brien, who played Father Duffy, the regiment's chaplain, with George Brent, a 1930s leading male actor, portraying Donovan. Although the unit was actually the 165th Infantry Regiment in the American Expeditionary Forces, the press and the public continued to refer to its historic unit identification, the 69th Infantry Regiment of the New York National Guard, whose lineage stretched back to the Union Army's famed "Irish Brigade" in the Civil War. Stephen L. Harris, *Duffy's War: Fr. Francis Duffy, Wild Bill Donovan, and the Irish Fighting 69th in World War I* (Washington, D.C.: Potomac Books, 2006).

6. In the 1928 Presidential campaign, Donovan was probably the most well known Catholic to support Herbert Hoover, and perhaps the only prominent Irish Catholic to oppose New York governor Al Smith, the first Roman Catholic Presidential nominee by a major party. Donovan's friend, Frank Knox, a Hearst newspaper executive, later told Secretary of the Interior Harold Ickes that Hoover had led Donovan to believe he would be appointed U.S. attorney general as a result. But although Hoover was elected, he did not appoint Donovan, who was bitterly disappointed. Diary entry of 23 December 1939 in Harold L. Ickes, *The Secret Diary of Harold L. Ickes* 3 vols. (New York: Simon and Schuster, 1954-55), 3: 88-89.

7. In April 1940, Donovan and his wife suffered the greatest personal tragedy of their lives when their 22-year-old daughter, Patricia, was killed when her automobile crashed on a rain slicked road near Fredericksburg, Virginia. Patricia's only sibling, David Donovan, named his daughter after his sister. Brown, *The Last Hero*, 78, 141-142.

8. Although Roosevelt's and Donovan's years at Columbia Law School had overlapped (Donovan from 1905-1908, Roosevelt from 1904-1906), they had not been friends there, nor did they associate much together before 1940. Even during the war, as head of OSS, Donovan never became a part of Roosevelt's inner circle. Bradley F. Smith, *The Shadow Warriors: O.S.S. and the Origins of the C.I.A.* (New York: Basic Books, 1983), 31; Robin W. Winks, *Cloak & Gown: Scholars in the Secret War, 1939-1961*, 2nd ed. (New Haven, Conn.: Yale University Press, 1996), 65-66. See also Elliott Roosevelt, ed., *F.D.R.: His Personal Letters*, 4 vols. (New York: Duell, Sloan and Pearce, 1947-1950), vol. 4, 975-76.

9. Thomas F. Troy, *Wild Bill and Intrepid: Donovan, Stephenson, and the Origin of the CIA* (New Haven, Conn.: Yale University Press, 1996), 19-30, 48-56; Diary Entry, 27 June 1940, *Secret Diary of Harold L. Ickes*, 3: 215.

10. Anthony Cave Brown, *"C": The Secret Life of Sir Stewart Graham Menzies, Spymaster to Winston Churchill* (New York: Macmillan, 1987). An official history of the British Secret Intelligence Service (MI6) was commissioned in 2005 to be written by Keith Jeffrey, an historian at Queen's University, Belfast, to be published to mark MI6's centenary in 2009. The official history of the British internal Security Service (MI5) was being written by Christopher Andrew, an historian at Cambridge University.

11. "Program—July 1940 Trip (Col. D)," and Admiral John H Godfrey to William J. Donovan, 28 July 1940, in William J. Donovan Papers, Box 81B, Vol. 34, U.S. Army Military History Institute, Carlisle, Pa., hereinafter Donovan Papers, USAMHI; for a full account of the trip, see Smith, *Shadow Warriors*, 33-37.

12. Roy Jenkins, *Churchill: A Biography* (New York: Farrar, Straus, 2001), 629-641; Edward Spiro, *Set Europe Ablaze* (New York: Crowell, 1967). What became known as the Special Operations Executive (SOE) was established 16 July 1940, its recruits were drawn from civilian society as well as the military.

13. Weinberg, *A World at Arms*, 150-51; David Stafford, *Roosevelt and Churchill: Men of Secrets* (Woodstock, N.Y.: Overlook Press, 2000), 40-44; Stafford, *Britain and the European Resistance, 1940-1945: A Survey of the Special Operations Executive with Documents* (London: Macmillan, 1980); Michael R.D. Foote, *SOE: An Outline History of the Special Operations Executive, 1940-46* (Frederick, Md.: University Publications of America, 1985). For an internal COI/OSS account of the formation of the British commandoes in 1940, see OSS Strategic Services Training Unit, "Commando Troops," pp. 1-6, a 200-page typescript dated 6 July 1942, copy in OSS Records (RG 226), Entry 136, Box 165, Folder 1804, National Archives II, College Park, Md., hereinafter OSS Records (RG 226), National Archives II.

14. William J. Donovan and Edgar Mowrer, *Fifth Column Lessons for America* (Washington, D.C., 1940); see also, Franklin D. Roosevelt's reference to a "war of nerves" in an informal radio talk, in *FDR's Fireside Chats*, Russell D. Buhite and David W. Levy, eds. (Norman: University of Oklahoma Press, 1992), 161.

15. John Whiteclay Chambers II, editor-in-chief, *The Oxford Companion to American Military History* (New York: Oxford University Press, 1999), see especially the following entries, Timothy J. Naftali, "Counterintelligence," 191-92; Mark M. Lowenthal, "Intelligence, Military and Political," 334-37; Owen Connelly, "Rangers, U.S. Army," 588-89; and Rod Pascall, "Special Operations Forces," 669-71.

16. Rhodri Jeffreys-Jones, *Cloak and Dollar: A History of American Secret Service*, 2nd ed. (New Haven, Conn.: Yale University Press, 2003), 22. This Secret Service fund could be spent without detailed accounting, and sometimes it was used for covert operations.

17. For overviews, see Charles D. Ameringer, *U.S. Foreign Intelligence: The Secret Side of American History* (Lexington, MA: D.C. Heath, 1990); Christopher Andrew, *For the President's Eyes Only: Secret Intelligence and the American Presidency from Washington to Bush* (New York: Harper Collins, 1995); Stephen Knott, *Secret and Sanctioned: Covert Operations and the American Presidency* (New York: Oxford University Press, 1996).

18. Russell F. Weigley, *History of the United States Army* (New York: Macmillan, 1967), 379-80; Marc B. Powe and Edward E. Wilson, *Evolution of American Military Intelligence* (Fort Huachuca, Ariz.: U.S. Army Intelligence Center and School, 1973); David Kahn, *The Codebreakers*, 2nd ed. (New York: Scribner, 1996); Curt Gentry, *J. Edgar Hoover: The Man and the Secrets* (New York: W.W. Norton, 1991).

19. David Kahn, *The Reader of Other People's Mail: Herbert O. Yardley and the Birth of American Codebreaking* (New Haven, Conn.: Yale University Press, 2004).

20. Stimson's remark is often misrepresented as having been made in 1929 when he closed the Cipher Bureau. Embittered, Yardley afterwards wrote a book revealing the operations of the "Black Chamber," an action that caused the Japan and other powers to change and increase the security of their ciphers, and the U.S. government to deny that the Cipher Bureau had ever existed. Herbert O. Yardley, *The American Black Chamber* (Indianapolis, Ind.: Bobbs Merrill, 1931); see also James Banford, *The Puzzle Palace* (Boston: Little, Brown, 1982). Due to Yardley's reckless action, Donovan refused to hire him for the OSS.

21. Other intelligence offices existed in the Treasury Department's Customs Service, Secret Service, and Bureau of Narcotics; the border patrol of the Labor Department's Immigration and Naturalization Service; and, in a different area, the Federal Communications Commission, which in addition to its regulatory duties, monitored nearly a million words a day of foreign radio broadcasts.

22. Ronald Lewin, *The American Magic: Codes, Ciphers and the Defeat of Japan* (New York: Farrar, Straus, Giroux, 1982).

23. As an example, they decided that the Army would decode Japanese diplomatic messages sent on even dates and the Navy on those bearing odd dates. The two services took turns delivering the decoded intercepts (the "Magic" decodes) to the President, the Army one month, the Navy the next.

24. Richard J. Aldrich, *Intelligence and the War against Japan: Britain, America and the Politics of Secret Service* (New York: Cambridge University Press, 2000), 73; Andrew, *For the President's Eyes Only*, 216-20. The greatest American intelligence and command failure was, of course, the inadequate preparation for a possible attacks that occurred against U.S. military facilities at Pearl Harbor, Hawaii, on December 7, 1941 and the next day on Luzon in the Philippines. American civilian and military leaders correctly viewed the Japanese invasion force as aimed at Malaya and the Dutch East Indies (now Indonesia), but they underestimated Japanese boldness and technological innovations that made possible the successful aerial attacks on the two American bases. Donovan's civilian organization was just in its infancy at the time, and neither the military nor the FBI had shared its intelligence readings of Japanese and German signals with the Office of the Coordinator of Information. On the intelligence and command failures at Pearl Harbor, see Gordon Prange, *At Dawn We Slept: The Untold Story of Pearl Harbor* (New York: McGraw-Hill, 1981); and Frederick Parker, *Pearl Harbor Revisited: U.S. Navy Communications Intelligence, 1924-1941* (Fort Meade, Md.: National Security Agency, 1994). In contrast, the U.S. Navy's greatest intelligence coup came a few months later in April 1942, when the Navy radio intelligence unit at Pearl Harbor broke the main Japanese naval operational code (labeled JN-25), in time to defeat decisively the numerically superior Imperial Fleet at the Battle of Midway in June 1942.

25. Troy, *Donovan and the CIA*, 36-40; Ford, *Donovan*, 94-106. The best primary source on Donovan's 1940-41 tour is the diary of his official guide, Lt. Colonel [later Brigadier] Vivian Dykes, "Personal Diary of Trip with Colonel William Donovan, 26 December 1940-3 March 1941," Tab 2, Exhibit III, Donovan Papers, USAMHI, Carlisle, Pa. This has been published as *Establishing the Anglo-American Alliance: The Second World War Diaries of Brigadier Vivian Dykes*, ed. Alex Danchev (London: Brassey's, 1990).

26. William J. Donovan to Frank Knox, 26 April 1940, reproduced as Appendix A in Troy, *Donovan and the CIA*, 417-18. For a discussion of the memorandum, see ibid., 56-57.

27. For example, in April 1941, Gen. Sherman Miles, Assistant Chief of Staff for Intelligence (G-2), warned Army Chief of Staff Gen. George C. Marshal: "In all confidence ONI tells me that there is considerable reason to believe that there is a movement on foot, fostered by Col[onel] Donovan, to establish a super agency controlling *all* intelligence. . . . From the point of view of the War Department, such a move would appear to be very disadvantageous, if not calamitous." Troy, *Donovan and the CIA*, 49. Equally hostile was J. Edgar Hoover, who had deployed the FBI for counterintelligence work in Latin America and was planning an expansion that would involve stationing his agents as legal attachés at U.S. diplomatic posts around the world. Brown, *The Last Hero*, 159; and Don Whitehead, *The F.B.I. Story* (New York: Random House, 1956), 228.

28. Michael Warner, *The Office of Strategic Services: America's First Intelligence Agency* (Washington, DC: Central Intelligence Agency, 2000), 2. Robert Dallek, *Franklin D. Roosevelt and American Foreign Policy, 1932-1943* (New York: Oxford University Press, 1979), 336.

29. Williamson Murray and Allan R. Millett, *A War to Be Won: Fighting the Second World War* (Cambridge, Mass.: Harvard University Press, 2000), 91-166; Weinberg, *A World at Arms*, 138-254. Chiang Kai-shek, as he was known in the West (but more correctly Jiang Jieshi) was head of the Nationalist (Kuomintang) government and commander of the Nationalist's armed forces in China from 1928 to 1949.

30. Mark A. Stoler, *Allies and Adversaries: The Joint Chiefs of Staff, the Grand Alliance, and U.S. Strategy in World War II* (Chapel Hill: University of North Carolina Press, 2000).

31. Theodore A. Wilson, *The First Summit: Roosevelt and Churchill at Placentia Bay, 1941*, rev. ed. (Lawrence: University Press of Kansas, 1999); Douglas Brinkley and David R. Facey-Crowther, eds., *The Atlantic Charter* (New York: St. Martin's Press, 1994).

32. War *Report of the OSS*, I, 7. Although this was also Donovan's explanation to military officials in September and November 1943 of what had happened, Troy, *Donovan and the CIA*, 55-56, does not see it as quite so simple. For Donovan's wartime explanations, see William J. Donovan to Maj. Gen. Walter Bedell Smith, 17 September 1943, Wash-Dir-Op 266, Folder 182; and Donovan, "Office of Strategic Services," lecture delivered at the Army and Navy Staff College, Washington, DC, 1 November 1943, typescript, Dir-Op-125, both in OSS Records (RG 226), National Archives II.

33. The final version of the memorandum is William J. Donovan, "Memorandum of Establishment of Service of Strategic Information," 3-4, 6, Memorandum to the President of the United States, 10 June 1941, six-page typescript with proposed organizational chart attached, copy in OSS Records (RG 226), Entry 139, Box 221, Folder 3096, National Archives II.

34. Troy, *Donovan and the CIA*, 59.

35. Bradley F. Smith, "Admiral Godfrey's Mission to America, June/July 1941," *INS*, 1. 3 (1986): 441-50. For Commander Ian Fleming's role, Andrew Lycett, *Ian Fleming* (London: Wiedenfeld & Nicolson, 1995), 129-30; and Cmdr. I.L.Fleming, Memorandum to Colonel Donovan, 27 June 1941, copy in Thomas Troy Files, CIA Records (RG 263), Box 2, Folder 19, National Archives II.

36. Adm. John H. Godfrey, "Naval Memoirs," pp. 132-33, John H. Godfrey Papers, Box 1, Folder 6, Churchill College Archive, Cambridge University, Cambridge, England, quoted in Jay Jakub, *Spies and Saboteurs: Anglo-American Collaboration and Rivalry in Human Intelligence Collection and Special Operations, 1940-45* (New York: St. Martin's, 1999), 26. Jakub also draws upon Godfrey's contemporary report of his 1941 visit located in Naval Intelligence Files, Admiralty Records, ADM 223/84, Public Records Office, Kew, London, United Kingdom.

37. Troy, *Donovan and the CIA*, 59-61.

38. Ibid., 63.

39. Roosevelt's 10 June 1941 handwritten note to John B. Blandford, Jr., Acting Director of the Bureau of the Budget, read: "Please set this up confidentially with Ben Cohen—Military—not O.E.M. FDR." As Thomas Troy explained, Roosevelt's use of the word "confidentially" probably referred to

the fact that the new agency would have the use of secret (unvouchered) funds and that such use should be purposely vague in descriptions of the purpose and functions of the new agency. Since the agency was initially a civilian one, the term "Military" presumably referred to a promise Roosevelt apparently made to give Donovan a military rank, presumably a promotion from colonel to brigadier or major general. "Not O.E.M." meant, Troy presumed, that unlike most of the other numerous new civilian war agencies, Donovan's organization would not be included under supervision of the new Office of Emergency Management. Donovan had extracted an agreement from Roosevelt that his agency would report directly to the President. Troy, *Donovan and the CIA*, 63, a photographic reproduction of Roosevelt's scrawled note is on ibid., 64.

40. All quotations from Stephenson's 18 June 1941 report are from his autobiography, William Stevenson, *A Man Called Intrepid* (New York: Harcourt, Brace Jovanovich, 1976), 249; or his earlier official biography by an aide, H. Montgomery Hyde, *The Quiet Canadian: The Secret Service Story of Sir William Stevenson* (London: Hamish Hamilton, 1962), 153.

41. Troy, *Donovan and the CIA*, 414; Aldrich, *Intelligence and the War against Japan*, 99.

42. Jakub, *Spies and Saboteur*, 23-28; see also Troy, *Wild Bill and Intrepid*.

43. One school, predominantly but not exclusively British and Canadian, has emphasized British tutelage of the Americans, including their allegedly decisive influence on Donovan's ideas on intelligence and special operations, his appointment, and indeed of the COI/OSS itself. See, for example, Hyde, *The Quiet Canadian*, 151-80; William Stevenson, *A Man Called Intrepid* (New York: Harcourt, Brace, Jovanovich, 1976), which emphasize Stevenson's role; and Donald McLachlan, *Room 39: A Study in Naval Intelligence* (New York: Atheneum, 1968), 224-39, with its emphasis on Adm. John Godfrey's influence. In contrast, another school of interpretation, composed predominantly American historians of the OSS and biographers of Donovan, tends to reject such claims as exaggerated. The Americans generally portray Donovan as too robust to be the malleable character depicted by the British and Canadians, and they emphasize the importance of U.S. history and the contemporary context, Donovan's own ideas developed over many years, his dynamic personal qualities, his political judgment, and a combination of good timing and highlevel connections, including Donovan's ability to keep the agency from being killed, before or after its birth, by its rivals among the American intelligence bureaucracies. See, for example, Lyman B. Kirkpatrick, *The Real CIA* (New York: Macmillan, 1968), 14-17; Ford, *Donovan of the OSS*, x; Smith, *OSS*, 2. For balanced assessments by an American, see Troy, *Wild Bill and Intrepid*, 186-211; and by a Briton, see Rhodri Jeffreys-Jones, "The Role of British Intelligence in the Mythologies Underpinning the OSS and Early CIA," in David Stafford and Rhodri Jeffreys-Jones, eds., *American-British-Canadian Intelligence Relations, 1939-2000* (London: Frank Cass, 2000), 5-19.

44. The order is reprinted in War *Report of the OSS*, I, 8; and also in its entirety as Appendix C in Troy, *Donovan and the CIA*, 423.

45. Harold D. Smith, Director, Bureau of the Budget, Memorandum for the President, 3 July 1941, Official File 4485, Box 1, OSS, 1940-41, Franklin D. Roosevelt Papers, Franklin D. Roosevelt Library, Hyde Park, N.Y., hereinafter Franklin D. Roosevelt Papers, Hyde Park, N.Y.

46. On the maneuvering in late June and early July 1941, especially the Army's successful efforts to limit the COI, see Troy, *Donovan and the CIA*, 63-70.

47. Maj. Gen. William J. Donovan to Adjutant General of the Army, 15 May 1945, subject: Recommendation for Promotion [of Col. Millard P. Goodfellow], p. 3, located in Millard Preston Goodfellow Papers, Box 2, Folder: Biographical Material, in the Millard Preston Goodfellow Papers, Hoover Institution, Stanford, Calif., herinafter, Goodfellow Papers, Stanford, Calif.

48. "White House Statement Announcing the President's Appointment of William J. Donovan as Coordinator of Information, July 11, 1941," in Franklin D. Roosevelt, *The Public Papers and Addresses of Franklin D. Roosevelt*, ed. Samuel I. Rosenman, 13 vols. (New York: Harper & Brothers, 1950), 10: 264. The original draft with some phrases crossed out because of objections by Press Secretary Steven Early and Budget Director Harold D. Smith, is attached to Harold D. Smith, Director, Bureau of the Budget, Memorandum for the President, 3 July 1941, Official File 4485, Box

1, OSS, 1940-41, Franklin D. Roosevelt Papers, Hyde Park, NY. On the press coverage, see *New York Times*, 12 July 1941, 5; "Job for Donovan," *Newsweek*, 21 July 1941, 15-16; "High Strategist," *Time*, 4 August 1941, 12.

49. John H. Waller, *The Unseen War in Europe: Espionage and Conspiracy in the Second World War* (New York: Random House, 1996), 148.

50. Lecture by "Captain [Francis P.] Miller on OSS [History] (Oct. 10, 1944)," p. 2, OSS Records (RG 226), Entry 161, Box 8, Folder 91, National Archives II. Miller was a member of the OSS headquarters staff, who had been with the organization since it was the COI.

51. *War Report of the OSS*, I, 8.

52. Troy, *Donovan and the CIA*, 129, 152, 218-19.

53. Ford, *Donovan of the OSS*, 109.

54. Brown, *The Last Hero*, 222-24; Richard Grid Powers, *Secrecy and Power: The Life of J. Edgar Hoover* (New York: Free Press/Macmillan, 1987), 267-269.

55. Edwin J. Putzell, a member of COI/OSS headquarters secretariat, oral history interview, 11 April 1997, pp. 39-40, conducted by Tim Naftali, OSS Oral History Transcripts, Box 3, CIA Records (RG 263), National Archives II. Donovan's organization was also convinced that columnist Drew Pearson had planted an informant in the organization (or was getting information from Hoover's "mole") because on several occasions, deciphered coded messages were printed in Pearson's column.

56. Richard Hack, *Puppetmaster: The Secret Life of J. Edgar Hoover* (Beverly Hills, Calif.: New Millennium Press, 2004), 239-242; Brown, *Last Hero*, 222-230.

57. *War Report of the O.S.S.*, 13; Ford, *Donovan of OSS*, 121; Troy, *Donovan and the OSS*, 77. A list of OSS buildings and personnel there in the District of Columbia and neighboring area in 1944, showed that, OSS occupied 340,000 square feet and had 3,400 personnel in the District. Federal Works Agency, Public Buildings Administration, Office of Planning and Space Control, "Bureau Space Record, Federal Space in the District of Columbia, Office of Strategic Services,"1 September 1944, Entry 132, Box 8, Folder 60, OSS Records (RG 226), National Archives II.

58. Arthur M. Schlesinger, Jr., Oral History Interview, 9 June 1997, p. 4, conducted by Petra Marquardt-Bigman and Christof Mauch, OSS Oral History Transcripts, CIA Records (RG 263), Box 4, National Archives II.

59. The proposed OSS insignia featured a gold spearhead on a black oval, and it is widely believed to have been the official emblem of the OSS and worn by its personnel. In fact, however, it fell victim to the antagonism of some professional officers of the regular armed forces toward what they considered Donovan's quasi-civilian, quasi-military organization of amateurs. Donovan had asked the Army's Quartermaster Corps to design an OSS insignia. On 16 June 1943, he approved the Quartermaster Corps' gold and black design and initiated a process for acquiring shoulder patches and collar devices, but the Joint Chiefs of Staff rejected his proposal, leaving the OSS with 195 cloth shoulder sleeve insignia patches and 200 metal collar devices with the proposed design that it no authority to use. Consequently, the insignia were not distributed and apparently not worn by OSS personnel. Les Hughes, "Insignia of the OSS," *The Trading Post* (Spring 1993); and Hughes, "The OSS Spearhead Insignia," http://www.insigne.org/OSS-Spearhead.htm, Accessed 14 March 2005.

60. Troy, *Donovan and the CIA*, 78.

61. Ford, *Donovan of OSS*, 134-35.

62. *War Report of the O.S.S.*, 11; Winks, *Cloak & Gown*, 60-114.

63. Joseph McBride, *Searching for John Ford: A Life* (New York: St. Martin's Press, 2001); Mark Cotta Vaz, *Living Dangerously: The Adventures of Merian C. Cooper, Creator of King Kong* (New York: Villard, 2005); Harriet Hyman Alonso, *Robert E. Sherwood: The Playwright in Peace and War* (Amherst: University of Massachusetts Press, 2007), 235-38.

64. "FDR" to "Dear Winston," 24 October 1941, Official File 4485 OSS, Box 1, OSS 1940-41, Franklin D. Roosevelt Papers, Hyde Park, NY.

65. *War Report of the O.S.S.*, 9-11; Troy, *Donovan and the CIA*, 77. 80; W.O. Hall, Bureau of the Budget, 16 July 1941, report on his conversation with Donovan, COI/BOB file, excerpts in Thomas Troy File, CIA Records (RG 263), Box 2, Folder 19, National Archives II; for the November figures, see Harold R. Smith, Director of the Bureau of the Budget, Memorandum for the President, 5 Nov. 1941, subject: Budget Request for the Coordinator of Information, and attachment, "Summary of 1942 Budget Request, Coordinator of Information," with Roosevelt's handwritten adjustments; plus "mp" to Secretary of the Treasury, 12-8-41, with the President's allocation of $3,162,786 to the Coordinator of Information. Both in Official File, 4485, OSS, Box 2, OSS Nov-Dec. 1941, Franklin D. Roosevelt Papers, Hyde Park, NY.

66. Warner, *Office of Strategic Services*, 4; *War Report of the O.S.S.*, 13-14, 84-85. Donovan, already wealthy, forsook any salary from the government. He received necessary transportation, subsistence, and other expenses connected to COI/OSS, but in fact, Donovan continued to spend a much of his own funds on his public service. Troy, *Donovan and the CIA*, 77.

67. Stanley P. Lovell, *Of Spies and Stratagems* (Englewood Cliffs, N.J.: Prentice-Hall, 1963), 111-13.

68. W.O. Hall, Bureau of the Budget, 16 July 1941 report on his conversation with Donovan, COI/BOB file, excerpts in Thomas Troy File, CIA Records (RG 263), Box 2, Folder 19, National Archives II

69. Troy, *Donovan and the CIA*, 107. Bruce replaced Wallace Banta Phillips, who Donovan had hired in October 1941 to organize the espionage side of COI's Special Activities. An enterprising American businessman who also ran an industrial spy service, Phillips, was working for the Office of Naval Intelligence in 1941, when he and their two dozen foreign agents were transferred to COI. Donovan asked him to prepare a plan for undercover intelligence activities, but he and Donovan had a falling out. Donovan fired him and replaced him with David Bruce. Joseph E. Persico, *Roosevelt's Secret War: FDR and World War II Espionage* (New York: Random House, 2002), 55-56, 112-13; Brown, *The Last Hero*, 175-78.

70. Obituaries, M.P. Goodfellow, *New York Times*, September 6, 1973, 40; Washington Post, Sept. 7, 1973, C-4; entry in *Who's Who in America*, Vol. 36; and the following documents, William J. Donovan to the Adjutant General, 15 May 1945, subject: Recommendation for Promotion [Colonel Millard P. Goodfellow]; and three-page handwritten biographical notes beginning "Goodfellow came over to OSS first in September of 1941—worked as liaison between G-2 and COI," both in the Millard Preston Goodfellow Papers, Box 4, Folder 1, Stanford, Calif.

71. Goodfellow and others at Army Intelligence had recommended Robert Solberg to Donovan for developing special operations. A Pole who had served as an officer in the Czar's Army in the First World War, Solberg had fled Russia after the Communist takeover. He subsequently may have done espionage work for the British while acting as an international businessman in the interwar period. Early in 1941, the U.S. Army's Military Intelligence Division gave him a commission as a lieutenant colonel, and in October 1941, Donovan was persuaded to bring him to COI to develop a program of special operations. A few weeks later, Solberg left for Great Britain for nearly three months to study the organization and methods of the British Special Operations Executive (SOE). Smith, *Shadow Warriors*, 58-59, 91-92; Troy, *Donovan and the CIA*, 82-83, 106-1108; *War Report of the O.S.S.*, 80; and also the biographical material in the Solberg File, Box 2 and Donovan order of 9 October 1941, OSS Memoranda, 1941-March 1942 File, Box 4, M. Preston Goodfellow Papers, Stanford, Calif.; plus Maj. M.P. Goodfellow to Lt. Col. Ralph C. Smith, 10 October 1941, subject: Lt. Col. Robert A. Solberg and points on liaison," copy in Thomas Troy Files, CIA Records (RG 263), Box 2, Folder 19, National Archives II.

72. Troy, *Donovan and the CIA*, 107.

73. *War Report of the O.S.S.*, 16.

CHAPTER 2

1. Roosevelt's quotation is from "Notes from WJD–April 5, '49," Donovan's notes on a postwar interview [presumably with William vanden Heuvel], cited in Thomas F. Troy, *Donovan and the CIA: A History of the Establishment of the Central Intelligence Agency* (Frederick, Md.: University Publications of America, 1981), 116. For the Polo Grounds episode, see the *New York Times*, 8 Dec. 1941, 32.

2. The Organization of Strategic Services (OSS) was created on 13 June 1942. By Executive Order, Roosevelt first moved the Foreign Information Service to the newly created Office of War Information (OWI). Then by military order, he transformed the reduced COI into the new Office of Strategic Services and placed it under the jurisdiction of the Joint Chiefs of Staff. He appointed Donovan as Director of the OSS. The Military Order of 13 June 1942, establishing the OSS is reprinted in Franklin D. Roosevelt, *The Public Papers and Addresses of Franklin D. Roosevelt*, comp. Samuel I. Rosenman, 13 vols. (New York: Harper & Brothers, 1950), 11: 283.

3. OSS, *War Report of the O.S.S.*, with new introduction by Kermit Roosevelt (New York: Walker and Co., 1976, orig., 1947), 18-27, 188-90. On the deleterious effects on counterespionage of the FBI rivalry with the COI/OSS/CIA, see Timothy Naftali, *Blind Spot: The Secret History of American Counterterrorism* (New York: Basic Books, 2005). Troy, *Donovan and the CIA*, 120-50; on the transfer of Robert Sherwood and the FIS to the OWI, see Harriet Hyman Alonso, *Robert E. Sherwood: The Playwright in Peace and War* (Amherst: University of Massachusetts Press, 2007), 245-50.

4. As the President told Sherwood in making the change, "I strongly felt that your work is essentially information and not espionage or subversive activity among individuals or groups in enemy nations. I know Bill Donovan does not agree with this, but the rest of the C.O.I., including himself, belongs under the Joint Chiefs of Staff." FDR to Dear Bob [Robert E. Sherwood], 13 June 1942, Official File 4485, OSS, Box 2, Folder OSS 1942-1945, Franklin D. Roosevelt Library, Hyde Park, N.Y., hereinafter, Franklin D. Roosevelt Papers, Hyde Park, N.Y.

5. On the creation of the U.S. Joint Chiefs of Staff (JCS) and strategic planning with the British Joint Chiefs of Staff, see Mark A. Stoler, *Allies and Adversaries: The Joint Chiefs of Staff, the Grand Alliance, and U.S. Strategy in World War II* (Chapel Hill: University of North Carolina Press, 2000), 64-67. For most of the war, the JCS included General George C. Marshall, U.S. Army; Admiral Ernest J. King, U.S. Navy; and General H.H. ("Hap") Arnold; U.S. Army Air Forces; together with Admiral William D. Leahy as the President's personal representative.

6. For interest by the Army's War Plans Division as early as February 1942 in active liaison with COI in regard to the Army's new mission of "subversive activities," see, "W.P.S., Draft Memorandum for General Gerow, Subject: Subversive Activities–Planning," n.d. [February ? 1942], and W.P.S. [of the War Plans Division], Memorandum to General Lee [probably Brig. Gen. Raymond E. Lee, chief of Army G-2, Intelligence], n.d. [February? 1942], Wash-OSS-Op-21 (COI Subversive Activities), photocopies in CIA Records (RG 263), Thomas F. Troy Files, Box 2, Folder 19, National Archives II.

7. Dwight D. Eisenhower, *The Papers of Dwight D. Eisenhower: The War Years*, ed. Alfred D. Chandler (Baltimore, Md.: Johns Hopkins Press, 1970), 253-54.

8. On the key negotiations were between Donovan and JCS secretary Brig. Gen. Walter Bedell ("Beatle") Smith, see Troy, *Donovan and the CIA*, 129-37, 143-50.

9. Troy, *Donovan and the CIA*, 173-91, 431-34.

10. JCS Directive 67 of 21 June 1942, approved 22 June 1942, recognized Donovan as having the military rank of colonel. For the dates of Donovan's promotions, 24 March 1943 and 10 November 1944, see Troy, *Donovan and the CIA*, 514n. See also Diary entry of 14 November 1942 in James Grafton Rogers, *Wartime Washington: The Secret OSS Journal of James Grafton Rogers, 1942-1943*, ed. Thomas F. Troy (Frederick, Md.: University Publications of America, 1987), 21.

11. William J. Donovan to Gen. Archibald Wavell, 6 July 1942, job 66-595, Box 1, Folder 48, Donovan Papers, U.S. Army Military History Institute, Carlisle, Pa. The JCS's Joint Psychological Warfare Committee reported sabotage and guerrilla as the OSS's military functions to be placed under theater commanders. See Joint Psychological Warfare Committee, memorandum to the Joint Chiefs of Staff, subject: Functions of the Office of Strategic Services in Relation to Secret Operations, 21 July 1942, and subsequent drafts of 24 and 28 July 1942, copies in OSS Schools and Training Branch; Schools, Functions of OSS in Relationship to Special Operations; Records of the Director's Office of OSS (RG 226), microfilm number 1642, roll 64, frames 952-968, National Archives II, College Park, Md.

12. OSS, *War Report of the O.S.S.*, 87.

13. Reported in James Grafton Rogers, diary entry of 6 September 1943, *Wartime Washington: The Secret OSS Journal of James Grafton Rogers*, 141.

14. A figure of 12,718 OSS personnel is provided in Lawrence H. McDonald, "The OSS and Its Records," in *The Secrets War: The Office of Strategic Services in World War II*, ed. George C. Chalou (Washington, DC: National Archives and Records Administration 1992), 81. See also Michael Warner, *The Office of Strategic Services: America's First Intelligence Agency* (Washington, D.C.: Central Intelligence Agency, 2000), 9. That apparently was the number of personnel at its peak. The total number of 21,600 people who had worked for COI or OSS at one time or another in civilian or military capacity is provided in Elizabeth P. McIntosh, *Sisterhood of Spies: The Women of the OSS* (Annapolis, Md.: Naval Institute Press, 1998), xi, 11. In August 2008, CIA released OSS personnel files suggesting 24,000 persons. Brett J. Blackledge and Randy Herschaft, Associated Press, "Newly Release Files Detail Early US Spy Network," 14 August 2008, http://www.washingtonpost.com/wp-dyn/content/article/2008/08/14/August; Spy Files Include a Justice, a Baker, and a Filmmaker," Newark (N.J.) *Star-Ledger*, 15 August 2008, A4.

15. Figures provided in Maj. Gen. William J. Donovan to Adjutant General of the Army, 15 May 1945, subject: Recommendation for Promotion [of Col. Millard P. Goodfellow], p. 3, located in Millard Preston Goodfellow Papers, Box 2, Folder: Biographical Material, in the Millard Preston Goodfellow Papers, Hoover Institution, Stanford, Calif.

16. Elizabeth P. McIntosh, *Sisterhood of Spies: The Women of the* OSS (Annapolis, Md.,: Naval Institute Press, 1998), xi, 11; Katherine Breaks, "The Ladies of the OSS: The Apron Strings of Intelligence in World War II," Senior Thesis in History, Yale University, 1991), cited by Robin W. Winks, "Getting the Right Stuff: FDR, Donovan, and the Quest for Professional Intelligence," in *The Secrets War: The Office of Strategic Services in World War II*, ed. George C. Chalou (Washington, D.C.: National Archives and Records Administration 1992), 24. But as Elizabeth McIntosh wrote in the late 1990s: "Discrimination against women in government service during the war was obvious. The women in X-2 [counter-espionage], for example, were as well educated as the men, they spoke the same number of foreign languages, on average were the same age (early thirties), and most had traveled abroad. But in X-2 they were generally secretaries, filing clerks, or translators. There was one decoder; two were listed as associate head and administrative assistant. None achieved executive positions in X-2." McIntosh, *Sisterhood of Spies*, 97-98.

17. Barbara Hans Waller, telephone interview with the author, 19 March 2005.

18. Barbara Hans Waller, telephone interview with the author,

19. March 2005. They were trained at the OSS Administration Building in Washington, D.C. She had no knowledge of the Communications Branch radio operation school at Area C. 19 Barbara Hans Waller quoted in McIntosh, *Sisterhood of Spies*, 215-16.

20. "OSS Personnel: By Branch," Table in McDonald, "The OSS and Its Records," *The Secrets War*, 96.

21. "OSS Personnel: By Theater" Table in McDonald, "The OSS and Its Records," *The Secrets War*, 92.

22. For a brief summary of the organization and operation of the Secret Intelligence Branch, see OSS, *OSS, War Report of the O.S.S.*, 179-87; and Warner, *Office of Strategic Services*, 21-24. David

K.E. Bruce headed SI until 1942 when he was appointed OSS chief in London and was succeeded by Whitney H. Shepardson, international lawyer and business executive, who headed SI until the end of the war. See David K. E. Bruce, *OSS against the Reich: The Wartime Diaries of Colonel David K.E. Bruce*, ed. Nelson Douglas Lankford (Kent, OH: Kent State University Press, 1991).

23. Lou Mumford, "Her Secret Life was Working with Spies. It Wasn't Like James Bond Movies, but World War II Era Job Was Intriguing," *South Bend [Indiana] Tribune*, 27 May 2005, reprinted in *OSS Society Digest* No. 1044, 28 May 2005, ossociety@yahoogroups.com, accessed 28 May 2005.

24. McIntosh, *Sisterhood of Spies*, 21-31.

25. Stephen Ambrose, *Ikes' Spies: Eisenhower and the Espionage Establishment* (Garden City, N.Y.; Doubleday, 1981), 39-56.

26. Peter Grose, *Gentleman Spy: The Life of Allen Dulles* (Boston: Houghton Mifflin, 1994); Christof Mauch, *The Shadow War against Hitler: The Covert Operations of America's Wartime Secret Intelligence Service* (New York: Columbia University Press, 2003). In January 2008, the CIA released in digitalized form nearly 8,000 formerly classified documents in the Allen W. Dulles Papers, 1939-1977, some of it heavily redacted. See, for example, information on Double Agent Lummy, in Switzerland and France, August 1943 through October 1944, in image 194308L70000029249; and reports from contacts about political and military developments in Germany's ally, Bulgaria, February 1943 through April 1944, image 19430210_0000029247; Sub-Series D, Correspondence General, English, 1942-1947, Allen W. Dulles Digital File Series, 1939-1977, Seeley Mudd Manuscript Library, Princeton University, Princeton, N.J.

27. William J. Casey, *The Secret War against Hitler* (Washington, DC: Regnery Gateway, 1988); Joseph E. Persico, *Casey: From the OSS to the CIA* (N.Y.: Viking, 1990).

28. OSS, *War Report of the O.S.S.*, 188-98; Warner, *Office of Strategic Services*, 8, 29. Caricatures of Angleton have appeared in numerous spy-thriller books and films, most recently, Robert DeNiro's film, *The Good Shepherd* (2007), starring Matt Damon as the Angleton character. On Angleton, see David C. Martin, *Wilderness of Mirrors* (New York: Harper & Row, 1980), 12, for his basic training at Area B; Timothy J. Naftali, "ARTIFICE: James Angleton and X-2 Operations in Italy," *The Secrets War: The Office of Strategic Services in World War II*, ed. George C. Chalou (Washington, D.C.: National Archives and Records Service, 1992), 218-45; for contrasting views of Angleton and the Cold War, see the favorable view of Robin W. Winks, *Cloak and Gown: Scholars in the Secret War, 1939-1961* 2nd ed. (New Haven, Conn.: Yale University Press, 1996), 322-437, 539 (p. 340 on Angleton's recruitment and training); and the more hostile view of Angleton as an ideologue in Tom Mangold, *Cold Warrior: James Angleton: The CIA's Master Spy Hunter* (New York: Simon & Schuster, 1991), passim (p. 42 for the Angleton's Legion of Merit Award from the U.S. Army in 1945 citing the capture of a thousand enemy agents).

29. Aline, Countess of Romanones, *The Spy Wore Red: My Adventures as an Undercover Agent in World War II* (New York: Random House, 1987), 18-44,56-57, 148-49, and passim; and *The Spy Went Dancing: My Further Adventures as an Undercover Agent* (New York: Putnam, 1990), 13-14.

30. McIntosh, *Sisterhood of Spies*, 13.

31. Wanda Di Giacomo quoted in McIntosh, *Sisterhood of Spies*, 17.

32. Elizabeth P. ("Betty") McIntosh, telephone interview with the author, 12 March 2005; on Di Giacomo and Lussier, see also McIntosh, *Sisterhood of Spies*, 16-17, 145-152.

33. Hélène Deschamps, *Spyglass: An Autobiography* (New York: Henry Holt, 1995); *OSS Society Digest*, Number 2078, 16 June 2008, osssociety@yahoogroups.com, accessed 16 June 2008; see also Margaret L. Rossiter, *Women in the Resistance* (New York: Praeger, 1986).

34. Judith L. Pearson, *Wolves at the Door: The True Story of America's First Female Spy* (Guilford, Conn.: Lyons Press, 2005); McIntosh, *Sisterhood of Spies*, 113-128. Despite the title of Pearson's book, Virginia Hall was hardly the first American female spy since their use dates back to at least the American Revolutionary War.

35. James Grafton Rogers, diary entry of 8 June 1943, *Wartime Washington: The Secret OSS Journal of James Grafton Rogers*, 107-8. See also entry of 20 February 1943, p. 57, for similar.

36. The SO branch went through several chiefs. See OSS-USA Organizational Chart November 1944, attached to H.C. Parton, Jr., Chief, Presentation Branch, to Lt. E.R. Kellogg, subject: Attachment, 2 Nov. 1944 in OSS Records (RG 226), Entry 133, Box 163, Folder 1384, National Archives II.

37. Bang-Bang Boys, phrase from "History of Schools and Training, OSS," p. 25, a 55-page-typescript, copy accompanied by a 7 January 1949 memorandum by Col. E.B. Whisner stating "Received this date from W[illiam]. J. Morgan the following report: History of Schools and Training, OSS, Part I thru Part VI," OSS Records (RG 226), Entry 176, Box 2, Folder 13, National Archives II.

38. One of Donovan's ideas had been to create American commando military units of foreign-speaking immigrants or second-generation Americans. This went so far as a proposal to the JCS for a Strategic Service Command and the formation of significant units in an independent corps. Although the JCS rejected the plan, something of it survived, via the broad authorization in JCS 155/4/D of 23 December 1942, in the full-fledged guerrilla companies of ethnic or foreign-speaking American troops, which became the OSS's different nationality Operational Groups. Wallace R. Deuel, "The History of the OSS," typescript, 1944, II:52, summarized in CIA Records (RG 263), Thomas Troy Files, Box 2, Folder, 19, National Archives II. Wallace R. Deuel, *Chicago Daily News* correspondent for Rome and Berlin in the 1930s, joined COI in 1941 as a special assistant to Donovan. He remained with OSS throughout the war and later worked for the CIA until his retirement in 1972. His typescript history of the OSS, written in 1944, has remained with the CIA. The present author requested its declassification in 2008.

39. The OSS Operational Groups (OGs) were created as 13 May 1943 as a separate tactical combat units under JCS Directive 155/7/D, 4 April 1943, Article 7, relating to Operational Nuclei for Guerrilla Warfare. The initial authorization was for 120 officers and 384 enlisted men. William J. Donovan, Special Order No. 21, issued 13 May 1943, effective, 4 May 1943; and Col. Ellery C. Huntington, Jr., C.O. Operational Groups, to Lt. Cmdr. R. Davis Halliwell, Chief of S.O., 22 June 1943, subject: Operational Groups, OSS— Organization and Functions, both in OSS Records (RG 226), Entry 136, Box 140, Folder 1460, National Archives II. Unlike SI, R&A, and even the Jedburgh teams of SO, there is, as of this writing, no overall history of the Operational Groups. A start has been made with a special OG section in *OSS Society Newsletter*, Winter 2007, 5-8; and the website www.ossog.org.

40. After the war, the legend emerged that the term "Jedburgh" came from a Scottish town where the teams trained. Although there is a town of Jedburgh in Scotland it is 250 miles north of Milton Hall in Peterborough, England, the main training site. A recent student of the subject has concluded that the name Jedburgh was given operation by being randomly selected from a list of town names in the U.K. Will Irwin, *The Jedburghs: The Secret History of the Allied Special Forces, France 1944* (New York: Public Affairs, 2005), 37-38.

41. OSS, *War Report of the OSS*, 225; Warner, *Office of Strategic Services*, 16-17.

42. "The Story of OSS OGs Worldwide," *OSS Society Newsletter*, Winter 2007, 5. Albert Materazzi, an OG veteran, believes there may have been 2,000 OGs. Albert Materazzi, "OGs get to tell their story," *OSS Society Digest* Number 1313, 26 March 2006, osssociety@yahoogroups.com, accessed 26 March 2006.

43. Harry Howe Ransom, *The Intelligence Establishment* (Cambridge, Mass.: Harvard University Press, 1970), 70.

44. Warner, *Office of Strategic Services*, 37-39. Richard Dunlop, *Behind Japanese Lines: With the OSS in Burma* (Chicago: Rand McNally, 1979); Roger Hilsman, *American Guerilla: My War Behind Japanese Lines* (Washington, DC: Brassey's, 1990).

45. E. Bruce Reynolds, *Thailand's Secret War: OSS, SOE, and the Free Thai Underground* (Cambridge, UK: Cambridge University Press, 2005); Dixee R. Bathrolomew-Feis, *The OSS and Ho*

Chi Minh: Unexpected Allies in the War against Japan (Lawrence: University Press of Kansas, 2006).

46. Maochun Yu, *OSS in China: Prelude to Cold War* (New Haven, Conn.: Yale University Press, 1996); Ronald H. Spector, *In the Ruins of Empire: The Japanese Surrender and the Battle for Postwar Asia* (New York: Random House, 2007), 9-21, 32-33, 44-47, 106-33.

47. OSS, *War Report of the OSS*, 225-28; John W. Brunner, *OSS Weapons*, 2nd ed. (Williamstown, NJ: Phillips Publications, 2005), 161-63; Francis Douglas Fane and Don Moore, *The Naked Warriors* (New York: Appleton-Century Crofts, 1956), 133; see also John B. Dwyer, *Seaborne Deception: The History of U.S. Beach Jumpers* (New York: Greenwood, 1992).

48. Barbara Lauwers Podaski, "Infiltrating Nazi Front Line with Morale Op's Disinformation" *OSS Society Newsletter*, Winter 2006, 3; and McIntosh, *Sisterhood of Spies*, 59-70.

49. McIntosh, *Sisterhood of Spies*, 233-234; and see her memoir, Elizabeth McDonald, *Undercover Girl* (New York: Macmillan, 1947).

50. Soley, *Radio Warfare*, 172-89.

51. Wink, *Cloak and Gown*, 114.

52. Bary M. Katz, *Foreign Intelligence: Research and Analysis in the Office of Strategic Services, 1942-1945* (Cambridge, MA: Harvard University Press, 1989).

53. OSS, *War Report of the OSS*, 48-69; Warner, *Office of Strategic Services*, 12. Examples of OSS R&A Reports include Report No. 1756, 27 January 1944: Development of German Pattern of Occupation; Report No. 232, 22 September 1944: South Germany: An Analysis of the Political and Social Organization, the Communications, Economic Controls, Agricultural and Food Supply, Mineral Resources, Manufacturing and Transportation Facilities of South Germany; Report No. 1844, 3 October 1944: Concentration Camps in Germany; Report No. 1999, 19 August 1944: The Belgian Underground; Report No. 2229, 15 June 1944: Burma: Enemy Shipping, October 1943-April 1944; Report No. 2993, 31 March, 1945: The Contributions of the Italian Partisans to the Allied War Effort; Current Intelligence Study No. 31, 20 July 1945: Japan's " Secret" Weapon: "Suicide."

54. R&A reports and estimates were used by Army intelligence in the Korean War and in Europe in the Cold War; they were "good and still valuable," according to Lt. Col. Elbert B. O'Keefe (U.S. Army-Ret.), former G-2 officer, conversation with the author at Catoctin Mountain Park annual dinner for park volunteers, 17 November 2007.

55. Edward Hymoff, *The OSS in World War II* (New York: Ballantine, 1972), 342,

56. OSS, *War Report of the O.S.S.*, 135.

57. Arthur Reinhardt, "Deciphering the Commo Branch," *OSS Society Newsletter*, Fall 2006, 6-7.

58. OSS, *War Report of the O.S.S.*, 90-91, 135-43. The new Communications Branch consolidated all the previous COI and OSS signal and traffic communications operations into one branch. There had been a difference of opinion over the desirability of lines of communication for SO and SI, but Donovan insisted that the organization be done through the headquarters unit. William J. Donovan to Colonel Buxton, Colonel Goodfellow and Major Bruce, Memorandum, 3 July 1942, OSS Records (RG 226), Director's Office Files, microfilm M1642, Roll 42, Frame 1122, National Archives II.

59. Lt. Cmdr. William H. Vanderbilt to Lawrence W. Lowman, 18 May 1942; Lt. Col. M. Preston Goodfellow to the Surgeon, Army Dispensary, 21 May 1942, subject: Physical Examination [of Lawrence Wise Lowman]; Lowman to Vanderbilt, 3 June 1942; "Check Slip," Security clearance 16 June 1942; all in OSS Records (RG 226), Entry 92A, Box 45, Folder 736, National Archives II.

60. On the detailed plans for such a Communications Branch, see Robert Cresswell to Major Bruce, 20 July 1942, subject: Communications Problem: Survey and Recommendation, attachment to David K.E. Bruce to William J. Donovan, 20 July 1942, in OSS Records (RG 226), Director's Office Files, microfilm M1642, Roll 42, Frames 1125-1131, National Archives II.

61. William J. Donovan to Assistant Chief of Staff, G-1 [Personnel], 10 June 1943, subject: Request for Priority to Procure Communications Personnel from the Signal Corps, OSS Records (RG 226), Director's Office Files, microfilm M1642, Roll 42, Frames 1319-1321, National Archives II.

466 | Endnotes: Chapter 2

62. The one-time pad had on the cardboard backing numerous alphabets, arranged in what seemed to be nonsensical order, especially when, as the agent would, use them in conjunction with the randomly generated key text printed on the pages of the one-time pad. The key was to know how to position the text (the clear text to be encoded or the enciphered text to be decoded) to the pad's key text using the Vigenere Square, referred to as "triads." Operators needed to memorize the 676 combinations possible in the square and then apply them. See, for example, W. Scudder Georgia, Jr., "It's All Greek to Me: And Other Stories," in James F. Ranney and Arthur L. Ranney, eds, *The OSS CommVets Papers*, 2nd ed. (Covington, Ky.: James F. Ranney, 2002), 158.

63. The figure is from Arthur Reinhardt, "Deciphering the Commo Branch," *OSS Society Newsletter*, Fall 2006, 6. Because of the temporary expansion of training in 1943, the Communications Branch also established an additional if smaller training facility in 1943 at a former Signal Corps Radar Training School at Camp McDowell, near Napierville, Illinois, which was designated Area M. OSS, *War Report of the O.S.S.*, 137.

64. OSS, *War Report of the O.S.S.*, 141.

65. Arthur Reinhardt, "Deciphering the Commo Branch," *OSS Society* Newsletter, Fall 2006, 6.

66. Warner, *Office of Strategic Services*, 32-33; for a detailed description, see John W. Brunner, *OSS Weapons*, 2nd ed. (Williamstown, NJ: Phillips Publications, 2005). While the transmitter used by the agent on the ground spread to about a 60 mile circle at 40,000 feet, the cone narrowed to only a few feet at ground level with little chance that the Germans would detect it. Marvin R. Edwards, "Joan and Eleanor: Radio Transmissions Aboard the Mossie [Mosquito bomber]," *OSS Society Newsletter*, Fall 2006, 8.

67. Reinhardt Krause, "Inventor of Portable Radio Developed Joan-Eleanor Project," *OSS Society Newsletter*, Fall 2006, 7.

68. Stanley P. Lovell, *Of Spies and Stratagems* (Englewood Cliffs, N.J.: Prentice-Hall, 1963), a memoir filled with details as extraordinary and often as harrowing as the lethal products of Lovell's laboratory.

69. John W. Brunner, *OSS Weapons*, 2nd ed. (Williamstown, NJ: Phillips Publications, 2005); H. Keith Melton, *The Ultimate Spy Book* (New York: DK Publishing, 1996), 28-41, 149, and passim. Many of these devices are on display at the International Spy Museum in Washington, D.C. and the Airborne and Special Operations Museum, Fayetteville, N.C., the latter which houses Brunner's extensive collection.

70. Harry J. Anslinger with J. Dennis Gregory, *The Protectors: Narcotics Agents, Citizens and Officials Against Organized Crime in America* (New York: Farrar, Straus, 1964), 76-80, one quarter of the FBN agents went into this wartime unit; Charles Siragusa, *The Trail of the Poppy: Behind the Mask of the Mafia* (Englewood Cliffs, N.J.: Prentice-Hall, 1966), vii, 58-59; John C. McWilliams, "Covert Connections: The FBN, the OSS, and the CIA," *The Historian*, 53/4 (Summer 1991): 660-72.

71. *History and Mission of the Counter Intelligence Corps in World War II* (n.p., n.d.), 1-5.

72. "Military Record of Garland H. Williams, Major, Infantry," attached to William J. Donovan to Secretary of War, 2 March 1942, [subject: promotion of Williams to lieutenant colonel], OSS Records (RG 226), Entry 92, Box 32, Folder 33, National Archives II.

73. William J. Donovan to Secretary of War, 31 December 1941 and 2 March 1942, the latter with threepage attachment, "Military Record of Garland H. Williams, Major, Infantry," OSS Records (RG 226), Entry 92, Box 32, Folder 33. Williams was promoted to lieutenant colonel in March 1942.

74. Maj. Garland H. Williams to Lt. Col. M.P. Goodfellow, 14 February 1942; William J. Donovan to Secretary of War, 14 February 1942; Donovan to Secretary of War, 17 February 1942; all in OSS Records (RG 226), Entry 136, Box 158, Folder 1721, National Archives II. On the difficulties in securing the allotment of 2,000 non-commissioned officers from the War Department, see OSS, *War Report of the O.S.S.*, 82, 100.

75. Tommy Davis, Report for the Special Operations Executive on visit to the United States, 15 Oct. 1941, Special Operations Executive Archives, Foreign and Commonwealth Office, London;

quoted in David Stafford, *Camp X* (New York: Dodd, Mead, 1987), 31. The decision for the SOE camp in Canada had been made at a meeting in William Stephenson's apartment in New York City on 6 September 1941, ibid. 17-20.

76. Lynn Philip Hodgson, *Inside Camp-X*, (Oakville, Ontario: L.P. Hodson, 1999); and Hodgson and Alan Paul Longfield, *Inside Camp-X, Part II*, forthcoming 2008). See also the Camp X Historical Society's website: www.campxhistoricalsociety.ca. The official name was Special Training School #103 (STS 103).

77. OSS, Strategic Services Training Unit, "[British] Commando Tactics" July 1942, typescript binder, p. 2, OSS Records (RG 226), Entry 136, Box 165, Folder 1804, National Archives II.

78. Stafford, *Camp X*, 81-82; George White's Diary, 1942, cited in McWilliams, "Covert Connections," 665 [the diary itself has subsequently been lost]; Anthony Moore, British Liaison, "Notes on Co-Operations between SOE and OSS," January 1945, pp. 1-2; OSS Records (RG 226), Entry 136, Box 158, Folder 1722, National Archives II.

79. The quotation is from one of Brooker's SOE colleagues, Bickham Sweet-Escott, *Baker Street Irregular* [SOE headquarters in London was on Baker Street] (London: Methuen, 1965), 143. On Brooker and the rest of the staff, see Stafford, *Camp X*, 10-12, 55-56.

80. Stafford, *Camp X*, 87.

81. Most of the training of Americans at Camp X occurred in 1942 and early 1943; thereafter, it trained mainly European guerrillas, until it closed as a secret agent training school in March 1944. www.campxhistoricalsociety.ca.

82. Anthony Moore, British Liaison, "Notes on Co-Operations between SOE and OSS," January 1945, pp. 1-6; OSS Records (RG 226), Entry 136, Box 158, Folder 1722, National Archives II. There was, of course, self-interest involved. In accepting more than a dozen OSS instructors for a month-long visit to SOE schools in Britain in the summer of 1942, SOE commended the idea of having even more OSS officers involved as "extremely desirable" because "this will not only serve as a means of training for the individuals concerned but will be a most powerful factor in insuring unity of doctrine and effort in the future operations of the two national S.O. organisations." "Record of Discussion regarding Collaboration between British and American S.O.E. [sic]," 23 June 1942, copy in papers of M. Preston Goodfellow, Box 4, Folder 3, Hoover Institution, Stanford, Calif.

83. *History of the Schools and Training Branch, Office of Strategic Services*, William L. Cassidy, ed. (San Francisco, Calif.: Kingfisher Press, 1983), 28-29, which is primarily a reproduction, without the extensive appendices, of the original 1945 typescript, "History of the Schools and Training Branch, OSS," completed in August 1945, probably by Maj. Kenneth P. Miller and most of the text of which was declassified by the CIA in 1981 as a result of a Freedom of Information Act request from William Cassidy (and fully declassified in 1985). Hereinafter cited as *History of Schools and Training Branch, OSS*. A copy of the original typescript is the OSS Records (RG 226), Entry 99, Box 78, among other locations in the OSS Records in the National Archives II. But for readers' convenience, citation in the present work will be to pages in the version published in 1983.

84. Maj. Garland H. Williams, "Training," eight-page, typed memorandum, undated [but January or February 1942 before Williams was promoted to lieutenant colonel in March] with attachments on details of proposed courses (the quotation is from page 2), located in OSS Records (RG 226), Entry 136, Box 161, Folder 1754, National Archives II.

85. Training films as well as films documenting OSS operations in the field were produced by the Field Photograph Branch, which included Hollywood director John Ford. Training films, included those on close combat techniques, clandestine communications, foreign weapons, uniforms, and insignia. OSS, *War Report of the O.S.S.*, 161-62. Many of these OSS films are located in the Visual Branch of National Archives II; copies of shooting scripts and narration are in the records of various OSS branches, for example, "Gutter Fighting," OSS Records (RG 226), Entry 133, Box 151, Folder 1258; "Short Range Intelligence," Entry 146, Box 220, Folder 3054, National Archives II. 86 Williams, "Training," 7.

87. Williams, "Training," 4. The draft of a memorandum in August 1942, indicates that Goodfellow's office was thinking of nearly a dozen different foreign guerrilla groups, each comprised of about sixty men and four officers who could be from the U.S. Army or foreign armies. The list of groups included "Norwegians, French, Italians, Austrians, German, Dutch, Hungarians, Spaniards, Poles, Czechs-Slovaks, Yugoslavs, Greeks, Chinese, Korean, Thai." Handwritten draft of a memorandum [written by M. Preston Goodfellow?], Joint Psychological Warfare Committee to Joint Chiefs of Staff, 31 August 1942, subject: Organization, Strategic Service Command [reference JCS 83/1 of 19 August 1942, further study as to training, type of men and organization of the Strategic Service Command," in M. Preston Goodfellow Papers, Box 4, Folder 4, Hoover Institute, Stanford, Calif.

88. Williams, "Training," 5-6.

89. Ibid.

90. For student capacity at the training areas, see Lt. Col. H. L. Robinson to Major Teilhet, 27 Nov. 1943, subject: Report on Training Areas, OSS Records (RG 226), Entry 146, Box 162, Folder 1757; for the station complement, see L.B. Shallcross, deputy Staff Training Branch/TRD, [Central Intelligence Agency], to John O'Gara, chief, Staff Training Branch, 1 February 1951, subject: Information on OSS Schools and Training Sites, OSS Records (RG 226), Entry 161, Box 7, Folder 76, both in National Archives II.

91. OSS, *War Report of the O.S.S.*, 81.

92. "History of Schools and Training Branch, OSS," pp. 7-8, attached to W[illiam] J. Morgan to Col. E.B. Whisner, 7 January 1949, pp. 10-11, OSS Records (RG 226), Entry 176, Box 2, Folder, 13, both in National Archives II.

93. Ibid., 10.

94. Ibid.

95. OSS, *War Report of the O.S.S.*, 81.

96. "History of Schools and Training Branch, OSS," p. 10, attached to W[illiam] J. Morgan to Col. E.B. Whisner, 7 January 1949, pp. 7-8, OSS Records (RG 226), Entry 176, Box 2, Folder, 13, both in National Archives II.

97. OSS, *War Report of the O.S.S.*, 81.

98. H. Preston Goodfellow, interviewed for the OSS official history in 1945, and quoted in *History of Schools and Training Branch*, OSS, 29.

99. Lt. Col. Garland H. Williams quoted in *History of the Schools and Training Branch*, OSS, 28; see also 31-32, 47. For similar, see Maj. Garland H. Williams, COI, to Joseph Green, supervising customs agent, Seattle, Washington, 12 January 1942, OSS Records (RG 226), Entry 92, Box 32, Folder 33, National Archives II.

100. Anonymous OSS recruiter, quoted in Russell Miller, *Behind the Lines: The Oral History of Special Operations [SOE and SO] in World War II* (New York: St. Martin's Press, 2002), 57.

101. OSS, *War Report of the O.S.S.*, 223-24.

102. OSS Security Office, "History of the Security Office," [1945], 4-5, 12, typescript, OSS Records (RG 226), Entry 99, Box 77, Folder 4, National Archives II.

103. OSS, *War Report of the O.S.S.*, 76-77, 241; *History of Schools and Training Branch*, OSS, 42-43, 47.

104. For location and the station complement, see L.B. Shallcross, deputy, Staff Training Branch/TRD, [Central Intelligence Agency], to John O'Gara, chief, Staff Training Branch, 1 February 1951, subject: Information on OSS Schools and Training Sites, OSS Records (RG 226), Entry 161, Box 7, Folder 76, National Archives II. "The Farm" of the OSS training was *not* the same site as "The Farm" later used by its successor, the CIA, which is located in southeastern Virginia.

105. *History of Schools and Training Branch*, OSS, 47-49.

106. Although it is clear that OSS Training Area D was on the east bank of the Potomac River, the precise location is still disputed, in part because there is are no remains of the facility and the OSS records regarding it are so sparse. Some veterans believe that it was at Indian Head, some contend Smallwood State Park, and other put it farther south. A contemporary reference by the Schools and Training chief in 1945, places it about 40 miles south of Washington, D.C., which would correspond to the identification of Smith's Point made in a 1951 CIA document which precisely and accurately identifies the sites of all the other OSS training schools. L.B. Shallcross, deputy Staff Training Branch/TRD, [Central Intelligence Agency], to John O'Gara, chief, Staff Training Branch, 1 February 1951, subject: Information [designation, location, type of training, and station complement] on OSS Schools and Training Sites, OSS Records (RG 226), Entry 161, Box 7, Folder 76, National Archives II.

107. John P. Spence, considered by the UDT-SEAL Association as the "First Frogman," to be one of the initial sailors assigned to Area D beginning in 1942 and who eventually served in OSS Maritime Unit L, telephone interview with the author, 28 January 2005.

108. Commander H. Woolley [Royal Navy] to Col. M.P. Goodfellow, 9 October 1942, subject: Guerrilla Notes—rough notes and suggestions [regarding use of Area D], M. Preston Goodfellow Papers, Box 4, Folder 4, Hoover Institution, Stanford, Calif.; *History of Schools and Training*, OSS, 135, which says portable buildings obtained from the Army; John P. Spence affidavit included in Tom Hawkins, "OSS Maritime," *The Blast*, (3rd Quarter 2000), 10-11, magazine of the UDT-SEAL Association; and layout of buildings in Area D, OSS Records (RG 226), Entry 85, Box 13, Folder 249, National Archives II.

109. OSS, *War Report of O.S.S.*, 225-227. The West Coast MU/SO training area at Tonyon Cove, 2-1/2 miles from Avalon on Catalina Island, reached its peak in 1944-1945 training teams for landings in the Pacific and the Far East.

110. "History of Schools and Training Branch, OSS," p. 10, attached to W[illiam] J. Morgan to Col. E.B. Whisner, 7 January 1949, pp. 7-8, OSS Records (RG 226), Entry 176, Box 2, Folder, 13, both in National Archives II.

111. The Area E facilities included E-1, the Inverness estate owned by Leslie Kieffer near Towson, which could handle 24 trainees; E-2, Oldfields School, Inc., owned by Maj. Watts Hill, near Towson, with a capacity of 77 trainees; and E-3, the Nolting estate, owned by Harry C. Gilbert, one miles east of Glencoe, Maryland, which had room for 64 trainees. For location and the station complement, see L.B. Shallcross, deputy Staff Training Branch/TRD, [Central Intelligence Agency], to John O'Gara, chief, Staff Training Branch, 1 February 1951, subject: Information [designation, location, type of training, and station complement] on OSS Schools and Training Sites, OSS Records (RG 226), Entry 161, Box 7, Folder 76; for trainee capacity, see layout of buildings, Areas E-2, E-2, E-3, in OSS Records (RG 226), Entry 85, Box 13, Folder 249, both in National Archives II.

112. Col. H.L. Robinson, Schools and Training Branch Order No. 1, issued 21 July 1944, effective 17 July 1944, OSS Basic Course, OSS Records (RG 226), Directors Office Files, Microfilm No. 1642, Roll 102, Frames 1120-1121, National Archives II.

113. Congressional Country Club, *The Congressional Country Club, 1924-1984* (n.p., n.d.; [Bethesda, Md., c. 1985]), 12, 26. I am grateful to Maxime D. Harvey, director of the club's membership services for supplying me with a copy of this volume.

114. "History of Schools and Training Branch, OSS," p. 10, attached to W[illiam] J. Morgan to Col. E.B. Whisner, 7 January 1949, pp. 12-13, OSS Records (RG 226), Entry 176, Box 2, Folder, 13, in National Archives II; the report explained also that additional factors leading to the decision to lease the club was its housing facilities, kitchen and dining room equipment, and grounds for tents or huts and the fact that a search had failed to obtain any other comparable facility available.

115. Ibid., 26, 33; see also A. William Asmuth, Jr., OSS Legal Division, to Col. Ainsworth Blogg, Schools and Training Branch, 26 April 1944, subject: Provisions of Lease with Congressional Country Club, Inc., OSS Records (RG 226), Entry 137, Box 3, Folder 24, National Archives II.

116. Joseph Kelley, an OSS veteran and member of the Congressional Country Club since 1949 quoted in "Society's Annual Meeting Held at Country Club Where OSS Trainees Once Blew Up the Greens," *OSS Society Newsletter*, Summer 2005,

117. Congressional Country Club, *The Congressional Country Club*, 27-31; Roger Hall, *You're Stepping on My Cloak and Dagger* (New York: Norton, 1957), 23-33; on the fatality on the bridge training exercise in July 1943, Albert Materazzi, veteran of the OSS Italian OG, telephone interview with the author, 26 January 2005.

118. Building and tent layout for Area F as of 5 Oct. 1943, in OSS Records (RG 226), Entry 85, Box 13, Folder 249, National Archives II.

119. Elizabeth P. ("Betty") McIntosh, oral history interview, 2 May 1997, pp. 5-6, conducted by Maochun Yu, OSS Oral History Transcripts, CIA Records (RG 263), Box 2, National Archives II.

120. Elizabeth P. ("Betty") McIntosh, telephone interview with the author, 12 March 2005. None of the male OSS veterans from Areas A, B, and C, in the two National Parks, whom the author interviewed, remembered ever seeing any OSS women, or any women at all, at those forested training areas, except for one dance held at Area C in 1945.

121. *History of Schools and Training Branch, OSS*, 29-31; see the dissatisfaction expressed in Lt. Col. Garland H. Williams to R.W. Billinghurst, 28 April 1942, Goodfellow Papers, Box 3, Folder OSS Correspondence, January to May 1942, Hoover Institution, Stanford, Calif.

122. After JCS was given supervision of the OSS in June 1942, a subcommittee of the Joint Psychological Warfare Committee of the JCS studied the issue of training saboteurs and commandos. It questioned SO training officers, observed the operation of one of the SO camps (presumably Area B), and compared OSS training with accounts of British SOE training and with a report on saboteur training in Germany based on interrogations of several captured Nazi saboteurs who had been landed by U-boat on the East Coast and quickly captured. The subcommittee reported that OSS training seemed suitable for saboteurs and their organizers, but given the type of training and missions involved, Army and Navy instructors should be used in the military and technical fields. Enclosure, "Functions of the Office of Strategic Services," JPWC 21/2, pp. 2-4, enclosed with Lt. Col. A.H. Onthank, secretary, to the subcommittee of the Joint Psychological Warfare Committee of the JCS, 28 July 1942, OSS Records (RG 226), Directors Office Files, Microfilm No. 1642, Role 64, Frames 952-958, National Archives II.

123. See *History of Schools and Training Branch, OSS*, 34-35, 48-49; Anthony Moore, British Liaison, "Notes on Co-Operations between SOE and OSS," January 1945, p. 2; OSS Records (RG 226), Entry 136, Box 158, Folder 1722; and a highly critical report by the newly named acting chief training officer, Capt. George E. Brewer, Jr., to Lt. Col. E. C. Huntington, Jr., 31 August 1942, subject: Problems Confronting the Training Unit, six typed pages, copy in OSS Records (RG 226), Entry 136, Box 161, Folder 1754, National Archives II.

124. On the reorganization, see memoranda from Colonel Donovan to All Branch Heads, 10 Aug. 1942; for Colonels Buxton, Goodfellow, Huntington, and Captain Doering, 17 Aug. 1942; and for Colonel Buxton, 23 Aug. 1942, all in OSS Records, Washington Directors Office-Op. 266, No. 519 (.S.O.); and on Goodfellow's naiveté in sometimes misjudging character, see William J. Donovan to "Dear Preston," 28 April 1942, Washington Director's Office Op-266, file No. 656; copies in CIA Records (RG 263), Thomas Troy Files, Box 2, Folder 19, National Archives II.

125. Wallace R. Deuel, "History of the OSS," typescript, 1944, Vol. 2, chapt. 30, p. 52ff. Deuel stated that the OSS guerrilla operations, particularly the OGs, outgrew the limitations set forth in JCS 155/4/D, 23 Dec. 1942 because theater commanders began in mid-1943 to ask SO chief Ellery Huntington to provide not merely organizers, fomenters and operational nuclei, which the JCS directive had authorized, but also full-fledged guerrilla companies. Plus a Norwegian general wanted Norwegian-speaking American solders for his forces. Deuel worked as one of Donovan's special assistants. This section of his typescript history was summarized by CIA historian Thomas Troy in the CIA Records (RG 263), Thomas Troy Files, Box 2, Folder 19, National Archives II. Deuel's typescript history remained classified by the CIA at the time the present study was completed. The

present author utilized the Deuel Papers at the Library of Congress and filed a request in 2008 for its declassification.

126. *History of Schools and Training, OSS,* 51. One of them, James Grafton Rogers, a former Assistant Secretary of State then chief of the OSS's Planning Group for covert action, belittled Goodfellow in his diary, as "a sort of promoter character, who ... is as irresponsible as a blue-bottle fly." James Grafton Rogers, diary entry of 3 August 1942, *Wartime Washington: The Secret OSS Journal of James Grafton Rogers, 1942-1943*, ed. Thomas F. Troy (Frederick, Md.: University Publications of America, 1987), 9.

127. John C. McWilliams, *The Protectors: Harry J. Anslinger and the Federal Bureau of Narcotics, 1930-1962* (Newark: University of Delaware Press, 1990), 161, 214n.

128. *History of Schools and Training Branch, OSS,* 35. Lt. Col. Garland H. Williams left OSS and was assigned to the U.S. Army's Airborne Command on 11 August 1942.

129. *History of Schools and Training Branch, OSS,* 51.

130. Robert E. Mattingly, *Herringbone Cloak–GI Dagger: Marines of the OSS,* Washington, D.C.: U.S. Marine Corps Headquarters, History and Museum Division, 1989, 109-11. Mattingly ignores Williams.

131. On Huntington's background, see McIntosh, *Sisterhood of Spies,* 21; Rogers, *Wartime Washington,* 14n.

132. Edwin Putzell, Oral History Interview, 11 April 1997, p. 13, CIA Records (RG 263), OSS Oral History Transcripts, Box 3, National Archives II.

133. Mattingly, *Herringbone Cloak–GI Dagger: Marines of the OSS,* 112-13.

134. The Training Directorate thus included Baker from SI, Huntington from SO, Lawrence W. Lowman from Communications Branch, and Brooker from SOE. *History of Schools and Training Branch, OSS,* 55-56. Baker had immense faith in Brooker's approach to training. Sweet-Escott, *Baker Street Irregular,* 143.

135. *History of Schools and Training Branch, OSS,* 53.

136. Lt. John A Bross, "Notes of Meeting with Geographic Desks, dated September 3, 1942," two page memorandum, 7 September 1942, OSS Records (RG 226), Entry 136, Box 161, Folder 1754, National Archives II.

137. Lt. Louis D. Cohen to Kenneth Baker, 18 December 1942, in response to "Student" to Baker, 8 December 1942, both in Appendix III, "Part Two of the History," OSS Records (RG 226), Entry 136, Box 158, Folder 1722, National Archives II.

138. The OGs were to be combat units behind enemy lines. Obtained from the Army, the men already had basic training. They were to be assembled primarily according to foreign language groups, then they would spend four to six weeks being trained as a unit by SO/OG instructors and by the officers who would lead them. OSS, Operational *Groups Field Manual–Strategic Services (Provisional), Strategic Services Field Manual No. 6,* issued 25 April 1944, pp. 7-13, in OSS Records (RG 226), Entry 136, Box 140, Folder 1465, National Archives II.

139. *History of Schools and Training Branch, OSS,* 66-68.

140. OSS General Order No. 9, establishing the Schools and Training Branch, was issued on 3 January 1943; *History of Schools and Training Branch, OSS,* 62.

141. Ibid., 62-63; see also Minutes of the OSS Executive Committee, June 23, 1945, p. 5, in black bound volume, "Minutes of Executive Committee, OSS, 19th meeting, May 11, 1943 through 43rd meeting, Oct. 19, 1943," in OSS Records (RG 226), Entry 139, Box 221, Folder 3097, National Archives II. Taking an indefinite leave from OSS, Baker was reassigned to the Army's Command and General Staff School at Fort Leavenworth, Kansas. He subsequently returned to OSS duty and wound up in southern France in late 1944, heading Special Forces Unit 4, the advance unit of Special Project Operations Center (SPOC), Algiers, and representing SPOC with the U.S. Seventh Army and the 6th Army Group. Erasmus H. Kloman, *Assignment Algiers: With the OSS in the Mediterranean Theater* (Annapolis, Md.: Naval Institute Press, 2005), 34-35.

472 | Endnotes: Chapter 2

142. Possibly due in part to that stress, Knox Chandler, a former journalist and college professor and the chief instructor at RTU-11, "The Farm," the SI finishing school, committed suicide in July. Chandler had sustained a head injury, and it may have contributed to his action. *History of Schools and Training Branch, OSS*, 63.

143. On the investigations of the OSS training schools see, for example, Minutes of the OSS Executive Committee, June 21, 1945, p. 2, and July 1, 1943, p. 4, in bound volume, "Minutes of Executive Committee, OSS, 19th meeting, May 11, 1943 through 43rd meeting, Oct. 19, 1943," in OSS Records (RG 226), Entry 139, Box 221, Folder 3097, National Archives II.

144. Table of organization for OSS Schools and Training Branch and OSS, in Lt. E.R. Kellogg to Chief, Presentation Branch, 2 November 1944, subject: Attachment, OSS Records (RG 226), Entry 133, Box 163, Folder 1384, National Archives II. See also *History of Schools and Training Branch, OSS*, 62-63, 85.

145. *History of Schools and Training Branch, OSS*, 85.

146. Blogg had also attended SOE schools in Britain in the late summer and early fall of 1942. On his background, see Ainsworth Blogg, Personnel File, CIA Records (RG 263), Accession #92-745, Box 10, National Archives II; Col. H.L. Robinson to Director, Strategic Services Unit (successor to OSS), 20 February 1946, subject: Recommendation for Citation [for Ainsworth Blogg], OSS Records (RG 226), Entry 133, Folder 814; Maj. John J. Sullivan to Col. William J. Donovan, 24 April 1942 [recommendation for Ainsworth Blogg], in OSS Records (RG 226), Director's Office Files, microfilm roll 36, frames 325-27, National Archives II; and Mrs. Dorothea ("Dodie") Dow, widow of Capt. Arden Dow, one of the instructors at Areas B and A, letter to the author, 6 August 2005.

147. Col. H.L. Robinson to Director, Strategic Services Unit, 20 Feb. 1946, subject: recommendation for citation [for Lt. Col. Philip K. (sic) Allen], OSS Records (RG 226), Entry 133, Box 105, Folder 814; and "History of Schools and Training Branch, OSS," pp. 7-8, attached to W. J. Morgan to Col. E.B. Whisner, 7 January 1949, OSS Records (RG 226), Entry 176, Box 2, Folder, 13, both in National Archives II.

148. *History of Schools and Training Branch, OSS*, 4, 77-78, 100, including the criticism from Joseph Anthony Kloman, former chief instructor at Area E, who had been moved to Area A-3.

149. In October 1943, OSS had 5,000 members and asked for 6,000 more. James Grafton Rogers, diary entry of 6 September 1943, *Wartime Washington*, 141.

150. *History of Schools and Training Branch, OSS*, 83; see also Kenneth H. Baker to Col. Edward Buxton, 13 March 1943, subject: Curriculum of Basic Training Course, reprinted as Exhibit A in ibid., 169-71. For an example, see *Operational Groups Field Manual—Strategic Services (Provisional)*, Strategic Services Field Manual No. 6, April 1944, Section IV, Training, pp. 10-13, OSS Records (RG 226), Entry 136, Box 140, Folder 1465, National Archives II.

151. John Waller, Oral History Interview, 27 January 1997, pp. 4, 6, CIA Records (RG 263), OSS Oral History Transcripts, Box 4, National Archives. Waller retired from the CIA as inspector general.

152. Maj. Gen. John K. Singlaub, US Army (Ret.), with Malcolm McConnell, *Hazardous Duty: An American Soldier in the Twentieth Century* (New York: Summit/Simon & Schuster, 1991), 32-33; also John K. Singlaub, telephone interview with the author, 11 December 2004.

153. Albert Materazzi, telephone interview with the author, 26 January 2005.

154. Frank Mills, Oral History Interview, 19 November 1996, p. 2, CIA Records (RG 263), OSS Oral History Transcripts, Box 3, National Archives II; on his background and deployment, see Francis B. Mills with John W. Brunner, *OSS Special Operations in China* (Williamstown, NJ: Phillips Publications, 2002).

155. Raymond Brittenham, Oral History Interview, 27 February 1997, pp. 13-14, 16, CIA Records (RG 263), Oral History Transcripts, Box 1, National Archives II.

156. Elizabeth McIntosh quoted in Miller, *Behind the Lines*, 59-60; see also Elizabeth McIntos telephone interview with the author, 12 March 2005.

157. "History of Schools and Training Branch, OSS," p. 35, attached to W[illiam] J. Morgan to Col. E.B. Whisner, 7 January 1949, OSS Records (RG 226), Entry 176, Box 2, Folder, 13, National Archives II.

158. A 1943 memorandum, for example, cited reports from China that "we have had at least eight men, who for various quirks in their make-up, have to be pulled from the field... should never have been sent to the field....Others simply won't fit anywhere. One was definitely a psychiatric case." OSS Assessment Staff, *Assessment of Men: Selection of Personnel for the Office of Strategic Services* (New York: Rinehart & Co., 1948), 4, 12-13.

159. W[illiam] J. Morgan, *The O.S.S. and I* (New York: Norton, 1957), 20-23, asserts that in developing these Selection Assessment Boards at the beginning of the war, the British combined the best features of German psychology, especially the techniques of a *Wehrmacht* psychologist named Simoneit, with the scientific, psychometric approach of American psychology. The Americans then developed this assessment system even further. A Yale trained psychologist, Morgan was sent by OSS in the summer of 1943 to observe the British SOE apply it to its agents, which he did for a year before parachuting into France in an SO mission in August 1944. See also OSS, *War Report of the O.S.S.*, 238-39.

160. "A Good Man is Hard to Find: The O.S.S. Learned How, with New Selection Methods that May Well Serve Industry," *Fortune*, 1946, 92-95, 218-19, 223. Station S was run first by Dr. Henry A. Murray of the Harvard Psychological Clinic, and subsequently by Dr. Donald W. MacKinnon, a Harvard trained psychologist from Bryn Mawr College. On the faculty at Station S in 1944 was a 31-year-old psychologist with a Ph.D. from Berkeley, John W. Gardner, later President of the Carnegie Corporation, Secretary of Heath, Education, and Welfare, and head of the consumer advocacy organization, Common Cause.

161. "OSS Assessment Program," pp. 15-16, Appendix IX, Part Three, Section F, of the "History [of the OSS]," typescript in OSS Records (RG 226), Entry 136, Box 158, Folder 1727, National Archives II.

162. Raymond Brittenham, Oral History Interview, 27 February 1997, pp. 14-15, CIA Records (RG 263), OSS Oral History Transcripts, Box 1, National Archives II.

163. "History of Schools and Training Branch, OSS,"pp. 39-41, attached to W[illiam] J. Morgan to Col. E.B. Whisner, 7 January 1949, OSS Records (RG 226), Entry 176, Box 2, Folder, 13; and John Waller, Oral History Interview, 27 January 1997, p. 4, CIA Records (RG 263), OSS Oral History Transcripts, Box 4, both in National Archives II.

164. "A Good Man is Hard to Find," *Fortune*, 1946, 94-95. As an OSS recruit, Sidney Harrow spent ten weeks at Area A in 1943 before being assigned to Station S. Mrs. May K. Harrow, Sidney Harrow's widow, telephone interviews with the author, October 17 and 27, 2005.

165. "History of Schools and Training Branch, OSS,"p. 39, 41-42, attached to W. J. Morgan to Col. E.B. Whisner, 7 January 1949, OSS Records (RG 226), Entry 176, Box 2, Folder, 13, National Archives II. The report indicated that at these three day assessments, "a surprising number of psycho-neurotics was found."

166. "History of Schools and Training Branch, OSS,"p. 42, attached to W[illiam] J. Morgan to Col. E.B. Whisner, 7 January 1949, OSS Records (RG 226), Entry 176, Box 2, Folder, 13, National Archives II. 167 OSS Assessment Staff, *Assessment of Men: Selection of Personnel for the Office of Strategic Services* (New York: Rinehart & Co., 1948), 58-202.

168. *War Report of the OSS*, 240-41; "The Test at Station S," *Time*, 21 Jan. 1946; "History of Schools and Training Branch, OSS,"p. 42, attached to W[illiam] J. Morgan to Col. E.B. Whisner, 7 January 1949, OSS Records (RG 226), Entry 176, Box 2, Folder, 13; and OSS Assessment Program," pp. 18-68, Appendix IX, Part Three, Section F, of the History [of the OSS], typescript in OSS Records (RG 226), Entry 136, Box 158, Folder 1727, National Archives II.

169. Edward Hymoff, *The OSS in World War II* (New York: Ballantine Books, 1972), 83. Hymoff, a Corps of Engineers' private from Boston, served briefly with the OSS in Italy and Yugoslavia, ibid., 10-13, and later became a newspaper reporter and author of several books on military history.

170. "A Good Man is Hard to Find," *Fortune*, 1946, 220-23.

171. OSS Assessment Staff, *Assessment of Men: Selection of Personnel for the Office of Strategic Services* (New York: Rinehart & Co., 1948); reprinted in paperback as Donald W. Fiske, Eugenia Hanfmann, Donald W. MacKinnon, James G. Miller, Henry A. Murray [OSS assessment professionals], *Selection of Personnel for Clandestine Operations: Assessment of Men* (Walnut Creek, Calif.: Aegean Park Press, 1996). For its continued impact, see Alfred H. Paddock, Jr., *U.S. Army Special Warfare, Its Origins: Psychological and Unconventional Warfare, 1941-1952* (Washington, D.C./Fort McNair: National Defense University Press, 1982), and Mattingly, *Herringbone Cloak— GI Dagger: Marines of the OSS*, 107.

172. *History of Schools and Training Branch, OSS*, 84; Col. H.L. Robinson, chief, Schools and Training Branch Order No. 1, issued 21 July 1944, effective 17 July 1944, subject: OSS Basic Course, OSS Records (RG 226), Entry 133, Box 163, Folder 1401, National Archives II.

173. Minutes, 3rd Meeting of [SI Advisory] Training Committee, 11 May 1944, p. 1; Minutes, 6th Meeting of the Training Committee, 29 May 1944, p. 2; see also Maj. Louis D. Cohen, Chief, SI Training, to Mr. Whitney H. Shepardson, S.I., 24 July 1944, subject: Training Program Washington; all in OSS Records (RG 226), Entry 146, Box 229, Folder 3239, National Archives II. In its defense, Schools and Training Branch complained that the operating branches retained their own ideas and programs for training their agents, resisted common training, and often failed to inform Schools and Training of their precise needs or how its training programs related to changing conditions and performance in the field overseas. *History of Schools and Training Branch, OSS*, 81-86.

174. Minutes of the [Eighth Meeting] of the [SI Advisory] Training Committee, 14 June 1944, p. 2, OSS Records (RG 226), Entry 146, Misc. Washington Files, Box 229, Folder 3239, "Advisory Training Committee," National Archives II.

175. Draft of Memorandum from SI, [June 1944], OSS Records (RG 226), Entry 146, Box 229, Folder 3239, National Archives II. Only three pages from the middle of this unsigned and undated draft memorandum remain in this folder; they are located between documents dated 24 May 1944 and 20 July 1944. The memo itself refers to the 14 June committee meeting. It is not clear that this memorandum was ever sent; nevertheless, it is certainly indicative of the existence of bitter condemnation of Schools and Training Branch at least in SI. Given the report from the conference on SI, SO, and X-2 representatives referred to in the Minutes of the Eighth Meeting of the SI Training Advisory Committee, 14 June [1944], it reflected similar sentiments in SO and X-2 as well. For additional evidence of the friction between the branches and Schools and Training, but also the ability of the branches to achieve at least a compromise from S&T, see Minutes of the 11th Meeting of the [SI Advisory] Training Committee, 6 July 1944, "a joint meeting of representatives of S.I., S.O., M.O., X-2 and personnel of S & T to discuss the [operating branches'] proposed changes in training," in ibid.

176. James L. McConaughy to Col. [G. Edward] Buxton, 20 July 1944, subject: Report of Mr. O'Gara, 15 July [a 10-page critical analysis of Schools and Training's program, by J.E. O'Gara of OSS Secret Intelligence Branch]. Both are in OSS Records (RG 226), Entry 146, Box 229, Folder 3239, National Archives II. John E. O'Gara was deputy director of OSS in charge of personnel. In civilian life, O'Gara had been general manager of Macy's Department Stores.

177. R. Boulton, vice chairman, S.I. Training Advisory Committee, to chief, S.I., 17 July 1944, subject: OSS Basic Two Weeks Course, OSS Records (RG 226), Entry 146, Box 229, Folder 3239, National Archives II.

178. R. Boulton, S.I., for the Training Representatives of SI, X-2, SO and MO Branches, to Col. H. L. Robinson [Schools and Training], 7 July 1944, subject: Meeting with Schools and Training Personnel, 6 July 1944; Training Board Meeting, 7 July 1944, "Notes on Discussion Regarding Area 'E' S.I., X-2, Basic Course Changes," OSS Records (RG 226), Entry 146, Box 229, Folder 3239, National Archives II.

179. *History of Schools and Training Branch, OSS*, 157-58. The Washington headquarters of

Schools and Training Branch had initially occupied part of Q Building in late 1943, but in early 1944, it moved to the North Building and finally to the first floor and basement of the Toner Building, a former public school in the District of Columbia, which it shared with the OSS Headquarters and Headquarters Detachment.

180. A number of OSS units were excluded from S&T's jurisdiction over training at least until late 1944. These included highly specialized technical units like the Communications Branch, the Maritime Unit, Counter-Espionage Branch (X-2), and the OSS Services Branch (reproduction, budget and procedures, procurement and supply), although some of them drew on S&T for supplies and school administration. By 1944 a number of them, such as Research and Analysis Branch, also sent their personnel for the OSS Basic Course, particularly when they were sending personnel overseas. *History of Schools and Training Branch, OSS*, 68-69.

181. OSS, *War Report of the O.S.S.*, 242-43; "History of Schools and Training Branch, OSS," pp. 44-50, attached to W[illiam] J. Morgan to Col. E.B. Whisner, 7 January 1949, OSS Records (RG 226), Entry 176, Box 2, Folder, 13, National Archives II.

182. On the West Coast, Area WP was set up at the Marine Base at Camp Pendleton near San Onofre north of San Diego. By the end of 1944, training Area WA on Santa Catalina Island off Los Angeles could handle 200 trainees (the sites on the island included Tonyon Cove for Maritime training, the largest facility; Howlands Landing for SO and MU; and 4th of July Cove for SO and MU training. An assessment station (WS) was opened at the Capistrano Beach Club in San Clemente on the mainland ; a West Coast Training Center (WN) responsible for maritime, Far Eastern Background, and assessment activities was set up in Newport Beach, connected through communications wire to the OSS West Coast headquarters in San Francisco. L.B. Shallcross, memorandum for John O'Gara, 1 Feburary 1951, subject: Information on OSS Schools and Training Sites, OSS Record (RG 226), Entry 161, Box 7, Folder 76, National Archives II.

183. "History of Schools and Training Branch, OSS,"pp. 31-32, attached to W[illiam] J. Morgan to Col. E.B. Whisner, 7 January 1949, OSS Records (RG 226), Entry 176, Box 2, Folder, 13, National Archives II.

184. In the one year of operation of the West Coast schools, nearly 1,000 men were given the Basic OSS Course, approximately 250 given Advanced SO training, 200 Advanced SI, and 100 Advanced MO. "History of Schools and Training Branch, OSS," pp. 31-32, 52, attached to W[illiam] J. Morgan to Col. E.B. Whisner, 7 January 1949, OSS Records (RG 226), Entry 176, Box 2, Folder, 13, National Archives II.

185. *History of Schools and Training Branch, OSS*, 81-82,84.

186. At the peak activity, six training areas and two assessment stations in California were turning out 100 students a month, including special groups of Chinese, Indonesians, Koreans, and a class of Japanese Americans (code named "Irish"). *History of Schools and Training Branch, OSS*, 13-24; *War Report of the OSS*, 251-53. In the one year of operation, the West Coast schools trained and dispatched 985 persons to the Far East. Col. H.L. Robinson to Director Strategic Services Unit, 20 Feb. 1946, Recommendation for Citation [for Philip K. Allen], p. 3, OSS Records (RG 226), Box 105, Folder 814, National Archives II. The Training Report for 23 January 1945 showed 132 enrolled in schools in the Washington, D.C. area and 199 enrolled in the West Coast training center as of 17 January 1945, reported in Minutes of the OSS Training Board Meeting of 25 January 1945, OSS Records (RG 226), Entry 133, Box 164, Folder 1414, National Archives II.

187. OSS Schools and Training Branch existed from January 1944 through the end of the war. During that period its Basic Espionage Schools graduated more than 1,800 trained men and women as operatives in gathering, analyzing and disseminating information. Its Paramilitary Schools, concerned with training saboteurs and guerrilla leaders, trained 1,027. "The Farm," which specialized in advanced intelligence training, graduated more than 800 men and women between May 1942 and December 1944. These figures cover only those trained for European operations, and they do not include specialized groups over which the Schools and Training Branch had little or at most divided control, such as Communications, Operational Groups, and the Maritime Unit.

During the operation of the West Coast training facilities in California, close to 1,000 personnel were given basic OSS training. Approximately 250 were given advanced SO training, some 200 had advanced SI training, and approximately 100 received Morale Operations training before being sent to the Far East. In addition to all of the above, countless other operatives, foreign or American, were given instruction in various OSS branch training programs in Europe, North Africa, or the Far East. OSS, *War Report of the O.S.S.*, 242-43.

188. History of Schools and Training, OSS," n.d. [1945-1947?], pp. 53-54, a 55-page-typescript, copy accompanied by a 7 January 1949 memorandum by Col. E.B. Whisner stating "Received this date from W[illiam]. J. Morgan [former OSS psychologist and SO team member] the following report: History of Schools and Training, OSS, Part I thru Part VI," OSS Records (RG 226), Entry 176, Box 2, Folder 13, National Archives II.

189. Ibid., 54, 7-8; for other examples, see OSS, *War Report of the O.S.S.*, 243.

190. History of Schools and Training, OSS," n.d. [1945-1947?], pp. 54-55, typescript accompanied by a 7 January 1949 memorandum by Col. E.B. Whisner stating "Received this date from W[illiam]. J. Morgan [former OSS psychologist and SO team member] the following report: History of Schools and Training, OSS, Part I thru Part VI," OSS Records (RG 226), Entry 176, Box 2, Folder 13, National Archives II.

CHAPTER 3

1. Maj. Garland H. Williams, "Training," memorandum, n.d. [January or February 1942], p. 6; located in OSS Records (RG 226), Entry 136, Box 161, Folder 1754, National Archives II, College Park, Md., hereinafter, National Archives II.

2. Lt. Col. H[enson]. L. Robinson to Col. Atherton Richards, OSS Planning Group, subject: Schools and Training Report [a 14-page, historical and geographical overview of the entire OSS training program]," 30 October 1943, p.1, OSS Records (RG 226), Entry 146, Box 162, Folder 1757, National Archives II.

3. Conrad L. Wirth, *Parks, Politics, and the People* (Norman: University of Oklahoma Press, 1980), 187, 189; see also a contemporary report on the program, U.S. Department of the Interior, National Park Service, "Emergency Conservation Work under National Park Supervision on State and County Parks, Metropolitan Sections of Municipal Parks and Resettlement Recreational Demonstration Projects," n.d. [c. 1937], in National Park Service Records (hereinafter NPS Records) (RG 79), Entry 54, Box 20, Folder "Virginia, R.S. 0-207, Reports," National Archives II.

4. On the origins of the U.S. National Park Service within the Department of the Interior, see Harland D. Unrau and G. Frank Williams, *Administrative History: Expansion of the National Park Service in the 1930s* (Denver, Colo.: NPS Denver Service Center Publication, 1983); Ronald A. Foresta, *America's National Parks and Their Keepers* (Washington, D.C.: Resources for the Future, Inc., 1984); Horace M. Albright as told to Robert Cahn, *The Birth of the National Park Service: The Founding Years, 1913-33* (Salt Lake City, Utah: Howe Bros., 1985); Richard West Sellars, *Preserving Nature in the National Parks: A History* (New Haven, Conn.: Yale University Press, 1997); U.S. National Park Service, *The National Parks: Shaping the System*, 3rd ed., rev. by Janet McDonnell (Washington, D.C. Govt. Printing Office, 2005).

5. Conrad Wirth was a landscape architect who worked with the National Capital Park and Planning Commission before he joined NPS in 1931. Sellars, *Preserving Nature in the National Parks*, 139-140. In 1934, Arno B. Cammerer, NPS Director, 1933-1940, established an NPS team to make a preliminary survey of recreational needs of the nation and their applicability to national, state, and local parks. Its members included George Wright, head of the Wildlife Division, Conrad Wirth, head of the Recreation, Land Planning and State Cooperation Division, and John Coffman, chief forester of the National Park Service. Coffman is a distant relative of the present author.

6. Wirth, *Parks, Politics, and the People*, 188-89. The estimated construction cost of $75,000 to $100,000 for each camp was in addition to the cost of acquiring the land, which NPS hoped to purchase for $5 an acre but which was later extended to an average of $10 per acre.

7. Wirth, *Parks, Politics, and the People*, 186. Wirth later served as NPS director, 1951-1963, initiating Mission 66, a large scale, long term project targeted at 1956-1966 for restoring and developing the national parks for post-war America.

8. Sellars, *Preserving Nature in the National Parks*, 140-142. NPS's emphasis on recreational expansion and management in the parks did generate criticism among some of it NPS's longtime supporters such as the Save the Redwoods League, the Wilderness Society and the National Parks Association, ibid., 142-145. Another area of expanded NPS responsibility in the 1930s was historic site management.

9. See the listing in Wirth, *Parks, Politics, and the People*, 187.

10. Conrad L. Wirth to Arthur E. Demaray, associate director of NPS, 22 April 1935, NPS Records (RG 79), Box 124, National Archives II.

11. "Chopawamic, Va., 'At the Small Lodge,' Preliminary Report," 28 January 1935, pp. 10-12; Arno B. Cammerer to C. Marshall Finnan, Supt., National Capital Parks, 2 February 1935; see also photographs and commentary in reports entitled "Observations and Photographs on Chopawamsic

Creek, December 17, 20, and on Quantico Creek, December 18, 24, 26, and 27, 1934; all in NPS Records (RG 79), E-47, Box 124, Folder 501, National Archives II.

12. Because of the multiple federal entities involved in the RDA project, it was actually the Resettlement Administration that acquired the land; NPS selected the site and would plan the facilities. Charles Gerner was the project manager and William R. Hall, assistant project manager from NPS for the Resettlement Administration. Hall acquired options to buy 12,400 acres, appraised at $181,000, for $139,000. Susan Cary Strickland, *Prince William Forest Park: An Administrative History* (Washington, D.C.: National Park Service, 1986), 9-10, 14-15; see also "Land Acquisition, Chopawamsic, Virginia, "Old Resettlement Administration File," closed 31 December 1937, NPS Records (RG 79), E-47, Box 126, Folder 630, National Archives II.

13. Strickland, *Prince William Forest Park*, 11-13.

14. Mary Edith Coulson, secretary of the Washington Council of Social Agencies, to William R. Hall, 13 July 1936, NPS Records (RG 79), RDA Files, Entry 47, Box 121, National Archives II.

15. The quotations are from the film, *Human Crop* (U.S. Dept. of the Interior, 1936), copy in Prince William Forest Park Nature Center; see also Strickland, *Prince William Forest Park*, 15-16, 19.

16. Reported in "14,000-Acre Park Approved by House," *Washington Post*, 6 August 1940, clipping in NPS Records (RG 79), E-47, Box 124, Folder 501,National Archives II. On the CCC, see John A. Salmond, *The Civilian Conservation Corps, 1933-1942: A New Deal Case Study* (Durham, N.C.: Duke University Press, 1967); see also John C. Paige, *The Civilian Conservation Corps and the National Park Service, 1933-1942: An Administrative History* (National Park Service, Department of the Interior, 1985). On the WPA, see Nick Taylor, *American-Made: The Enduring Legacy of the WPA: When FDR Put the Nation Back to Work* (New York: Bantam, 2008).

17. The three original CCC work camps (SP-22 near the current park headquarters; SP-26, near Camp 2, site of the Old Randal home off Route 626; and SP-26, near Camp 3) were abandoned in 1938-39. SP-26, near Camp 3, was re-established in March 1938. Two additional CCC camps were built in 1941 and which operated from June 1941 through February 1943. Their mission was primarily defense-related construction work at military facilities in the area, such as Fort Belvoir, the Corps of Engineers' main training facility. These two new CCC camps were designated NP-D-5 and NP-D-16, near Camp 3. John Gum was the superintendent for NPS of these two CCC camps, which came under the War Department. See the listings in Strickland, *Prince William Forest Park*, 17, 24. Most of the buildings in these CCC camps were demolished in 1959-1960.

18. Progress Report, 1 July 1936, NPS Records (RG 79), Entry 54, Box 121, and W.R. Hall, project manager, "Report on the First Open House Held at Chopawamsic on Wednesday, August 16, 1939," NPS Records (RG 79), E-47, Box 124, Folder 501, National Archives II.

19. Strickland, *Prince William Forest Park*, 1986), 7. Public Law 763, Chapter 663, 76th Cong., 3rd Sess., approved 13 August 1940.

20. Beginning in 1934, Lykes was assigned to the NPS office in Atlanta, Georgia, where he worked in various capacities on Recreational Demonstration Area projects in a number of southeastern states, returning to Richmond in 1937 to handle special services, including publication of the *Regional Review*. In 1939, he was at various times in charge of all Virginia Wayside development, Superintendent of North Carolina Beach Erosion Control Project (later Cape Hatteras National Seashore), Superintendent of Kings Mountain National Military Park, South Carolina, and NPS Assistant State Inspector for FloridaIra B. Lykes' biographical information obtained from various documents, including Temporary Appointment, Transfer, Reinstatement, or Promotion Form for US Civil Service Commission, 17 July 1941; Loyal Data Form, 17 November 1947; "Profiles: Ira B. Lykes," unidentified clipping; all in Ira B. Lykes, Personnel File, National Archives and Records Administration, National Personnel Records Center, St. Louis, Mo., obtained by the author, 26 April 2007.

21. Mrs. J. Atwood Maulding, Director of Personnel, Department of the Interior, to Ira B. Lykes, 24 August 1940 in Ira B. Lykes, Personnel File, National Archives and Records Administration, National Personnel Records Center, St. Louis, Mo.

22. Physical description of Ira B. Lykes in early 1940s provided by the widow of Capt. Arden Dow, chief instructor at OSS Training Area A-4 in Chopwamsic RDA, Mrs. Dorothea ("Dodie") D. Dow, Athens, Georgia, telephone interview with the author, 16 June 2005.

23. Ira B. Lykes, oral history interview, 23 November 1973, by S. Herbert Evison, pp. 30-31, tape number 261. I am indebted to David Nathanson, NPS, Harpers Ferry Center, Harpers Ferry, W. Va., with responding on 1 November 2005, to my request for a transcript of the Lykes' oral history interview.

24. "The Effects of Segregation on Park Management," chapter three in Strickland, *Prince William Forest Park*, 39-49.

25. For itemization of the racial identification and the uses of the five camps (plus the CCC camp), see the list in Strickland, *Prince William Forest Park*, 43, which is based on the Minutes of a meeting of National Capital Parks officials and representatives of the Social Service Agencies of Washington area, 11 December 1942. See also the Report on Camp Good Will and Camp Pleasant in Chopawamsic RDA for the summer of 1938 submitted to the NPS, 11 October 1938 by William H. Savin, Director of the Family Service Association of Washington, D.C.; and for appraisals and pictures of some of the camps, see Camp Appraisal Reports on the use of Camp 2-G for July and August in both 1938 and 1939 by the Jewish Community Center, Washington, D.C., and on Camp 2 for summer 1941 by the Arlington, Virginia, Council of Girls Scouts; in "Camp Appraisal Reports, Chopawamic, Virginia [1938-1941]," NPS Records (RG 79), E-47, Box 123, Folders 207-25 and Folder 314, National Archives II.

26. Ira B. Lykes, superintendent, 1939-1951 of Prince William Forest Park, oral history interview by S. Herbert Evison, 23 November 1973, Audiotape #261, transcript, pages 38-39, both located at NPS, Harpers Ferry Center Library, Harpers Ferry Center, W. Va.

27. Strickland, *Prince William Forest Park*, 42-47.

28. Ira B. Lykes, retired in Venice, Florida, interviewed 4 September 1985, by Susan Cary Strickland, and quoted in Strickland, *Prince William Forest Park*, 23.

29. Strickland, *Prince William Forest Park*, 23-24.

30. Arthur Demaray to Lt. Col. R. Valliant, 31 August 1938, NPS Records (RG 79), Entry 54, Box 121, National Archives II.

31. Ira Lykes, Monthly Report, May 1942, quoted in Strickland, *Prince William Forest Park*, 25-26.

32. Matt Huppuch to Conrad Wirth, 12 July 1934, NPS Records (RG 79), RDA Program Files, Box 57, National Archives II. See also Edmund F. Werhle, *Catoctin Mountain Park: An Historic Resource Study* (Washington, D.C.: National Park Service, 2000), 159-160.

33. Werhle, *Catoctin Mountain Park*, 160-161.

34. "Catoctin Park Plans Menaced as Owners Refuse to Sell Land," *Washington Post*, 25 June 1935, 1.

35. To avoid confusion, the present study will refer to the two Garland Williams by using either their full names with different middle initials or by reference to "Mike" Williams for the NPS project manager and Major, later Lieutenant Colonel Williams for the official in Donovan's organization. Lieutenant Colonel Williams left the OSS in July 1942; Mike Williams was the NPS manager at Catoctin from 22 January 1935 until his retirement on 31 August 1957.

36. Documents in the personnel file of Garland B. ("Mike") Williams from the National Archives and Records Administration, National Personnel Records Center, Civilian Personnel Records, St. Louis, Mo, obtained by the author, 8 February 2005.

37. See the appointment documents signed by H.C. Bryant, acting associate director, NPS, FERA Land Program, and Guy W. Numbers, acting chief, Division of Appointments, Mail, and Files of NPS; personnel file of Garland B. ("Mike") Williams National Archives and Records Administration, National Personnel Records Center, St. Louis, Mo.

38. Werhle, *Catoctin Mountain Park*, 163-166; Barbara M. Kirkconnell, *Catoctin Mountain Park: An Administrative History* (Washington, D.C.: National Park Service, 1988), 17.

39. *Catoctin Clarion*, 18 October 1935, 1.

40. Werhle, *Catoctin Mountain Park*, 168-171.

41. A.P. Bursley to Garland B. Williams, 29 May 1936, Catoctin Mountain Park Archives, Catoctin Mountain Park Office.

42. Wehrle, *Catoctin Mountain Park*, 176-178; Kirkconnell, *Administrative History of Catoctin Mountain Park*, 35-39.

43. Although Hi-Catoctin was planned as a boys' camp, it was used in the summers of 1939 to 1941 for family camping by the Federal Camp Council of Washington, D.C., a group that offered camping vacations for federal employees. Kirkconnell, *Administrative History of Catoctin Mountain Park*, 30, 41.

44. "Catoctin Recreational Demonstration Area, Maryland, General Information," material for an informational leaflet on Catoctin Recreational Demonstration Area including capacity numbers or each camp, included in E. M. Lisle, acting regional director, memorandum for the director, attn: editor-in-chief, 9 May 1939, NPS Records (RG 79), Entry 47, RDA Files, Box 59, Folder 501, National Archives II.

45. Although the park would benefit the public, many of the mountain people who had to give up their homes remained disgruntled even decades later. Werhle, *Catoctin Mountain Park*, 172-175.

46. CCC Company Number 1374, arrived from Quantico, Virginia where it had spent the previous four years constructing cabins and other facilities at Recreational Demonstration Areas like Chopawamsic RDA, in the area around Washington, D.C. John C. Paige, *The Civilian Conservation Corps and the National Park Service, 1933-1942: An Administrative History* (Washington, D.C. National Park Service, 1985), 50.

47. *Catoctin Clarion*, 28 April 1939.

48. That station was enlarged in 1964 and served as the visitor center and park administrative headquarters until 1973.

49. When the CCC unit departed, Mike Williams was left with only 30 WPA workers. On the CCC projects at Catoctin, see Werhle, *Catoctin Mountain Park*, 178, 185-186, 194-196; and Kirkconnell, *Administrative History of Catoctin Mountain Park*, 43-49, 71-75.

50. The crippled children temporarily occupied Camp Number 1 (Misty Mount) as soon as it was completed in 1937, but moved to the new special needs facility, Camp Number 2 (Greentop), when it opened in 1938.

51. Kirkconnell, *Administrative History of Catoctin Mountain Park*, 65-69. Racial segregation was still widespread in American society in the 1930s, reinforced by state laws and local ordinances in the South and often by custom in the North. Secretary of the Interior Harold Ickes supported the idea of making New Deal programs and benefits to all Americans, but he was also aware of the power of local customs and conditions. Thus black and white campgrounds, such as those at Chopawamsic RDA, were in practice segregated. However, Catoctin RDA did not provide for any black campgrounds, despite an effort by several prominent African Americans and black aid organizations in Baltimore in 1939 to have a camp for crippled black children constructed next to that for the white crippled children at Greentop. A 1940 NPS master plan recommended "a comprehensive development outline for Negro use." However, Catoctin RDA remained exclusively for white use in this period. Wehrle *Catoctin Mountain Park*, 179-80. Wehrle, Appendix 17, lists the organizations that did use Catoctin's cabin camp facilities in the prewar period.

52. Werhle, *Catoctin Mountain Park*, 194.

53. "Use of organized camps by British Sailors [Catoctin]," report, n.d., in NPS Records (RG 79), Entry 7, Central Classified File, 1939-1949, General, 201 National Defense-General, Box 73, Black Binder, "War Uses., C," National Archives II. Some 7,000 British sailors enjoyed outdoor relaxation in NPS Recreational Demonstration Areas in the summer of 1941; By 1942, some 22,000 British sailors had participated "British Seamen Rest and Romp in U.S. Camps," *New York Herald-Tribune*, 1 March 1942

54. Matt C. Huppuch, chief, recreational use and consulting service, NPS, memorandum for Conrad Wirth, 14 June 1941, subject: Cooperation with the Navy Department, NPS Records (RG 79), Entry 47, Box 57, Folder, 201; Garland B. Williams to Regional Director, report, 16 September to 1 November 1941, and statistical report to Conrad Wirth, 4 July 1944; all in NPS Records (RG 79), Entry 7 Central Classified Files, 1933-1949, 201 National Defense, C., Box 79, Folder 201, all in National Archives II. Lontime Thurmont resident, James H. Mackley, who was 12 in 1941, remembers that some of the British sailors visited the local school. "One of the sailors gave me a [British] penny. It was really big, the size of a half dollar." James H. Mackley, telephone interview with the author, 13 June 2006.

55. Formerly director of the Save the Redwoods League in California, Drury was the first NPS director not to have prior experience in National Park management. Sellars, *Preserving Nature in the National Parks*, 150; on Drury's pre-NPS career, see Susan R. Schrepfer, *The Fight to Save the Redwoods: A History of Environmental Reform, 1917-1978* (Madison, University of Wisconsin Press, 1983), 23-76. For his appointment, see diary entry of 23 June 1940, Harold L. Ickes, *Secret Diary of Harold L. Ickes*, 3 vols. (New York: Simon & Schuster, 1954), 3:213.

56. Newton B. Drury, "Former Directors Speak Out," *American Forests*, 82/6 (June 1976): 28; and Foresta, *America's National Parks*, 48-49.

57. Newton B. Drury to the Secretary of the Interior, 28 July 1950, NPS Records (RG 79), Entry 19, National Archives; for an earlier statement, see Drury, "What the War Is Doing to National Parks and Where They Will Be at Its Close," *Living Wilderness* (May 9, 1944): 11.

58. The number of permanent fulltime NPS employees was cut from 6,000 in November 1941 to 1,600 in June 1944. NPS headquarters was moved to Chicago for the duration of the war. The basic NPS budget (for administration, maintenance, and operations) remained stable at around $4.5 million a year, but appropriations for land acquisition, construction, and development were cut from $26 million in fiscal 1940 to $16 million by fiscal 1943, and then completely eliminated for the rest of the war. Charles W. Porter III, History Branch, NPS, "National Park Service War Work, December 7, 1941-June 30, 1944," pp. 47-48, a 63 page typescript report, n.d. c. 1944, in NPS Records (RG 79), Director Newton B. Drury's Files, Entry 19, Box 23, National Archives II.

59. Newton B. Drury to Secretary of the Interior, subject: War Times Uses of the National Park," 28 July 1950 [during the Korean War], NPS Records (RG 79), Papers of Newton B. Drury, Entry 19, Box 23, Folder, "Wartime Experience of N.P.S." Charles W. Porter, III, "Supplement Covering National Park Service War Work, June 30, 1944 to October 1, 1945," p. 13, n.d., c. 1945, 16 page typescript, accompanying "National Service War Work" NPS Records (RG 79), Papers of Newton B. Drury, Entry 19, Box 23, National Archives II.

60. Visitors to the national parks dropped from 21 million in fiscal 1941 to 10 million in 1942, and reached the lowest point of 7 million in 1943, but there 8 million in 1944, 10 million in 1945, and back to 21 million in 1946. During the war, a total of 8 million visitors were men and women in uniform. Newton B. Drury to Assistant Secretary of the Interior Doty, subject: War Times Uses of the National Parks," 2 August 1950 [during the Korean War], NPS Records (RG 79), Papers of Newton B. Drury, Entry 19, Box 23, "Wartime Experience of N.P.S." folder, National Archives II.

61. Newton B. Drury statement in the National Park Service, Annual Report to the Secretary of the Interior, 1942, reprinted in Charles W. Porter III, History Branch, NPS, "National Park Service War Work, December 7, 1941-June 30, 1944," p. 1. For similar during the war as NPS sought support from conservation groups, see Newton B. Drury, "The National Parks in Wartime," *American Forests*

(August 1943); and Carl P. Russell, National Park Service, "The National Parks in Wartime," *Outdoor America* (July-August 1943), clippings, which along with the above document are in NPS Records (RG 79), Papers of Newton B. Drury, Entry 19, Box 23, "Wartime Experience of N.P.S." folder, National Archives II.

62. Memorandum from NPS Director Newton B. Drury to Washington Office and all Field Offices, 27 November 1940, reprinted in "National Park Service War Work, December 7, 1941-June 30, 1944" pp. 2-3, NPS Records (RG 79), Director Newton B. Drury's Files, Entry 19, Box 23, National Archives II.

63. Newton B. Drury, Oral History Interview, "Parks and Redwoods, 1919-1971," conducted by Amelia Roberts Fry and Susan Schrepfer, 2 vols., Regional Oral History Office, Bancroft Library, University of California, Berkeley, 1972, 2:405-6.

64. In keeping with the NPS criteria for preserving the parks, Interior initially resisted the War Department's request for the properties. In mid-March 1942, Donovan's office had sent up the chain of command a letter stating that there was a "military necessity" for obtaining substantial parts of the two RDAs for a military training program. The classified "secret" letter of 11 March 1942 from the Coordinator of Information asserting the military necessity of acquiring parts of the parks, in this case Catoctin, is cited in Maj. Gen. Thomas H. Robins, assistant chief of engineers, to Commanding General of the Services of Supply, 19 March 1942, subject: Land Acquisition–Troop Housing, Vicinity of Washington, D.C. copy in World War II File, General Information Folder, Catoctin Mountain Park Archives, Catoctin Mountain Park, Thurmont, Md.

65. Secretary of War Henry L. Stimson to Secretary of the Interior Harold L. Ickes, 24 March 1942, NPS Records (RG 79), Central Classified File, 1933-1949, General, 201 National Defense, Box 73, Folder 201, National Archives II.

66. Secretary of the Interior to Secretary of War, 25 March 1942, copy in National Park Service Records (RG 79), Central Classified File, 1933-1949, General, 201 National Defense, Box 73, Folder 201, National Archives II. *Kirkconnell's Administrative History of Catoctin Mountain Park*, page 85, errs in listing the permit to 1 June 1943. Ickes' letter granted it only to 1 June 1942.

67. Ira B. Lykes to Conrad L. Wirth, Confidential memorandum, 30 March 1942, with copy of 19 April 1937 Rules and Regulations, in National Park Service Records (RG 79), Central Classified File, 1933-1949, General, 201 National Defense, Chopawamsic, Project CHO-IV-1&1A, Box 73, Folder 201, National Archives II. The two officers were probably Capt. William J. Hixson, the first commanding officer of Detachment A and Lt. Rex Applegate, an instructor in pistol shooting at Area B, both of whom returned to Chopawamsic on Tuesday, 31 March 1942 with Lt. Col. Garland H. Williams

68. A year later, an OSS report indicated that "we have utilized the property [at Catoctin] for the Office of Strategic Services since April 1, 1942...." "F.J.B." [Lt. F.J. Ball] to Major [Otto C.] Doering, Executive Officer OSS HQ, 30 March 1943, subject: Release of Area "B," OSS Records (RG 79), Entry 137, Box 3, Folder 24, National Archives II.

69. The *Baltimore American*, 5 April 1942; the columnist was J. Hammond Brown.

70. Karl E. Pfeiffer to Conrad L. Wirth, 6 April 1942 and Wirth to Pfeiffer, 13 April 1942, National Park Service Records (RG 79), Central Classified File, 1933-1949, General, 201 National Defense, Box 73, Folder 201, National Archives II.

71. "Catoctin Area Is Ordered Closed to the Public," *Catoctin Enterprise*, 10 April 1942, 1.

72. Garland B. Williams, Narrative Reports for months of April 1942 and June 1943; NPS Records (RG 79), Entry 47, RDA Files, Boxes 57 and 61, National Archives II.

73. Telephone interviews by the author with longtime residents of Thurmont, Maryland: John Kinnaird, Frank Long, 7 June 2006; James H. Mackley, 13 June 2006; and with longtime Triangle, Virginia resident, Robert A. Noile, 27 April 2007.

74. Ira B. Lykes, oral history interview by S. Herbert Evison, 23 November 1973, Audiotape #261, transcript, page 39, both located at NPS's Harpers Ferry Center Library, Harpers Ferry Center, W. Va.

75. Newton B. Drury to Assistant Secretary of the Interior Oscar L. Chapman, 15 April 1942; Stimson to Ickes, 22 May 1942, both in NPS Records (RG 79), Entry 47, Box 126, Folder 901; Herbert Evison, acting superintendent of land planning, memo to Garland B. (Mike) Williams, 11 June 1942; and G. B. Williams, memorandum to the director, attn: Mr. Wirth, 14 June 1942; both in (RG 79), Entry 47, RDA Files, Box 61, Folder, Catoctin; and Herbert Evison, NPS, memo for the manager of Chopawamsic RDA, 16 June 1942, NPS Records (RG 79), Entry 47, RDA Files, Box 124, Folder 600, all in National Archives II.

76. Secretary of the Interior Harold L. Ickes to Secretary of War Henry L. Stimson, 22 April 1942; Stimson to Ickes, 3 May 1942, NPS (RG 79), Central Classified File, 1933-1949, General, 201 National Defense, Box 73, Folder 201, National Archives II.

77. John J. Dempsey, Acting Secretary of the Interior to Secretary of War, 16 May 1942, and "Special Use Permit Authorizing Use of Land in the Catoctin Recreational Demonstration Area, Maryland, by the War Department for National Defense Purposes" signed by Dempsey, 16 May 1942 both in NPS Records (RG 79), Central Classified File, 1933-1949, General, 201 National Defense, Box 73, Folder 201. Special Use Permit for Chopawamsic, 16 May 1942, with identical provisions, is in (RG 79), Central Classified File, 1933-1949, General, 201 National Defense, Chopawamsic, Project CHO-IV-1&1A, Box 73, Folder 201, National Archives II. Dempsey's original letter of 16 May 1942 had accepted "for the duration of the presentation emergency," but the permit, pre-dated 16 May 1942 (apparently neither written nor delivered until mid-June), listed no expiration date, indicating instead that the permit was "revocable at will by the Secretary of the Interior," which was apparently as far as Ickes was willing to go.

78. By 12 March 1942, the Coordinator of Information's office had allocated funds for the acquisition and construction of training centers, and within a week after the Secretary of the Interior had signed the temporary permit on 25 March, a representative of the War Department's Division of [Land] Acquisition arrived at Chopawamsic RDA W.O. Hall, to Bernard L. Gladieux, 12 March 1942, subject: status of K & L Activities—COI, in Records of the Bureau of the Budget (RG 51), Series 39.19, Emergency and War Agencies, Folder, COI-General Administration. Hall reported that in regard to the K & L (unvouchered funds) activities, the Office of the Coordinator of Information had allocated around $2 million for the purchase of a vessel for secret missions, for airplanes for secret missions, and for the construction of training centers in the United States and England for enlisted men and volunteer personnel for special missions. A copy of the document is in CIA Records (RG 263), Thomas Troy File, Box 2, Folder 19, National Archives II.

79. Ira B. Lykes to Conrad L. Wirth, confidential memorandum, 30 March 1942 [a separate memorandum from the one of the same date cited above], in NPS Records (RG 79), Central Classified File, 1933-1949, General, 201 National Defense, Chopawamsic, Project CHO-II-1, Box 81, Folder 201, National Archives II. A longtime resident of the area, Gum supervised CCC camp ND-16.

80. Ira B. Lykes to Conrad L. Wirth, confidential memorandum, 31 March 1942, in NPS Records (RG 79), Central Classified File, 1933-1949, General, 201 National Defense, Chopawamsic, Project CHO-II-1, Box 81, Folder 201, National Archives II.

81. The 14,446 acre figure is from Ira B. Lykes to F. F. Gillen, December 1941, NPS Records (RG 79), File 1460/3, which in 1986 was at the Washington National Records Center, Suitland, Maryland, and was cited in Strickland, *Prince William Forest Park*, 28. Archivists believe that those NPS records, including Ira Lykes' correspondence and monthly reports cited by Strickland were later transferred to National Archives II in College Park, Maryland, but they cannot be found there. Telephone conservations by the author with Joseph N. Groomes, Washington National Records Center, 14 October 2005; Joseph D. Schwarz, archivist for NPS records at the National Archives II, 17 October 2005; and Michael Grimes and Rene Jaussaud of NPS's National Capital Region, 21 October 2005.

82. The transactions were handled by the Army Corps of Engineers' regional Real Estate and Land Acquisition Office in Baltimore. Capt. Stanley H. Lyson, S-4 (Supply), OSS Schools and

Training Division, to Lt. Warner, undated [c. February 1945], subject: Data on Detachment A, OSS Records (RG 226), Entry 136, Box 153, Folder 1658, National Archives II.

83. Kirkconnell, *Administrative History of Catoctin Mountain Park*, 92. For details, see War Department, Corps of Engineers, Real Estate Branch, "Map Showing Government Property, Catoctin Area," 26 May 1942, NPS Records (RG 79), Entry 21, Office of the Chief Counsel, Legislature File, 1932-1950, Box 76, Folder HR 3807; and "Report Furnished Director in Response to Memorandum of August 30, 1943," "County of Frederick," "County of Washington," NPS Records (RG 79), Entry 47, RDAs, Box 60, Folder 601, both in National Archives II.

84. Map showing Government Property Catoctin Area, prepared by the War Department, Corps of Engineers, Real Estate Branch, originally 26 May 1942, but updated, in NPS Records (RG 79), Entry 21, Office of the Chief Counsel, Legislative File, 1932-1950, Box 76, Folder "HR 3807, National Archives II. 85 M. Preston Goodfellow, interviewed for the OSS official history in 1945, and quoted in *History of Schools and Training Branch*, OSS, 29.

CHAPTER 4

1. F.J. B., Jr. [Lt. F. J. Ball, Jr.], OSS, to Major [Otto C.] Doering [Donovan's Executive Officer at OSS HQ], 30 March 1944, subject: Release of Area "B," OSS Records (RG 226), Entry 137, Box 3, Folder 24, National Archives II, College Park, Md.

2. Maj. Ainsworth Blogg, Area B, commanding officer, Area B, April 1942 to May 1943, OSS personnel file, CIA Records (RG 263), National Archives II Reference to his previous employment is from Mrs. Dorothea ("Dodie") Dow, widow of Capt. Arden W. Dow, an instructor at Area B, telephone interview with the author, 15 May 2005.

3. Charles M. Parkin, OSS personnel file, CIA Records (RG 263), National Archives II. Blogg was the only other person from the Office of the Coordinator of Information at Area B when Parkin arrived. Charles M. Parkin, telephone interview with the author, 10 May 2005.

4. William R. Peers and Dean Brelis, *Behind the Burma Road: The Story of America's Most Successful Guerrilla Force* (Boston: Little, Brown, 1963), 27, 30-32, and passim. Troy J. Sacquety, a civilian historian with the U.S. Army Special Operations Command at Fort Bragg, N.C., and a specialist on OSS Operational Groups, is currently writing a history of Detachment 101 based on his doctoral dissertation from Texas A&M University. Troy Sacquety, telephone interview with the author, 25 April 2008.

5. G. B. ("Mike") Williams to Capt. Charles M. Parkins [sic; Parkin], [Acting] Commanding Officer, Area B-2, 2 October 1942, World War II Files, Correspondence Folder, Catoctin Mountain Park Archives, Catoctin Mountain Park, Thurmont, Md.

6. G.B. ("Mike") Williams, Memorandum for the Commanding Officer, Camp No. 1, USMC, 13 February 1943, p. 1, World War II Files, Correspondence Folder, Catoctin Mountain Park Archives, Thurmont, Md.

7. G.B. ("Mike") Williams, Confidential Memorandum for the Director [of NPS], 28 January 1943, p. 2, copy in World War II Files, Correspondence Folder, Catoctin Mountain Park Archives, Thurmont, Md. James H. Mackley, a longtime resident of Thurmont and a friend of one of Williams' sons, Aulick ("Ookie"), said the Williams family lived in a house on east Main Street during the war. James H. Mackley, Thurmont, Md., telephone interview with the author, 13 June 2006. The original residence for the Park Manager, located behind the Visitor Center, burned down in 1945 and was rebuilt in 1947. Mackley may have been thinking of the 1945-47 period.

8. R.P. Tenney to J[oseph]. R. Hayden [political scientist, member of COI Board of Analysts, and first head of training for the Secret Intelligence Branch], 8 June 1942, interoffice memo, subject: "Area B," OSS Records (RG 226), Entry 136, Box 158, Folder, 1721, National Archives II.

9. Barbara M. Kirkconnell, *Catoctin Mountain Park: An Administrative History* (National Park Service, 1988), 86, and note 58, lists the date of the announcement of the closing of the Foxville-Deerfield Road as both 22 June 1942 and 22 June 1943.

10. J. Mel Poole, Superintendent, and James Voigt, Resource Manager, Catoctin Mountain Park, interviewed by the author during a tour of the park with three OSS veterans. Mel Poole said the park's original address was Lantz rather than Thurmont as it is today. The northern entrance and the route to B-2 via B-5 (instead of via Cabin Camp 1 as it currently is from the main entrance on Route 77 on the south) was remembered and retraced in a tour of the park and the training camp arranged by the author for OSS veterans, Charles M. Parkin, Frank A. Gleason, and Reginald G. Spear, 18 May 2005.

11. Jerry Sage, *Sage* (Wayne, Pa.: Miles Standish Press, 1985), 12.

12. None of the three OSS veterans taken on a tour of the park and training site on 18 May 2005 mentioned a swimming pool; nor is it referred to in reports by visitors or staff during the war.

13. R.P. Tenney to J[oseph]. R. Hayden [political scientist, member of COI Board of Analysts, and first head of training for the Secret Intelligence Branch], 8 June 1942, interoffice memo, subject: "Area B," OSS Records (RG 226), Entry 136, Box 158, Folder, 1721, National Archives II.

14. The OSS Visual Presentation Branch drew up charts of building layout and complement size for each training area in the fall of 1943. See "Area B-2," map of buildings, including total capacity, 5 October 1943, in OSS Records (RG 226), Entry 85, Box 13, Folder 249, National Archives II.

15. Comments of Frank A. Gleason, OSS instructor at B-2, June 1942 to March 1943, on a tour of the park, arranged by the author, 18 May 2005.

16. Charles M. Parkin, OSS executive officer and instructor at Area B-2, comments on a tour of the park arranged by the author, 18 May 2005, and in telephone interview with the author, 10 May 2005. For similar, see the memoir of another instructor, Jerry Sage, in Sage, *Sage*, 25-26.

17. Reginald G. ("Reg") Spear, who was a Special Operations trainee at Area B-2 in early 1944, comments made during a tour of the park arranged by the author, 18 May 2005.

18. "List of Improvements to Property by War Department...Restoration Survey, Catoctin Training Center, Thurmont, Maryland," 31 October 1945, 1, 3, included with survey by G.B. ("Mike") Williams, memorandum for Regional Director, NPS Region One, 7 March 1946, WWII Files, General Information Folder, Catoctin Mountain Park Archives, Catoctin Moutain Park, Thurmont, Md.

19. G. B. ("Mike") Williams, custodian, to Regional Director, 30 January 1944, NPS Records (RG 79), Entry 7, Central Classified File, 1933-1949, General, 201 National Defense, C [Catoctin], Box 79, Folder 201, National Archives II.

20. J. Mel Poole, Superintendent, Catoctin Mountain Park, on the tour of the park with three OSS veterans arranged by the author, 18 May 2005. The veterans touring with the author and park officials on 18 May 2005 did not remember so many large buildings on the site as there are now. The large structure near the current dining hall, for example, is the comfort station for the area, and was built after the camp was put on a septic system in the late 1940s and early 1950s.

21. Reginald G. ("Reg") Spear, on a tour of the park, tape recorded by the author, 18 May 2005.

22. OSS veterans, Frank A. Gleason and Reginald G. ("Reg") Spear, on a tour of the park, tape recorded by the author, 18 May 2005.

23. Observations of Frank A. Gleason, instructor at B-2, June 1942 to March 1943, on a tour of the park tape recorded by the author, 18 May 2005.

24. J. Mel Poole, Superintendent, Catoctin Mountain Park, on the tour of the park with three OSS veterans, tape recorded by the author, 18 May 2005.

25. Observations of Frank A. Gleason, instructor at B-2,on a tour of the park with the author, 18 May 2005.

26. J. Mel Poole, Superintendent, Catoctin Mountain Park, on the tour of the park with three OSS veterans, arranged by the author, 18 May 2005. The Job Corps was one of President Lyndon Johnson's "War on Poverty" programs adopted by Congress in 1964 and implemented by the Office of Equal Opportunity. It was a program of work training especially for urban minority youths in outdoor settings reminiscent of Roosevelt's Civilian Conservation Corps.

27. On 25 September 1942, OSS Detachment B temporarily vacated Camp No. 2 and occupied Area B-5, an the abandoned CCC Camp MD-NP-3, adjacent to Round Meadow Area. Camp MD-NP-3 was transferred on 1 October 1942 under authorization of the Director of the CCC to the U.S. Army, Commanding Officer, Headquarters, Office of Strategic Services, which thereby took over jurisdiction of the CCC camp. Thomas J. Allen, Regional Director, Region One, memorandum for the Director, 29 December 1945, in NPS Records (RG 79), Entry 47 RDAs, Box 60, Folder 601.03.2, National Archives II.

28. Frank A. Gleason on a tour of Catoctin Mountain Park with three OSS veterans arranged by the author, 18 May 2005.

29. "List of Improvements to Property by War Department... Restoration Survey, Catoctin Training Center, Thurmont, Maryland," 31 October 1945, 5-7, included with survey by G.B. ("Mike") Williams, memorandum for Regional Director, NPS Region One, 7 March 1946; World War II Files, General Information Folder, Catoctin Mountain Park Archives, Catoctin Moutain Park, Thurmont, Md. The old concrete septic tank, measuring 8x18x10 feet or the 25,000 gallon water reservoir may be the buried concrete object discovered by the park staff several years ago.

30. "Area B-5," map of buildings, including total capacity, 5 October 1943, located in OSS Records (RG 226), Entry 85, Box 13, Folder 249, National Archives II.

31. G.B. ("Mike") Williams to Conrad L. Wirth, 5 November 1942, NPS Records (RG 79), Entry 47 RDAs, Box 57, Folder 201, National Archives II.

32. Albert R. Guay (pronounced "Gay"), diary entry, Tuesday, 27 October 1942, photocopies of the diary pages during his tour of duty at B-5, 21 October to 19 December 1942, have been generously deposited by Mr. Guay at the request of the author in the World War II Archives of Catoctin Mountain Park, Thurmont, Md. The second quotation is from a telephone interview with Mr. Guay with the author, 24 October 2005. Guay said the barracks at B-5 were pretty good sized, but not as big as regular two-story military barracks. The B-5 barracks were all on one floor and held about twenty men in rows of double-decked bunk beds in each building, at least for the enlisted men, like Private Guay. He also noted that he and the company clerk worked in one small office where the personnel records were. The camp commander, Major Ainsworth Blogg, had a separate office in the camp headquarters elsewhere in B-5.

33. Albert R. Guay, telephone interview with the author, 24 October 2005.

34. Albert R. Guay, telephone interview with the author, 24 October 2005. On 12 October 2005, Guay returned to Catoctin Mountain Park for the first time since 1942 and toured the old Area B-5 with Sally E. Griffin, supervisory park ranger. The PX was the only building he recognized. It stood in what was the center of the old area.

35. J. Mel Poole, Superintendent of Catoctin Mountain Park, on a tour of Catoctin Mountain Park with three OSS veterans arranged by the author, 18 May 2005.

36. G.B. Williams, memorandum for Regional Director, NPS Region One, 7 March 1946; WWII Files, General Information Folder, Catoctin Mountain Park Archives, Catoctin Moutain Park, Thurmont, Md.

37. Frank A. Gleason, instructor at B-2, through March 1943, telephone interview with the author, 1 May 2006.

38. R.P. Tenney to J[oseph]. R. Hayden [political scientist, Far East expert, and an original member of COI's and then OSS's Board of Analysts], 8 June 1942, interoffice memo, subject: "Area B," OSS Records (RG 226), Entry 136, Box 158, Folder, 1721, National Archives.

39. Charles M. Parkin, one of three OSS veterans of Area B-2, on a tour of Catoctin Mountain Park arranged by the author, 18 May 2005.

40. The estimated construction cost of the pistol house, or "house of horrors," was $6,000, a considerable expense at the time G.B. ("Mike") Williams, memorandum for Regional Director, NPS Region One, 7 March 1946, WWII Files, General Information Folder, Catoctin Mountain Park Archives, Catoctin Moutain Park, Thurmont, Md.

41. R.P. Tenney to J[oseph]. R. Hayden, 8 June 1942, interoffice memo, subject: "Area B," OSS Records (G 226), Entry 136, Box 158, Folder, 1721, National Archives II.

42. Richard Dunlop, *Behind Japanese Lines: With the OSS in Burma* (Chicago: Rand McNally, 1979), 86; see also, the description of the mystery house reported in the *Baltimore Sun*, 26 July 1948.

43. Lt. Col. Corey Ford and Maj. Alastair MacBain, *Cloak and Dagger: The Secret Story of OSS* (New York: Random House, 1945), 24-25.

44. The 300-yard rifle range was due for completion in September 1943. 1st Lt. Montague Mead, commanding officer, Detachment B, to Maj. Wickham, Training Branch, OSS, Washington, D.C.,

14 August 1943, OSS Records (G 226), Entry 136, Box 158, Folder, 1721, National Archives II. For an examples of standard construction with drawings and photographs, see "Completion Report of Constructing Concrete Target Butts at Lorton Rifle Range, Fort Humphreys, Va.," U.S. Army Corps of Engineers, Office of the Chief of Engineers Records (RG 77), Entry 391, Construction Reports, 1917-1943, Box 30, Fort Belvoir, Folder/Book 2, National Archives II.

45. Reginald G. ("Reg") Spear, on a tour of the park, arranged by the author, 18 May 2005, and telephone interview with the author, 11 May 2006.

46. The location of these areas is difficult to resolve. The area 1,000 feet east of Round Meadow, where the Fire Cache is today, shows evidence of previous clearing work. On the other hand, Corps of Engineers maps indicate that an area 8,000 feet north, northeast of Round Meadow was used for a rocket, grenade, mortar and demolition area. Current park personnel have not been able to locate any evidence of clearing within that area. Mel Poole, Park Superintendent, email to the author, 6 June 2008.

47. G.B. ("Mike") Williams, memorandum for Regional Director, NPS Region One, 7 March 1946; and "Restoration Survey, Catoctin Training Center, Thurmont, Md.," 31 October 1945, p. 7, WWII Files, General Information Folder, Catoctin Mountain Park Archives, Catoctin Moutain Park, Thurmont, Md.

48. Conclusion of Superintendent Mel Poole based on the directions given by former OSS instructor Frank Gleason on a tour of the park by three OSS veterans arranged by the author, 18 May 2005.

49. Reginald ("Reg") Spear, telephone interview with the author, 11 May 2006. The first chief instructor, Capt. Charles Parkin, called the initial earthen cave for storing munitions, a "go-down." Charles M. Parkin, on a tour of the park, arranged by the author, 18 May 2005.

50. Albert R. Guay, diary entry, Tuesday, 3 November 1942, copies now in WWII Archives of Catoctin Mountain Park, Thurmont, Md., plus Albert Guay, telephone interview with the author, 24 October 2005.

51. G.B. ("Mike") Williams, memorandum for Regional Director, NPS Region One, 7 March 1946, WWII Files, General Information Folder, Catoctin Mountain Park Archives, Catoctin Mountain Park, Thurmont, Md.

52. See, for example, the dimensions, construction information and photographs of sample Army magazines of the period in "Completion Report, Construction of Magazines at Fort Humphreys, Virginia," completed 24 March 1934, U.S. Army Corps of Engineers, Office of the Chief of Engineers Records (RG 77), Entry 391, Construction Reports, 1917-1943, Box 30, Fort Belvoir, Folder/Book 2, National Archives II.

53. R.P. Tenney to J[oseph]. R. Hayden, 8 June 1942, interoffice memo, subject: "Area B," OSS Records (RG 226), Entry 136, Box 158, Folder, 1721, National Archives II.

54. Ibid.

55. Reginald G. Spear, one of three OSS veterans of Area B-2, on a tour of Catoctin Mountain Park arranged by the author, 18 May 2005. A contemporary reference in 1946 by Superintendent Garland B. ("Mike") Williams incorrectly referred to these as "target pits," which suggests that OSS may not have wanted the ranger to see all of the kinds of exercises they were conducting. G.B. ("Mike") Williams, memorandum for Regional Director, NPS Region One, 7 March 1946, WWII Files, General Information Folder, Catoctin Mountain Park Archives, Catoctin Mountain Park, Thurmont, Md.

56. The actual location of the Trainazium was right about where the current pump house (a black building) is located next to the swimming pool at Camp 1 (Greentop). In fact, Reginald Spear and Franklin Gleason agreed that the Trainazium was just on the north side of the fence that currently encloses the swimming pool. The area there now, they said, had been plowed and graded since the 1940s. And the two men did not remember a swimming pool being there. Reginald Spear, OSS trainee at B-2 in early 1944, and Frank Gleason, instructor at B-2 from June 1942 to March

1943, were two of three OSS veterans of Area B-2, on a tour of Catoctin Mountain Park arranged by the author, 18 May 2005. The other OSS veteran on the tour, instructor Charles Parkin, concurred. A contemporary document locates the "log obstacle course" as "erected 100 feet west of the pistol house." The pistol house or "mystery house," is described below. G.B. ("Mike") Williams, memorandum for Regional Director, NPS Region One, 7 March 1946, WWII Files, General Information Folder, Catoctin Mountain Park Archives, Catoctin Mountain Park, Thurmont, Md.

57. Frank A. Gleason, telephone interviews with the author, 31 January 2005, 1 May 2006, 9 February 2007.

58. Frank A. Gleason, letter to the author, 2 May 2006, in response to author to Gleason, et al., 27 April 2006. Gleason's letter includes a drawing he made of the Trainazium as a result of a request by the author. The pronunciation of the structure, Gleason said in an interview, was "Train-A-ZEE-um."

59. Frank A. Gleason, telephone interviews with the author, 31 January 2005, 1 May 2006, 9 February 2007. "The logs were notched, and we also used spikes in holding them together."

60. Frank A. Gleason, A, on a tour of Catoctin Mountain Park arranged by the author, 18 May 2005.

61. Charles M. Parkin, chief instructor at B-2 from April 1942 to March 1943, one of three OSS veterans of Area B-2, on a tour of Catoctin Mountain Park arranged by the author, 18 May 2005.

62. Richard Dunlop, *Behind Japanese Lines: With the OSS in Burma* (Chicago: Rand McNally, 1979), 86.

63. Frank A. Gleason, instructor at Area B-2, June 1942 to March 1943, telephone interview with the author, 1 May 2006.

64. Frank Gleason, telephone interview with the author, 31 January 2005; and Joseph Lazarsky, telephone interview with the author 14 March 2005. This may be why Casey in his memoir about his role in the war, did not include any reference to his training but skipped directly from his enlistment in the OSS in early spring 1943 to his arrival in London in October 1943. See William Casey, *The Secret War against Hitler* (Washington, D.C.: Regnery Gateway, 1988), 3-5, 22.

65. Roosevelt had first gone to Warm Springs, Georgia, in 1924, seeking to recover from the paralysis inflicted by polio. He soon purchased the dilapidated mineral springs resort, turned it into a nonprofit foundation for treatment of polio victims, and built a cabin there. During his presidency, he used it so often that it came to be known as the "Little White House." He died there of a stroke on 12 April 1945. Turnley Walker, *Roosevelt and the Warm Springs Story* (New York: A.A. Wyn, 1953). In Washington, air conditioning equipment had been installed in the White House in 1933, but the Presidential retreats served as respites from the White House workplace as well as from Washington weather.

66. *A Brief History of Camp Hoover* (Washington, D.C.: Department of the Interior, 1987). Jimmy Carter, a fly fisherman like Hoover, was apparently the only other President to use Camp Hoover. W. Dale Nelson, *The President Is at Camp David* (Syracuse, N.Y.: Syracuse University Press, 1995), 4.

67. Conrad L. Wirth, *Parks, Politics, and People* (Norman: University of Oklahoma Press, 1980), 200-201. Earlier in March, Roosevelt, through Secretary of the Interior Ickes had sought to obtain a site on 1,300 foot-high Sugarloaf Mountain located some 30 miles north of Washington, but the owner, Gordon Strong, a Chicago banker, heir to a railroad fortune, and an ardent Republican, refused his request. Chris Lampton, "One Man and His Mountain," *Maryland Magazine*, Autumn 1980, 2-5.

68. Drury's suggested sites were Comers Deadening at 3,400 feet in Shenandoah National Park in Virginia, about a 2 ½ hour drive from the White House; Cactoctin Mountain at approximately 1,700 feet in Catoctin Recreational Demonstration Area in Maryland, about a two hour drive; and Furnace Mountain, Virginia, at about 900 feet, across the river from Point of Rocks, Maryland, a 90-minute drive from Washington. All were equal in terms of seclusion, Drury noted, but he

recommended the Shenandoah site first, if the travel time was not deemed to length, mainly because he thought the wartime Presidential retreat if not used by subsequent Presidents would be of more use to the public in Shenandoah. He ranked Catoctin second and Furnace Mountain third, but he recommended that the President visit Shenandoah and Catoctin before making a final decision. Newton B. Drury to Secretary of the Interior, 31 March 1942, memorandum, National Park Service Records (RG 79), Entry 19, Records of Newton B. Drury, 1940-1951, Box 4, Folder "Catoctin," National Archives II. Nelson, *The President Is at Camp David*, 5-6, relying upon Wirth, *Parks, Politics, and People*, 201, errs in listing the three suggested sites as Comer's Deadening, and Camps 3 and 4 at Catoctin. Neither source apparently had access to Drury's 31 March 1942 memorandum, which clearly lists Furnace Mountain as the third possible site and does not distinguish different sites within Catoctin.

69. Author's deduction from references to these two campsites in Wirth, *Parks, Politics, and People*, 201.

70. Conrad L. Wirth, Supervisor of Recreation and Land Planning, to the Director of NPS, 16 April 1942, memorandum [on Reilly's inspection of 15 April], National Park Service Records (RG 79), Entry 19, Records of Newton B. Drury, 1940-1951, Box 4, Folder "Catoctin," National Archives II.

71. Newton B. Drury to the Secretary of the Interior, 23 April 1942, memorandum, National Park Service Records (RG 79), Entry 19, Records of Newton B. Drury, 1940-1951, Box 4, Folder "Catoctin," National Archives II, which includes the President's sketches. See also Grace G. Tully, President's private secretary, to Newton B. Drury, 15 August 1942; Drury to Tully, 18 August 1942, in President's Secretary's File, Subject File, Box 164, Folder "Shangri-La," and Secretary of the Interior Harold L. Ickes to President Franklin D. Roosevelt, 19 February 1943, enclosing a commemorative album prepared by Conrad L. Wirth, "'Shangri-La,' the President's Lodge, 1942," in President's Personal File 3650, "Shangri-La," all in Franklin D. Roosevelt Papers, Franklin D. Roosevelt Library, Hyde Park, NY.

72. Nelson, *The President Is at Camp David*, 6.

73. The Arcadian idyll in a mysterious valley in the Himalaya Mountains of Tibet, depicted in *Lost Horizon* (Columbia Pictures, 1937), directed by Frank Capra and staring Ronald Coleman, was based on James Hilton's novel, *Lost Horizon* (New York: W. Morrow, 1933).

74. Nelson, *The President Is at Camp David*, 1.

75. OSS veterans Charles Parkin, Frank Gleason, and Reginald Spear, on a tour of Catoctin Mountain Park arranged by the author, 18 May 2005. The wartime extent of the OSS camp was confirmed by Catoctin Superintendent Mel Poole and Resource Manager James Voigt, who accompanied the tour.

76. Conrad L. Wirth to the Director [of NPS], 25 April 1942, memorandum with enclosure, Cost Estimates Hi-Catoctin Lodge, 24 April 1942, in National Park Service Records (RG 79), Entry 19, Records of Newton B. Drury, 1940-1951, Box 4, Folder "Catoctin," National Archives II.

77. Newton B. Drury to Mike Riley [sic], 25 April 1942, memorandum, National Park Service Records (RG 79), Entry 19, Records of Newton B. Drury, 1940-1951, Box 4, Folder "Catoctin," National Archives II.

78. Wirth, *Parks, Politics, and the People*, 202. Cost for physical labor and equipment provided by WPA and the National Park Service came in at the $15,000 that Roosevelt had mandated. But the total cost for the retreat and especially the security measures required far exceeded that figure, and the excess was provided by the Emergency Fund for the President, National Defense, 1942. Additional equipment and supplies were provided by the Army, Navy, and Marine Corps. "A Summary of the Development of 'Shangri-La,'" typescript, Franklin D. Roosevelt Library, Hyde Park, N.Y.

79. Newton B. Drury to Secretary of the Interior, 2 May 1942, National Park Service Records (RG 79), Entry 19, Records of Newton B. Drury, 1940-1951, Box 4, Folder "Catoctin; " and Hillory A.

Tolson, acting associate director of NPS, to G.B. ("Mike') Williams, 7 May 1942, National Park Service Records (RG 79), Entry 54 Box 57, both in National Archives II.

80. Frank A. Gleason, telephone interview by the author, 31 January 2005. Account confirmed by Charles M. Parkin during a tour of Catoctin Mountain Park, 18 May 2005.

81. Newton B. Drury to the President, 8 June 1942, memorandum, NPS Records (NG 79), Entry 19, Records of Newton B. Drury, 1940-1951, Box 4, Folder, "Catoctin," National Archives II.

82. Log Book of the "U.S. S. Shangri-La," President's Naval Collection, Franklin D. Roosevelt Library, Hyde Park, NY. On 8-11 August 1942, Roosevelt formally "accepted" the mythical yacht, "U.S.S. Shangri-La," into the U.S. Navy, just as the real yacht, the U.S.S. *Potomac*, was a naval vessel. The Presidential retreat was then officially transferred from the responsibility of the National Park Service to the Department of the Navy.

83. Charles M. Parkin, transcript of interview by the author during tour of Catoctin Mountain Park, 18 May 2005; confirmed by Frank A. Gleason, Jr., that same day; and independently by Joseph Lazarsky, telephone interview with the author, 14 March 2005.

84. Charles M. Parkin, and Cheryl Parkin Evans, his daughter, interview with the author during tour of Catoctin Mountain Park, 18 May 2005.

85. Charles M. Parkin, transcript of interview by the author during tour of Catoctin Mountain Park, 18 May 2005.

86. List of "FDR's Trips to Shangri-La (now Camp David)," typed list of dates of visits, Vertical File, "Shangri-La;" see also "FDR: Day by Day" chronology prepared by filmmaker Pare Lorentz; both in the Franklin D. Roosevelt Library, Hyde Park, N.Y.

87. Nelson, *The President Is at Camp David*, 16-17.

88. Fairbairn is sometimes erroneously referred to as having been the chief of the Shanghai police, for example, in William R. Peers and Dean Brelis, *Behind the Burma Road: The Story of America's Most Successful Guerrilla Force* (Boston: Little, Brown, 1963), 31. In fact, Fairbairn's highest law enforcement position had been as Assistant Commissioner of the international Shanghai Municipal Police in charge of their Armed and Training Reserve, which was the special riot duty unit and its training force. See William L. Cassidy, "Fairbairn in Shanghai," *Soldier of Fortune*, 4 (September 1979): 66-71.

89. Lt. Col. William E. Fairbairn, British Army, to Director of Office of Strategic Services, 23 March 1945, memorandum subject: Resume of service of Lt. Col. William E. Fairbairn, p.3, in OSS Records (RG 226), Director's Office Files, Microfilm 1642, Roll 46, Frames 19-20, National Archives II.

90. James F. Byrnes, *All in One Lifetime* (New York: Harper, 1958), 195.

91. Nelson, *The President Is at Camp David*, 15.

92. Franklin D. Roosevelt to Winston Churchill, 1 July 1943, President's Personal File 7683, in unnumbered box, Franklin D. Roosevelt Library, Hyde Park, N.Y.

93. Joseph Lazarsky, telephone interviews with the author, 14 March 2005, and 2 January 2008.

94. Sharon Carstairs, "Mary Ellen Smith, First Woman Cabinet Minister in Commonwealth [sic]," *Debates of the Canadian Senate (Hansard)*, 1st Session, 36th Parliament, Vol. 137, Issue 84, Oct. 21, 1998.

95. Reginald G. Spear, telephone conversations with the author, 21 December 2004, 6 January 2005, 7 January 2005, 25 March, 2005, 27 May 2005, 24 June 2005; as well as statements made during a tour of Catoctin Mountain Park and personal discussions while driving roundtrip from McLean, Virginia to Catoctin Mountain Park, Maryland, 18 May 2005.

96. Reginald G. Spear, telephone interview with the author, 27 May 2005.

97. Michael F. ("Mike") Reilly, weekly report, quoted in Chief, U.S. Secret Service to Assistant Secretary [of the Treasury] Gaston, 1 June 1942, memorandum, U.S. Secret Service Records, File

103A-3 "Catoctin Mountain," Box 16, Franklin D. Roosevelt Library, Hyde Park, NY. Reilly misspelled the name of instructor William E. Fairbairn as "C.T. Farber."

98. Diary entry of 29 August 1942, in William D. Hassett, *Off the Record with F.D.R., 1942-1945* (New Brunswick, N.J.: Rutgers University Press, 1958), 113.

99. "The camp," in W.J.H. [Presidential secretary William D. Hassett] to Mr. Latta, 12 May 1944; President's Personal File 8086; Marines "on the hill" (quotation marks in original), Lt. Col. Charles Brooks to Capt. John L. McCrea, 8 July 1942, in President's Secretary's File, Subject File, Box 164, Folder "Shangri-La," both in Franklin D. Roosevelt Library, Hyde Park, NY.

100. That was the experience of Private Albert R. Guay, diary entry of 15 November 1942, and telephone interview with the author, 24 October 2005. Private Guay, served as assistant company clerk at B-5, from October to December 1942.

101. Information from an exhibit of Presidential armored limousines at the William Jefferson Clinton Presidential Library at the time of a visit by the author during the Annual Meeting of the Oral History Association, held in Little Rock, Ark., October 26-28, 2006.

102. William A. Willhide, Thurmont, Md., telephone interview with the author, 13 June 2006.

103. James H. Mackley, Thurmont, Md., telephone interview with the author, 13 June 2006.

104. Nelson, *The President Is at Camp David*, 19.

105. Frank A. Gleason, Jr., telephone interview with the author, 31 January 2005. Emphasis in the original conversation.

106. Frank A. Gleason and Charles M. Parkin on a tour of Catoctin Mountain Park arranged by the author of 18 May 2005.

107. Brian Albrecht, "WWII Vet Recalls the 'Stuff that Stays with You," *Cleveland Plain Dealer*, 27 May 2007, reprinted in *OSS Society Digest*, Number 1760, 30 May 2007, osssociety@yahoogroups.com, accessed 30 May 2007.

108. Albert R. Materazzi, email to the author, 24 January 2005, and telephone interview with the author, 23 September 2005.

109. Roger Hall, *You're Stepping on My Cloak and Dagger* (Annapolis, Md.: Naval Institute Press, 1957), 34-35.

110. Robert E. Carter, telephone interview with the author, 28 August 2008.

111. Caesar J. Civitella, telephone interview with the author, 18 April 2008.

112. G.B. ("Mike") Williams to Conrad L. Wirth, 5 November 1942, p.5, a seven-page report on developments at the park since the completion of the main lodge for the President, photocopy in World War II Archives, Catoctin Mountain Park, Thurmont, Md.

113. G.B. {("Mike") Williams to [Marine] Commander, Camp No. 1, 5 March 1943, memorandum, World War II Archives, Catoctin Mountain Park Archives at Catoctin Mountain Park Headquarters, Thurmont, Md.

114. Col. Paul J. McDonnell, OSS HQ Detachment, to Col. William J. Donovan, 5 March 1943, subject: Retention of Detachment "B" in its Present Status [more accurately, a recommendation not to retain it], in OSS Records (RG 226), Entry 137, Box 3, Folder 24, National Archives II.

115. G. Edward Buxton, memorandum for the Secretariat, 8 May 1943; Lt. F. J. Ball to Major Doering, 30 March 1943, subject: Release of Area "B," all in OSS Records (RG 226), Entry 137, Box 3, Folder 24, National Archives II.

116. Area B was conspicuously missing from the list of SI-SO training camps prepared by the newly established OSS Training Directorate in January 1943; see Training Directorate to All Geographic Desks and Area Operations Officers, 1 January 1943, subject: The Training Areas, OSS Records (RG 226), Entry 137, Box 3, Folder 24, National Archives II.

117. In order to keep its options open in case it needed the site again, OSS headquarters decided to indicate to the War Department that it did not need the use of the Catoctin area "at the present

time" and that it was agreeable to having the area made available to Camp Ritchie "on a revocable permit basis," thus being able to "keep a 'string' on the property and re-enter at such time as the premises are again needed by us." A. William Asmuth, Jr. office of the counsel, OSS, memorandum for the files, 26 April 1944, subject: Arrangement with the War Department for Temporary Disposition of Area B, OSS Records (RG 226), Entry 137, Box 3, Folder 24, National Archives II.

118. Col. H. L. Robinson, Executive Officer, Schools and Training, OSS, to Col. Sherman I. Strong, 11 March 1944, subject: Proposed Improvements at Area B, in OSS Records (RG 226), Entry 136, Box 163, Folder 1772, National Archives II.

119. Col. Sherman I. Strong, Commanding Officer, OSS HQ and HQ Detachment, to Capt. Montague Mead, 11 May 1944, subject: Discontinuance of Area B, in OSS Records (RG 226), Entry 136, Box 163, Folder 1772, National Archives II.

120. "Fort Ritchie," in Robert B. Roberts, *Encyclopedia of Historic Forts* (New York: Macmillan, 1988), 388. After the war, Camp Ritchie was reactivated in 1951 by the Army, renamed Fort Ritchie, and designated to support the Alternate Joint Communications Center. From 1964 until the closure of the base in 2005, Fort Ritchie served as headquarters of the U.S. Army's Communications Command (under various titles) responsible for the management of communications and electronics functions of more than one hundred military agencies and activities in the United States. Camp Ritchie was located half a dozen miles north of Thurmont, Maryland on Route 550.

121. "Military Intelligence Training Center at Camp Ritchie, Md.," National Army Security Agency Association. http://www.nassa-home.org/history/schools.htm, accessed 14 April 2007. The MITC was formally activated 19 June 1942, but its cadre began to arrive earlier. For accounts by individual trainees at the MITC on their experiences in the classrooms, firing ranges, and field exercises in and around Camp Ritchie, see, for example, Richard Warren Seltzer, Sr, account posted at http://www.samizdat.com/gen/seltzer/lifeand times.html, accessed 14 April 2007; and Lewis Bloom, Military Intelligence Service veteran, oral history, Rutgers Oral History Archives of World War II, http://oralhistory.rutgers.edu, and also telephone interview with the author, 17 January 2007.

122. On Brig. Gen. Charles Y. Banfill, U.S. Army Air Forces, see official biography in http://www.af.mil/bios/bio.asp?bioID=4597 accessed 14 April 2007.

123. Hillory A. Tolson, acting director, memorandum, 23 December 1942; Col. Charles Y. Banfill to Regional Director, Region One, 1 April 1943; and G.B. ("Mike") Williams to the Director, 8 February 1943, and Williams to the Acting Regional Director, 6 April 1943; Special Use Permit... by the Military Intelligence Training Center for war training purposes," 5 May 1943, signed by G.B. Willliams, all in NPS Records (RG 79), Entry 7, Central Classified File, 1933-1949, General 201, National Defense "C" Box 79, Folder 201; and see also support from Fred T. Johnson, acting regional director of NPS Region One, memorandum for the director, 10 April 1943, in NPS Records (RG 79), Entry 47, RDAs, Box 60, Folder 601 Lands, Catoctin, all in the National Archives II.

124. For secrecy purposes, the NPS never used the term OSS, but referred only vaguely to "the military unit occupying Areas B-2 and B-5," and the like in its reports.

125. G. B. ("Mike") Williams to Col. Charles Y. Banfill, 20 October 1943, in NPS Records (RG 79), Entry 7, Central Classified File, 1933-1949, General 201, National Defense "C" Box 79, Folder 201, National Archives II.

126. Undated report "Use of northern portion of area [Catoctin Recreational Demonstration Area] for special field maneuvers [1942-1943]," located in NPS Records (RG 79), Entry 7, Central Classified File, 1933-1949, General 201, National Defense-General, Box 73, Black Binder, "War Uses, C," and A. Van Beuren to Mr. Weston Howland, 24 September 1943, subject: Areas B and B-2, OSS Records (RG 226), Entry 146, Box 223, Folder 3106, both in National Archives II.

127. In addition to troops from Camp Ritchie, some units from the Ordnance School at Aberdeen Proving Ground in Maryland were authorized to use Catoctin Mountain Park. During the spring of 1944, NPS granted permits on 16 March and 27 April 1944 for overnight bivouacking of troops at the West Picnic Area, south of Route 77, for troops from the Ordnance Bomb Disposal School at

Aberdeen Proving Ground, Md. Fifty troops participated in March and a convoy of eleven vehicles carrying 70 troops came in April Mike Williams reported that the grounds remained in excellent condition. G. B. ("Mike") Williams to the Director, 22 April 1944 and 30 June 1944, both in NPS Records (RG 79), Entry 7, Central Classified File, 1933-1949, General 201, National Defense "C" Box 79, Folder 201, National Archives II.

128. Acting Secretary of the Interior Abe Fortas to Secretary of War Henry L. Stimson, 31 May 1944, and attached Special Use Permit Authorizing Use of Land in the Catoctin Recreational Demonstration Area, Maryland, by the War Department, for war purposes, 31 May 1944, plus attached map, OSS Records (RG 226), Entry 137, Box 3, Folder 23, National Archives II; accepted by Acting Secretary of War Robert P. Patterson, 15 July 1944, in NPS Records (RG 79), Entry 7, Box 79, Folder 201, National Archives II. There is a copy in Special Use Permits Folder, Catoctin Mountain Park Archives, World War II, located at Catoctin Mountain Park Headquarters, Thurmont, Md.

129. Capt. Montague Mead, OSS, to Brig. Gen. Charles Y. Banfill, commandant, Camp Ritchie, 7 June 1944, and Brig. Gen. Charles Y. Banfill to Capt. Montague Mead, 13 June 1944, in OSS Records (RG 226), Entry 136, Box 163, Folder 1772, National Archives II.

130. Because the OSS ended its use of the park, including Cabin Camp 2 (and the old CCC camp which had reverted to Army), and because the Marine Corps continued to use Cabin Camp 1, the Department of the Interior in May 1944 cancelled the use permit issued to the War Department on 16 May 1942. The new use permit for the War Department, issued by Assistant Secretary Fortas on 31 May 1944, effective 1 June 1944, was "revocable at will" by the Secretary of the Interior. For coverage, see undated report "Use of Camp No. 2, abandoned CCC Camp MD-NP-3, and northwesterly undeveloped portion of Catoctin Area consisting in all of approximately 2,000 acres," located in NPS Records (RG 79), Entry 7, Box 73, Black Binder, "War Uses, C," National Archives II. The authorized area appears smaller than that granted for use in 1942. In 1944, it was 2,000 acres including Cabin Camp 2 and the area northwest of it. See A.E. Demaray, Associate Director to Director, 16 Jan. 1945 [sic] in file below Entry 7, Box 79, Folder 201, National Archives II. The 1944 permit granted to the War Department included a number of conditions that had not been in the 1942 permit. As OSS counsel explained after talking with Col. Ainsworth Blogg, the original commander of Area B, "It is our understanding that shortly after our original entrance into the Area, certain verbal restrictions were imposed upon our use of the premises by the Department of the Interior, and that the revised permit simply reflects these limitations." A. William Asmuth, Jr., office of the general counsel of OSS, to Charles S. Cheston, acting assistant director of OSS, 26 June 1944, subject: Transfer of Area "B" to Camp Ritchie, OSS Records (RG 226), Entry 137, Box 3, Folder 24, National Archives II. The provisions added to the 1944 permit included the following: NPS employees should not be interfered with in use of the road between the abandoned CCC camp and the NPS Central Service Group. Precaution must be taken to preserve and protect geological and historical objects, and wherever possible structures, roads, trees, shrubs, and other natural terrain features should remain unmolested. Precaution should be taken against fire and vandalism and if fire occurred, War Department personnel and equipment should be made available for fire suppression within the area. The pumping station that supplied water to Group Camp 2 (Greentop) would be operated by NPS, but minor maintenance and repair of the station would be performed by the War Department. The Permittee (the War Department) would bear the cost of electrical energy consumed by Permittee by reimbursing the NPS upon proper billing. All garbage and flammable rubbish would be disposed of at a location designated by the manger of the Catoctin Recreational Demonstration Area, and all garbage would be buried and all rubbish burned at least once every five days. Only pistol ammunition could be used in Camp 2 and no live ammunition of any description could be used in the area east of the Foxville-Deerfield Road except rifle or pistol ammunition which could be used on the range situated approximately 400 yards west of the abandoned CCC MD-NP-3. The permittee was granted permission to erect additional housing facilities if and when necessity arose, the exact location of such structure to be determined by the manager of the Catoctin RDA and the proper Army officials, and the plans for the structures to be approved by the NPS. At the termination of the use of the area, all buildings, except those of a strictly military technical nature, were to be transferred to the Department of the Interior or, like those of

strictly military nature, removed by the permittee and the site restored as nearly as possible to its condition at the time of the issuance of the permit. "Special Use permit... by the War Department," 31 May 1944, effective 1 June 1944, Map Attached; Abe Fortas, Acting Secretary of the Interior; accepted 15 July 1944, by Robert P. Patterson, Acting Secretary of War, NPS Records (RG 79), Entry 7, Box 79, Folder 201, National Archives II.

131. Col. Donald J. Kendall, USMC, to Garland B. ("Mike") Williams, NPS, 3 July 1945; and G.B. Williams to Director NPS Region One, 6 July 1945, memorandum; in Correspondence Folder, Catoctin Mountain Park Archives, World War II, located at Catoctin Mountain Park Headquarters, Thurmont, Md.

132. The Marine Corps' use of Cabin Camp 2 (Greentop), 1946-1947, is covered in a subsequent chapter on the postwar period.

CHAPTER 5

1. Ira B. Lykes manager Chopawamsic RDA, to Conrad L. Wirth, Assistant Director of NPS, confidential memorandum, 31 March 1942, in National Park Service Records (RG 79), Central Classified File, 1933-1949, General, 201 National Defense, Chopawamsic, Project CHO-IV-1&1A, Box 73, Folder 201, National Archives II, College Park, Md.

2. Ibid.

3. Ira B. Lykes, interview with Susan Cary Strickland, historian for NPS, 4 September 1985, quoted in Susan Cary Strickland, *Prince William Forest Park: An Administrative History* (Washington, D.C.: National Park Service, 1986), 27.

4. In addition to the guard houses at the two main gates, there were 8 by 8-foot guard houses erected at A-2, the initial headquarters for the OSS camp, and at A-4, a former CCC work camp, which ultimately served as the headquarters for the entire Area A from fall 1942 until the closure of Area A in January 1945. Lt. Col. Ainsworth Blogg, executive officer, schools and training branch, Strategic Services Unit [formerly OSS], to General Counsel Office, SSU, 16 October 1945, subject: Liquidation of Areas "A" and "C," in OSS Records (RG 226), Entry 146, Box 8, Folder 199, National Archives II. In addition to the main gate at Area C-1, for example, there was also a guard house at the entrance to Area C-4.

5. Lee Lansing, historian for the town of Dumfries, Va., interview with Susan Cary Strickland, 15 July 1985, in Strickland, *Prince William Forest Park*, 27. Robert A. Noile, 68, who grew up on Joplin Road near the park headquarters, said his father told him that that the perimeter of the park had been patrolled by men on horseback. Noile who, as a youth played in the park, at least the part north across Joplin Road, said he never saw any fences. Perhaps wire fences were erected on the Batestown side on the east. Robert A. Noile, Triangle, Va., telephone interview with the author, 27 April 2007.

6. The CCC operated in the park until February 1943. Strickland, *Prince William Forest Park*, 17.

7. NPS had for some time sought to have the supervisors of Prince William County officially abandon a number of the old dirt roads, overgrown with brush, within the park that were considered unsafe and often impassable. William R. Hall, project manager, Chopawamsic RDA, to Board of Supervisors, Prince William County, 25 November 1939, and clipping, "Roads May Be Closed," [Manassas?] *Star*, 12 November 1939, both in NPS Records (RG 79), Entry 47, Box 126, Folder 630, National Archives II.

8. Conrad L. Wirth, Memorandum for the Departmental Representative on the Advisory Board, CCC, 24 April 1942, and referring to Major General James A. Ulio's letter of 11 April to the Director of CCC, requesting the allocation of a national defense project on the Chopawamsic RDA, in NPS Records (RG 79), Entry 47, Box 126, Folder 621, National Archives II. General Ulio, was the War Department representative on the Advisory Council of the Civilian Conservation Corps from 1940-1943.

9. Stan Cohen, *The Tree Army: A Pictorial History of the Civilian Conservation Corps, 1933-1942* (Missoula, Mont.: Pictorial Histories, 1980), 129; for Kent and Company's work since 1942, see Lt. Col. Ainsworth Blogg to Charles S. Cheston, via A. William Asmuth, Jr.,Legal Division, 5 May 1944, subject: new sewer field at Area A-2, OSS Records (RG 226), Entry 137, Box 3, Folder 24, National Archives II.

10. On the role of John T. Gum [whose name was sometimes misspelled as John E. Gum or Gun], as manager of the CCC facilities at Chopawamsic RDA before the during the war, see Strickland, *Prince William Forest Park*, 17, although she misspells Gum's name as Gun.

11. Robert A. Noile, Triangle, Va., telephone interview with the author, 27 April 2007.

12. For John T. Gum's supervision of construction, position as sergeant in Detachment A, and eventual transfer to Detachment C, see Capt. Stanley H. Lyson, S-4 [Supply], OSS Schools and

Training Branch, to Lt. Warner, undated [February 1945], subject: Data on Detachment A, OSS Records (RG 226), Entry 136, Box 153, Folder 1658, National Archives II.

13. Ira B. Lykes to Conrad L. Wirth, confidential memorandum, 30 March 1942 [a separate memo from the one of the same date cited above], in NPS Records (RG 79), Central Classified File, 1933-1949, General, 201 National Defense, Chopawamsic, Project CHO-II-1, Box 81, Folder 201, National Archives II.

14. Ibid.

15. Patricia Parker, *The Hinterland: An Overview of the Prehistory and History of Prince William Forest Park, Virginia* (Washington, D.C: National Park Service, 1986); and Arvilla Payne-Jackson and Sue Ann Taylor, *Prince William Forest Park: The African-American Experience* (Washington, D.C.: National Park Service, 2000).

16. For example, in 1942-43, the Richard W. Wheat family did not want to sell their 20 acres for construction of a new entrance road to the Chopawamsic RDA. For a reference to their resistance and to negotiations being "acrimonious," see Olinus Smith, engineer, memorandum for the Chief Counsel of NPS, 19 October 1942, NPS Records (RG 79), Entry 47, Box 126, Folder 630, National Archives II.

17. Robert A. Noile, Triangle, Va., telephone interview with the author, 27 April 2007. Mr. Noile's mother taught at the Joplin school.

18. Ira B. Lykes to F.F. Gillen, December 1941, NPS Records (RG 79), File Number 1460/3, National Record Center, Suitland, Md., cited in Susan Cary Strickland, *Prince William Forest Park: An Administrative History* (Washington, D.C.: National Park Service, 1986), 28, n.99. These records are believed to have been subsequently transferred to the National Archives, but cannot be found.

19. Lt. Col. Robert H. Fabian, Army Service Forces, Corps of Engineers, for the Chief of Engineers, completed form for Declaration of Surplus Real Property, 23 January 1946, Recommendation for Disposal to the U.S. Department of the Interior, located in the Park Archives, Prince William Forest Park, Triangle, Va.; see also Lt. Col. C.J. Blair, Jr., Chief, Real Estate Division, Middle Atlantic Division, US Army Corps of Engineers, to National Park Service, 3 December 1945, subject: Chopawamsic Recreational Demonstration Area, NPS Records (RG 79), Central Classified File, 1933-1949, General, 201, National Defense "C," Box 73, Folder 201, National Archives II.

20. Edgar Prichard, OSS veteran, SO Branch and in the postwar an Arlington, Virginia, lawyer, "Address to Historic Prince William, Inc.," 16 January 1991 [sic, actually 1992], typescript and accompanying newspaper clipping, Denyse Tannenbaum, "Veteran Recalls Learning to Spy in the Park," *Potomac News*, 18 January 1992, both in White Binder, OSS, Park Archives, Prince William Forest Park, Triangle, Va.

21. On George White, see John C. McWilliams, "Covert Connections: The FBN, the OSS, and the CIA," *The Historian*, 53/4 (Summer 1991): 665-69; Harry J. Anslinger with J. Dennis Gregory, *The Protectors: Narcotics Agents, Citizens and Officials against Organized Crime in America* (New York: Farrar, Straus,1964), 79-80; William R. Peers and Dean Brelis, *Behind the Burma Road: The Story of America's Most Successful Guerrilla Force* (Boston: Little, Brown, 1963), 30; Nicol Smith and Blake Clark, *Into Siam: Underground Kingdom* (Indianapolis, Ind.: Bobbs-Merrill, 1946), 49-50; David Stafford, *Camp X* (New York: Dodd, Mead, 1987), 82. White kept a diary, which included his service in the OSS. It is part of the Douglas Perham Collection, one of the largest collections of electronic related historical materials in the United States. Unfortunately, the diary cannot be located at present because the Perham Collection is currently in two hundred, unsorted, unmarked boxes as a result of its transfer from Foothill College, Mountain View, California to History San Jose in San Jose, California in 2003. Telephone interview James Reed, archivist of the Perham Collection, San Jose, Calif., with the author, 4 October 2005.

22. The quotation is from the director of OSS Research and Development, Stanley P. Lovell, *Of Spies and Stratagems* (Englewood Cliffs, NJ: Prentice-Hall, 1963), 57-59.

23. Lovell, *Of Spies and Stratagems*, 57; McWilliams, "Covert Connections: The FBN, the OSS, and the CIA," 665-69.

24. Capt. Stanley H. Lyson, S-4, Schools and Training, to Lt. Warner, undated [February 1945], subject: Data on Detachment A, OSS Records (RG 226), Entry 136, Box 153, Folder 1658, ; and Estimate #3, Estimated Prices Prepared by T. Sgt. John E. [sic] Gum, both attachments to Col. W.L. Rehm, Acting Assistant Director for Services, Strategic Services Unit (SSU), for the Director, to Commanding General, Army Service Forces, Mobilization Division, 19 October 1945, subject: Chopwamsic National Recreational Demonstration Area, Attachments #1, #2, and #3 in OSS Records (RG 226), Entry 146, Box 8, Folder 199, both in National Archives II.

25. It is not absolutely certain which CCC work camp in Prince William Forest Park became Area A-4. By location, the evidence (CCC flagpole, parade ground and CCC era buildings currently numbered 3-32, 3-35, 3-36, 3-36A, and 3-39) suggests that it may have been the CCC work camp located at what is currently the Maintenance Area of the Orenda Historic District, about one to two miles from current Cabin Camp Number 3. That was CCC work camp S.P. 26. Judy Volonoski, museum technician at PWFP, email correspondence with the author, 20 March and 7 June 2006, the latter enclosing Map of Chopawamsic RDA showing CCC Camp Locations, c. 1935-39 [but not 1941-43], source: Civilian Conservation Corps Activities in NPS National Capital Region, HABS NO. DC-858, p. 146.

26. Statistics computed from the figures listed on the maps for each sub-camp in Area A, undated [October 1943, date on similar map for Area B], in OSS Records (RG 226), Entry 85, Box 13, National Archives II.

27. Ira B. Lykes manager Chopawamsic RDA, to Conrad L. Wirth, Assistant Director of NPS, confidential memorandum, 31 March 1942, in NPS Records (RG 79), Central Classified File, 1933-1949, General, 201 National Defense, Chopawamsic, Project CHO-IV-1&1A, Box 73, Folder 201, National Archives II.

28. Edgar Prichard, OSS veteran, SO Branch and in the postwar an Arlington, Virginia, lawyer, "Address to Historic Prince William, Inc.," 16 January 1991 [sic, actually 1992], typescript and accompanying newspaper clipping, Denyse Tannenbaum, "Veteran Recalls Learning to Spy in the Park," *Potomac News*, 18 January 1992, both in White Binder, OSS, Park Archives, Prince William Forest Park, Triangle, Va.

29. Course A-5 (Specialist Course) SO Finishing Course, undated, unsigned document, located in White Binder, entitled, "OSS," Park Archives, Prince William Forest Park, Triangle, Va.; plus Lt. Col. Henson L. Robinson, "Schools and Training," report, October 1943, p. 1, OSS Records (RG 226), Entry 136, Box 158, Folder 1723, both in National Archives II.

30. Capt. Stanley H. Lyson, S-4 [Supply], Schools and Training, to Lt. Warner, undated [February 1945], subject: Data on Detachment A, OSS Records (RG 226), Entry 136, Box 153, Folder 1658; and Lt. Col. Ainsworth Blogg, executive officer, Schools and Training Branch, Strategic Services Unit [formerly OSS], to General Counsel Office, SSU, 16 October 1945, subject: Liquidation of Areas "A" and "C," in OSS Records (RG 226), Entry 146, Box 8, Folder 1999, National Archives II.

31. "Area A-5," map and capacity list, undated [October 1943, date given on similar maps for Area B] located in OSS Records (RG 226), Entry 85, Box 13, Folder 249, National Archives II.

32. Capt. Don R. Callahan, commanding officer, Detachment A, to Col. H.L. Robinson, executive, Schools and Training Branch, OSS, 13 March 1944, subject: Housing and Training Facilities, Detachment A, in OSS Records (RG 226), Entry 136, Box 153, Folder 1658, National Archives II.

33. Capt. Don R. Callahan, commanding officer, Detachment A, to Col. H.L. Robinson, executive, Schools and Training Branch, OSS, 13 March 1944, subject: Housing and Training Facilities, Detachment A, in OSS Records (RG 226), Entry 136, Box 153, Folder 1658, National Archives II.

34. Arthur F. ("Art") Reinhardt, OSS Communication Branch veteran, who was at Area A for a few days and then trained at Area C-4 for ten weeks from April to early July 1944 before being sent to China. Mr. Reinhardt, toured Cabin Camp 4 (the former OSS Training Area C-4), with the author and Prince William Forest Park Museum Technician Judy Volonoski on 13 December 2004.

The author tape recorded and transcribed Mr. Reinhardt's commentary and the current identification by Ms. Volonoski of the building numbers and usage today at Prince William Forest Park.

35. The location of the Boat House is given variously as at A-2 and at A-5; however, A-5 seems the most probable. Strickland, *Prince William Forest Park: An Administrative History*, 36, states that the Army Corps of Engineers relocated the Boat House from Camp 5 to Camp 2. The Boat House location at A-2 and its dimensions are given in Capt. Stanley H. Lyson, S-4 [Supply], Schools and Training, to Lt. Warner, undated [February 1945], subject: Data on Detachment A, OSS Records (RG 226), Entry 136, Box 153, Folder 1658. But it is cited at A-5 in Estimate #1, Estimated Prices Prepared by Quartermaster Depot Office, Ft. Belvoir, Va., and again at A-5 in Estimate #3, Estimated Prices Prepared by T. Sgt. John E. [sic] Gum, both attachments to Col. W.L. Rehm, Acting Assistant Director for Services, Strategic Services Unit [formerly OSS], for the Director, to Commanding General, Army Service Forces, Mobilization Division, 19 October 1945, subject: Chopwamsic National Recreational Demonstration Area, Attachments #1, #2, and #3 in OSS Records (RG 226), Entry 146, Box 8, Folder 199, all in National Archives II.

36. Capt. Don R. Callahan, commanding officer, Detachment A, to Col. H.L. Robinson, executive, Schools and Training Branch, OSS, 13 March 1944, subject: Housing and Training Facilities, Detachment A, in OSS Records (RG 226), Entry 136, Box 153, Folder 1658, National Archives II. Callahan reported that at least in March 1944 the Boat House was at Area A-5.

37. Photographs of Louis J. Gonzalez, Bill Dyck, Bill Messina, and Jerry Lapid at a lake in Chopawamsic in July 1942, provided by his son, Ron Gonzalez, to the author, 16 January 2008.

38. Lt. Col. Henson L. Robinson, "Schools and Training," report, October 1943, p. 1, OSS Records (RG 226), Entry 136, Box 158, Folder 1723, National Archives II.

39. Material on OSS building usage and construction at A-2 is primarily from two documents: Capt. Stanley H. Lyson, S-4, Schools and Training, to Lt. Warner, undated [February 1945], subject: Data on Detachment A, OSS Records (RG 226), Entry 136, Box 153, Folder 1658 ; and Lt. Col. Ainsworth Blogg, executive officer, Schools and Training Branch, Strategic Services Unit [formerly OSS], to General Counsel Office, SSU, 16 October 1945, subject: Liquidation of Areas "A" and "C," in OSS Records (RG 226), Entry 146, Box 8, Folder 1999. In March 1945, Fairbairn remembered his "indoor mystery house" as having been at A-3, but the above cited reports indicate that it was at A-2. It is possible that Fairbairn misremembered; a less likely possibility is that the structure was moved between A-2 and A-3. Lt. Col. William E. Fairbairn, British Army, to Director of Office of Strategic Services, 23 March 1945, memorandum, subject: Resume of service of Lt. Col. William E. Fairbairn, p. 3, in OSS Records (RG 226), Director's Office Files, Microfilm 1642, Roll 46, Frames 19-20, all in National Archives II.

40. Lt. Col. Ainsworth Blogg to Charles S. Cheston, via A. William Asmuth, Jr. of the Legal Division, 5 May 1944, subject: new sewer field at Area A-2, OSS Records (RG 226), Entry 137, Box 3, Folder 24, National Archives II

41. "Area A-2," map and capacity list, undated [October 1943, the date given on similar maps for Area B] located in OSS Records (RG 226), Entry 85, Box 13, Folder 249, National Archives II.

42. Capt. Don R. Callahan, commanding officer, Detachment A, to Col. H.L. Robinson, executive, Schools and Training Branch, OSS, 13 March 1944, subject: Housing and Training Facilities, Detachment A, in OSS Records (RG 226), Entry 136, Box 153, Folder 1658, National Archives II.

43. Harry F. Belfry, "Office of Strategic Services (OSS)," in *U.S.A. Airborne: 50th Anniversary: A Commemorative History* (Atlanta, Ga.: Turner Pub. Co., 1997), 350.

44. Jerry Sage, *Sage* (Wayne, Pa.: Miles Standish Press, 1985), 53, 55.

45. Ibid., 53.

46. "History of the Schools and Training Branch, Office of Strategic Services," typescript history prepared in 1945, but not declassified until the 1980s; for handiest reference, see the published

version, William L. Cassidy, ed., *History of the Schools & Training Branch, Office of Strategic Services* (San Francisco: Kingfisher Press, 1983), 33.

47. Frank A. Gleason, telephone interview with the author, 9 May 2005; and interviews with Gleason and Charles M. Parkin, veteran instructors from Area B, interview during tour of Catoctin Mountain Park arranged by the author, 18 May 2005.

48. Col. W.L. Rehm, 13 June 1944, notation at the bottom of Lt. Col. Lucius O. Rucker, Jr., to Col. W.L. Rehm, 12 June 1944; as well as other documents n OSS Records (RG 226), Entry 169A, Washington-SFBPersonnel, Box 23, Folder 1014, National Archives.

49. Lt. Col. Henson L. Robinson, "Schools and Training," report, October 1943, pp. 1-2, OSS Records (RG 226), Entry 136, Box 158, Folder 1723, both in National Archives II.

50. Capt. Stanley H. Lyson, S-4 [Supply], Schools and Training, to Lt. Warner, undated [February 1945], subject: Data on Detachment A, OSS Records (RG 226), Entry 136, Box 153, Folder 1658; and Lt. Col. Ainsworth Blogg, executive officer, Schools and Training Branch, Strategic Services Unit [formerly OSS], to General Counsel Office, SSU, 16 October 1945, subject: Liquidation of Areas "A" and "C," in OSS Records (RG 226), Entry 146, Box 8, Folder 1999, both in National Archives II.

51. Ibid., both Lyson's and Blogg's letters. The Demolition Area near A-3 is also referred to in Estimate #3, Estimated Prices Prepared by T. Sgt. John E. [sic] Gum, attachment to Col. W.L. Rehm, Acting Assistant Director for Services, Strategic Services Unit (SSU), for the Director, to Commanding General, Army Service Forces, Mobilization Division, 19 October 1945, subject: Chopawamsic National Recreational Demonstration Area, in OSS Records (RG 226), Entry 146, Box 8, Folder 199, National Archives II.

52. "Area A-3," map and capacity list, undated [October 1943, the date given on similar maps for Area B] located in OSS Records (RG 226), Entry 85, Box 13, Folder 249; another report indicated that the classes of 33 men each were staggered so that a new class arrived each week. Lt. Col. Henson L. Robinson, "Schools and Training," report, October 1943, p. 2, OSS Records (RG 226), Entry 136, Box 158, Folder 1723, both in National Archives II.

53. Jerry *Sage*, 30-31.

54. Serge Obolensky, *One Man in His Time: The Memoirs of Serge Obolensky* (New York: McDowell, Obloensky, 1958), 342-343, and 352-391.

55. "Major Obolensky's Training Schedule, Outline of the Strategic Services Command," OSS Records (RG 226), Entry 61, Box 11, Folder 121, National Archives II.

56. "Types of Training," p. 4, document attached to L.B. Shallcross, deputy, STB/TRD [CIA] to John O'Gara, 1 February 1951, subject: information on OSS schools and training sites, OSS Records (RG 226), Entry 61, Box 7, Folder 76, National Archives II.

57. Estimate #3, Estimated Prices Prepared by T. Sgt. John E. [sic] Gum, attachment to Col. W.L. Rehm, Acting Assistant Director for Services, Strategic Services Unit (SSU), for the Director, to Commanding General, Army Service Forces, Mobilization Division, 19 October 1945, subject: Chopawamsic National Recreational Demonstration Area, in OSS Records (RG 226), Entry 146, Box 8, Folder 199, National Archives II.

58. Material on OSS building usage and construction at A-4 is derived primarily from Capt. Stanley H. Lyson, S-4, Schools and Training, to Lt. Warner, undated [February 1945], subject: Data on Detachment A, OSS Records (RG 226), Entry 136, Box 153, Folder 1658; and Lt. Col. Ainsworth Blogg, executive officer, Schools and Training Branch, Strategic Services Unit [formerly OSS], to General Counsel Office, SSU, 16 October 1945, subject: Liquidation of Areas "A" and "C," in OSS Records (RG 226), Entry 146, Box 8, Folder 1999, National Archives II.

59. "Area A-4," map and capacity list, undated [October 1943, the date given on similar maps for Area B] located in OSS Records (RG 226), Entry 85, Box 13, Folder 249; plus Lt. Col. Henson L. Robinson, "Schools and Training," report, October 1943, p. 1, OSS Records (RG 226), Entry 136, Box 158, Folder 1723, both in National Archives II.

60. Capt. Stanley H. Lyson, S-4, Schools and Training, to Lt. Warner, undated [February 1945], subject: Data on Detachment A, OSS Records (RG 226), Entry 136, Box 153, Folder 1658, National Archives II.

61. Capt. Don R. Callahan, commanding officer, Detachment A, to Col. H.L. Robinson, executive, Schools and Training Branch, OSS, 13 March 1944, subject: Housing and Training Facilities, Detachment A, in OSS Records (RG 226), Entry 136, Box 153, Folder 1658, National Archives II.

62. Capt. Don R. Callahan, CO at Detachment A, to Col. H.L. Robinson, Executive, Training, OSS, 13 March 1944, subject: Housing and Training Facilities Detachment A, OSS Records (RG 226), Entry 136 Schools and Training Branch, Box 153, Folder 1658, National Archives.

63. Capt. Don R. Callahan, commanding officer, Detachment A, to Col. H.L. Robinson, executive, Schools and Training Branch, OSS, 13 March 1944, subject: Housing and Training Facilities, Detachment A, in OSS Records (RG 226), Entry 136, Box 153, Folder 1658; and Maj. Robert C. Wright, Schools and Training Branch OSS, to Commanding Officer, Detachment C, 29 January 1945, subject: memorandum from Military District of Washington, SPWTD 471.6, in OSS Records (RG 226), Entry 136, Box 153, Folder 1658, both in National Archives II.

64. Strickland, *Prince William Forest Park*, 30, citing "6.5-64, 1946 source," and "6.5-56, 56A, 1946 source."

65. Worth H. Storke, clerk, Minute from the Prince William County Board of Supervisors Meeting of 6 April 1944, copy attached to Lt. Col. Ainsworth Blogg to A. William Asmuth, Jr., Legal Division, 4 May 1944, subject: closing of roads and highways, Chopawamsic Area, OSS Records (RG 226), Entry 137, Box 3, Folder 24, National Archives II.

66. See maps of number of field exercises/problems located in White Binder entitled "OSS," in the Park Archives, Prince William Forest Park, Triangle, Va.

67. On the number and functions of the pistol houses (he reported three pistol houses, some other sources indicate two) and the one mystery house in Area A, see Lt. Col. Ainsworth Blogg, executive officer, schools and training branch, Strategic Services Unit [formerly OSS] to General Counsel Office, SSU, 16 October 1945, subject: Liquidation of Areas "A" and "C," in OSS Records (RG 226), Entry 146, Box 8, Folder 199, National Archives II.

68. Lykes may be confusing two different houses here, or he may be describing a cluster of houses. In a 1985 interview with historian Susan Carey Strickland, the retired park manager told her that a "Little Tokyo" was built in the woods and regularly assaulted in training practice, which sounds more like a Problem House. Strickland, *Prince William Forest Park*, 27-28, n. 96. But in the 1973 interview quoted here with Herbert Evison, his former supervisor at NPS, Lykes although using the term "Little Tokyo," spends the rest of his statement taking about what appears to have been one of Fairbairn's Mystery Houses with dummies representing German soldiers and officers inside. Ira B. Lykes, oral history interview, 23 November 1973, by S. Herbert Evison, page. 42 of transcript made from Tape Number 261, at NPS's Harpers Ferry Center Library, Harpers Ferry, W. Va.

69. Ira B. Lykes, oral history interview, 23 November 1973, by S. Herbert Evison, his supervisor at NPS, page. 42 of transcript made from Tape Number 261, at NPS's Harpers Ferry Center Library, Harpers Ferry, W. Va. I am indebted to David Nathanson there for responding to my request for a copy of the transcript.

70. Lt. Col. Corey Ford and Maj. Alastair MacBain, *Cloak and Dagger: The Secret Story of OSS* (New York: Random House, 1945), 24-25.

71. Capt. Eliot N. Vestner to Director of Training, OSS, 28 June 1943, subject: demolition of unacquired property, with attached hand drawn map of "General Area Used for Training," OSS Records (RG 226), Entry 137, Box 3, Folder 24, National Archives II.

72. Ira B. Lykes to Irving Root, NPS, 20 October 1942, NPS Records (RG 79), File 1460/3, Federal Records Center, Suitland, Md., cited in Strickland, *Prince William Forest Park*, 28, n.100. These records have disappeared.

73. Ira B. Lykes, Monthly Report, July 1942, in ibid., 28, n. 101.

74. Lt. Col. Ainsworth Blogg to Charles S. Cheston, through William Asmuth, OSS legal division, 5 May 1944, subject: repair of roads—Areas "A" and "C," OSS Records (RG 226), Entry 137, Box 3, Folder 24, National Archives II.

75. Ira B. Lykes, Monthly Report, August [1945], NPS Records (RG 79), File 1460/3, Federal Records Center, Suitland, Md., cited in Strickland, *Prince William Forest Park*, 29, n. 102. These records have disappeared.

76. Col. H.L. Robinson, chief, Schools and Training Branch, OSS to Lt. Col. James A. Hewitt, Post Engineer, Fort Belvoir, Va., 6 January 1945, subject: Closing of Detachment A, OSS in OSS Records (RG 226), Entry 136, Box 153, Folder 1658, National Archives II.

77. On the use of the lake near A-5 and also the dock at Quantico, see Capt. Don R. Callahan, commanding officer, Detachment A, to Col. H.L. Robinson, executive, Schools and Training Branch, OSS, 13 March 1944, subject: Housing and Training Facilities, Detachment A, in OSS Records (RG 226), Entry 136, Box 153, Folder 1658, National Archives II. On the establishment of a maritime school at Area D, see *War Report of the OSS (Office of Strategic Services)*, with a new introduction by Kermit Roosevelt (New York: Walker and Co., 1976), 81, 225-26, a commercially published, declassified version of the original "War Report of the OSS," prepared by the History Project, Strategic Services Unit, Office of the Assistant Secretary of War, 5 September 1947 (hereinafter cited as, OSS, *War Report of the O.S.S.*).

78. For the Smith's Point, Charles County, Maryland location, see L.B. Shallcross, deputy, STB/TRD [CIA] to Mr. John O'Gara, 1 February 1951, subject: Information on OSS Schools and Training Sites, OSS Records (RG 226), Entry 61, Box 7, Folder 76, National Archives II. This document has proven accurate in the location of every other OSS training and assessment site. All sources agree that it was located on the Potomac River a comparatively short boat ride from the U.S. Marine Corps Base at Quantico. Some place it down river. The President of the UDT-SEAL Association seeking to commemorate the founding of the American frogmen wrote in 2000 that Area D, "was located south of Quantico, VA on the Potomac River, and is discussed extensively in the MU history to be presented. We have not yet located it on the map." Tom Hawkins, "OSS Maritime: We Have Looked into the Past, and It is Us!," *The Blast*, (2000), 10-11. Frank A. Gleason, who was at Area D recalled it as north more than 10 to 20 miles south of Quantico on the eastern shore of the Potomac River. Frank A. Gleason, telephone interview with the author, 3 April 2008. Some think it was upriver from Quantico. Jerry Sage, *Sage*, 50-51, called the location Indian Head, Maryland, which is today the location of the U.S. Navy's Naval Surface Warfare Center and was then the Naval Ordnance Station. OSS radio operator James Ranney located it in what is now Smallwood State Park, Maryland, just south of Indian Head; James F. Ranney, "OSS Training Areas," in James F. Ranney and Arthur L. Ranney, eds., *The OSS CommVet Papers*, 2nd ed. (Covington, Ky.: James F. Ranney, 2002), 119. OSS, *War Report of the O.S.S.*, 80, says only that it was on "1,400 acres of wooded terrain on the Potomac River across from Quantico." The most contemporaneous document indicating location of Area D is by the then new head of Schools and Training Branch who stated only that it was "located on the Potomac River about 40 miles from Washington [which would put it around Smith's Point]. It is a CCC type camp, situated on the river bank." is Lt. Col. Henson L. Robinson, "Schools and Training," report, October 1943, p. 4, OSS Records (RG 226), Entry 136, Box 158, Folder 1723. A typescript history of the Schools and Training Branch apparently written at the end of the war, indicates only that Area D was "located on the Potomac River in an isolated section comprising fourteen hundred acres. This area, like Areas A, B, and C, had formerly been a summer recreation area." "History of Schools and Training Branch, OSS," p. 12, received by Col. E.B. Whisner, 7 January 1949 from W.J. Morgan [a former OSS officer], OSS Records (RG 226), Entry 176, Box 2, Folder 13, both in National Archives II. In an attempt to find any indication of OSS Area D located at a summer recreation area that might have become one of the Maryland state parks along the Potomac, or a CCC camp, most of whose buildings were easily portable and were moved from place to place as needed, the present author conducted telephone interviews with the following individuals. From the Maryland Park Service: Ross Kimmell, 22 May

2007; William Moffat, 5 June 2007; William Farrall, 21 June 2007; from the Maryland Forest Service, Craig Henderson, 1 April 2008; a letter to Frank Owens, local historian, 3 April 2008, which went unanswered; and telephone interviews with two OSS veterans of Area D, Jonathan Spence, former MU frogman, 28 January 2005; and Frank A. Gleason, former SO instructor, 3 April 2008; plus a search of the CCC records for Charles County, Maryland, by Eugene Morris, archivist, National Archives II, 14 April 2008. The result was inconclusive in regard to locations at a state park or CCC camp. It should be noted that Purse State Park is located just north of Smith's Point, and also that one of the typescript histories of the Schools and Training Branch, the one edited by William L. Cassidy, indicates that the temporary housing for Area D was obtained from the Army, but it does not reveal whether this housing had been from CCC camps supervised by the U.S. Army or directly from the Army. The mystery remains unresolved.

79. OSS, *War Report of the O.S.S.*, 80.

80. Frank A. Gleason, telephone interview with the author, 3 April 2008.

81. *History of the Schools and Training Branch: Office of Strategic Services*, ed. William L. Cassidy (San Francisco: Kingfisher Press, 1983), 135. This is largely a reprint of the typescript official history of the S&T Branch of the OSS prepared by Schools and Training, probably Maj. Kenneth P. Miller, in 1945 and obtained by William Cassidy in 1983 from the CIA through Freedom of Information Act requests. On the possibility that they were CCC structures, see note 77 above.

82. Map of layout of buildings with statistics on capacity of Area D, undated [October 1943, the date given on similar maps for Area B] located in OSS Records (RG 226), Entry 85, Box 13, National Archives II.

83. *War Report of the OSS*, 81, 226; and Commander H. Woolley [Royal Navy] to Colonel M.P. Goodfellow, 9 October 1942, subject: Guerrilla Notes–Rough Notes and Suggestions, M. Preston Goodfellow Papers, Box 4, Folder 4, Hoover Institute, Stanford, Calif.

84. Sage, *Sage*, 50, 55.

85. Ibid., 51, 443.

86. Frank A. Gleason, telephone interviews with the author, 16 December 2007, and 3 April 2008.

87. John Pitt Spence, telephone interview with the author, 28 January 2005. Spence remembered the name of the yacht at the *Maybelle*, but Commander Woolly listed it in October 1942 as the *Marsyl*. Commander H. Woolley [Royal Navy] to Colonel M.P. Goodfellow, 9 October 1942, subject: Guerrilla Notes–Rough Notes and Suggestions, p. 2, M. Preston Goodfellow Papers, Box 4, Folder 4, Hoover Institute, Stanford, Calif. In 2005, the 85-year-old Spence recalled that a group of men went to Quantico Marine Base by boat every Sunday to get the newspapers; he recalled that Area D was on the east side of the Potomac, but he could not remember whether it was north or south of Quantico.

88. Tom Hawkins, "OSS Maritime: We Have Looked into the Past and It Is Us!" *The Blast*, newsletter of the UDT [Underwater Demolition Team]-SEAL Association, (Third Quarter, 2000), p. 10; plus materials on the OSS combat swimmers reunion at the U.S. Naval Academy, Annapolis, Md., 27-29 March 2001, and the memorial plaques to the OSS MU operational swimming units at Fort Bragg, N.C., and the UDT-Seal Museum at Fort Pierce, Fla., sent to the author by John Pitt Spence, 5 February 2005.

89. OSS, *War Report of the OSS*, 226-27; Kenneth Finlayson, "Key West: Home of ARSOF Underwater Operations," *Veritas: Journal of Army Special Operations History*, 3/1 (2007): 3-4; Francis Douglas Fane and Don Moore, *The Naked Warriors* (New York: Appleton-Century Crofts, 1956), an account by Commander Fane of the origins and accomplishments of the Underwater Demolition Teams of the U.S. Navy, and the OSS, in World War II.

90. "Chief Instructors by Training Area," a list, 1942-1945, included in Lt. Col. Henson L. Robinson, "Schools and Training," report, October 1943, p. 15, OSS Records (RG 226), Entry 136, Box 158, Folder 1723; on Area D, see also Gregory Gluba, NWDT2, National Archives to Christina

J. Snyder, Prince William Forest Park Visitor Center, 16 October 1997, located in the PWFP Library. He indicated that his information was from the OSS Records (RG 226), Entry 136, Box 174, both in National Archives II.

91. Lawrence L. Hollander, "A Brief History of Area C," in Ranney and Ranney, eds., *The OSS CommVet Papers*, 2nd ed., 11. Hollander, like James Ranney, was a veteran of the COI/OSS Communications Branch and a member of the CommVet alumni association. Hollander indicated that Area C was established in mid-1942; the biography of R. Dean Cortright, p. 190, declared it had been set up by COI. COI was succeeded by OSS in mid-June 1942.

92. The Training Directorate to All Geographic Desks and Administrative Officers, SI-SO, n.d., [winter 1942-43?], subject: Training Procedures, OSS Records (RG 226), Entry 146, Misc. Washington Files, Security Office Files, Box 223, Folder 3106, Schools and Training, National Archives.

93. OSS, *War Report of the OSS*, 136. For the background to this decision, see Col. William J. Donovan to Colonel Buxton, Colonel Goodfellow, and Major Bruce, 3 July 1942, subject: a study of the question of communications. OSS Records (RG 226), Records of the Director's Office, Microfilm M-1642, Roll 42, Frame 1122. For recommendation that for security purposes, SI eschew joint communication operations at least at the agent to supervisor level, but cooperate with a central communications branch at higher levels, see Robert Cresswell to Major Bruce, 20 July 1942, subject: communications problem: survey and recommendation, a six-page report attached to David Bruce to William J. Donovan, 20 July 1942, ibid., Frames 1125-1131. Under the prodding of Donovan and C.W. Horn, his technical adviser for communications, Bruce at SI and Goodfellow at SO agreed, apparently in late July 1942 to the creation of a central communications officer; Lawrence Lowman was chosen to head the new branch when it was created in September. C.W. Horn to Colonel Donovan, 23 July 1942, subject: Communication System for the OSS, ibid., Frame 1132, all in National Archives II.

94. On Lowman's recruitment, see exchange of letters, 18 May to 22 June, 1942 between Lowman and Lt. Cmdr. William H. Vanderbilt, great-grandson of the railroad tycoon and himself a former governor of Rhode Island who was one of Donovan's main advisers and who handled Lowman's recruitment. OSS Records (RG 226), Entry 92A, Box 45, Folder 736, National Archives II.

95. OSS, *War Report of the OSS*, 136.

96. From B.M. Brooker, Kenneth H. Baker, and L.W. Lowman to Maj. David Bruce, Lt. Col. E.C. Huntington, memorandum, 16 November 1942, subject: proposed change for communications school, OSS Records (RG 226), Records of the Director's Office, Microfilm M-1642, Roll 42, Frames 1175-1176, National Archives II

97. Arthur Reinhardt, "Deciphering the Commo Branch," *OSS Society Newsletter*, Fall 2006), 6.

98. "Richard Dean Cortright, W9IRY/5," in Ranney and Ranney, eds., *The OSS CommVet Papers*, 2nd ed., 190.

99. These were Thai students in the United States who, through the Thai Embassy, had offered their services to Donovan's organization. The first group of 21 Thais, from what was then called Siam, were trained at B-2, A-2, and other areas, including Area C, before being sent to the China-Burma-India Theater in the spring of 1943. They were led there by SO officer and former Far Eastern travel writer, Lt. Nicol Smith, see Nicol Smith and Blake Clark, *Into Siam: Underground Mission* (Indianapolis, Ind.: Bobbs-Merrill, 1946), 1-28, 44-45, 49.

100. R. Dean Cortright, "Early Days of Area C," in Ranney and Ranney, eds., *The OSS CommVet Papers*, 2nd, 9.

101. Ibid.

102. *History of the Schools and Training Branch: Office of Strategic Services*, ed. William L. Cassidy (San Francisco: Kingfisher Press, 1983), 47-49.

103. Allen R. Richter, telephone interview with the author, 25 March 2005.

104. James F. Ranney, telephone interview with the author, 8 January 2005.

105. Marvin S. Flisser, telephone interview with the author, 27 January 2005. Similarly, John W. Brunner, who admittedly was there only for one week in October 1944, said he only saw two officers and those very briefly, and he never saw Major Jenkins. Like Flisser, Brunner emphasized the informality of the OSS. "Everything that I experienced was very informal. No military routine. No marching or parading. We simply got up when we were told. An enlisted man would come down to our hut to wake us up, and we went to the mess hall. Then we went out to our assignment for the day. There was no marching there. We just walked." John W. Brunner, telephone interview with the author, 21 March 2005.

106. Timothy Marsh, telephone interview with the author, 18 January 2006.

107. "Lawrence L. Hollander, N4JYV," in Ranney and Ranney, eds., *The OSS CommVet Papers*, 2nd ed., 196.

108. Lawrence L. Hollander, "A Brief History of Area C," in Ranney and Ranney, eds., *The OSS CommVet Papers*, 2nd ed., 11.

109. Arthur F. ("Art") Reinhardt, during a tour of former OSS Training Area C-4 arranged by the author, 13 December 2004.

110. Roger L. Belanger, letter to the author, 28 December 2004.

111. Col. W.L. Rehm, Acting Assistant Director for Services, Strategic Services Unit (SSU), for the Director, to Commanding General, Army Service Forces, Mobilization Division, 19 October 1945, subject: Chopwamsic National Recreational Demonstration Area, Estimates #1, #2 and #3, in OSS Records (RG 226), Entry 146, Box 8, Folder 199, National Archives II.

112. Lt. Col. Ainsworth Blogg, executive officer, OSS headquarters, to Charles S. Cheston, through A. William Asmuth, Jr., legal division, 6 May 1944, subject: floor covering for kitchen at Detachment "C," OSS Records (RG 226), Entry 137, Box 3, Folder 24, National Archives II.

113. Capacity listed on building layout map, "Area C," undated [October 1943, when similar map of Area B was prepared], OSS Records (RG 226), Entry 85, Box 13, Folder 249, National Archives II.

114. Charles S. Cheston, acting assistant director OSS to chief, schools and training branch, 25 May 1944, subject: repairs and improvements at Areas A and C; Cheston to chief, schools and training branch, 5 June 1944, subject: swimming facilities at Area C; and Cheston to Col. H.L. Robinson, chief, schools and training branch, 19 June 1944, subject: construction at Area A and E, all in OSS Records (RG 226) Entry 137, Box 3, Folder 24, National Archives II.

115. Col. W.L. Rehm, Acting Assistant Director for Services, Strategic Services Unit (SSU), for the Director, to Commanding General, Army Service Forces, Mobilization Division, 19 October 1945, subject: Chopwamsic National Recreational Demonstration Area, Estimates #1, #2 and #3, in OSS Records (RG 226), Entry 146, Box 8, Folder 199, National Archives II.

116. Map of "Area C," OSS Records (RG 226), Entry 85, Box 13, Folder 249, National Archives II.

117. Arthur F. ("Art") Reinhardt, OSS Communication Branch veteran, who trained at C-4 from April to early July 1944 before being sent to China, interviewed during a tour of Area C with the author and Judy Volonoski, museum technician at Prince William Forest Park, on 13 December 2004. The author recorded and transcribed Mr. Reinhardt's commentary and the identification by Ms. Volonoski of the numbers of the buildings and their current usage at Prince William Forest Park.

118. Arthur F. Reinhardt, during a tour of former Area C, Prince William Forest Park, 13 December 2001.

119. Capt. Duncan C. Lee to Brig. Gen. William J. Donovan, 11 and 15 June 1943, subject: new building for Area C; Lee to Commanding Officer Hq. and Hq. Detachment, 16 June 1944, subject: erection of multipurpose building at Area C, 16 June 1944, all in OSS Records (RG 226), Entry 137, Box 3, Folder 24, National Archives II.

120. Arthur F. Reinhardt, during a tour of former Area C, Prince William Forest Park, 13 December 2001.

121. Col. W.L. Rehm, Acting Assistant Director for Services, Strategic Services Unit [formerly OSS], for the Director, to Commanding General, Army Service Forces, Mobilization Division, 19 October 1945, subject: Chopwamsic National Recreational Demonstration Area, Estimates #1, #2 and #3, in OSS Records (RG 226), Entry 146, Box 8, Folder 199, National Archives II. A Quonset hut was a pre-fabricated metal structure in which a semi-circular, corrugated steel roof curved down to form the main walls. Portable and quickly installed, it was widely used by the Army and Navy in World War II.

122. OSS, *War Report of the O.S.S.*, 137.

123. Col. W.L. Rehm, Acting Assistant Director for Services, Strategic Services Unit [formerly OSS], for the Director, to Commanding General, Army Service Forces, Mobilization Division, 19 October 1945, subject: Chopawamsic National Recreational Demonstration Area, Estimates #1, #2 and #3, in OSS Records (RG 226), Entry 146, Box 8, Folder 199, National Archives II.

124. Map of "Area C," OSS Records (RG 226), Entry 85, Box 13, Folder 249, National Archives II.

125. Ibid., plus Marvin S. Flisser, telephone interview with the author 27 January 2005.

126. Marvin S. Flisser, telephone interview with the author, 27 January 2005; Arthur F. Reinhardt, during a tour of Prince William Forest Park, 13 December 2004.

127. Col. W.L. Rehm, Acting Assistant Director for Services, Strategic Services Unit [formerly OSS], for the Director, to Commanding General, Army Service Forces, Mobilization Division, 19 October 1945, subject: Chopwamsic National Recreational Demonstration Area, Estimates #1, #2 and #3, in OSS Records (RG 226), Entry 146, Box 8, Folder 199, National Archives II.

128. Map of "Area C," OSS Records (RG 226), Entry 85, Box 13, Folder 249, National Archives II.

129. Arthur F. Reinhardt, on a tour of former Area C, Prince William Forest Park, arranged by the author, 13 December 2004. Shooters probably had to pickup the spent brass cartridges and turn them in to get a new box of shells, because, as Mr. Reinhardt remembers, the camp was kept pristine.

130. "Harry N. Neben, W9QB," in Ranney and Ranney, eds., *The OSS CommVet Papers*, 2nd ed., 202.

131. "Notes from the 1996 OSS CommVets Reunion," in "OSS," white binder, Park Archives, Prince William Forest Park, Triangle, Va.

132. R. Dean Cortright, "Early Days of Area C," in Ranney and Ranney, eds., *The OSS CommVet Papers*, 2nd ed., 10.

133. Listing in Headquarters Detachment A, "Standard Operating Procedures," 13 April 1944, p. 5; and ibid., "Pass and Furlough Policy," 24 April 1944, p. 2, which lists Gum as a "S/Sgt," a Staff Sergeant, which is higher than Technical Sergeant; both documents in OSS Records (RG 226), Entry 136, Box 153, Folder 1658, National Archives. Gum lived in Dumfries with his wife, young son and two daughters. Robert A. Noile, Triangle, Va., who knew the family well, telephone interviews with the author, 27 April and 16 May 2007.

134. Joseph J. Tully, telephone interview with the author, 23 March 2005.

135. The Communications Branch also supervised the preliminary radio and code training given by SO, SI, MO, and MU branches of OSS for their agents at Areas A, B, E, D, F, and RTU-11 and elsewhere. Areas C and M were where CB trained its own personnel.

136. OSS, *War Report of the OSS*, 137; Lt. Col. Henson L. Robinson, "Schools and Training," report, October 1943, p.4, OSS Records (RG 226), Entry 136, Box 158, Folder 1723, National Archives II.

137. Biographical sketch of "Lawrence L. Hollander, N4JYV," in Ranney and Ranney, eds., *The OSS CommVet Papers*, 2nd ed., 196. In addition, the OSS communications base station from Mackinac Island was transferred to Area M.

138. Biographical sketch of "Lawrence L. Hollander, N4JYV," in Ranney and Ranney, eds., *The OSS CommVet Papers*, 2nd ed., 196, indicates that the Japanese-American radio operators were sent

to Italy. But it seems unlikely that the OSS would train radio operators for the Nisei Regimental Combat Team in the regular Army. Since the OSS was indeed training Asians and Asian Americans and deploying them in the China-Burma-India Theater, it seems more probable that the Far East was the destination of these Japanese Americans working for OSS.

139. OSS, *War Report of the O.S.S.*, 137.

140. Ibid, 137, 241.

141. Strickland, *Prince William Forest Park*, 53.

142. Strickland, *Prince William Forest Park*, 54; Robert A. Noile, Triangle, Va., telephone interview with the author, 27 April 2007.

143. On the development of the Marine base at Quantico, see Allan R. Millett, *Semper Fidelis: The History of the United States Marine Corps*, rev. ed. (New York: Free Press, 1991), 284, 291,323-24.

144. Lt. Col. K.E. Rockey, USMC, to Charles H. Gerner, NPS, 24 June 1935, pp. 2-3, NPS Records (RG 79), Central Classified File, 1933-1949, General 201 National Defense, "C," Box 81, Folder 201, National Archives II.

145. Acting Secretary of the Interior E.K. Burlew to Secretary of the Navy, 27 May 1941, copy in the Park Archives, Prince William Forest Park, Triangle, Va.

146. Arthur Demaray, National Park Service, to Lt. Col. R. Valliant, 31 August 1938, NPS Records (RG 79), Box 121, Folder 110, National Archives II.

147. Ira Lykes, Monthly Report, May 1942, NPS Records (RG 79), File Number 1460/3, Federal Records Center, Suitland, Md., cited in Strickland, *Prince William Forest Park*, 25, n. 84. These records have disappeared.

148. Ira Lykes, Monthly Report, 1941, NPS Records (RG 79), File Number 1460/3, Federal Records Center, Suitland, Md., cited in Strickland, *Prince William Forest Park*, 26, n. 87. These records have disappeared.

149. Ira Lykes, Monthly Report, November 1942, cited in Strickland, *Prince William Forest Park*, 25-26, n. 86. These records have disappeared.

150. Ira B. Lykes, oral history interview, 23 November 1973, by S. Herbert Evison [who had been Lykes' last superior at NPS], pp. 42-43, tape number 261, NPS's Harper's Ferry Center, Harper's Ferry, W.Va.

151. Lykes said General Torrey's offer was made in January. Lykes' NPS personnel file indicates that he was assigned to active duty with the USMC on 8 April 1943; he was relieved from active duty with the Marines on 3 December 1945, put on leave and was honorably discharged on 21 January 1946. Documents in the Ira B. Lykes, Personnel File, National Archives and Records Administration, National Personnel Records Center, St. Louis, Mo., obtained by the author, 26 April 2007.

152. Ira B. Lykes, oral history interview with S. Herbert Evison, 23 November 1973, pages 42-43, transcript in NPS's Harpers Ferry Center Library, Harpers Ferry, W.Va.

153. Robert A. Noile, Triangle, Va., telephone interview with the author, 27 April 2008.

154. Frank Gartside, Assistant Superintendent, National Capital Parks, National Park Service, to Ira B. Lykes, 13 September 1943, subject: Furlough (military duty), copy in Ira B. Lykes, Personnel File, obtained by the author 26 April 2007 from the National Personnel Records Center, St. Louis, Mo.

155. Ira B. Lykes to Susan Cary Strickland, 8 September 1985, quoted in Strickland, *Prince William Forest Park*, 26.

156. Susan Cary Strickland interview with Thelma Williams Hebda, 15 July 1985 (Williams had married another park employee, Joseph Hebda); also Ira B. Lykes to Frank Gartside, 31 May 1943, both cited in Strickland, *Prince William Forest Park*, 26-27.

157. "We made arrangements with the Marines at Quantico to go shopping from their food supplies." R. Dean Cortright, "Early Days of Area C," in Ranney and Ranney, eds., *The OSS CommVet Papers*, 2nd ed., 9. 158 John Pitt Spence, telephone interview with the author, 28 January 2005.

159. Frank A. Gleason and Charles M. Parkin, interview at Catoctin Mountain Park, Thurmont, Md., on a tour of the park with the author, 18 May 2005, and Gleason, telephone interview with the author, 3 April 2008.

160. Acting Secretary of the Interior E.K. Burlew to Secretary of the Navy, 27 May 1941, copy in the Park Archives of Prince William Forest Park, Triangle, Va.

161. "Marines to Add 50,000 Acres to Quantico Reservation," *Washington Evening Star*, 6 October 1942, clipping with Conrad L. Wirth, NPS, to Ira B. Lykes, 7 October 1942, in NPS Records (RG 79), Central Classified File, 1933-1949, General 201 National Defense, "C," Box 81, Folder 201 National Archives II.

162. Ira B. Lykes to Conrad L. Wirth, 9 October 1942, with map attached, NPS Records (RG 79), Central Classified File, 1933-1949, General 201 National Defense, "C," Box 81, Folder 201, National Archives II.

163. Ira B. Lykes to Superintendent National Capital Parks [Irving Root], 4 February 1943, Memorandum, National Park Service Records (RG 79), Central Classified File, 1933-1949, General 201 National Defense, "C," Box 81, Folder 201, National Archives II.

164. Ira B. Lykes, oral history interview, 23 November 1973, by S. Herbert Evison, p. 44, tape number 261, NPS's Harper's Ferry Center, Harper's Ferry, W.Va.

165. Newton B. Drury, Director NPS, to Mr. Demaray, 28 June 1943, Memorandum; Acting Secretary of the Interior Abe Fortas to Acting Secretary of the Navy James Forrestal, 21 June 1943; Forrestal to Secretary of the Interior, 12 June 1943, requesting transfer of jurisdiction of 4, 862 acres in the park lying south of Road No. 626 and west of road No. 619 to Navy Department "for the period of the present national emergency and for a period of six months thereafter," all in NPS Records (RG 79), Central Classified File, 1933-1949, General 201 National Defense, "C," Box 81, Folder 201, National Archives II.

166. Ira B. Lykes, oral history interview, 23 November 1973, by S. Herbert Evison, p. 43, tape number 261, NPS's Harper's Ferry Center, Harper's Ferry, W.Va.

167. Ira B. Lykes to Superintendent, 12 April 1946, NPS Records (RG 79), File Number 1460/4, Federal Records Center, Suitland, Md., cited in Strickland, *Prince William Forest Park*, 57, n. 164. These records have disappeared.

168. Secretary of the Interior Harold Ickes to Acting Secretary of the Navy James Forrestal, 5 August 1944, and Forrestal to Ickes, 18 September 1944; on other hand some Marine officers thought that the Quantico base might be expanded to a takeover of the entire Chopawamsic Recreational Demonstration Area after the war, Lt. Col. M. B. Twining, USMC, executive officer, to Post Quartermaster, 4 December 1944; all in NPS Records (RG 79), File Number 1460/4, Federal Records Center, Suitland, Md., cited in Strickland, *Prince William Forest Park*, 58, n. 165-167. These records have disappeared. 169 The negotiations and final resolution in 2003 are described in Chapter 10 of the present work.

CHAPTER 6

1. Edgar Prichard, "Address to Historical Prince William, Inc.," 16 January 1991 [actually 1992], p. 1, typescript of a talk and clipping, Deynse Tannenbaum, "Veteran Recals Learning to Spy in Park," *Potomac News*, 18 January 1992, in the Park Archives of Prince William Forest Park, Triangle, Va.

2. Ainsworth Blogg, personnel file, CIA Records (RG 263), OSS Personnel Records, Box 10, National Archives II; plus telephone interviews by the author with Dorothea Dean Dow, Athens, widow of Arden W. Dow, 15 May and 6 September 2005; Albert R. Guay, 24 October 2005; and Frank A. Gleason, 26 December 2007. People's ranks changed as they were promoted. I shall use the rank they had at the time mentioned.

3. R.P. Tenney to Joseph R. Hayden [COI's SI Branch], memorandum, 8 June 1942, subject: "Area B," OSS Records (RG 226), Entry 136, Box 158, Folder 1721, National Archives II, College Park, Md., hereinafter National Archives II; Jerry Sage, *Sage* (Wayne, Pa.: Miles Standish Press, 1985), 22.; Joseph J. Tully, telephone interview with the author, 29 December 2007; and Dr. Jonathan Clemente, email to the author, 26 December 2007.

4. Albert R. Guay, diary entry 6 December 1942, and telephone interview with the author, 24 October 2005. At the author's request, Mr. Guay agreed to deposit photocopies of the OSS related parts of his diary in the Park Archives at Catoctin Mountain Park, Thurmont, Md. Guay's last name stemmed from his French-Canadian ancestry and is pronounced "Gay."

5. Albert R. Guay, telephone interview with the author, 24 October 2005. In 1943, he left the OSS, went to Officers' Candidate School and became a personnel officer in the Army Air Forces.

6. Albert R. Guay, diary entries of 1, 2, 3, 4, 7, 13 December 1942, and telephone interview with the author, 24 October 2005.

7. Albert R. Guay, diary entry, 5 December 1942; telephone interview with the author, 24 October 2005.

8. Ainsworth Blogg file, CIA Records (RG 263), OSS Personnel Files, Box 10, National Archives II.

9. "History of Schools and Training, OSS, Part I thru Part VI Inclusive," pp. 3-4, typescript history, n.d. [before 1948 from internal evidence], included with cover note, by Col. E.B. Whisner, 7 January 1949, "Received this date from W.J. Morgan the following report." OSS Records (RG 226), Entry 176, Box 2, Folder 13, National Archives II. Dr. William J. Morgan had been an OSS psychologist sent to England in 1943 to assess training there; he was later a member of an SO team that parachuted into central France in August 1944. See William James Morgan, *The O.S.S. and I* (New York: Norton, 1957), 1-104, 164.

10. "Chief Instructors [and Commanding Officers] by Training Area," included in "Appendix IV, Part Three of the History [of Schools and Training Branch], OSS Records (RG 226), Entry 136, Box 158, Folder 1723, National Archives II; for Cohen's inclusion in the first group of Americans at Camp X in Canada (but spelling his first name as Lewis), see "History of the Schools and Training Branch, Office of Strategic Services," typescript history prepared in 1945, but not declassified until the 1980s; for handiest reference, see the published version, William L. Cassidy, ed., *History of the Schools & Training Branch, Office of Strategic Services* (San Francisco: Kingfisher Press, 1983), 28.

11. Sage, *Sage*, 21.

12. Sage, *Sage*, 1-5, 83-423.

13. Ibid., 10.

14. Ibid., 10-11.

15. Ibid., 20.

16. Ibid., 20-21.

17. Background information from Dorothea Dean Dow (Mrs. Arden W. Dow), telephone interview with the author, 15 May 2005; Joseph Lazarsky, telephone interview with the author, 14 March 2005; see also R.P. Tenney to Joseph R. Hayden, memorandum, 8 June 1942, subject: "Area B," OSS Records (RG 226), Entry 136, Box 158, Folder 1721, National Archives II.

18. Robert E. Mattingly, *Herringbone Cloak—GI Dagger: Marines of the OSS* (Washington, D.C.: History and Museums Division, Headquarters, U.S. Marine Corps, 1989), 173; Dorothea Dean Dow (Mrs. Arden W. Dow), telephone interview with the author, 15 May, 9, 16 June 2005; and Sage, *Sage,*, 22.

19. Dorothea Dean Dow (Mrs. Arden W. Dow), telephone interviews with the author, 15 May and 9 and 15 June 2005. For Rumburg's recruitment by the OSS as an instructor at Area B-2, see Sage, *Sage*, 21, 430.

20. R.P. Tenney to Joseph R. Hayden, memorandum, 8 June 1942, subject: "Area B," OSS Records (RG 226), Entry 136, Box 158, Folder 1721, National Archives II.

21. "Battle Firing: For Those Who Want to Live," in Lt. Col. Serge Obolensky and Capt. Joseph E. Alderdice, "Outline of Training Programs at Areas 'A' and 'F' for Operational Groups, Office of Strategic Services," typescript, n.d., [August 1943], OSS Records (RG 226), Entry 136, Box 140; Folder 1466, National Archives II.

22. Rex Applegate, *Kill or Get Killed* (Harrisburg, Pa.: Military Service Publishing Co., 1943); Rex Applegate, *Kill or Get Killed: Manhandling Techniques for Police and the Military*, 3rd ed. (Harrisburg, Military Service Publishing Co., 1956); Rex Applegate and Michael D. Janich, *Bullseyes Don't Shoot Back: The Complete Textbook of Point Shooting for Close Quarters Combat* (Boulder, Colo.: Paladin Press, 1998); Rex Applegate and Chuck Melson, *The Close-Combat Files of Colonel Rex Applegate* (Boulder, Colo.: Paladin Press, 1998).

23. For celebratory accounts, see "Rex Applegate," American Combatives, www.americancombatives.com/appplegate.php; "Military Expert Dies: Rex Applegate," www.fas.org/irp/news/ 1998, accessed 22 may 2007.

24. Dorothea ("Dodie") Dean Dow (Mrs. Arden W. Dow), telephone interview with the author, 15 May 2005. Mrs. Dow could remember the first names of only Rita Collart and Betty Harris.

25. John C. McWilliams, "Covert Connections: The FBN, the OSS, and the CIA," *The Historian*, 53/4 (Summer 1991): 665-69; Harry J. Anslinger with J. Dennis Gregory, *The Protectors: Narcotics Agents, Citizens and Officials against Organized Crime in America* (New York: Farrar, Straus, 1964), 79-80.

26. William R. Peers and Dean Brelis, *Behind the Burma Road: The Story of America's Most Successful Guerrilla Force* (Boston: Little, Brown, 1963), 30; Nicol Smith and Blake Clark, *Into Siam: Underground Kingdom* (Indianapolis, Ind.: Bobbs-Merrill, 1946), 49-50.

27. Charles M. Parkin, telephone interview with the author, 10 May 2005; and discussion with the author and other OSS veterans after a tour of Catoctin Mountain Park, 18 May 2005.

28. Charles M. Parkin, conversation with the author and other veterans after a tour of Catoctin Moujntain Park, 18 May 2005. A brief history of Parkin's military service is included in Lt. Col. C.M. Parkin to Assistant Chief of Staff, G-1, War Department, General Staff, 6 June 1945, subject: request to enter Army and Navy Staff College, in Charles M. Parkin, personnel file, OSS Records (RG 226), Entry 92A, Box 47, Folder 785, National Archives II.

29. See for example, comments by Private Albert Guay, diary entry of 5 November 1942; Lt. Jerry Sage in *Sage*, 22; and assessment of Gleason's personality by interviewer, 27 April 1945, in CIA Records (RG 263), OSS Personnel Files, Box 22, National Archives II.

30. Joseph Lazarsky, telephone interview with the author, 2 January 2008.

31. Frank A. Gleason, Atlanta, Ga., telephone interviews with the author, 31 January 2005, 9 May 2005.

32. Charles M. Parkin and Frank A. Gleason on a tour of Catoctin Mountain Park, arranged by the author, 18 May 2005, and reference to the "Three Skis" in Peers and Brelis, *Behind the Burma Road*, 30.

33. Charles M. Parkin, Frank A. Gleason and Reginald Spear, OSS veterans, interviewed during and after a tour of Catoctin Mountain Park arranged by the author, 18 May 2005.

34. Sage, *Sage*, 22.

35. "Record of Discussion regarding collaboration between British and American S.O.E.," 23 June 1942, a British document, in M. Preston Goodfellow Papers, Box 4, Folder 3, Hoover Institution, Stanford, Calif.

36. Frank A. Gleason, telephone interview with the author, 31 January 2005; and Gleason and Charles M. Parkin, during a tour of Catoctin Mountain Park, 18 May 2005. Those sent to the SOE schools in Britain in staggered, six-week periods between July and October 1942 included a least Ainsworth Blogg, Charles Parkin, Frank Gleason, Jerry Stage, Joseph Collart and Arden Dow.

37. Sage, *Sage*, 34; and R.P. Tenney to Joseph R. Hayden [COI's SI Branch], memorandum, 8 June 1942, subject: "Area B," OSS Records (RG 226), Entry 136, Box 158, Folder 1721, National Archives II.

38. Charles M. Parkin, interview during a tour of Cactoctin Mountain Park, 18 May 2005. Princess Märtha, born a Swedish princess, had married to Crown Prince Olav of Norway in 1929 and became a Norwegian princess. After the Germans invaded Norway she fled to Sweden and then the United States.

39. Biographical material on Fairbairn's early years from Peter Robbins, *The Legend of W.E. Fairbairn, Gentleman and Warrior: The Shanghai Years* (London: CQB Publications, 2005); William L. Cassidy, "Fairbairn in Shanghai," *Soldier of Fortune*, September 1979; Fred Wakeman, Jr., *Policing Shanghai, 1927-1937* (Berkeley: University of California Press, 1955). Fairbairn recommended such riot squads for maintaining order in occupied Germany and Japan; see Maj. W.E. Fairbairn, "Shock and Riot Police: Proposal Regarding Establishment and Maintenance of Order in Enemy Territory after Successful Allied Invasion," June 1943, OSS Records (RG 226), Entry 1936, Box 164, Folder 1785, National Archives II.

40. Charles M. Parkin, telephone conversation with the author, 10 May 2005.

41. W.E. Fairbairn, *Defendu (Scientific Self-Defence): The Official Textbook of the Shanghai Municipal Police, Hong Kong Police, and Singapore Police* (Shanghai: North-China Daily News & Herald, 1926); W. E. Fairbairn, *Get Tough: How to Win in Hand-to-Hand Combat* (New York : Appleton, 1942).

42. W.E. Fairbairn and E. A. Sykes, *Shooting to Live with the One-Hand Gun* (Edinburgh: Oliver, 1942).

43. John W. Brunner, *OSS Weapons*, 2nd ed. (Williamstown, N.J.: Phillips Publications, 2005), 68-72.

44. Lt. Col. William E. Fairbairn, British Army, to Director of Office of Strategic Services, 23 March 1945, memorandum, subject: resumé of service of Lt. Col. William E. Fairbairn, in OSS Records (RG 226), Director's Office Files, Microfilm 1642, Roll 46, Frames 18-23. A typical week for Fairbairn in 1943 included demonstrations at A-3 on Monday, A-4 on Tuesday, Area E on Wednesday, and Area B on Thursday and Friday. "W.E.F., Itinerary, 15th November-21st November 1943," OSS Records (RG 226), Entry 136, Box 163, Folder 1784, National Archives II.

45. Maj. W.E. Fairbairn, script for OSS training film, "Gutter Fighting," Part I, "The Fighting Knife," 11 February 1944, p. 2, OSS Records (RG 226), Entry 133, Box 151, Folder 1258, National Archives II.

46. Lt. Col. William E. Fairbairn, British Army, to Director of Office of Strategic Services, 23 March 1945, memorandum subject: resumé of service of Lt. Col. William E. Fairbairn, pp. 1-2, in OSS Records (RG 226), Director's Office Files, Microfilm 1642, Roll 46, Frames 19-20, National Archives II.

512 | Endnotes: Chapter 6

47. Marvin S. Flisser, telephone interview with the author, 27 January 2005.

48. Lt. Col. William E. Fairbairn, British Army, to Director of Office of Strategic Services, 23 March 1945, memorandum subject: resumé of service of Lt. Col. William E. Fairbairn, pp. 1-2, in OSS Records (RG 226), Director's Office Files, Microfilm 1642, Roll 46, Frames 19-20, National Archives II.

49. There is a discrepancy in the records over whether Fairbairn's "indoor mystery house" at Area A was at A-2 or A-3. Contemporary records indicate A-2, but Fairbairn remembered it as at A-3. It is possible that Fairbairn misremembered; a less likely possibility is that the structure was moved between A-2 and A-3.

50. Lt. Col. William E. Fairbairn, British Army, to Director of Office of Strategic Services, 23 March 1945, memorandum subject: resumé of service of Lt. Col. William E. Fairbairn, pp. 2-3, in OSS Records (RG 226), Director's Office Files, Microfilm 1642, Roll 46, Frames 19-20, National Archives II.

51. "History of the Schools and Training Branch, Office of Strategic Services," prepared in 1945, but not declassified until 1980s; published version in Cassidy, ed., *History of the Schools & Training Branch*, 88.

52. OSS, *War Report of the OSS*, 69-70, 161. OSS training films transferred to videotape are in OSS Records (RG 226), Motion Picture Branch of the National Archives II; films, scripts and continuity sheets are in OSS Records (RG 226), e.g., Entry 133, Box 151, Folder 1258; and Box 154, Folder 1299, National Archives II.

53. "The OSS Training Group," a 27-minute film, a videotape of which Rex Applegate donated to the CIA Museum in January 1996, and which Applegate stated was filmed by Ford at Catoctin in 1942. Ford's filming there was confirmed by former Area B instructors Frank A. Gleason and Charles M. Parkin in an interview during a tour of Catoctin Mountain Park 18 May 2005; Sage, *Sage*, 37-38; and James W. Voigt, Resource Manager, Catoctin Mountain Park, after looking at videotaped excerpts from "OSS Basic Training," confirmed the location as Cabin Camp 2 (Area B-2). James W. Voigt, email to the author, 21 August 2007. Videotape copies at Catoctin Mountain Park and Prince William Forest Park. Similar scenes are described in the script for "OSS Basic Training," Entry 133, Box 116, Folder 937, National Archives II

54. For nearly 40 still photographs of OSS Training Areas A, B, C, F, and RTU-11, see OSS Records (RG 226), Entry 136, Box 159, Folder 1730, National Archives II.

55. H.E. Miner to DZ [David Zablodowsky], untitled Memo, n.d. [October 1943], p. 1, OSS Records (RG 226), Entry 85, Box 13, Folder 225, National Archives II.

56. Col. Lawrence W. Lowman, chief, communications branch, to Lt. Kellogg, 16 May 1944, subject: production of communications film, SSTR-1, OSS Records (RG 226), Entry 90, Box 18, Folder 244, National Archives II.

57. William R. Peers and Dean Brelis, *Behind the Burma Road: The Story of America's Most Successful Guerrilla Forces* (Boston: Little, Brown, 1963), 31-32.

58. William L. Mudge, Jr. to J.M. Scribner, Chief, SSO, 1 January 1943, subject: plan and operation of SO scheme, OSS Records (RG 226), Entry 136, Box 164, Folder 1791, National Archives II.

59. Charles M. Parkin and Frank A. Gleason, interview during tour of Catoctin Mountain Park, 18 May 2005; Sage, *Sage*, 38-49. B-2 was largely shut down from July 1942 until spring 1943. B-5 remained in reduced operation for the cadre. Albert R. Guay, diary entries, 21 October to 19 December 1942.

60. Maj. Garland H. Williams, "Training," p. 4, n.d. [January-February 1942], in OSS Records (RG 226), Entry 136, Box 161, Folder 1754, National Archives II.

61. Frank A. Gleason, telephone interview with the author, 31 January 2005, and letter to the author 9 August 2005. The Allies feared that Franco might ally with Hitler instead of remaining

neutral. Despite helping to supply Germany, Franco ultimately rejected Hitler's overtures for an alliance.

62. Office of Strategic Services, *Operational Group Command*, OSS, December 1944. I am indebted to Caesar J. Civitella, St. Petersburg, Fla., for sending me a copy of this booklet, 25 April 2008.

63. Troy J. Sacquety, "OSS (Office of Strategic Services)," *Veritas: Journal of Army Special Operations History*, 3/4 (2007): 40-41; and OG website, OSS Society. http://www.ossorg.org/ accessed 14 May 2008.

64. Jacques F. Snyder, "Cloak and Dagger Days," typescript memoir, n.d., late 1990s, pp. 1-2, provided to the present author by James Snyder, his son, 16 January 2008.

65. "History of the Schools and Training Branch, Office of Strategic Services," prepared in 1945, but not declassified until the 1980s, published version in Cassidy, ed., *History of the Schools & Training Branch*, 87, 138-39, 7-24; Lt. Col. H.L. Robinson, "Schools and Training," October 1943, p. 3, plus "Outlines of the Courses at All of the U.S. Training Areas, December 1943, both in "Appendix IV, Part Three of the History [of the Schools and Training Branch of OSS], in OSS Records (RG 226), Entry 136, Box 258, Folder 1723; also *Mobilization Training Program Mobilization Training Programs for Officers and Enlisted Personnel Required for Special Overseas Assignments* (Washington, D.C.: Office of Strategic Services, n.d. [in compliance with JCS 155/11/D of 27 October 1943], p. 8, in OSS Records (RG 226), Entry 146, Box 223, Folder 3106, all in National Archives II.

66. In August 1944, there were 246 OSS enlisted men at Detachment A, 233 at Detachment C, 162 at Detachment F, 51 at E and none at B which had been closed down. Chief Warrant Officer Stanley S. Stokwitz to Files, 7 August 1944, subject: strength return for purpose of determining welfare funds dividends, OSS Records (RG 226) Entry 136, Box 177, Folder 1885, National Archives II.

67. "Close Combat," in Lt. Col. Serge Obolensky and Capt. Joseph E. Alderdice, "Outline of Training Programs at Areas 'A' and 'F' for Operational Groups, Office of Strategic Services," n.p., typescript, n.d., [August 1943], OSS Records (RG 226), Entry 136, Box 140; Folder 1466, National Archives II.

68. "Close Combat," typed lecture, December 1943, included in "Syllabus of Lectures," February 1944, in OSS Records (RG 226), Entry 136, Box 158, Folder 1717, National Archives II. Emphasis in the original.

69. Ibid., Emphasis in the original document.

70. Ellsworth ("Al") Johnson, telephone interview with the author, 27 June 2008.

71. "Operational Groups Training, Preliminary Course F and Final Course B," December 1943, in Appendix IV, Part Three of the History [of Schools and Training Branch], typescript in OSS Records (RG 226), Entry 136, Box 158, Folder 1723, National Archives II.

72. Ian Sutherland, "The OSS Operational Groups: Origin of the Army Special Forces," *Special Warfare* 15 (June 2002).

73. Henrik Krüger, biography of Hans Tofte (in Danish, *Hans V. Tofte: Den danske Kriegshelt, der kom til top i CIA* (Copenhagen: Ascheoug, 2005), pages 25-31. Krüger email to the author, 5 September 2006.

74. For Tofte on the instructional staff, see Capt. Don R. Callahan, commanding officer, Detachment A, to Col. H.L. Robinson, executive officer, S&T Branch, 13 March 1944, subject: housing and training facilities, Detachment A, OSS Records (RG 226), Entry 136, Box 153, Folder 1658, National Archives II.

75. William B. Dreux, *No Bridges Blown* (Notre Dame, Ind.: University of Notre Dame Press, 1971), 12-13.

76. Krüger, *Hans V. Tofte: Den danske Kriegshelt*, 25-31.

77. John K. Singlaub, telephone interview with the author, 11 December 2004; plus interview by Maochun Yu and Christof Mauch, 31 October 1996, pp. 1-8, OSS Oral History Transcripts, CIA Records (RG 2673), Box 4, National Archives II.

78. John K. Singlaub, *Hazardous Duty: An American Soldier in the Twentieth Century* (New York: Summit Books, 1991), 30-31, 69; Will Irwin, *The Jedburghs The Secret History of the Allied Special Forces, France 1944* (New York: Public Affairs Press, 2005), 258-59.

79. Jonathan Clemente, email to the author, 26 December 2007; Frank A. Gleason, telephone interview with the author, 26 December 2007; Joseph J. Tully, telephone interview with the author, 29 December 2007.

80. White was no longer at Area B, when R.P. Tenney reported to Joseph R. Hayden, memorandum, 8 June 1942, subject: "Area B," OSS Records (RG 226), Entry 136, Box 158, Folder 1721, National Archives II.

81. Dorothea Dean Dow (Mrs. Arden W. Dow), telephone interview with author, 15 May 2005.

82. Charles M. Parkin, during a tour of Catoctin Mountain Park arranged by the author, 18 May 2005.

83. Frank A. Gleason, telephone interview with the author, 9 May 2005.

84. Ibid., and also "Training Program at Pact Rowdy," typescript report, 1 January 1944, [by Arden W. Dow], in OSS Records (RG 226), Entry 146, Box 256, Folder 3550, National Archives II.

85. Arden W. Dow, file in CIA Records (RG 263), OSS Personnel Files, Box 16; and Lt. Col. C.M. Parkin, to Assistant Chief of Staff, G-1, War Department, General Staff, 6 June 1945, subject: request to enter Army and Navy Staff College, in Charles M. Parkin, personnel file, OSS Records (RG 226), Entry 92A, Box 47, Folder 785, National Archives II.

86. The following information on Area A sub-camps is derived from several sources, including, but not limited to Training Directorate to All Geographic Desks and Area Operations Officers, 1 January 1943, subject: The Training Areas, OSS Records (RG 226), Entry 146, Box 223, Folder 3106; Lt. Col. H.L. Robinson, executive, S&T, "Schools and Training," October 1943, pp. 1-2, report on status of all S&T's camps and programs; plus "Outlines of the Courses at All of the U.S. Training Areas, December 1943, both in "Appendix IV, Part Three of the History [of the Schools and Training Branch of OSS], in OSS Records (RG 226), Entry 136, Box 258, Folder 1723; *Mobilization Training Program Mobilization Training Programs for Officers and Enlisted Personnel Required for Special Overseas Assignments* (Washington, D.C.: Office of Strategic Services, n.d. [probably January 1944], pp. 4-6, in OSS Records (RG 226), Entry 146, Box 223, Folder 3106; and Capt. Don R. Callahan, C.O., Detachment A, to Lt. Col. H.L. Robinson, executive, S&T Branch, 13 March 1944, subject: housing and training facilities, Detachment A, OSS Records (RG 226), Entry 136, Box 153, Folder 1658; Chief Warrant Officer Stanley Stokwitz to Files, 7 August 1944, subject: strength return for purpose of determining welfare funds dividends, OSS Records (RG 226) Entry 136, Box 177, Folder 1885, National Archives II.

87. J.R. Brown to J.R. Hayden, 7 July 1942, [report on the course given at Area A (A-2), 21 June to 3 July 1942], OSS Records (RG 226), Entry 136, Box 161, Folder 1754. The amount of ammunition for practice at the SO training camps far exceeded the limited amounts available for recruits in US Army basic training where, at least at the beginning of the war, recruits were given a dozen rounds of rifle ammunition for target practice. See 1st Lt. James E. Rodgers, supply officer, report on ammunition supplies expended, 6 September 1943, in OSS Records (RG 226), Entry 136, Box 153, Folder 1658, National Archives II.

88. Beginning April 1944, a four-week basic training course was instituted at A-2 to take the place of the course at A-4. All Army enlisted personnel who were in OSS, or who transferred to OSS, were required to take this course at A-2 or Area C. OSS Training Board, Minutes of Meeting of 14 April 1944; and Maj. Philip K. Allen, for Col. H. L. Robinson, to "All Concerned," n.d. [April 1944], subject: OSS Basic Training Course, both in OSS Records (RG 226), Entry 146, Box 223, Folder 3106, National Archives II.

89. Mattingly, *Herringbone Cloak—GI Dagger: Marines of the OSS*, 173-74.

90. "History of the Schools and Training Branch, Office of Strategic Services," prepared in 1945, but not declassified until the 1980s, published in Cassidy, ed., *History of the Schools & Training Branch*, 91-92.

91. Ibid., 100. Kloman was the cousin of Erasmus ("Ras") Kloman, SO veteran who trained at Area B as well as other areas in winter 1943-44 before being sent to the Mediterranean Theater. Erasmus Kloman said that his cousin's name had been Joseph Kloman, but he changed it to Anthony Kloman for unknown reasons. Erasmus Kloman, telephone interview with the author, 24 January 2005.

92. Instruction Staff, A-3 to Commanding Officer, Detachment A, 15 February 1944, subject: weekly report on training & instruction, OSS Records (RG 226), Entry 136, Box 153, Folder 1661, National Archives II.

93. "History of the Schools and Training Branch, Office of Strategic Services,"prepared in 1945, but not declassified until the 1980s, published version in Cassidy, ed., *History of the Schools & Training Branch*, 101-102. Final chief instructor, November to December 1944, at A-3 was Lt. Compton N. Crook, formerly training executive at Area E. "Chief Instructors [and Commanding Officers] by Training Area," in "Appendix IV, Part Three of the History [of Schools and Training Branch], OSS Records (RG 226), Entry 136, Box 158, Folder 1723, National Archives II.

94. David Kenney, telephone interview with the author, 11 April 2005.

95. Mattingly, *Herringbone Cloak—GI Dagger: Marines in the OSS*, 114-31; Francis L. Coolidge, "Jedburghs," 30 November 1944, pp. 2-3, Interviews with Returned Men, OSS Records (RG 226), Entry 161, Box 2, Folder 31, National Archives II.

96. "History of the Schools and Training Branch, Office of Strategic Services," prepared 1945, but not declassified until 1980s, published version, Cassidy, ed., *History of the Schools & Training Branch*, 66-68.

97. Lt. Col. H. L. Robinson, executive S&T branch to Directorate, et al., 29 January 1944, subject: enlarged course at Area A-4, OSS Records (RG 226), Entry 137, Box 3, Folder 24, National Archives II.

98. "Paramilitary Course, A-4," "December 1944 [in pencil]," unsigned document in White Binder, "OSS," in Park Archives, Prince William Forest Park, Triangle, Va.

99 "Communications Course, A-4," OSS Records (RG 226), Entry 136, Box 177, Folder 1883, National Archives II.

100 "History of the Schools and Training Branch, Office of Strategic Services," prepared 1945, but not declassified until 1980s, published as Cassidy, ed., *History of the Schools & Training Branch*, 88, 91.

101. "History of the Schools and Training Branch, Office of Strategic Services," prepared 1945, but not declassified until 1980s, published version, Cassidy, ed., *History of the Schools & Training Branch*, 90.

102. "A Good Joe Named George," 9 December 1944, p. 1, Interviews with Returned Men, Schools and Training Branch, OSS Records (RG 226), Entry 161, Box 2, File 31, National Archives II. Although "George's" last name is redacted on the typescript, the tab attached to it indicates "Wuchinich." For the account of the training of a recruit named "George," see "History of the Schools and Training Branch, Office of Strategic Services," prepared 1945, but not declassified until 1980s, published as Cassidy, ed., *History of the Schools & Training Branch*, 30-31.

103. "More About George," 29 January 1945, pp. 4-5, Interviews with Returned Men, Schools and Training Branch, OSS Records (RG 226), Entry 161, Box 2, File 31, National Archives II.

104. "History of the Schools and Training Branch, Office of Strategic Services," prepared 1945, but not declassified until 1980s, published as Cassidy, ed., *History of the Schools & Training Branch*, 31-32.

105. Ibid.

106. Ibid., 33. On the OSS in Yugoslavia, see Strategic Services Unit, Office of the Assistant Secretary of War, *War Report of the OSS*, Volume II: *Overseas Targets* (New York: Walker and Company, 1976, original classified version, 1947), 127-132; and Patrick K. O'Donnell, *Operatives*,

516 | Endnotes: Chapter 6

Spies, and Saboteurs: The Unknown Story of the Men and Women of World War II's OSS (New York: Free Press, 2004), 79-101.

107. Lt. Col. Garland H. Williams quoted in "History of the Schools and Training Branch, Office of Strategic Services," prepared 1945, but not declassified until 1980s, published as Cassidy, ed., *History of the Schools & Training Branch*, 33.

108. Agent/trainee "George," quoted in "History of the Schools and Training Branch, Office of Strategic Services," prepared 1945, but not declassified until 1980s, published as Cassidy, ed., *History of the Schools & Training Branch*, 33.

109. Ibid., 30, 33.

110. "Chief Instructors [and Commanding Officers] by Training Area," included in "Appendix IV, Part Three of the History [of Schools and Training Branch], OSS Records (RG 226), Entry 136, Box 158, Folder 1723, National Archives II.

111. Capt. Joseph J. Grant, Jr., to Capt. Eliot N. Vestner, 10 May 1943, subject: failure to comply with order governing forest fire precautions; and Grant to Vestner, 12 May 1943, copy to Col. Paul J. McDonnel and Lt. Col. Kenneth Baker; OSS Records (RG 226), Entry 136, Box 153, Folder 1658, National Archives II.

112. Capt. Eliot N. Vestner to Capt. Joseph J. Grant, 13 May 1943, memo, and attachment, Vestner to Director of Training, OSS HQ, 12 May 1943, subject: report of incident, both in OSS Records (RG 226), Entry 136, Box 153, Folder 1658, National Archives II.

113. See memoranda by Capt. Joseph J. Grant, 12, 13, 20, 22, 26 May, 4, 8 June, 1943, in OSS Records (RG 226), Entry 136, Box 153, Folder 1658, National Archives II.

114. Vestner to Baker, 8 June 1943; Baker to Grant, 9 June 1943, subject: issuance of orders, Area A; and ibid., 11 June 1943, subject: orders restricting activities of training personnel; Lt. Don R. Callahan by order of the Camp Commander, to Chief Instructor, Detachment A, 12 June 1943, subject: enlisted quarters, training staff; Grant to Baker, 15 June 1943; all in OSS Records (RG 226), Entry 136, Box 153, Folder 1658, National Archives II.

115. Lt. Col. Kenneth H. Baker to Commander of Troops [Detachment A], 17 June 1943, subject: treatment of civilian instructors in OSS Records (RG 226), Entry 136, Box 153, Folder 1658, National Archives II.

116. J.W. [Capt. John W. Williams?], "Notes on Interviewing Martin on the Quarters and Messing Status at "A" of himself and Ed," 16 June 1943, report [to Kenneth Baker, director of training], in OSS Records (RG 226), Entry 136, Box 153, Folder 1658, National Archives II.

117. "Chief Instructors [and some Commanding Officers] by Training Area," included in "Appendix IV, Part Three of the History [of Schools and Training Branch], OSS Records (RG 226), Entry 136, Box 158, Folder 1723, National Archives II. Lieutenant Gordon B. Hartzfeld and then Lieutenant John T. Handy, each served at chief instructor at Area A for three-month periods during the winter of 1943-1944.

118. "History of the Schools and Training Branch, Office of Strategic Services," prepared 1945, but not declassified until 1980s; published as Cassidy, ed., *History of the Schools & Training Branch*, 72.

119. Ibid., 72-73. After Brooker left, Skilbeck worked with OSS Schools and Training in July and August 1943, and Dehn, whose specialty was propaganda, assisted as a general instructor at Area A during those two months and also worked with MO as an instructor and adviser in training officers. Anthony Moore, "Notes on Co-Operation between SOE and OSS," January 1945, p. 6; OSS Records (RG 226), Entry 136, Box 158, Folder 1722, National Archives II.

120. Ibid., 73.

121. Ibid.

122. "Operations," in Lt. Col. Serge Obolensky and Capt. Joseph E. Alderdice, "Outline of Training Programs at Areas 'A' and 'F' for Operational Groups, Office of Strategic Services, " n.p.," typescript, n.d., [August 1943], OSS Records (RG 226), Entry 136, Box 140; Folder 1466, National Archives II.

123. "Lecture: Guerrilla Warfare," in Ibid.

124. "Two Weeks Training Course, Area 'A-2'" in Ibid.

125. Aaron Bank, *From OSS to Green Berets: The Birth of the Special Forces* (Novato, Calif.: Presidio Pres, 1986), 5-6; "Col. Aaron Bank, U.S. Special Forces Founder, Dies," *OSS Society Newsletter*, Summer 2004, 13.

126. Capt. Don R. Callahan, C.O., Detachment A, to Col. H.L. Robinson, Executive, Training, OSS, 13 March 1944, subject: Housing and Training Facilities Detachment A, OSS Records (RG 226), Entry 136, Box 153, Folder 1658, National Archives II.

127. "Instructor's Manuscript: Explosives & Demolitions, Tactical Problem, Lesson No. 8," in Lt. Col. Serge Obolensky and Capt. Joseph E. Alderdice, "Outline of Training Programs at Areas 'A' and 'F' for Operational Groups, Office of Strategic Services," n.p. typescript, n.d., [August 1943], OSS Records (RG 226), Entry 136, Box 140; Folder 1466, National Archives II. Bland's Ford Bridge may have been an old wooden bridge across the Occoquan River at Woodbridge. The bridge washed out in 1972.

128. "Problem No. 2: Clearing Buildings of Personnel," in Lt. Col. Serge Obolensky and Capt. Joseph E. Alderdice, "Outline of Training Programs at Areas 'A' and 'F' for Operational Groups, Office of Strategic Services, n.p. typescript, n.d., [August 1943], OSS Records (RG 226), Entry 136 Washington & Field Station File, Wash-OG-AD-1-1; Box 140; Folder 1466, National Archives II. The other night field problems in the folder included mock attacks on a hydroelectric dam and on a power plant.

129. Kermit Roosevelt, "Introduction," to OSS, *War Report of the OSS: Overseas Targets*, xii.

130. Unidentified SO student [only the first two pages of the report remain] to Lt. Col. P[hilip]. G. Strong, 15 August 1942, subject: operation of the SA/G [SO] Schools, p. 2, OSS Records (RG 226), Entry 136, Box 161, Folder 1754, National Archives II.

131. Interviews with Returned Men, "A Good Joe Named George," [George Wuchinich, his name is on the tab], 9 December 1944, p. 1, OSS Records (RG 226) Entry 161, Box 2, Folder 31, National Archives II. K.P. was an abbreviation for "Kitchen Police," referring to duty peeling potatoes and washing dishes.

132. Lawrence L. Hollander, N4JYV, "A Brief History of Area C," in James J. Ranney and Arthur L. Ranney, eds., *The OSS CommVet Papers*, 2nd ed. (Covington, Ky.: James F. Ranney, 2002), 11.

133. "History of the Schools and Training Branch, Office of Strategic Services," history prepared in 1945, but not declassified until 1980s, published as Cassidy, ed., *History of the Schools & Training Branch*, 134.

134. "This camp is part of America's war effort, and the enemy would like to know who is trained her, how they are trained, when they leave, where they go, and many other facts which may seem small and unimportant but which may cost American lives or delay the winning of the war. Therefore: In talking to anyone in this camp— If you are a student or trainee, do not tell anyone anything about yourself—your personal history or the mission you are being prepared for....In talking or writing to anyone outside this camp— DO NOT TELL ANYONE: 1. The location of this or other O.S.S. camps, or anything which would indicate their location— such as the names of nearby towns, or the characteristics of the surrounding countryside. 2. Names of persons here, or other information about them or their arrivals at and departures from 'C'. 3. The methods and equipment of the Communications School or other information about it. Don't try to maintain secrecy by telling lies. In talking or writing to your family, friends or other people—if you are an enlisted man or officer—just tell them you are in a military camp where you get the usual military training—such as drill and marksmanship with weapons. Then you won't arouse curiosity or get tangled up with a story you can't make 'stick'.... [Instructions on mail, which went through P.O. Box 2601, Washington, 13, D.C.; proper response to armed guards who patrolled Area C during hours of darkness; prohibition of cameras or taking pictures; the posting of Communications School Training schedule for each week; the posting each day of schedules for physical training, weaponry, close combat, etc. The training schedules were posted on the bulletin boards at the Mess

Hall, Code Room, and School Headquarters] You are held responsible for being there at the right time, and no claim that you did not notice your name or section number of this schedule will excuse you....Maps, showing camp C-1 and C-4 and the ranges are posted in the mess halls and elsewhere, and you can also get information from men who have been here long enough to know their way around." Albert H. Jenkins, Major, USMCR, Commanding, "Information and Regulations for All Personnel at Detachment 'C,'" 18 August 1943, copy in White Binder, "OSS," Park Archives, Prince William Forest Park, Triangle, Va.

135. Joseph J. Tully, telephone interview with the author, 23 March 2005.

136. Ibid., and Roger L. Belanger, telephone interview with the author, 11 April 2005.

137. Betty Bullard Manning, widow of Howard Manning, telephone interview with the author, 4 March 2005.

138. Howard E. Manning, file, and Col. H.L. Robinson to Chief, Personnel Procurement Branch, OSS, 3 January 1945, subject: request for procurement of military personnel, both in CIA Records (RG 263), OSS Personnel Files, Box 37, National Archives II.

139. Ibid.

140. David Kenney, telephone interview with the author, 11 April 2005.

141. Joseph J. Tully, telephone interview with the author, 23 March 2005.

142. Col. Sherman I. Strong to Deputy Director, Administrative Services, 14 June 1945, subject: Capt. Howard E. Manning, and the formal letter from Strong to C.O., Hq. & Hq. Detachment, OSS, 14 June 1945, both in OSS Records (RG 226), Entry 132, Box 10, National Archives II.

143. Betty Bullard Manning, widow of Howard Manning, telephone interview with the author, 4 March 2005.

144. Roger L. Belanger, telephone interview with the author, 11 April 2005.

145. Obie L. Etheridge, telephone interview with the author, 14 January 2008. The episode was also remembered by Joseph Tully, telephone interview with the author, 17 January 2008. Neither recalled the officer's name or position.

146. Lostfogel and Putnam by Joseph J. Tully, interview with the author, 23 March 2005.

147. R. Dean Cortright, W9IRY, "Early Days of Area C," in Ranney and Ranney, eds., *The OSS CommVet Papers*, 2nd ed., 10.

148. Joseph J. Tully, telephone interview with the author, 23 March 2005; similarly Arthur F. Reinhardt, interview with the author at Prince William Forest Park, 14 December 2004.

149. Marvin S. Flisser, telephone interview with the author, 27 January 2005.

150. Table of Organization, "Area C," two typed pages, undated, from the National Archives, copies in White Binder, "OSS," Park Archives, Prince William Forest Park, Triangle, Va.

151. Daily Morning Report 23 August 1943, in Major [Frederick] Willis to General Donovan, 25 August 1943, OSS Records (RG 226), Director's Office Records, Microfilm M 1642, Reel 42, Frame 1330. By November 1943, the Communications School at Area C had 220 CB trainees, plus 12 students from other branches, a total of 232 students. Lt. Col. H.L. Robinson to Major Teilart, 17 November 1943, subject: Report on Training Areas, OSS Records (RG 226), Entry 146, Box 162, Folder 1757, National Archives II.

152. Timothy Marsh, telephone interview with the author, 18 January 2006.

153. On Lieutenant Lethgo's comment, Joseph J. Tully, telephone interview with the author, 23 March 2005. Lethgo is listed on the Area C instructional staff for March 1943 in "Schools and Training Branch, Monthly Report, March 1943," p. 2, Appendix III, Part Two of the History [of Schools and Training Branch], OSS Records (RG 226), Entry 136, Box 158, Folder 1722, National Archives II

154. Arthur F. Reinhardt, interview with the author at Prince William Forest Park, 14 December 2004.

155. Ibid.

156. Timothy Marsh, telephone interview with the author, 18 January 2006.

157. Ibid.

158. James F., Ranney, telephone interview with the author, 8 January 2005.

159. Ibid., see also James F. Ranney, W4KFR, "OSS radio station DMX–Bari Italy," OSS radio station JCYX–Cairo, Egypt," and "OSS Radio Station WLUR–Chihkiang, China," in Ranney and Ranney, eds., *The OSS CommVet Papers*, 2nd ed., 29-34.

160. Maj. Garland H. Williams, "Training," p. 3, n.d. [c. February or March 1942], in OSS Records (RG 226), Entry 136 Box 161, Folder 1754, National Archives II.

161. Short-waves are electromagnetic radio waves whose frequencies range from about 3 to 25 megahertz (MHz), or 3 to 25 millions of cycles of waves per second. This is roughly similar to the high-frequency (actually super high frequency, SHF) band or range. It is, much higher than the normal AM commercial radio station band, which is medium frequency (MF) and ranges from 540 to 1,800 kilohertz (kHz), thousands of cycles per second. When short waves strike certain layers of the ionosphere, they are usually reflected back to earth. Through such "bouncing" off the ionosphere, such short waves of high frequency radio transmission can be sent and received at long distances. To direct them to specific places over such distances requires a combination of transmission power, appropriate antenna, skillful manipulation of the antenna and good fortune with atmospheric conditions.

162. Lt. Col. H.L. Robinson, "Schools and Training," October 1943, in OSS Records (RG 226), Entry 136, Box 158, Folder 1723, National Archives II.

163. OSS, *War Report of the O.S.S.*, 136.

164. William J. Donovan to Assistant Chief of Staff (G-1 [personnel]), 19 June 1943, subject: request for priority to procure communications personnel from the Signal Corps, OSS Records (RG 226), Director's Office Records, microfilm 1642, Reel 42, Frame 1319, National Archives II.

165. The morning report for 24 August showed nearly 350 persons at Area C (a station complement of 87, 27 instructors, 210 trainees, plus 3 officers and 24 graduates awaiting orders). Daily Morning Report, Communications School, Detachment "C," 14 August 1943, OSS Records (RG 226), Washington Director's Office Records, microfilm 1642, Reel 42, Frame 1334, National Archives II.

166. Major [Frederick] Willis to Officers of the Communications Branch, 23 August 1943, subject: raising our sights on equipment and manpower, OSS Records (RG 226), Director's Office Records, microfilm, 1642, Reel 42, Frames 1330-1332, National Archives II.

167. Information on major, later lieutenant colonel, Peter G.S. Mero derived from several sources, including Communications Branch veterans Arthur Reinhardt, Great Falls, Va., email to the author, 23 September 2005; Peter M.F. Sichel, email to the author, 20 September 2005; Allen Richter, telephone interview with the author, 16 March 2007; SO veteran, Albert Materazzi, email to the author, 19 September 2005; the obituary of Mero's widow, Sarah M. Sabow Mero, Brenda Warner Rotzoll, "Sarah M. Mero: Handled Secret WWII Transmissions," *Chicago Sun-Times*, 25 November 2003; and various first-hand references in James J. Ranney and Arthur L. Ranney, eds., *The OSS CommVet Papers*, 2nd ed. (Covington, Ky.: James F. Ranney, 2002).

168. W. Scudder Georgia, Jr., KD3P, "It's All Greek to Me and Other Stories," in Ranney and Ranney, eds., *The OSS CommVet Papers*, 2nd ed., 151. The meeting was probably in late June 1943.

169. Steven Huston, son of Frank V. Huston, telephone interview with the author 24 September 2005; and Maj. C.A. Porter to Lt. Frank Huston, 22 February 1944, subject: message center, Bari, copy provided by Steven Huston to the author, 14 October 2005; W. Scudder Georgia, Jr., KD3P, "It's All Greek to Me and Other Stories," in Ranney & Ranney, eds., *The OSS CommVet Papers*, 151-164.

170. Roger L. Belanger, telephone interview with the author, 11 April 2005.

171. "Ed Nicholas, Jr., WB4TJJ," in *XBLCD: Newsletter of the OSS CommVets*, VI.6 (March-April 1995), 1; and Army personnel record of Edward E. Nicholas, Jr. Ned Nicholas, son of Edward N. Nicholas, Jr., telephone interview 17 July 2005 and letter and packet to the author, 18 July 2005.

172. "Vincent L. Gonzalez, Jr., W2JHU," Ranney & Ranney, eds., *The OSS CommVet Papers*, 2nd ed., 195.

173. Spyridon George Kapponnis (Spiro Cappony), telephone interviews with the author, 16 September and 4 October 2006, plus material in Spiro Cappony, packet to the author, 31 October 2006.

174. Arthur F. Reinhardt, interview with the author at Prince William Forest Park, 14 December 2004, and additional information in Arthur F. Reinhardt, email to the author, 24 February 2007.

175. As described by the Communications Branch in October 1943, the Communications School curriculum for OSS radio operators and technicians included the following totals during ten weeks of training: Introduction and training objectives 1 hour; proficiency tests (code aptitude test) 3 hours; physical training 60 hours; weapons (pistol and M1 rifle) 32 hours; dismounted drill 60 hours; International Morse Code, receiving and sending—including code, table nets, procedure, visual communications (wigwag and blinker), general security and transmission security—114 hours; radio material: including fundamentals of electricity, receiver and transmitter—96 hours; cryptography, including cryptographic security—22 hours; camouflage -2 hours; hygiene and camp sanitation—4 hours; practical communications work (watches in base station, 8 hour shifts; watches in mobile sub-base, 8 hour shifts; field work with special field equipment)—96 hours. Course total: 490 hours of instruction and practice in ten weeks. *Mobilization Training Program: Mobilization Training Programs for Officers and Enlisted Personnel Required for Special Overseas Assignments* (Washington, D.C.: Office of Strategic Services, n.d. [January 1944?]), 9-10, pamphlet in OSS Records (RG 226), Entry 146, Box 223, Folder 3106, National Archives II.

176. OSS, *War Report of the O.S.S.*, 136-140.

177. R. Dean Cortright, W9IRY, "Early Days of Area C," in Ranney and Ranney, eds., *The OSS CommVet Papers*, 2nd ed., 9.

178. James F. Ranney, telephone interview with the author, 8 January 2005; Timothy Marsh, telephone interview with the author, 18 January 2006.

179. "Codes and Ciphers," typed lecture, September 1943, included in "Syllabus of Lectures," February 1944, binder, in OSS Records (RG 226), Entry 136, Box 158, Folder 1717, National Archives II.

180. Gail F. Donnalley, telephone interview with the author, 30 April 2005.

181. John W. Brunner, telephone interview with the author, 21 March 2005.

182. Arthur Reinhardt, interview with the author at Prince William Forest Park, 14 December 2004.

183. David Kahn, *The Codebreakers: The Story of Secret Writing* (New York: Simon & Schuster, 1999). The man popularly if unfairly, credited with devising it ia Blaise de Vignenére (Veez-ih-nair) in France in the 1580s.

184. Arthur Reinhardt, "Deciphering the Commo Branch," *OSS Society Newsletter* (Fall 2006): 6; and Arthur Reinhardt, interview with the author at Prince William Forest Park, 14 December 2004.

185. John W. Brunner, *OSS Weapons*, 2nd ed. (Williamstown, N.J.: Phillips Pub., 2005), 240-43, which includes photographs of One-Time-Pads and training charts.

186. John W. Brunner, telephone interview by the author, 21 March 2005.

187. OSS, *War Report of the O.S.S.*, 136.

188. R. Dean Cortright, W9IRY, "Early Days of Area C," in Ranney and Ranney, eds., *The OSS CommVet Papers*, 2nd ed., 9; Marvin S. Flisser, telephone interview with the author, 27 January 2005.

189. Marvin S. Flisser, telephone interview with the author, 27 January 2005.

190. Arthur Reinhardt, "Deciphering the Common Branch," *OSS Society Newsletter*, Fall 2006, 6.

191. Lt. Col. H.L. Robinson, chief, schools and training branch, to director OSS, 15 May 1944, subject: Basic Training for Military Enlisted Personnel destined for Overseas Service, OSS Records (RG 226), Entry 136, Box 153, National Archives II.

192. "History of the Schools and Training Branch, Office of Strategic Services," history prepared 1945, but not declassified until 1980s, published as Cassidy, ed., *History of the Schools & Training Branch*, 23, 142.

193. Joseph J. Tully, telephone interview with the author, 23 March 2005.

194. Ibid.

195. Kenneth H. Baker to Lt. Col. Lane Rehm, 18 November 1942, subject: report on trainees at present date, OSS Records (RG 226), Entry 146, Box 223, Folder 3106, National Archives II. Dean Courtright remembered a few Thais being trained as agent-operators at C in late 1942; R. Dean Cortright, W9IRY, "Early Days of Area C," in Ranney and Ranney, eds., *The OSS CommVet Papers*, 2nd ed., 9.

196. "R.L. (Bob) Scriven, K5WFL," in Ranney and Ranney, eds., *The OSS CommVet Papers*, 2nd ed., 212; Col. Lawrence Lowman, chief, CB, to Lt. Kellogg, 16 May 1944, subject: production of communications film, SSTR-1, OSS Records (RG 226), Entry 90, Box 18, Folder 244, National Archives II.

197. John W. Brunner, telephone interview with the author, 21 March 2005.

198. Timothy Marsh, telephone interview with the author, 18 January 2006.

199. Roger L. Belanger, telephone interview with the author, 11 April 2005.

200. "History of the Schools and Training Branch, Office of Strategic Services," history prepared in 1945, but not declassified until 1980s, published as Cassidy, ed., *History of the Schools & Training Branch*, 142.

201. Joseph J. Tully, telephone interview with the author, 23 March 2005.

202. Betty Bullard Manning, widow of Howard Manning, telephone interview with the author, 4 March 2005.

203. As late as autumn 1943, some OG students complained that the guerrilla warfare lectures at Area F were often given by senior officers who had never seen such operations or combat and were simply teaching it from the manual. One of the students, Robert Farley, who been in combat with the International Brigade in the Spanish Civil War, was outspoken in his criticism, contending that from what he had seen and experienced in Spain, some of the instructors' assertions about guerrilla warfare were erroneous. William B. Dreux, *No Bridges Blown* (Notre Dame, Ind.: University of Notre Dame Press, 1971), 13-15.

204. See Theodore A. Wilson, *Building Warriors: Selection and Training of U.S. Ground Combat Troops in World War II* (Lawrence: University Press of Kansas, forthcoming, 2009).

205. Report by Capt. John Tyson to Col. Charles Vanderblue, chief, SO Branch, 30 July 1943, *History of the London Office of the OSS, 1942-1945: War Diary*, SO Branch, OSS London, Vol. 9: Training, iii-iv, quoted in Charles H. Briscoe, "Major Herbert R. Brucker: SF Pioneer, Part III: SOE Training & 'Team Hermit' into France," *Veritas: Journal of Army Special Operations History* 3:1 (2007): 72.

206. *History of the London Office of the OSS, 1942-1945: War Diary*, SO Branch, OSS London, Vol. 9: Training, viii, quoted in ibid.

207. Ellsworth ("Al") Johnson, telephone interview with the author, 27 June 2008.

CHAPTER 7

1. "Richard P. Scott, W3EFZ," in James F. Ranney and Arthur L. Ranney, eds, *The OSS CommVets Papers*, 2nd ed. (Covington, Ky.: James F. Ranney, 2002), 211.

2. Caesar J. Civitella, telephone interview with the author, 18 April 2008.

3. Maj. Gen. John K. Singlaub (USA-Ret.), transcript of interview by Maochun Yu and Christof Mauch, 31 October 1996, p. 9, OSS Oral History Transcripts, CIA Records (RG 263), Box 4, National Archives II, College Park, Md., hereinafter, National Archives II.

4. William B. Dreux, *No Bridges Blown* (Notre Dame, Ind.: University of Notre Dame Press, 1971), 1-2.

5. Transcript of an undated interview with OSS veteran, Lt. Rafael Hirtz posted on Library of Congress Veterans History Project, 12 February 2004, accessed 10 January 2007.

6. James Snyder, son of Jacques L. Snyder, email correspondence with the author, 16 January and 13 February 2008.

7. Edgar Prichard, "Address to Historical Prince William, Inc.," 16 January 1991, p. 1, typescript of a talk, located, along with a newspaper clipping of the speech, Park Archives, Prince William Forest Park, Triangle, Va.

8. Robert R. Kehoe, "An Allied Team with the French Resistance in 1944," *Studies in Intelligence: Journal of the American Intelligence Professional*, June 2002, 103, originally written in 1997

9. James F. Ranney, telephone interview with the author, 8 January 2005.

10. Albert Materazzi, Bethesda, Md., *OSS Society Digest*, Number 992, 31 March 2005, osssociety@yahoogroups.com, accessed 31 March 2005.

11. Joe Holley, "Albert Materazzi, 92, Served in OSS during World War II," *Washington Post*, 17 April 2008, B7.

12. Arthur Reinhardt, interview with the author, during a tour of Prince William Forest Park, Triangle, Va. 14 December 2004.

13. Ibid.

14. Francis ("Frank") Mills in Russell Miller, *Behind the Lines: The Oral History of Special Operations in World War II* (New York: St. Martin's, 2002), 60.

15. Ralph Tibbett, memoir in *OSS CommVets Newsletter*, Spring 1997, paraphrased in Dorothy Ringlesbach, *OSS: Stories That Can Now Be Told* (Bloomington, Ind.: Author House, 2005), 33.

16. Roger Hall, *You're Stepping on My Cloak and Dagger* (New York: W.W. Norton, 1957), 13-14.

17. Dreux, *No Bridges Blown*, 11-12.

18. Ibid., 13, 15, 17.

19. William Casey, *The Secret War against Hitler* (Washington, D.C.: Regnery Gateway, 1988), 3-5.

20. Richard Helms interview by Christof Mauch, 21 April 1997, CIA Records (RG 263), OSS Oral Histories, Box 2, National Archives II.

21. David Kenney, telephone interview with the author, 11 April 2005.

22. Arthur Reinhardt, "Deciphering the Commo Branch," *OSS Society Newsletter*, Fall 2006,: 6-7.

23. Arthur Reinhardt, interview with the author during a tour of Prince William Forest Park, Triangle, Va., 14 December 2004.

24. John K. Singlaub, *Hazardous Duty: An American Soldier in the Twentieth Century* (New York: Summit Books, 1991), 31.

25. John W. Brunner, telephone interview with the author, 21 March 2005.

26. Edgar Prichard, "Address to Historic Prince William, Inc., 16 January 1991, typescript, p. 4, copy in Park Archives, Prince William Forest Park, Triangle, Va.

27. Robert R. Kehoe, "An Allied Team with the French Resistance in 1944," *Studies in Intelligence: Journal of the American Intelligence Professional*, June 2002, 104.

28. Marvin S. Flisser, telephone interview with the author, 27 January 2005.

29. David Kenney, telephone interview with the author, 11 April 2005; and Kenney email letter to the author, 15 March 2005.

30. Ibid.

31. For such lists of "school names" for particular trainees at Area A, B, C, E, F, and RTU-11, including women trainees such as "Sallie" and "Helen" at the SI training school at RTU-11, as well as identification of the particular class, such as A-33 or E-44 or C-204, see Kenneth H. Baker to Lt. Col. Lane Rehm, 18 November 1942, subject: report on trainees at present date; and Schools and Training Branch to Mr. R.H.I. Goddard, subject: report on trainees at present date, 29 September 1943, these and others in OSS Records (RG 226), Entry 146, Box 223, Folder 3106, National Archives II.

32. Stephen J. Capestro, interviewed by G. Kurt Piehler, 17 August 1994, oral history transcript, p. 20, in files of Rutgers Archives of World War II, Rutgers University, New Brunswick, N.J.

33. Headquarters, Detachment A, "Standard Operating Procedure," 13 April 1944, p. 2, OSS Records (RG 226), Entry 136, Box 174, Folder 1841, National Archives II. Emphasis in the original.

34. Frank A. Gleason, telephone interview with the author, 31 January 2005. When Erasmus H. Kloman arrived at A-4, his cousin, Anthony Kloman, an instructor there, quickly informed him not to give any indication that they were related or even knew each other. Since the trainees had fictitious names, the family relationship remained a secret. Erasmus H. Kloman, telephone interview with the author, 24 January 2005.

35. Charles H. Briscoe, "Herbert R. Brucker, SF Pioneer: Part II, Pre-WWII-OSS Training 1943," *Veritas: Journal of Army Special Operations History*, 2:3 (2006): 26-35. In a July 2006 interview for the article, Brucker referred to his SO training near Phoenix, north of Baltimore, Maryland, as being at "the Farm," but what OSS itself referred to as "the Farm," was the advanced SI training school, RTU-11, located at Clinton, Maryland, south of Washington, D.C. It is possible that some of the trainees and others at Area E-3, which was a country estate with manor house, may have referred to it also as "the Farm."

36. Stephen J. Capestro, interviewed by G. Kurt Piehler, 17 August 1994, oral history transcript, p. 20, in files of Rutgers Archives of World War II, Rutgers University, New Brunswick, N.J.

37. Reginald G. Spear, telephone interview with the author, 25 March 2005.

38. James L. Boals, III email to the author, 20 February 2007. Whether headquarters had fabricated the story of the Nazi spy to encourage security among the trainees or whether there actually had been a Nazi agent among the trainees remains unclear. The 1947 Hollywood feature film, *13 Rue Madeleine*, starring James Cagney as the OSS instructor and Richard Conte as the German spy portrayed it. Certainly, it was something that the OSS was concerned about, and one of the missions of OSS X-2, counter-intelligence branch was to ferret out enemy agents in Allied countries and in Allied intelligence organizations.

39. "MTP [Military Training Program]-OSS, Training Schedule for Class III, week of 21 to 27 May 1944, in OSS Records (RG 226), Entry 136, Box 174, Folder 1840; see also Headquarters, Detachment A, "Standard Operating Procedures," 13 April 1944, pp. 1, 4, ibid., Folder 1841, both in National Archives II.

40. "Terry Samaras, W2QFP," and "Harry M. Neben, W9QB," in James F. Ranney and Arthur L. Ranney, eds, *The OSS CommVets Papers*, 2nd ed. (Covington, Ky.: James F. Ranney, 2002), 201, 208; Arthur Reinhardt, interview with author at Prince William Forest Park, Va., 14 December 2004.

41. John W. Brunner, telephone interview with the author, 21 March 2005.

42. Arthur Reinhardt, interview with the author at Prince William Forest Park, Va., 14 December 2004.

43. William R. Peers and Dean Brelis, *Behind the Burma Road: The Story of America's Most Successful Guerrilla Force* (Boston: Little, Brown, 1963), 30.

44. R.P. Tenney to J.R. Hayden [schools and training in Secret Intelligence Branch of OSS], interoffice memo, subject: Area B, 8 June 1942, located in OSS Records (RG 226), Entry 136, Box 158, Folder 1721, National Archives II.

45. "Area 'B' Training Course," one-page, typed schedule for instruction 16 May to 13 June 1942, attached to J.R. Brown to Mr. J.R. Hayden [schools and training in Secret Intelligence Branch of OSS], 14 June 1942, forwarded by Hayden to Dr. [Kenneth H.] Baker, 7 July 1942, in OSS Records (RG 226), Entry 136, Box 161, Folder 1754. I have also integrated comments from "Bill's" oral report midway through the course reported in R.P. Tenney to J.R. Hayden, interoffice memo, subject: Area B, 8 June 1942, Appendix II to Part One of History of the Schools and Training Office, located in OSS Records (RG 226), Entry 136, Box 158, Folder 1721, both documents in the National Archives II.

46. Arthur Reinhardt, interview with the author at Prince William Forest Park, Va., 14 December 2004.

47. Obie L. Etheridge, telephone interview with the author, 14 January 2008.

48. Albert R. Guay, telephone interview with the author, 24 October 2005, and diary entries for 28 October; 1, 4, 18, 20, 29, 30 November; 2 December 1942.

49. A. Van Beuren to Mr. Weston Howland [forward to H.L. Robinson] 24 September 1943, subject: Areas B[-5] and B-2, OSS Records (RG 226), Entry 146, Box 223, Folder 3106, National Archives II.

50. Arthur Reinhardt, interview with the author at Prince William Forest Park, Va., 14 December 2004. Both Sterling Hayden and Douglas Fairbanks, Jr. had joined the OSS, but Hayden went on dangerous Special Operations missions behind enemy lines in Yugoslavia before returning to the United States. Donovan rejected Errol Flynn's inquiry about joining the OSS.

51. Jacques L. Snyder, "Cloak and Dagger Days," typescript memoir, n.d., [late 1990s], p. 1, provided by his son, James Snyder, to the present author, 16 January 2008.

52. Some accounts refer erroneously to Fairbairn as former chief of the Shanghai police, for example, Peers and Brelis, *Behind the Burma Road*, 31. In fact, Fairbairn's highest law enforcement position had been as Assistant Commissioner of the international Shanghai Municipal Police. He was in charge of the Armed and Training Reserve, which was the special riot duty unit and its training force. See William L. Cassidy, "Fairbairn in Shanghai," *Soldier of Fortune*, 4 (September 1979): 66-71.

53. Edgar Prichard, "Address to Historical Prince William, Inc.," 16 January 1991, pp. 2, 3, typescript of a talk, located, along with a newspaper clipping of the speech, in the Park Archives of Prince William Forest Park, Triangle, Va. Prichard trained at both Area B and Area A.

54. Lt. Col. Corey Ford and Maj. Alastair MacBain, *Cloak and Dagger: The Secret Story of OSS* (New York: Random House, 1945), 26.

55. Jerry Sage, *Sage* (Wayne, Pa.: Miles Standish Press, 1985), 20.

56. Edward E. Nicholas, "How We Won the Great War," typescript memoir, n.d., [c. 1994], p. 5. Copy sent by Nicholas's son, Ned Nicholas, to the author, 18 July 2005.

57. Joseph Lazarsky, telephone interview with the author, 14 March 2005.

58. Dreux, *No Bridges Blown*, 17-18. It is possible that Sergeant Bolinksi was the same demolition instructor that Jacques L. Snyder had at Area A in late 1943. An old timer with years of experience with explosives, he had only two fingers remaining on his right hand and still used his teeth to crimp the detonating caps on primer cord, a practice as dangerous as Russian roulette and one which he told the students never to do. Jacques L. Snyder, "Cloak and Dagger Days," typescript memoir, n.d., [late 1990s], p. 3, provided by his son, James Snyder, to the present author, 16 January 2008.

59. Spiro Cappony, telephone interview with the author, 16 September 2006.

60. Reginald Spear, telephone interview with author, 25 March 2005 and 24 June 2005.

61. John W. Brunner, reading from and elaborating on his diary entry for Friday, 27 October 1944, in a telephone interview with the author, 21 March 2005. At Area A, Lt. William Dreux remembered the thrill of firing pistols, rifles, carbines, and submachine guns at moving and bobbing cardboard targets. Dreux, *No Bridges Blown*, 18.

62. John W. Brunner, reading from and elaborating on his diary entry for 24 October 1944, in a telephone interview with the author, 21 March 2005.

63. Reginald G. Spear, telephone conversation with the author, 25 March 2005.

64. Maj. Gen. John K. Singlaub with Malcolm McConnell, *Hazardous Duty: An American Soldier in the Twentieth Century* (New York: Summit/Simon & Schuster, 1991), 32.

65. Jacques L. Snyder, "Cloak and Dagger Days," typescript memoir, n.d., [late 1990s], p. 4, provided by his son, James Snyder, to the present author, 16 January 2008.

66. John W. Brunner, reading from and elaborating on his diary entry for 24 October 1944, in a telephone interview with the author, 21 March 2005.

67. Edgar Prichard, Arlington, Va., "Address to Historical Prince William, Inc.," 16 January 1991, p. 1, typescript of a talk and newspaper clipping of the speech, in Park Archives of Prince William Forest Park, Triangle, Va. Prichard's reference in this case was to Area B.

68. Ibid. Prichard trained at both Area A and Area B. A similar account of Area B was given by Spiro Cappony, telephone interview with the author, 16 September 2006.

69. Arne [I. Herstad] to his fiancé, Andi, 1 August 1943, *OSS Society Digest*, Number 2064, 28 May 2008, osssociety@yahoogroups.com, accessed 29 May 2008. For the NORSO group at Area B-2. A, see Van Beuren to Weston Howland, 24 September 1943, subject: Areas B and B-2, OSS Records (RG 226), Entry 146, Box 223, Folder 3106, National Archives II.

70. Singlaub with McConnell, *Hazardous Duty*, 32-33.

71. Ellsworth ("Al") Johnson, telephone interview with the author, 27 June 2008.

72. John W. Brunner, reading from and elaborating on his diary entry for Wednesday, 25 October 1944, in a telephone interview with the author, 21 March 2005.

73. Edgar Prichard, "Address to Historical Prince William, Inc.," 16 January 1991, p. 5, typescript of a talk, in the Park Archives of Prince William Forest Park, Triangle, Va.

74. Arne Herstad to Andi, 25 August 1943, in *OSS Society Digest*, Number 2088, 1 July 2008, osssociety@yahoogroups.com, accessed 1 July 2008.

75. Erasmus H. Kloman, *Assignment Algiers: With the OSS in the Mediterranean Theater* (Annapolis, Md.: Naval Institute Press, 2005), 13; and Erasmus Kloman, telephone interview with the author, 24 January 2005.

76. Albert R. Guay, diary entry for 16 December 1942, and telephone interview with the author, 24 October 2005.

77. Spiro Cappony, telephone interview with the author, 16 September 2006.

78. Jacques L. Snyder, "Cloak and Dagger Days," typescript memoir, n.d., [late 1990s], p. 2, provided by his son, James Snyder, to the author, 16 January 2008. Similarly, one of the Italian OGs in Fall 1943 remembered raiding the nearby farm of columnist Drew Pearson, author of hostile columns about the OSS. Caesar Civitella, telephone interview with the author, 18 April 2008.

79. Jacques L. Snyder, "Cloak and Dagger Days," typescript memoir, n.d., [late 1990s], p. 3, provided by his son, James Snyder, to the present author, 16 January 2008.

80. Albert Materazzi, telephone interview with the author, 26 January 2005.

81. Edgar Prichard, "Address to Historical Prince William, Inc.," 16 January 1991, pp. 5-6, typescript of a talk, in the Park Archives of Prince William Forest Park, Triangle, Va. The five SO

agents were sent to Nigeria in August 1942 to work with British SOE and set up chains of agents in case the German offensive in North Africa drove down the Atlantic coast of West Africa. After the U.S. landings in North Africa in November, the team was recalled and re-assigned. OSS, *War Report of the OSS, Overseas Targets*, 40.

82. Kloman, *Assignment Algiers*, 4, 7-9, 11, 111n.

83. Ibid., 11-12, and passim.

84. R. Dean Cortwright, W9IRY, "Reflections Concerning the SSTR-1," and James F. Ranney, W4KFR, "The SST-1 Transmitter," both in Ranney and Ranney, eds, *The OSS CommVets Papers*, 2nd ed., 61-65; John W. Brunner, *OSS Weapons*, 2nd ed. (Williamstown, NJ: Phillips Publications, 2005), with pictures, 232-235.

85. Arthur Reinhardt, interview with the author at Prince William Forest Park, Va., 14 December 2004; see also Arthur Reinhardt, "Deciphering the Commo Branch," *OSS Society Newsletter* (Fall 2006): 6.

86. Marvin S. Flisser, telephone interview with the author, 27 January 2005.

87. Arthur Reinhardt, interview with the author at Prince William Forest Park, Virginia, 14 December 2004.

88. "Harry M. Neben, W9QB," in Ranney and Ranney, eds, *The OSS CommVets Papers*, 2nd ed., 201.

89. Arthur Reinhardt, interview with the author at Prince William Forest Park, Virginia, 14 December 2004.

90. James F. Ranney, telephone interview with the author, 8 January 2005.

91. Marvin S. Flisser, telephone interview with the author, 27 January 2005.

92. Timothy Marsh, telephone interview with the author, 18 January 2006.

93. John W. Brunner, telephone interview with the author, 21 March 2005; similarly Obie L. Etheridge, telephone interview with the author, 14 January 2008.

94. Arthur Reinhardt, interview with the author at Prince William Forest Park, Va., 14 December 2004.

95. James F. Ranney, telephone interview with the author, 8 January 2005.

96. Ibid.

97. Frank A. Gleason, telephone interview with the author, 26 December 2007.

98. Obie L. Etheridge, telephone interview with the author, 14 January 2008. Etheridge witnessed the accident.

99. Edgar Prichard, "Address to Historical Prince William, Inc.," 16 January 1991, p. 5, typescript of a talk in the Park Archives of Prince William Forest Park, Triangle, Va.

100. Joseph Lazarsky, telephone interview with the author, 14 March 2005.

101. Frank A. Gleason, telephone interview with the author, 31 January 2005.

102. Congressional Country Club, *Congressional Country Club, 1924-1984*, n.p., n.d. (Bethesda, Md.: Congressional Country Club, c. 1985), 29. I am indebted to Maxine D. Harvey of the club for this volume.

103. John Pitt Spence, telephone interview with the author, 28 January 2005.

104. Albert Materazzi, telephone interview with the author, 26 January 2005.

105. Albert Materazzi, Close Combat, *OSS Society Digest*, Number 1123, 18 August 2005, osssociety@yahoogroups.com, accessed 23 August 2005.

106. Spiro Cappony, telephone interview with the author, 16 September 2006.

107. Ibid.

108. I am indebted to the grandson of Joseph ("Jumping Joe") Savoldi for this information. Emails from J.G. Savoldi to the author, 26 May 2006. See also, "Information Sought on OSSer 'Jumping Joe' Savoldi," OSS Society *Newsletter* (Summer 2005), 16.

109. There are two references to Joseph ("Jumping Joe") Savoldi at OSS Training Area B in the spring of 1942. He was identified by name, but incorrectly described as an instructor, by Edgar Pritchard, a 71-yearold Arlington, Virginia, lawyer and OSS/SO veteran in a talk to an historical society in Prince William County, Virginia, "Address to Historic Prince William, Inc.," 16 January 1991, page 2, typescript of talk in Park Archives at Prince William Forest Park, Triangle, Va. A 1942 contemporaneous, internal report on OSS Training Area B does not give Savoldi's name, but states that one of the students there in early June 1942 was "a professional wrestler," presumably a reference to Joseph Savoldi. R.P. Tenney to J.R. Hayden, interoffice memo, subject: Area B, 8 June 1942, in OSS Records (RG 226), Entry 136, Box 158, Folder 1721, National Archives II.

110. Jerry Sage, *Sage* (Wayne, Pa.: Miles Standish Press, 1985), 36.

111. Michael Burke, *Outrageous Good Fortune* (Boston: Little Brown, 1984), 94. See also Corey Ford and Alastair MacBain, *Cloak and Dagger: The Secret History of OSS* (New York: Random House, 1945), 149-176; and Max Corvo, *The O.S.S. in Italy, 1942-1945: A Personal Memoir* (New York: Praeger, 1990), 83, 86, 113, 134. All of the above authors had served as officers in the OSS.

112. Ford and Mac Bain, *Cloak and Dagger*, 173-74.

113. Moe Berg's fascinating life has been sketched in several biographies, the fullest being Nicholas Dawidoff, *The Catcher was a Spy: The Mysterious Life of Moe Berg* (New York: Pantheon, 1994); but see also Louis Kauffman, Barbara Fitzgerald, and Tom Sewell, *Moe Berg: Athlete, Scholar, Spy* (Boston: Little, Brown, 1975); and a family memoir/document collection by his sister, Ethel Berg, *My Brother Morris Berg: The Real Moe* (Newark, N.J.: E. Berg, 1976), copy in Special Collections, Alexander Library, Rutgers University, New Brunswick, N.J. Another, if lesser known professional baseball player who trained at Catoctin Mountain Park was Richard W. ("Bobo") Breck, an OSS Special Operations agent who served in Italy. Breck had played baseball at preparatory school and at Harvard and after the war, was a pitcher for the Pawtuckett Slaters, a farm team for the Boston Braves in 1946 and 1947 until he broke his leg. He subsequently worked as a production manager for Raytheon on radar for missile and other defense systems. C.G. Lynch, obituary, "Richard W. Breck, 83, was Veteran of World War II, Raytheon Manager," *Patriot Ledger*, Quincy, Mass., 17 March 2005, p. 25, reproduced in *OSSSociety Digest*, Number 984, 18 March 2005, osssociety@yahoogroups.com, accessed 18 March 2005.

114. Dawidoff, *The Catcher was a Spy*, 150, 152-153.

115. Alex Flaster, Chicago, email to the author, 24 February 2006; Flaster is a Chicago based documentary filmmaker who was in 2006 producing a feature length documentary film on Moe Berg. At the infiltration test at the end of Berg's training, he was apprehended in his attempt to penetrate the Glenn Martin aircraft factory in Baltimore. The publicity from his arrest caused a minor scandal in Washington, because there already had been some concern that the OSS might be used for domestic spying on Americans.

116. "I was in charge of General Bill Donovan's OSS Balkan Desk from August 1943 until I left for England in May 1944." "Moe Berg," photocopy of a summary in Berg's handwriting, undated; and Mr. Morris Berg, Acting Area Operations Officer, ME-SO, to Mr. Thomas Damberg, Special Relations, 12 November 1943, subject: future correspondence, both documents in Berg, *My Brother Morris Berg*, 200, 224.

117. Dawidoff, *The Catcher was a Spy*, 169-217. For some unknown reason, Berg declined to accept the award in 1946. His sister accepted it for him after Moe Berg's death in 1972.

118. Serge Obolensky, *One Man in His Time: The Memoirs of Serge Obolensky* (New York: McDowell, Obolensky, 1958), 342. Obolensky wrote in 1958 that Goodfellow sent him immediately "to a commando school in Virginia," but he undoubtedly confused this with Maryland, as Lt. Jerry Sage remembers training him at B-2.

119. Ibid., 342-343.

120. Sage, *Sage*, 30-31.

121. Ibid., 31-32.

122. Obolensky, *One Man in His Time*, 343.

123. Ibid., 347-348.

124. Ibid., 343, 349, 404, although Obolensky spelled it Alderdyce. See "Lecture: Guerrilla Warfare," in Lt. Col. Serge Obolensky and Capt. Joseph E. Alderdice, "Outline of Training Programs at Areas 'A' and 'F' for Operational Groups, Office of Strategic Services, n.p., typescript, n.d., [August 1943], OSS Records (RG 226), Entry 136 Box 140; Folder 1466, National Archives II.

125. Kloman, *Assignment Algier*, 34.

126. Obolensky, *One Man in His Time*, 353-371; Albert Garland and Howard Smyth, *Sicily and the Surrender of Italy* (Washington, D.C.: U.S. Army Center for Military History, 1965), 258-261.

127. Sterling Hayden, *Wanderer* (New York: Knopf, 1963), 310.

128. On Hayden the actor, see, for example, the entries on him in David Thomson, *The New Biographical Dictionary of Film*, exp. and rev. (New York: Knopf, 2004); and Karen Burroughs Hannsberry, *Bad Boys: The Actors of Film Noir* (Jefferson, N.C.: McFarland, 2003), 288-299.

129. Spiro Cappony, telephone interview with the author, 16 September 2006. Mr. Cappony recounted much of his entire OSS experience at home and abroad in a letter to a friend in 2004. At the author's request, Mr. Cappony has made a recording on 4 October 2006 of his reading that letter to the author. The recording has been donated by Mr. Cappony to Catoctin Mountain Park, Thurmont, Md.

130. Ibid.

131. Ibid.

132. Albert R. Guay, diary entries of 11 and 30 November 1942.

133. Albert R. Guay, diary entry of 28 October 1942.

134. "Richard P. Scott, W3EFZ," in Ranney and Ranney, eds, *The OSS CommVets Papers*, 2nd ed., 211.

135. Timothy Marsh, telephone interview with the author, 18 January 2006.

136. Frank A. Gleason, telephone interview with the author, 9 May 2005.

137. Charles M. Parkin, transcript of interview on a tour of Catoctin Mountain Park, 18 May 2005.

138. Frank A. Gleason, telephone interview with the author, 31 January 2005; and interview on tour of tour of Catoctin Mountain Park, Thurmont, Md., 18 May 2005.

139. Arne [I. Herstad] to Dearest Andi, 6 October 1943, *OSS Society Digest*, Number 2062, 26 May 2008, osssociety@yahoogroups.com, accessed 26 May 2008. They were selected from nearly 100 men under Captain William F. Larsen at Area B-2. A, see Van Beuren to Weston Howland, 24 September 1943, subject: Areas B and B-2, OSS Records (RG 226), Entry 146, Box 223, Folder 3106, National Archives II. Larsen later led his Norsos on the Percy Red OG Mission in Central France in early August 1944.

140. Singlaub with McConnell, *Hazardous Duty*, 33.

141. Dreux, *No Bridges Blown*, 18.

142. Marvin S. Flisser, telephone interview with the author, 27 January 2005.

143. Arthur Reinhardt, interview with the author while touring Prince William Forest Park, 14 December 2004.

144. Caesar Civitella, telephone interview with the author, 18 April 2008.

145. Timothy Marsh, telephone interview with the author, 18 January 2006.

146. Albert R. Guay, telephone interview with the author, 24 October 2005, and diary entry for 1 November 1942.

147. Frank Long, Thurmont, Md., 5 June 2006; William A. Willhide, Thurmont, Md., 13 January 2006; and James H. Mackley, Thurmont, Md., interviewed by the author, 13 June 2006.

148. William A. Willhide, Thurmont, Maryland, telephone interview with the author, 13 June 2006.

149. A. van Beuren to Weston Howland, 24 September 1943, subject: Areas B and B-2, p. 1, OSS Records (RG 226), Entry 146, Box 223, Folder 3106, National Archives II.

150. Albert R. Guay, diary entries for 12 and 28 November, and 9 and 11 December.

151. Albert R. Guay, telephone interview with the author, 24 October 2005.

152. Albert R. Guay, diary entries of 12 and 28 November and 9, 11, and 12 December 1942.

153. Albert R. Guay, telephone interview with the author, 24 October 2005, and diary entry for 23 October 1942.

154. Albert R. Guay, telephone interview with the author, 24 October 2005.

155. Albert R. Guay, diary entry for 29 October 1942.

156. Ibid., diary entry for 31 October 1942.

157. Ibid., diary entries for 10, 12, 14, 19, 21 November and 1, 3, 10, 17, 18 December 1942.

158. Spiro Cappony, telephone interview with the author, 16 September 2006.

159. Ibid.

160. Hall, *You're Stepping on My Cloak and Dagger*, 34.

161. Ibid., 36.

162. Albert R. Guay, diary entry for 5 November 1942.

163. Ibid., diary entry for 15 December 1942.

164. James F. Ranney, telephone interview with the author, 8 January 2005.

165. Headquarters, Detachment A, "Standard Operating Procedure," 13 April 1944, p. 2, OSS Records (RG 226), Entry 136, Box 174, Folder 1841, National Archives II.

166. James F. Ranney, telephone interview with the author, 8 January 2005. Civilian instructor Timothy Marsh never went to the Marine base at Quantico. Timothy Marsh, telephone interview with the author, 18 January 2006.

167. Marvin S. Flisser, telephone interview with the author, 27 January 2005. The men from Area C were in uniform, but as far as the dance hall patrons knew, "we were [just] regular soldiers."

168. Arthur Reinhardt, interview with the author at Prince William Forest Park, Va., 14 December 2004; also James F. Ranney, telephone interview with author, 8 January 2005.

169. Timothy Marsh, telephone interview with the author, 18 January 2006.

170. Ibid.

171. Timothy Marsh, telephone interview with the author, 18 January 2006.

172. Lee Lansing, town historian, Dumfries, Va., interviewed by Susan Cary Strickland, 15 July 1985, in Susan Cary Strickland, *Prince William Forest Park: An Administrative History* (Washington, D.C.: National Park Service, 1986), 27.

173. Robert A. Noile, Triangle, Va., telephone interview with the author, 27 April 2007. Noile, who grew up on Joplin Road across from the park, reported on what his father told him soon after World War II.

174. John Hammond Moore, "Hitler's Wehrmacht in Virginia, 1943-1946," *Virginia Magazine of History and Biography*, 85 (July 1977); on a wider scale, see Judith M. Gansberg, *Stalag USA: The Remarkable Story of German POWs in America* (New York: Crowell, 1977). The most super secret camp for German prisoners of war was actually an interrogation center at Fort Hunt on the Potomac River down near Mount Vernon in Fairfax County, Virginia. Known only as P.O. Box 1142, this facility was an Army/Navy installation at which 3,400 captured German submariners, airmen,

soldiers, and scientists were held incognito for a few weeks or several months while German speaking interrogators questioned them about plans, equipment, and inventions, before they were sent to regular POW camps. Fort Hunt is today a park on the George Washington Memorial Parkway administered by the National Park Service. Petula Dvorak, "A Covert Chapter Opens for Fort Hunt Veterans; As files on Nazi POWs Are Declassified, Their Interrogators Break Their Silence," *Washington Post*, 20 August 2006, A1.

175. Dorothea ("Dodie") Dean Dow, telephone interviews with the author, 15 May 2005, and 9 and 16 June 2005.

176. Later in the war, when Detachment A was officially closed effective 11 January 1945, the mail drop for the OSS personnel at Prince William Forest Park was shifted from Washington, D.C. to the Marine Base at Quantico. The new mailing address was Detachment A OSS, Box 1000 Marine Barracks, Quantico, Va.; see Col. H.L. Robinson, chief, Schools and Training Branch, OSS to Lt. Col. James A. Hewitt, Post Engineer, Fort Belvoir, Virginia, 6 January 1945, subject: Closing of Detachment A, OSS in OSS Records (RG 226), Entry 136, Box 153, Folder 1658, National Archives II.

177. Dorothea ("Dodie") D. Dow, telephone interviews with the author, 15 May 2005, and 9 and 16 June 2005.

178. Ibid., She remembered the name of Park Manager Ira Lykes, when the author mentioned it to her but not the other NPS supervisor, John Gum.

179. Ibid., 15 May 2005, and 9 and 16 June 2005.

180. Ibid., 1 January 2008.

181. Ibid., 15 May 2005, and 9 and 16 June 2005.

182. Ibid., 15 May 2005.

183. Frank A. Gleason, telephone interview with the author, 31 January 2005; and letter to the author 9 August 2005.

184. Frank A. Gleason, interview during a tour arranged by the author of Catoctin Mountain Park, Thurmont, Md., 18 May 2005.

185. Albert R. Guay, diary entries for 22 and 23 November 1942.

186. Ibid., diary entries for 29 November and 6 December 1942.

187. Charles M. Parkin, interview during a tour arranged by the author of Catoctin Mountain Park, Thurmont, Md., 18 May 2005.

188. Frank A. Gleason, telephone interview by the author, 31 January 2005.

189. Albert R. Guay, telephone interview with the author, 24 October 2005, and diary entries for 3 and 4 November 1942.

190. Albert R. Guay, telephone interview with the author, 24 October 2005, and diary entry for 15 December 1942.

191. Timothy Marsh, telephone interview with the author, 18 January 2006.

192. Ira B. Lykes, oral history interview by S. Herbert Evison, 23 November 1973, tape number 261, transcription, p. 45, National Park Service, Harpers Ferry Center Library, Harpers Ferry, W. Va. I am indebted to David Nathanson of the Harpers Ferry Center for responding to my request for a copy the relevant parts of Ira Lykes' oral history interview. David Nathanson to the author, 10 November 2005.

193. Ira B. Lykes, interview with Susan Cary Strickland, 4 September 1985, in Susan Cary Strickland, *Prince William Forest Park: An Administrative History* (Washington, D.C.: National Park Service, 1986), 27.

194. Robert A. Noile, Triangle, Va., telephone interviews with the author 27 April and 16 May 2007. Born in 1939, on the south side of Joplin Road diagonally across from the park, Noile knew both the Lykes and Gum families and played with his friends in the park beginning when the War

Department returned it to the Park Service in 1946. He added that the park office was near the road to the old Wilson place.

195. Lykes was provost marshal, security officer, for 94,000 acres of Marine Base at Quantico, which included 13 firing ranges, plus what he called "little Tokyo's" makeup houses for simulated urban fighting. The Marine base during the war included 8,000 to 10,000 acres of the park south of Joplin Road which was known as the "Guadalcanal Area" and was used for training. By the end of the war, Lykes was a major. He was mustered out in January 1946 and returned immediately to his job as project manager of the park. Ira B. Lykes, oral history interview by S. Herbert Evison, 23 November 1973, tape number 261, transcription, pp. 42-43, National Park Service, Harpers Ferry Center Library, Harpers Ferry, W.Va.

196. Strickland, *Prince William Forest Park*, 26.

197. Thelma Williams Hebda, Dumfries, Va., interview with Susan Cary Strickland, 15 July 1985 cited in Strickland, *Prince William Forest Park*, 26. Ms. Williams later married Joseph Hebda, who became an NPS employee at the park after the war. Robert A. Noile says other than Superintendent Ira Lykes, the only civilian allowed into the OSS area of the park north of Joplin Road during the war was Epp Williams, who did maintenance work. Robert A. Noile, Triangle, Va., telephone interviews with the present author 27 April and 16 May 2007.

198. Ira B. Lykes to Frank T. Gartside, superintendent, NPS National Capital Region, 31 May 1943, Records of the National Park Service (RG 79), File Number 1460/4, which in the 1980s was at the Federal Records Center, Suitland, Md, as quoted in Strickland, *Prince William Forest Park*, 27. These records have disappeared.

199. Robert A. Noile, Triangle, Va., telephone interviews with the author 27 April and 16 May 2007.

200. Ira B. Lykes, oral history interview by S. Herbert Evison, 23 November 1973, tape number 261, transcription, p. 44, National Park Service, Harpers Ferry Center Library, Harpers Ferry, W.Va.

201. Ralph Tibbett, memoir in *OSS CommVets Newsletter* (March 1997), paraphrased in Dorothy Ringlesbach, *OSS: Stories That Can Now Be Told* (Bloomington, Ind.: Author House, 2005), 34.

202. Ira B. Lykes, oral history interview by S. Herbert Evison, 23 November 1973, tape number 261, transcription, p. 42, National Park Service, Harpers Ferry Center Library, Harpers Ferry, W.Va.

203. Ibid., 39-42.

204. The author has not heard such stories from any of the former OSS Special Operations, Operational Group, Secret Intelligence, or Communications personnel that he has interviewed.

205. Author's note: The author has not found any other evidence of this restriction and the continuous close supervision by an instructor, except perhaps for certain groups of foreign nationals and their own leaders training at OSS camps.

206. Another's note: William E. ("Dan") Fairbairn, former Associate Commissioner of the international Municipal Police of Shanghai.

207. Ira B. Lykes, oral history interview by S. Herbert Evison, 23 November 1973, tape number 261, transcription, pp. 39-42, National Park Service, Harpers Ferry Center Library, Harpers Ferry, W.Va. I am indebted to David Nathanson of the Harpers Ferry Center for responding to my request for a copy the transcript of Ira Lykes' oral history interview. David Nathanson to the author, 10 November 2005.

208. Ira B. Lykes, Venice, Fla., interview with Susan Cary Strickland, 4 September 1984, quoted in Strickland, *Prince William Forest Park*, 28.

209. Ibid., 26.

CHAPTER 8

1. "OSS Organization and Function," (June 1945), OSS Records (RG 226), Entry 141, Box 4; and "History," ibid., Entry 99, Box 75, National Archives II, College Park, Md., hereinafter, National Archives II.

2. Gen. Dwight D. Eisenhower to the Director, OSS, UK base, 31 May 1945, reprinted in Office of Strategic Services, *War Report of the OSS*, vol. 2: *The Overseas Targets* (New York: Walker & Co., 1976), 222. A similar letter was sent to the Executive Director of SOE, British Maj. Gen. Colin Gubbins.

3. These figures were given by Geoffrey M.T. Jones, an OSS veteran and then President of the Veterans of the O.S.S. Association, at an international historical conference on the topic, "The Americans and the War of Liberation in Italy: Office of Strategic Services and the Resistance," held in Venice, Italy, 17-18 October 1994, proceedings published in Italian and English, as *Gli Americani e la Guerra de Liberazione in Italia: Office of Strategic Service(O.S.S.) e la Resistenza/ The Americans and the War of Liberation in Italy: Office of Strategic Services(O.S.S.) and the Resistance* (Venice: Institute of the History of the Resistance, 1995), 202.

4. Patrick K. O'Donnell, *Operatives, Spies, and Saboteurs: The Unknown Story of the Men and Women of World War II's OSS* (New York: Free Press, 2004), 12, which deals with the European and Mediterranean Theaters but not the Far East.

5. The official OSS report on the agency's overseas activities runs 460 pages. Office of Strategic Services, *War Report of the OSS*, Vol. 2, *The Overseas Targets*, with new introduction by Kermit Roosevelt (New York: Walker & Co., 1976). This report, declassified and published in 1976, was originally prepared by the Strategic Services Unit, successor to the OSS, in 1947.

6. OSS, *War Report of the OSS*, Overseas Targets, 11-16. See also, "Certain Accomplishments of the Office of Strategic Services," p. 1; attached to William J. Donovan to W.B. Kantack, OSS Reports Officer, 14 November 1944, "Accomplishments of OSS, 15654, copy in CIA Records (RG 263), Thomas F. Troy Files, Box 12, Folder 98, National Archives II.

7. H. Montgomery Hyde, *Cynthia: The Most Seductive Secret Weapon in the Arsenal of the Man Called Intrepid* (New York: Ballantine, 1965); Elizabeth P. McIntosh, *Sisterhood of Spies: The Women of the OSS* (Annapolis, Md.: Naval Institute Press, 1998), 26-31. OSS also conducted such a "black bag" break-in at the embassy of Spain, which was then officially neutral but actually pro-Axis under Franco.

8. "Certain Accomplishments of the Office of Strategic Services," p. 1; attached to William J. Donovan to W.B. Kantack, Reports Officer, 14 November 1944, "Accomplishments of OSS, 15654, copy in CIA Records (RG 263), Thomas F. Troy Files, Box 12, Folder 98, National Archives II.

9. OSS, *War Report of the OSS*, Overseas Targets, 16-18.

10. The price in casualties for French North Africa was 1,200 suffered by the Americans, 700 by the British, and 1,300 by the French. On the invasion, see Rick Atkinson, *An Army at Dawn: The War in North Africa, Theater in World War II* (New York: Farrar, Straus and Giroux, 2004).

11. Bradley F. Smith, *The Shadow Warriors: O.S.S. and the Origins of the C.I.A.* (New York: Basic Books, 1983), 156, for reference to the Army's assessment; OSS, *War Report of the OSS, Overseas Targets*, 18, for Marshall's December 1942 letter to William J. Donovan noting the important role of the OSS.

12. Carleton S. Coon, *A North African Story: An Anthropologist as OSS Agent* (Ipswich, Mass.: Gambit, 1980).

13. *An Army at Dawn*, 361; OSS, *War Report of the OSS*, Overseas Targets, 20-21.

14. Robert E. Mattingly, *Herringbone Cloak—GI Dagger: Marines of the OSS* (Washington, D.C.: History and Museums Division, HQ, U.S. Marine Corps, 1989), 174.

15. Donald C. Downes, a graduate of Phillips Exeter and Yale who had taught at a boy's preparatory school, in Cheshire, Connecticut, had worked before U.S. entry into the war as an amateur agent for Office of Naval Intelligence in Turkey and the Middle East, then joined and became a rising star in the OSS. See Downes' memoir, *The Scarlet Thread: Adventures in Wartime Espionage* (London: Verschoyle, 1953). After the German counteroffensive was defeated, Downes began to establish a clandestine network in Spain. However, a key Spanish spy was caught, betrayed the network, and all thirteen of the spies that Downes had sent into Spain were arrested and executed by Franco's government. It was a major disaster for the OSS. Bradley Smith, *OSS: The Secret History of America's First Central Intelligence Agency* (Berkeley: University of California Press, 1972), 75-82.

16. Jerry Sage, *Sage* (Wayne, Pa.: Miles Standish Press, 1985), 69-70. Another African American, elderly jazz band leader Henry Perkins, was briefly recruited as a spy by OSS operative Waller Booth in Tangiers in late 1942. O'Donnell, *Operatives, Spies, and Saboteurs*, 45.

17. Maj. Jerry Sage, interview, 30 March 1945, p. 1; Schools and Training Branch, "Interviews with Returned Men," OSS Records (RG 226), Entry 136, Box 159, Folder 1729, National Archives II.

18. Ibid., 2; see also Sage, *Sage*, 83-88; and OSS, *War Report of the OSS, Overseas Targets*, 20-21.

19. "Irving Goff," interview in Studs Terkel, *"The Good War": An Oral History of World War Two* (New York: Ballantine Books, 1984), 494. Terkel spelled the name as Feldsen; Sage spelled it Felsen.

20. Maj. Jerry Sage, interview, 30 March 1945, p. 3, Schools and Training Branch, "Interviews with Returned Men," OSS Records (RG 226), Entry 136, Box 159, Folder 1729, National Archives II; and Sage, *Sage*, 58, 83-88.

21. Ibid., 1-10, Schools and Training Branch, "Interviews with Returned Men," OSS Records (RG 226), Entry 136, Box 159, Folder 1729, National Archives II. On Sage's many escape attempts, see also Walter Wager, "Slippery Giant of the OSS," *Men*, July 1961, 32; and Lt. Cmdr. Richard M. Kelly. USNR, "He Never Stropped Trying [to Escape]," *Blue Book: The Magazine of Adventure for Men* (September 1946), 42; Richard Kelly had head of OSS's Maritime Unit in the Adriatic, 1943-45.

22. Sage, *Sage*, 305.

23. Sage, *Sage*, 207, 327, 431. Upon his return to Washington, Sage told his old mentor, "Dan" Fairbairn "that his training really worked, including the sentry-kill and everything else he had taught."

24. Ibid., 277-304; the 1963 film was based on Paul Brickhill, *The Great Escape* (London: Farber & Farber, 1951). For details of the escape attempt, see also Alan Burgess, *The Longest Tunnel: The True Story of World War II's Greatest Escape Tunnel* (New York; G. Weidenfeld, 1990), and Tim Carroll, *The Great Escape from Stalag Luft III* (New York: Pocket Books, 2005).

25. Sage, *Sage*, 396. "Dagger" was Sage's code name.

26. Sage, *Sage*, 431; for documentation, see Lt. Col. John W. Williams to Capt. David C. Crockett, 24 March 1945, subject: Major Jerry Sage, OSS Records (RG 226), Entry 160A, Box 23, Folder 1028, National Archives II.

27. Their stories are recounted in Robert E. Mattingly, *Herringbone Cloak—GI Dagger: Marines of the OSS* (Washington, D.C.: History and Museums Division, HQ, U.S. Marine Corps, 1989).

28. Ibid., 40-47; and "Peter Julien Ortiz," in W. Thomas Smith, Jr., *Encyclopedia of the Central Intelligence Agency* (New York: Facts on File, 2003), 180-181. Surprisingly, Ortiz is barely mentioned in the official history of the Marine Corps and then only in the POW section. Benis M. Frank and Henry I. Shaw, *History of Marine Operations in World War II* (Washington, D.C.: Historical Branch, G-3 Division, Headquarters, U.S. Marine Corps, 1968), 5: 748. *Operation Secret* (Warner Bros., 1952), starring Cornel Wilde, was very loosely based on Ortiz.

29. OSS, *War Report of the OSS, Overseas Targets*, 57, 166-69.

30. Obituary, "Richard W. Breck, 83, Was Veteran of World War II, Raytheon Manager," *Patriot Ledger* (Quincy, Mass.), *OSS Society Digest* Number 984, 18 March 2005, at osssociety@yahoogroups.com.

534 | Endnotes: Chapter 8

31. Jack Hemingway, *Misadventures of a Fly Fisherman: My Life With and Without Papa* (Dallas, Tex.: Taylor Pub. Co., 1986), 127-134. Quotation is on page 131.

32. Stephen J. Capestro, oral history interview with G. Kurt Piehler, 17 August 1994, p. 29, typescript in the office of the Rutgers Oral History Archives of World War II, New Brunswick, N.J. Capestro's back was broken in a military vehicle accident in North Africa in which the driver was killed. Capestro was returned to the USA for treatment.

33. Obituary, "Sarah M. Mero: Handled Secret WWII Transmissions," *Chicago Sun Times*, 25 November 2003.

34. OSS, *War Report of the OSS, Overseas Targets*, 170-71.

35. James F. Ranney, "OSS Radio Station JCYX–Cairo, Egypt," in James F. Ranney and Arthur L. Ranney, eds., *The OSS CommVets Papers*, 2nd ed. (Covington, Ky.: James F. Ranney, 2002), 31.

36. James F. Ranney, "OSS Radio Station DMX–Bari, Italy," in Ranney and Ranney, eds., *The OSS CommVets Papers*, 2nd ed., 29-30.

37. Frank V. Hutson's personnel file, supplied by his son Steve Huston to the author, 24 September 2005. For Max Corvo's SI side of the dispute with Mero over relocating the radio station, see Max Corvo, *The O.S.S. in Italy, 1942-1945* (New York: Praeger, 1990), 127, 138-40, 161.

38. Gail F. Donnalley, telephone interview with the author, 30 April 2005. See also the chronology in Guy H. Nicholason, "Operation Sunrise," Ranney and Ranney, eds., *The OSS CommVets Papers*, 2nd ed., 43-47.

39. Richard Dunlap, *Donovan: America's Master Spy* (Chicago: Rand McNally, 1982), 399, quoting his own interview with Paul Gale.

40. "Irving Goff," interview in Terkel, *"The Good War,"* 494.

41. The chief of the SO unit in Sicily, Lt. Col. Guido Pantaleoni, was captured leading a team through enemy lines. OSS, *War Report of the OSS, Overseas Targets*, 55.

42. "Minutes of the meeting of the Joint Chiefs of Staff, 108th meeting, 19 August 1943, CCS 334 (8-7-43x), item 9, copy made by CIA historian Thomas F. Troy from a copy in the files of Edward P. Lilly, Ph.D., a JCS historian; CIA Records (RG 263), Thomas F. Troy Files, Box 4, Folder 30, National Archives II.

43. Ibid.; correspondence including Marshall to Eisenhower, 21 August 1943; Eisenhower (actually via his assistant, W. Bedell Smith) to Marshall, 22 August 1943, Marshall to Eisenhower, 23 August 1943, and Eisenhower to Marshall, 28 August 1943, all in Dwight D. Eisenhower, *The Papers of Dwight David Eisenhower*, 21 Vols. (Baltimore, Md.: Johns Hopkins University Press, 1970–), 2: 1211-12.

44. Smith, *OSS*, 16-18; Serge Obolensky, *One Man in His Time: The Memoirs of Serge Obolensky* (New York: McDowell, Obolensky, 1958), 340-351.

45. Obolensky, *One Man in His Time*, 354-359.

46. Quotations from ibid., 360-363. See also Lt. Col. Serge Obolensky, "Report of Sardinia Operation," typed copy of Obolensky's report in History of OSS OGs in Italy, n.d. [1945] typescript, pp. 24-25, OSS Records (RG 226), Entry 143, Box 11, Folder 1, National Archives II.

47. OSS, *War Report of the OSS, Overseas Targets*, 61-62; Albert Garland and Howard Smyth, *Sicily and the Surrender of Italy* (Washington, D.C.: Center of Military History, 1965), 258-61.

48. Mattingly, *Herringbone Cloak–GI Dagger: Marines of the OSS*, 174-76.

49. History of OSS OGs in Italy, n.d. [1945] typescript, p. 14, OSS Records (RG 226), Entry 143, Box 11, Folder 1, National Archives II. See also OSS, *War Report of the OSS, Overseas Targets*, 59-61.

50. "Certain Accomplishments of the Office of Strategic Services," p. 2; attached to William J. Donovan to W.B. Kantack, Reports Officer, 14 November 1944, "Accomplishments of OSS, 15654," copy in CIA Records (RG 263), Thomas F. Troy Files, Box 12, Folder 98, National Archives II.

51. Douglas Porch, *The Path to Victory: The Mediterranean Theater in World War II* (New York: Farrar, Straus and Giroux, 2004); Rick Atkinson, *The Day of Battle: The War in Sicily and Italy, 1943-1944* (New York: Henry Holt, 2007).

52. "Louis Joseph Gonzalez," biographical information provided by his son, Ronald Gonzalez, letter to the author, 16 January 2008.

53. History of OSS OGs in Italy, p. 14, n.d. [1945] typescript, OSS Records (RG 226), Entry 143, Box 11, Folder 1, National Archives II.

54. Information on the McGregor Operation (the spelling of the project's name varies.) is in Lt. Col. Corey Ford and Maj. Alastair MacBain, *Cloak and Dagger: The Secret Story of the OSS* (New York: Random House, 1946), 149-176; Michael Burke's memoir, *Outrageous Good Fortune* (New York: Random House, 1984), 94-100; Corvo, *OSS in Italy*, 83-86, 95-97, 110-13, 134; and in the records of the McGreggor [sic] Mission, OSS Records (RG 226), Entry 179, Box 286, National Archives II.

55. On Savoldi's recruitment and mission, see Lt. Col. Carroll T. Harris to Lt. Cmdr. William H. Vanderbilt, 23 July 1942; and 1st Lt. W.R. Mansfield to Mr. William Mudge, 19 May 1943, subject: prospective Italian recruits, p. 2, and Maj. Charles J. Eubank to Mr. Joe Savoldi, 9 December 1944, all in OSS Records (RG 226), Entry 92A, COI/OSS Central Files, Box 42, Folder 687, National Archives II. I am indebted to archivist Larry McDonald for bringing this document to my attention. On the mission, see also O'Donnell, *Operatives, Spies, and Saboteurs*, 53-56, who does not mention Savoldi; and lifeline chronology, photographs and clippings as well as copies of falsified Italian Army identification cards and other wartime documents, and a postwar undated, unidentified U.S. newspaper clipping, William P. Moloney, "Bare Savoldi's War Mission." Included in J.G. Savoldi, grandson of Joseph Savoldi, Jr., email messages to the author, 26 and 30 May 2006, and packet to the author, 30 May 2006.

56. OSS, *War Report of the OSS, Overseas Targets*, 67, 81.

57. Peter Tompkins, *A Spy in Rome* (New York: Simon and Schuster, 1962), 29-72. Credit to OSS Rome's Radio Vittoria for saving the Anzio beachhead was given by the G-2 of 6th Corps, a Colonel Langevin, is in "Accomplishments of the OSS, 15654," p. 2, attached to William J. Donovan to W.B. Kantack, Reports Officer, 14 November 1944, copy in CIA Records (RG 263), Thomas F. Troy Files, Box 12, Folder 98, National Archives II; for similar see Martin Blumenson, *Salerno to Cassino* (Washington, D.C.: Center of Military History, 1969), 393. For Tompkins' critiques, see Peter Tompkins, "Are Human Spies Superfluous?" in George C. Chalou, ed., *The Secrets War: The Office of Strategic Services in World War II* (Washington, D.C.: National Archives and Records Administration, 1992), 129-39; and Peter Tompkins, "Operative in Rome," 4 October 1944, Interviews with Returned Men, Schools and Training Branch, OSS Records (RG 226), Entry 161, Box 2, Folder 31, National Archives II.

58. The controversy over OSS SI operations in Italy, 1942-45, is reflected not least in the negative assessments by British author Anthony Cave Brown, *The Last Hero: Wild Bill Donovan* (New York: Times Books, 1982), 484-507; and the positive explanations provided by the operations officer in the Italian Secret Intelligence Section of the OSS, Max Corvo, *The O.S.S. in Italy*. Marine Lt. Walter W. Taylor, 29-year-old Yale graduate, Harvard Ph.D. and neighbor of Donvan's, who trained at SO schools at Areas A and B in 1943, wound up infiltrating Italian SI agents from Corsica to the mainland in PT boats and rubber dinghies. He earned a Bronze Star for his heroism in one such expedition in June 1944 near Genoa. Mattingly, *Herringbone Cloak—GI Dagger: Marines of the OSS*, 177-82.

59. David C. Martin, *Wilderness of Mirrors* (New York: Harper & Row, 1980), 12-13.

60. Caricatures of Angleton, particularly in his role as CIA's counter-intelligence chief, have appeared in numerous spy-thriller books and films, most recently, Robert DeNiro's film, *The Good Shepherd* (2007), starring Matt Damon. For contrasting views of Angleton and the Cold War, see the favorable view of Robin W. Winks, *Cloak and Gown: Scholars in the Secret War, 1939-1961* 2nd ed. (New Haven, Conn.: Yale University Press, 1996), 322-437, 539; and the hostile view of Tom Mangold, *Cold Warrior: James Angleton: The CIA's Master Spy Hunter* (New York: Simon & Schuster, 1991), 42 and passim.

61. "Operational Group 'A,'" chronology, no page number but following page 31, in History of OSS OGs in Italy, n.d. [1945] typescript, OSS Records (RG 226), Entry 143, Box 11, Folder 1, National Archives II.

62. Albert Materazzi, telephone interview with the author, 23 September 2005; and Caesar Civitella, telephone interview with the author, 18 April 2008. Both contended that the criticism of some of the Italian American OGs during their training made by Roger Hall, an instructor at Areas F and B, in his acerbic account of his OSS experience, *You're Stepping on My Cloak and Dagger* (New York: Norton, 1957), 39-42, was unfair and unwarranted. For support of the positions of Materazzi and Civitella, see Emilio T. Caruso, "Italian-American Operational Groups of the Office of Strategic Services," *The Americans and the War of Liberation in Italy*, 219-220. On the other hand, Area B commander Capt. Montague Mead made a stark contrast to a visiting inspector from the Security Branch, praising the Norwegian OG then at the camp and denigrating the recently departed Italian OG. A. van Beuren to Weston Howland, 24 September 1943, subject: Areas B and B-2, p. 2, OSS Records (RG 226), Entry 146, Box 223, Folder 3106, National Archives II.

63. The most important of the islands were Gorgona and Capraia. OSS, *War Report of the OSS, Overseas Targets*, 77-79. Max Corvo, "The OSS and the Italian Campaign," Chalou, ed., *The Secrets*, 192.

64. History of OSS OGs in Italy, n.d. [1945] typescript, p. 15, OSS Records (RG 226), Entry 143, Box 11, Folder 1, National Archives II.

65. Albert Materazzi interview in O'Donnell, *Operatives, Spies and Saboteurs*, 59-60.

66. Joseph Squatrito, *Code Name: Ginny* (Staten Island, NY: Forever Free Publishing, 2001).

67. History of OSS OGs in Italy, n.d. [1945] typescript, p. 17, OSS Records (RG 226), Entry 143, Box 11, Folder 1, National Archives II; Squatrito, *Code Name: Ginny*. See also Brown, *Last Hero*, 475-83; OSS, *War Report of the OSS, Overseas Targets*, 79; O'Donnell, *Operatives, Spies, and Saboteurs*, 59-61; Corvo, *OSS in Italy*, 162.

68. On Hitler's instructions, see "Commando Order," I.C.B. Dear and M.R.D. Foot, eds., *The Oxford Companion to World War II* (Oxford: Oxford University Press, 1995), 257-58; also Brown, *Last Hero*, 473-74.

69. Brown, *Last Hero*, 479-83.; and Squatrito, *Code Name: Ginny*; Capt. Albert R. Materazzi testified against Dostler at his trial in Rome in November 1945. "OSS Raiders' Story is Told," *New York Sun*, 18 November 1945, clipping in packet sent by Albert R. Materazzi to the author, 31 January 2005.

70. Peter Tompkins, *A Spy in Rome* (New York: Simon and Schuster, 1962), 18-19.

71. History of OSS OGs in Italy, n.d. [1945] typescript, p. 17, OSS Records (RG 226), Entry 143, Box 11, Folder 1, National Archives II.

72. Under an agreement signed 7 December 1944, the CLN agreed to put 90,000 partisans into the field and the American SO and British SOE agreed to provide a monthly expense per man of 1,500 lire, a total of 80,000,000 lire a month, to be repaid by the Italian government after the war. OSS, *War Report of the OSS: Overseas Targets*, 109.

73. Albert Materazzi, telephone interview with the author, 23 September 2005; and Max Corvo, "The O.S.S. and the Italian Campaign," *The Americans and the War of Liberation in Italy*, 35, albeit most were from the south of Italy and their dialect was often difficult for northerners to understand.

74. The following is based on Piero Boni, Rome, Italy, in-person interview with the author, 15 July 2007. Silvia Boni, his daughter and a good friend of the author, served as interpreter; plus Piero Boni, "Relations between the Americans and the Italian Resistance," *The Americans and the War of Liberation in Italy: Office of Strategic Services (O.S.S.) and the Resistance* (Venice: Institute of the History of the Resistance, 1995), 214-217. I am indebted to Mr. Boni for the interview and for a copy of this volume. For concurrence about the differences between the American and British attitudes toward the partisans, see Mario Fiorentini, "The "Dingo" Mission: Operating in the Genoa-

Piancenza-Parma Triangle,"ibid., 248; Tullio Lussi, "The O.R.I. and the O.S.S. Collaboration," ibid., 272-273; and Ennio Tassinari, "My Four Missions with the Allies: From Initial Distrust to an Effective Co-operation," ibid., 314.

75. Piero Boni, Rome, Italy, in-person interview with the author, 15 July 2007.

76. Ibid. In addition to the bazooka, Boni preferred, for his own personal weapons, an Italian Beretta pistol and submachine gun. He said the latter had a longer range than the British Sten submachine gun. He also liked the Colt .45 automatic pistol for its impact. In the field, he carried both a Beretta and a Colt .45.

77. Mattingly, *Herringbone Cloak—GI Dagger: Marines of the OSS*, 176.

78. See for example, Lossowski, Ops [Operations], cable to Renata [Mission], 14 October 1944, OSS Records (RG 226), Entry 139, Box 48, Folder 446, National Archives II.

79. Piero Boni, Rome, Italy, in-person interview with the author, 15 July 2007.

80. Ibid.

81. History of OSS OGs in Italy, n.d. [1945], Operations, p. 5, OSS Records (RG 226), Entry 143, Box 11, Folder 1, National Archives II.

82. "Concerning the Relationship between Rochester and Cayuga Missions," memorandum prepared by Piero Boni (code-named Coletti) for the author and translated and emailed to the author by his son-in-law, Professor Ugo Rubeo of the University of Rome, 11 September 2007. A detailed history of the Renata, Rochester, and Cayuga missions may be found in Piero Boni, *Giorni a Compiano* (Days in Compiano) (Compiano, Parma, Italy: Compiano Arte Storia, 1984); and the reports on OSS OG website, ossog.org.

83. See incoming messages from the Rochester Mission to OSS headquarters in Siena and then Florence, 24 November 1944 through 16 March 1945, succeeded by the Dodgers Mission (when new radio crystals had arrived) to OSS HQ, 16 March through 5 April 1945, all in OSS Records (RG 226), Entry 139, Box 51, Folder 474, National Archives II.

84. Outgoing message from Fajans, Ops, Florence to Rochester, 22 February 1945, OSS Records (RG 226), Entry 139, Box 51, Folder 474, National Archives II.

85. "Irving Goff," in Terkel, *"The Good War*, 496.

86. "Report of the Cayuga Mission," typed copy of the original report by Capt. Michael Formichelli in History of OSS OGs in Italy, n.d. [1945], OSS Records, Entry 143, Box 11, Folder 1, National Archives II.

87. Outgoing message, Ops 2 to Dodgers, 4 April 1945, OSS Records (RG 226), Entry 139, Box 51, Folder 474, National Archives II.

88. Report of the Cayuga Mission," p. 4 by Capt. Michael Formichelli,, in History of OSS OGs in Italy, n.d. [1945], OSS Records, Entry 143, Box 11, Folder 1, National Archives II.

89. "Report of the Cayuga Mission," typed, undated copy of the original report by Capt. Michael Formichelli, p. 6, in History of OSS OGs in Italy, n.d. [1945], OSS Records, Entry 143, Box 11, Folder 1, National Archives II.

90. Information on Howard W. Chappell from his son, Jack R. Chappell, email to the author, 5 October 2007, telephone interview, 13 October 2007, and Howard Chappell's military records and other documents supplied to the author, 17 October 2007; confirmation of the nickname from Caesar J. Civitella, telephone interview with the author, 18 April 2008.

91. Albert R. Materazzi, telephone interview with the author, 30 August 2007.

92. Dewey Linze, "Howard Chappell Recalls Christmas Flight Behind Enemy Lines in War," *Los Angeles Times*, 10 December 1961, E-6; William L. White, "Some Affairs of Honor," *Reader's Digest*, December 1945, 137.

93. "Report on the Tacoma Mission," p. 1, typed, copy of the original report by Capt. Howard Chappell, in History of OSS OGs in Italy, n.d. [1945], OSS Records, Entry 143, Box 11, Folder 1, National Archives II.

538 | Endnotes: Chapter 8

94. "Report on the Tacoma Mission," p. 6, typed copy of the original report by Capt. Howard Chappell, in History of OSS OGs in Italy, n.d. [1945], OSS Records, Entry 143, Box 11, Folder 1; and Lt. Cmdr. Richard M. Kelly, "Torture Preferred! [an article on Salvadore Fabrega in the Tacoma Mission]" *Blue Book: Magazine of Adventure for Men* (June 1946), 60-61; clipping in OSS Records (RG 226), Entry 161, Box 7, Folder 7, National Archives II. See also Summary of the Tacoma Mission on OSS Operational Groups website, www.ossog.org/italy/tacoma.html, accessioned 23 September 2007.

95. Kelly, "Torture Preferred!" *Blue Book*, 62; clipping in OSS Records (RG 226), Entry 161, Box 7, Folder 7, National Archives II.

96. "Report on the Tacoma Mission," pp.6-7, typed copy of the original report by Capt. Howard Chappell, in History of OSS OGs in Italy, n.d. [1945], OSS Records, Entry 143, Box 11, Folder 1, National Archives II; and Kelly, "Torture Preferred!" *Blue Book*, 62-63; clipping in OSS Records (RG 226), Entry 161, Box 7, Folder 7, National Archives II.

97. "Report on the Tacoma Mission," p. 8-10, and Report of the Tacoma Mission by Cpl. Oliver M. Silsby, pp. 4-5, attached to typed copy of the original report by Capt. Howard Chappell, in History of OSS OGs in Italy, n.d. [1945], OSS Records, Entry 143, Box 11, Folder 1; and Kelly, "Torture Preferred! *Blue Book*, 64; clipping in OSS Records (RG 226), Entry 161, Box 7, Folder 7, National Archives II.

98. Kelly, "Torture Preferred" *Blue Book*, 66-67, clipping in OSS Records (RG 226), Entry 161, Box 7, Folder 7, National Archives II.

99. William L. White, "Some Affairs of Honor," *Reader's Digest* (December 1945), 136-54; Kelly, "Torture Preferred," *Blue Book*, 67-70 in OSS Records (RG 226), Entry 161, Box 7, Folder 7, National Archives II.

100. "Report on the Tacoma Mission," pp. 15-16, copy of report by Capt. Howard Chappell, in History of OSS OGs in Italy, [1945], OSS Records, Entry 143, Box 11, Folder 1, National Archives II. Italics added.

101. Ibid., 17; plus Albert R. Materazzi, telephone interview with the author, 30 August 2007.

102. "Report on the Tacoma Mission," by Capt. Howard Chappell, p. 17; and Report of the Tacoma Mission by Cpl. Oliver M. Silsby, pp. 6-8, attached to ibid., both in History of OSS OGs in Italy, n.d. [1945], OSS Records, Entry 143, Box 11, Folder 1, National Archives II. In the ranking of military awards, the Distinguished Service Cross, is the second highest in the Army, after the Medal of Honor. The Distinguished Service Medal is third, the Silver Star is fourth, and the Bronze Star is fifth in precedence in the awards for bravery in combat.

103. See the after action reports on the OSS Operational Groups website, www.ossog.org.

104. Mattingly, *Herringbone Cloak—GI Dagger*, 162-73.

105. Summary of Spokane and Sewanee Missions, OSS Operational Groups website, www.ossog.org/italy/spokane-sewanee.html, accessed 10 June 2008.

106. Caesar J. Civitella, telephone interview with the author, 18 April 2008.

107. See Jochen von Lang, *Top Nazi: SS General Karl Wolff: The Man between Hitler and Himmler* (New York: Enigma Books, 2005).

108. On 26 February 1945, Kesselring, telegraphed the following to SS, police, and Army commanders in northern Italy: "Activity of partisan bands... particularly in the areas of Modena, Reggio and Parma... has spread like wildfire in the last ten days. The concentration of partisan groups of varying political tendencies into one Organization, as ordered by the Allied High Command, is beginning to show clear results. The execution of partisan operations shows considerably more commanding leadership. Up to now it has been possible for us, with a few exceptions, to keep our vital rear lines of communications open by means of our slight protective forces, but this situation threatens to change considerably for the worse in the immediate future. Speedy and radical counter measures anticipate this development. It is clear to me that our only remedy, and one which is unavoidably necessary to meet the situation, is the concentration of all

available forces, even if this means temporary weakening in other places [i.e., the front lines]." Reprinted in OSS, *War Report of the OSS, Overseas Targets*, 114.

109. OSS, *War Report of the OSS, Overseas Targets*, 115.

110. Porch, *Path to Victory*, 641; David Travis, "Communism and Resistance in Italy, 1943-1948" in Tony Judt, ed., *Resistance and Revolution in Mediterranean Europe, 1939-1948* (London: Routledge, 1989), 91.

111. Porch, *Path to Victor*, 640; Albert Kesselring, *Kesselring: A Soldier's Record* (New York: William Morrow, 1954), 272.

112. Figures on OSS OG casualties compiled by the author from the chronology of Operational Group "A" in History of OSS OGs in Italy, OSS Records (RG 226), Entry 143, Box 11, Folder 1, National Archives II.

113. A member of a prominent New Hampshire family, who attended, without graduating, both Harvard and Yale, Hall had often skied in the Italian Alps in the prewar period. Joining the Army and then the OSS, Lieutenant Hall received SO training at Areas F and A, and was sent to Italy. In August 1944, Hall was a member of an SO team parachuted into northern Italy near the Austria Alps, to aid the partisans and hamper German lines of communication and supply. The team split up. Hall's group blew up a number of railroad bridges, but Hall, a loner, left Lt. Joseph Luckitsch in November 1944 and set out alone toward Ampezzo, where he was convinced the Germans had been building a secret highway which he hoped to destroy. After the war, an investigation revealed that he had been captured in Ampezzo by the OVRA, Fascist secret police, who tortured and finally killed him. Stewart Alsop and Thomas Braden, *Sub Rosa: The O.S.S. and American Espionage* (New York; Reynal & Hitchcock, 1946), 200-214. Hall had trained with Hall at Areas F and A with Jacques F. Snyder, author of "Cloak and Dagger Days," typescript memoir, n.d. [1990s], p. 23, provided by his son, James Snyder, to the author, 16 January 2008, plus information he obtained for the author from his father's training. James Snyder, email to the author, 13 February 2008.

114. Holohan, a Wall Street lawyer, former staff member of the Securities and Exchange Commission, and an Irish Catholic close to Donovan in age, was selected by Donovan to lead a delicate mission in the fall of 1943 to reduce internecine strife among the political partisan groups in northern Italy. Both his personality and policies in dealing with the partisans, angered at least two of the other five members of his team assembled from SI, SO, and OG personnel. He mysteriously disappeared 27 September 1943. In the 1950s, the Italian government found Holohan's body in Lake of Orta north of Milan and with testimony from witnesses, convicted two members of Holohan's team, Lt. Aldo Icardi, a Pittsburgh lawyer who had been his intelligence chief, and Sgt. Carl LoDolce, an engineer in Rochester, New York, of his murder. But their 1953 conviction in an Italian court had been *in absentia* and its sentences of death for Icardi and 17 years in prison for LoDolce were never carried out. The U.S. government did not proceed against the two men, who proclaimed their innocence, and their lawyers successfully resisted all efforts to extradite them. Anthony Cave Brown, *The Last Hero: Wild Bill Donovan* (New York: Times Books, 1982), 721-27, 804-18.

115. Porch, *Path to Victor*, 638; Paul Ginsburg, *A History of Contemporary Italy: Society and Politics, 1943-1988* (London: Penguin, 1990), 70.

116. Gen. Mark W. Clark, HQ, 15th Army Group, to Commanding Officer, 2671st Special Reconnaissance Battalion, Separate (Provisional) [the final designation of the OSS OG headquarters attached to 5th Army], May 1945, letter of commendation, plus other letters, and Presidential award provided in General Order 72, War Department, 18 July 1946, reprinted in *The Americans and the War of Liberation in Italy*, 223-224.

117. "Donovan's Devils" referred especially to the OGs. Al Materazzi, "Italian-American OGs Attend Memorial Ceremony in Ameglia," *OSS Society Newsletter*, Summer 2004, 10.

118. *Maquis* was the term for the thick, hampering and resilient underbrush native to French Corsica, and the predominantly young men and women of the French Resistance adopted the term to apply to themselves.

119. H.R. Kedward, *Resistance in Vichy France* (Oxford, U.K.: Oxford University Press, 1978); and Kedward, *In Search of the Maquis* (Oxford: Oxford University Press, 1993). For examples of this rivalry in operation and its effects on some of the OSS operations, see Hélène Deschamps Adams, "Behind Enemy Lines in France," in Chalou, ed., *The Secrets War*, 140-64.

120. Neilson MacPherson, *American Intelligence in War-Time London: The Story of the OSS* (London: Frank Cass, 2003), 82.

121. OSS, War *Report of the OSS, Overseas Targets*, 192.

122. Ibid., 192-93.

123. On such casualties, see, for example, the "Stockbroker" circuit that operated in Eastern France near the Swiss border. OSS, *War Report of the OSS, Overseas Targets*, 193.

124. OSS, *War Report of the OSS, Overseas Targets*, 177-79.

125. Susan Ottaway, *Violette Szabo* (Annapolis, Md.: Naval Institute Press, 2002), 97-112, 115-116, 147-148. A list of the SOE women agents executed by the Nazis is on pages 178-179.

126. Gestapo report quoted in O'Donnell, *Operatives, Spies, and Saboteurs*, 173.

127. Judith L. Pearson, *Wolves at the Door: The True Story of America's First Female Spy* (Guilford, Conn.: Lyons, 2005); see also "Mission Heckler, Activity Report of Virginia Hall (DIANE)," OSS Records (RG 226), Entry 190, Box 741, National Archives II.

128. Col. Francis B. Mills with John W. Brunner, *OSS Special Operations in China* (Williamsburg, N.J.: Phillips Publications, 2002), 435-437, quote on 437; obituary, "Francis Byron Mills," *Washington Post*, 1 October 2005. A more detailed account of the OSS off Omaha and Utah beaches was provided in a memoir by David Doyle, an OSS radio operator on board the same vessel as Mills. David W. Doyle, *True Men and Traitors: From the OSS to the CIA: My Life in the Shadows* (New York: John Wiley, 2001), 34-35.

129. Anecdote by David K.E. Bruce at a speech at the annual dinner meeting of OSS veterans, Washington, D.C., 26 May, 1971, quoted in R. Harris Smith, *OSS: The Secret History of America's First Central Intelligence Agency* (Berkeley: University of California Press, 1972), 184-85. A more extended version is in Richard Dunlop, *Donovan: America's Master Spy* (Chicago: Rand McNally, 1982), 439-440. Bruce may have embellished the story a bit when he told it in 1971. In a 1958 account to a former colleague, Bruce emphasized their search for their lethal pills but did not have Donovan saying he would use his pistol on both of them. David Bruce to Whitney H. Shephardson, 16 August 1958 and attachment, copy on Wallace R. Deuel Papers, Box 61, Folder 5, Library of Congress.

130. Will Irwin, *The Jedburghs: The Secret History of the Allied Special Forces, France 1944* (New York: Public Affairs, 2005), 36-38, who rejects the prevalent story that the name came from a training camp near Jedburgh, Scotland, as he notes that the Jedburgh training camp in Britain was 250 miles away at Milton Hall near Peterborough, in east, central England not too far from Cambridge. His persuasive argument is that the term was simply a name selected from the approved list of British code names. On the training in the United States OSS camps of future Jedburghs, see pp. 43-50. See also Colin Beavan, *Operation Jedburgh* (New York: Viking/Penguin, 2006).

131. OSS, *War Report of the OSS, Overseas Targets*, 199. There were a few Dutch and Belgian Jeds; they served on the teams sent into those countries.

132. Irwin, *The Jedburghs*, 114.

133. The squadrons, first the 36th and 406th, later joined by the 788th and 850th Bomb Squadrons, were part of the 801st Heavy Bombardment Group, 8th U.S. Air Force. The nickname "Carpetbaggers" came from the original codename for the operation. During the war, the Carpetbaggers flew more than 3,000 missions and delivered 556 agents and 4,500 tons of supplies to Resistance forces, but in the process, 208 members of the aircrews were lost to German nightfighter planes, antiaircraft guns, and crashes. OSS and other agents were also airdropped by British planes as well. Ben Parnell, *Carpetbaggers: America's Secret War in Europe*, rev. ed. (Austin, Tex.: Eakin Press, 1993); Ron Clarke, "The Carpetbaggers," illustrated history on website of the

Carpetbagger Aviation Museum, Harrington, Northamptonshire, England, http://harringtonmuseum.org.uk/CarpetbaggerMuseumHomePage.htm, accessed 17 July 2006.

134. Eugene Polinsky, Grand-View-on-Hudson, New York, presentation at the author's seminar on World War II at Rutgers University, New Brunswick, N.J., 21 March 2007. His earlier oral history interview is on the website of the Rutgers Oral History Archives of World War II. http://oralhistory.rutgers.edu.

135. Robert R. Kehoe, "Jed Team Frederick: 1944: An Allied Team with the French Resistance," *Studies in Intelligence: Journal of the American Intelligence Professional* (June 2002), Office of Strategic Services, 60th Anniversary, Special Edition, p. 104.

136. Kehoe, "Jed Team Frederick: 1944," p. 127.

137. Ibid., 121.

138. Bernard Knox, *Essays, Ancient and Modern* (Baltimore, Md.,: The Johns Hopkins University Press, 1989), xi-xxvii. See also, "Nice Article from Jed Bernard Knox," [transcript of a 1998 lecture by Knox at New York University], *OSS Society Digest*, Number 2102, 18 July 2008, osssociety@yahoogroups.com.

139. O'Donnell, *Operatives, Spies, and Saboteurs*, 177-78; Irwin, *Jedburghs*, xii-xvii; Jed Team "Giles" Report, OSS Records (RG 226), Entry 103, Box 1, National Archives II. In 1945, Knox was sent to northern Italy where he helped Italian partisans harass the retreating Germans. His unit was subsequently scheduled to assist in the invasion of Japan. Knox, *Essays*, xxix-xxxiii.

140. Kehoe, "Jed Team Frederick: 1944," 130.

141. Ibid., 130.

142. Ibid., 130-32. The fact that the Germans were able to hold out for many months in the major ports in Brittany was a result of the Allied command's underestimation of the German defenses there.

143. John K. Singlaub with Malcolm McConnell, *Hazardous Duty: An American Soldier in the Twentieth Century* (New York: Summit Books/Simon & Schuster, 1991), 68.

144. Irwin, *Jedburghs*, 121-130, 260; Captain P.C. [Paul Cyr], "Jedburghs in Action," 16 December 1944, pp. 1-11, "Interviews with Returned Men," Schools and Training Branch, OSS Records (RG 226), Entry 161, Box 2, Folder 31; see also an account of Capt. George G. Thomson, radioman, Sgt. John White of Cambridge, Massachusetts, and their French teammate on Jedburgh Team Alec in Lt. Cmdr. Richard M. Kelly, "Guarding Patton's Flank," *Blue Book Magazine* (January 1947), 62-71, copy in OSS Records (RG 226), Entry 161, Box 7, Folder 7, National Archives II. Cyr and his teammates parachuted in again in the "George II" mission into the Loire region of western France on 7-8 September 1944.

145. William B. Dreux, *No Bridges Blown* (Notre Dome, IN: University of Notre Dame Press, 1971), 229-57, the quotation is on xii. For more on Dreux, see a memoir by his friend and fellow trainee at Areas F and A, Aaron Bank, *From OSS to Green Berets: The Birth of the Special Forces* (Novato, Calif.: Presidio Press, 1986), 4-7, 65-68.

146. Dreux, *No Bridges Blown*, 314.

147. Singlaub, *Hazardous Duty*, 34-41; see also John K. Singlaub, interviewed by Maochun Yu and Christof Mauch, 31 October 1996, CIA Records (RG 263), OSS Oral History Transcripts, Box 4, National Archives II; and John K. Singlaub, Arlington, Va., telephone interview with the author, 11 December 2004 and 6 January 2005.

148. Singlaub, *Hazardous Duty*, 42-43.

149. Max Hastings, *Das Reich: The March of the 2nd SS Panzer Division through France* (New York: Holt, Rinehart, 1981), 161-79.

150. Singlaub, *Hazardous Duty*, 45-69; Maj. H.A. Murray, "Interviewer's Report," on John K. Singlaub, 1 February 1945, CIA Records (RG 263), OSS Personnel Records, Accession # 61-574, Box 52, National Archives II.

151. "Interview: William Colby, Former Director, Central Intelligence Agency," *Special Warfare*, 7:2 (April 1994); 40-41; Will Irwin, *The Jedburghs: The Secret History of the Allied Special Forces, France 1944* (New York: Public Affairs Press, 2005), 132-134.

152. Colby, *Honorable Men*, 23-26, 38-44; see also Iwin, *Jedburghs*, 132-151.

153. Richard Goldstein, "Col. Aaron Bank, Who Was 'Father of Special Forces,' Dies at 101," *New York Times*, 6 April 2004; "Col. Aaron Bank, U.S. Special Forces Founder, Dies," *OSS Society Newsletter* (Summer 2004), 13; "'Father of Special Forces,' Dies at Age 101 in California Home," *Special Warfare*, 16:4 (May 2004): 48-49.

154. Aaron Bank, *From OSS to Green Berets: The Birth of Special Forces* (Navato, CA: Presidio, 1986), 13-62, quotation on 59. Erwin, Jedburghs, 244, 274-75. Captain Bank was later selected to head Operation Cross (often erroneously referred to as Operation Iron Cross), a plan authorized by General Eisenhower in March for snatching or killing Adolph Hitler and key members of his inner circle should they retreat to a "national redoubt" in the Bavarian Alps. Bank recruited 100 young anti-Nazi Germans, mainly communists, from POW camps, trained them for two months as commandos. In early May 1945, posing as German mountain infantry, with *Wehrmacht* uniforms and weapons, they were ready to board the planes which would drop them into the target area, when Bank received word that the mission had been aborted. Hitler had not left Berlin, there was no Alpine Redoubt, and the war in Europe was nearing its end. "'Father of Special Forces,' Dies at Age 101 in California Home," *Special Warfare*, 16:4 (May 2004): 48-49; Bank, *From OSS to Green Berets*, 72-99; Christof Mauch, *The Shadow War Against Hitler: The Covert Operations of America's Wartime Intelligence Service* (New York: Columbia University Press, 2003), 178, 185-97.

155. Erwin, *Jedburghs*, 280-71; Bart Barnes, "Lucien E. Conein Dies at 79: Fabled Agent for OSS and CIA," *Washington Post*, 6 June 1998, B6; Tim Weiner, "Lucien Conein, 79, Legendary Cold War Spy," *New York Times*, 7 June 1998; Conein's name was pronounced co-NEEN.

156. Alsop and Braden, *Sub Rosa: The OSS and American Espionage*, partially a history and partially a memoir including a recounting of Alsop's own experiences as a Jed but under the pseudonym "Bill Wheeler," 136-84; see also Smith, *OSS*, 190.

157. Mattingly, *Herringbone Cloak–GI Dagger: Marines of the OSS*, 132-46.

158. Mattingly, *Herringbone Cloak–GI Dagger: Marines of the OSS*, 114-118; Benis Frank, "Colonel Peter Julien Ortiz," OSS Marine, Actor, Californian, http://www.militarymuseum.org/Ortiz.html, accessed 3 September 2007.

159. Francis L. Coolidge, summary of interview by Schools and Training Branch, 30 November 1944, p. 3, Interviews with Returned Men," OSS Records (RG 226), Entry 161, Box 2, Folder 31, National Archives II.

160. Merritt Binns, Fannitsburg, Pa., telephone interviews with the author, 3 and 4 April 2007.

161. John P. Bodnar, 82, at the commemoration ceremony in Centron, France, 1 August 2004 of the 60th anniversary of the team's airdrop into France, quoted in Master Sgt. Phil Mehringer, "Operation Union II: Marines Land in France 60 Years Ago," http://www.leatherneck.com/forums/archive/index.php/t-16379.thml, accessed 3 September 2007. Jack Rislar, 83, the only other survivor still alive in 2004 was also at the ceremony. A slightly different rendition of Bodnar's response to Ortiz's order for the two sergeants to try to escape was given by Ortiz himself in his after-action report more than a year later. Ortiz wrote that "Sergeant Bodnar was next to me and I explained the situation to him and what I intended to do. He looked me in the eye and replied, 'Major, we are Marines, what you think is right goes for me too.'" Maj. Peter J. Ortiz, USMR, "Chronological Report of the Capture and Subsequent Captivity of Members of the Mission Union," 12 May 1945, quoted in Mattingly, *Herringbone Cloak–GI Dagger: Marines of the OSS*, 122.

162. Master Sgt. Phil Mehringer, "Operation Union II: Marines Land in France 60 Years Ago," http://www.leatherneck.com/forums/archive/index.php/t-16379.thml, accessed 3 September 2007; "Peter Julien Ortiz," in W. Thomas Smith, Jr., *Encyclopedia of the Central Intelligence Agency* (New York: Facts on File, 2003), 180-181.

163. Charles H. Briscoe, "Herbert R. Brucker, SF Pioneer: Part II Pre-WWII-OSS Training 1943," *Veritas: Journal of Army Special Operations History* 2:3 (2006): 26-35.

164. Herbert Brucker interviewed by Charles H. Briscoe, 8 March 2007, Fayetteville, N.C., quoted in Charles H. Briscoe, "Major Herbert R. Brucker: SF Pioneer, Part III: SOE Training & 'Team Hermit' into France," *Veritas: Journal of Army Special Operations History* 3:1 (2007): 73.

165. Herbert Brucker quoted in O'Donnell, *Operatives, Spies, and Saboteurs*, 173.

166. Charles H. Briscoe, "Major (R) Herbert R. Brucker, DSC: Special Forces Pioneer: SOE France, OSS Burma and China, 10th SFG, SF Instructor, 77th SFB, Laos, and Vietnam," *Veritas: Journal of Army Special Operations History* 2:2 (2006): 33-35.

167. Lt. William B. Macomber, Jr., Activity Report, Circuit: Freelance, OSS (SO), W.E. Section, 18 September 1944, quoted in Mattingly, *Herringbone Cloak—GI Dagger: Marines of the OSS*, 159-60.

168. Jacques L. Snyder, "Cloak and Dagger Days," typescript memoir, n.d. [1990s], pp. 13-28, provided by his son, James Snyder, to the author, 16 January 2008, plus information he obtained for the author from his father about Areas F and A, James Snyder, email to the author, 13 February 2008.

169. Later, as an American university professor of Spanish, Roberto Esquenazi-Mayo wrote an account entitled of his military service, memories of a student solider, *Memorias de un Estudiante Soldado* (Habana: Dirección de Cultura, Ministerio de Educación, 1951). A briefer version for use in Spanish language classes, edited, with introduction, notes, exercises, and vocabulary by George Thomas Cushman was published by W.W. Norton in New York in 1954.

170. Roger Hall, *You're Stepping on My Cloak and Dagger* (New York: W.W. Norton, 1957), 163-66.

171. William E. Colby quoted in Henry Buningham, "Ex-CIA Chief Salutes Army Green Berets," *Fayetteville [N.C.] Observer-Times*, 11 September 1993, 15. I am indebted to Caesar J. Civitella, for this article. In an attempt to enhance understanding of the OGs, an OSS Operational Group website was constructed in 2005. See http://www.ossog.org.

172. Ian Sutherland, "The OS Operational Groups: Origin of Army Special Forces," *Special Warfare*, 15:2 (June 2002): 2. The hometowns, or more accurately permanent addresses given in 1945, by members of the OSS French OGs and OSS Chinese OGs were listed in Lt. Col. Alfred T. Cox, HQ OG Command, SSUK, to All Personnel Concerned, 8 November 1945, subject: Permanent Addresses, French and Chinese OGs, a six-page typed list attached, provided in "OG French and Chinese Address List," *OSS Society Digest* Number 1132, 4 September 2005, osssociety@yahoogroups.com, accessed 4 September 2005.

173. "Office of Strategic Services, Operational Groups, French Operational Group," www.ossog.org/france.html. accessed 12 June 2008; Troy J. Sacquety, "The OSS... Operational Groups," *Veritas: Journal of Army Special Operations History*, 3:4 (2007): 40.

174. OSS, *War Report of the OSS, Overseas Targets*, vii.

175. Edward E. Nicholas, III, "How We Won the Great War," typescript memoir notes, p. 7. I am indebted to his son, Ned Nicholas, for providing me with a copy of this document.

176. For OSS role, see Richard S. Friedman, KK4XX, "OSS Operations in Anvil/Dragoon," in Ranney and Ranney, eds., *The OSS CommVets Papers*, 2nd ed., 49-54; and "Excerpts—OSS Activities with 7th Army," 5-page typescript in Donovan Papers, "Black Book," CIA Records (RG 263), Thomas F. Troy File, Box 4, Folder 30, National Archives II.

177. Ian Sutherland, "The OSS Operational Groups: Origin of Army Special Forces," *Special Warfare*, 15:2 (June 2002): 2-13.

178. "Office of Strategic Services, Operational Groups, Percy Red," www.ossog.org/france/percyred.html, accessed 24 June 2008.

179. Rafael Hirtz, transcript of interview, 12 February 2004, Veterans History Project, Library

of Congress, http://1cweb2.loc.gov/cocoon/vhp-stories/loc.natlib.afc2001001.00094, accessed 10 January 2007; and www.ossog.org./france/donald.html, accessed 24 June 2008.

180. On the death of 1st Lt. W. Larson, see "Office of Strategic Services, Operational Groups, French OG, Section Christopher," www.ossog.org./france/chrisotpher.html. Accessed 12 June 2008.

181. Obolensky, *One Man in His Time*, 380-88.

182. Emmett F. McNamara, telephone interview with the author, 2 September 2008; see also OSS, French Operational Group, U.K.-to-France, Lindsey Section," www.ossog.org.

183. Operational Report, Company B, 267th Special Reconnaissance Battalion (Provisional), Narrative History of the French OGs, OSS Records (RG 226), Entry 99, Box 44, National Archives II.

184. Erasmus H. Kloman, *Assignment Algiers: With the OSS in the Mediterranean Theater* (Annapolis, Md.: Naval Institute Press, 2005), 36-41.

185. OSS, *War Report of the OSS, Overseas Targets*, 207.

186. "Office of Strategic Services, Operational Groups, French OG, Section Lehigh," www.ossog.org./france/lehigh.html. Accessed 12 June 2008.

187. O'Donnell, *Operatives, Spies, and Saboteurs*, 189-93; OSS, *War Report of the OSS; Overseas Targets*, 205-6. "Office of Strategic Services, Operational Groups, French OG, Section Louise," www.ossog.org./france/louise.html. Accessed 12 June 2008.

188. "Office of Strategic Services, Operational Groups, French OG, Section Lafayette," www.ossog.org./france/lafayette.html. Accessed 12 June 2008; and Caesar J. Civitella, telephone interview with the author, 18 April 2008.

189. OSS, *War Report of the OSS, Overseas Targets*, 219.

190. OSS, *War Report of the OSS, Overseas Targets*, 198-99. These supplies included 27,000 containers, each holding up to 220 pounds, which were parachuted in, and 10,000 packages, holding up to 100 pounds of non-breakable items, such as clothing and packaged food rations, which were dropped free.

191. OSS, *War Report of the OSS, Overseas Targets*, 220-21.

192. Ibid., 192.

193. Ibid., 216-19, 238-39.

194. Ibid., 248.

195. Hemingway, *Misadventures of a Fly Fisherman*, 135-138.

196. Ibid., 147.

197. Peter M.F. Sichel, interview with the author, 9 July 2008; and messages in the *OSS Society Digest*, Number 1262, 29 January 2006, and Number 1263, 30 January 2006, osssociety@yahoogroups.com, both accessed 31 January 2006; other accounts of the episode are in Hemingway, *Misadventures of a Fly Fisherman*, 170-175; and Lt. Cdr. Richard M. Kelly, USNR, "Spy Work Ahead," *Blue Book Magazine*, August 1947.

198. Eric Homberger, "Obituary: Jack Hemingway," *Guardian*, 4 December 2000, http://www.guardian.co.uk./Archive, accessed 17 September 2007; and Hemingway, *Misadventures of a Fly Fisherman*.

199. Peter M.F. Sichel, telephone interview with the author, 9 July 2008; Mattingly, *Herringbone Cloak—GI Dagger: Marines of the OSS*, 183-85/

200. Joseph E. Persico, *Piecing the Third Reich: The Penetration of Nazi Germany by American Secret Agents during World War II* (New York: Viking, 1979), 109; OSS, *War Report of the OSS; Overseas Targets*, 294.

201. Mattingly, *Herringbone Cloak—GI Dagger, Marines of the OSS*, 186-90.

202. Peter M.F. Sichel, telephone interview with the author, 9 July 2008. Although poor weather conditions in December 1944 prevented photo-reconnaissance and the Germans had imposed radio

silence at the tactical level, "Ultra" intelligence, as well as the German POW OSS agents revealed the massing of German forces, at Allied commanders at the highest level concluded, mistakenly, that they were being assembled to resist the next Allied offensive and possibly to try to counterattack it after the Allies had launched the next attack on either side of the Ardennes Forest. Charles B. Macdonald, *A Time for Trumpets: The Untold Story of the Battle of the Bulge* (New York: Morrow, 1984).

203. OSS War, *Report of the OSS, Overseas Targets*, 305; O'Donnell, *Operatives, Spies and Saboteurs*, 255.

204. Tom Moon, *This Grim and Savage Game: OSS and the Beginning of U.S. Covert Operations in World War II* (New York: Da Capo Press, 2000), 237-38. Beginning in January 1945, William Casey, who had become SI chief in London, began parachuting anti-Nazi German POWs, converted by OSS into agents, into Germany. Thirty-four teams were dropped safely due especially to the increasing chaos in Germany in 1945, but for a variety of reasons, only seven ever established direct communication. Joseph E. Persico, *Piecing the Third Reich*; OSS, *War Report of the OSS; Overseas Targets*, 305.

205. OSS, *War Report of the OSS, Overseas Targets*, xii, 220.

206. Ibid. Of the roughly 280 Jedburghs of all nationalities, a total of 21 were killed, a death rate of almost 8 percent. John K. Singlaub with Malcolm McConnell, *Hazardous Duty: An American Soldier in the Twentieth Century* (New York: Summit Books/Simon & Schuster, 1991), 69, 530n. At least three British-led Jedburgh teams were ambushed shortly after landing and most of their members killed. Two Jedburgh teams were entirely wiped out. In the Vosges Mountains in mid-August 1944, the French lieutenant of Team Jacob was killed in a firefight and the two other members captured; the commanding officer, a British major, was executed and the radio operator, a British sergeant imprisoned. Iwin, *Jedburghs*, 175.

207. Irwin, *Jedburghs*, Appendix, pp. 248-279, a list of Jedburgh Teams in France, including casualties; on Manierre, see also Singlaub with McConnell, *Hazardous Duty*, 69, 451.

208. Singlaub with McConnell, *Hazardous Duty*, 69; Irwin, *Jedburghs*, 258-259; John K. Singlaub, telephone interview with the author, December 11, 2004. Although a trainee, Singlaub assisted Larry Swank in demolitions instruction at Catoctin Mountain Park in November 1943.

209. Irwin, *Jedburghs*, 155-164.

210. OSS/London, OSS/London: SO Branch War Diary, Volume 4, Book 6, Report on Team Augustus, p. 4; and SFHQ G-3 Periodic Report Number 84 (27 August 1944), p. 5, both cited in Irwin, *Jedburghs*, 164.

211. Irwin, *Jedburghs*, 164-176.

212. OSS, *War Report of the OSS, Overseas Targets*, xii, 204-5, 220. Nearly half of the injuries resulted from the initial parachute jump.

213. Newspaper clipping, "KIA-France," by Corporal. M.L., undated [November-December 1944] and unidentified [OSS Area F Newsletter, *Attention Please*], in OSS Records (RG 226), Entry 146, Box 223, Folder 3105, National Archives II. See also OSS, *War Report of the OSS, Overseas Targets*, 206.

214. Lt. Col. Alfred T. Cox to Director [of the OSS], 20 December 1944, subject: Visit to Families of Operational Group Personnel Killed in Action, OSS Records (RG 226), Entry 143, Box 12, Folder 149, National Archives II.

215. Newspaper clipping, "KIA-France," by Corporal. M.L., undated [November-December 1944] and unidentified [OSS Area F Newsletter, *Attention Please*], in OSS Records (RG 226), Entry 146, Box 223, Folder 3105; General Orders Number 70, 17 April 1945, Award of Distinguished Service Cross (Posthumous), "Paul A. Swank," and General Orders Number 71, 18 April 1945, Award of Silver Star, "Nolan J. Frickey," from Maj. Gen. George D. Pence, chief of staff, by command of Gen. McNarney, OSS Records (RG 226), Entry 143, Box 12, Folder 149, National Archives II.

216. General Orders Number 71, 18 April 1945, Award of Silver Star (Posthumous), "Raymond Bisson," "Bernard F. Gautier," from Maj. Gen. George D. Pence, chief of staff, by command of Gen. McNarney; and Col. Alfred T. Cox to Director [of the OSS], 20 December 1944, subject: Visit to Families of Operational Group Personnel Killed in Action; and letters from Maj. Gen. William J. Donovan to Mrs. Agnes Bisson, and the next of kin of other OG members killed in southern France, 23 December 1944, all in OSS Records (RG 226), Entry 143, Box 12, Folder 149, National Archives II.

217. Jason Krump, "Stories that Live Forever, Part III: The Epitome of Courage [Ira Christopher Rumburg]," Washington States University: Official Site of Cougar Athletics, wsucougars.com, http://www.cstv.com/printable/schools/wast/genrel/052207aae.html. Accessed 13 June 2008.

218. On Chris Rumburg in the OSS and at Areas A and B, see Sage, *Sage*, 21, 430; Dorothea Dean Dow (Mrs. Arden W. Dow), telephone interviews with the author, 15 May and 9 and 15 June 2005; J.R. Brown to J.R. Hayden, 7 July 1942, [report on course at Area A-2, 21 June to 3 July 1942], OSS Records (RG 226), Entry 136, Box 161, Folder 1754, National Archives II.

219. Krump, "Stores that Live Forever, Part III: The Epitome of Courage."

220. See Allan Andrade, *S.S. Leopoldville Disaster, December 24, 1944* (New York: Tern Book Co., 1997).

221. Krump, "Stores that Live Forever, Part III: The Epitome of Courage."

222. Capt. Robert Campbell, 66th Infantry Division, to J. Fred ("Doc") Bohler, WSC Athletic Director, 9 August 1945, excerpted in Krump, "Stores that Live Forever, Part III: The Epitome of Courage."

223. "Cover-up: The Story of the S.S. Leopoldville," The History Channel, www.history.com; see also Andrade, S.S. Leopoldville Disaster, and Andrade's website, www.msnusers.com/ssleopoldville; Krump, "Stores that Live Forever, Part III: The Epitome of Courage;" Dennis Hevesi, "A Fight to Honor Forgotten Men of the Leopoldville," *New York Times*, 8 May 1995. Of British, American and Belgian governments, the first to begin declassification of information on the Leopoldville disaster was Great Britain in 1996.

224. Krump, "Stores that Live Forever, Part III: The Epitome of Courage;" www.history.com; www.infantry.Army.mi/musuem/outside-tour/monuments/66th_inf_div.htm, accessed 14 June 2008.

225. Mark Mazower, *Inside Hitler's Greece: The Experience of Occupation, 19141-1944* (New Haven, Conn.: Yale University Press, 1993).

226. Lt. Cmdr. Richard M. Kelly, USNR, "Mission to Greece," *Blue Book: Magazine of Adventure for Men* (November 1946), 76-78, copy in OSS Records (RG 226), Entry 161, Box 7, Folder 7, National Archives II. Kelly was head of the OSS Maritime Unit in the eastern Mediterranean, 1943-1945.

227. Kelly, "Mission to Greece," 76-78, and Spiro Cappony, telephone interview with the author, 16 September 2006. In his memoirs, Hayden does not mention Cappony by name but refers only to "an enlisted man from the Navy who was fluent in Greek, telegraphy, and cipher," Sterling Hayden, *Wanderer* (New York: Knopf, 1970), 310.

228. Spiro Cappony, telephone interview with the author, 16 September 2006.

229. Christin Nance Lazerus, "Greece Honors Spiro Cappony," *Post-Tribune*, 23 November 2007, reprinted in *OSS Society Digest*, Number 1911, 25 November 2007, osssociety@yahoogroups.com, accessed 25 November 2007.

230. Kelly, "Mission to Greece," 79-82.

231. Arthur E. Mielke, "Gunny Curtis," *Leatherneck Magazine*, January 1946, 22; Mattingly, *Herringbone Cloak—GI Dagger: Marines of the OSS*, 99.

232. Kelly, "Mission to Greece," 83.

233. Report of the Evros [River] Mission, as quoted in OSS, *War Report of the OSS, Overseas Targets*, 123.

234. Mielke, "Gunny Curtis," 22.

235. Kelly, "Mission to Greece," 83-86.

236. Maj. James Kellis Report, except reprinted in Kelly, "Mission to Greece," 86.

237. OSS, *War Report of the OSS, Overseas Targets*, 123-24.

238. "Office of Strategic Services, Operational Groups, Greek Operations," www.ossog.org/greek. html. Accessed 16 June 2008.

239. Ted Russell, interview in O'Donnell, *Operatives, Spies and Saboteurs*, 110-11. The "Eisenhower jacket" adopted in 1943 was a trim, waist-length jacket, popularized by General Dwight D. Eisenhower.

240. "Office of Strategic Services, Operational Groups, Greek Operations," www.ossog.org/greek. html, accessed 16 June 2008.

241. MTO USA, General Orders No. 71, 18 April 1945, Award of Silver Star, Maj. Gen. George D. Pence, chief of staff, p. 4; OSS Records (RG 226), Entry 143, Box 12, Folder 149, National Archives II.

242. Patrick Kane, "As the Tide Turned in World War II, Keeping Nazi Troops Away," Petersburg, Va., *Progress-Index*, 16 May 2008, A1, reprinted in *OSS Society Digest*, Number 2063, 27 May 2008, osssociety@yahoogroups.com, accessed 27 May 2008.

243. OSS, *War Report of the OSS, Overseas Targets*, 124; O'Donnell, *Operatives, Spies, and Saboteurs*, 110-23.

244. Report filed with OSS Headquarters, 24 December 1944 of Greek/US OG Operations in Greece, 1944. I am indebted to Robert E. Perdue, Ph.D., for telling me about this document from the OSS Records (RG 226), telephone conversations, 31 August and 5 September 2007. This report deals only with the Greek/US OGs. There were other casualties in other OSS branches in Greece.

245. O'Donnell, *Operatives, Spies, and Saboteurs*, 118; "WWII O.S.S. Volunteer Unit to Be Awarded Bronze Star for Valor in Nazi-Occupied Greece," *OSS Society Digest*, Number 2043, 29 April 2008, osssociety@yahoogroups.com, accessed 29 April 2008; Greek Operations, Group II, "Office of Strategic Services, Operational Groups, Greek Operations," www.ossog.org/greek. html, accessed 16 June 2008.

246. David Hendrix, "Under Deep Cover [Spiro Cappony]," undated [1996] and unidentified newspaper clipping, p. C-2, provided to the author by Spiro Cappony, 31 October 2006. Spiro Cappony, telephone interview with the author, 16 September 2006. Because of the wound in his right arm, Cappony had to telegraph his messages with his left hand. The base station operator noticed the difference and questioned whether it was really Cappony or whether the set had been captured. Through prearranged signal, Cappony was able to convince the base station that it was he. Cappony was awarded a Bronze Star.

247. The term Chetniks was derived from *cheta*, the traditional Serbian name for the armed bands who had fought against the Ottoman Turks. As the rival Yugoslav movements fought each other as much as the Germans and took no prisoners, most of the 1.2 million Yugoslavs who were killed in the war may have died at the hands of other Yugoslavs. Walter R. Roberts, *Tito, Mihailovic and the Allies, 1941-1945*, 2nd ed. (Durham, NC: Duke University Press, 1987).

248. Mattingly, *Herringbone Cloak—GI Dagger: Marines of the OSS*, 73-75.

249. Walter R. Mansfield, "Marine with the Chetniks," *Marine Corps Gazette* (January-February 1946): 4.

250. Walter R. Mansfield, "Is There a Case for Mihailovic?" *American Mercury* (June 1946): 716; Albert B. Seitz, *Mihailovich: Hoax or Hero* (Columbus, Oh.: Leigh House, 1953).

251. British policy initially supported Mihailovic, but by the summer of 1943, exaggerated reports from British SOE/SIS officers in Yugoslavia contrasting Mihailovic as inactive and Tito more aggressive led Churchill to reassess and break with Mihailovic in favor of Tito. F. H. Hinsley, et al., *British Intelligence in the Second World War*, vol. 3, *Its Influence on Strategy and Operations*,

part I (London: H.M. Stationers, 1984), 137-72. On the controversy, see Roberts, *Tito, Mihailovic and the Allies*; David Martin, *The Web of Disinformation: Churchill's Yugoslav Blunder* (New York: Harcourt, Brace, Jovanovich, 1990); and Franklin Lindsay, *Beacons in the Night: With the OSS and Tito's Partisans in Wartime Yugoslavia* (Stanford, Calif.: Stanford University Press, 1993).

252. Martin Gilbert, *Winston S. Churchill*, Volume VII, *Road to Victory, 1941-1945* (Boston: Houghton Mifflin, 1986), 317-22, 571, 614, 640-41, 739-40, 755, 893.

253. Even though the OSS removed its official mission to Mihailovic, it placed some SI officers with him for "infiltrating agents into Austria and Germany." William J. Donovan to Dwight D. Eisenhower, March 31, 1944; Eisenhower to Donovan, 9 April 1944, in Eisenhower, *Papers of Dwight D. Eisenhower*, 3: 1815.

254. James Grafton Rodgers, *Wartime Washington: The Secret OSS Journal of James Grafton Rogers, 1942-1943*, Thomas F. Troy, ed., (Frederick, Md.: University Publications of America, 1987), 32n. After the war, Weil became President of Macy's flagship store in New York City in 1949.

255. "A Good Joe Named George," 9 December 1944, p. 1, Interviews with Returned Men, Schools and Training Branch, OSS Records (RG 226), Entry 161, Box 2, File 31, National Archives II. Although "George's" last name is redacted on the typescript, the tab attached to it indicates "Wuchinich."

256. Ibid., 1-10.

257. Ibid., 11-19.

258. O'Donnell, *Operatives, Spies, and Saboteurs*, 84.

259. Raids were staged from Vis against the islands of Hvar, Solta, Korcula, Mjle, and Brac, and the Peljesac Peninsula. These, together with a list of the members of the Yugoslavian OGs (including some members of Greek and other OGs) are in "Office of Strategic Services, Operational Groups, Yugoslavian Operations," www.ossog.org/yugoslavian.html. accessed 16 June 2008.

260. Brian Albrecht, "WWII Vet Recalls the 'Stuff that Stays with You," *Cleveland Plain Dealer*, 27 May 2007, reprinted in *OSS Society Digest*, Number 1760, 30 May 2007, osssociety@yahoogroups.com, accessed 30 May 2007. The one quarter casualty rate may have included the more than a dozen Americans from the Dawes Mission to Czechoslovakia killed in January 1945.

261. Hayden, *Wanderer*, 299-302; see also Mattingly, *Herringbone Cloak—GI Dagger: Marines of the OSS*, 80-86.

262. Hayden, *Wanderer*, 313-14; John Hamilton [Sterling Hayden], "Liaison Officer with the Partisans," 10 November 1944, p. 2, Interviews with Returned Men, Schools and Training Branch, OSS Records (RG 226), Entry 161, Box 2, File 31, National Archives II.

263. Ibid., 3.

264. Ibid., 2. See also the account in his memoirs, which is even more forthcoming about his admiration for the tough, communist partisans. Hayden, *Wanderer*, 315-320.

265. Henrik Krüger, Hans V. Tofte: *Den danske Krigshelt, der kom til tops i CIA* (Copenhagen: Aschelhoug, 2005). I am indebted to Henrik Krüger for this information from his book; Henrik Krüger emails to the author, 4 and 5 August 2006.

266. Hayden, *Wanderer* 320-333; Mattingly, *Herringbone Cloak—GI Dagger: Marines of the OSS*, 86.

267. OSS, *War Report of the OSS, Overseas Targets*, 132; Gregory A. Freeman, *The Forgotten 500: The Untold Story of the Men Who Risked All for the Greatest Rescue Mission of World War II* (New York: Nal Caliber, 2007), of an OSS mission with the Chetniks that rescued 512 downed American airmen in 1944.

268. Brian Albrecht, "WWII Vet Recalls the 'Stuff that Stays with You," *Cleveland Plain Dealer*, 27 May 2007, reprinted in *OSS Society Digest*, Number 1760, 30 May 2007, osssociety@yahoogroups.com.

269. OSS, Schools and Training Branch, "Office of Strategic Services (OSS): Organization and Functions," June 1945, p. 18, OSS Records (RG 226), Entry 141, Box 4, Folder 36, National Archives II.

270. The 15 German divisions in Yugoslavia would otherwise have been a substantial help to the 26 German divisions fighting the Allies in Italy, for example. OSS, *War Report of the OSS, Overseas Targets*, 127.

271. After the German surrender in May 1945, Tito ended the coalition government in Yugoslavia, consolidated his and the Communist Party's power and took revenge on his enemies. Mihailovic was denounced as a traitor and executed in July 1946.

272. The Communist guerrilla leader, Enver Hoxha, remained dictator of Albania until his death in 1983. The OSS mission to Albania beginning in November 1943, was led by Army Captain Thomas E. Stefan from Laconia, New Hampshire, the son of Albanian immigrants, recruited by the OSS after Military Intelligence Service training at Camp Ritchie, and Marine Sergeant (and later lieutenant) Nick R. ("Cooky") Kukich, 27, a coal miner from Ohio and the son of Serbian immigrants, who received OSS paramilitary training at Area B and radio training at Area C. Peter Lucas, "A Marine Has Landed—Albania, 1943," *Leatherneck Magazine*, July 2004, 46-49; and Mattingly, *Herringbone Cloak—GI Dagger: Marines of the OSS*, 87-89, an otherwise careful study, which mistakes Kukich's nickname for his surname, i.e., "Lieutenant Cooky." In a special mission in December 1943, Major Lloyd Smith of SI, with the aid of the guerrillas, successfully rescued 13 female U.S. Army nurses, 13 male medics, and four crewmen, who had been trapped in the Albanian mountains when their plane went off course in a storm and crashed in early November. See the memoir of one of the nurses, Agnes Jensen Mangerich as told to Evelyn M. Monahan and Rosemary L. Neidel, *Albanian Escape: The True Story of U.S. Army Nurses Behind Enemy Lines* (Lexington: University Press of Kentucky, 1999); and Peter Lucas, *The OSS in World War II Albania: Covert Operations and Collaboration with Communist Partisans* (Jefferson, N.C.: McFarland & Co., 2007), 52-53.

273. Jim Downs, *World War II: OSS Tragedy in Slovakia* (Oceanside, Calif.: Liefrinck, 2002), 23-46. After Germany absorbed Czechoslovakia in 1938, the eastern part, Slovakia, was made a separate state, albeit a vassal state, and the collaborationist leaders who were put in power adopted a Nazi style political system.

274. Ibid., Edward Hymoff, *The OSS in World War II* (New York: Ballantine, 1972), 184-85, and "Allied Help of the USA and Participation of the Americans in the SNU [Slovak National Uprising] in the Year 1944," Slovak National Uprising Museum, Banská Bystrica, Slovkia, www.muzeumsnp.sk.

275. Downs, *World War II: OSS Tragedy in Slovakia*, 48-55.

276. Hymoff, *OSS in World War II*, 186.

277. Hymoff, *OSS in World War II*, 186-89; Hymoff interviewed Schwartz, Catlos and Dunlevy in 1965, and reprints their report, ibid., 189-221, 378n; Downs, *World War II: OSS Tragedy in Slovakia*, 11-18, 47-55, 78-84; and "Short Report on the American Mission to the Czechoslovak Forces of the Interior (CFI) at Banská-Bystrica; Sept.-Oct.1994; and "Dawes Military Mission to the Czechoslovak Forces of the Interior (C.F.I.), Banská-Bystrica, Slovakia; September-October, 1944," both in OSS Records (RG 226), Entry 143, Box 12, Folder 149, National Archives II.

278. OSS, *War Report of the OSS: Overseas Targets*, 134. The decision of the officers in Bari, Italy in charge of the "Dawes" mission to send in the second detachment despite Holt Green's advice led to a bitter dispute after the war. The official OSS evaluation did not acknowledge any responsibility although noting that there was "incomplete planning" in regard to the Slovak Resistance, and it blamed the failure of the uprising on lack of support from the Soviet Union and Great Britain. Downs, *World War II: OSS Tragedy in Slovakia*, 305-306.

279. O'Donnell, *Operatives, Spies, and Saboteurs*, 218.

280. O'Donnell, *Operatives, Spies, and Saboteurs*, 221, citing a postwar report by Lt. Stefan Zenopian, one of the British SOE agents, "Zenopian Report," OSS Records (RG 226), Entry 210, Box

295, National Archives II; Downs, *World War II: OSS Tragedy in Slovakia*, 205-236; Sonya N. Jason, *Maria Gulovich, OSS Heroine of World War II: The Schoolteacher Who Saved American Lives in Slovakia* (Jefferson, N.C.: McFarland, 2008).

281. The Nazi regime publicly announced that 17 British and American "spies" had been executed on 24 January 1944; however, two of the witnesses, SS Officer Josef Niedermayer and Mauthausen prisoner Wilhelm Ornstein, later testified that only five to seven were killed on that date and the rest were executed subsequently. Downs, *World War II: OSS Tragedy in Slovakia*, passim, especially 285n; see also O'Donnell, *Operatives, Spies, and Saboteurs*, 224-225; "Report on Progress, 'Dawes' Case," OSS Records (RG 226), Entry 146, Box 36. Daniel Pavletich was captured in Piešt'any and was executed either at Zvolen or possibly at Mauthausen; Emil Tomes was captured at Polana, escaped and was killed on 5 May 1945 in an uprising against the Germans. List of members with dates of capture and execution (or not) in Dawes Military Mission to the Czechoslovak Forces of the Interior (C.F.I.), Banská-Bystrica, Slovakia; September-October, 1944," OSS Records (RG 226), Entry 143, Box 12, Folder 149, National Archives II.

282. Downs, *World War II: OSS Tragedy in Slovakia*, 247-325; O'Donnell, *Operatives, Spies, and Saboteurs*, 225-227, citing "Gulovich Report to OSS HQ," OSS Records (RG 226), Entry 108, Box 84; and "Experiences of Sgt. Catlos and Pvt.[sic] Dunlevy, Members of an OSS Mission to Slovakia," ibid., Entry 210, Box 295, both in National Archives II.

283. Downs, *World War II: OSS Tragedy in Slovakia*, 150, 248, 271, 276, 322; Peter R. Black, *Ernst Kaltenbrunner: Ideological Soldier of the Third Reich* (Princeton, N.J.: Princeton University Press, 1984).

284. OSS, *War Report of the OSS, Overseas Targets*, 146; OSS, Schools and Training Branch, "Office of Strategic Services (OSS): Organization and Functions," June 1945, p. 18, OSS Records (RG 226), Entry 141, Box 4, Folder 36; "Summary of Westfield Activities," [Swedish-based OSS Special Operations sabotage missions in Norway] OSS Records (RG 226), Entry 210, Box 30, National Archives II.

285. "Office of Strategic Services, Operational Groups, Norwegian Operational Group," www.ossog.org/norway.html. Accessed 18 June 2008.

286. Corporal Arne Herstad to Dearest Andi, 6 October 1943, *OSS Society Digest*, Number 2062, 26 May 2008, osssociety@yahoogroups.com, accessed 26 May 2008.

287. Irwin, *Jedburghs*, 132-54.

288. OSS, *War Report of the OSS, Overseas Targets*, 146.

289. "Office of Strategic Services, Operational Groups, Norwegian Operational Group," www.ossog.org/norway.html. Accessed 18 June 2008.

290. William E. Colby, "OSS Operations in Norway: Skis and Daggers," *Studies in Intelligence* (June 2002): 145-149. This mission report was originally written by Major Colby in 1945. See also William E. Colby and Peter Forbath, *Honorable Men: My Life in the CIA* (New York: Simon & Schuster, 1978), 26-29, 44-50; and "Office of Strategic Services, Operational Groups, Norwegian Operational Group, Norso I," www.ossog.org/norway/norso_01.html. Accessed 18 June 2008.

291. Colby, "OSS Operations in Norway, 149.

292. Ibid., 149-52; OSS, *War Report of the OSS, Overseas Targets*, 139.

293. "Office of Strategic Services, Operational Groups, Norwegian Operational Group, Norso II," www.ossog.org/norway/norso_02.html. Accessed 18 June 2008. Hall, *You're Stepping on My Cloak and Dagger*, 196-213.

294. William E. Colby quoted in Lt. Cmdr. Richard M. Kelly, "The Norso Mission," *Blue Book: Magazine of Adventure for Men* (December 1946), 43; clipping in OSS Records (RG 226), Entry 161, Box 7, Folder 7, National Archives II.

295. Maj. Alfred T. Cox, Commanding Officer's Report, Operational Report, Company B, 2671st Special Reconnaissance Battalion, Separate (Provisional), Grenoble, France, 20 September 1944, OSS Records (RG 226), Entry 190, Box 741, National Archives II.

296. Caesar J. Civitella, telephone interview with the author, 29 April 2008; see also "OSS Operational Groups: Italian Operational Group, Spokane/Sewanee Missions," and French Operational Group, Algiers-to France Sections, Nancy Mission," www.ossog.org.

297. Ian Sutherland, "The OSS Operational Groups: Origin of Army Special Forces," *Special Warfare* 15:2 (June 2002): 11.

298. "Interview: William Colby, Former Director, Central Intelligence Agency," *Special Warfare*, 7:2 (April 1994): 42.

299. OSS, *War Report of the OSS: Overseas Targets*, xii, 220.

300. Maj. Gen. William J. Donovan to Joint Chiefs of Staff, February 1945, reprinted in Francis B. Mills with John W. Brunner, *OSS Special Operations in China* (Williamstown, NJ: Phillips Publications, 2002), 442-446. Mills served in a SO team in France before being sent to China.

301. Brigadier General Benjamin Franklin Caffey, Jr., while serving as a young officer, had earned a law degree at the University of Michigan in 1916, been a G-3 operations officer with the 1st Division in France in World War I, met and served with Eisenhower in the Philippines in the late 1930s. In World War II, Colonel Caffey had led the 39th Regimental Combat Team at the capture of Algiers, then was made brigadier general and assistant commander of the 34th ("Red Bull") Infantry Division in Italy. Serving in its advance in Italy, he was twice injured and except for those disabling occurrences would have become a major general and a division commander. The second injury, badly frozen feet, led to his assignment in an important staff position in the Allied Force Headquarters, Mediterranean in early 1945, which is probably when he wrote this report. Subsequently, he was returned to the United States where he expedited demobilization. Subsequently, when Eisenhower became chief of staff, he appointed Caffey to the Operations Division of the General Staff. Later, Caffey served as U.S. military attaché to Switzerland before retiring from the Army and becoming first a professor at a university in Orlando, and then obtaining positions in the Veterans' Administration and the Federal Civil Defense Administration during the Eisenhower Administration. Quotation and most of the background information from Dwight D. Eisenhower, Army Chief of Staff, to Director, Personnel and Administration, U.S. Army, 7 Feb. 1948, in Dwight D. Eisenhower, *Papers of Dwight David Eisenhower*, ed. Alfred D. Chandler, et al., 21 vols. (Baltimore, Md.: The Johns Hopkins University Press, 1970-2007), 11:237; see also 2:687n; 2: 1293; and 14:430n.

302. Brig. Gen. B.F. Caffey, U.S.Army., chief, SP Ops S/Sec., G-3, [Special Projects Operations Center, G-3, Special Operations, Allied Force Headquarters, Mediterranean, under which the OSS SO and OGs operated in the Mediterranean theater] "Resistance Movements in Occupied Countries," OPD 334.8 TS, Case #285, "top secret," 5-page typescript report, undated [c. early 1945, before the end of the war in Europe], copy from the papers of Dr. E.P. Lilly, former JCS historian, located in the CIA Records (RG 263), Thomas F. Troy Files, Box 4, Folder 30, National Archives II.

303. Ibid.

304. OSS, *War Report of the OSS: Overseas Targets*, vii, 239; Irwin, *The Jedburghs*, 236-239. Ralph Ingersoll, one of the planning officers on the staff of Gen. Omar Bradley, commander of the U.S. Forces in the invasion of Normandy, wrote later that half a dozen German divisions had been used to contend with the French Resistance, divisions, which otherwise would have been fighting the Allies in Normandy. "It is a military fact, that the French were worth at least a score of divisions to us, maybe more." Ralph Ingersoll, *Top Secret* (New York: Harcourt, Brace, 1946), 181-183. In contrast, John Keegan, *The Second World War* (New York: Viking, 1990), belittled the impact of the French Resistance and suggested that their entire contribution was barely equal to that of a single Allied division.

305. On von Rundstedt and Model, see Irwin, *The Jedburgh*, 240-241; on Kesselring, see Kesselring, *Kesselring: A Soldier's Record*, 272; and OSS, *War Report of OSS: Overseas Targets*, 114.

306. Quoted in Blake Ehrlich, *Resistance: France, 1940-1945* (Boston: Little, Brown, 1965), 194.

307. Gen. Dwight D. Eisenhower to Director, OSS, UK Base, ETO USA, 31 May 1945, reprinted in OSS, *War Report of the OSS, Overseas Targets*, 222. For a recent scholarly assessment that OSS's

SI and SO operations made "a useful contribution to the Normandy campaign," see Nelson MacPherson, *American Intelligence in War-time London* (London: Frank Cass, 2003), 91; and an even more glowing assessment of the achievements of the multinational teams is provided in Irwin, *The Jedburghs*, 235-245.

308. Dwight D. Eisenhower, *Crusade in Europe* (New York: Doubleday, 1948), 296.

CHAPTER 9

1. OSS, *War Report of the OSS*, Volume II, *Overseas Targets*; originally written 1947 by the OSS's immediate successor, the Strategic Services Unit of the War Department, declassified and with a new introduction by Kermit Roosevelt (New York: Walker and Company, 1976), 365. Hereinafter OSS, *War Report of the OSS: Overseas Targets*.

2. For the representative's memoir, see Esson Gale, *Salt for the Dragon* (East Lansing: Michigan State College Press, 1953), 215.

3. Many of Donovan's supporters remain convinced that MacArthur never became reconciled to the fact that Donovan and not MacArthur, his division commander in World War I, had been awarded the Medal of Honor for heroism in France. Robert Harris Smith, *OSS: The Secret History of America's First Central Intelligence Agency* (Berkeley: University of California Press, 1972), 250-51; Richard Dunlop, *Donovan: America's Master Spy* (Chicago: Rand McNally, 1982), 402-414; Corey Ford, *Donovan of OSS* (Boston: Little, Brown, 1970), 253. MacArthur's defenders support his judgment: Charles A. Willoughby and John Chamberlain, *MacArthur, 1941-1951* (New York: McGraw Hill, 1954), 144-145; D. Clayton James, *The Years of MacArthur*, Vol. II, *1941-1945* (Boston: Houghton Mifflin, 1975), 510-11; and William B. Breuer, *MacArthur's Undercover War: Spies, Saboteurs, Guerrillas, and Secret Missions* (New York: John Wiley, 1995), 32-35, 226-227, who asserts that MacArthur was influenced by rumors from Washington that the OSS was filled with left-leaning liberals, fuzzy-thinking amateurs, and eager "cowboys" playing "cops and robbers."

4. William F. Halsey, *Admiral Halsey's Story* (New York: McGraw Hill, 1947), 170; and for a different account, Richard Dunlop, *Donovan: America's Master Spy* (Chicago: Rand McNally, 1982), 403.

5. Lillian Cox, "San Diego Paper Reports on OSS American Indian in Philippines," *OSS Society Newsletter*, Fall 2005, 16.

6. OSS, *War Report of the OSS, Overseas Targets*, 365-366. Earlier, Nimitz allowed John Ford and the OSS Field Photographic unit to photograph the damage to the naval base at Pearl Harbor in December 1941 and later to accompany the fleet for a prize-winning wartime documentary, *The Battle of Midway* (1942).

7. Francis Douglas Fane and Don Moore, *The Naked Warriors* (New York: Appleton-Century Crofts, 1956), 133; and John B. Dwyer, *Seaborne Deception: The History of U.S. Beach Jumpers* (New York: Greenwood, 1992).

8. John P. Spence, telephone interview with the author, 28 January 2005; and supporting material, including a commemoration of Spence, from which the quotations were taken, at a conference, "Naval Forces Under the Sea: Yesterday, Today and Tomorrow," U.S. Naval Academy, 29-29 March 2001, mailed by Spence to the author, 4 Feb. 2005.

9. Robert E. Mattingly, *Herringbone Cloak—GI Dagger: Marines of the OSS* (Washington, D.C.: History and Museum Division, Headquarters, U.S. Marine Corps, 1989), 176.

10. Fane and Moore, *Naked Warriors*, 132; Kenneth Finlayson, "Key West: Home of ARSOF Underwater Operations," *Veritas: Journal of Army Special Operations History*, 3:1 (2007): 3.

11. OSS, *War Report of the OSS, Overseas Targets*, 366; Fane and Moore, *Naked Warriors*, 133-37, and the declassified report of 24 August 1944 from the commanding officer of the USS *Burrfish* to the commander of submarine force, Pacific fleet, http://www.missingaircrew.com/yap2.asp, accessed 26 July 2006.

12. Carole LaMond, "Parade Marshal, WWII Vet, A Quiet Hero," unidentified newspaper story, Sudbury [Massachusetts], 22 May 2008, reprinted in *OSS Society Digest*, Number 2063, 27 May 2008, www.osssociety@yahoogroups.com, accessed 27 May 2008.

13. Fane and Moore, *Naked Warriors*, 168-69; OSS, *War Report of the OSS, Overseas Targets*, 366.

14. Official U.S. Navy SEAL Information Web Site, http://www.sealchallenge. Navy/seal/introduction.aspx, accessioned 24 September 2007.

15. Reginald G. Spear's paternal grandfather, Sir Richard Spear, owned coal mines in Newcastle. In the late 19th century, he immigrated to Nanaimo, Victoria Island, British Columbia, Canada, and started exporting coal from there. Reginald Spear's aunt, Mary Ellen Smith, was the first woman government minister in the British Empire. His father, an officer in a British Columbia regiment, fought and lost an arm at the Battle of Vimy Ridge in France in World War I. Spear's mother was a volunteer nurse with the Canadian Expeditionary Force. After the war, the Spears moved to California, where Reginald G. Spear was born in January 1924. Reginald G. Spear, telephone conversation with the author, 27 May 2005.

16. Ibid., telephone interviews with the author, 21 December 2004, 6 January and 25 March 2005.

17. Reginald G. Spear, interviewed by the author during an OSS veterans' tour of Area B. Catoctin Mountain Park, arranged by the author 18 May 2005, and telephone interview with the author, 27 May 2005.

18. Reginald G. Spear, telephone interviews with the author, 26 and 27 May 2005.

19. Ibid., 24 June 2005.

20. Ibid., 7 January 2005.

21. Reginald G. Spear conversation with the author while returning to Tyson's Corner, Vienna, Va. from an OSS veterans' tour arranged by the author of Catoctin Mountain Park, Md., 18 May 2005; plus telephone interview 3 July 2008.

22. Alan R. Millett, *Semper Fidelis: The History of the United States Marine Corps*, rev. ed. (New York: Free Press, 1991), 419-423.

23. Reginald G. Spear conversation with the author while driving back to Tyson's Corner, Vienna, Va. from an OSS veterans' tour arranged by the author of Catoctin Mountain Park, 18 May 2005; and telephone interview 3 July 2008.

24. Gavin Daws, *Prisoners of the Japanese* (New York: William Morrow, 1995). Eleven American prisoners escaped, eventually reached American lines and told their story to Army intelligence officers.

25. Reginald Spear, telephone interview with the author, 27 May 2005. Robert Ross Smith, *Triumph in the Philippines* (Washington, D.C.: U.S. Army Center of Military History, 1963); William B. Breuer, *Retaking the Philippines: America's Return to Corregidor and Bataan, October 1944-March 1945* (New York: St. Martin's Press, 1986).

26. Reginald Spear, telephone interview with the author, 25 March 2005.

27. Ibid., 27 May 2005.

28. Ibid.

29. Ibid., 27 May 2005, and 2 July 2008. Mr. Spear delivered a videotaped talk on his experience in the Philippines at a 60th anniversary commemorations held in February 2005 in San Diego and Las Vegas of the liberation of the Japanese prison camp at Santo Tomas in Manila. For accounts of the rescues at the Luzon camps at Cabanatuan, Santo Tomas, Biblibad, and Los Banos, see Gavin Daws, *Prisoners of the Japanese* (New York: William Morrow, 1995); Anthony Arthur, *Deliverance at Los Banos* (New York: St. Martin's Press, 1985); and Hampton Slides, *Ghost Soldiers: The Epic Account of World War II's Greatest Rescue Mission* (New York: Doubleday, 2001).

30. Reginald Spear, telephone interview with the author, 2 July 2008.

31. Richard J. Aldrich, *Intelligence and the War against Japan: Britain, America and the Politics of Secret Service* (New York: Cambridge University Press, 2000), 129, 409 note 72; David Stafford, *Roosevelt and Churchill: Men of Secrets* (Woodstock, NY: Overlook Press, 1999), 261-262.

32. Reginald Spear, telephone interview with the author, 27 May 2005.

33. Ibid., telephone interview with the author, 2 July 2008.

34. Ibid., telephone interview with the author, 3 July 2008.

35. Louis Morton, "Germany First," in Kent Roberts Greenfield, ed., *Command Decisions* (Washington, D.C.: U.S. Government Printing Office, 1960); Michael Schaller, *The U.S. Crusade in China, 1938-1945* (New York: Columbia University Press, 1979).

36. Hsi-Sheng Ch'i, *Nationalist China at War* (Ann Arbor: University of Michigan Press, 1984).

37. Merion and Susie Harries, *Soldiers of the Sun: The Rise and Fall of the Imperial Japanese Army* (New York Random House, 1991).

38. Mao Zedong in current spelling. The present work will use the World War II spellings concerning China.

39. OSS, *War Report of the OSS: Overseas Targets*, 357.

40. Barbara Tuchman, *Sand against the Wind: Stilwell and the American Experience in China, 1911-1945* (New York: Macmillan, 1971); and David Rooney, *Stilwell the Patriot: Vinegar Joe, the Brits, and Chiang Kai-shek* (London: Greenhill, 2005).

41. Troy J. Sacquety, "The OSS," *Veritas: Journal of Army Special Operations History*, 3:4 (2007); 48. Dr. Sacquety is currently writing a history of Detachment 101. The first OSS Special Operations force sent overseas was called Detachment 101, because Donovan's headquarters decided that calling it Detachment 1 did not suggest enough weight or experience William R. Peers and Dean Brelis, *Behind the Burma Road: The Story of America's Most Successful Guerilla Force* (Boston: Little Brown, 1963), 27.

42. For example, James R. Ward, "The Activities of Detachment 101 of the OSS," in George C. Chalou, ed, *The Secrets War: The Office of Strategic Services in World War II* (Washington, D.C.: National Archives and Records Administration, 1992), 318; and Detachment 101veterans' website, http://www.oss-101.com.

43. "'Deadliest Colonel' Dies at Age 95," *U.S. Customs Today*, August 2002, http:/www.cbp.gov/xp/CustomsToday/2002/August/other/colonel.xml, accessed 30 December 2007; Thomas N. Moon and Carl F. Eifler, *The Deadliest Colonel* (New York: Vantage Press, 1975), 1-25; Eifler was not Donovan's first choice, Maochun Yu, *OSS in China: Prelude to Cold War* (New Haven, Conn.: Yale University Press, 1996), 24-25.

44. Tom Moon, *This Grim and Savage Game: OSS and the beginning of U.S. Covert Operations in World War II* (New York: Da Capo Press, 2000), 44-46, 51.

45. Ibid., 50-51; and Peers and Brelis, *Behind the Burma Road*), 26-34, 63, 109-10, 188. Hemming, the demolition expert, was later injured in Burma when his jeep rolled over on him; Moree was the photo expert; Pamplin and Eng had some communications expertise.

46. OSS, *War Report of the OSS, Overseas Targets*, 369; Otha C. Spencer, *Flying the Hump: Memories of an Air War* (College Station: Texas A&M University Press, 1994). Kunming, pronounced "Cue-en-MING."

47. Barbara Tuchman, *Stilwell and the American Experience in China, 1911-1945* (New York: Macmillan, 1972), 340. Stilwell's alleged remark appears in various versions in different works. Tuchman cites as her source, "China-Burma-India Theater History, OSS Narrative, Annex B, Section 2." Other sources cite Carl Eifler to M. Preston Goodfellow, OSS headquarters, 28 Sept. 1942, Goodfellow Papers, Hoover Institution, Stanford, Calif. The present author has been unable to locate this document in the Goodfellow Papers. The memoir by Moon and Eifler, *The Deadliest Colonel*, 60-61, recounts Stilwell's orders to Eifler, but has "booms" remark coming from Eifler as a way of summarizing what Stilwell wanted to hear. Ray Peers, who was there, does not use Stilwell's quote at all in his own memoir, *Behind the Burma Road*, 42-43.

48. Kermit Roosevelt, "Introduction to the 1976 Edition," OSS, *War Report of the OSS; Overseas Targets*, xvii.; Troy J. Sacquety, "The Failures of Detachment 101 and Its Evolution into a Combined Arms Team," *Veritas: Journal of Army Special Operations History*, 3 (2006).

49. "Reports from 101," Maj. Archie Chun-Ming [medical officer], Schools & Training Branch, "Interviews with Returned Men," OSS Records (RG 226) Entry 161, Box 2, Folder 31, National Archives II.

50. Nicol Smith and Blake Clark, *Into Siam: Underground Kingdom* (Indianapolis, Ind.: Bobbs-Merrill, 1946), 46-47.

51. Allen Richter, W4PHL, "Burma Homebrew," James F. Ranney and Arthur L. Ranney, eds., *The OSS CommVets Papers*, 2nd ed. (Covington, Ky.: James F. Ranney, 2002), 95.

52. Ibid., 96.

53. Allen R. Richter, telephone interview with the author, 25 March 2005.

54. Jack Pamplin, [& Eifler] "Two Report from 101," [1944], p. 4, Schools & Training Branch, "Interviews with Returned Men," OSS Records (RG 226), Entry 161, Box 2, Folder 31, National Archives II.

55. Colonel Lowman to General Donovan, 14 November 1944, subject: OSS Communications, pp. 3-4, copy in CIA Records (RG 263), Thomas Troy Files, Box 12, Folder 98, National Archives II.

56. OSS, *War Report of the OSS, Overseas Targets*, 375-76.

57. "O.S.S. Detachment 101: A Brief History of the Detachment for NCAC Records," p. 2, OSS Records, Wash-Dir-Off, OP-108 (Det. 101), April 1945, copy in CIA Records (RG 263), Thomas F. Troy Files, Box 4, Folder 30, National Archives II.

58. Ibid.

59. Lt. Col. W.R. Peers to Brig. Gen. William J. Donovan, 29 February 1944, subject: report covering period 1 February to 29 February 1944, inclusive, pp. 5-6, OSS Records (RG 226), Entry 99, Box 66, Folder 402, National Archives II.

60. "Reports from 101," Lt. Jack Pamplin, [1944], Schools and Training Branch, "Interviews with Returned Men," OSS Records (RG 226), Entry 161, Box 2, Folder 31, National Archives II.

61. "Interviews with Colonel Eifler, Some of Colonel Eifler's View on Training; Based on Talks en route to and at Areas A-4, E, and F, July 1944," p. 4, OSS Records (RG 226), Entry 161, Box 2, Folder 28, National Archives II.

62. Moon and Eifler, *The Deadliest Colonel*, Appendix, 312. Moon served under Eifler in Burma, and related his own experiences in *This Grim and Savage Game* (New York: Burning Gate Press, 1991).

63. Roger Hilsman, *American Guerrilla: My War Behind Japanese Lines* (Washington, D.C.: Brassey's, 1990), 124. The more aggressive Kachin tribes also had a tradition of torturing prisoners before killing them, and they resisted American attempts to stop this practice.

64. Donovan Webster, *The Burma Road: The Epic Story of the China-Burma-India Theater in World War II* (New York: Farrar, Straus and Giroux, 2003), 159-60; Peers and Brelis, *Behind the Burma Road*, 95.

65. Col. John C. Hooker, Jr., (U.S. Army-Retired), "Biography," typescript, 1945 Chapter. I am grateful to Colonel Hooker for sending me a copy of his manuscript. John C. Hooker, email to the author, 27 June 2008.

66. Moon and Eifler, *Deadliest Colonel*, 170-72; Richard Dunlop, *Behind Japanese Lines: With the OSS in Burma* (Chicago: Rand McNally, 1979), 255-61, provides a longer account by a 101 veteran. A straight forward narrative without quotations is in Peers and Brelis, *Behind the Burma Road*, 130-32.

67. Richard Dunlop, *Donovan: America's Master Spy* (Chicago: Rand McNally, 1982), 423. The L (for lethal) pill was a capsule filled with deadly potassium cyanide. Agents were directed to put the pill under their tongue if they were in danger of immediate torture. The hard shell of the capsule was insoluble and would not dissolve in the body, but if the agent could no long stand the pain of torture, he (or she) was told to chew the pill, which would then break open, release the cyanide, and cause almost instantaneous death.

68. Smith and Clark, *Into Siam*, 56. The account by Smith, who was there, differs somewhat from Dunlop's. Smith indicates that when Donovan arrived along with director John Ford, the general had hardly stepped out of the plane when he announced "I'm going behind Jap lines." Unlike Detachment 101 Executive Officer, John Coughlin, Smith said that with that trip into danger, Donovan "earned the lifelong loyalty of everyone at 101."

69. Dunlop, *Behind Japanese Lines*, 257.

70. Ibid., 260.

71. For some time, Eifler had shown signs of physical and emotional fatigue, probably the result of a head injury suffered on an amphibious operation on the south Burmese coast in May 1943. Moon and Eifler, *Deadliest Colonel*, 118-20; and Peers and Brelis, *Behind the Burma Road*, 132. John Patton Davies of the U.S. Foreign Service and chief political adviser to Stilwell reported that although Eifler was still "outstanding" as a "lusty killer and saboteur," he was beginning to display a "lack of mental and emotional stability." John P. Davies to William J. Donovan, 6 October 1943, OSS Records (RG 226), Entry 139, File 2548, Box 193, National Archives II.

72. Moon, *This Grim and Savage Game*, 258-70.

73. Troy J. Sacquety, "A Special Forces Model: OSS Detachment 101 in the Myitkyina Campaign, Part I," *Veritas: Journal of Army Special Operations History*, 4:1 (2008): 30-47.

74. Lt. Col. W. R. Peers to Brig. Gen. W.J. Donovan, "Report Covering Period 1 April to 30 April, 1944, inclusive," 30 April 1944, OSS Records (RG 226), Entry 190, Box 40, Folder 54, National Archives II.

75. Joseph Lazarsky, telephone interview with the author, 11 February 2007; in fact, Lazarsky's guerillas repeatedly ambushed Colonel Maruyama and his troops as they evacuated Myitkyina airfield; only a small number of them reached Bhamo. Peers and Brelis, *Behind the Burma Road*, 30, 168.

76. Sacquety, "A Special Forces Model: OSS Detachment 101 in the Myitkyina Campaign." 46. While Stilwell's Chinese and American units pressed south, British General William Slim's Indian, British and other troops fought up the Burmese peninsula, defeated the Japanese and took the capital at Rangoon.

77. Peers and Brelis, *Behind the Burma Road*, 184-85. Joseph E. Lazarsky, telephone interview with the author, 9 July 2008, who noted that Peers had misidentified Lazarsky's unit, it was the 1st Kachin Battalion not the 3rd Kachin Battalion.

78. Ibid., 205-7; Roger Hilsman, *American Guerrilla: My War Behind Japanese Lines* (Washington, D.C.: Brassey's, 1990), 49, 65-92, 119-127.

79. Peers and Brelis, *Behind the Burma Road*, 204-6. See also Hilsman, *American Guerrilla*, 153-95.

80. "O.S.S. Detachment 101: A Brief History of the Detachment for NCAC Records," pp. 2, 9, April 1945, copy in CIA Records (RG 263), Thomas F. Troy Files, Box 4, Folder 30, National Archives II, which quotes high praise from the chief of staff and head of G-2 of the U.S. Northern Area Combat Command headed by Stilwell and then by Gen. Daniel Sultan. See also 101 Veteran James S. Fletcher, "Kachin Rangers: Fighting with Burma's Guerrilla Warriors," *Special Warfare*, 1:2 (July 1988): 19-27; and Charles H. Briscoe, "Kachin Rangers: Allied Guerrillas in World War II Burma," ibid., 15:4 (December 2002): 35-43. Although the Allied victory in Burma was, in sheer manpower, due primarily to the Indian Army under the leadership of General Sir William Slim and other British commanders, the victory was a coalition achievement involving troops from throughout the British Empire, was well as Americans, Chinese, and indigenous peoples. Louis Allen, *Burma: The Longest War, 1941-1945* (New York: St. Martin's, 1984).

81. For the statistics, see OSS, *War Report of the OSS, Overseas Targets*, 391-92; Peers and Brelis, *Behind the Burma Road*, 217-19; and "O.S.S. Detachment 101: A Brief History of the Detachment for NCAC Records," pp. 7-8, copy in CIA Records (RG 263), Thomas F. Troy Files, Box 4, Folder 30, National Archives II. In addition, the detachment's agents provided 75 percent of all the targeting

intelligence used by the US 10th Air Force and 85 per cent of intelligence received by the U.S. Northern Combat Area Command. Between 200 and 400 downed Allied airmen were rescued by Detachment 101.

82. Peers and Brelis, *Behind the Burma Road*, 220. On page 126, Peers states that about 50 per cent of all the agent personnel who were lost were Anglo-Burmese, who Peers praises as extraordinarily brave. OSS, *War Report of the OSS, Overseas Targets*, 391, written in 1947 lists the number of Americans killed at 15. A newspaper account of a reunion of Kachins and Americans from Detachment 101 at Lake Arrowhead, California in 2005, gave the strength as 1,000 Americans and 10,000 Kachin Rangers, the results as 5477 Japanese killed and 10,000 missing, at a cost of 18 Americans and 184 Kachins dead. Joe Vargo, "Burmese Remember American Help in WWII," *The Press-Enterprise*, 30 May 2005, www.pe.com, reprinted *in OSS Society Digest*, 31 May 2005, www.osssociety@yahoogroups.com.

83. William Ray Peers, June 1944 report quoted and Carl Eifler quoted in "Interviews with Colonel Eifler, Some of Colonel Eifler's View on Training; Based on Talks en route to and at Areas A-4, E, and F, July 1944," pp. 1-3, OSS Records (RG 226), Entry 161, Box 2, Folder 28, National Archives II.

84. Gen. Dwight D. Eisenhower, U.S. Army Chief of Staff, General Orders, 17 January 1946, The Presidential Distinguished Unit Citation is reprinted in Peers and Brelis, *Behind the Burma Road*, 208-209.

85. Troy J. Sacquety, "The OSS...Detachment 404: 1944-1945," *Veritas: Journal of Army Special Operations History*, 3.4 (2007): 49.

86. Elizabeth P. McIntosh, *Sisterhood of Spies: The Women of the OSS* (Annapolis, Md.: Naval Institute Press, 1998), 34-5, 212-3.

87. Col. John C. Hooker, Jr., (U.S. Army-Retired), "Biography," typescript, chapter for 1944. I am grateful to Colonel Hooker for sending me a copy of his manuscript. John C. Hooker, email to the author, 27 June 2008.

88. Col. John C. Hooker, Jr., (U.S. Army-Retired), "Biography," typescript, chapter for 1944. I am grateful to Colonel Hooker for sending me a copy of his manuscript. John C. Hooker, email to the author, 27 June 2008.

89. Judith A. Stowe, *Siam Becomes Thailand: A Story of Intrigue* (London: Hurst, 1991).

90. E. Bruce Reynolds, *Thailand and Japan's Southern Advance, 1940-1945* (New York: St. Martins, 1994).

91. Aldrich, *Intelligence and the War against Japan*, 194-200, 424n, from a British perspective.

92. William J. Donovan to U.S. Joint Chiefs of Staff, 18 August 1942, memorandum [regarding the Free Thai contingent], OSS Records (RG 226), Entry 146, Box 256, Folder 3553, National Archives II.

93. The Royal Thai Legation, Washington, D.C., "Project of the Free Thai Movement," undated [summer 1942], typescript, p. 7, OSS Records (RG 226), Entry 146, Box 256, Folder 3553, National Archives II.

94. Smith, who had traveled extensively in the Far East, had written a book on the Burma Road. For Smith's OSS training, see Smith and Clark, *Into Siam*, 13-15, 48-50; Peers and Brelis, *Behind the Burma Road*, 31.

95. "OSS in Thailand," "Major S_" [Nicol Smith], 24 April 1945, p. 1, Schools and Training Branch, "Interviews with Returned Men," OSS Records (RG 226), Entry 136, Box 159, Folder 1729; and Donovan to JCS, 18 August 1942; Adm. William D. Leahy, for the JCS to Director, Office of Strategic Services, 11 September 1942, subject: Mission of Thai Nationals for Far East, both in OSS Records (RG 226), Entry 146, Box 256, Folder 3553, all in National Archives II.

96. Although contemporary documents are more precise, Smith's postwar account is vague as to which of what he calls, the "alphabet resorts" the 21 Thais went to for their OSS training. He is

clear that they began their training at Area B in early 1942 but confuses the location of their holding area when they had completed all their training, stating that by mid-January 1943, "the twenty-one Thais were isolated at Area D, thirty miles from Washington in the Virginia woods, to wait for the date of departure." Since Area D was in Maryland, across the Potomac, the holding area in the Virginia woods south of Washington must have been Area A or C in Prince William Forest Park. Smith and Clark, *Into Siam*, 25, 28, 45, 48.

97. "OSS in Thailand," "Major S__" [Nicol Smith], 24 April 1945, p. 1, Schools and Training Branch, "Interviews with Returned Men," OSS Records (RG 226), Entry 136, Box 159, Folder 1729, National Archives II; plus Frank A. Gleason, telephone interview with the author, 31 January 2005; and Joseph Lazarsky, telephone interview with the author, 11 February 2007.

98. Smith and Clark, *Into Siam*, 69; "OSS in Thailand," "Major S__" [Nicol Smith], 24 April 1945, p. 1, Schools and Training Branch, "Interviews with Returned Men," OSS Records (RG 226), Entry 136, Box 159, Folder 1729, National Archives II.

99. E. Bruce Reynolds, "Opening Wedge: The OSS in Thailand," in George C. Chalou, ed, *The Secrets War: The Office of Strategic Services in World War II* (Washington, D.C.: National Archives and Records Administration, 1992), 331-35.

100. The OSS officers were John Wester, a businessman in Thailand before he joined OSS/SI and Richard Greenlee, a tax lawyer in Donovan's firm before the war. Greenlee was Chief of SO in Detachment 404, and OSS unit with headquarters in Ceylon, whose area of responsibility extended from India to Indochina.

101. Memorandum, "Background of Thailand Operations for OPD [Operations and Plans Division of the U.S. Joint Chiefs of Staff]," n.d., OSS Records (RG 226), Entry 154, File 2297, Box 131, National Archives II.

102. Mattingly, *Heringbone Cloak—GI Dagger: Marines of the OSS*, 160, although Mattingly apparently erred in indicating Macomber was with Peers' Detachment 101 in these operations. Peers and Brelis, *Behind the Burma Road*, 228-29. does not include Macomber in his roster of Detachment 101. The area was instead under the jurisdiction of Detachment 404.

103. "Behind Enemy Lines by Steve Sysko," [diary excerpts], *OSS Society Newsletter*, Summer 2004, 9; Bruce Edwards, "With the OSS, Steve Sysko Helped Liberate Thailand from the Japanese," *Rutland [Vermont] Herald*, 11 August 2008, reprinted in *OSS Society Digest*, Number 2125, 12 August 2008, osssociety@yahoogroups.com, accessed 12 August 2008.

104. Bob Bergin, "Pearl Harbor Payback: The AVG's Surprise Raid on Chiang Mai," *Flight Journal* (April 2005); see also http://thaiaviation.com. The American Volunteer Group's surprise attack and destruction of 15 to 30 Japanese planes on the ground provided an uplift in a dismal period and was headlined by the *New York Times*. McGarry parachuted out; 2nd Squadron leader and ace John ("Scarsdale Jack") Newkirk was killed when his plane was shot down and crashed with Newkirk aboard after strafing a Japanese convoy.

105. Bob Bergin, "Claire Chennault and the OSS: A Favor Done—and Returned," *OSS Society Newsletter*, Winter 2004-225, 2; and "OSS in Thailand," "Major S__" [Nicol Smith], 24 April 1945, pp. 3-4, Schools and Training Branch, "Interviews with Returned Men," OSS Records (RG 226), Entry 136, Box 159, Folder 1729, National Archives II.

106. "OSS in Thailand," "Major S__" [Nicol Smith], 24 April 1945, p. 5, Schools and Training Branch, "Interviews with Returned Men," OSS Records (RG 226), Entry 136, Box 159, Folder 1729, National Archives II. A similar report about the need to treat the indigenous people working for OSS with respect came from Ray F. Kauffman, an OSS SO training and operations officer in Detachment 404, who dealt mainly with indigenous people on Sumatra, formerly part of the Dutch East Indies (now Indonesia), which was occupied by the Japanese in World War II. Scion of an affluent family in Des Moines, Iowa, Kauffman, like many members of the OSS, had already demonstrated initiative and daring as a civilian. In his case at age 28 in 1935, he began a three-year circumnavigation of the world on a 46-foot ketch, *Hurricane*, sometimes single-handed, sometimes accompanied by a friend from Des Moines, and sometimes with hired natives, a story he told in a

1940 book, *Hurricane's Wake*. Ray F. Kauffman, *Hurricane's Wake* (New York: Macmillan, 1940). Joining the OSS at age 35, Kauffman, attended a number of the training schools in 1942-1943, including Area D, the Maritime School; Area C, where he took the full communication course at Prince William Forest Park; and RTU-11 ("the Farm"), where he studied Secret Intelligence. He worked in conjunction with a group of Malayan trainees at those camps, and in the fall of 1943, Kauffman and the Malayan agents set out of Detachment 404 headquarters in Kandy, Ceylon. On Ceylon, Kauffman was assigned as a training instructor for Maritime operations out of the port of Trincomalee on the Bay of Bengal in northeast Ceylon. He also led two seaborne penetration operations himself. Back in Washington in November 1944, Kauffman told Schools and Training Branch that ingenuity and common sense were keys to successful training and operations. In all his work, Kauffman said he kept to "one Golden Rule for handling natives...'live with them and become their friend.' Already, he believes, that this policy has proved its efficiency in the Far East." Unlike Carl Eifler, Kauffman believed that loyalty was a more important motivation than money. "The formula for making successful operatives out of reluctant natives," Kauffman said, "is 90% friendship and 10% money." What would motivate a local person to take life and death chances "for our side," Kauffman asked. His answer was the "friendship policy." Instead of ordering people to do things in training or in the field, Kauffman sought ways to establish interesting goals in which the training would be incidental, for example, sending them on a pigeon hunt as a way of teaching them compass use to plot a course or a fishing expedition as a way of learning how to launch small boats in the surf while keeping the equipment dry. Kauffman emphasized consulting with the natives about problems. Whether field stripping a weapon or repairing a radio set, he asked the native his opinion. Sometimes the trainee would prove to be an able mechanic, but whether he was or not, Kauffman had won the respect of his students by making them feel that he respected them. "Training the Native," Ray Kauffman, 11 November 1944, p. 1, Schools and Training Branch, "Interviews with Returned Men," OSS Records (RG 226), Entry 161, Box 2, Folder 31, National Archives II.

107. OSS, Schools and Training Branch, "Office of Strategic Services (OSS), Organization and Functions," June 1945, p. 28, OSS Records (RG 226), Entry 141, Box 4, Folder 36, National Archives II.

108. Reynolds, "Opening Wedge," 335-43; OSS, *War Report of the OSS, Overseas Targets*, 407-14; Aldrich, *Intelligence and the War against Japan*, 323-30.

109. Mark Atwood Lawrence, *Assuming the Burden: Europe and the American Commitment to War in Vietnam* (Berkeley: University of California Press, 2005), 19-26. On Churchill's resistance to trusteeship and decolonization, see Peter Clarke, *The Last Thousand Days of the British Empire: Churchill, Roosevelt and the Birth of Pax Americana* (London: Bloomsbury Press, 2008).

110. Lawrence, *Assuming the Burden*, 74; Scott L. Bills, *Empire and Cold War: The Roots of U.S.-Third World Antagonism, 1945-1947* (New York: Macmillan, 1990).

111. Ronald H. Spector, *Advice and support: The Early Years of the U.S. Army in Vietnam, 1941-1960* (New York: Free Press, 1985); Dixee R. Bartholomew-Feis, *The OSS and Ho Chi Minh: Unexpected Allies in the War against Japan* (Lawrence: University Press of Kansas, 2006), 1-33.

112. For differing views of Ho's motivation in trying to enlist the Americans in his cause, see Jean Lacoutre, *Ho Chi Minh: A Political Biography* (New York: Random House, 1968); Peter M. Dunn, *The First Vietnam War* (London: Hurst, 1985), 50; Charles Fenn, *Ho Chi Minh: A Biographical Introduction* (New York: Scribners, 1973), 74-76, Archimedes L.A. Patti, *Why Viet Nam? Prelude to America's Albatross* (Berkeley: University of California Press, 1980), 43-47; and William J. Duiker, *Ho Chi Minh* (New York: Hyperion, 2000). On the OSS relationship with Ho and he Viet Minh, see most recently, Bartholomew-Feis, *The OSS and Ho Chi Minh*, which provides a detailed account generally sympathetic to the OSS.

113. Bartholomew-Feis, *The OSS and Ho Chi Minh*, 300-320;

114. Charles Fenn, *At the Dragon's Gate: With the OSS in the Far East* (Annapolis, Md.,: Naval Institute Press, 2004), 4-11; Fenn, *Ho Chi Minh*, 78-82; Bartholomew-Feis, *OSS and Ho Chi Minh*, 154-155; Aldrich, *Intelligence and the War against Japan*, 292-93.

115. Patti, *Why Viet Nam?*, 6-7, 28-31, 83-88

116. Bartholomew-Feis, *The OSS and Ho Chi Minh*, 188-89; and Henry Prunier, telephone interview with the author, 15 August 2008.

117. René Défourneaux, email to the author, 7 August 2008. Défourneaux was trained by SOE in Britain and then by OSS SO on Catalina Island, California. Henry Prunier spent a day or two at Area F, then was trained at Catalina Island and Newport Beach. Henry Prunier, telephone interview with the author, 15 August 2008.

118. Sergeant William Zielski, Distinguished Service Cross Citation, March 10, 1945, OSS Records (RG 226), Entry 92A, Box 118, Folder 2553, National Archives II.

119. Allison Thomas, "'Welcome to Our American Friends,'" in *Strange Ground: Americans in Vietnam, 1945-1975: An Oral History*, ed. Harry Mauer (New York: Henry Holt, 1989), 29-30, 33.

120. René Défourneaux, "A Secret Encounter with Ho Chi Minh," *Look* magazine, 9 August 1966, 32-33; William Broyles, Jr., *Brothers in Arms: A Journey form War to Peace* (New York: Knopf, 1986), 104.

121. Bartholomew-Feis, *The OSS and Ho Chi Minh*, 160, 176-188, 193-208; Patti, *Why Viet Nam?*, 29-30.

122. Thomas, "'Welcome to Our American Friends,'" 35.

123. Bartholomew-Feis, *The OSS and Ho Chi Minh*, 209-215.

124. Ibid., 179, 204-6, 218, 368n; Will Irwin, *The Jedburghs: The Secret History of the Allied Special Forces in France, 1944* (New York: Public Affairs Press, 2005), 252.

125. Aaron Bank, *OSS to Green Berets: The Birth of Special Forces* (Novato, Calif.: Presidio Press, 1986), 100-104.

126. Ibid., 105-14; Patti, *Why Vietnam?*, 566. Bank located more than 150 French civilian internees.

127. Henry Prunier, telephone interview with the author, 15 August 2008; Patti, *Why Viet Nam?*

128. Bartholomew-Feis, *The OSS and Ho Chi Minh*, 218. See also Thomas's "Report on the Deer Mission," September 1945, OSS Records (RG 226), Entry 154, Box 199, Folder 2277, National Archives II.

129. Bank, *From OSS to Green Berets*, 116-29.

130. Obituary, Tim Weiner, "Lucien Conein, 79, Legendary Cold War Spy," *New York Times*, 87 June 1998; Bartholomew-Feis, *The OSS and Ho Chi Minh*, 245, 251, 377n.

131. Lawrence, *Assuming the Burden*, 81-82, 90-96, 134, 144; Bartholomew-Feis, *The OSS and Ho Chi Minh*, 312-320; Maochun Yu, *OSS in China: Prelude to Cold War* (New Haven, Conn.: Yale University Press, 1996), 232-233.

132. Ronald H. Spector, *In the Ruins of Empire: The Japanese Surrender and the Battle for Postwar Asia* (New York: Random House, 2007), 124; Patti, *Why Viet Nam?*, 480-81. Dewey's actual rank was major, but he elevated himself to lieutenant colonel to have more influence with those with whom he dealt.

133. Charles M. Parkin, interview after a tour arranged by the author of former OSS Training Area B, Catoctin Mountain Park, Thurmont, Md., 18 May 2005.

134. Bartholomew-Feis, *The OSS and Ho Chi Minh*, 268-291; R. Harris Smith, *OSS: The Secret History of America's First Central Intelligence Agency* (Berkeley: University of California Press, 1972), 337-338; "Captain Albert Peter Dewey, AC [Air Corps]," OSS Records (RG 226), Entry 92A, Box 29, Folder 421, National Archives II.

135. Patti, *Why Viet Nam?*, 320-23 Bartholomew-Feis, *The OSS and Ho Chi Minh*, 265-299, Ho Chi Minh reference on p. 298; Aldrich, *Intelligence and the War against Japan*, 346-49; Bills, *Empire and the Cold War*, 124-26; The official OSS history barely mentions French Indochina. OSS, *War Report of the OSS, Overseas Targets*, 414. Nancy Dewey Hoppin, Peter Dewey's daughter, who was

an infant at the time of her father's death, visited Vietnam in 2005 and was told that her father had been mistaken for a French officer and that his body, which was never found, had been thrown into a river at Go Vap near Saigon. Seymour Topping, "Vietnamese Historian Recalls Untold Story of Tragic Murder of Peter Dewey," *OSS Society Newsletter*, Summer 2005, 3-4.

136. Peter Dewey quoted in Patti, *Why Viet Nam?*, 320.

137. Barbara Tuchman called Dai Li, "China's combination of Himmler and J. Edgar Hoover," *Stilwell and the American Experience*, 334; for an attempt at a balanced view, see Yu, *OSS in China*, 31-32, 286n, and Frederick Wakeman, Jr., *Spymaster: Dai Li and the Chinese Secret Service* (Berkeley: University of California Press, 2003), xiii-xv, 1-11, 355-358, 365. There are various English spellings of Dai Li's name (e.g. Tai Li). I have used Wakeman's spelling.

138. Wakeman, *Spymaster*, 290-91.

139. Milton Miles, *A Different Kind of War: The Little Known Story of the Combined Guerrilla Forces Created in China by the U.S. Navy and the Chinese During World War II* (Garden City, N.Y.: Doubleday, 1967). His nickname "Mary," resulted from a Hollywood star, Mary Miles, having the same last name.

140. Dai Li's decision to establish a partnership with Miles and the Navy may have been caused by recent bureaucratic setbacks he had suffered domestically over cryptography and the discovery that his intelligence center was penetrated by a half dozen communist agents who were reporting directly to Mao's Chinese Communist Party headquarters in Yenan. Yu, *OSS in China*, 38-58.

141. In October 1942, for example, OSS had deposited $50,000 in half a dozen Asian banks for Miles, at titular head of OSS in China, to draw upon in helping to obtain agreement. OSS also purchased and distributed 500 gold wristwatches as gestures of good will. Yu, *OSS in China*, 84.

142. Yu, *OSS in China*, 77-83, 90, 296n. In his recommendation after his China visit, Lusey confided to Donovan., "none of the [SO] men sent to china as instructors [for Dai Li's guerrillas] should be told they will eventually be used as SI men, or that they will be used for anything excerpt instructing the Chinese SO people; when the time comes for them to do SI work, we will work them in." "This would have to be handled very carefully, as our whole show would blow up if the Chinese ever found out we were doing anything like this." Alghan R. Lusey to William J. Donovan, 14 September 1942, Secret Memorandum, OSS Records (RG 226), Entry 139, Box 267, Folder 3934, National Archives II.

143. Wakeman, *Spymaster*, 290-293. OSS, *War Report of the OSS, Overseas Targets*, 424-27.

144. Yu, *OSS in China*, 94-98.

145. Wakeman, *Spymaster*, 294-295.

146. Miles, *A Different Kind of War*, 149-50; Wakeman, *Spymaster*, 379.

147. Description of duties in China in Lt. Col. Charles M. Parkin, to Assistant Chief of Staff, G-1, WD General Staff (through channels), 6 June 1945, subject: Request to Enter Army and Navy Staff College, OSS Records (RG 226), Entry 92A, Box 47, Folder 785, National Archives II. It is possible that Parkin was the commander of the instructional staff rather than the commander of the unit staff, as he listed himself as former commanding officer of Area B, when he had been the chief instructor. Frederick Wakeman lists Maj. John. H. ("Bud") Masters, USMC, as commander of the U. S. personnel at SACO Unit 1; but like Parkin, Wakeman reports that the trainees conducted several sabotage raids including wrecking trains. Wakeman, *Spymaster*, 379.

148. Charles M. Parkin, telephone interview with the author, 10 May 2005;

149. Lt. Col. Jacque B. deSibour, to Commanding General, Rear Echelon, HQ, USF, China Theater, 10 February 1945, subject: Promotion of Office [Major Charles M. Parkin to Lieutenant Colonel], OSS Records (RG 226), Entry 92A, Box 47, Folder 785, National Archives II. The report provides dates of service and describes Parkin as a "young, energetic" officer of "superior" quality and in the upper third of all officers of his relative rank. See also Parkin's OSS personnel file transferred to CIA Records (RG 263), Box 31, National Archives II.

150. Wakeman, *Spymaster*, 379.

151. Ibid.

152. Lt. Col. Charles M. Parkin, to Assistant Chief of Staff, G-1, WD General Staff (through channels), 6 June 1945, subject: Request to Enter Army and Navy Staff College, OSS Records (RG 226), Entry 92A, Box 47, Folder 785, National Archives II.

153. Miles, *A Different Kind of War*, 157; Wakeman, Spymaster, 380.

154. Joseph Lazarsky, telephone interview with the author, 11 February 2007.

155. Yu, *OSS in China*, 84, who cites Capt. Jeffrey C. Metzel to Lt. Cmdr. Milton Miles, 7 February 1943, Records of the Chief of Naval Intelligence, Foreign Intelligence, the Far East Desk (RG 38), Miles Papers, Box 1, Folder 1, National Archives II.

156. Frank A. Gleason, telephone interview with the author, 9 May 2005.

157. Ibid., Miles, *A Different Kind of War*, 158. On Dow's name for the camp, "Pact Rowdy," see Maj. Arden W. Dow to M.E. Miles, Captain, USN, U.S. Naval Observer, 8 January 1943 [1944], subject: Complete Training Program at S.A.C.O. Number 3, Pact Rowdy, a 14-page report, OSS Records (RG 226), Entry 146, Box 256, Folder 3550, National Archives II.

158. Maj. Frank A. Gleason, "Summary of the Activities of Major Frank A. Gleason with the Office of Strategic Services in the China Theater from March 16, 1943 thru and including March 20, 1945," p. 1, tenpage typed report accompanying, Gleason to Director of OSS (through Chief, SO Branch), 7 June 1945, subject: Report of Activities of Major Frank A. Gleason in China Theater, OSS Records (RG 226), Entry 146, Box 256, Folder 3550, National Archives II. Emphasis added.

159. Maj. Arden W. Dow to M.E. Miles, Captain, USN, U.S. Naval Observer, 8 January 1943 [1944], subject: Complete Training Program at S.A.C.O. Number 3, Pact Rowdy, a 14-page report, OSS Records (RG 226), Entry 146, Box 256, Folder 3550, National Archives II.

160. Wakeman,, *Spymaster*, 380-381.

161. Ibid., 294-298.

162. Frank A. Gleason, telephone interview with the author, 9 February 2007.

163. Maj. Arden Dow to Director of OSS, CBI Theater, Report, 20 March 1944, pp. 9, 19, in OSS Records, quoted in Wakeman, 500n 12, 17.

164. Wakeman, *Spymaster*, 299, citing a memoir published in 1981 of four months experience at the Linru camp by one of the Chinese political instructors there, Zhong Xiangbai.

165. Maj. Frank A. Gleason, "Summary of the Activities of Major Frank A. Gleason with the Office of Strategic Services in the China Theater from March 16, 1943 thru and including March 20, 1945," p. 4, tenpage typed report accompanying, Gleason to Director of OSS (through Chief, SO Branch), 7 June 1945, subject: Report of Activities of Major Frank A. Gleason in China Theater, OSS Records (RG 226), Entry 146, Box 256, Folder 3550, National Archives II.

166. Maj. Frank G.___ [Gleason], interview, 30 April 1945, p. 1, Schools and Training Branch, "Interviews with Returned Men," OSS Records (RG 226), Entry 136, Box 159, Folder 1729, National Archives II.

167. At a banquet on Donovan's arrival in Chungking, 2 December 1943, Donovan bluntly told Dai Li that if he would not cooperate, OSS would establish itself in China on its own. The Chinese spymaster bridled and threatened to kill any OSS agents operating outside of SACO. Donovan shouted, "For every one of our agents you kill, we will kill one of your generals!" The two men shouted at each other for a minute before regaining their composure. The next day, when Donovan met with Chiang Kai-shek, the generalissimo reminded him that China was an ally and a sovereign country and Donovan as a high U.S. representative was expected to remember that and act accordingly. "You do not expect a secret service from another country to go into the United States and start operations," Chiang explained. "You would object seriously." Milton E. Miles, "On SSU Organization in China," postwar report dated, 17 May 1946, in Milton E. Miles Personal Papers, Box 3, Folders 1-2, Hoover Institution, Stanford, Calif.

168. Claire L. Chennault, *Way of a Fighter: The Memoirs of Claire Lee Chennault*, ed. Robert Hotz (New York: G.P. Putnam's, 1949); Martha Bird, *Chennault: Giving Wings to the Tiger* (Tuscaloosa: University of Alabama Press, 1987).

169. Wakeman, *Spymaster*, 319; Dunlop, *Donovan: America's Master Spy*, 427; OSS, *War Report of the OSS, Overseas Targets*, 430-37, which includes a copy of the OSS agreement with the 14th Air Force.

170. Col. John G. Coughlin [head of OSS/ SO in China] to Maj. Arden W. Dow, 3 April 1944, subject: Special Mission, pp. 2-3, OSS Records (RG 226), Entry 92A, Box 39, Folder 611, National Archives II.

171. Smith, *OSS*, 258-62; Ford, *Donovan*, 267-68; Yu, *OSS in China*, 153-157; OSS, *War Report of the OSS, Overseas Targets*, 418.

172. In April 1945, AGFRTS and all of its original members (Air Force and OSS) came under jurisdiction of the OSS Zhijian (Chihkiang) Field Unit, which still under the leadership of Air Force Lt. Col. Wilfrey Smith, became responsible for OSS SO operations in the vast area between the Yangtze and West rivers. Yu, *OSS in China*, 156-67.

173. Francis B. Mills with John W. Brunner, *OSS Special Operations in China* (Williamstown, NJ: Phillips Publications, 2002), 468.

174. Elizabeth P. McIntosh, *Sisterhood of Spies: The Women of the OSS* (Annapolis, Md.: Naval Institute Press, 1998), 197-198, 204-207, 216-222, 233-236.

175. Lloyd E. Eastman, *Seeds of Destruction: Nationalist China in War and Revolution, 1937-1945* (Stanford, Calif.: Stanford University Press, 1984); David P. Barrett and Larry N. Shyu, *China in the Anti-Japanese War, 1937-1945* (New York: Peter Lang, 2001). *Ichi-Go* translates as "Operation One."

176. OSS, *War Report of the OSS, Overseas Targets*, 440-43.

177. Ford, *Donovan of OSS*, 273.

178. Maj. Frank G.___ [Gleason], interview, 30 April 1945, p. 1, Schools and Training Branch, "Interviews with Returned Men," OSS Records (RG 226), Entry 136, Box 159, Folder 1729, National Archives II.

179. Maj. Frank A. Gleason, "Summary of the Activities of Major Frank A. Gleason with the Office of Strategic Services in the China Theater from March 16, 1943 thru and including March 20, 1945," pp. 8-9, ten-page typed report accompanying, Gleason to Director of OSS (through Chief, SO Branch), 7 June 1945, subject: Report of Activities of Major Frank A. Gleason in China Theater, OSS Records (RG 226), Entry 146, Box 256, Folder 3550, National Archives II.

180. Ibid., 9.

181. Maj. Frank G.___ [Gleason], interview, 30 April 1945, p. 4, Schools and Training Branch, "Interviews with Returned Men," OSS Records (RG 226), Entry 136, Box 159, Folder 1729, National Archives II.

182. Frank A. Gleason, telephone interview with the author, 31 January 2005. In destroying the first nine bridges, Gleason used 1,000-poud naval mines obtained from the Navy in Luichow, and which had been destined for harbors along the coast to destroy enemy shipping. email to the author, 28 May 2005.

183. Theodore H. White and Annalee Jacoby, *Thunder Out of China* (New York: William Sloane, 1946), 195-196. White's original account appeared in *Time Magazine*, 8 January 1945, 57-58.

184. Lt. George C. Demas to Lt. Col. O.C. Doering, Jr., et al., 23 January 1945, subject: Excerpt from Time Magazine of 8 January 1945, pp. 57-58, OSS Records (RG 226), Entry 139, Box 195, Folder 2584; see also Maj. Frank G.___ [Gleason], interview, 30 April 1945, p. 3, Schools and Training Branch, "Interviews with Returned Men," OSS Records (RG 226), Entry 136, Box 159, Folder 1729, National Archives II. Frank A. Gleason, telephone interview with the author, 15 July 2008.

185. Interviewer's Report on Maj. Frank A. Gleason, 27 April 1945, Washington, D.C., Frank A. Gleason, OSS Personnel File, CIA Records (RG 263), Box 22, National Archives II.

186. Frank A. Gleason, telephone interview with the author, 31 January 2005.

187. White and Jacoby, *Thunder Out of China*, 196.

188. Maj. Frank G.__ [Gleason], interview, 30 April 1945, pp. 6-7, Schools and Training Branch, "Interviews with Returned Men," OSS Records (RG 226), Entry 136, Box 159, Folder 1729, National Archives II.

189. Lt. George C. Demas to Lt. Col. O.C. Doering, Jr., et al., 23 January 1945, subject: Excerpt from *Time* Magazine of 8 January 1945, pp. 57-58, OSS Records (RG 226), Entry 139, Box 195, Folder 2584, National Archives II.

190. Frank A. Gleason, email to the author, 28 May 2005.

191. Ibid.

192. White and Jacoby, *Thunder Out of China*, 196-97.

193. Lt. George C. Demas to Lt. Col. O.C. Doering, Jr., et al., 23 January 1945, subject: Excerpt from Time Magazine of 8 January 1945, pp. 57-58, OSS Records (RG 226), Entry 139, Box 195, Folder 2584, National Archives II.

194. Frank Gleason, email to the author, 28 May 2005; Theodore H. White and Annalee Jacoby, *Thunder Out of China* (New York: William Sloane, 1946), 195-196; Theodore H. White, *The Mountain Road* (New York: W. Sloan, 1958); and *The Mountain Road* (US: Columbia Pictures, 1960), 102 minutes, starring Jimmy Stewart as "Major Baldwin" and directed by Daniel Mann. Gleason says *Thunder Out of China* provides an accurate account of his team's activities.

195. OSS Schools and Training Branch, "Office of Strategic Services (OSS): Organization and Functions," June 1945, pp. 32-33, OSS Records (RG 226), Entry 141, Box 4, Folder 36, National Archives II.

196. Excerpts from Monthly Report of OSS, SU, Detachment 202, CBI, 1 November 1944, from Ensign Gunnar G. Mykland, executive officer, located in Donovan's "Black Book" copy in CIA Records (RG 263), Thomas Troy Files, Box 4, Folder 30, National Archives II.

197. Colonel Lowman to General Donovan, 14 November 1944, subject: OSS Communications, pp. 3-4, OSS Records, Wash-Dir-Op-27, "Accomplishments of OSS," copy in CIA Records (RG 263), Thomas Troy Files, Box 12, Folder 98, National Archives II.

198. Kenneth E. Shewmaker, *Americans and Chinese Communists, 1927-1945: A Persuading Encounter* (Ithaca, NY: Cornell University Press, 1971).

199. David D. Barrett, *Dixie Mission: U.S. Army Observer Group in Yenan, 1944* (Berkeley: University of California Press, 1970); Carolle J. Carter, "Mission to Yenan: The OSS and the Dixie Mission," in George C. Chalou, ed., *The Secrets War: The Office of Strategic Services in World War II* (Washington, D.C.: National Archives and Records Service, 1992), 303-4.

200. Carter, "Mission to Yenan, 305.

201. Ibid., 305-6.

202. Yu, *OSS in China*, 166-67, 224-25. The Dixie mission of the U.S. Army and the OSS became controversial after the war, when Republicans blamed the "loss of China," the defeat of the Nationalists by the Communists in 1949, on allegedly Communist sympathizers in the Roosevelt and Truman Administrations, the State Department, the OSS, and the Democratic Party. It was one of the issues Republicans used to end nearly twenty years of Democratic control of the White House in 1952. But although some OSS planners saw American operations in Yenan as a possible basis for major victories against the Japanese and a way to expand the role of the OSS in North China, these operations were not a major concern of the OSS outside of China. Furthermore, the OSS's reputation with the Communist leaders in Yenan was not good, partly due to mutual suspicion and mistrust and partly because most of the promised arms and supplies were not delivered. Most OSS operations behind enemy lines in Communist infiltrated areas in China were seen by the Communists as hostile to their aims and interests. The main American criticism during the war against the U.S. Army and OSS for "working with the Communists" came from Donovan's

rival there, Lt. Cmdr. Milton Miles, who was closely linked to Nationalist spy chief, Gen. Dai Li. Most American leaders in China urged cooperation with the Communists as well as the Nationalists in 1944-1945. In the spring of 1945, irregardless of their later postwar anti-Communist rhetoric, American civilian and military leaders in China, including Ambassador Patrick Hurley and Generals Albert Wedemeyer, Claire Chennault, and Curtis LeMay, praised the Chinese Communists for their alleged help against the Japanese. Carter, "Mission to Yenan, 312-13. The official history, OSS, *War Report of the OSS, Overseas Targets*, 419, 437-438, devotes only three pages to the Dixie Mission.

203. Yu, *OSS in China*, 221-22; Maj. Frank L. Coolidge to Colonel Heppner, 13 September 1945, Report on the Spaniel Mission, OSS Records (RG 226) Entry 148, Box 6, Folder 87, National Archives II.

204. John W. Brunner, telephone interview with the author, 21 March 2005.

205. Ibid.

206. James F. Ranney, telephone interview with the author, 8 January 2005. At Chihkiang, Maj. James O. Swenson was the commanding officer; Capt. Richard S. Buchholz was in charge of communications, Lt. Ranney was his second in command. PFC Elmer Schubert, helped operate the transmitter station; David Kenney helped operate the receiver station.

207. David Kenney, telephone interview with the author, 11 April 2005.

208. James F. Ranney, W4KFR, "OSS Radio Station WLUR–Chihkiang, China," in James F. Ranney and Arthur L. Ranney, eds., *The OSS CommVets Papers*, 2nd ed. (Covington, KY: James F. Ranney, 2002), 33-34.

209. "R.L. ("Bob") Scriven, K5WFL," in *ibid.*, 212.

210. Arthur Reinhardt, interview with the author at Prince William Forest Park, Triangle, Va., 14 December 2004, with additions made to the transcript by Mr. Reinhardt on 23 February 2007.

211. Ibid., for a published account of running a network of coast watchers, see Dan Pinck, *Journey to Peking: A Secret Agent in Wartime* China (Annapolis, Md.: Naval Institute Press, 2003). Pinck's youngest son, Charles T. Pinck, was President of the OSS Society in 2008 when the present volume went to press.

212. Francis B. Mills with John W. Brunner, *OSS Special Operations in China* (Williamstown, NJ: Phillips Publications, 2002), 20.

213. Yu, *OSS in China*, 185, 210; OSS, *War Report of the OSS: Overseas Targets*, 440-45.

214. Figures on OSS strength in China are from Yu, *OSS in China* (1996), 226, who cites Capt. J.W. Kruissink, adjutant officer, to HQ investigation boards, US Forces, China Theater, 9 October 1945, memorandum, subject: Strength Report of OSS taken from Morning Reports, in Records of Allied and US Army Commands in the China-Burma-India Theater of Operations (World War II) (RG 493), Records of the China Theater of Operations, U.S. Army, Records of the General Staff, G-5 (Civil Affairs) Section, formerly classified report–special agencies in China, Section V, OSS, Box 61, National Archives and Federal Records Center, Suitland, Md.

215. In addition to Heppner as OSS chief in China, Detachment 202 in early1945 included, Lieutenant Colonel Willis Bird as Deputy Chief; Colonel William P. Davis as Operations Officer; Colonel Paul L. E. Helliwell, chief of SI Branch; Lieutenant Colonel Nicholas W. Willis, chief of SO Branch; Lieutenant Colonel Charles A. Porter, chief of the Communications Branch; Captain Eldon Nehring, formerly from Area A in Prince William Forest Park, chief of the Schools and Training Branch; plus others in charge of the Morale Operations, Research and Analysis, Counterespionage, Field Photographic, and Administrative branches Yu, *OSS in China* (1996), 226, citing Investigating Boards, US Forces, China Theater, undated, subject: Report on OSS, in Records of Allied and US Army Commands in the China-Burma-India Theater of Operations (World War II) (RG 493), Records of the China Theater of Operations, U.S. Army (CT or Wedemeyer), Record of the General Staff, G-5 (Civil Affairs) Section, formerly classified report–special agencies in China, Section V, OSS, Box 61, National Archives and Federal Records Center, Suitland, Md.

216. Francis B. Mills with John W. Brunner, *OSS Special Operations in China* (Williamstown, N.J.: Phillips Publications, 2002), 38-39.

217. Yu, *OSS in China*, 215-16.

218. Yu, *OSS in China*, 216-19, 227, 315n. Much assistance was provided by the extraordinary network of Chinese Roman Catholics operated by Bishop Thomas Megan, a 44-year-old, Irish-American clergyman, known as the "Fighting Bishop," who inherited them from the legendary Father Vincent Lebbe, a Belgian priest who arrived in China in 1895 and in the 1920s and 1930s, organized Catholic Christian groups as alternatives to Communism and Confucianism. For that and aiding wounded Chinese soldiers and providing intelligence information to the guerrillas, Lebbe was kidnapped by Chinese Communists in March 1940. The Communists tortured him so badly that he died soon after being released.

219. Yu, *OSS in China*, 227, 241; contrast with Wuchinich's own later account in Smith, *OSS*, 281.

220. On Operation Carbonado and the Akron Mission, see OSS, *War Report of the OSS, Overseas Targets*, 444-445, 455-456.

221. Charles M. Parkin, telephone interview with the author, 10 May 2005.

222. Charles M. Parkin, discussion with the author at dinner following a tour of former Area B in Catoctin Mountain Park, Thurmont, Md., 18 May 2005. Colonel Parkin told the author that it was a three-man team—himself, a photographer and an interpreter. He must have forgotten the radio operator. OSS Communications Branch cipher clerk John W. Brunner told the author subsequently that he knew Parkin's radio operator, who roomed with Brunner in Kunming when the radioman was sent back on leave after the Akron mission. John W. Brunner, email to the author, 14 June 2005.

223. Description of duties in China in Lt. Col. Charles M. Parkin, Jr. to Assistant Chief of Staff, G-1, WD General Staff (through channels), 6 June 1945, subject: Request to Enter Army and Navy Staff College, OSS Records (RG 226), Entry 92A, Box 47, Folder 785, National Archives II.

224. OSS, *War Report of the OSS, Overseas Targets*, 418.

225. Charles M. Parkin, discussion with the author at dinner following a tour of former Area B in Catoctin Mountain Park, Thurmont, Md., 18 May 2005. The August invasion envisioned in Carbonado was never implemented. By the summer of 1945, it had become clear that the operational emphasis had shifted from China to the Japanese home islands, and the original plan was reduced merely to seizing the Hong Kong-Canton port area for use as a base against the Japanese. Even that plan was abandoned with the sudden end of the war in mid-August 1945. OSS, *War Report of the OSS, Overseas Targets*, 455-56.

226. Mills with Brunner, *OSS Special Operations in China*, 14-15.

227. Ibid., 435.

228. Obituary, Francis Byron Mills, *Washington Post*, 1 October 2005.

229. Mills with Brunner, *OSS Special Operations in China*, 17.

230. Ibid., 27.

231. Ibid., 40, 43.

232. Ibid., 20.

233. Charles H. Briscoe, "Major Herbert R. Brucker, SF Pioneer, Part II: SOE Training & 'Team Hermit' into France," *Veritas: Journal of Army Special Operations History*, 2:3 (2007):72-85.

234. Charles H. Briscoe, "Major (R) Herbert R. Brucker, DSC, Special Forces Pioneer," *Veritas: Journal of Army Special Operations History*, 2:2 (2006): 33; Banks, *From OSS to Green Berets*, between 62-63, and 104. The spelling of the name of the commander of Team Ibex is spelled variously as Demers or de Meis.

235. The author's father, John McCausland Chambers, U.S. Army Medical Corps, 1944-1946, served in Chungking for most of 1945.

236. Mills with Brunner, *OSS Special Operations in China*, 29.

237. Ibid., 38-40.

238. Ibid., 31.

239. Ibid., 46-47.

240. Ibid., 45.

241. Ibid., 49.

242. Ibid., 49-66.

243. Ibid., 49-66, 187-190, 197.

244. Paul Cyr quoted in Ibid.,193.

245. Ibid., 193-203; see also OSS, *War Report of the OSS; Overseas Targets*, 447.

246. Mattingly, *Herringbone Cloak—GI Dagger: Marines of the OSS*, 73-75.

247. Field Report of Capt. Walter R. Mansfield, 22 May 1945, reprinted in Mills with Brunner, *OSS Special Operations in China*, 321.

248. Ibid., 275-277.

249. OSS, *War Report of the OSS, Overseas Targets*, 446.

250. Mills with Brunner, *OSS Special Operations in China*, 346, which also recognizes Team Elephant as "the most effective SO unit in the field."

251. Ibid., 345-346.

252. Field Report of Capt. Walter C. Hanna, Jr., 25 September 1945, reprinted in Mills with Brunner, *OSS Special Operations in China*, 351-352.

253. Field Report of Capt. Walter C. Hanna, Jr., 25 September 1945, reprinted in Mills with Brunner, *OSS Special Operations in China*, 347; see also Hanna's bitter messages to headquarters, 27 May, 11 June, 23 June 1945, pages 348-356.

254. Field Report of Capt. Walter C. Hanna, Jr., 25 September 1945, reprinted in Mills with Brunner, *OSS Special Operations in China*, 358-359.

255. Mills with Brunner, *OSS Special Operations in China*, 359.

256. Lt. Cmdr. Richard Kelly, "Operation Dormouse," *Blue Book Magazine* (October 1945), 64, copy in OSS Records (RG 226), Entry 161, Box 7, Folder 7, National Archives II.

257. Quoted in Ibid., 64.

258. Ibid., 64, 66; "Team Dormouse," in Mills with Brunner, *OSS Special Operations in China*, 361-364.

259. Capt. Raymond Moore quoted in Lt. Cmdr. Richard Kelly, "Operation Dormouse," *Blue Book Magazine* (October 1945), 69-70, copy in OSS Records (RG 226), Entry 161, Box 7, Folder 7, National Archives II.

260. Lt. Cmdr. Richard Kelly, "Operation Dormouse," *Blue Book Magazine* (October 1945), 70, copy in OSS Records (RG 226), Entry 161, Box 7, Folder 7, National Archives II.

261. Ibid., 71-74.

262. Citation for Silver Star Medal for Capt. Raymond E. Moore, reprinted in Lt. Cmdr. Richard Kelly, "Operation Dormouse," *Blue Book Magazine* (October 1945), 76, copy in OSS Records (RG 226), Entry 161, Box 7, Folder 7, National Archives II.

263. Capt. Raymond E. Moore quoted in Lt. Cmdr. Richard Kelly, "Operation Dormouse," *Blue Book Magazine* (October 1945), 74, in OSS Records (RG 226), Entry 161, Box 7, Folder 7, National Archives II.

264. Capt. Raymond E. Moore, quoted in Kelly, "Operation Dormouse," 74.

265. Kelly, "Operation Dormouse," 75.

266. By the time the war ended, the number of Chinese being trained as commandos had grown to 3,000 and the American personnel with them had reached 390. The plan was for each commando unit to include 154 Chinese soldiers, 8 interpreters, and 19 Americans. Each had three rifle sections, a 60 mm mortar section, a light machine gun section, and a demolition section, plus an advance team of an SI officer and a radio operator who would precede the unit into the field and prepare for its arrival. "Chinese Operational Group[s]," http://www.ossog.org/china.html, accessed 23 September 2007. OSS, *War Report of the OSS, Overseas Targets*, 454-55.

267. For a list of American personnel in the Commando Units of the Chinese Operational Groups, see "Office of Strategic Services Operational Groups," "Personnel Chinese Operational Group," http://www.ossog.org/personnel.html. Accessed 23 September 2007.

268. Serge Obolensky, *One Man in His Time: The Memoirs of Serge Obolensky* (New York: McDowell, Obolensky, 1958), 347-349.

269. Operational Report, Company B, 267th Special Reconnaissance Battalion (Provisional), Narrative History of the French OGs, OSS Records (RG 226), Entry 99, Box 44, National Archives II.

270. Arthur P. Frizzell, Excerpt of Remarks at 1997 Chinese OG Reunion, http://www.ossog.org/china/blackberry_frizzell.html, accessed 23 June 2008.

271. Emmet F. McNamara, Excerpt of Remarks at 1997 Chinese OG Reunion, http://www.ossog.org/china/apple_mcnamara.html, accessed 23 June 2008. 272 Mattingly, *Herringbone Cloak—GI Dagger*, 176.

273. Ford, *Donovan of OSS*, 272-73.

274. [Ellsworth] Al Johnson, "One Small Part," typescript recollection, of France and China, http://www.ossog.org/france/patrick.html, accessed 21 June 2008. See also, OSS, *War Report of the OSS: Overseas Targets*, 454.

275. Col. John C. Hooker, Jr., (U.S. Army-Retired), "Biography," typescript, chapter for 1945. I am grateful to Colonel Hooker for sending me a copy of his manuscript. John C. Hooker, email to the author, 27 June 2008.

276. [Ellsworth] "Al" Johnson, "One Small Part," typescript recollection, of France and China, http://www.ossog.org/france/patrick.html, accessed 21 June 2008.

277. Ellsworth ("Al") Johnson, telephone interview with the author, 17 June 2008. I am indebted to Nancy Moseler, Mr. Johnson's daughter, for putting me in touch with her father and helping with his materials.

278. [Ellsworth] "Al" Johnson, "One Small Part," typescript recollection, of France and China, http://www.ossog.org/france/patrick.html, accessed 21 June 2008.

279. Ford, *Donovan of OSS*, 272-73; "Chinese Operational Group[s]," http://www.ossog.org/china.html. Accessed 23 September 2007.

280. Arthur P. Frizzell, Excerpt of Remarks at 1997 Chinese OG Reunion, http://www.ossog.org/china/blackberry_frizzell.html, accessed 23 June 2008.

281. "Chinese Operational Group, Commando Unit Apple, Summary of Lt. Col. Cox," http://www.ossog.org/china/apple.html. Accessed 23 September 2007; also OSS, *War Report of the OSS, Overseas Targets*, 456.

282. Emmett F. McNamara, telephone interview with the author, 2 September 2008.

283. John C. Hooker, email to the author, 3 July 2008. As an OG, he would have been trained at Area F and then Area B or A, but B was closed by that time.

284. Col. John C. Hooker, Jr., (U.S. Army-Retired), "Biography," typescript, chapter for 1945. I am grateful to Colonel Hooker for sending me a copy of his manuscript. John C. Hooker, email to the author, 27 June 2008. He provided further details in an email to the author of 3 July 2008.

285. "Chinese Operational Group, Commando Unit Blueberry, Summary of Lt. Col. Cox," http://www.ossog.org/china/blueberry.html. Accessed 23 September 2007; and OSS, *War Report of the OSS, Overseas Targets*, 456.

286. Rickerson quoted in [Ellsworth] "Al" Johnson, "One Small Part," typescript recollection, of France and China, http://www.ossog.org/france/patrick.html, accessed 21 June 2008.

287. Ellsworth ("Al") Johnson, telephone interview with the author, 27 June 2008.

288. [Ellsworth] "Al" Johnson, telephone interview with the author, 27 June 2008. He pronounced his name "Gal-ANT."

289. [Ellsworth] "Al" Johnson, "One Small Part," typescript recollection, of France and China, http://www.ossog.org/france/patrick.html, accessed 21 June 2008.

290. [Ellsworth] "Al" Johnson, "One Small Part," typescript recollection, of France and China, http://www.ossog.org/france/patrick.html, accessed 21 June 2008; and Cook quote in Remarks of Ellsworth ("Al") Johnson, Excerpts from 1997 Chinese OG Reunion [in New Hampshire]," http://www.ossog.org./china/chineseog_reunion_1997.html, accessed 23 September 2007.\

291. James E. Cook quoted in "Chinese Operational Group, Commando Unit Blueberry, Summary of Lt. Col. Cox," http://www.ossog.org/china/blueberry.html. Accessed 23 September 2007. For a brief, diplomatic summary of the engagement, see OSS, *War Report of the OSS, Overseas Targets*, 456-57.

292. [Ellsworth] "Al" Johnson, "One Small Part," typescript recollection, of France and China, http://www.ossog.org/france/patrick.html, accessed 21 June 2008.

293. Ellsworth ("Al") Johnson, telephone interview with the author, 27 June 2008.

294. Arne I. Herstad to "My dearest wife," 15 September 1945, *OSS Society Digest*, Number 2061, 25 May 2008, osssociety@yahoogroups.com, accessed 25 May 2008.

295. Col. John C. Hooker, Jr., (U.S. Army-Retired), "Biography," typescript, chapter for 1945; John C. Hooker, email to the author, 27 June 2008.

296. [Ellsworth] "Al" Johnson, "One Small Part," typescript recollection, of France and China, http://www.ossog.org/france/patrick.html, accessed 21 June 2008.

297. Yu, *OSS in China*, 236.

298. Ibid., 236-237.

299. Ibid., 15-17; and for a fuller background, Allan R. Millett, *The War for Korea, 1945-1950: A House Burning* (Lawrence, University Press of Kansas, 2005), 16-38.

300. On Rhee's first interview with Goodfellow in early 1942, see M. Preston Goodfellow, "Personal Experience Article," undated [after 1950] typescript in M. Preston Goodfellow Papers, Box 2, Folder "Biographical Material," Hoover Institution Archives, Stanford, Calif.; on the Chinese rejection, see Smith, *Shadow Warriors*, 130.

301. Moon and Eifler, *The Deadliest Colonel*, 48.

302. Bickham Sweet-Escott, Baker Street Irregular (London: Methuen & Co., 1965), 142. Sweet-Escott was an SOE representative in Washington.

303. See correspondence between Goodfellow and Syngman Rhee, 1943-1945, in M. Preston Goodfellow Papers, Box 4, Syngman Rhee, Subject file, Hoover Institution Archives, Stanford, Calif.

304. Peter F.M. Sichel, telephone interview with the author, 9 July 2008.

305. Moon and Eifler, *The Deadliest Colonel*, 48-49, 215-227, 232-33, 323.

306. Robert E. Carter, telephone interview with the author, 28 August 2008.

307. Joseph J. Tully, telephone interview with the author, 23 March 2005, who said in June and July 1945, there were about thirty to fifty Korean Army officers given two months training in special operations, intelligence, and radio operation at Area C-1 while he was there. R. Harris Smith, *OSS*, 26, says most of the Korean agents selected by Rhee and trained by the OSS were "never used."

Smith probably refers to their nonuse during the war. The widow of Maj. Howard Manning, the last commander of Area C, said that her husband told her that some of the Koreans he trained at Area C later served as high government officials in the Republic of Korea. Betty Bullard (Mrs. Howard) Manning, telephone interview with the author, 4 March 2005.

308. Yu, *OSS in China* (1996), 15-17, who cites "History of the SI Branch, Office of Strategic Services, China Theater," October 1945, OSS Records (RG 226), Entry 154, Folder 3333, National Archives II. The Soviet Union also trained some Korean expatriates during the 1940s in military tactics and Marxist-Leninist ideology for possible deployment in Korea at a later date. By 1945, the main leader of that group was a partisan named Kim Il Sung who would later become the head of the Communist regime in North Korea. Millett, *The War for Korea*, 39.

309. E. Howard Hunt, *Undercover: Memoirs of an American Secret Agent* (New York: G.P. Putnam's Sons, 1974), 32-45; Tim Weiner, "E. Howard Hunt, Agent Who Organized Botched Watergate Break-In, Dies at 88," *New York Times*, 24 January 2007, C13. Hunt's memoir has been likened to the novels he wrote.

310. Chester L. Cooper, "Remembering 109—Recollections of OSSers," *OSS Society Newsletter*, Summer 2005, 11; see also Chester L. Cooper's memoir, *In the Shadows of History: Fifty Years Behind the Scenes of Cold War Diplomacy* (Amherst, NY: Prometheus Books, 2005). Cooper later became a consummate government insider, serving in key positions with the CIA, the State Department, and the National Security Council. Joe Holley, Obituary, "Diplomatic Insider, Chester L. Cooper, *Washington Post*, 3 November 2005, B8.

311. Moon, *This Grim and Savage Game*, 233, 259-70, 295-96; Greg Schulte, "Were Five OSS Agents Hanged in Japan in 1944?," *OSS Society Digest*, Number 1251, 17 January 2006, osssociety@yahoogroups.com, accessed 17 January 2006; and Chong Lee, Korean Broadcasting System, Seoul, Korea, "Media Inquiry—Napko Project," *OSS Society Digest*, Number 1973, 6 February 2008, osssociety@yahoogroups.com, accessed 6 February 2008.

312. The agents allegedly parachuted in during a B-29 mission. With transmitters, weapons, rations, and currency, they soon began to report on shipping, particularly troop transports, which were then sunk by the Americans. Japanese search teams eventually caught five of the agents, took them to Sugamo Prison in Tokyo, interrogated and executed them in November 1944. The sixth agent was never caught and allegedly became part of the U.S. occupation forces after the war. The author provides no sources for this assertion, and there has been no collaboration of the account John L. Ginn, *Sugamo Prison, Tokyo: An Account of the Trial and Sentencing of Japanese War Criminals in 1948, by a U.S. Participant* (Jefferson, N.C.: McFarland & Co., 1992), 4; see also Greg Schulte, "Were Five OSS Agents Hanged in Japan in 1944?," *OSS Society Digest*, Number 1251, 17 January 2006, osssociety@yahoogroups.com.

313. Moon and Eifler, *The Deadliest Colonel*, 232-33, 323.

314. Howard Chappell, remarks at reunion of OSS members in Washington, D.C., quoted in Donnie Radcliffe, "Intelligence Gathering, Bush Honored at OSS Reunion," *Washington Post*, 24 October 1991, C1, C13; Albert R. Materazzi, "Training at Catalina," *OSS Society Digest*, Number 991, 31 March 2005, osssociet@yahoogroups.com, accesses 31 March 2005. His son indicated that he believed that his father at some period had trained on Catalina Island. Jack R. Chappell, email to the author, 5 October 2007. Robert E. Carter, in charge of two camps on Catalina Island where Eifler's Koreans were being trained, said he never saw Chappell on the island in 1945, but he added that he and Chappell and the German OGs had trained there in 1944 after being at Areas F and B. Robert E. Carter, telephone interview with the author, 28 August 2008.

315. Yu, *OSS in China*, 230, citing William J. Donovan to Albert C. Wedemeyer, 10 August 1945, OSS Records (RG 226), Entry 154, Box 192, Folder 3285 "OSS Wash/ Donovan Trip, August 1945," National Archives II.

316. Yu, *OSS in China*, 231, citing Richard Heppner to R. Davis Halliwell, 10 August 1945, Urgent cable, OSS Records (RG 226), Entry 154, Box 192, Folder 3285, "OSS Wash/Donovan Trip, August 1945," National Archives II.

317. Yu, *OSS in China*, 231-32, citing Richard Heppner to W.P. Davis, 10 August 1945, Top Secret Cable, OSS Records (RG 226), Entry 90, Box 3, Folder 30, "Jap Surrender—in and out, Aug. 1945," National Archives II.

318. Yu, *OSS in China*, 232.

319. Mills with Brunner, *OSS Special Operations in China*, 391-393.

320. There was concern everywhere about whether Japanese soldiers would react to the surrender with revenge on their captives or simply withdraw and leave the prisoners and internees to perish without food or medical supplies. There were some 20,000 American and other Allied POWs and some 15,000 internees held in Japanese-run camps throughout Asia and the Pacific in late summer of 1945. Ford, *Donovan of OSS*, 297.

321. William Craig, *The Fall of Japan* (New York: Dial Press, 1967); E. Barlett Kerr, *Surrender and Survival: The Experience of American POWs in the Pacific, 1941-1945* (New York: William Morrow, 1985); and Gavin Daws, *Prisoners of the Japanese* (New York: William Morrow, 1995);

322. Spector, *In the Ruins of Empire*, 8-9.

323. OSS, *War Report of the OSS, Overseas Targets*, 457-58.

324. Ford, *Donovan of OSS*, 298. Although OSS was able to rescue some of the Doolittle fliers, the Japanese had after the 1942 raid executed four of the fliers and deliberately starved another to death.

325. Frank A. Gleason, telephone interview with the author, 15 July 2008.

326. Spector, *In the Ruins of Empire*, 16, citing an OWI News Dispatch, attached to Gustav Krause to Richard Heppner, 22 Aug. 1945.

327. OSS, *War Report of the OSS, Overseas Targets*, 458; and John K. Singlaub, *Hazardous Duty: An American Soldier in the Twentieth Century* (New York: Summit/ Simon & Schuster, 1991), 71-101.

328. Singlaub with McConnell, *Hazardous Duty*, 89.

329. Ibid., 91; for a detailed account of the treatment of the Allied POWs on Hainan Island, see Courtney T. Harrison, *Ambon, Island of Mist, 2/21st Battalion AIF (Gull Force), Prisoners of War, 1941-45* (North Geelong, Australia: T.W. and C.T. Harrison, 1988), 186-259.

330. Singlaub with McConnell, *Hazardous Duty*, 100-101; see also Lawrence J. Hickey, *Warpath across the Pacific: The Illustrated History of the 345th Bombardment Group during World War II*, 2nd ed., rev. (Boulder, Colo.: International Research and Publishing Co., 1982), 289-300.

331. Yu, *OSS in China*, 242, citing Richard Heppner to Gustav Krause, for information John Magruder and Whitney Shephardson, 10 August 1945, cable, OSS Records (RG 226), Entry 90, Box 3, Folder 30, "Jap Surrender—in and out, August 1945," National Archives II.

332. Hilsman, *American Guerrilla*, 232. Hilsman arrived on a second plane two days later.

333. Yu, *OSS in China*, 243; quoting a letter from Gustav Krause to Betty MacDonald [later MacIntosh], 6 November 1946. Colonel Krause was head of the OSS base in Hsian [Xian], where Wainwright was brought from Manchuria. See also Hal Leith, *POWs of Japanese Rescued!: General J.M. Wainwright* (Trafford Publishing, 2004). Leith, then a sergeant, was a member of the rescue team.

334. Hilsman, *American Guerrilla*, 234. In 1990, Hilsman wrote that his father recognized him right away and walked over an embraced him. This is in sharp contrast to the account in Corey Ford, *Donovan of OSS*, 300, who reported that young Hilsman found his father lying on a wooden cot amidst filth and suffering, and looking up at him blankly, and to convince his father that he was not delirious, Hilsman began recounting his own experiences in Burma and said his Kachin guerrillas had killed 300 Japanese without one casualty. "When Dad sat up and contradicted me," Ford has young Hilsman saying, "I knew he was all right."

335. Spector, *In the Ruins of Empire*, 20. Having rescued the Allied POWs, the OSS began covert operations to establish a intelligence network in Manchuria, parachuting in a large number of agents, many of them Chinese members of Bishop Megan's Catholic network. OSS reported the

secret entrance of thousands of Chinese Communist soldiers into Mukden on 7 September, the capture of Henry Pu Yi, the puppet head of Manchukuo and the last emperor of China, as well as the Soviet seizure of not just Japanese but also western properties such as the British-American Tobacco Company. The Soviets barely accepted the American rescue mission in Manchuria, which they saw as their sphere of influence occupied by the Red Army, but when the POWs and internees had been evacuated by mid-September, the Soviet's commander in Mukden ordered the OSS team out of Manchuria by 5 October, an order they followed under protest. Yu, *OSS in China*, 242-47.

336. Figures for December 1943 and February 1945 from OSS, War Report of the OSS: Overseas Targets, 442-443; for April 1945 from OSS Theater Reports, S&T Branch Excerpts, China, 30 April 1945, p. 13, OSS Records (RG 226), Entry 136, Box 159, Folder 1729, National Archives II; for August 1945, General Orders No. 27, Headquarters, U.S. Forces, China Theater, by order of General Wedemeyer, 1 February 1946, reprinted General Orders No. 27, Headquarters, U.S. Forces, China Theater, by order of General Wedemeyer, 1 February 1946, reprinted in Mills with Brunner, *OSS Special Operations in China*, 432-433.

337. John W. Brunner, telephone interview with the author, 21 March 2005.

338. Maj. Leonard Clark, a noted explorer and author of a book about Hainan Island before the war, was assigned to Secret Intelligence in China. In October 1945, after returning to Canton from an intelligence mission in Taiwan, he became involved in an argument with two other OSS officers and killed them with his pistol. Clark was imprisoned by Chinese authorities and only released in 1949 with Donovan's personal intervention. Ever an adventurer, Clark resumed his explorations looking for gold in Nicaragua, a mountain higher than Everest in Nepal, and a treasure in the Amazon where his travels came to an end with two native arrows in his back. Mills with Brunner, *OSS Special Operations in China*, 18-19.

339. Mills with Brunner, *OSS Special Operations in China*, 15, 426.

340. Mills with Brunner, *OSS Special Operations in China*, 409-410, 426, 453, 473, 484-486; the two others who died fighting the Japanese were Specialist First Class Dean A. Cline, U.S. Navy Reserve; and a Flight Officer Evans, U.S. Army Air Force. David Kenney, an OSS Communications man, was at Chihkiang when the falling out of the plane incident occurred. He remembered that one of the men, about 30 years old, had become so excited that the Air Force would let him go along on the mission. David Kenney, telephone interview with the author, 11 April 2005.

341. Yu, *OSS in China*, 235-36; Smith, *OSS*, 281-82.

342. The Americans included Captain Birch, Lt. Laird M. Ogle, Morale Operations; Sgt. Albert C. Meyes, Communications Branch, and civilian Albert C. Grimes, Counter-Espionage Branch, reported in H. Ben Smith, chief, intelligence division, Strategic Services Unit [successor of the OSS], China Theater, to Commanding General, AAF/CT, 7 November 1945, subject: Additional Report on Death of Capt. John Birch, Alfred Wedemeyer Papers, Box 87, Folder 87.2, the Hoover Institution, Stanford, Calif.

343. Details on John Birch's murder from Yu, *OSS in China*, 235-40; Aldrich, *Intelligence and the War against Japan*, 372; Smith, *OSS*, 280-81; Testimony of Tung Chin-sheng, a.k.a. Tung Fu Kuan, Entry 148, Box 16, Folder 225; OSS Investigation report, "Account of the Death of Captain John Birch," 14 September 1945, by John S. Thomson to Headquarters Central Command, OSS, Entry 148, Box 6, Folder 87; William Miller's report to General Alfred Wedemeyer, 1 September 1945, Entry 168, Box 16, Folder 225, "Death of Capt. John Birch," all in OSS Records (RG 226), National Archives II. Lt. Col. Jeremiah J. O'Connor, Deputy Theater Commander Judge Advocate, China Theater, to Chief of Staff, 13 November 1945, subject: Death of Captain John Birch, copy in Alfred Wedemeyer Papers, Box 87, Folder 87.2, the Hoover Institution, Stanford, Calif.

344. Yu, *OSS in China*, 240-41.

345. Years later, an ultra-conservative California businessman, Robert H. W. Welch, Jr., formed a right-wing political organization in the United States and seized upon Birch's slaying by the Chinese Communists for the name of his group. He called it the John Birch Society, and in 1954, Welch published a book, *The Life of John Birch: In the Story of One American Boy, the Ordeal of His*

Age (Los Angeles, Calif.: Western Islands Press, 1954), which although ostensibly a biography is primarily a political diatribe. There is no evidence, however, that John Birch himself had such an extremist anti-Communist ideology as Robert Welch enunciated. Yu, *OSS in China*, 235; Ford, *Donovan of OSS*, 300, Smith, *OSS*, 280-81.

346. OSS, *War Report of the OSS, Overseas Targets*, 391-392.

347. OSS Schools and Training Branch, "Office of Strategic Services (OSS): Organization and Functions," June 1945, p. 31, in a 37-page booklet designed to supplement lectures on OSS organization; OSS Records (RG 226), Entry 141, Box 4, Folder 36, National Archives II.

348. OSS Schools and Training Branch, "Office of Strategic Services (OSS): Organization and Functions," June 1945, p. 31, in a 37-page booklet designed to supplement lectures on OSS organization; OSS Records (RG 226), Entry 141, Box 4, Folder 36, National Archives II.

349. Totals, 2,000 and 12,348 are from General Orders No. 27, Headquarters, U.S. Forces, China Theater, by order of General Wedemeyer, 1 February 1946, reprinted in Mills with Brunner, *OSS Special Operations in China*, 431-432; see also subtotal figures by branch and area in OSS, *War Report of the OSS: Overseas Targets*, 440-447.

350. General Orders No. 27, Headquarters, U.S. Forces, China Theater, by order of General Wedemeyer, 1 February 1946, reprinted in Mills with Brunner, *OSS Special Operations in China*, 432-433.

351. [Ellsworth] "Al" Johnson, "One Small Part," typescript recollection, of France and China, p.70, http://www.ossog.org/france/patrick.html, accessed 21 June 2008.

CHAPTER 10

1. Michael Warner, *The Office of Strategic Services: America's First Intelligence Agency* (Washington, D.C.: Central Intelligence Agency, 2002), 42; Anthony Cave Brown, *The Last Hero: Wild Bill Donovan* (New York: Times Books, 1982), 508-515; R. Harris Smith, *OSS: The Secret History of America's First Central Intelligence Agency* (Berkeley: University of California Press, 1972), 5; Corey Ford, *Donovan of OSS* (Boston: Little, Brown, 1970), 239.

2. Edwin J. Putzell, deputy and subsequent successor to Otto C. Doering, Jr., executive officer of the OSS, recalled an office rebellion by a group of four or five of the OSS's top administrators who sent a bluntly worded memo to Donovan in early 1944 "suggesting that he was a poor administrator and that the agency needed better top leadership in its day-to-day operation" and calling for major reorganization. Donovan felt betrayed and rejected their proposal. Edwin Putzell, interviewed by Tim Naftali, 11 April 1997, pp. 2, 15, OSS Oral History Transcripts, CIA Records (RG 263), Box 3, National Archives II, College Park, Md.

3. William J. Donovan quoted in Robert Alcorn, *No Banners, No Bands* (New York: Donald McKay, 1965), 182; Tom Moon, *This Grim and Savage Game: OSS and the Beginning of U.S. Covert Operations in World War II* (New York: Da Capo press, 2000), 249.

4. On Donovan's 18 November 1944 plan for a permanent central intelligence agency, see Richard Dunlop, *Donovan: America's Master Spy* (Chicago: Rand McNally, 1982), 458-459; Smith, *OSS: The Secret History*, 363, both of whom claim Roosevelt had requested it. But the much more detailed account in Thomas F. Troy, *Donovan and the CIA: A History of the Establishment of the Central Intelligence Agency* (Frederick, Md.: University Publications of America, 1981), 218-222, indicates that the impetus came from Donovan rather than the President. Donovan's first recorded statement of his goal of a permanent OSS is in a question and answer period following a talk to an audience of Army officers in Washington in May 1943; he later put such a recommendation in a memorandum to General Eisenhower on 17 September 1943.

5. William J. Donovan, Memorandum for the President, 18 November 1944, and draft Executive Order for "Substantive Authority Necessary in Establishment of a Central Intelligence Service," Donovan Papers, "OSS Reports to the White House, Nov.-Dec. 1944," reprinted as Appendix M in Troy, *Donovan and the CIA*, 445-47.

6. Ibid., plus Troy, *Donovan and the CIA*, 231-34. The only restrictions Donovan put on his proposed intelligence organization were that it would be prohibited from any police or law enforcement functions at home or abroad and that in wartime, the agency and its members would be coordinated or controlled by Joint Chiefs of Staff or by the military Theater Commander where they operated.

7. William J. Donovan, Memorandum for the President, 18 November 1944, Donovan Papers, "OSS Reports to the White House, Nov.-Dec. 1944," reprinted as Appendix M in Troy, *Donovan and the CIA*, 445-47.

8. Walter Trohan, "Donovan Proposes Super-spy System for Postwar New Deal: Would Take over FBI, Secret Service, ONI and G-2 to Watch Home, Abroad," *Washington Times-Herald*, 9 February 1945, 1.

9. "Roosevelt Plans Post-War Global Secret Service; Donovan Maps New Agency to Keep U.S. Alert to Threat of a New War," *New York Herald-Tribune*, 10 February 1945; see also *New York Times*, 12 February 1945; *Chicago Sun*, 15 February 1945, and *Washington Post*, 16 February 1945, in Troy, *Donovan and the CIA*, 256-57.

10. Donovan was furious with the leak and slanderous attack. The OSS traced the leak of one of the "secret" numbered copies of the memorandum to the Office of Naval Intelligence and the FBI. Troy, *Donovan and the CIA*, 258. OSS executive officer, Otto C. Doering, told Donovan that "J. Edgar Hoover had personally handed the memorandum to Trohan." Dunlop, *Donovan: America's Master*

Spy, 464, who received the story from Doering himself. Dunlop added that other sources implicated Army G-2 in the leak. This was the traditional interpretation when CIA historian Thomas F. Troy originally completed his classified, manuscript history of Donovan, the OSS, and the origins of the CIA in 1975, but afterwards, when Troy finally was able to get in touch with Walter Trohan in the late 1970s, Trohan told me that the documents had been leaked to him by none other than Steve early, Roosevelt's executive secretary, who told Trohan, that "FDR wanted the story out." The truth of such an allegation and the purposes that Roosevelt would have had if true remain matters of speculation. Troy, *Donovan and the CIA*, vi, 255-60.

11. "F.D.R." to Major General Donovan, Memorandum, 5 April 1945; Donovan immediately did so, see William J. Donovan to Secretary of War, et al., Memorandum, 6 April 1945, both in CIA Records (RG 263), Thomas Troy Files, Box 6, Folder 46, National Archives II.

12. Harry S Truman had no middle name. His parents simply gave him a middle initial; thus there is no period after it.

13. Harry S Truman, *Memoirs*, 2 vols. (Garden City, N.Y.: Doubleday, 1955), 1:58; see also Troy, *Donovan and the CIA*, 266-267.

14. Ibid., 270-271.

15. Margaret Truman, *Harry S. Truman* (New York; William Morrow, 1973), 250, quoting from her father's appointment book.

16. Dunlop, *Donovan*, 468.

17. Edwin J. Putzell, executive officer of the OSS, interviewed by Tim Naftali, 11 April 1997, p. 74, 76; OSS Oral History Transcripts, CIA Records (RG 263), Box 3, National Archives II.

18. Richard Dunlop, a former OSS officer selected and aided by Donovan in writing the general's biography, contended, in what may have been Donovan's own view, that "as for Truman, it was not at all definite that he wanted a central intelligence agency, and certainly not one with Bill Donovan as director. Donovan was among other things a Republican, and a dangerous one at that, who might still become a candidate for the presidency." Dunlop, *Donovan*, 467. Anthony Cave Brown, *The Last Hero*, 790-92, cited another possible personal reason for Truman's antipathy toward Donovan, claiming that in World War I, Lt. Col. Donovan complained about Capt. Harry S Truman's Missouri National Guard artillery battery for alleged inadequate support of the attack by Donovan's unit. A similar story circulated within the OSS. Rafael Hirtz, OSS SO officer in France and China, interview posted 12 February 2004, on the website of the Library of Congress, Veterans History Project, accessed 10 January 2007.

19. On the Park report, see Troy, *Donovan and the CIA*, 282; Christopher Andrew, *For the President's Eyes Only: Secret Intelligence and the American Presidency from Washington to Bush* (New York: Harper/Collins, 1995), 156; and Michael Warner, "The Creation of the Central Intelligence Group," *Studies in Intelligence: Journal of the American Intelligence Professional*, 39/5 (Fall 1995), unclassified and published in expanded form in 1996, and reprinted in *Studies in Intelligence, Office of Strategic Services, 60th Anniversary, Special Edition* (Spring 2002): 170-71.

20. Col. Richard Park, Jr., GSC, Memorandum for the President, n.d., "Top Secret," in the files of President Harry S. Truman's secretary, Rose A. Conway, Box 15, "OSS/Donovan" folder, Harry S. Truman Papers, Truman Library, Independence, Mo. There is a copy of the Park Report in CIA Records (RG 263), Thomas Troy Files, Box 6, Folder 44, National Archives II. The present author consulted the copy in the National Archives II. Although the report is undated, Park apparently completed it in mid-March 1945, before Roosevelt's death, because he showed it to Maj. Gen. Clayton Bissell, GSC, Acting Chief of Staff of Army G-2, on 12 March 1945. Maj. Gen. Clayton Bissell, Memorandum for the Record, 12 March 1945, subject: Colonel Park's Comments on OSS, Military Intelligence Division Records (RG 165), WFRC, Folder 334, Officer of Strategic Services, copy attached to the Park Report in the Thomas Troy Files, cited above.

21. Col. Richard Park, Jr., GSC, Memorandum for the President, n.d., "Top Secret," Part II, p. 1, CIA Records (RG 263), Thomas Troy Files, Box 6, Folder 44, National Archives II.

22. On the communists and OSS issue, see, for example, Patrick K. O'Donnell, *Operatives, Spies, and Saboteurs: The Unknown Story of the Men and Women of World War II's OSS* (New York: Free Press, 2004), 310-11; Robin W. Winks, *Cloak & Gown: Scholars in the Secret War, 1939-1961*, 2nd ed. (New Haven, Conn.: Yale University Press, 1987), 356-59; Charles T. Pinck, President, OSS Society, "Communists at the OSS," letter to the editor, *Washington Post*, 8 January 2008, rejoinder to column by Robert D. Novak, Subverting Bush at Langley," in ibid., 24 December 2007, and dismissing Novak's assertion that the OSS was "infiltrated by communists."

23. Bickham Sweet-Escott, *Baker Street Irregular* [SOE headquarters in London was on Baker Street] (London: Methuen, 1965), 155.

24. Col. Richard Park, Jr., GSC, Memorandum for the President, n.d., "Top Secret," Part II, p. 2, CIA Records (RG 263), Thomas Troy Files, Box 6, Folder 44, National Archives II. Park, of course, was working fast and loose with figures as well as innuendos. OSS records show that regular funds were closely monitored, but the unvouchered funds posed a more difficult problem, particularly as they were distributed to gain information and cooperation from anonymous indigenous informants and guerrillas. Edwin J. Putzell, executive officer of the OSS, recalled in speaking of W. Lane Rehm, former financier and OSS chief financial officer, "Well, you know, the use of money and gold was one of the big problems, always, when you are in this kind of operation. Rehm, being a conservative type and fiscally oriented and here we were using funds for which we didn't have to make an accounting bothered him a good bit. So there was pulling and hauling as to how much and whether to do it." Putzell interviewed by Tim Naftali, 11 April 1997, p. 74, 76; OSS Oral History Transcripts, CIA Records (RG 263), Box 3, National Archives II.

25. To identify just a few of the inaccuracies in the Park Report: Park reported (Appendix I, p. 15) that in China "it is generally known that a tie-up has been made by the O.S.S. with Chiang Kai-shek's own intelligence and are integrated with them as they are with the British." In fact, OSS and Chiang's intelligence service under Dai-Li were rivals and highly suspicious of each other, and their cooperative arrangement, forced upon them by the U.S. Navy mission in China, was evaded by each. Park asserted that "Donovan tries to control completely every activity" (ibid., p. 20). In fact, Donovan was anything but a micromanager. Park sought to discredit Donovan and the OSS by indicating that many appointments came from the Social Register. "The social director of the St. Regis Hotel, where General Donovan resides when in New York, attained the rank of Lt. Colonel in the O.S.S." (ibid., p. 31). This half-truth avoids mentioning that Serge Obolensky had been an officer in the Czarist Army and fought guerrilla war against the Red Army, had training at a British commando school, was an officer in the New York National Guard, attended OSS training camps, helped write the OSS manual for Operational Groups, and conducted daring and successful OSS missions into Sardinia and France. Park contended (ibid., p. 39), that "A completely reliable source stated that O.S.S. intends to represent to the American public that it sends its own members across the lines into European enemy territory although this is not actually the case since, due to many blunders committed by that service, SHAEF, about May 1944, instructed the O.S.S. Espionage Section to refrain from dispatching any more agents into enemy territory. The exclusive right to dispatch such agents was given to the British." Virtually nothing in that assertion was true. OSS had been praised by General Eisenhower, the head of SHAEF, and it continued to send espionage agents behind enemy lines in Western Europe and in Germany itself until the spring of 1945.

26. Col. Richard Park, Jr., GSC, Memorandum for the President, n.d., "Top Secret," Part II, p. 2; Appendix I, pp. 36-38, CIA Records (RG 263), Thomas Troy Files, Box 6, Folder 44, National Archives II.

27. Ibid., Part II, p. 2; Appendix I, pp. 27-28. Park also made the extraordinary claim that an unnamed businessman said he was told by a member of the OSS at a dinner party in June 1943, "that the O.S.S. planned to enter the field of domestic investigation and would be sort of a U.S. Gestapo with power to penetrate every government agency, trade union, large corporation, etc." Again, untrue.

28. Troy *Donovan and the CIA*, 282 lines up similar sections of the Trohan and Park accounts side by side to confirm their virtually identical nature.

29. Col. Richard Park, Jr., GSC, Memorandum for the President, n.d., "Top Secret," Part III, p. 1; Appendix III, pp. 1-2; CIA Records (RG 263), Thomas Troy Files, Box 6, Folder 44, National Archives II.

30. See, for example, Bradley F. Smith, *The Ultra-Magic Deals and the Most Secret Special Relationship* [between the United States and Great Britain] (Novato, Calif.: Presidio Press, 1993).

31. Troy, *Donovan and the CIA*, 283-284. The OSS request for FY 1946 for $45 million was cut to $24 million; in contrast the FBI request for $49 million was only slightly reduced to $43 million.

32. Troy, *Donovan and the CIA*, 284.

33. Brig. Gen. John Magruder to William J. Donovan, memorandum, 2 May 1945, and Donovan to Secretary of War Henry L. Stimson, 16 May 1945, both cited in Troy, *Donovan and the CIA*, 270-71.

34. William J. Donovan to Samuel I. Rosenman, accompanying a copy of Donovan to Budget Director Harold Smith, 25 August 1945, quoted in OSS, *War Report of the OSS*, 119. This impolitic letter seems to have been ignored by Donovan's sympathetic biographers.

35. Roger Hall, *You're Stepping on My Cloak and Dagger* (New York: Norton, 1957), 216-217; Hall, who had been a Special Operations instructor at Areas F and B and later parachuted into France and Norway, returned to the United States in summer 1945 and worked on this project under John M. ("Deadline Johnny") Shaheen, a former publicity man from Chicago, who headed the new Reports Declassification Section.

36. Troy, *Donovan and the CIA*, 284, 291-292. Although the OSS publicity office ceased to exist after 1945, some celebratory stories and films about the exploits of Donovan's organization continued to appear for several years. With the active cooperation of former OSS personnel, Hollywood produced three major feature films involving the OSS between 1946 and 1947. Paramount's 1946 thriller, *O.S.S.*, starred Alan Ladd. Republic's *Cloak and Dagger*, directed the same year by Fritz Lang, featured Gary Cooper. In 1947, James Cagney starred in 20th Century Fox's *13 Rue Madeleine*, a film about an OSS agent sacrificing his own life to avoid revealing the identities of other Allied agents or leaders of the French Resistance.

37. The key decision, apparently on 27 or 28 August 1945, was made by Truman or his main advisory committee on demobilization and reconversion, a committee composed of Budget Director Harold D. Smith, Special Counsel Samuel Rosenman, and Director of War Mobilization John W. Snyder. OSS learned about the decision on 29 August 1945. Warner, "Creation of the Central Intelligence Group," 171.

38. Harold Smith, Diary entry, 13 September 1945, quoted in Troy, *Donovan and the CIA*, 296.

39. Executive Order 9621, 20 September 1945, "Termination of the Office of Strategic Services and Disposition of Its Functions," is reprinted in Troy, *Donovan and the CIA*, Appendix S, 461-62.

40. Harry S. Truman to William J. Donovan, 20 September 1945, in U.S. President, *Papers of the Presidents of the United States. Harry S. Truman. 1945* (Washington, D.C.: Government Printing Office, 1961), 330. Neither President Truman nor Budget Director Smith wanted to confront Donovan directly with the news that would be so disappointing to him. Truman told Smith to do it, but Smith sent his assistant, Donald Stone, to deliver the document to Donovan at OSS headquarters. Stone later reported that "when I delivered the document, Donovan took it with a kind of stoic grace. He knew it was coming, but he gave no outward indication of the personal hurt he felt by the manner in which he was informed." Richard Dunlop, *Donovan: America's Master Spy* (Chicago: Rand McNally, 1982), 473.

41. Harold Smith, Diary Entry, 20 September 1945, Smith Papers, "Conferences with President Truman, 1945," in Troy, *Donovan and the CIA*, 302.

42. In an apparent oversight, the original termination date of 1 October 1945, which had been inserted in a earlier draft of the Executive Order, was not altered when the final version of the Executive Order was prepared for the President's signature.

43. Donovan, his secretary, and a handful of people, with the help of Ray Kellogg from the Photographic Unit, microfilmed the records of the director's office to preserve them. Edwin J. Putzell, deputy and subsequent successor to Otto C. Doering, Jr., executive officer of the OSS, interviewed by Tim Naftali, 11 April 1997, pp. 72-73; OSS Oral History Transcripts, CIA Records (RG 263), Box 3, National Archives II.

44. An abbreviated text of Donovan's farewell speech of 28 September 1945 is reprinted in Ford, *Donovan of OSS*, 344, Appendix F. A fuller version is in Dunlop, *Donovan*, 473-474.

45. "General Donovan Welcomes Holdees, Sends a Message," *Attention Please* [Area F newspaper, 1944-1945] 9 April 1945, 1; OSS Records (RG 226), Entry 85, Box 27, Folder 449, National Archives II.

46. Donovan's directive is cited in A. van Buren, Security Officer, to Deputy Director, Schools and Training Branch, 23 December 1944, subject: Security Policy at Area F [under the director's new mission for Area F], OSS Records (RG 226), Entry 146, Box 223, Folder 3105, National Archives II.

47. Roger L. Belanger, letter to the author, 4 February 2005, and telephone interview with the author, 11 April 2005.

48. A. van Beuren, chief of security branch, to Charles S. Cheston, Assistant Director, 1 June 1945, subject: Area Recreation, OSS Records (RG 226), Entry 146, Box 223, Folder 3104, National Archives II.

49. Staff Sergeant E.D., "Overseas Veterans Amazed at Transformation of Area F," "Meet the Staff," "Dance Tomorrow in Main Lounge," *Attention Please*, 19 February 1944, 1, 4; OSS Records (RG 226), Entry 85, Box 27, Folder 449, National Archives II.

50. "Softball Season Starts; Area Team Sensational in League," *Attention Please*, 31 May 1944; "Defeat Area 'C' 4-2," *Attention Please*, 9 July 1945, 3; "Holdees Win Ball Game; Beat Sta. Complement; Area F Victor in League," *Attention Please*, 19 July 1945, 1; "Area F Team Lags in OSS Softball League," *Attention Please*, 31 August 1945, 1, 3; OSS Records (RG 226), Entry 85, Box 27, Folder 449, National Archives II.

51. Pfc. Vernon Taylor, "New Program Set Up at Area; Tours and Postwar Info Stressed," *Attention Please*, 31 August 1945, 1; OSS Records (RG 226), Entry 85, Box 27, Folder 449, National Archives II.

52. David Kenney, telephone interview with the author, 11 April 2005.

53. Spiro ("Gus") Cappony, telephone interview with the author, 16 September 2006.

54. Arthur ("Art") Reinhardt, interview with the author following a tour arranged by the author of Prince William Forest Park, Triangle, Va., 22 February 2007.

55. Smith, *OSS*, 238-39. The prosecution team also included former OSS General Counsel James B. Donovan (no relation to "Wild Bill" Donovan); James Donovan, *Challenges* (New York: Atheneum, 1967).

56. Dunlop, *Donovan*, 483.

57. Telford Taylor, *The Anatomy of the Nuremberg Trials: A Personal Memoir* (New York: Knopf, 1992), 148.

58. Dunlop, *Donovan*, 484.

59. Troy, *Donovan and the CIA*, 303.

60. Kai Bird, *The Chairman: John J. McCloy, the Making of the American Establishment* (New York: Simon and Schuster, 1992), 129-30.

61. William W. Quinn, *Buffalo Bill Remembers: Truth and Courage* (Fowlerville, Mich.: Wilderness Adventure Books, 1991), 240.

62. Richard Helms with William Hood, *A Look Over My Shoulder: A Life in the Central Intelligence Agency* (New York: Random House, 2003), 72.

63. Dunlop, *Donovan*, 485.

64. Michael Warner, "The Creation of the Central Intelligence Group," *Studies in Intelligence: Journal of the American Intelligence Professional,*, 39/5 (Fall 1995), unclassified and published in expanded form in1996, and reprinted in *Studies in Intelligence, Office of Strategic Services, 60th Anniversary, Special Edition* (Spring 2002): 169-78.

65. Dunlop, *Donovan*, 486.

66. Warner, "The Creation of the Central Intelligence Group," 174-78.

67. Troy, *Donovan and the CIA*, 377, 382-384, 402-410; Dunlop, *Donovan*, 491.

68. Troy, *Donovan and the CIA*, 277-410, and Appendices W, X, Y, 467-72; Arthur Darling, *The Central Intelligence Agency: An Instrument of Government to 1950* (University Park: Pennsylvania State University Press, 1990).

69. Peter M.F. Sichel, telephone interview with the author, 9 July 2008.

70. For histories of CIA's covert operations and the controversies they entailed, see, for example, Rhodri Jeffreys-Jones, *Cloak and Dollar: A History of American Secret Intelligence*, 2d ed. (New Haven, Conn.: Yale University Press, 2003), 255-88; Charles D. Ameringer, *U.S. Foreign Intelligence: The Secret Side of American History* (Lexington, MA: D.C. Heath, 1990), 201-405; and the report of a Senate committee chaired by Senator Frank Church (Dem.-Idaho), U.S. Congress, Senate Select Committee to Study Government Operations with Respect to Intelligence Activities, *Final Report*, 94th Cong. 2nd Sess., Senate Report No. 94-755, 6 vols. (Washington, D.C.: Govt. Printing Office, 1976).

71. Ibid., 130-131; Evan Thomas, *The Very Best Men: Four Who Dared: The Early Years of the CIA* (New York: Simon & Schuster, 1995), 28-30; Wiener, *Legacy of Ashes*, 32-34.

72. Caesar J. Civitella, telephone interview with the author, 25 April 2008.

73. Robert Harris Smith, *O.S.S.: The Secret History of America's First Central Intelligence Agency* (Berkeley: University of California Press, 1972), 361-62.

74. A point made by many commentators, but in regard to the CIA leadership, see in particular, Evan Thomas, *The Very Best Men: Four Who Dared: The Early Years of the CIA* (New York: Simon & Schuster, 1995), 10-12, 340-341, and passim. The four officials in the CIA's Directorate of Plans, its operational arm, who Thomas wrote about were Frank Wisner, Desmond FitzGerald, Tracy Barnes, and Richard Bissell. Thomas notes that Bissell, who was in no small measure responsible for the Marshall Plan and the U-2 spy plane, was also responsible in large part for the assassination plots against Fidel Castro and the abortive invasion at the Bay of Pigs.

75. As evident by comments from some OSS veterans (other OSS veterans disagree) on contemporary actions and issues in the OSS Society's electronic bulletin board and chat room on the internet, osssociety@yahoogroups.com. This site is open to OSS veterans, their families and descendants and to other persons interested in the OSS, but applicants must be cleared to have access. The author has had access since he began his research on this subject in 2005. See also the public writings of some former OSSers, for example, John K. Singlaub with Malcolm McConnell, *Hazardous Duty: An American Soldier in the Twentieth Century* (New York: Summit/Simon & Schuster, 1991), 445-46, 525-26.

76. Smith, *O.S.S.*, 363, 375, 381-383.

77. Rafael D. Hitz, Transcript of Interview, n.d. [1980s], Rafael Hirtz Collection (AFC/2001/001/94), Veterans History Project, American Folklife Center, Library of Congress, http://lcweb2.loc.gov/diglib/vhp/bib/loc.natlib.afc2001001.00094, accessed 10 January 2007.

78. Peter M.F. Sichel, telephone interview with the author, 9 July 2008.

79. As evident by comments by some OSS veterans on contemporary actions and issues in the OSS Society's electronic bulletin board and chat room on the internet, osssociety@yahoogroups.com.

80. Smith, *O.S.S.*, 381-383. After what he called "a very brief, uneventful, and undistinguished association with the most misunderstood bureaucracy of the American government," Smith left

the CIA in May 1968 and wrote this history of the OSS primarily with the cooperation of some 200 OSS and State Department veterans who provided recollections and recent observations.

81. Tim Weiner, *Legacy of Ashes: The History of the CIA* (New York: Doubleday, 2007), 513-514. In 2007, as a result of the terrorist attacks on the United States on 11 September 2001, Congress gave oversight over all 15 agencies in the U.S. intelligence community to a newly-created position, the Director of National Intelligence, who would report directly to the President. Ibid., 504-12.

82. The U.S. Army's official lineage for the Special Forces omits the OSS as a direct predecessor because it was a civilian organization and, therefore, outside of the official chain of command. Officially recognized as predecessor units are Airborne Divisions, Ranger Units, and Joint U.S.-Canadian 1st Special Services Unit (the "Devil's Brigade"). The issue of lineage of the Army's Special Forces has been made more complicated and confused by the various reorganizations and designations in the U.S. Army as well as the decision, unjustified in the opinion of many Special Forces veterans, by the Army's Center of Military History to designate the U.S.-Canadian 1st Special Force (the "Devil's Brigade"), which served in Italy and southern France, as the forerunner of Army Special Forces. www.Army.mil/cmh//lineage/branches/sf. A number of authoritative sources disagree and cite OSS SO and OG as direct ancestor of the Army's Special Forces, for example, Alfred H. Paddock, Jr., *US Army Special Warfare: Its Origins: Psychological and Unconventional Warfare, 1941-1952* (Washington, D.C.: National Defense University Press, 1982; rev. ed., University Press of Kansas, 2005), 23-25; Aaron Bank, *From OSS to Green Berets: The Birth of the Special Forces* (Novato, Calif.: Presidio Press, 1986), 205-206; Gordon L. Rottman, *The U.S. Army Special Forces, 1952-1984* (Osprey, 1985); Charles Simpson, *Inside the Green Berets: The First Thirty Years: A History of the U.S. Army's Special Forces* (Presidio Press, 1983); and Mike Yard, "OGs Get to Tell Their Story," *OSS Society Digest*, Number 1317, 30 March 2006, accessed 30 March 2006. Paddock and Bank credit OSS as more representative than most of those units of the range of actions conducted by today's Special Forces. Furthermore, the Special Forces themselves have informally adopted OSS and its units such as Special Operations, Operational Groups, and Detachment 101 in Burma, as among the predecessors of today's Special Forces. See, for example, the websites of the U.S. Army's John F. Kennedy Special Warfare Museum at Fort Bragg, Fayetteville, North Carolina at www.soc.mil/swcs/museum, and the Airborne and Special Operations Museum, Fayetteville, North Carolina at www.asomf.org/museum. On some of the OSS insignia being incorporated, see story on the opening of an OSS Exhibit at the latter in January 2008, reported in "Retired General [John K. Singlaub] to Speak on Career," *Fayetteville Observer*, 24 January 2008, reprinted in *OSS Society Digest* Number 1962, 26 January 2008, accessed 26 January 2008. For example, the Army has authorized the wearing of the Army Special Forces arc on the upper left sleeve by those who have served in World War II at least 120 consecutive days in OSS Detachment 101, OSS Jedburgh Detachments, OSS Operational Groups, and the OSS Maritime Unit. http://www.tioh.hquda.pentagon.mil/Tab/SpecialForcesTab.htm, accessed 11 August 2008; and see also the various issues of *Veritas: Journal of Army Special Operations History*.

83. Alfred H. Paddock, Jr., *U.S. Army Special Warfare, Its Origins: Psychological and Unconventional Warfare, 1941-1952* (Washington, D.C.: National Defense University Press, 1982), 156. The semi-official Airborne and Special Operations Museum in Fayetteville, North Carolina agrees. It credits Donovan's OSS, along with the airborne divisions, Rangers, and the Joint US-Canadian 1st Special Services Unit, as the predecessors of the Army's Special Forces. http://www.asopf.org/museum_exhibits_wwii.htm, accessed 2 January 2008.

84. Paddock, *U.S. Army Special Warfare*, 69-70. This despite the advice of Army Colonel William R. ("Ray") Peers, former head of Detachment 101.

85. Ibid., 130-32.

86. Brig. Gen. Robert A. McClure, 55, had been U.S. military attaché in London before being appointed by Eisenhower as chief of intelligence for the European Theater of Operations in 1942. In 1944, he was appointed head of the newly created Psychological Warfare Division of SHAEF. Since Special Forces work was classified they originally received much less publicity than the Psychological Warfare department in the new center. 86 Paddock, *U.S. Army Special Warfare*, 142-

146. The location was "Smoke Bomb Hill," a collection of old wooden barracks just across the street from where the current Special Forces Center is located. Caesar J. Civitella, telephone conversation with the author, 29 April 2008.

87. Aaron Bank, *From OSS to Green Berets: The Birth of Special Forces* (Novato, Calif.; Presidio Press, 1986), 131, 139-67; Paddock., *U.S. Army Special Warfare*, 158; Obituary, "Col. Aaron Bank, Who Was 'Father of Special Forces,' Dies at 101," *New York Times*, 6 April 2004; "Col. Aaron Bank, U.S. Special Forces Founder, Dies," *OSS Society Newsletter*, Summer 2004, 13.

88. Caesar J. Civitella, "U.S. Army Special Forces Originals," *The Drop* [magazine of the U.S. Army Special Forces Association], Summer 1998, 55; I am indebted to Major Civitella for sending me a copy of this article and the OG pamphlet, OSS, *Operational Group Command* (OSS: December 1944), Caesar J. Civitella, to the author, 25 April 2008. See also "Orientation Material for Use in Connection with Selection of Volunteers for Special Forces," 24 September 1952, quoted in Paddock, *U.S. Army Special Warfare*, 148; and Bank, *From OSS to Green Berets*, 186-87.

89. Troy J. Sacquety, U.S. Army Special Operations Command History Office, Fort Bragg, N.C., email to the author, 8 August 2008; see also Peers and Brelis, *Behind the Burma Road*, 230.

90. Caesar J. Civitella, telephone interview with the author, 18 April 2008

91. Ibid., 25 April 2008; and Charles H. Briscoe, "Major (R) Herbert R. Brucker, DSC, Special Forces Pioneer," *Veritas: Journal of the U.S. Army Special Forces*, 2:2 (2006): 34-35.

92. Jack Hemingway, *Misadventures of a Fly Fisherman: My Life With and Without Papa* (Dallas, Tex.: Taylor Pub. Co., 1986), 257.

93. Caesar J. Civitella, telephone interview with the author, 25 and 29 April 2008; and Civitella, "U.S. Army Special Forces Originals," *The Drop* [magazine of the U.S. Army Special Forces Association], Summer 1998, 55.

94. Bank, *From OSS to Green Berets*, 175-176. Caesar J. Civitella, telephone interview with the author, 25 April 2008

95. Maj. Gen. John K. Singlaub quoted in "SWCS Dedicates Bank Hall: Building Named for 'Father of Special Forces,'" *Special Warfare* 19:1 (January-February 2006): 7.

96. Camp Mackall is located in the North Carolina sand hills, 40 miles southwest of Fort Bragg. Several airborne divisions trained there in World War II. Robert W. Jones, Jr., "Camp Mackall: A History of Training," *Veritas: Journal of Army Special Operations History*, 3:4 (2007): 26-27; and http://www.bragg.Armymil/18abn/CampMackall.htm.

97. Bank, *From OSS to Green Berets*, 179-185; Paddock, *U.S. Army Special Warfare*, 150-51.

98. Bank, *From OSS to Green Berets*, 186-187; see also Dunlop, *Donovan*, 499-500.

99. Bank, *From OSS to Green Berets*, 187-188; Paddock, *U.S. Army Special Warfare*, 149-150.

100. Caesar J. Civitella, telephone interviews with the author, 25 and 29 April 2008. Among other things, Bank took the organization and statement of mission of Special Forces in1952 verbatim from the organization and statement of mission in the OSS Operational Group Command in World War II.

101. Susan L. Marquis, *Unconventional Warfare: Rebuilding U.S. Special Forces* (Washington, D.C.: Brookings, 1997).

102. "Col. Aaron Bank, U.S. Special Forces Founder, Dies," *OSS Society Newsletter*, Summer 2004, 13.

103. Aaron Bank quoted in Dunlop, *Donovan*, 500.

104. Col. H.L. Robinson to Director [Brig. Gen. John Magruder], Strategic Services Unit, War Department, 4 October 1945, subject: Report of Schools & Training Activities as of 1 October 1945, OSS Records (RG 226), Records of the Director's Office, Microfilm 1642, Roll 102, Frames 561-64, National Archives II.

105. On the West Coast, the OSS S&T schools on Catalina Island and the Headquarters Detachment at Newport Beach, California were in the process of being moved to the East Coast for installation in the camps of Area A at Prince William Forest Park, Virginia, when V-J Day occurred on 14 August 1945. Ibid.

106. Capt. James M., Rodgers to Liaison Officer, Schools and Training Branch, OSS, 9 January 1945, subject: Authorization to Destroy Training Property, and authorization from Lt. Leonard Karsakov, 9 January 1945, OSS Records (RG 226), Entry 136, Box 153, Folder 1658, National Archives II.

107. On the role of John Gum, as wartime superintendent of the CCC camps at Chopawamsic, see Susan Carey Strickland, *Prince William Forest Park: An Administrative History* (Washington, D.C.: National Park Service, 1986), 17, although she misspells Gum's name as Gun.

108. Capt. Stanley H. Lyson, S-4, Schools and Training, to Lt. Warner, undated [February 1945], subject: Data on Detachment A, OSS Records (RG 226), Entry 136, Box 153, Folder 1658, National Archives II.

109. Ibid.

110. Estimate #3, Estimated Prices Prepared by T. Sgt. John E. [sic] Gum, attachment to Col. W.L. Rehm, Acting Assistant Director for Services, Strategic Services Unit (SSU), for the Director, to Commanding General, Army Service Forces, Mobilization Division, 19 October 1945, subject: Chopwamsic National Recreational Demonstration Area, in OSS Records (RG 226), Entry 146, Box 8, Folder 199, National Archives II.

111. Capt. Stanley H. Lyson, S-4, Schools and Training, to Lt. Warner, undated [February 1945], subject: Data on Detachment A, OSS Records (RG 226), Entry 136, Box 153, Folder 1658; and Estimate #3, Estimated Prices Prepared by T. Sgt. John E. [sic] Gum, both attachments to Col. W.L. Rehm, Acting Assistant Director for Services, Strategic Services Unit (SSU), for the Director, to Commanding General, Army Service Forces, Mobilization Division, 19 October 1945, subject: Chopawamsic National Recreational Demonstration Area, Attachments #1, #2, and #3 in OSS Records (RG 226), Entry 146, Box 8, Folder 199, National Archives II.

112. The location of the Boat House is given variously as at A-2 and at A-5, but A-5 seems the most probable. Strickland, *Prince William Forest Park*, 36, states that the Army Corps of Engineers relocated the Boat House from Camp 5 to Camp 2. The Boat House location at A-2 and its dimensions are given in Capt. Stanley H. Lyson, S-4 [Supply], Schools and Training, to Lt. Warner, undated [February 1945], subject: Data on Detachment A, OSS Records (RG 226), Entry 136, Box 153, Folder 1658, National Archives II. But it is cited at A-5 and its construction cost of $300 is given in Estimate #1, Estimated Prices Prepared by Quartermaster Depot Office, Ft. Belvoir, Va., and again at A-5 in Estimate #3, Estimated Prices Prepared by T. Sgt. John E. [sic] Gum, both attachments to Col. W.L. Rehm, Acting Assistant Director for Services, Strategic Services Unit (SSU), for the Director, to Commanding General, Army Service Forces, Mobilization Division, 19 October 1945, subject: Chopawamsic National Recreational Demonstration Area, Attachments #1, #2, and #3 in OSS Records (RG 226), Entry 146, Box 8, Folder 199, National Archives II.

113. Capt. Stanley H. Lyson, S-4, Schools and Training, to Lt. Warner, undated [February 1945], subject: Data on Detachment A, OSS Records (RG 226), Entry 136, Box 153, Folder 1658, National Archives II.

114. Estimate #3, Estimated Prices Prepared by T. Sgt. John E. [sic] Gum, attachment to Col. W.L. Rehm, Acting Assistant Director for Services, Strategic Services Unit (SSU), for the Director, to Commanding General, Army Service Forces, Mobilization Division, 19 October 1945, subject: Chopawamsic National Recreational Demonstration Area, OSS Records (RG 226), Entry 146, Box 8, Folder 199, National Archives II.

115. Col. H.L. Robinson to Director, Strategic Services Unit, War Department [Brig. Gen. John Magruder], 4 October 1945, subject: Report of Schools & Training Activities as of 1 October 1945, OSS Records (RG 226), Records of the Director's Office, Microfilm 1642, Roll 102, Frames 561-64, National Archives II.

116. Joseph J. Tully, telephone interview with the author, 29 December 2007.

117. Col. W.L. Rehm, Acting Assistant Director for Services, Strategic Services Unit (SSU), for the Director, to Commanding General, Army Service Forces, Mobilization Division, 19 October 1945, subject: Chopawamsic National Recreational Demonstration Area, Estimates #1, #2 and #3, in OSS Records (RG 226), Entry 146, Box 8, Folder 199, National Archives II.

118. Lt. Col. Ainsworth Blogg to General Counsel Office, SSU, 16 October 1945, subject: Liquidation of Areas "A" and "C," OSS Records (RG 226), Entry 136, Wash S&T OP89, Box 174, Folder 1841, National Archives II. Blogg noted that the labor involved in removing the equipment as well as restoring the buildings and property to prewar condition would offset the limited salvage value of the equipment itself.

119. Col. W.L. Rehm, Acting Assistant Director for Services, Strategic Services Unit (SSU), for the Director, to Commanding General, Army Service Forces, Mobilization Division, 19 October 1945, subject: Chopawamsic National Recreational Demonstration Area, OSS Records (RG 226), Entry 146, Box 8, Folder 199, National Archives II.

120. Lt. Col. Ainsworth Blogg to General Counsel Office, SSU, 16 October 1945, subject: Liquidation of Areas "A" and "C," OSS Records (RG 226), Entry 136, Box 174, Folder 1841, National Archives II.

121. Capt. James E. Rogers, commander, Detachment A, to Commanding General, Fort Belvoir, 19 October 1944, subject: Range Clearance; and Lt. Stanley H. Lyson, to Commanding Officer, HQ Detachment, OSS, 22 December 1944, subject: Request for Clearance of Dud [Mortar Shell] Field, Detachment "A" Mortar Range; and Lyson to Col. H.L. Robinson, 6 January 1945, subject: Dud Field Mortar Range Detachment A; and Maj. Robert C. Wright, memorandum, 29 January 1945; all in OSS Records (RG 226), Entry 136, Box 153, Folder 1658, National Archives II.

122. Col. H.L. Robinson to Capt. Howard J. Preston, 10 January 1946, subject: Chopawamsic National Recreational Demonstration Area, OSS Records (RG 226), Entry 146 General Counsel's Office, Box 8, Folder 199, National Archives II.

123. Colonel Blair to Frank T. Gartside, Superintendent, National Capital Parks, 21 August 1946, National Park Service Records (RG 79), File Number 1460/5, cited in Strickland, *Prince William Forest Park*, 29, as being in the Federal Records Center, Suitland, Md., but which seems to be no longer there.

124. In the summer of 1985, for example, James Fugate, Chief of Maintenance at Prince William Forest Park, found a dead launching grenade in the roof of one of the cabins in Camp 3, Strickland, *Prince William Forest Park*, 30. In 2002, the empty propellant tip of an anti-tank missile, probably a bazooka rocket, was found on a ridge in the root mound of a recently fallen tree near Parking Lot "I." Prince William Forest Park Archives, Cultural Resources, Catalog item number PRWI 13462.

125. Robert A. Noile, Triangle, Virginia, telephone interview with the author, 27 April 2007. His name is pronounced "No-LEE."

126. National Park Service, "Fort Hunt—the Forgotten Story," http://www.nps.gov/archive/gwump/fohu/fortgotten.thm, accessed 15 August 2007; Petula Dvorak, "Fort Hunt's Quiet Men Break Silence on WWII," *Washington Post*, 6 October 2007, A1. I am indebted to Vincent Santucci, chief ranger at the NPS's George Washington Memorial Parkway, for informing me of the material he and his rangers have uncovered concerning this long secret operation. Vincent Santucci, Arlington, Va,, telephone interview with the author, 22 August 2007.

127. Ira B. Lykes, oral history interview with S. Herbert Evison, 23 November 1973, page 45, transcript in the National Park Service's Harpers Ferry Center Library, Harpers Ferry, W.Va. I am indebted to David Nathanson of the library for responding to my request for a copy of this transcript.

128. Ibid., 45-50. In this 1973 interview, Lykes stated that original appropriation was $35,000. A decade later on 4 September 1985, in an interview with Susan Cary Strickland, an historian, Lykes remembered the amount as $25,000. Strickland, *Prince William Forest Park*, 34, 35.

129. Ira Lykes to Irving Root, 20 February 1948, in Strickland, *Prince William Forest Park*, 36.

130. Ira B. Lykes, oral history interview with S. Herbert Evison, 23 November 1973, pages 51-52, transcript in the National Park Service's Harpers Ferry Center Library, Harpers Ferry, W. Va. In this 1973 interview, Lykes stated that he had calculated it one day and the National Park Service had gotten "over a million dollars worth of development" in the cooperative project with the Corps of Engineers. A decade later, on 4 September 1985, in an interview with Susan Cary Strickland, an historian, Lykes remembered the amount as $2 million. Strickland, *Prince William Forest Park*, 35. An official commendation in 1965, states that over a two year period, the cooperation with the Corps of Engineers, "saved the Government almost $500,000." Ira B. Lykes, Citation for Distinguished Service, signed by Secretary of the Interior Stewart L. Udall, 15 July 1965, copy in Ira B. Lykes, Personnel File, National Personnel Records Center, St. Louis, Mo., obtained by the author, 26 April 2007.

131. Ira B. Lykes, oral history interview with S. Herbert Evison, 23 November 1973, pages 53-54, transcript in the National Park Service's Harpers Ferry Center Library, Harpers Ferry, W. Va.

132. Strickland, *Prince William Forest Park*, 47.

133. Ibid., 100, note 110.

134. Allan R. Millett, *Semper Fidelis: The History of the United States Marine Corps*, rev. ed. (New York: Free Press, 1991), 284, 291,323-24.

135. Lt. Col. Keller E. Rockey, USMC, to Charles H. Gerner, Assistant Superintendent, National Park Service, 24 June 1935, NPS Records (RG 79), National Archives II, copy in Prince William Forest Park archive. In World War II, Major General Rockey commanded the 5th Marine Division at Iwo Jima.

136. Strickland, *Prince William Forest Park*, 53-54.

137. Millett, *Semper Fidelis*, 344.

138. Arthur Demaray to Lt. Col. R. Valliant, 31 August 1938, NPS Records (RG 79), Box 121, National Archives II.

139. Ira Lykes, Monthly Report, May 1942, quoted in Strickland, *Prince William Forest Park: An Administrative History* (Washington, D.C.: National Park Service, 1986), 25.

140. Ira Lykes, Monthly Report, 1941, cited in Strickland, *Prince William Forest Park*, 26.

141. Ira Lykes, Monthly Report, November 1942, cited in Strickland, *Prince William Forest Park*, 25-26.

142. Ira B. Lykes, oral history interview with S. Herbert Evison [one of his former supervisors], 23 November 1973, pages 42-43, transcript in the National Park Service's Harpers Ferry Center Library, Harpers Ferry, West Virginia. Lykes apparently misremembered the number of acres south of Joplin Road, it was approximately 5,000 acres. Bob Hickman, Superintendent, Prince William Forest Park, telephone conversation with the author, 11 August 2008.

143. Robert A. Noile, Triangle, Va., telephone interview with the author, 27 April 2008. Born in 1939, Noile grew up in a house on old Joplin Road 300 yards from the park.

144. Frank Gartside, Assistant Superintendent, National Capital Parks, National Park Service, to Ira B. Lykes, 13 September 1943, subject: Furlough (military duty), copy in Ira B. Lykes, Personnel File, obtained by the author 26 April 2007 from the National Personnel Records Center, St. Louis, Mo.

145. Ira B. Lykes to Susan C. Strickland, 8 September 1985, quoted in Strickland, *Prince William Forest Park*, 26.

146. Susan C. Strickland interview with Thelma Williams Hebda, 15 Juy 1985, and Ira B. Lykes to Frank Gartside, 31 May 1943, both cited in Strickland, *Prince William Forest Park*, 26-27. Thelma Williams later married a park employee, Joseph Hebda.

147. Irving C. Root, Superintendent, National Capital Parks, National Park Service, to Ira B. Lykes, 6 December 1945, copy in Ira B. Lykes, Personnel File, obtained by the author 26 April 2007 from the National Personnel Records Center, St. Louis, Mo.

148. "Marine Corps Base Quantico," http://en.wikipedia.org/wiki/Marine_Corps_Base_Quantico. Accessed 5 January 2008.

149. Strickland, *Prince William Forest Park*, 55-54.

150. Acting Secretary of the Interior E.K. Burlew to Secretary of the Navy, 27 May 1941, copy in the Prince William Forest Park archives, Triangle, Va.

151. Secretary of the Navy James Forrestal to Acting Secretary of the Interior Abe Fortas, 12 June 1943, cited in Strickland, *Prince William Forest Park*, 57.

152. Strickland, *Prince William Forest Park*, 58-59.

153. Public Law 736, June 11, 1948, 80th Congress, 2nd Session. It originated as H.R. 6246 sponsored by Rep. Howard Smith. Copy of the act as well as well as House and Senate Reports and interdepartmental correspondence relating to it from White House Bill File, Harry S Truman Papers, Harry S Truman Presidential Library, Independence, Mo. I am indebted to Robert Hickman, superintendent at Prince William Forest Park, for sending me copies of these documents, 11 January 2008.

154. Strickland, *Prince William Forest Park*, 60-61.

155. Documents including, A.E. Damaray, director NPS, to Ira B. Lykes, 27 July 1951, subject: Promotion and Transfer [to Shiloh National Military Park] and [name unclear] Acting Personnel Director to Ira B. Lykes, 8 August 1968, subject: Resignation, copies in Ira B. Lykes, Personnel File, obtained by the author 26 April 2007 from the National Personnel Records Center, St. Louis, Mo.

156. George B. Hartzog, Jr., *Battling for the National Parks* (Mt. Kisco, N.Y.: Moyer Bell Ltd., 1988), 134-137.

157. Robert Hickman, biographical sketch, copy in Prince William Forest Park archives. I am indebted to George Liffert, Assistant Superintendent of the park, for providing me with this information. George Liffert email to the author, 7 January 2008.

158. Robert Hickman, Superintendent of Prince William Forest Park, telephone interview with the author, 24 January 2008, and email to the author, 25 January 2008. for a photograph of the signing by Hickman and Wilson in Lejeune Hall on the Marine Corps base, see "Coming to an Understanding," *Quantico Sentry*, 13 March 1998, clipping in the Archives of Prince William Forest Park, Triangle, Va.

159. "Memorandum of Understanding: A Memorandum of Understanding between Prince William Forest Park and Marine Corps Base, Quantico," 10 March 1998, reprinted in "Prince William Forest Park, General Management Plan, February 1999," pp. 37-40, copy in the Archives of Prince William Forest Park, Triangle, Va.

160. Robert Hickman, Superintendent of Prince William Forest Park, Triangle, Va., telephone interview with the author, 24 January 2008, and email to the author, 25 January 2008.

161. Public Law 107-314 (H.R. 4546), Section 2835, signed into law by President George W. Bush, Dec. 2, 2002.

162. "Agreement to Transfer Administrative Jurisdiction of Land," effective 22 September 2003, signed on 18 August 2003 by Robert S. Hickman and Terry R. Carlstrom for the NPS and on 22 August 2003 by Brig. Gen. Joseph Composto for the Marine Corps. Signed copies of the agreement included with Col. J.M. Lowe, USMC, to Robert Hickman, NPS, 21 November 2003, copy in the Archives of Prince William Forest Park, Triangle, Va. In effect, the final agreement provided for the Marine base to transfer jurisdiction to the park of 352 acres of land that the Marines had purchased from private owners but would now be in the park area. The park transferred jurisdiction to the Marine base of 3,398 acres south and west of the park that the Marines had been using for decades. The special use permit for the Marines to operate on park land was nullified, and 1,346 acres of the park south of old Joplin Road, which the Marines had occupied since World War II, were relinquished by the Marines and recognized to be part of the park.

163. Robert Hickman, superintendent of Prince William Forest Park, Triangle, Va., telephone interview with the author, 24 January 2008, and email to the author, 25 January 2008. Hickman

added that throughout the negotiations that led to the final transfer, he had the active support of the directors NPS's National Capital Region, sequentially, Bob Stanton, Terry Carlstrom, and Joe Lawler, and they had worked with several commanding officers at the Marine Base, including not simply Generals Martin R. Steele, Edwin Kelley, and Frances Wilson but also Brig. Gen. Leif Hendrickson and Brig. Gen. Joseph Composto. Providing continuity and cooperation throughout from the Marine Corps were attorney, Penny Clark, and base community relations officer, Ken Oliver.

164. Strickland, *Prince William Forest Park*, 83-92.

165. Acting Secretary of the Interior Abe Fortas to Secretary of War Henry L. Stimson, 31 May 1944; and to Secretary of War Robert L. Patterson, 4 October 1945, in "World War II Correspondence," Catoctin Mountain Park Archives, Catoctin Mountain Park Headquarters, Thurmont, Md.

166. Acting Secretary of the Interior Abe Fortas, Special Use Permit Authorizing Use of Land in the Catoctin Recreational Demonstration Area, Maryland, by the War Department, for war purposes, 31 May 1944; accepted by Acting Secretary of War Robert P. Patterson, 15 July 1944, Special Use Permits Folder, Catoctin Mountain Park Archives, World War II, located at Catoctin Mountain Park Headquarters, Thurmont, Md.

167. G.B. Williams, custodian, Catoctin RDA, to Director, NPS Region One, memorandum, 19 October 1945, "World War II Correspondence," Catoctin Mountain Park Archives, Catoctin Mountain Park Headquarters, Thurmont, Md.

168. Garland B. Williams was born December 17, 1893, photostatic copy of birth certificate, plus other documents related to Willliams' continuing role as custodian of Catoctin Recreational Demonstration Area during the war, all in Garland B. Williams Personnel File, National Personnel Records Center, St. Louis, Mo., obtained by the author, 8 February 2005.

169. Handwritten note [by historian Barbara Kirkconnell?], "21 Oct. '45" located in World War II Records, Correspondence Folder, Catoctin Mountain Park Archives, Catoctin Mountain Park Headquarters, Thurmont, Md.

170. James H. Mackley, Thurmont, Md., telephone interview with the author, 13 June 2006. Mackley, a lifelong resident of Thurmont, was born there in 1929.

171. The Visitor Center was originally built in 1941.

172. U.S. Army Corps of Engineers, Real Estate Branch, map 26 May 1942, "Government Property, Catoctin Area," National Park Service Records (RG 79), Office of the Chief Counsel, Legislative Files, 1932-1950, Box 76, National Archives II. This report indicated that the War Department had acquired 275 acres, perhaps not including the acreage leased from the Church of the Brethren.

173. G.B. Williams, custodian, Catoctin RDA, to Director, NPS Region One, memorandum, 19 October 1945, and 7 March 1946 memo from Williams to Director Region One, both located in World War II Records, Correspondence Folder, Catoctin Mountain Park Archives, Catoctin Mountain Park Headquarters, Thurmont, Md.

174. *Catoctin Enterprise*, 28 January 1944, and 2 June 1944, cited in Edmund F. Wehrle, *Catoctin Mountain Park: A Historic Resource Study* (Washington, D.C.: National Park Service, 2000), 209-210.

175. Post Engineer's report of 25 September 1945, in Lt. Col. F.B. Grider, Army Services Forces, Third Service Command, to Division Engineer, Middle Atlantic Division, 11 October 1945, subject: Excess Off-Port Facility: Catoctin Recreational Demonstration Area, located in World War II Records, General Information Folder, Catoctin Mountain Park Archives, Catoctin Mountain Park Headquarters, Thurmont, Md.

176. Lt. Charles E. Spears, Detachment 6 of the 9800th TSU-CE Bomb & Shell Disposal Team, Blue Ridge Summit, Pa., to Real Estate Officer, Middle Atlantic Division, Corps of Engineers, 4 March 1946 and G.B. Williams to NPS Regional Director, 5 March 1946, both in World War II File,

General Information Folder, Catoctin Mountain Park Archives, Catoctin Mountain Park Headquarters, Thurmont, Md.

177. Mel Poole, Superintendent, Catoctin Mountain Park, to the author, February 2008.

178. War Department, [name of bureau unclear in photocopy, United States Remainder Office?], "Restoration Survey Catoctin Training Center, Thurmont, Maryland," 31 October 1945, located in World War II Records, General Information Folder, Catoctin Mountain Park Archives, Catoctin Mountain Park Headquarters, Thurmont, Md.

179. G.B. Williams, custodian, Catoctin RDA, to Director, NPS Region One, memorandum, 19 October 1945, located in World War II Records, Correspondence Folder, Catoctin Mountain Park Archives, Catoctin Mountain Park Headquarters, Thurmont, Md.

180. War Department, [name of bureau unclear in photocopy, United States Remainder Office?], "Restoration Survey Catoctin Training Center, Thurmont, Maryland," 31 October 1945, located in World War II Records, General Information Folder, Catoctin Mountain Park Archives, Catoctin Mountain Park Headquarters, Thurmont, Md.

181. Ibid.

182. Ibid.

183. The Marines suffered 45,000 casualties on the two islands. Allan R. Millett, *Semper Fidelis: The History of the United States Marine Corps*, rev. ed. (New York: Free Press, 1991), 431, 438. Quotation from the request of Col. Donald J. Kendall, USMC, to Garland B. Williams, NPS, 3 July 1945; and G.B. Williams to Director NPS Region One, 6 July 1945, memorandum; in Correspondence Folder, Catoctin Mountain Park Archives, World War II, located at Catoctin Mountain Park Headquarters, Thurmont, Md.

184. G.B. Williams, custodian, Catoctin RDA, to Director, NPS Region One, memorandum, 19 October 1945, located in World War II Records, Correspondence Folder, Catoctin Mountain Park Archives, Catoctin Mountain Park Headquarters, Thurmont, Md; A.E. Demaray, associate director, NPS to Abe Fortas, Acting Secretary of the Interior, 20 September 1945; Acting Secretary of the Interior Abe Fortas to Secretary of War Robert L. Patterson, 4 October 1945; NPS Records (RG 79), Entry 7, Central Classified File, 1933-1949, General, 201 National Defense, C [Catoctin], Box 79, Folder 201, National Archives II.

185. Fortas, Special Use Permit Authorizing Use of Land in the Catoctin Recreational Demonstration Area, Maryland, by the Commandant, U.S. Marine Corps, for rehabilitation and security purposes, 4 October 1945; accepted by Maj. Gen. W.P.T. Hill, Quartermaster General, USMC, 19 October 1945; World War II Records, Special Use Permits Folder, Catoctin Mountain Park Archives, located at Catoctin Mountain Park Headquarters, Thurmont, Md.

186. G.B. Williams to NPS Region One Director, 7 January 1946, memorandum, handwritten draft and typed final version, located in World War II Records, Correspondence Folder, Catoctin Mountain Park Archives, Catoctin Mountain Park Headquarters, Thurmont, Md. The is some discrepancy between the handwritten draft of the 7 January 1946 memorandum, which states clearly that it was the Amy that renovated the buildings before the Marines moved into Camp No. 2 and the typed version of the memo which indicates that the renovation between November and January may have been done by the Marine Corps, an assertion reinforced in Williams to Harry T. Thompson, Asst. Supt., National Capital Parks, 8 February 1946, in ibid. Thus it would appear that Kirkconnell, *Catoctin Mountain Park*, 88, may have erred in attributing this winter renovation to the Army instead of the Marines.

187. G.B. Williams to Director, National Park Service, Region One, 7 March 1946; in World War II Records, Correspondence Folder, Catoctin Mountain Park Archives, Catoctin Mountain Park Headquarters, Thurmont, Md.; Lt. Col. C.J. Blair, Jr., chief, Real Estate Division, Middle Atlantic Division, Corps of Engineers, to Department of the Interior, 3 December 1945, World War II, Correspondence Folder, Catoctin Mountain Park Archives, Catoctin Mountain Park Headquarters, Thurmont, Md.; G.B. Williams, custodian, to Regional Director, 7 January 1946; and A.E. Demaray, associate director NPS, to Lt. Col. C.J. Blair, Jr., chief, real estate division, Corps of

Engineers, Office of the Division Engineer, War Department, Baltimore, MD, 23 January 1946, all in NPS Records (RG 79), Entry 7, Central Classified File, 1933-1949, General, 201 National Defense, C [Catoctin], Box 79, Folder 201, National Archives II.

188. Lt. Charles E. Spears, Detachment 6 of the 9800th TSU-CE Bomb & Shell Disposal Team, Blue Ridge Summit, Pa., to Real Estate Officer, Middle Atlantic Division, Corps of Engineers, 4 March 1946 and G.B. Williams to NPS Regional Director, 5 March 1946, both in World War II File, General Information Folder, Catoctin Mountain Park Archives; Lt. Col. William C. Ready, Asst. Division Engineer, Middle Atlantic Division, Corps of Engineers, to A. E. Demaray, National Park Service, 30 January 1946; Demaray to Director, NPS Region One, 15 February 1946; and G.B. Williams to Director NPS Region One, 7 March 1946; all in World War II Records, Correspondence Folder, Catoctin Mountain Park Archives, Catoctin Mountain Park Headquarters, Thurmont, Md.

189. Transfer of Surplus Property Agreement, Catoctin Recreational Demonstration Area, 14 May 1946, signed by representatives of the Corps of Engineers Washington District and the U.S. Department of the Interior; G.B. Williams to Harry T. Thompson, Acting Supt., National Capital Parks, 17 May 1946, both in World War II Records, Correspondence Folder, Catoctin Mountain Park Archives, Catoctin Mountain Park Headquarters, Thurmont, Md.

190. G.B. Williams to Maj. W.J. Dickinson, 2 October 1946, in World War II Records, Correspondence Folder, Catoctin Mountain Park Archives, Catoctin Mountain Park Headquarters, Thurmont, Md.

191. Irving C. Root, Supt. NPS National Capital Parks, to Commandant U.S. Marine Corps, memorandum, 17 February 1947; and G.B. Williams, park custodian, memorandum certifying receipt of equipment from USMC, 18 March 1947. both in World War II Records, Correspondence Folder, Catoctin Mountain Park Archives, Catoctin Mountain Park Headquarters, Thurmont, Md.

192. Kirkconnell, *Catoctin Mountain Park*, 92, n. 82, citing her telephone interview with Garland Williams, Jr., 1 March 1987; see also W. Dale Nelson, *The President Is at Camp David* (Syracuse, N.Y.: Syracuse University Press, 1995), 26.

193. "Site Survey Summary Sheet for DERP-FUDS Site No. CO3MDO346," Catoctin Recreational Demonstration Area, 11 September 1995, pp. 4-5, in World War II Records, General Information Folder, Catoctin Mountain Park Archives, Catoctin Mountain Park Headquarters, Thurmont, Md.

194. G.B. Williams to Director, NPS Region One, memorandum, 7 January 1946; Elbert Cox, Acting Director, NPS Region One, to Arthur A. Demaray, NPS Associate Director, 21 January 1946, in World War II Records, Correspondence Folder, Catoctin Mountain Park Archives, Catoctin Mountain Park Headquarters, Thurmont, Md.

195. Since the Marines had abandoned the Misty Mount cabin camp to move to Greentop, the National Park Service offered Cabin Camp No. 1 (Misty Mount) as an alternative to the Maryland League for Crippled Children, but the League rejected that site as too rugged and steep for its handicapped youngsters. Secretary of Agriculture Clinton Anderson to Matthew J. Connelly, 20 April 1946; Secretary of the Interior Julius A. Krug to Matthew J. Connelly, 15 May 1946, memorandum, both in the Papers of Presidential naval aide William M. Rigdon at the Truman Presidential Library and cited in Nelson, *The President is at Camp David*, 22-25; E. M. Lisle, asst. regional director, NPS Region One, Memorandum to the Director, 21 August 1946, NPS Records (RG 79), Entry 54, Box 60, National Archives II.

196. When the Marines had vacated Misty Mount to move to Greentop, the National Park Service, at Williams recommendation, granted the Salvation Army a permit to use Cabin Camp No. 1 during the summer of 1946 as it had before the war. But the permit was subsequently cancelled, and Misty Mount remained unoccupied until 1947. E.M. Lisle, assistant regional director, Region One, to Director, NPS, 21 August 1946, memorandum, and Arthur E. Demaray, Associate Director of NPS, 25 November 1946, "Confidential" memorandum, both in NPS Records (RG 79), RDA Program Files, Entry 54, Box 60, National Archives II.

197. "Shangri-La Revealed," *Baltimore Sun*, 16 September 1945; "Mountain Top White House: F.D. and Winnie Met at Maryland Hideout," Washington, D.C. *Times-Herald*, 18 September 1942, p. 12;

clippings, and Newton B. Drury to International News Photos, Inc., 17 September 1945; all in National Park Service Records (NG 79), Entry 19, Records of Newton B. Drury, 1940-1951, Box 4, Folder, "Catoctin," National Archives II. As the late President had predicted, his critics accused him of extravagance. The McCormickowned, isolationist, anti-New Deal *Chicago Tribune* complained that the costs had run over $100,000, which was true, but also incorrectly that the swimming pool had been constructed especially for the President and that the whole facility was quite pretentious. *Chicago Tribune*, 21 September 1945, p. 1. Two weeks later, reporters and photographers were allowed inside the camp and the media presented illustrated accounts of the late President's mountain hideaway. "Shangri-La," *Baltimore Sun*, 7 October 1945, Photographic Section, 1; "Roosevelt Hideaway: The Late President Had a Secret Retreat in Maryland's Mountains," *Life*, 15 October 1945, 101-104, clippings in Vertical File, "Shangri-La," Franklin D. Roosevelt Library, Hyde Park, NY.

198. Nelson, *The President Is at Camp David*, 22-50.

199. Maryland League for Crippled Children to President Truman, 1 August 1947, Truman Presidential Library, cited in Nelson, *The President is at Camp David*, 22-23, 25. For the Presidential visit with the crippled children, see also the account in the *Catoctin Enterprise*, 8 August 1947.

200. James H. Mackley, Thurmont, Md., telephone interview with the author, 13 June 2006.

201. Federal responsibility within the park was divided by agreement in 1948. Although the land at the Presidential camp remained within the jurisdiction of the National Park Service, which also maintained the access roads, the Department of the Navy accepted responsibility for administration, protection, operation and improvements at the Presidential Retreat. Agreement between the National Park Service and the Department of the Navy's Bureau of Yards and Docks, 1 November 1948, Catoctin Mountain Park Archives, Catoctin Mountain Park Headquarters, Thurmont, Md. The agreement was subsequently modified to provide concurrent protection by the two agencies. See Kirkconnell, *Administrative History of Catoctin Mountain Park*, 214-217; and Edmund F. Wehrle, *Catoctin Mountain Park: An Historic Resource Study* (Washington, D.C.: National Park Service, 2000), 211-12.

202. League of Maryland Sportsmen, "A Resolution Calling upon the Federal Government to Return the Catoctin National Recreation Area to the State of Maryland," attached to Sen. Millard E. Tydings to Newton B. Drury, director of NPS, 1 July 1944; Joseph F. Kaylor, Maryland State Forester, to Hillory A. Tolson, NPS, 1 June 1945; Records of NPS (RG 79), RDA Program Files, Box 60, National Archives II.

203. Harry S Truman to Herbert R. O'Connor, Governor of Maryland, December 1945, letter in the Harry S Truman Papers, Truman Presidential Library, Independence, Mo. Executive Order 7496 of 12 November 1936 had made Catoctin RDA part of the National Park System, but the federal status of the RDA's was generally considered temporary, with the original goal being federal acquisition and development and then transfer to the states. However, in the summer of 1942 when President Roosevelt had signed legislation which would turn the RDAs over to the states, under certain conditions, he explicitly excluded Catoctin RDA, where he had established his Presidential Retreat, as well as half a dozen other RDAs. The legislation being pursued immediately after the war was to make the park, or at least part of it, a *permanent* part of the National Park System. Wehrle, *Catoctin Mountain Park*, 211-212, 214.

204. Wehrle, *Catoctin Mountain Park*, 211-17; Kirkconnell, *Administrative History of Catoctin Mountain Park*, 92-99.

205. The property of the Presidential Retreat at Shangri-La/Camp David, like the White House, is the responsibility of the National Park Service.

206. Kirkconnell, *Administrative History of Catoctin Mountain Park*, 101-102; Wehrle, *Catoctin Mountain Park*, 217; see also Maryland Department of Natural Resources, Cunningham Falls State Park website, http://www.dnr.state.md.us/publiclands/western/cunninghamfalls.html. 30 September 2006.

207. Garland B. ("Mike") Williams retired on 31 August 1957. Garland B. Williams, Sr., "Citation for Commendable Service" signed by Conrad L. Wirth; "Personnel History Record to Accompany Honor Award Nominations;" and G.B. Williams, Application for Retirement, 12 August 1957; Garland B. Williams, "Notice of Personnel Action, Retirement–Optional," signed by Maurice K. Green, personnel officer, 16 September 1957, Garland B. Williams Personnel File, National Personnel Records Center, St. Louis, Mo., obtained by the author, 8 February 2005.

208. The following summary of recent and current uses of Catoctin Mountain Park was provided by Superintendent Mel Poole and his staff at the park to the author in January 2008.

209. Mel Poole, Superintendent, Catoctin Mountain Park, telephone interview with the author, 13 August 2008.

210. The above summary of recent and current uses of Catoctin Mountain Park was provided by Superintendent Mel Poole and his staff at the park to the author in February 2008.

211. *Congressional Country Club, 1923-1984* (Baltimore, Md.: Wolk Press, n.d.), 33-35. I am indebted to Maxine D. Harvey, director of member services at the Congressional Country Club, for providing me with a copy of this volume, 1 October 2007.

212. John W. Brunner, telephone interview with the author, 21 March 2005.

213. Ibid. Brunner, who had trained as a cipher clerk in the Communications Branch at OSS headquarters and Area C, arrived in China in early 1945. He spent more than a year there, first in Kunming, and then, after the Japanese surrender, he was dispatched to Shanghai and Tientsin. There he did code work, but because he had learned Chinese, he was subsequently used as an interpreter for X-2, the Counterintelligence Branch. OSS's successor, the Strategic Services Unit, kept him on in its Secret Intelligence Branch, until he was mustered out of the Army in March 1946. Returning to Ursinus College in Pennsylvania under the GI Bill, Brunner graduated with proficiency in Greek, Latin, German and Chinese. Subsequently, he earned a doctorate in German at Columbia University, and beginning in 1954 taught that language and later served as head of the German department at Muhlenberg College in Allentown, Pennsylvania, until his retirement in 1989. In addition to dozens of scholarly articles in his field, he published several books on OSS weaponry, including, all of them published by Phillips Publishing of Williamstown, New Jersey. They include *The OSS Crossbows* (1990), *The Colt Glock Pocket Hammerless Automatic Pistols* (1996), *The OSS Weapons* (1994, 2nd edition, 2005), and with Francis B. Mills, *OSS Special Operations in China* (2002).

214. Will Irwin, *The Jedburghs* (New York: Public Affairs Press, 2005), 225, 262-63; Homer, *The Illiad*, Robert Fagles, trans., introduction and notes by Bernard Knox (New York: Viking, 1990); and similarly, Homer, *The Odyssey* (Viking 1996), and Virgil, *The Aeneid* (Viking 2006); see also Bernard Knox, *Essays: Ancient and Modern* (Baltimore, Md.; The Johns Hopkins University Press, 1989), xxxii-xxxiii.

215. Irwin, *The Jedburghs*, 231-32.

216. Others among the U.S. ambassadors to more than twenty countries, who had served in the OSS include George Garrett, ambassador to Ireland, 1947-53; Edwin Martin, ambassador to Argentina, 1964-68, and Thomas Beale, ambassador to Jamaica, 1965-68. Smith, *OSS*, 22.

217. Francis B. Mills and John W. Brunner, *OSS Special Operations in China* (Williamstown, N.J.: Phillips Publications, 2002), 321.

218. Betty Bullard Manning (Mrs. Howard Manning), telephone interview with the author, 4 March 2005.

219. Smith, *OSS*, 23-24.

220. Lynn Philip Hodgson, author of *Inside Camp-X*, email to the author, 5 March 2008.

221. Fleet Marine Force Reference Publication (FMFRP) 12-80, *Kill or Get Killed* (1976).

222. "Rex Applegate," International Close Combat Instructors Association, www.clscbtassc.com; tribute in www.americancombatives.com/applegate.php; and www.paladin-press.com, for Paladin Press, which published many of Applegate's books and videos.

223. John C. McWilliams, "Covert Connections: The FBN, the OSS, and the CIA," *The Historian*, 53, no. 4 (June 1991): 663-665.

224. Ibid., 665-671; Kathryn Meyer and Terry Parssinen, *Webs of Smoke: Smugglers, Warlords, Spies and the History of the International Drug Trade* (Lanham, Md.: Rowman and Littlefield, 1998), 246-247.

225. Howard W. Chappell, resumé; letter, Garland Williams, Head, Intelligence Division, [Federal Bureau of Narcotics], to Howard Chappell, 23 March 1953; both from the papers of Howard Chappell. I am indebted to Howard Chappell's son, Jack Chappell, letter to the author 17 October 2007, for providing me with copies of this and other documents from his late father's files.

226. Thomas N. Moon and Carl F. Eifler, *The Deadliest Colonel* (New York: Vantage, 1975), 247-266.

227. "'Deadliest Colonel,' dies at age 95," *U.S. Customs Today*, August 2002, www.cbp.gov/xp/CustomsToday/200/August/other/colonel.xml, accessed 30 December 2007.

228. These films included *My Darling Clementine* (1947), *Fort Apache* (1948), *She Wore a Yellow Ribbon* (1950), and later and with greater complexity, *The Searchers* (1957).

229. His film noir movies of the 1950s, included most memorably *The Asphalt Jungle* (1950), *Johnny Guitar* (1953), and *The Killing* (1956).

230. The films about the OSS were *13 Rue Madeleine* (1946) starring James Cagney, and *Operating Secret* (1952) starring Cornel Wilde. Ortiz played small parts most famously in Twelve O'Clock High (1949), *What Price Glory* (1952), *Retreat Hell!* (1952), and *Wings of Eagles* (1957).

231. Entries on "Joe Savoldi," and "Houston Harris" in Wikipedia, http://en.wikipedia.org/wik; and "Bobo Brazil dies at age 74," *Slam! Sports Wrestling*, 23 January 1998, http://slam.canoe.ca/SlamWrestlingArchive/jan23_for.html; each accessed 30 December 2007.

232. "Joe Savoldi," Wikipedia, http://en.wikpedia.org/wik; and James Gregory Savoldi, a grandson of "Jumping Joe" Savoldi, email to the author, 29 December 2007. I am indebted to J.G. Savoldi for providing me with photographs and other materials about his grandfather.

233. One account indicates that Berg worked as a contract employee for the CIA in 1951, advising and consulting, and that "several years later, he served on the staff of NATO's Advisory Group for Aeronautical Research and Development." "Morris ('Moe') Berg," G.J.A. O'Toole, *The Encyclopedia of American Intelligence and Espionage* (New York: Facts on File, 1988), 65.

234. "In Memoriam," *OSS Society Newsletter*, Spring 2004, 9.

235. "London Times Obituary: Peter Tompkins," *OSS Society Digest* Number 1575, 25 February 2007, osssociety@yahoogroups.com, accessed 25 February 2007.

236. Reginald Spear, telephone interview with the author 15 August 2008.

237. Serge Obolensky, *One Man in His Time: The Memoirs of Serge Obolensky* (New York: McDowell, Obolensky, 1958), 395-396; K. Michell Moran, "The Prince and Princess of the Pointes: Estate Auction Recalls Colorful Lives of Late Russian Prince and his Wife, a Grosse Pointe Native," unidentified newspaper article reprinted in *OSS Society Digest* Number 1954, 18 January 2008, osssociety@yahoogroups.com, accessed 18 January 2008.

238. Smith, *OSS*, 26. See also, correspondence between Goodfellow and Rhee, 1944-45, in M. Preston Goodfellow Papers, Box 4, Subject File: Syngman Rhee, Hoover Institution Archives, Stanford, Calif.

239. Biographical information and unidentified newspaper clipping, "Name Goodfellow Adviser in Korea," all in Goodfellow Papers, Box 2, Folder: Biographical Materials, Hoover Institution Archives, Stanford, Calif.

240. Dun & Bradstreet Report on Overseas Reconstruction Corporation, 19 September 1960, p. 1, copy in Goodfellow Papers, Box 2, Folder: Biographical Material, Hoover Institution Archives, Stanford, Calif.

241. Obituary, "M. Preston Goodfellow", *New York Times*, 6 September 1973, 40; "M.P. Goodfellow, 81, Publisher, Dies," *Washington Post*, 7 September 1973, C4.

242. Robert E. Mattingly, *Herringbone Cloak—GI Dagger: Marines of the OSS* (Washington, D.C.: History and Museums Division, Headquarters, U.S. Marine Corps, 1989), 182.

243. Ellsworth ("Al") Johnson, telephone interview with the author, 27 June 2008.

244. Rolf Herstad, email to the author, 7 August 2008. I am indebted to Rolf Herstad for supplying me with much information and illustrative material about his father's service in the OSS.

245. Albert ("Al") Materazzi, telephone interviews with the author, 26 January 2005, 23 September 2005, "Biography," sent to the author, 31 January 2005; Al Materazzi, "Remembering 109 [William J. Donovan]—Recollections of OSSers," *OSS Society Newsletter*, Winter 2004-05, p. 7; email to the author, 6 March 2007.

246. Obituary, Richard W. Breck, Quincy, Mass. *Patriot-Ledger*, 17 March 2005, reprinted in *OSS Society Digest* Number 984, 18 March 2005, osssociety@yahoo.groups.com, accessed 18 March 2005.

247. Allen R. Richter, telephone interview with the author, 16 March 2007.

248. Arthur Reinhardt, email to the author, 25 July 2007, attributing this information to John W. Coffey.

249. Arthur Reinhardt, email to the author, 23 September 2005; Brenda Warner Rotzoll, obituary "Sarah M. Mero: Handled Secret WWII Transmissions," *Chicago Sun-Times*, 25 November 2003.

250. Allen R. Richter, telephone interview with the author, 16 March 2007.

251. Ibid.

252. Ibid., 25 March 2005.

253. James F. Ranney, telephone interview with the author, 8 January 2005; James F. Ranney and Arthur L. Ranney, eds., *The OSS CommVets Papers*, 2nd ed. (Covington, Ky.: James F. Ranney, 2002).

254. Timothy R. Marsh, telephone interview with the author, 18 January 2006.

255. Roger L. Belanger, telephone interview with the author, 11 April 2005.

256. Joseph J. Tully, telephone interview with the author, 29 December 2007.

257. David Kenney, telephone interview with the author, 11 April 2005.

258. Spiro Cappony, telephone conversation with the author, 16 September 2006, and David Hendrix, "Under Deep Cover," *Times*, pp. C-1, C-2, unidentified, undated clipping of a news story about him sent to the author by Spiro Cappony, 31 October 2006.

259. Marvin Flisser, telephone interview with the author, 27 January 2005.

260. "In Memoriam: Jack Kilby," *OSS Society Newsletter*, Summer 2005, 16, and Spring 2006, 17; John Markoff, "Jack S. Kilby, an Inventor of the Microchip, Is Dead at 81," *New York Times*, 22 June 2005; T.R. Reid, "Jack Kilby, Touch Lives on Micro and Macro Scales," *Washington Post*, 22 June 2005, C1.

261. "Peers, William Raymond," W. Thomas Smith, Jr., *Encyclopedia of the Central Intelligence Agency* (New York: Facts on File, 2003), 184; William R. Peers and Dean Brelis, *Behind the Burma Road: The Story of America's Most Successful Guerrilla Force* (Boston: Little, Brown, 1963); William R. Peers, *The Peers Report* (1970); "Army Values: Integrity: Ray Peers," *Special Warfare*, 13:2 (Spring 2000): 21.

262. "Gen. John K. Singlaub New Society Chairman," *OSS Society Newsletter*, Spring 2005, 1; John K. Singlaub, with Malcom McConnell, *Hazardous Duty: An American Soldier in the Twentieth Century* (New York: Simon and Schuster, 1991).

263. It was the 878th Airborne Engineer Battalion, Lt. Col. Joseph H. Collart, http://www.ixengineer command.com/listmen.php. Accessed 3 January 2008.

264. Maj. Gen. Robert R. Ploger, *Vietnam Studies: U.S. Army Engineers, 1965-1970* (Washington, D.C.: U.S. Army, Center of Military History, 1974), 63-64, 72.

265. Frank A. Gleason, telephone interview with the author, 26 December 2007; Charles M. Parkin, ed., *The Rocket Handbook for Amateurs* (New York: John Day, 1959); and Francis E. Cross and Charles M. Parkin, *Captain Gray in the Pacific Northwest* (Bend, Oregon: Maverick Publications, 1987). Located in the coastal fishing town of Garibaldi, in Tillamook County, the Garibaldi Maritime Museum opened to the public in 2004. Judy Fleagle, "Captain Gray Takes a Stand," *Oregon Coast* magazine, May/June 2005, 10.

266. Frank A. Gleason, telephone interview with the author, 26 December 2007. Joseph Lazarsky, who was with the CIA in the Far East in those years, doubts whether Leo Kawarski "spent more than 1% of his time" with the CIA. Joseph Lazarsky, telephone interview with the author, 27 December 2007.

267. Frank A. Gleason, email to the author, 28 January 2005; and telephone interviews with the author, 31 January 2005, and 26 December 2007.

268. Dorothea ("Dodie") Dow (Mrs. Arden W. Dow), telephone interview with the author, 15 May 2005.

269. John C. Hooker, resumé, John C. Hooker, email to the author, 27 June 2008.

270. Jerry Sage, *Sage* (Wayne, Pa.: Miles Standish Press, 1985), 467-470.

271. Irwin, *The Jedburghs*, 228; some of Cyr's postwar problems are recounted in Joseph C. Goulden with Alexander W. Raffio, *The Death Merchant: The Rise and Fall of Edwin P. Wilson* (New York: Simon and Schuster, 1984), 69-72, 237-38, 313, 433.

272. Joseph Lazarsky, telephone interviews with the author, 14 March 2005, 11 February and 27 December 2007.

273. Obituary, "Oliver Mowrer Silsby, Jr., CIA Agent," *Washington Post*, 29 May, 2006.

274. The Col. Arthur D. ("Bull") Simons Award presented to Caesar Civitella at the USSOCOM HQ, MacDill AFB, Tampa, Fla. Troy J. Sacquety, "Caesar J. Civitella: Bull Simons Award 2008," *Veritas: Journal of Army Special Operations History*, 4:1 (2008): 77.

275. Tim Weiner, "Lucien Conein, 79, Legendary Cold War Spy," *New York Times*, 7 June 1998.

276. Erasmus H. Kloman, *Assignment Algiers: With the OSS in the Mediterranean Theater* (Annapolis, Md.: Naval Institute Press, 2005), 127.

277. "Louis Joseph Gonzalez," biographical statement, provided by his son, Ronald Gonzalez, San Mateo, to the author, 22 January 2008. I am indebted to Ronald Gonzalez for the material about his father.

278. Information from James Snyder, son of Jacques F. Snyder, email to the author, 8 August 2008. I am indebted to James Snyder for the information and material about his father.

279. Elizabeth P. MacIntosh, *Undercover Girl* (New York: Macmillan, 1947, reprinted 1995); she also wrote two books for children.

280. Elizabeth P. MacIntosh, *Sisterhood of Spies* (Annapolis, Md.: Naval Institute Press, 1998), 242, for the statistics on CIA employment by gender.

281. Ibid., 169. Aline, Countess of Romanones, *The Spy Wore Red: My Adventures as an Undercover Agent in World War II* (New York: Random House, 1987).

282. Aline, Countess of Romanones, *The Spy Wore Red*; and *The Spy Went Dancing* (New York: G. Putnam's Sons, 1990), 11-14, and passim.

283. Ibid., 244.

284. Douglas Martin, "Hélène Deschamps Adams, Wartime Hero, Dies at 85," *New York Times*, 24 September 2006, A4. Her books were *The Secret War* (New York: W.H. Allen, 1980) and *Spyglass: An Autobiography* (New York: Henry Holt, 1995).

285. MacIntosh, *Sisterhood of Spies*, 68-69; and Interview with Barbara Lauwers Podoski conducted by Christof Mauch, 4 September 1996, OSS Oral History Transcripts, CIA Records (RG 263), Box 3, National Archives II.

286. MacIntosh, *Sisterhood of Spies*, 167; Jim Downs, *WWII OSS Tragedy in Slovakia* (Oceanside, Calif.: Liefrinck Pub., 2002), 314, 324.

287. Laura Shapiro, *Julia Child* (New York: Viking, 2007).

288. Irwin, *The Jedburghs*, 226.

289. Gail F. Donnalley, telephone interview with the author, 30 April 2005.

290. Arthur ("Art") Reinhardt, interview with the author and Judy Volonoski, museum technician, following a tour at Prince William Forest Park, Triangle, Va., 22 February 2007.

291. William J. Donovan, "American Foreign Policy Must Be Based on Facts," *Vital Speeches*, 1 May 1946, 446-48; Donovan, "Intelligence: Key to Defense," *Life*, 30 September 1946, 108-10.

292. For example, "Donovan Points to Russians in Defense Appeal," *New York Herald Tribune*, 10 January 1947.

293. Anthony Cave Brown, *The Last Hero: Wild Bill Donovan* (New York: Times Books, 1982), 795-800. Brown indicates that Donovan, who was not a stockholder, did not receive much income from the company; his main income was his salary from his law firm.

294. Dunlop, *Donovan*, 486-488, 490-491, 495, 497, 499.

295. Ibid., 488.

296. Anthony Cave Brown cites several possible reasons for Truman's antipathy toward Donovan. In World War I, Donovan's units, attacking the Hindenburg Line in 1918 were deprived of artillery support and suffered major casualties. Donovan complained and Capt. Harry S Truman's artillery battery, which had been directed to provide support for that section of the line the night before Donovan's mission, was withdrawn from that mission soon after Donovan's complaint. Later as a U.S. Senator, Truman became close friends with Democratic Senator Burton K. Wheeler of Montana, who had been prosecuted by Donovan as assistant U.S. attorney general for crimes related to the Teapot Dome scandal, but who had been found innocent at his trial; subsequently, Wheeler formed an anti-Donovan clique in the Senate. Cave Brown, *The Last Hero*, 790-792.

297. Dunlop, *Donovan*, 491, 499. Donovan did not think Allen Dulles was the best choice to manage a large intelligence organization. At the end of World War II, despite Allen Dulles's achievements in the Bern office of the OSS and recommendations by a number of OSSers, Donovan had refused to appoint him to succeed David Bruce as chief of OSS European operations. Instead, he put Dulles in charge of OSS in a particular country: Germany. "I thought Allen was a fine operative, but I did not think he had the organizational skill to handle all of Europe," Donovan said after the war. Some thought it was because of Dulles's supposed mishandling of the German bid for a separate peace in Italy, Operation Sunrise, but Donovan denied that. Donovan's friends thought that Donovan's bid for DCI was undermined by his old nemesis, J. Edgar Hoover. Cave Brown, *The Last Hero*), 821.

298. This was at the urging of Gen. Walter Bedell Smith. Smith had been Eisenhower's wartime chief of staff, and served as Director of Central Intelligence from October 1952 to February 1953. As DCI, he frequently consulted with Donovan. In May 1953, as Deputy Secretary of State, he summoned Donovan to accept the post in Thailand as a frontline in the fight against communist expansion. Ibid., 822.

299. Dunlop, *Donovan*, 488.

300. Ibid., 500.

301. Dunlop, *Donovan*, 502. During Donovan's travels around Thailand, the men trailing him turned out to be not Soviet KGB agents, as he had first suspected, but members of the CIA sent to report his activities to the suspicious CIA chief Allen Dulles. The SEATO nations included Australia, Great Britain, France, New Zealand, Pakistan, the Philippines, Thailand, and the United States. In a separate protocol, SEATO protection was extended to Cambodia, Laos, and South Vietnam, which had been barred by the Geneva Agreements of 1954 from joining any military alliance.

302. Dunlop, *Donovan*, 504-505.

303. Cave Brown, *The Last Hero*, 830-833.

304. Otto C. Doering, New York City, interview with Anthony Cave Brown, 1 November 1977, quoted in Cave Brown, *The Last Hero*, 833.

CHAPTER 11

1. Brett J. Blackledge and Randy Herschaft, Associated Press, "Newly Release Files Detail Early US Spy Network," 14 August 2008, http://www.washingtonpost.com/wp-dyn/content/article/2008/08/14/August; Spy Files Include a Justice, a Baker, and a Filmmaker," Newark (NJ) *Star-Ledger*, 15 August 2008, A4. The 750,000 newly declassified documents also seem to suggest that OSS had a total of 24,000 members rather than the 13,000 previously believed, but their status, whether permanent, temporary, member or consultant, American or foreigner, remains to be determined. The release of these three-quarters of a million documents occurred as the present study was going to press, and they have not been included in it.

2. Quoted in "History of Schools and Training, OSS, Part I: Chronology and Administration, June 1942 – October 1945," p. 25, typescript, n.d. [apparently written in 1947], copy delivered by W.J. Morgan, who had been with OSS Schools and Training Branch during World War II, to Col. E.B. Whisner, 7 January 1949, OSS Records (RG 226), Entry 176, Box 2, Folder 12, National Archives II, College Park, Md.

3. "History of Schools and Training, OSS, Part I: Chronology and Administration, June 1942 – October 1945," p. 17, typescript, n.d. [apparently written in 1947], copy delivered by W.J. Morgan, who had been with OSS Schools and Training Branch during World War II, to Col. E.B. Whisner, 7 January 1949, OSS Records (RG 226), Entry 176, Box 2, Folder 12, National Archives II.

4. "Excerpts from History of Schools and Training, OSS," attached to L.B. Shallcross, Deputy, STB/TRD, to John O'Gara, Chief, Staff Training Branch [of CIA], 1 February 1951, subject: information on OSS Schools and Training Sites, OSS Records (RG 226), Entry 161, Box 7, Folder 76, National Archives II.

5. However, as Schools and Training Branch acknowledged after the war, too often OSS men were sent overseas without any military training, because it was assumed they would continue to work in purely service and support functions, such as Research and Analysis or Administrative Services, but once overseas were transferred to operational or other duties. "History of Schools and Training, OSS, Part I: Chronology and Administration, June 1942 – October 1945," p. 29, typescript, n.d. [apparently written in 1947], copy delivered by W.J. Morgan, who had been with OSS Schools and Training Branch during World War II, to Col. E.B. Whisner, 7 January 1949, OSS Records (RG 226), Entry 176, Box 2, Folder 12, National Archives II.

6. George H. White, Diary, 1942, quoted in John C. McWilliams, "Covert Connections: The FBN, the OSS, and the CIA," *The Historian*, 53.4 (Summer 1991): 665.

7. H. Stuart Hughes, oral history interview conducted by Barry Katz, 10 May 1997, p. 13, in OSS Oral History Transcripts, CIA Records (RG 263), Box 2, National Archives II.

8. "Outlines of Courses at All U.S. Training Areas," December 1943, OSS Records (RG 226), Entry 136, Box 158, Folder 1723, National Archives II.

9. "Outlines of Courses at All U.S. Training Areas," December 1943, OSS Records (RG 226), Entry 136, Box 158, Folder 1723, National Archives II.

10. George Maddock, newspaper interview in 2007, reprinted in *OSS Society Digest*, Number 1918, 2 December 2007, osssociety@yahoogroups.com, accessed 2 December 2007.

11. Unidentified trainee [signature page missing] to Lt. Col. P.G. Strong, 15 August 1942, subject: operation of SA/G [SO] schools, OSS Records (RG 226), Entry 136, Box 161, Folder 1754, National Archives II.

12. Arthur ("Art") Reinhardt, interview at Prince William Forest Park, Triangle, Va., by the author 14 December 2004.

13. "History of Schools and Training, OSS, Part I: Chronology and Administration, June 1942 – October 1945," p. 42, typescript, n.d. [apparently written in 1947], copy delivered by W.J. Morgan,

who had been with OSS Schools and Training Branch during World War II, to Col. E.B. Whisner, 7 January 1949, OSS Records (RG 226), Entry 176, Box 2, Folder 12, National Archives II. The report, p. 39, stated that a surprising number of "psycho-neurotics" was found.

14. "Spotting the Talent," *Financial Express*, 16 October 2005, reprinted in *OSS Society Digest*, Number 1172, 17 October 2005, osssociety@yahoogroups.com, accessed 17 October 2005; Office of Strategic Services, *War Report of the OSS* (New York: Walker and Co., 1976), 238-241; reprint of the original typescript report prepared by the Strategic Services Unit of the War Department in 1947 (hereinafter OSS, *War Report of the OSS*); "A Good Man Is Hard to Find: The O.S.S. Learned How, with New Selection Methods that May Well Serve Industry," *Fortune* (1946): 92-95, 217-223; OSS Assessment Staff, *Assessment of Men: Selection of Personnel for the Office of Strategic Services* (New York: Rinehart, 1948). 15 Francis ("Frank") Mills, oral history interview conducted by Maochun Yu, 19 November 1996, p. 2, OSS Oral History Transcripts, CIA Records (RG 263), Box 3, National Archives II.

16. Erasmus H. Kloman, *Assignment Algiers: With the OSS in the Mediterranean Theater* (Annapolis, Md.: Naval Institute Press, 2005), 10, 14.

17. Ibid., 13; Erasmus Kloman, telephone interview with the author, 24 January 2005. In fact, his assignment was changed, and he was never sent to Yugoslavia.

18. Kloman, *Assignment Algiers*, 13-14.

19. George Maddock, newspaper interview 2007, reprinted in OSS Society Digest, Number 1918, 2 December 2007, osssociety@yahoogroups.com, accessed 2 December 2007.

20. See, for example, the complaint by an unidentified trainee [signature page missing] to Lt. Col. P[hilip].G. Strong, [head of Special Operations Branch],15 August 1942, subject: operation of SA/G [SO] schools, p. 2, OSS Records (RG 226), Entry 136, Box 161, Folder 1754, National Archives II.

21. Frank A. Gleason, telephone interview with the author, 31 January 2005.

22. "One afternoon of tremendous benefit to the group [of students] was that spent with a man recently returned from the field, who reported on his own activities and personal experiences interestingly and in detail. His talk and his answers to their questions were of real value in helping the men to picture the situations they may encounter, and the operations they may undertake. The group was enthusiastic over the opportunity to hear him, and their reaction certainly was evidence of the desirability of bringing in a man with actual field experience whenever possible." "Student" to Kenneth Baker [chief, Schools and Training Branch], 8 December 1942, p. 2, a three-page typed report by an anonymous student, in Appendix III, Part Two of the History [of Schools and Training Branch], OSS Records (RG 226), Entry 136, Box 158, Folder 1722, National Archives II.

23. Capt. W.B. Kantack, reports officer, SO, to Lt. Bane, 16 March 1945, subject: reports requirements of Schools and Training [in response to request from Mr. William R. Stewart, Assistant Intelligence Officer, S&T], OSS Records (RG 226), Directors Office Files, microfilm M1642, Roll 63, Frames 663-664, National Archives II.

24. See the recommendations of Maj. Arthur Goldberg, head of the SI Labor Desk in Europe, "Report on an Hour with Major Goldberg," pp. 3-4, in "Interviews with Returned Men," OSS Records (RG 226), Entry 161, Box 2, Folder 31. In April 1945 S&T in China made a similar suggestion, ending, "It cannot be said that advanced training in the U.S. is a waste of time, but there is little doubt but that the U.S. training staffs could accomplish infinitely more if operating here." OSS Theater Reports, S & T Branch Excerpts, China Theater, 30 April 1945, p. 10, OSS Records (RG 226), Entry 136, Box 159, Folder 1729, both in National Archives II.

25. Allen R. Richter, telephone interview with the author, 25 March 2005.

26. Lt. Col. H.L. Robinson, Executive, Schools and Training Branch, October 1943, "Schools and Training," p. 12, a 14-page typed report, included in Appendix IV, Part Three of the History [of Schools and Training Branch], OSS Records (RG 226), Entry 136, Box 158, Folder 1723, National Archives II.

27. Ibid.

28. Ibid., 13.

29. OSS, Schools and Training Branch, "Interviews with Returned Men," fall 1944 to spring 1945, in OSS Records (RG 226), Entry 161, Box 2, Folder 31; and Lt. Arthur Simon, Area "K" [England], to Maj. Ezra Shine, Chief, S&T Branch [in England], 24 May 1945, subject: summary and analysis of deprocessing interviews [of men from the 13 teams parachuted into Germany between January and April 1945 and their assessments and recommendations regarding their training and its relationship to their missions], plus the transcripts of the interviews with the team members, all in OSS Records (RG 226), Entry 136, Box 158, Folder 26, National Archives II.

30. Maj. Peter Dewey quoted in Deane W. Starrett, Chief, Training Materials and Research Section [of CIA] to Col. [E.B.] Whisner, Deputy Chief, TRS, 16 May 1949, subject: wartime recommendations for the training of personnel in OSS, p. 16, OSS Records (RG 226), Entry 161, Box 2, Folder 32, National Archives II. A Major Caskey from SI service in Greece and Turkey, p. 13, echoed that and contended that too much stress on security even at RTU-11 tended to make the student overly conscious of his cover." On arriving in the field, he would be conspicuous by his secrecy." Lt. George Demas from the Far East, p. 22, agreed that "being too tight-lipped, etc. tends to attract attention." The summaries in the 1949 report were made by Donald C. Baker at the request of W.J. Morgan, who had served in OSS S&T Branch in WWII, and were taken from a series of interviews by Sgt. J.C. Gibbs of S&T with returning field personnel from SI, SO, MO, and MU during 1944 and 1945.

31. Quotation from an unidentified Greek SI officer, ibid., p. 21; for similar complaints about the need for more training on observation and reporting, see "John_" SI, Greece, p. 16, "Dorothy C._" SI, Turkey, p. 17, and Huntington Bliss, cable officer, Bari, Italy, p. 20.

32. Quotations from Ray F. Kauffman, SO Ceylon, ibid., p. 17; and "Mr. X" stationed in a "neutral country," p. 8; and "chief organizer of a sabotage group," p. 9. Emphasis in the original.

33. Maj. Gen. John K. Singlaub (USA-Ret.), transcript of interview by Maochun Yu and Christof Mauch, 31 October 1996, pp. 8, 13-14, OSS Oral History Transcripts, CIA Records (RG 263), Box 4, National Archives II. Asked if there were any areas of training that proved inadequate, General Singlaub replied in 1996, "I can't think of any area [of training] that showed up at being deficient. There may be some, but I can't think of any."

34. Robert R. Kehoe, "1944: An Allied Team with the French Resistance," *Studies in Intelligence: Journal of the American Intelligence Profession*, OSS 60th Anniversary Issue (June 2002): 104.

35. Jerry Sage, *Sage* (Wayne, Pa.: Miles Standish Press, 1985), 207-208.

36. Ibid., 305, 327, 431.

37. Frank A. Gleason, Jr., telephone interview with the author, 9 February 2007.

38. Joseph Lazarsky, telephone interview with the author, 11 February 2007.

39. "History of Schools and Training, OSS, Part I: Chronology and Administration, June 1942 – October 1945," p. 53, typescript, n.d. [apparently written in 1947], copy delivered by W.J. Morgan, who had been with OSS Schools and Training Branch during World War II, to Col. E.B. Whisner, 7 January 1949, OSS Records (RG 226), Entry 176, Box 2, Folder 12, National Archives II.

40. "OSS Training Branch, Chapter VI (History)," p. 1, typescript n.d. [1946-1947?], recommendations for "the Training Section of a secret intelligence agency in time of war," OSS Records (RG 226), Entry 176, Box 2, Folder 14, National Archives II.

41. "Strategic Services, Office of (OSS)," W. Thomas Smith, Jr., *Encyclopedia of the Central Intelligence Agency* (New York: Facts on File, 2003), 220. Smith provides a rounded off figure of 13,000 as the 1944 peak strength of the OSS. The precise peak figure of 12,974 in December 1944 is listed and compared with other months since October, October 1944 being, according to the report, "the first month in which total figures were available" for "OSS the world over...." Louis M. Ream [OSS deputy director for administrative services, including personnel] to Col. G. Edward Buxton [deputy director of OSS and Donovan's right hand man], memorandum, 18 January 1945, [no

subject line, but the topic is Ream's concern with the fact that although on 23 October 1944 Donovan had promised Harold D. Smith, head of the Bureau of the Budget, an immediate 5 per cent reduction in the total OSS personnel complement, resulting in a reduction of about 600 persons, in fact there had been an increase rather than a decrease in OSS personnel from 12,740 in October to 12,974 in December, due primarily to delays in terminating military and civilian personnel from overseas, but, as Ream emphasized, "it is very necessary that General Donovan's promise to Mr. Smith be kept."OSS Wash-Dir-Op-266, #55, "Liquidation of OSS," copy in CIA Records (RG 263), Thomas Troy Files, Box 6, Folder 46, National Archives II.

42. Louis M. Ream, deputy director, administrative services, to Charles S. Cheston [second assistant director of OSS since March 1943], memorandum, 29 January 1945, pp. 2-3 [no subject line, but Ream provides some statistical analyses of OSS personnel strength between October and December 1944, and also a functional distribution by percentage and sometimes with actual numbers of personnel in various categories]. For a breakdown of personnel by geographical theaters of operation as well as branches and other functional categories, see the tables for March and April 1945 entitled simply "Summary" [presumably April 1945], also in OSS Wash-Dir-Op-266, #55, "Liquidation of OSS," copy in CIA Records (RG 263), Thomas Troy Files, Box 6, Folder 46, National Archives II.

43. The figures given in May 1945 were 12,816 OSS personnel at that time, of which 8,939 were numbers of the armed forces. Of a total of 2,593 officers, 2, 192 had commissions in the Army or Army Air Corps; of a total of 6,346 enlisted personnel, 5,817 were serving on detached duty from the Army. A relatively few service members were from the Navy or Marines. Maj. Gen. William J. Donovan to The Adjutant General, War Department, 15 May 1945, subject: recommendation for promotion [of Col. Millard P. Goodfellow], p. 3, copy in Papers of M. Preston Goodfellow, Box 2, Biographical Material Folder, Hoover Institution Archives, Stanford, Calif.

44. Quoted in "History of Schools and Training, OSS, Part I: Chronology and Administration, June 1942 – October 1945," p. 52, typescript, n.d. [apparently written in 1947], copy delivered by W.J. Morgan, who had been with OSS Schools and Training Branch during World War II, to Col. E.B. Whisner, 7 January 1949, OSS Records (RG 226), Entry 176, Box 2, Folder 12, National Archives II.

45. Quoted in ibid., 51, emphasis in the original.

46. Joseph Lazarsky, telephone interview with the author, 11 February 2007. Frank Gleason concurred. "They [the CIA] had to because of the sound training we had." Frank A. Gleason, Jr., telephone interview with the author, 9 February 2007.

47. Richard Harris Smith, *OSS: The Secret History of America's First Central Intelligence Agency* (Berkeley: University of California Press, 1972), 265n. Tom Weiner, *Legacy of Ashes: The History of the CIA* (New York: Doubleday, 2007), 60, reports that Peers ran Western Enterprises, a CIA front organization in Taiwan, designed to help subvert Mao Tse-tung's regime.

48. Ray S. Cline, *Secrets, Spies, and Scholars: Blueprint of the Essential CIA* (Washington, D.C.: Acropolis Books, 1976), 109, 111, quotation on 122.

49. "Peary, Camp," W. Thomas Smith, Jr., *Encyclopedia of the Central Intelligence Agency* (New York: Facts on File, 2003), 183.

50. Valerie Plame Wilson, *Fair Game* (New York: Simon & Schuster, 2007), 1-2, 315.

51. Her training courses continued in intelligence gathering and analysis before she was given her first assignment. Ibid., 12-27, quotation at 12.

52. Joseph Lazarsky, telephone interview with the author, 11 February 2007.

53. Cline, *Secret Spies and Scholars*, 67. It could be added that Donovan's decision was validated as there were no known compromises of the OSS Communications Branch system, in contrast to the penetration of numerous other coding and communications systems. Arthur Reinhardt, email to the author, 27 June 2007.

54. Cline, *Secret Spies and Scholars*, 67-68.

55. Gen. Dwight D. Eisenhower to War Department, 26 May 1945, Records of the Joint Chiefs of Staff (RG 218), Admiral Leahy's Files, Box 9, Folder 54, quoted in Bradley F. Smith, *The Shadow Warriors: O.S.S. and the Origins of the C.I.A.* (New York: Basic Books, 1993), 307.

56. "SWCS Dedicates Bank Hall: Building Named for 'Father of Special Forces,'" press release from the John F. Kennedy Special Warfare Center and School, reprinted in *OSS Society Digest*, Number 1291, 1 March 2006, osssociety@yahoogroups.com, accessed, 1 March 2006. On November 21, 2005, the JFK Special Warfare Center and School at Fort Bragg, North Carolina, dedicated the former Special Operations Academic Facility as Colonel Aaron Bank Hall after a man known as the "father of Special Forces."

57. Aaron Bank, *From OSS to Green Berets: The Birth of Special Forces* (Novato, Calif.; Presidio Press, 1986), 172-174.

58. Ibid., 175-185; and Alfred H. Paddock, Jr., *U.S. Army Special Warfare, Its Origins: Psychological and Unconventional Warfare, 1941-1952* (Washington, D.C.: National Defense University Press, 1982), 150-151.

59. The issue of lineage of the Army's Special Forces has been made more complicated and confusing by the various reorganizations and designations in the U.S. Army as well as the decision, unjustified in the opinion of many Special Forces veterans, by the Army's Center of Military History to designate the U.S.-Canadian 1st Special Force (the "Devil's Brigade"), which served in Italy and southern France, as the forerunner of Army Special Forces. www.Army.mil/cmh//lineage/branches/sf. A number of authoritative sources disagree with the decision of the Army's Center of Military History and cite OSS's SO and OG combat units as direct ancestor of the Army's Special Forces. See, for example, Alfred H. Paddock, Jr., *U.S. Army Special Warfare: Its Origins: Psychological and Unconventional Warfare, 1941-1952* (Washington, D.C.: National Defense University Press, 1982; rev. ed., University Press of Kansas, 2005), 23-25; Aaron Bank, *From OSS to Green Berets: The Birth of the Special Forces* (Novato, Calif.: Presidio Press, 1986), 205-206; Gordon L. Rottman, *The U.S. Army Special Forces, 1952-1984* (Osprey, 1985); Charles Simpson, *Inside the Green Berets: The First Thirty Years: A History of the U.S. Army's Special Forces* (Novato, Calif.: Presidio Press, 1983); and Mike Yard, "OGs Get to Tell Their Story," *OSS Society Digest*, Number 1317, 30 March 2006, accessed 30 March 2006. Paddock and Bank credit OSS as more representative than ranger, paratrooper or the "Devil's Brigade," of the kind and range of actions conducted by today's Special Forces. Furthermore, the Special Forces themselves have informally adopted OSS and its combat units such as Special Operations, Operational Groups, Detachment 101, and Maritime Unit, as among the predecessors of today's Special Forces. See, for example, the websites of the U.S. Army's John F. Kennedy Special Warfare Museum at Fort Bragg, Fayetteville, N.C., at www.soc.mil/swcs/museum, and the semi-official Airborne and Special Operations Museum, Fayetteville, N.C., at www.asomf.org/museum. On some of the OSS insignia being incorporated into today's Special Forces and the U.S. Special Operations Command, see the story on the opening of an OSS Exhibit at the Airborne and Special Operations Museum in January 2008, reported in "Retired General [John K. Singlaub] to Speak on Career," *Fayetteville Observer*, 24 January 2008, reprinted in *OSS Society Digest*, Number 1962, 26 January 2008, accessed 26 January 2008; and Troy Sacquety, "The Special Forces Patch: History and Origins, *Veritas: Journal of Army Special Operations History*, 3.3 (2007): 59-63.

60. Otto C. Doering, New York City, interview with Anthony Cave Brown, 1 November 1977, quoted in Anthony Cave Brown, *The Last Hero: Wild Bill Donovan* (New York: Times Books, 1982), 833.

61. The romantic cult of Donovan and the daring men and women of the OSS, although it largely ignored the laborious research and analysis which is such a major part of any major intelligence organization, was maintained by the CIA among others as it contributed to that patrimony contributed to the legitimacy and mystique of the postwar central intelligence and covert operations agency. Significantly, periods of major public criticism and assaults upon the CIA in America such as in the 1970s, 1980s, and the first decade of the twenty-first century, often included disputes over the nature of its predecessor, the OSS, as well as the linkage in personnel and policies between the

two. See, for example, the defense of the OSS amidst the critiques of the CIA in the 1970s by Richard Harris Smith, *OSS: The Secret History of America's First Central Intelligence Agency* (Berkeley: University of California Press, 1972); Ray S. Cline, *Secret Spies and Scholars: Blueprint of the Essential CIA* (Washington, D.C.: Acropolis Books, 1976); and Thomas F. Troy, *Donovan and the CIA: A History of the Establishment of the Central Intelligence Agency* (Frederick, Md.: University Publications of America, 1981, orig. ms. Classified secret, 1975), esp. v-vii, 402-415. For the evidence of the dichotomy amidst the controversy over the CIA today, see, for example, critiques of OSS in Rhodri Jeffreys-Jones, *Cloak and Dollar: A History of American Secret Intelligence*, 2nd ed. (New Haven, Conn.: Yale University Press, 2003, orig. ed., 2002) 130-153; and Tom Weiner, *Legacy of Ashes: The History of the CIA* (New York: Doubleday, 2007), 3-8; and a championing of the OSS by Patrick K. O'Donnell, *Operatives, Spies, and Saboteurs: The Unknown Story of the Men and Women of World War II's OSS* (New York: Free Press, 2004), 311-314.

62. Elbert G. ("Al") O'Keefe, a retired lieutenant colonel from U.S. Army intelligence (G-2), told the author that both in Europe and in the Korean War, Army G-2 used estimates compiled by OSS's Research and Analysis Branch. "They were good and still valuable," declared O'Keefe, who as a professor soldier, said he did not otherwise think much of Donovan and his organization. Elbert G. ("Al") O'Keefe, conversation with the author after a presentation on the OSS given by the author at Catoctin Mountain Park, Thurmont, Md., 17 November 2007.

63. In December 1942, General George C. Marshall, Army Chief of Staff, paid special notice to the importance of OSS's assistance in Operation Torch, the invasion of North Africa, OSS, *War Report of the OSS (Office of Strategic Services)*, Vol. 2, *The Overseas Targets* (New York: Walker Pub. Co., 1976; declassified version of the 1947 official history of the OSS), 18.

64. Ludwell Lee Montague, "The Origins of the National Intelligence Estimate," p. 66, offprint of an article from an unidentified journal, indicating that the text was an address given by "the late Dr. Montague, a retired member of the Board of National Estimates, at the first meeting of the Intelligence Forum, 11 May 1971." Copy in the CIA Records (RG 263), Thomas Troy Files, Box 12, Folder 99, National Archives II.

65. "Office of Strategic Services," George J.A. O'Toole, *The Encyclopedia of American Intelligence and Espionage: From the Revolutionary War to the Present* (New York: Facts on File, 1988), 339.

66. Richard Helms with William Hood, *A Look Over My Shoulder: A Life in the Central Intelligence Agency* (New York: Random House, 2003), 37.

67. John W. Brunner, *OSS Weapons*, 2nd ed. (Williamstown, N.J.: Phillips Publications, 2005), 2-12, and Toni L. Hiley, curator, CIA Museum, interview with the author, 14 January 2005.

68. As related by Donovan the next day to Stanley P. Lovell, head of OSS's Research and Development Branch, and included in Lovell's memoir, *Of Spies & Stratagems* (Englewood Cliffs, N.J.: Prentice Hall, 1963), 40-41.

69. Cline, *Secrets, Spies, and Scholars*, 67.

70. Francis B. Mills with John W. Brunner, *OSS Special Operations in China* (Williamstown, N.J.: Phillips Publications, 2002), 20.

71. Maj. Gen. William J. Donovan to Joint Chiefs of Staff, February 1945, reprinted in Mills with Brunner, *OSS Special Operations in China*, 442-446.

72. See, for example, regarding OSS/SO in Italy, "Company D [2677th Regiment OSS (Provisional)] – Semi Monthly Reports," September 1944 through February 1945; and Reports of SO under HQ Company D, 5th Army Detachment; 8th Army Detachment; and SO Maritime Detachment [all in Italy], April to May 1945, all in OSS Records (RG 226), Entry 136, Box 177, Folder 1886, and decoded copies of W/T messages to and from agents in the field, in Folder 1884, National Archives II. For SO in France, see Will Irwin, *The Jedburghs: The Secret History of the Allied Special Forces, France 1944* (New York: Public Affairs Press, 2005); and Colin Beavan, *Operation Jedburgh: D-Day and America's First Shadow War* (New York: Viking, 2006).

73. Quoted in "Office of Strategic Services," George J.A. O'Toole, *The Encyclopedia of American Intelligence and Espionage: From the Revolutionary War to the Present* (New York: Facts on File, 1988), 338.

74. This problem with the Air Force and lack of OSS's own planes was also a problem in OSS activities in Italy and other parts of the Mediterranean Theater, see Geoffrey M.T. Jones, comments in *Gli Americani e la Guerra de Liberazione in Italia: Office of Strategic Service(O.S.S.) e la Resistenza/ The Americans and the War of Liberation in Italy: Office of Strategic Services(O.S.S.) and the Resistance* (Venice: Institute of the History of the Resistance, 1995), p. 203. I am indebted to one of the Italian participants in the conference from which this book is drawn (and a former member of the Resistance and OSS's Cayuga and Rochester Missions), Piero Boni of Rome, father of my good friend, Silvia Boni, for providing me with a copy of this book In Italy, the OSS did not have the use of a specially designated unit of the U.S. Army Air Force, unlike the OSS in northern Europe which was served by a unit of the 8th U.S. Air Force known as the "Carpetbaggers" which flew black-painted, specially modified B-24 Liberators delivering OSS supplies and personnel. However, even in northern Europe, OSS teams often had difficulty obtaining sorties of fighter-bombers to attack targets they identified. In China, combat radio operator Arthur Reinhardt acknowledged similar difficulties because the calls had to go through the Chinese Army headquarters and then the 14th U.S. Air Force with its own scarce resources, by which time it was "hoped the target hadn't vanished." According to Reinhardt, "Initially the close support missions were flown by P-40's. The pilots were gutsy, flew in low, sometimes with flaps down and bombed and strafed. Later on (although I observed little), the P-40's were replaced by P-51's, which were faster and better armed, but the pilots were a different breed and did not get down low." Arthur Reinhardt, email to the author, 24 February 2007.

75. Brig. Gen. B.F. Caffey, U.S.A., chief, SP[Special Operations] Ops S/Sec., G-3, "Resistance Movements in Occupied Countries," OPD 334.8 TS, Case #325, "top secret," pp. 2, 5, in a 5-page typescript report, undated [c. early1945, before the end of the war in Europe in May 1945], copy from the papers of Dr. E.P. Lilly, former JCS historian, located in the CIA Records (RG 263), Thomas F. Troy Files, Box 4, Folder 30, National Archives II. Brigadier General Benjamin Franklin Caffey, Jr., was a graduate of the University of Michigan Law School, who became an Army officer in World War I, served in France, and became a lifelong friend of Dwight Eisenhower's beginning in the 1930s in the Philippines. In World War II, Colonel Caffey had led the 39th Regimental Combat Team in the capture of Algiers, then was made brigadier general and assistant commander of the 34th ("Red Bull") Infantry Division in Italy. Serving in its advance through Italy, he was twice injured and except for those disabling occurrences, Caffey would have become, according to Eisenhower, a major general and a division commander. The second injury, badly frozen feet, led to his assignment in an important staff position in the Allied Force Headquarters, Mediterranean in early 1945, which is probably when he wrote this report. After the war, Eisenhower appointed Caffey to the Operations Division of the General Staff. See Dwight D. Eisenhower, *Papers of Dwight David Eisenhower*, ed. Alfred D. Chandler, et al., 21 vols. (Baltimore, Md.: The Johns Hopkins University Press, 1970-2007), 2:687n; 2: 1293; 11:237; and 14:430n.

76. Ralph Ingersoll, *Top Secret* (New York: Harcourt, Brace, 1946), 181-183. In contrast, John Keegan, *The Second World War* (New York: Viking, 1990), belittles the impact of the French Resistance and contends that their entire contribution was barely equal to a single Allied division. A 1991 Army study of the relationship of 11 Jedburgh Teams to Omar Bradley's 12th Army Group in its rapid drive across eastern France from Paris to Nancy in August 1944, concluded that despite the courage and daring of the Jed teams, they were "only marginally significant" in that Army's offensive that month largely because the Army was advancing so rapidly, its staffs were overwhelmed with data, and the commanders were not familiar with nor receptive to such groups operating behind enemy lines. At the same time, the author acknowledges that the Jedburgh special operations teams in 1944 were an innovative idea but one that because of technical problems and lack of receptiveness by Army commanders was an idea ahead of its time. Samuel J. Lewis, *Jedburgh Team Operations in Support of the 12th Amy Group, August 1944* (Ft. Leavenworth, Kans.: Combat Studies Institute, U.S. Army Command and General Staff College, 1991), 65-66. On the other hand,

Stewart King, son of Donald King of the OSS, who served with the 36th Infantry Division during attacks in the Vosges (south of the 12th Army Group) in November 1944, "while there is a lot of information in the declassified OSS files [e.g. RG 226, Entry 190], much of the communication between the SSS forward teams that my father served on and the intelligence sections of the divisions (and later armies) they served with was informal and thus only recorded elliptically. OSS was one of many sources that intelligence staffs used." Stewart King, "Re: How Did OSS Help U.S. Soldiers in Combat?" *OSS Society Digest* Number 1539, 13 January 2007, ossociety@yahoogroups.com, accessed 13 January 2007. Concerning another action by OSS, see Robert G. Gutjahr, *The Role of Jedburgh Teams in Operation Market Garden* (Ft. Leavenworth, Kans.: Combat Studies Institute, U.S. Army Command and General Staff College, 1990), which examines what Gutjahr considers the understandable failure of the two Jed teams to blow up the assigned bridge as simply part of the overall failure of British Geneneral Bernard Law Montgomery's complex plan, Operation Market Garden, to get across the lower Rhine in September 1944.

77. On German Field Marshals Gerd von Rundstedt and Walther Model in France, see Irwin, *The Jedburghs*, 240-241; on Field Marshal Albert Kesselring in Italy, see, *Kesselring: A Soldier's Record* (New York: William Morrow, 1954), 272.

78. Gen. George C. Marshall quoted in Blake Ehrlich, *Resistance: France, 1940-1945* (Boston: Little, Brown, 1965), 194.

79. Dwight D. Eisenhower, *Crusade in Europe* (New York: Doubleday, 1948), 296.

80. William R. Peers and Dean Brelis, *Behind the Burma Road: The Story of America's Most Successful Guerrilla Force* (Boston: Little, Brown, 1963), 220.

81. In a letter to the head of the U.S. Army Air Forces in April 1945, Donovan said more than 200 downed American aviators had been rescued from the jungles by OSS personnel and that the 10th U.S. Air Force had received 90 per cent of its target intelligence in 1944 from OSS units. William J. Donovan to General [H.H. "Hap"] Arnold, 6 April 1945, pp. 1-2, OSS Records (RG 226), Directors Office Files, Microfilm M1642, Roll 21, Frames 107-108, National Archives II.

82. In twelve months in 1944-1945, Detachment 101 was responsible for the deaths of an average of nearly 200 Japanese soldiers per month in jungle ambushes in Burma. Donovan reported in April 1945 that an analysis of enemy casualties inflicted in the Northern Combat Area Command (NCAC) between 15 October 1944 and 15 January 1945, by all ground troops in that area "reveals that Det. 101, with less than one per cent of total strength in NCAC, has inflicted 29 per cent of all casualties." William J. Donovan to Gen. Marshall [Gen. George C. Marshall, US Army Chief of Staff], 6 April 1945, p. 2, in OSS Records (RG 226), Directors Office Files, Microfilm M1642, Roll 21, Frames 109-110, National Archives II.

83. Toni L. Hiley, curator of the CIA Museum, Langley, Va., interview with the author while providing a personal tour of the exhibits, 14 January 2005. Indeed, the first American casualty in the war against Al Qaeda and the Taliban regime in Afghanistan in 2001 was Mike Spann, a CIA Special Activities Division, Ground Branch operative, who was working with U.S. Army troops there when he was killed.

84. General Orders, War Department, Washington, D.C., 17 January 1946, by order of Gen. Dwight D. Eisenhower, Army Chief of Staff, reprinted in Peers and Brelis, *Behind the Burma Road*, 208-209.

85. "Recent rescues have raised to well over 200 the number of Air Force personnel brought to safety by OSS units in Burma." William J. Donovan to General Arnold [Gen. H.H. Arnold, Chief of Staff of U.S. Army Air Forces], 6 April 1945, p. 1, in OSS Records (RG 226), Directors Office Files, Microfilm M1642, Roll 21, Frames 107-108, National Archives II.

86. John W. Brunner, telephone interview with the author, 21 March 2005. The figures of 2,000 OSS personnel in China in 1945 being credited with direct responsibility for the killing of more than 12,000 Japanese troops, are in General Orders No. 27, Headquarters, U.S. Forces, China Theater, by Order of General [Albert] Wedemeyer, 1 February 1946, reprinted in Mills with Brunner, *OSS Special Operations in China*, 431-432.

87. General Orders No. 27, Headquarters, U.S. Forces, China Theater, by Order of General [Albert] Wedemeyer, 1 February 1946, reprinted in Mills with Brunner, *OSS Special Operations in China*, 431-432.

88. See, for example, Ray S. Cline, *Secrets, Spies, and Scholars: Blueprint of the Essential CIA* (Washington, D. C.: Acropolis Books, 1976), 76.

89. Edward Hymoff, who worked for OSS and later became a war correspondent and author, notes that even calculating the numbers of people who served in or with OSS is difficult if not impossible cause it would have to take into account those who were assigned full-time, those assigned temporarily, as well as, even more difficult, those foreign partisans, guerrillas, and intelligence gatherers who were for various periods of time directly under the command of OSS officers of branches such as SO, OG, SI, MO, and X-2. Edward Hymoff, *The OSS in World War II* (New York: Ballantine Books, 1972), 340.

90. Casualty figures from the following: OSS, *War Report of the OSS, Overseas Targets*, 220, which lists a 10 per cent casualty rate as an average for SO/France and 7 per cent for SI/France; see also Cave Brown, *The Last Her)*, 787-788; Robin Winks, *Cloak and Gown: Scholars in the Secret War, 1939-1961* (New York: Morrow, 1987), 203; Donnell, *Operatives, Spies, and Saboteurs*, 15.

91. Brig. Gen. B.F. Caffey, U.S.A., chief, SP[Special Operations] Ops S/Sec., G-3, "Resistance Movements in Occupied Countries," OPD 334.8 TS, Case #325, "top secret," p. 3, in a 5-page typescript report, undated [c. early1945, before the end of the war in Europe in May 1945], copy from the papers of Dr. E.P. Lilly, former JCS historian, located in the CIA Records (RG 263), Thomas F. Troy Files, Box 4, Folder 30, National Archives II.

92. Jim Downs, *World War II: OSS Tragedy in Slovakia* (Oceanside, Calif.: Liefrinck, 2002), 247-325.

93. See, for example, OSS-101 Association at www.oss-101.com; OSS ComVets Association; Association of Foreign Intelligence Officers, www.afio.com; and the OSS Society at www.osssociety.org. The Operational Groups have their own website at www.ossog.org. The author wishes to thank especially, Arthur Reinhardt of the OSS Society and the ComVets Association and Charles T. Pinck, President of the OSS Society, for their assistance.

94. Websites of the International Spy Museum at www.spymuseum.org ; and U.S. Army's John F. Kennedy Special Warfare Museum at Fort Bragg, Fayetteville, N.C. at www.soc.mil/swcs/museum, and the Airborne and Special Operations Museum, Fayetteville, N.C. at www.asomf.org/museum. The latter, which became part of the U.S. Army's museum system in 2005, includes references to the OSS in its regular exhibits, but established a special exhibit on the OSS in January 2008, which included materials from its own collection, plus artifacts loaned by the North Carolina Museum of History from the collection of George Hill, a member of a prominent North Carolina family, who served in the OSS's Research and Development Branch. Exhibit reported in "Retired General [John K. Singlaub] to Speak on Career," *Fayetteville Observer*, 24 January 2008, reprinted in *OSS Society Digest*, Number 1962, 26 January 2008, accessed 26 January 2008.

95. "Naval Special Warfare Heritage: Navy Seal History," The National Navy UDT-Seal Museum, Fort Pierce, Florida, www.Navysealmuseum .com/heritage/history.php, accessed 26 January 2008.

96. National Museum of the Marine Corps, World War II Gallery, "Global Developments," www.usmcmuseum.org, accessed 26 January 2008.

97. The author expresses his appreciation to Ms. Toni L. Hiley, curator of the CIA museum, for a personal tour 14 January 2005 of the displays, including the OSS Gallery, which opened in 2001 on the 60th anniversary of the establishment of Donovan's organization, the Cold War Gallery, Directorate of Intelligence Gallery and Directorate of Science and Technology Gallery. Although the CIA museum inside the agency's compound in Langley, Va., is closed to the public, the agency has an electronic virtual museum at www.cia.gov/about_cia/cia-museum/index-html, which provides a glimpse of the galleries.

98. See, for example, R. Harris Smith, *OSS: The Secret History of America's First Central Intelligence Agency* (Berkeley: University of California Press, 1972), 367-383. The point about the different strategic situation was made forcefully by Ray S. Cline, veteran of OSS's R&A Branch, who later served in high positions in both the intelligence and clandestine services divisions of the CIA, in *Secrets, Spies, and Scholars*, 75-76. An Army report in 1945 had emphasized the necessity of support for OSS covert operations teams by the local civilian population and indicated that where that had been lacking, Austria, Hungary, Bulgaria, and southern Serbia, OSS missions there had failed. Brig. Gen. B.F. Caffey, U.S.A., chief, SP[Special Operations] Ops S/Sec., G-3, "Resistance Movements in Occupied Countries," OPD 334.8 TS, Case #325, "top secret," p. 3, typescript report, undated [c. early1945, before the end of the war in Europe in May 1945], copy from the papers of Dr. E.P. Lilly, former JCS historian, located in the CIA Records (RG 263), Thomas F. Troy Files, Box 4, Folder 30, National Archives II.

99. As evidenced by comments by some OSS veterans on contemporary actions and issues in the OSS Society's electronic bulletin board and chat room on the internet, osssociety@yahoogroups.com. This site is open to OSS veterans, their families and descendants and to other persons interested in the OSS, but applicants must be approved to have access. The author has had access since he began his research on this history in 2005.

100. Max Boot, testimony before a House committee, reported in Jefferson Morris, "Analyst Says SOCOM Not Well Suited for Terror War," *Aerospace Daily & Defense Report*, 30 June, 2006, p. 5, reprinted in *OSS Society Digest*, Number 1401, 16 July 2006, accessed 16 July 2006. Boot is the author, among other works, of *The Savage Wars of Peace: Small Wars and the Rise of American Power* (New York: Basic Books, 2002). Two candidates for the Republican nomination for President in 2008, Senator John McCain and Mitt Romney, also proposed recreating an OSS-like intelligence and operations agency. OSS Reborn by Charles and Dan Pinck, "Glorious Amateurs: A New OSS," www.ossreborn.com, accessed 25 January 2008.

101. Max Boot, "Why the OSS Succeeded and the CIA is Failing," *Los Angles Times*, 22 July 2004. Boot argued that "Donovan's high-powered recruits did impressive work, often utilizing connections that no humdrum bureaucrat could possibly have cultivated."

BIBLIOGRAPHY

UNPUBLISHED SOURCES

Archives of Organizational Records

College Park, Maryland, National Archives II, Records of the following agencies:
  Central Intelligence Agency (Record Group 263)
  National Park Service (Record Group 79)
  Office of Strategic Services (Record Group 226)
U.S. Army Corps of Engineers (Record Group 77)
Hyde Park, New York, Franklin D. Roosevelt Presidential Library
  Franklin D. Roosevelt Papers
Independence, Missouri, Harry S Truman Presidential Library
  Harry S Truman Papers
Thurmont, Maryland, Catoctin Mountain Park Archives
Triangle, Virginia, Prince William Forest Park Archives
St. Louis, Missouri, National Archives and Records Service, National Personnel Records Center:
  Lykes, Ira B., Personnel Records
  Williams, Garland B. ("Mike"), Personnel Records
Suitland, Maryland, Federal Records Center, National Park Service Records (RG 79)

Papers of Individuals

Carlisle, Pennsylvania, U.S. Army Historical Research Institute
  Donovan, William J., Papers
Princeton, New Jersey, Princeton University Libraries
  Dulles, Allan W. Papers
Stanford, California, The Hoover Institution
  Goodfellow, M. Preston, Papers
  Miles, Milton E., Papers
  Wedemeyer, Albert, Papers
Thurmont, Md., Catoctin Mountain Park
  Guay, Albert R., Diary, October-December 1942

Washington, D.C., Library of Congress
  Deuel, Wallace R., Papers
  Goldberg, Arthur J., Papers

Interviews conducted by the author

OSS Veterans interviewed, their OSS branch, and Training Areas.

Belanger, Roger L, interviews and correspondence, 21 January 2005 through 17 January 2008 (OSS CB; Area C)

Binns, Merritt, interviews 3, 4 April 2007 (OSS SO; Area A)

Boals, James L. interview, correspondence, 20, 25 February 2007 (OSS CB, Area C)

Boni, Piero, interview 15 July 2007 (OSS SO, SI; Training Area Naples, Italy)

Brunner, John W., interview 21 March 2005, correspondence through July 2008 (OSS CB; HQ and Area C)

Cappony, Spiro ("Gus"), interview 16 September 2006 (OSS SO, Areas C, B)

Carter, Robert E., interview 28 August 2008 (OSS OG, SO, S&T; Areas F, B, W-A)

Christensen, Hal S., interview, 26 August 2008 (OSS CB, Areas C, B)

Civitella, Caesar J., interviews 18, 25, 29 April 2008 (OSS OG; Areas F, B)

Di Giacomo, Wanda, interview, 21 March 2005 (OSS Personnel, SI; HQ & Roslyn, Va.)

Défourneaux, René. correspondence, 7, 8 August 2008 (OSS SO; Area WS and SOE)

Donnally, Gail F., interview, 30 April 2005 (OSS CB; Area C)

Etheridge, Obie L., interview 14 January 2008 (OSS SI, CB; Areas F and C)

Flisser, Marvin S., interview 27 January 2005 (OSS, CB, R&A, Area C)

Gleason, Frank A., interviews 31 January 2005 through 15 August 2008 (OSS SO; Area B)

Guay, Albert R. ("Al"), interview 24 October 2005 (OSS SO, Area B)

Hess, Robert O., interview, 27 August 2008 (OSS CB, X-2, Area F and SOE)

Hooker, John C., Jr., interview and correspondence 23 June through 14 August 2008 (OSS MU, SO, OG, Areas F, A)

Johnson, Ellsworth ("Al"), interview 27 June 2008 (OSS OG, Areas F, B)

Kenney, David A., interview 11 April 2005 (OSS CB, Areas A, C)

Kloman, Erasmus H. ("Ras"), interview 24 January 2005 (OSS SO, OG; Areas F, A, B, C, D)

Lazarsky, Joseph E., interviews 14 March 2005 through 9 July 2008 (OSS SO, Area B)

Loomis, Vader ("Jim"), interview, 10 January 2008 (OSS CB, MC; Areas B, C)

McNamara, Emmett, interview, 2 September 2008 (OSS OG; Areas B, F.)

Marsh, Timothy R., interview 18 January 2006 (OSS S&T, Area C)

Materazzi, Albert ("Al"), interview 26 January 2005, correspondence 23 January 2005 through 6 March 2007 (OSS OG; Areas A, F, B)

McIntosh, Elizabeth P. ("Betty"), interview 12 March 2005 (OSS MO; OSS HQ)
Nelson, Charles, interview 27, 2008, (OSS/SSU, Med.Corps, Kunming)
Parkin, Charles M., interviews 10, 15, 18 May 2005 (OSS SO, SI, Area B)
Prunier, Henry, interview 25 August 2008 (OSS SO, Area F, W-A)
Ranney, James F, interview 8 January 2005 (OSS CB, Areas B, C)
Richter, Allen R. interview 25 March 2005, 16 March 2007 (OSS CB, Area C)
Reinhardt, Arthur F. ("Art"), interview 13 December 2004 and correspondence through 19 August 2008 (OSS CB; Areas A, C)
Sichel, Peter F.M., interview, 9 July 2008 (OSS SI; Area B)
Singlaub, John K. ("Jack"), interview 11 December 2004, 6 January 2005 (OSS SO; Areas F, B)
Spear, Reginald G. ("Reg"), interviews 25 March 2005 through 16 August 2008 (OSS SO, SI, Areas F, B)
Spence, John, interview 28 January 2005 (OSS, SO, MU, Area D)
Tully, Joseph J., interviews 23 March 2005, 29, 30 December 2007 (OSS CB; Area C)
Waller, Barbara Hans, interview, 19 March 2005 (OSS CB; OSS HQ)

CORRESPONDENCE (OR INTERVIEWS) WITH RELATIVES OF DECEASED OSS VETERANS

Dow, Dorothea D. ("Dodie"), widow or Arden W. Dow, interviews 15 May 2005, 1 January 2008 (OSS SO, Areas B and A)
Chappell, Jack R., son of Howard W. Chappell, 4, 17 October 2007 (OSS OG, Areas F and B)
Gonzalez, Ronald, son of Louis J. Gonzalez, 14, 16, 23 January 2008 (OSS R&A, Area A)
Harrow, Mary, widow of Sidney L. Harrow, interview 17 October 2005 (OSS Assessment, Area A)
Herstad, Rolf, son of Arne I. Herstad, 24 June through 9 August 2008 (OSS OG, Areas A, B)
Huston, Steven A., son of Frank V. Huston, interview 24 September 2005, and correspondence, 14 October 2005 (OSS CB; Area C)
Kloman, Erasmus H., cousin of Joseph Anthony Kloman, interview 6 January 2005 (OSS SO; Area A)
Manning, Betty, widow of Howard Manning, interview 4 March 2005 (OSS S&T, Area C)
Nicholas, Ned, son of Edward E. Nicholas, III, 18 July 2008 (OSS CB, Area C)
Savoldi, J.G., grandson of Joseph ("Jumping Joe") Savoldi, 9 May 2006 through 29 December 2007 (OSS SO; Area B)
Snyder, James, son of Jacques L. Snyder, 15 January through 8 August 2008 (OSS OG, SO, SI, X-2, Areas F, A)

Other interviews by the author

Bloom, Lewis, 17 January 2007, former Army Intelligence Officer, trained Camp Ritchie
Farrall, William, 21 June 2007, employee, Smallwood State Park, Maryland
Henderson, Craig, 1 April 2008, Ranger, Maryland State Forest Service
Hickman Robert, 24 January 2008, Superintendent, Prince William Forest Park
Hiley, Toni L., 14 January 2005, Curator, CIA museum
Kimmel, Ross, 22 May 2007, Historian, Maryland Park Service
Kinnaird, John, 7 June 2006, longtime resident of Thurmont, Maryland
Long, Frank, 7 June 2006, longtime resident of Thurmont, Maryland
Mackley, James H., 13 June 2006, longtime resident of Thurmont, Maryland
Moffatt, William, 5 June 2007, Associate in Maryland State Park Service
Noile, Robert A., 27 April, 16 May 2007, longtime resident of Triangle, Virginia
O'Keefe, Elbert B., 17 November 2007, Retired Army Intelligence Officer (G-2)
Polinsky, Eugene J., 21 March 2007, former navigator B-24 "Carpetbagger" unit.
Poole, J. Mel, 18 May 2005, Superintendent, Catoctin Mountain Park
Santucci, Vincent, 22 August 2007, Chief Ranger, George Washington Parkway
Voigt, James W., 18 May 2005, Resource Manager, Catoctin Mountain Park
Volonoski, Judy, 13 December 2004, Museum Technician, Prince William Forest Park

Oral History Interviews Conducted by Others

Berkeley, California, Regional History Office, Bancroft Library, University of California.
    Newton B. Drury, Oral History Interview, "Parks and Redwoods, 1919-1971," conducted by Amelia Roberts Fry and Susan Schrepfer, transcript, 2 vols., 1972.

College Park, Maryland, National Archives II, Transcripts of Oral History interviews of OSS veterans, CIA Records, OSS Oral History Project, interviews with
    Raymond Brittenham, 27 February 1997 by Siegfried Beer
    Samuel Halpern, 16 June 1997 by Maochun Yu
    August Heckscher, 25 March 1997 by Siegfried Beer
    Richard Helms, 21 April 1997 by Christof Mauch
    Robert Francis Houlihan, 16 September 1996 by Christof Mauch
    H. Stuart Hughes, 10 May 1997 by Barry Katz
    Jacob J. Kaplan, 6 November 1997 by Siegfried Beer
    Elizabeth McIntosh, 2 May 1997 by Maochun Yu
    Eloise Page, 31 March 1997 by Timothy Naftali

Barbara Lauwers Podoski, 4 September 1997 by Christof Mauch
Edwin Putzell, 11 April 1997 by Timothy Naftali
Walt W. Rostow, 3 February 1997 by Barry Katz
Arthur M. Schlesinger, Jr. 9 June 1997 by Petra Marquardt-Bigman and Christof Mauch
John K. Singlaub, 31 October 1996 by Maochun Yu and Christof Mauch
Turner T. Smith 28 March 1997 by Timothy Naftali
John Waller, 17 January 1997 by Timothy Naftali

Harpers Ferry, West Virginia, Harpers Ferry Center, National Park Service
Ira B. Lykes, Oral History Interview 23 November 1973 by S. Herbert Evison

New Brunswick, New Jersey, Rutgers Oral History Archives of World War II
Transcripts of interviews with
Lewis Bloom, 21 June 1994 by G. Kurt Piehler
Stephen J. Capestro, 17 August 1994 by G. Kurt Piehler
Eugene Polinsky 11 December 2001 by Shaun Illingworth and Lauren O'Gara

UNPUBLISHED SOURCES

Books

Albright, Horace M. as told to Robert Cahn, *The Birth of the National Park Service: The Founding Years, 1913-33*. Salt Lake City, Utah: Howe Bros., 1985.

Alcorn, Robert H. *No Banners, No Bands: More Tales from the OSS*. New York: Donald McKay, 1965.

———. *No Bugles for Spies: Tales of the OSS*. New York: Donald McKay, 1962.

Aldrich, Richard J. *Intelligence and the War against Japan: Britain, America and the Politics of Secret Service*. New York: Cambridge University Press, 2000.

Aline, Countess of Romanones. *The Spy Went Dancing: My Further Adventures as an Undercover Agent*. New York: Putnam, 1990.

———. *The Spy Wore Red: My Adventures as an Undercover Agent in World War II*. New York: Random House, 1987.

Alsop, Stewart and Thomas Braden. *Sub Rosa: The O.S.S. and American Espionage*. New York: Reynal and Hitchcock, 1946.

Ambrose, Stephen E. *Ikes' Spies: Eisenhower and the Espionage Establishment*. Garden City, N.Y.; Doubleday, 1981.

Ameringer, Charles D. *U.S. Foreign Intelligence: The Secret Side of American History*. Lexington, Mass.: D.C. Heath, 1990.

Andrew, Christopher. *For the President's Eyes Only: Secret Intelligence and the American Presidency from Washington to Bush*. New York: Harper/Collins, 1995.

Andrade, Allan. *S.S. Leopoldville Disaster, December 24, 1944*. New York: Tern Book Co., 1997.

Andrew, Christopher. *For the President's Eyes Only: Secret Intelligence and the American Presidency from Washington to Bush*. New York: Harper Collins, 1995.

Applegate, Rex. *Kill or Get Killed*. Harrisburg, Pa.: Military Publishing Co., 1943.

———, and Chuck Melson. *The Close-Combat Files of Colonel Rex Applegate*. Boulder, Colo.: Paladin Press, 1998.

Atkinson, Rick. *An Army at Dawn: The War in North Africa, Theater in World War II* (New York: Farrar, Straus and Giroux, 2004.

———. *The Day of Battle: The War in Sicily and Italy, 1943-1944*. New York: Henry Holt, 2007.

Bathrolomew-Feis, Dixee R. *The OSS and Ho Chi Minh: Unexpected Allies in the War against Japan*. Lawrence: University Press of Kansas, 2006.

Bank, Aaron. *From OSS to Green Berets: The Birth of the Special Forces*. Novato, Calif.: Presidio Press, 1986.

Barrett, David D. *Dixie Mission: U.S. Army Observer Group in Yenan, 1944*. Berkeley: University of California Press, 1970.

Barrett, David P. and Larry N. Shyu. Eds. *China in the Anti-Japanese War, 1937-1945: Politics, Culture and Society*. New York: Peter Lang, 2001.

Beavan, Colin. *Operation Jedburgh: D-Day and America's First Shadow War*. New York: Viking, 2006.

Breuer, William B. *MacArthur's Undercover War: Spies, Saboteurs, Guerrillas, and Secret Missions*. New York: John Wiley, 1995.

Bruce, David K.E. *OSS against the Reich: The World War II Diaries of Colonel David K.E. Bruce*. Ed. Nelson Douglas Lankford. Kent. Oh.: Kent State University Press, 1991.

Brunner, John W. *OSS Weapons*, 2nd ed. Williamstown, N.J.: Phillips Publications, 2005.

Burke, Michael. *Outrageous Good Fortune*. Boston: Little, Brown, 1984.

Casey, William J. *The Secret War against Hitler*. Washington, DC: Regnery Gateway, 1988.

Cassidy, William L. Ed. *History of the Schools and Training Branch of the OSS*. San Francisco: Kingfisher Press, 1983.

Cave Brown, Anthony. *The Last Hero: Wild Bill Donovan*. New York: Times Books, 1982.

Chalou, George C. Ed. *The Secrets War: The Office of Strategic Services in World War II*. Washington, DC: National Archives and Records Administration, 1992.

Cline, Ray S. *Secrets, Spies, and Scholars: Blueprint of the Essential CIA*. Washington, D.C.: Acropolis Books, 1976.

Colby. William and Peter Forbath. *Honorable Men: My Life in the CIA*. New York: Simon & Schuster, 1978.

Coon, Carleton S. *A North African Story: An Anthropologist as OSS Agent*. Ipswich, Mass.: Gambit, 1980.

Congressional Country Club. *The Congressional Country Club, 1924-1984*. [Bethesda, Md.: Congressional Country Club, c. 1985].

Cooper, Chester L. *In the Shadows of History: Fifty Years Behind the Scenes of Cold War Diplomacy*. Amherst, N.Y.: Prometheus Books, 2005.

Corvo, Max. *The O.S.S. in Italy, 1942-1945: A Personal Memoir*. New York: Praeger, 1990.

Darling, Arthur. *The Central Intelligence Agency: An Instrument of Government to 1950*. University Park: Pennsylvania State University Press, 1990.

Dallek, Robert. *Franklin D. Roosevelt and American Foreign Policy, 1932-1943*. New York: Oxford University Press, 1979.

Dawidoff, Nicholas. *The Catcher was a Spy: The Mysterious Life of Moe Berg*. New York: Pantheon, 1994.

Daws, Gavin. *Prisoners of the Japanese*. New York: William Morrow, 1995.

Delattre, Lucas. *A Spy at the Heart of the Third Reich: The Extraordinary Story of Fritz Kolbe, America's Most Important Spy in World War II*. Trans. George A. Holoch, Jr. New York: Atlantic Monthly Press, 2005.

Deschamps, Hélène. *Spyglass: An Autobiography*. New York: Henry Holt, 1995.

Downs, Jim. *World War II: OSS Tragedy in Slovakia*. Oceanside Calif.: Liefrinck, 2002.

Doyle, David W. *True Men and Traitors: From the OSS to the CIA: My Life in the Shadows*. New York: John Wiley, 2001.

Dreux, William B. *No Bridges Blown*. Notre Dame, Ind.: University of Notre Dame Press, 1971.

Dulles, Allen W. *The Secret Surrender*. New York: Harper and Row, 1966.

Dunlop, Richard. *Behind Japanese Lines: With the OSS in Burma*. Chicago: Rand McNally, 1979.

____. *Donovan: America's Master Spy*. Chicago: Rand McNally, 1982.

Ehrlich, Blake. *Resistance: France, 1940-1945*. Boston: Little, Brown, 1965.

____. *Donovan: America's Master Spy*. New York: Rand McNally, 1982.

Eisenhower, Dwight D. *Crusade in Europe*. New York: Doubleday, 1948.

____. *The Papers of Dwight David Eisenhower*. Ed. Alfred D. Chandler, et al. 21 vols. Baltimore, Md.: The Johns Hopkins University Press, 1970-2007.

Fairbairn, William E. *Get Tough: How to Win in Hand-to-Hand Combat*. New York: Appleton, 1942.

Fane, Francis Douglas and Don Moore. *The Naked Warriors*. New York: Appleton-Century Crofts, 1956.

Fenn, Charles. *At the Dragon's Gate: With the OSS in the Far East*. Annapolis, Md. Naval Institute Press, 2004.

Fondazione Corpo Volontari Della Liberta (F.V.L.), Veterans Association of O.S.S. *Gli Americani e la Guerra de Liberazione in Italia: Office of Strategic Service(O.S.S.) e la Resistenza / The Americans and the War of Liberation in Italy: Office of Strategic Services(O.S.S.) and the Resistance*. Venice: Institute of the History of the Resistance, 1995.

Foote, Michael R.D. *SOE: An Outline History of the Special Operations Executive, 1940-46*. London: British Broadcasting Corporation, 1984.

Ford, Corey. *Donovan of OSS*. Boston: Little, Brown, 1970.

____, and Alistair McBain, *Cloak and Dagger*. New York: Random House, 1945.

Ford, Kirk. *OSS and the Yugoslav Resistance, 1943-1945*. College Station: Texas A&M University Press, 1992.

Ford, Roger. *Steel from the Sky: The Jedburgh Raiders, France 1944*. London: George Weidenfeld and Nicholson, 2004.

Foresta, Ronald A. *America's National Parks and Their Keepers*. Washington, D.C.: Resources for the Future, Inc., 1984.

Franks, Lucinda. *My Father's Secret War: A Memoir*. New York: Miramax, 2007.

Goulden, Joseph C. *The Best Years, 1945-1950*. New York: Atheneum, 1976.

____, with Alexander W. Raffio. *The Death Merchant: The Rise and Fall of Edwin P. Wilson*. New York: Simon and Schuster, 1984.

Grose, Peter. *Gentleman Spy: The Life of Allen Dulles*. Boston: Houghton Mifflin, 1994.

Gutjahr, Robert G. *The Role of Jedburgh Teams in Operation Market Garden*. Fort Leavenworth, Kans.: Combat Studies Institute, U.S. Army Command and General Staff College, 1990.

Hall, Roger. *You're Stepping on My Cloak and Dagger*. New York: Norton, 1957.

Harris, Stephen L. *Duffy's War: Fr. Francis Duffy, Wild Bill Donovan, and the Irish Fighting 69th in World War I*. Washington, D.C.: Potomac Books, 2006.

Hayden, Sterling. *Wanderer*. New York: Knof, 1970.

Helms, Richard with William Hood. *Look Over My Shoulder: A Life in the Central Intelligence Agency*. New York: Random House, 2003.

Hemingway, Jack. *Misadventures of a Fly Fisherman: My Life With and Without Papa*. Dallas, Tex.: Taylor Pub. Co.,1986.

Hilsman, Roger. *American Guerrilla: My War Behind Japanese Lines*. Washington, DC: Brassey's, 1990.

Hodgson, Lynn Philip. *Inside Camp-X*. Oakville, Ontario: L.P. Hodgson, 1999.

Hunt, E. Howard. *Undercover: Memoirs of an American Secret Agent*. New York: G.P. Putnam's Sons, 1974.-

Hyde, H. Montgomery. *Cynthia: The Most Seductive Secret Weapon in the Arsenal of the Man Called Intrepid*. New York: Ballantine, 1965.

____. *Secret Intelligence Agent*. New York: St. Martin's Press, 1983.

Hymoff, Edward. *The OSS in WWII*. New York: Ballantine, 1972.

Ickes, Harold L. *The Secret Diary of Harold L. Ickes.* 3 vols. New York: Simon and Schuster, 1953-54.

Ingersoll, Ralph. *Top Secret.* New York: Harcourt, Brace, 1946.

Irwin, Will. *The Jedburghs: The Secret History of the Allied Special Forces, France 1944* New York: Public Affairs/Perseus, 2005.

Jakub, Jay. *Spies and Saboteurs: Anglo-American Collaboration and Rivalry in Human Intelligence Collection and Special Operations, 1940-1945.* New York: St. Martin's Press, 1999.

Jason, Sonya N. *Maria Gulovich, OSS Heroine of World War II: The Schoolteacher Who Saved American Lives in Slovakia.* Jefferson, N.C.: McFarland, 2008.

Jeffreys-Jones, Rhodri. *Cloak and Dollar: A History of American Secret Service*, 2nd ed. New Haven, Conn.: Yale University Press, 2003.

Jenkins, Roy. *Churchill: A Biography.* New York: Farrar, Straus, 2001.

Judt, Tony. Ed. *Resistance and Revolution in Mediterranean Europe, 1939-1948.* London: Routledge, 1989.

Kahn, David. *The Codebreakers: The History of Secret Writing.* New York: Simon and Schuster, 1999.

Katz, Barry M. *Foreign Intelligence: Research and Analysis in the Office of Strategic Services, 1940-1945.* New York: St. Martin's, 1999.

Kauffman, Louis, Barbara Fitzgerald, and Tom Sewell. *Moe Berg: Athlete, Scholar, Spy.* Boston: Little, Brown, 1975.

Kedward, H.R. *In Search of the Maquis.* Oxford: Oxford University Press, 1993.

Keegan, John. *Intelligence in War: Knowledge of the Enemy from Napoleon to Al-Qaeda.* New York: Knopf, 2003.

_____. *The Second World War.* New York: Viking, 1990.

Kesselring, Albert. *Kesselring: A Soldier's Record.* New York: William Morrow, 1954.

Kloman, Erasmus H. *Assignment Algiers: With the OSS in the Mediterranean Theater* Annapolis, Md.: Naval Institute Press, 2005.

Knott, Stephen. *Secret and Sanctioned: Covert Operations and the American Presidency.* New York: Oxford University Press, 1996.

Knox, Bernard. *Essays: Ancient and Modern.* Baltimore, Md.: The Johns Hopkins University Press, 1989.

Lawrence, Mark Atwood. *Assuming the Burden: Europe and the American Commitment to War in Vietnam.* Berkeley: University of California Press, 2005.

Lee, David. *Up Close and Personal: The Reality of Close-Quarter Fighting in World War II.* Annapolis, Md.: Naval Institute Press, 2006.

Leith, Hal. *POWs of Japanese Rescued!: General J.M. Wainwright.* Trafford Publishing, 2004.

Lewis, Samuel J. *Jedburgh Team Operations in Support of the 12th Army Group, August 1944.* Ft. Leavenworth, Kans.: Combat Studies Institute, U.S. Army Command and General Staff College, 1991.

Lindsay, Franklin A. *Beacons in the Night: With the OSS and Tito's Partisans in Wartime Yugoslavia.* Stanford. Calif.: Stanford University Press, 1995.

Lovell, Stanley P. *Of Spies and Stratagems.* Englewood Cliffs, N.J.: Prentice-Hall, 1963.

Lucas, Peter. *The OSS in World War II Albania: Covert Operations and Collaboration with Communist Partisans.* Jefferson, N.C.: McFarland & Co., 2007.

MacDonald [later Elizabeth MacDonald McIntosh], Elizabeth. *Undercover Girl* (New York: Macmillan, 1947.

MacPherson, Nelson. *American Intelligence in WarTime London: The Story of the OSS.* London: Frank Cass, 2003.

McIntosh, Elizabeth P. *Sisterhood of Spies: The Women of the OSS.* Annapolis, Md.: Naval Institute Press, 1998.

Mangold, Tom. *Cold Warrior: James Angleton: The CIA's Master Spy Hunter.* New York: Simon & Schuster, 1991.

Marquis, Susan L. *Unconventional Warfare: Rebuilding U.S. Special Forces.* Washington, D.C.; Brookings Institute, 1997.

Mauch, Christof. *The Shadow War against Hitler: The Covert Operations of America's Wartime Secret Intelligence Service.* New York: Columbia University Press, 2003.

Melton, H. Keith. *OSS Special Weapons and Equipment: Spy Devices of World War II.* New York: Sterling, 1991.

Miles, Milton. *A Different Kind of War: The Little Known Story of the Combined Guerrilla Forces Created in China by the U.S. Navy and the Chinese during World War II.* Garden City, N.Y.: Doubleday, 1967.

Miller, Russell. *Behind the Lines: The Oral History of Special Operations in World War II.* New York: St. Martin's, 2002.

Millett, Allan. R. *Semper Fidelis: The History of the United States Marine Corps.* Rev. ed. New York: Free Press, 1991.

Millett, Allan R. and Williamson Murray. *A War to Be Won: Fighting the Second World War.* Cambridge, Mass.: Harvard University Press, 2000.

Mills, Francis B. with John W. Brunner, *OSS Special Operations in China.* Williamstown, NJ: Phillips Publications, 2002.

Montague, Ludwell Lee. *General Walter Bedell Smith as Director of Central Intelligence, October 1950-February 1953.* University Park: Pennsylvania State University Press, 1992.

Moon, Thomas N. *This Grim and Savage Game: OSS and the Beginning of Covert Operations in World War II.* New York: Da Capo Press, 2000.

____, and Carl F. Eifler. *The Deadliest Colonel.* New York: Vantage, 1975.

Morgan, William J. *The O.S.S. and I.* New York: Norton, 1957.

Nelson, W. Dale. *The President is at Camp David.* Syracuse, N.Y.: Syracuse University Press, 1995.

Obolensky, Serge. *One Man in His Time: The Memoirs of Serge Obolensky.* New York: McDowell, Obolensky, 1958.

O'Donnell, Patrick K. *Operatives, Spies, and Saboteurs: The Unknown Men and Women of World War II's OSS.* New York: Free Press/ Simon & Schuster, 2004.

Office of Strategic Services. *War Report of the OSS (Office of Strategic Services)*, 2 vols. New York: Walker and Co., 1976.

OSS Assessment Staff. *Assessment of Men: Selection of Personnel for the Office of Strategic Services.* New York: Rinehart & Co., 1948.

O'Toole, George J.A. Ed. *The Encyclopedia of American Intelligence and Espionage: From the Revolutionary War to the Present.* New York: Facts on File, 1988.

Ottaway, Susan. *Violette Szabo.* Annapolis, Md.: Naval Institute Press, 2002.

Paddock, Jr., Alfred H. *US Army Special Warfare: Its Origins: Psychological and Unconventional Warfare, 1941-1952.* Washington, D.C.: National Defense University Press, 1982. Rev. Ed. Lawrence: University Press of Kansas, 2005.

Parnell, Ben. *Carpetbaggers: America's Secret War in Europe.* Rev. ed. Austin, Tex.: Eakin Press, 1993.

Patti, Archimedes L.A. *Why Viet Nam? Prelude to America's Albatross.* Berkeley: University of California Press, 1980.

Pearson, Judith L. *Wolves at the Door: The True Story of America's First Female Spy.* Guilford, Conn.: Lyons Press, 2005.

Peers, William R. and Dean Brelis. *Behind the Burma Road: The Story of America's Most Successful Guerrilla Forces.* Boston: Little, Brown, 1963.

Persico, Joseph E. *Casey: From the OSS to the CIA.* NY: Viking, 1990.

_____. *Piercing the Reich: The Penetration of Nazi Germany by American Secret Agents during World War II.* New York: Viking, 1979.

_____. *Roosevelt's Secret War: FDR and World War II Espionage.* New York: Random House, 2001.

Pinck, Dan C. *Journey to Peking: A Secret Agent in Wartime China.* Annapolis, Md: Naval Institute Press, 2003.

_____, with Goeffrey M.T. Jones and Charles T. Pinck. *Stalking the History of the Office of Strategic Services: An OSS Bibliography.* Boston: OSS/Donovan Press, 2000.

Porch, Douglas. The Path to Victory: The Mediterranean Theater in World War II. New York: Farrar, Strauss and Giroux, 2004.

Ranney, James F. and Arthur L. Ranney, eds. *The OSS CommVets Papers*, 2nd ed. Covington, Ky.: James F. Ranney, 2002.

Ransom, Harry Howe. *The Intelligence Establishment.* Cambridge, Mass.: Harvard University Press, 1970.

Reynolds, E. Bruce. *Thailand's Secret War: OSS, SOE and the Free Thai Underground during World War II.* Cambridge, U.K.: Cambridge University Press, 2004.

Ringlesbach, Dorothy. *OSS: Stories That Can Now Be Told.* Bloomington, Ind.: Author House, 2005.

Robbins, Peter. *The Legend of W.E. Fairbairn, Gentleman and Warrior: The Shanghai Years.* London: CQB Publications, 2005.

Roberts, Walter R. *Tito, Michailovic and the Allies, 1941-1945.* 2nd Ed. Durham, N.C.: Duke University Press, 1987.

Rogers, James Grafton. *Wartime Washington: The Secret OSS Journal of James Grafton Rogers, 1942-1943.* Ed. Thomas F. Troy. Frederick, Md.: University Publications of America, 1987.

Rooney, David. *Stilwell the Patriot: Vinegar Joe, the Brits, and Chiang Kai-shek.* London: Greenhill, 2005.

Roosevelt, Franklin D. *The Public Papers and Addresses of Franklin D. Roosevelt.* Ed. Samuel I. Rosenman. 13 Vols. New York: Harper & Brothers, 1950.

Rossiter, Margaret L. *Women in the Resistance.* New York: Praeger, 1986.

Rottman, Gordon L. *The U.S. Army Special Forces, 1952-1984.* Osprey, 1985.

Sage, Jerry. *Sage.* Wayne, Pa.: Miles Standish Press, 1985.

Salmond, John A. *Civilian Conservation Corps, 1933-1942: New Deal Case Study* Durham, NC: Duke University Press, 1967.

Sellars, Richard West. *Preserving Nature in the National Parks: A History* (New Haven, Conn.: Yale University Press, 1997);

Shapiro, Laura. *Julia Child.* New York: Viking, 2007.

Simpson, Charles. *Inside the Green Berets: The First Thirty Years: A History of the U.S. Army's Special Forces.* Novato, Calif.: Presidio Press, 1983.

Singlaub, John K. with Malcolm McConnell. *Hazardous Duty: An American Soldier in the Twentieth Century.* New York: Summit Books, 1991.

Smith, Bradley F. *The Shadow Warriors: OSS and the Origins of the CIA.* New York: Basic Books, 1983.

____. *The Ultra-Magic Deals and the Most Secret Special Relationship.* Novato, Calif.: Presidio Press, 1993.

Smith, Nicol and Blake Clark. *Into Siam: Underground Kingdom.* Indianapolis, Ind.: Bobbs-Merrill, 1946.

Smith, Robert Harris. *OSS: The Secret History of America's First Central Intelligence Agency.* Berkeley: University of California Press, 1972.

Smith, W. Thomas, Jr. *Encyclopedia of the Central Intelligence Agency.* New York: Facts on File, 2003.

Soley, Lawrence C. *Radio Warfare: OSS and CIA Subversive Propaganda.* New York: Praeger, 1989.

Spector, Ronald H. *In the Ruins of Empire: The Japanese Surrender and the Battle for Postwar Asia.* New York: Random House, 2007.

Squatrito, Joseph. *Code Name: Ginny.* Staten Island, N.Y.: Forever Free Publishing, 2001.

Stafford, David. *Britain and the European Resistance, 1940-1945: A Survey of the Special Operations Executive with Documents.* London: Macmillan, 1980.

____. *Camp X.* New York: Dodd, Mead, 1986.

____. *Roosevelt and Churchill: Men of Secrets.* Woodstock, N.Y.: Overlook Press, 2000.

____, and Rhodri Jeffreys-Jones. Eds. *American-British-Canadian Intelligence Relations, 1939-2000*. London: Frank Cass, 2000.

Stevenson, William. *A Man Called Intrepid*. NY: Harcourt, Brace, Jovanovich, 1976.

Stoler, Mark A. *Allies and Adversaries: The Joint Chiefs of Staff, the Grand Alliance, and U.S. Strategy in World War II*. Chapel Hill: University of North Carolina Press, 2000.

Sweet-Escott, Bickham. *Baker Street Irregular*. London: Methuen and Co., 1965.

Taylor, Nick. *American-Made: The Enduring Legacy of the WPA: When FDR Put the Nation Back to Work* (New York: Bantam, 2008).

Terkel, Studs. *"The Good War": An Oral History of World War Two*. New York: Ballantine Books, 1984.

Thomas, Evan. *The very Best Men: Four Who Dared: The Early Years of the CIA*. New York: Simon and Schuster, 1995.

Tompkins, Peter. *A Spy in Rome*. New York: Simon and Schuster, 1962.

Troy, Thomas F. *Donovan and the CIA: A History of the Establishment of the Central Intelligence Agency*. Frederick, Md.: University Publications of America, 1981.

____. *Wild Bill and Intrepid: Donovan, Stevenson and the Origin of the CIA*. New Haven, Conn.: Yale University Press, 1996.

Truman, Harry S. *Memoirs*. 2 vols. Garden City, N.Y.: Doubleday, 1955.

Truman, Margaret. *Harry S Truman*. New York: William Morrow, 1973.

Tuchman, Barbara. *Stilwell and the American Experience in China, 1911-1945*. New York: Macmillan, 1972.

Unrau, Harland D. and G. Frank Williams, *Administrative History: Expansion of the National Park Service in the 1930s*. Denver, Colo.: NPS Denver Service Center Publication, 1983.

U.S. National Park Service, *The National Parks: Shaping the System*, 3rd ed., rev. by Janet McDonnell. Washington, D.C. Govt. Printing Office, 2005. U.S. President, *Papers of the Presidents of the United States, Harry S Truman, 1945*. Washington, D.C.: Government Printing Office, 1961.

Wakeman, Frederick, Jr. Spymaster: Dai Li and the Chinese Secret Service. Berkeley: University of California Press, 2003.

Waller, John H. *The Unseen War in Europe: Espionage and Conspiracy in the Second World War*. New York: Random House, 1996.

Warner, Michael. *The Office of Strategic Services: America's First Intelligence Agency*. Washington, D.C.: Central Intelligence Agency, 2000.

Weinberg, Gerhard L. *A World at Arms: A Global History of World War II*. New York: Cambridge University Press, 1994.

Wiener, Tim. *Legacy of Ashes: The History of the CIA*. New York: Doubleday, 2007.

Wilson, Valerie Plame. *Fair Game*. New York: Simon and Schuster, 2007.

White, Theodore H. and Annalee Jacoby. *Thunder Out of China*. New York: William Sloane, 1946.

Winks, Robin. *Cloak and Gown: Scholars in the Secret War, 1939-1961.* New York: William Morrow, 1987.

Wirth, Conrad L. *Parks, Politics, and the People.* Norman: University of Oklahoma Press, 1980.

Yu, Maochun. *OSS in China: Prelude to the Cold War.* New Haven, Conn.: Yale University Press, 1996.

Articles

"A Good Man is Hard to Find: The O.S.S. Learned How, with New Selection Methods that May Well Serve Industry," *Fortune*, 1946, 92-95, 218-19, 223.

Aline, Countess of Romanones, "The OSS in Spain during World War II." *The Secrets War: The Office of Strategic Services in World War II*, George C. Chalou, ed. Washington, DC: National Archives and Records Administration, 1992: 122-28.

Allison, Thomas. "Welcome to Our American Friends." Harry Mauer, Ed. *Strange Ground: Americans in Vietnam, 1945-1975: An Oral History.* New York: Henry Holt, 1989, 28-46.

Belfry, Harry F. "Office of Strategic Services (OSS)." *U.S.A. Airborne: 50th Anniversary: A Commemorative History.* Atlanta, Ga.: Turner Pub. Co., 1997: 35.

Briscoe, Charles H. "Major (R) Herbert R. Brucker, DSC: Special Forces Pioneer." *Veritas: Journal of Army Special Operations History.* 2.2 (2006): 33-35; 3.1 (2007): 72-86.

Cassidy, William L. "Fairbairn in Shanghai." *Soldier of Fortune*, 4 (September 1979): 66-71.

Chamberlain, John. "OSS." *Life* 19 November 1945: 119-24.

Défourneaux, René. "A Secret Encounter with Ho Chi Minh." *Look Magazine.* 9 August 1966: 32-33.

Drury, Newton B. "What the War Is Doing to National Parks and Where They Will Be at Its Close," *Living Wilderness* (May 9, 1944): 11.

Fischer, Joseph R. "Cut from a Different Cloth: The Origins of U.S. Army Special Forces." *Special Warfare.* 8.2 (April 1995): 28-39.

"Fort Ritchie," *Encyclopedia of Historic Forts.* Ed. Robert B. Roberts. New York: Macmillan, 1988: 388.

Funk, Arthur L. "Churchill, Eisenhower, and the French Resistance." *Military Affairs.* 45 (February 1981): 29-33.

Haines, Gerald K. "Virginia Hall Goillot: Career Intelligence Officer." *Prologue* 26 (Winter 1994): 249-59.

"Interview: William Colby, Former Director, Central Intelligence Agency." *Special Warfare.* 7.2 (April 1994): 42-44.

Kehoe, Robert R. "An Allied Team with the French Resistance in 1944." *Studies in Intelligence: Journal of the American Intelligence Professional* (June 2002): 101-36.

Lupyak, Joseph. "The Evolution of Special Forces Training: Maintaining High Standards." *Special Warfare.* 16.2 (August 2003): 2-5.

McDonald, Lawrence H. "The OSS and Its Records," *The Secrets War: The Office of Strategic Services in World War II,* George C. Chalou, ed. Washington, DC: National Archives and Records Administration, 1992: 78-102.

McWilliams, John C. "Covert Connections: The FBN, the OSS, and the CIA." *The Historian,* 53.4 (Summer 1991): 660-72.

Morris, Jefferson. "Analyst Says SOCOM Not Well Suited for Terror War." *Aerospace Daily & Defense Report.* 30 June 2006: 5.

"OSS: 60th Anniversary Issue," *Studies in Intelligence: Journal of the American Intelligence Professional* (June 2002): 1-196.

Peers, William R. "Guerrilla Operations in Northern Burma." *Military Review* (June 1948): 10-16; and (July 1948): 12-20.

Sacquety, Troy J. "A Special Forces Model: OSS Detachment 101 in the Myitkyina Campaign, Part I." *Veritas: Journal of Army Special Operations History.* 4.1 (2008): 30-47.

_____. "Caesar J. Civitella: Bull Simmons Award 2008. *Veritas: Journal of Army Special Operations History.* 4.1 (2008):77.

_____. "OSS (Office of Strategic Services)." *Veritas: Journal of Army Special Operations History.* 3.4 (2007): 34-51.

_____. "The Special Forces Patch: History and Origins, *Veritas: Journal of Army Special Operations History,* 3.3 (2007): 59-63.

Sutherland, Ian. "The OSS Operational Groups: Origin of the Army Special Forces." *Special Warfare.* 15 (June 2002): 11-15.

Warner, Michael. The Creation of the Central Intelligence Group." *Studies in Intelligence: Journal of the American Intelligence Professional.* (June 2002): 169-78.

Government Documents

Mattingly, Robert E. *Herringbone Cloak–GI Dagger: Marines of the OSS,* Washington, D.C.: U.S. Marine Corps Headquarters, History and Museum Division, 1989.

Paige, John C. "The Civilian Conservation Corps and the National Park Service, 1933-1942: An Administrative History." National Park Service 1985.

Parker, Patricia. *The Hinterland: An Overview of the Prehistory and History of Prince William Forest Park, Virginia.* National Park Service, 1986.

Payne-Jackson, Arvilla and Sue Ann Taylor, *Prince William Forest Park: The African-American Experience.* National Park Service, 2000.

Strickland, Susan Cary. *Prince William Forest Park: An Administrative History.* National Park Service, January 1986.

U.S. Congress, Senate Select Committee to Study Government Operations with Respect to Intelligence Activities, *Final Report,* 94th Cong., 2nd Sess., Senate Report No. 94-755 [the Church Report], 6 vols. Washington, D.C.: GPO, 1976.

Werhle, Edmund F. *Catoctin Mountain Park: An Historic Resource Study*. National Park Service, March 2000.

Periodicals

Attention Please. Newsletter of OSS Area F
Blue Book: The Magazine of Adventure for Men
Journal of Military History
Leatherneck Magazine
Military Affairs
OSS Society Newsletter
Saturday Evening Post
Special Warfare (Army's Special Forces)
Studies in Intelligence: Journal of the American Intelligence Profession
Veritas: Journal of Army Special Operations History

Internet Sources

Airborne and Special Operations Museum, Fayetteville, N.C.: www.asomf.org/museum_exhibits_wwii.htm

Camp X Historical Society, Toronto, Canada: www.campxhistoricalsociety.ca

Catoctin Mountain Park, Thurmont, Md.: www.nps.gov/cato

Central Intelligence Agency, Langley, Va.: www.cia.gov/about_cia/cia-museum/index

National Museum of the Marine Corps, Quantico, Va.: www.usmcmuseum.org

National Navy UDT-SEAL Museum, Fort Pierce, Fla.: www.navy.sealmuseum.com/heritage/history.php

National Park Service, Washington, D.C.: www.nps.gov/history; www.cr.nps.gov/histroy/online_books

OSS Operational Groups: www.ossog.org

OSS Society discussion group (limited access): osssociety@yahoogroups.com

OSS Society website: www.osssociety.org

Prince William Forest Park, Triangle, Va.: www.nps.gov/prwi/historyculture/oss.htm

U.S. Army, Center for Military History, Washington, D.C.: www.army.mil/cmh/lineage/branches/sf

U.S. Army, John F. Kennedy Special Warfare Museum, Ft. Bragg, Fayetteville, N.C.: www.soc.mil/swcs/museum

THE AUTHOR

John Whiteclay Chambers II is Distinguished Professor and former Chair of the History Department at Rutgers University, New Brunswick, New Jersey. He earned his Ph.D. at Columbia University and taught there for ten years before joining the Rutgers University faculty in 1982. His books, several of which have won prizes, include *To Raise an Army: The Draft Comes to Modern America* (Free Press/Macmillan, 1987); *The Tyranny of Change: America in the Progressive Era, 1890-1920* (St. Martin's Press, 1980; 2nd edition, revised, Rutgers University Press, 2000); *World War II, Film and History* (Oxford, 1996); *Major Problems in American Military History* (Houghton Mifflin, 1999), and *The Oxford Companion to American Military History* (Oxford, 1999). He helped to found the Rutgers Oral History Archives of World War II and still serves as Chair of its Academic Advisory Committee. He has authored many scholarly articles and book reviews, and he regularly reviews books on military history for the *Washington Post*.

Among his professional awards are Outstanding Teacher Awards at Columbia University and Rutgers University, a Rockefeller Fellowship, a Fulbright Fellowship to the University of Rome, a Visiting Lectureship at the University of Tokyo, and a Fellowship at the Institute for Advanced Study in Princeton.

ILLUSTRATIONS

Major General William J. Donovan, portrait of the Director of the Office of Strategic Services (OSS). Courtesy of the Central Intelligence Agency.

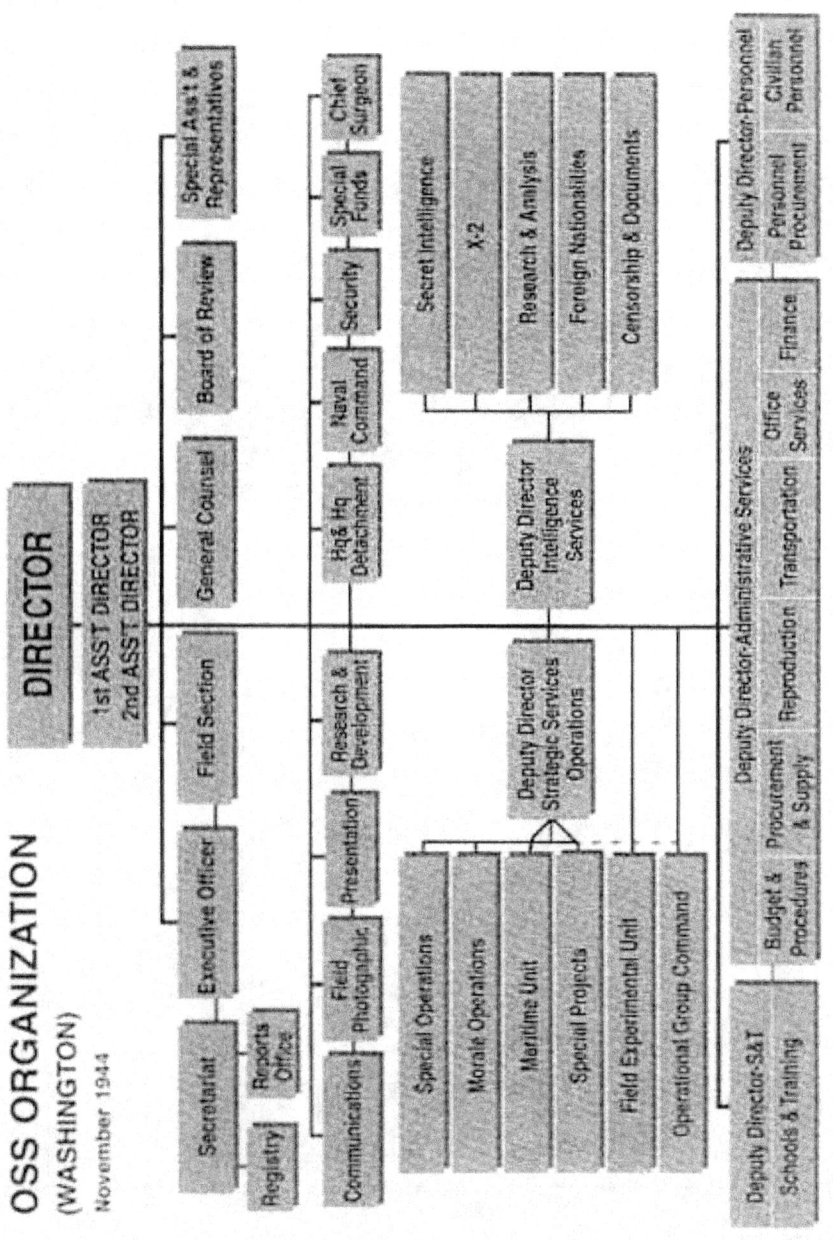

628 | Illustrations

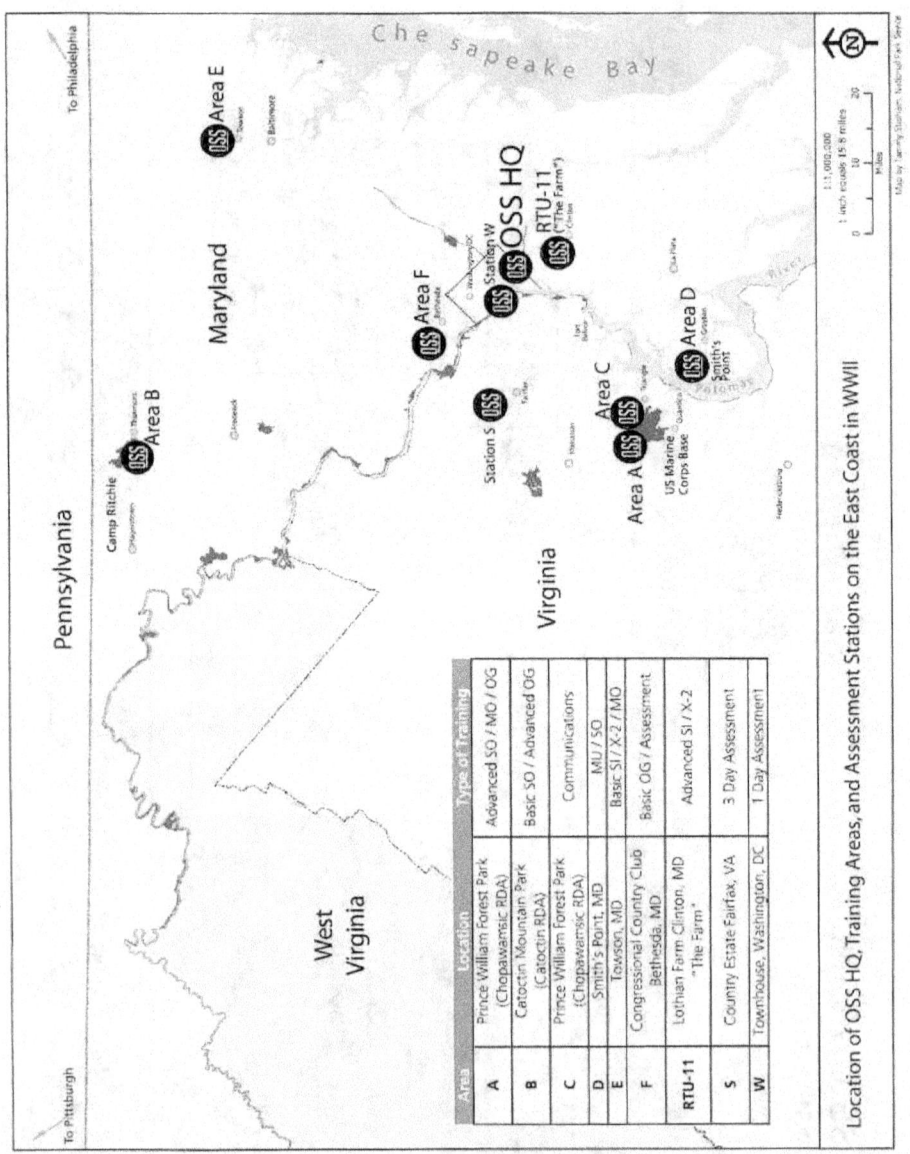

Location of OSS HQ, Training Areas, and Assessment Stations on the East Coast in WWII

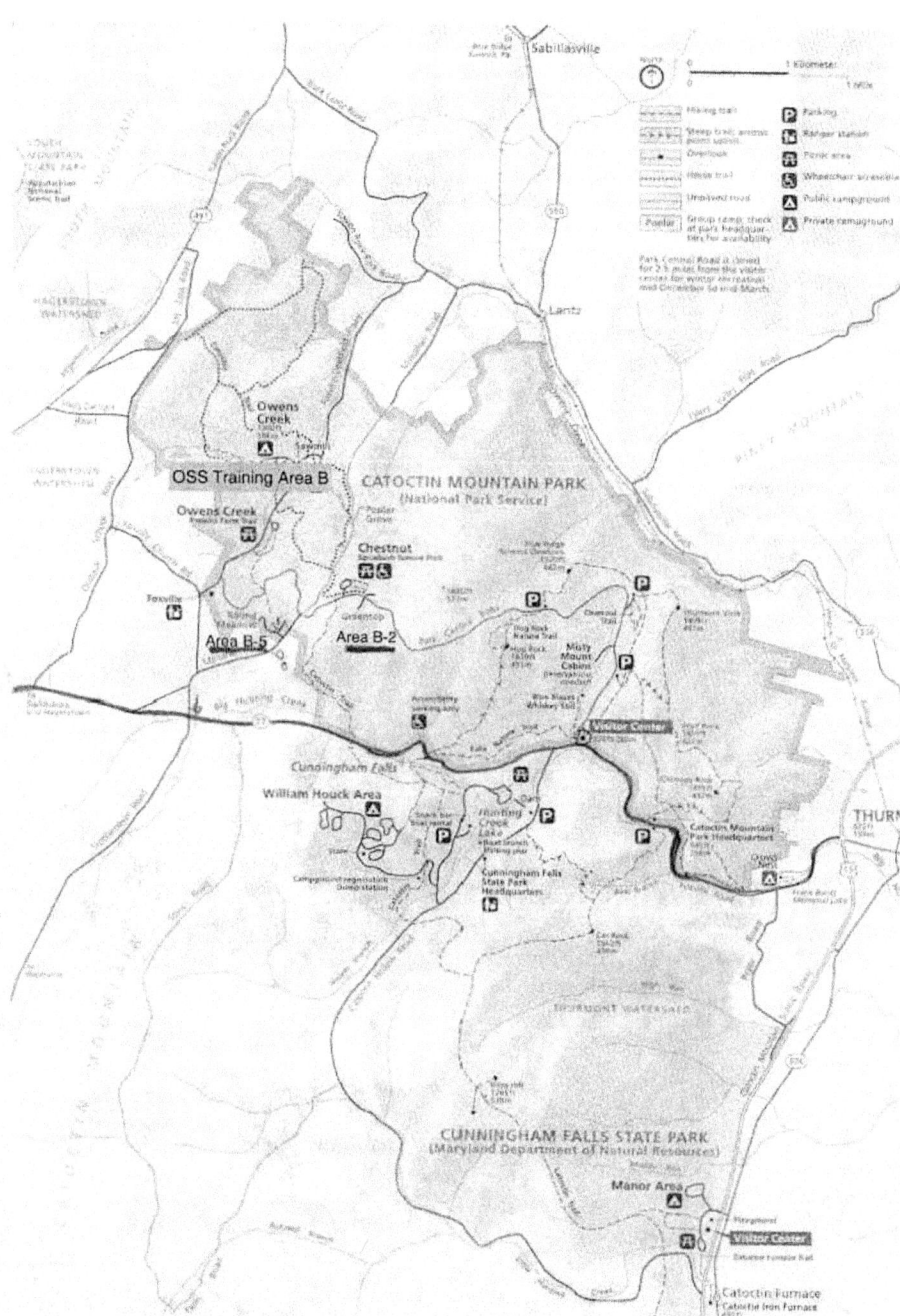

Current Map of Catoctin Mountain Park with locations of OSS Training Area B and Sub-Camps Area B-2 and Area B-5 superimposed
Catoctin Mountain Park

Area F, the Former Congressional Country Club, OSS Records, National Archives

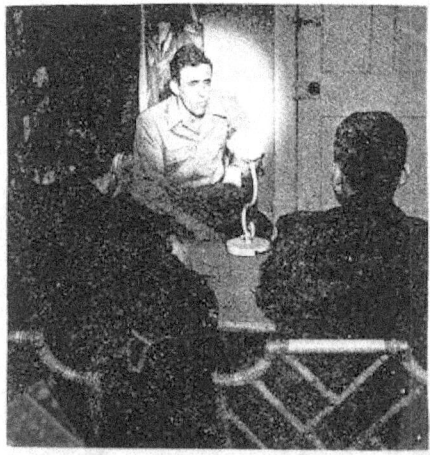

Assessing trainee's ability to
withstand hostile interrogation. OSS Photo.

Right: Learning to build and use
rope bridge crossings. OSS Photo.

Left: Lt. Erasmus H. Kloman practice firing
a German submachine gun on the beach.
Courtesy of Erasmus H. Kloman.

William E. ("Dan") Fairbairn on the left, and Hans Tofte, instructors in close combat fighting. Courtesy of the Congressional Country Club.

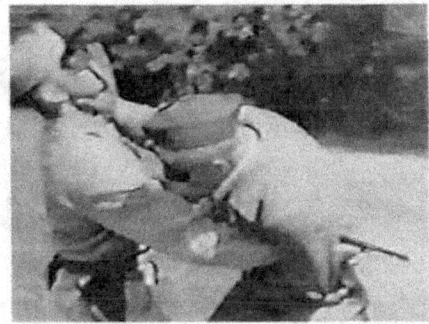

William E. Fairbairn demonstrated a disarming technique at Area B. This is a still from an OSS training film shot under John Ford's direction at Area B in 1942. Both students and instructors wore masks during the filming to preclude identification. Courtesy of Lynn Philip Hodgson

Left: Fairbairn, on right, accompanying a trainee firing at a mock-up German in one of Fairbairn's "Mystery Houses," or "House of Horrors." Courtesy of Lynn Philip Hodgson.

Area C served as the OSS Communications School. Civilian instructors like John Balsamo, front left, formerly a Wall Street telegrapher, and Timothy Marsh, front right, helped OSS "Commo" operators reach a minimum, accurate speed of at least 25 words per minute in OSS coded International Morse Code in classrooms like this. OSS Records.

Communications Branch trainees at Area C also learned how to set up various types of radio antennas for clandestine or base station wireless telegraphy transmitters and receivers. OSS Records, National Archives.

Instructors like Lt. F. Ralph Ward, left, and John Balsamo, right, demonstrated how the small, portable, suitcase-size SSTR-1, transmitter and receiver could be quickly set up by clandestine operators in the woods, its antenna wire, passing over Ward's shoulder, hung from a nearby tree branch. OSS Records, National Archives.

The SSTR-1 wireless telegraphy, short-wave radio transmitter and receiver in its suitcase carrying case complete with battery on the right and wires for connection to the battery and a plug for use with an electrical outlet. The battery was recharged by hand crank. Courtesy of John W. Brunner.

Maj. Albert Jenkins, USMCR, commanding officer of Area C for most of the war, sometimes personally instructed trainees in the use of the .45 caliber automatic pistol as he did here in December 1943. On the extreme left are Martin Lubowsky from Pittsburgh, half in the picture, James Bisaccia from the Bronx, holding up a pistol, and Corporal Mulford, weapons instructor. Identifications by Roger L. Belanger. OSS Photo.

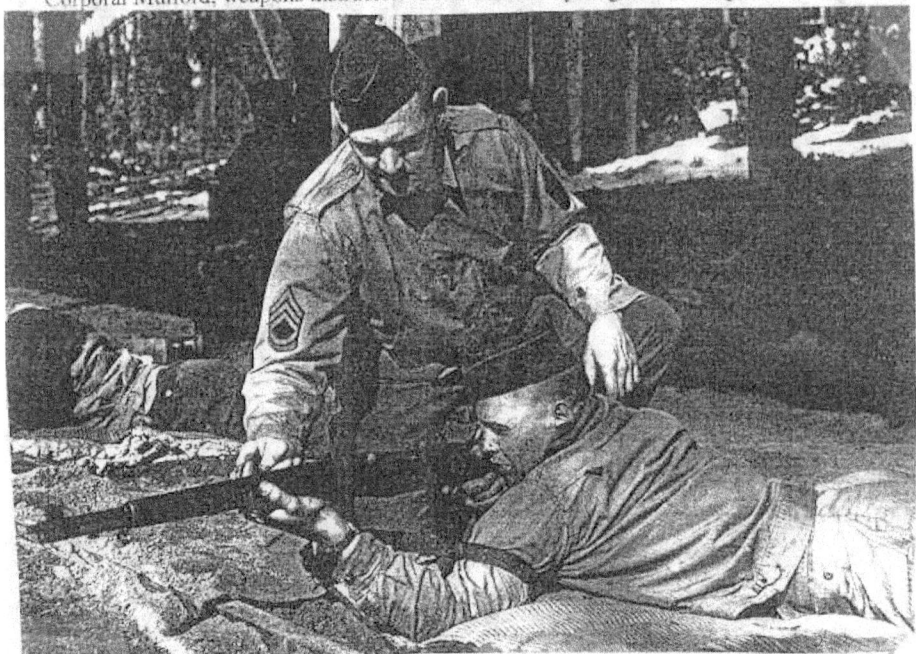

A trainee is instructed in the use of the M-1 Garand semi-automatic rifle at the rifle target range at Area C. OSS Records, National Archives.

Although instruction was rigorous at the Training Camps, there was usually some time for outdoor recreation, softball, basketball, or volleyball. Here a volleyball game is held in front of the Recreation Building in Area C. OSS Records, National Archives.

Since so much of the paramilitary training was so demanding physically, OSS tried to ensure that the quality of the food was better than standard Army chow. In Training Area C, former restaurant owner Capt. John Navarro obtained first class cooks. OSS Records.

Maj. Gen. William Donovan, OSS Director, center, at a breakfast meeting in winter of 1944-1945 with Maj. Howard Manning, left, who succeeded Albert Jenkins as head of Area C, and Manning's adjutant, Lt. Edward Katszka. Photo courtesy of Betty Manning.

General Donovan addresses an audience of trainees, probably Special Operations or Operational Group members, as they are in military uniform with shoulder patches and the paratroopers among them are wearing jump boots. The setting is probably one of the recreation halls in Area A or Area C as the pool tables have been moved to the front of the room. A dummy used for bayonet, knife, or unarmed combat exercise has been moved to the left rear corner of the room. Photo courtesy of Betty Manning.

OSS in the European and Mediterranean Theaters of Operation

Specially modified, black-painted B-24 "Liberator" bombers flown by the U.S. Army Air Force's "Carpetbaggers," the 801st/492nd Bombardment Group of the Eighth Air Force, for low level airdrops of OSS personnel and supplies in enemy occupied territory in Europe. Courtesy of Rolf Herstad

An unidentified woman and man, members of OSS Secret Intelligence Branch, being suited up in British SOE jumpsuits and parachutes in preparation for airborne infiltration from England to France. Courtesy of John W. Brunner.

Members of an OSS Operational Group Mission, Percy Red, are trucked to a waiting plane to parachute them into central France in August 1944. Courtesy of Rolf Herstad.

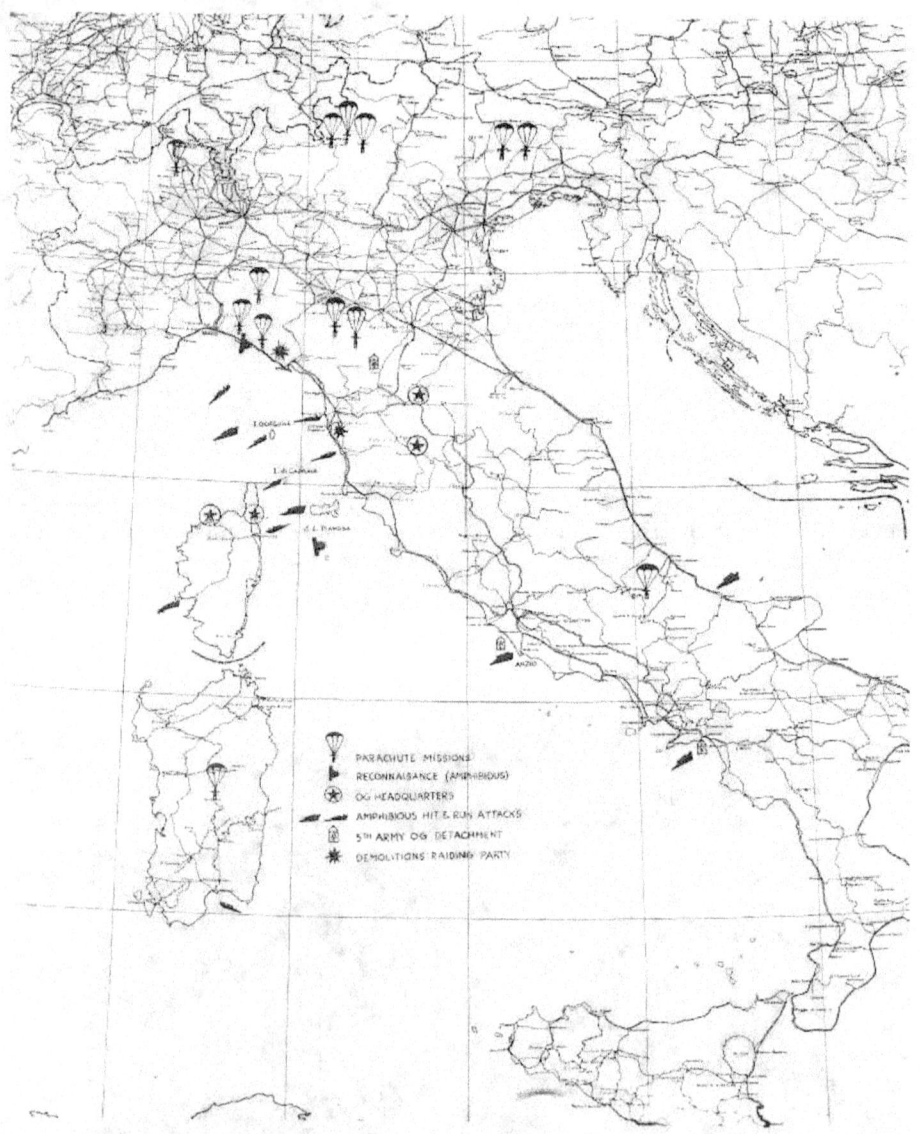

Map of OSS Italian Operational Group operations in Italy 1943 to 1945. OSS Records. National Archives.

Lt. Col. Serge Obolensky, Russian prince, New York socialite, OSS officer, who trained at Areas B, A, and F, parachuted into Sardinia in 1943, becoming at age 52, the oldest combat paratrooper in the Army. In 1944, he led OSS French OG Mission Patrick into central France, fighting for a month behind enemy lines.

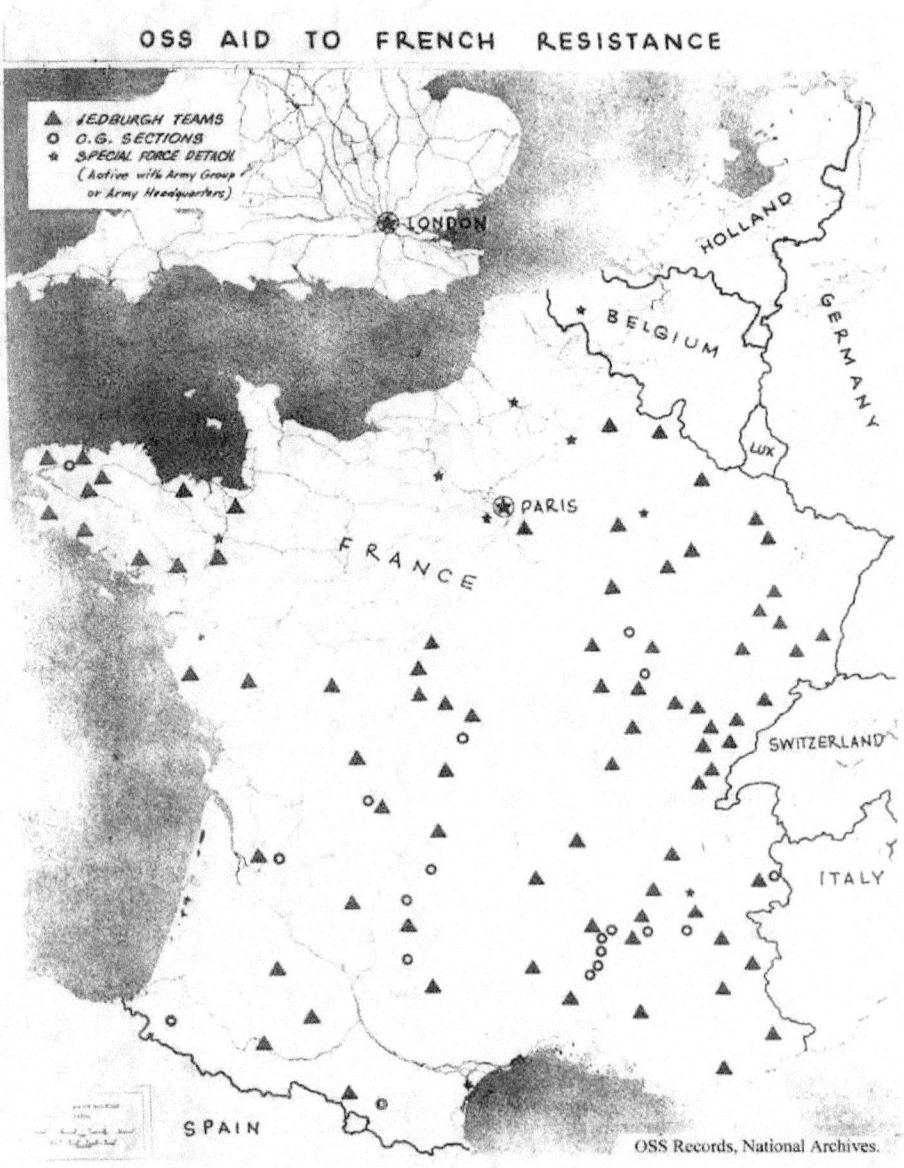

OSS Jedburgh Team "James" with Lt. Jack Singlaub, second from left with hands on hip, interrogating a German soldier captured by *maquis* guerrillas directed by the OSS team. Courtesy of John K. Singlaub.

OSS men help celebrate the liberation of Castres, France, September 1944. Left to right: T/5 Raymond Picard, Cpl. Herbert Shapiro, and T/5 Marcel Landry armed with Marlin submachine guns and .45 caliber pistols. Courtesy of John W. Brunner.

Left: Two members of the OSS Norway OG, who trained at Areas F and B, enjoy an evening out before leaving for Europe. Left to right: Leif Eide and his girlfriend, Margy; Andi Kindem and her fiancé, Arne I. Herstad.

Right: A group of buddies from the OSS Norway OG, who called themselves "The Unholy Four." (l. to r.) Cpl. Odd Aanonsen, Cpl. Kai Johansen, Pvt. Leif Eide, and T/5 Arne I. Herstad. Photo on "Percy Red" Mission in France 1944.

Above: A section of OSS Norway OG team being flown back to England in a C-47 transport plane after the successful "Percy Red" Mission in central France in 1944. All three pictures courtesy of Rolf Herstad.

Major Peter J. Ortiz and the others in OSS SO's Union II Mission team a few days after parachuting onto a plateau in the French Alps, Aug. 1, 1944, (l. to r.) Sgt. John Bodnar, Maj. Ortiz, Sgts. Robert Lasalle and Fred Brunner, Capt. Frank Coolidge, Sgt. Jack Risler, all Marines except Coolidge. Courtesy of the California State Military Museum.

A week after his arrival near Albertville, Ortiz, in his Marine Corps service uniform, inspected guerrilla members of the French Forces of the Interior. The young man with glasses and beret, *Captaine* Bulle the local FFI commander and already a hero, was later killed in action liberating Albertville. Courtesy of the California State Military Museum.

Sgt. Charles R. Perry, USMC, a member of OSS's Union II Mission, was killed when his steel parachute cable snapped during the team's airborne infiltration into France on Aug. 1, 1944. He was buried with full military honors provided by guerrillas of the French Forces of the Interior and the surviving team members of the Union II Mission in a ceremony held in the drop zone itself. Courtesy of the California State Military Museum.

Chris Rumburg: An American Hero

Lieutenant Colonel Ira "Chris" Rumburg

Ira Christopher ("Chris") Rumburg, who had been captain of the Washington State University football team, a champion wrestler, and president of the student body, had as a young lieutenant in OSS's SO Branch been an instructor at Area A in 1942-1943.

By 1944, Rumburg was a lieutenant colonel assigned to a regimental headquarters in the 66th ("Black Panther") Infantry Division, possibly as an OSS adviser. On Christmas Eve, Rumburg and 2,200 soldiers boarded a transport ship, the *SS Leopoldville*, which was to take them from Southampton, England to Cherbourg, France for the trip to the front. The ship was torpedoed six miles from France. Before the ship sank and he drowned, Rumburg was credited with saving the lives of more than one hundred of his fellow soldiers. He was one of the nearly 800 American soldiers who perished in the disaster. Rumburg was posthumously awarded the Bronze Star and Purple Heart, and commemorated by a Memorial Fund by the Washington State University.
Photos courtesy of Washington State University Athletics.

Tragedy in Slovakia

In the fall of 1944, in an attempt to assist a Nationalist Uprising in Slovakia against Germany and also to establish information networks in central Europe, OSS flew two teams, including nearly two dozen Special Operations and Secret Intelligence personnel, into site of the uprising in central Slovakia. The SO "Dawes" Mission to aid the Slovak Resistance and rescue downed Allied fliers was headed by Navy Lieutenant J. Holt Green from Charleston and his deputy, Navy Lieutenant James Harwey Gaul from Pittsburgh. Many of the team members were Czech or Slovak immigrants, such as Jaroslav ("Jerry") G. Mican, the oldest team member at 42, a native of Prague, who had emigrated to Chicago in the 1920s and become a school teacher. Army Air Force Captain Edward V. Baranski, an ethnic Slovak from Illinois, headed the OSS SI "Day" Group, whose mission was to establish a ring of local spies in Czechoslovakia. B-17 bombers flew the teams into Slovakia in September 1944 and in early October, half a dozen B-17s landed on a grass-covered airfield at Tri Duby near Banská Bystrica with additional planeloads of supplies for the 1st Czechoslovakian Army and the Slovak Resistance. But the uprising was crushed by the Germans. Although the OSS team, as well as British SOE members accompanying them, fled into the Tatra Mountains, most of them were captured, tortured and executed, many of them at Malthausen Concentration Camp near Linz, Austria. Of the twenty OSS team members, fourteen were killed, and only six survived. Most of the SOE team members were also captured and killed. Two of the OSS team members evaded capture with the help of 24-year-old Maria Gulovich, a Slovakian schoolteacher and partisan, who had been hired as an interpreter and guide by the OSS.

Slovakia, showing location of Banská Bystrica, the focal point of the Slovak National Uprising and site of OSS landings in nearby Tri Duby airfield, as well as the Tatra Mountains where most members of the OSS and SOE teams were captured by the Nazis.

Unless otherwise noted, all of the following illustrations are courtesy of the *Múzeum Slovenského národného povstania*, Museum of the Slovak National Uprising Banská Bystrica, Slovakia (SNU Museum, Banská Bystrica, Slovakia).

Above: Navy Lieutenant, senior grade, J. Holt Green, Commander of Special Operations' "Dawes" Mission to Slovakia. Below: Navy Lieutenant James H. Gaul, his deputy commander. Both were among those later captured and executed by the Nazis. (SNU Museum, Banská Bystrica, Slovakia).

Army Air Force Captain Edward V. Baranski, Commander of SI's "Day" Group.
Also captured and executed by the Nazis.
U.S. National Archives

Six U.S. B-17G "Flying Fortress" Bombers from the 15th U.S. Army Air Force from Bari, Italy at the insurgent-held, Tri Duby Airfield near Banská Bystrica, on Oct. 7, 1944. (SNU Museum, Banská Bystrica, Slovakia).

Bazookas and other weapons as well as munitions, explosives, medical and other supplies for the Slovakian National Uprising unloaded from a B-17 by Slovakian soldiers and members of the OSS. (SNU Museum, Banská Bystrica, Slovakia).

During the air supply operation, Navy Lieutenant Holt Green, left, conferred with Sergeant Jerry Mican and British Major John Sehmer, right, commander of the British SOE intelligence mission. All three were later captured and executed by the Germans. (SNU Museum, Banská Bystrica, Slovakia).

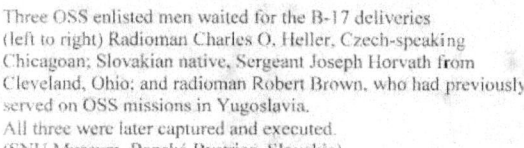

Three OSS enlisted men waited for the B-17 deliveries (left to right) Radioman Charles O. Heller, Czech-speaking Chicagoan; Slovakian native, Sergeant Joseph Horvath from Cleveland, Ohio; and radioman Robert Brown, who had previously served on OSS missions in Yugoslavia.
All three were later captured and executed.
(SNU Museum, Banská Bystrica, Slovakia).

In the shade of a B-17 wing, war correspondent Joseph Morton of the Associated Press typed a story on his portable typewriter perched atop a box of supplies. He too was executed by the Nazis. (SNU Museum, Banská Bystrica, Slovakia).

Above: OSS Photographer, Seaman Nelson Paris, was among those executed. U.S. National Archives.

A month later fleeing into the Lower Tatra Mountains, the main OSS group stopped for a break on November 11, 1944. Team photographer, Navy Seaman Nelson Paris took a picture of Joe Morton, left, Lt. Holt Green center, and Sgt. Jerry Mican, right. They were captured December 26, 1944 and executed at Mauthausen Concentration Camp in early 1945.
(SNU Museum, Banská Bystrica, Slovakia).

Above: Margita Kocková, a 30-year-old Slovak-American, was assigned by the Czechoslovak Army HQ as an interpreter with the British SOE team in Slovakia headed by Major John Sehmer. Like him, she was captured and executed by the Nazis. (SNU Museum, Banská Bystrica, Slovakia).

Left: Maria Gulovich, 24, Slovakian schoolteacher, partisan, and interpreter for the "Dawes" Mission, led two OSS enlisted men to safety. Her Slovakian identity card, included her signature, Mária Gulovičová in Slovak, scrawled over part of her face. Following the escape, she continued to work for the OSS and was later brought to the United States where she received the Bronze Star Medal. (SNU Museum, Banská Bystrica, Slovakia).

Mauthausen Concentration Camp

Most of interrogations, torture and execution of at least seventeen members of the OSS and SOE Missions to Slovakia in 1944-1945 took place at Mauthausen Concentration Camp near Linz, Austria, one of the largest, oldest and most brutal of the Nazi concentration camps (at least 70,000 persons, mainly Jews perished there during the war). The camp commander at the time the OSS and SOE members were executed was SS Colonel Franz Ziereis, who personally supervised the torture and execution of the captives, often aided by deputy commandant Georg Bachmayer.

In 1941, after Ziereis had become the new commandant of Mauthausen, Nazi SS Chief Heinrich Himmler and his main subordinate Ernst Kaltenbrunner, who oversaw the SS, the secret police, and the management and the methods of liquidation in the camps, made an inspection tour of the Mauthausen camp.

The top photograph, shows, left to right in the front row, Kaltenbrunner, Ziereis, Himmler and August Eigruber, Nazi *Gauleiter* (administrative leader) for Upper Austria. At the far right in the second row is Georg Bachmayer.

The bottom photograph shows Himmler, Ziereis, and Kaltenbrunner.
Photos courtesy of the *KZ-Gedenkstätte Mauthausen*, the Mauthausen Concentration Camp Memorial, Mauthausen, Austria.

OSS Detachment 101 operated in northern Burma from its headquarters in Nazira, India; Detachment 202 in China and northern French Indochina from Chungking and Kunming; Detachment 404 operated in southern Burma, Thailand, Malaya, and southern French Indochina from its main headquarters in Kandy, Ceylon (Sri Lanka). John W. Chambers.

One of the handful of OSS officers engaged in undercover operations in the Pacific was Lieutenant Reginald G. ("Reg") Spear, center, who trained at Area B, shown here in Hawaii with Admiral Chester Nimitz, right, commander-in-chief of the U.S. Pacific Fleet, and an aide, Commander Howell A. ("Hall") Lamar. Courtesy of Reginald G. Spear.

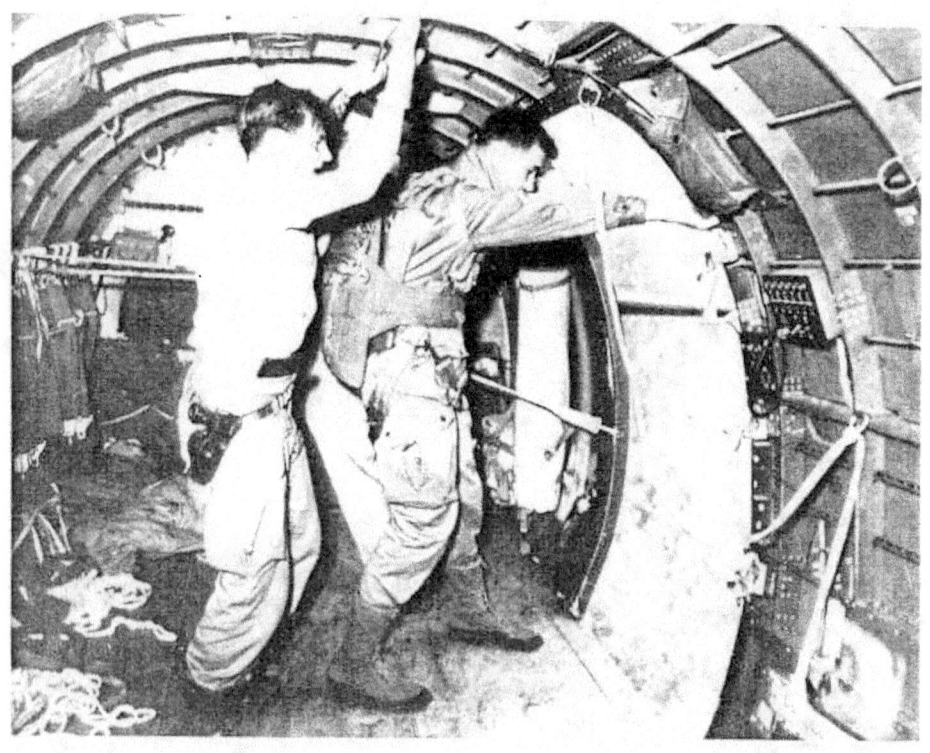

Lt. Richard T. Walsh, Detachment 101, prepares to jump into Japanese occupied Burma. Photo courtesy of John W. Brunner.

Col. Carl Eifler, left, commander of Detachment 101, prepares to take Gen. William Donovan on a flight behind Japanese lines in Burma, Dec. 1943, while film director John Ford, with sun glasses, and Nicol Smith, who trained at Area B, watch.

Lt. Joseph Lazarsky, former Instructor at Area B, in Burma, where he commanded the 1st Kachin Ranger Battalion, 1944

OSS on the offensive in the jungle of northern Burma, 1944. Lt. Joe Lazarsky, right, CO of 1st Kachin Ranger Battalion, his deputy, a British officer, second from left, and two Anglo-Burmese agents accompanying the battalion and its elephant supply train south to Lashio before heading to Rangoon. Photo courtesy of Joseph Lazarsky

Lt. Roger Hilsman, West Point graduate, and former member of Merrill's Marauders, joined the OSS in 1944, serving with OSS Detachment 101 in Burma, where this picture was taken of him carrying a pistol and an OSS silenced M3 submachine gun. Later, serving with OSS Detachment 202, he would later help rescue his father and General Wainwright from a Japanese POW camp in China. Photo courtesy of John W. Brunner.

OSS Detachment 404 "War Room" in Kandy, Ceylon (now Sri Lanka) with, left to right, British Admiral Lord Louis Mountbatten, Supreme Allied Commander of the Southeast Asia Command; Cora du Bois, chief of the Research & Analysis unit for Detachment 404; and Colonel John Coughlin, former deputy commander of Detachment 101, later head of OSS Detachment 202 in China in 1944. U.S. National Archives.

Major Lloyd E. Peddicord, left, OG section chief, and Captain George H. Bright, planning operations along the southern Burma coast in December 1944. Both officers, and the entire OG unit, trained at Areas F and A in 1944. Photo courtesy of Dr. Christian J. Lambertsen

Arakan Field Unit of Detachment 404 and later Detachment 101, was composed of OG, SI, and MU personnel, and engaged in operations along Burma's southern Arakan coast, inlets and rivers. They used Maritime Unit fastboats, including the P-564, shown here skippered by First Lieutenant Walter L. Mess of Arlington, Virginia. These U.S. Army Air-Sea Rescue Boats were similar to Navy PT (Patrol Torpedo) boats but shorter and without the torpedo tubes. Photo courtesy of Troy Sacquety.

OSS Map of China. Arrows show Japanese Offensive's penetration (May 1, 1945). Dotted lines and Roman Numerals indicate Chinese War Zones and the black dots denote War Zone Headquarters. OSS Records, National Archives.

OSS China Theater Headquarters Compound north of Kunming.
Photo courtesy of John W. Brunner

Left: OSSer Julia McWilliams later married Paul Child whom she met while she headed the OSS Registry for secret documents, first in Kandy, Ceylon and then in Kunming, China. Later, as his wife, Julia Child, she accompanied him to France, famously mastering the art of French cooking. Photo courtesy of the OSS Society.

Above: Elizabeth ("Betty") P. MacDonald (later McIntosh) former war correspondent, joined OSS, trained at Area F, and was sent to Ceylon and then China, to write material for the Morale Operations Branch.

Left: When the MO print shop in Kunming, China, was flooded in 1945, Betty P. MacDonald was carried to dry land by noted artist William A. Smith, also in MO, accompanied by Tong Ting, a famous China cartoonist. Photos courtesy of Elizabeth ("Betty") P. McIntosh.

Lt. Col. Charles M. Parkin in front of a pagoda in one of the joint Sino-American training camps in which American instructors trained Chinese soldiers to become Special Operations guerrillas. Parkin was chief instructor and eventually commanding officer of this camp in the mountains five miles south of Huizhou (Shexian), about 200 miles west of Shanghai. Later Parkin led a major reconnaissance mission, the "Akron" Mission, behind Japanese lines along the south China coast. Courtesy of Charles M. Parkin.

In China, Lt. Col. Arden W. Dow, right, former instructor at Area B and chief instructor at Area A, and Maj. Frank A. Gleason, left, former instructor at Area B, were commander and deputy commander respectively of a joint Sino-American training camp sixty miles southeast of Loyang to prepare Chinese Special Operations guerrilla teams to operate behind enemy lines. Courtesy of Frank A. Gleason.

Maj. Frank Gleason, OSS Special Operations and Army Corps of Engineers, instructs Chinese Special Operations trainees on the necessary amount and proper placement of modern explosives to destroy an industrial steel boiler. Courtesy of Frank A. Gleason.

Frank A. Gleason, 24, points to his jeep's windshield which was smashed by rock fragments from an explosive charge set by his Special Operations Team as they set out in late summer 1944 to impede the Japanese Army's advance Courtesy Frank A. Gleason.

In late summer and early winter 1944, in an attempt to impede the Japanese Army's *Ichi-Gō* offensive toward Kweilin and Liuchow in southern China, Maj. Frank Gleason led an OSS Special Operations unit of nearly a dozen Americans to destroy the area's transportation net.

Above: Maj. Frank Gleason stands atop the ruins of a stone bridge blown up by his team.

Left: A river ferry burns near Pinglo, set fire by Gleason's men.

Right: Villagers survey the remains of a wooden bridge with stone supports destroyed by Gleason's team, which set off a 1,000 pound bomb there to delay the Japanese Army.

Photos from OSS Records, National Archives.

Right: Capt. Leopold ("Leo") Karwaski, on left, former instructor at Area B, and a Marine Major Benson, right, both members of Maj. Gleason's SO team, take a roadside break near Kweilin, south China, late fall 1944. OSS Records, National Archives.

Below: Destruction achieved by Maj. Frank Gleason's OSS team helping to stop the Japanese Army's *Ichi-Gō* campaign toward Liuchow, Kweilin, and north toward Kunming and Chungking. The Japanese were halted. OSS Records, National Archives.

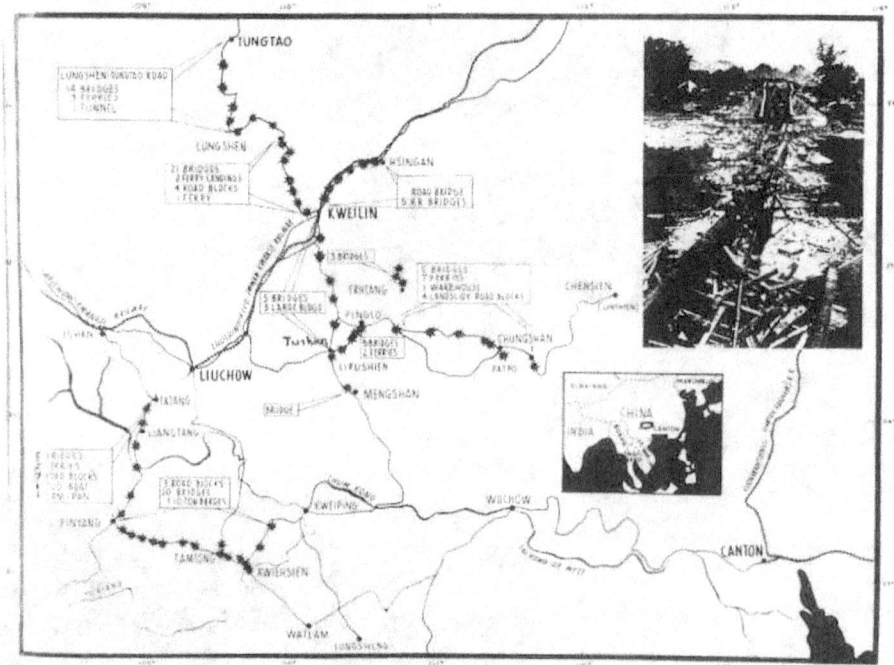

Corporal Conrad D. Beck, radio operator of Lieutenant Colonel Charles Parkin's "Akron" Mission to survey the south China coast behind Japanese lines, transmitting a message from the village of Tinting back to base headquarters, April 24, 1945, with an SSTR-1 to the fascination of Nationalist Chinese Army bodyguards and some local residents. Beck trained with Arthur Reinhardt at Area C. Photo courtesy John W. Brunner

The OSS communications room in Tientsin (Tianjin), China, southeast of Peking (Beijing) in 1945. John W. Brunner, left, who trained at Area C as well at the OSS coding center in Washington, D.C., encodes a message, while Guy E. Webb, right, transmits another message. Photo courtesy of John W. Brunner.

Two leaders of OSS SO Team "Ibex" in North China in 1945. Lt. Herbert R. Brucker, left, trained at Area E and probably C. An OSS radioman in France in 1944, Brucker had been promoted to lieutenant. In China, Brucker, with his Alsatian accent and beret, quickly became known as "Frenchy." He was executive officer of SO Team "Ibex" commanded by Capt. Leon Demers, right, which trained and led Chinese guerrillas behind Japanese lines in North China in 1945. Courtesy of Troy J. Sacquety.

From Trainees to Combat Veterans: A French OG from Europe to Asia

Portrait of part of a sizable part of the OSS French Operational Group in their Class-A uniforms at Brockhall training estate in England. They had previously trained at Areas F and B. In the summer of 1944, they would be parachuted behind German lines. After the liberation of France, these French OGs volunteered for service in Asia in 1945.

A section of this same OSS French Operational Group after their combat success in France and before being returned to England and then transferred to China. Rear row, left to right: Grant B. Hill, medic Ellsworth ("Al") Johnson, radioman Thomas F. McGuire, two unidentified OSSers, then Robert Anderson, and Capt. James E. Cook. Front row: two unidentified OSSers on the left, then J.L. DuBois, and with the cap, Roy Gallant Photos courtesy of Ellsworth ("Al") Johnson.

OSS Rescue Mission

Members of the OSS "Pigeon" Mission arrive at a Japanese camp for Allied prisoners of war on Hainan Island off the south China coast after parachuting in on 27 August 1945 with news of Tokyo's surrender. Maj. John K. ("Jack") Singlaub, center, commander of the mission, a Jedburgh in France who had trained at Areas F and B and in England, directs a Japanese Army captain to take them to the colonel in charge, as they hasten through the village where were the guards resided toward the camp where the POWs were incarcerated.

Left to right: Lt. Ralph Yempuku, Japanese-American Nisei from Hawaii, the interpreter, with bandaged chin from the parachute drop; unidentified Japanese Army captain; Lt. John Bradley, Special Operations Branch; Major John K. Singlaub, SO, commander; Cpl. Jim Healey, medic; and Lt. Arnold Breakey, Secret Intelligence Branch. Photo courtesy of John K. Singlaub.

Starved Allied POWs

Some of the Allied prisoners of war rescued by the OSS "Pigeon" Mission on Hainan Island. The team rescued more than 500 in all. This group of Australian officers with their colonel, back row center, posed for an OSS photographer. Malnutrition, disease and brutal treatment had led to the deaths of 200 Australian and Dutch prisoners at the Hainan POW camp since their capture in 1942, but 500 were still alive. Eight American airmen had been captured in 1945, but only one was still alive. Aided by Singlaub's men and the medical team that arrived afterwards, the sick and starving prisoners were treated and put aboard ships for transportation to Allied hospitals. Photo courtesy of John K. Singlaub.

On the way home to the United States.

Five members of the OSS French Operational Group, who had trained at Areas F and B and then in England, had fought behind German lines in France in 1944 and Japanese lines in China in 1945, were on their way home from China via Calcutta, India, in September 1945. They were Nicholas Burke, from Chicago, sitting; others, from the left, Emil Roy from Kansas, Ellsworth ("Al") Johnson from Grand Rapids, Michigan; Gus (full name unknown), Robert Vernon from Idaho. Courtesy of Ellsworth ("Al") Johnson.

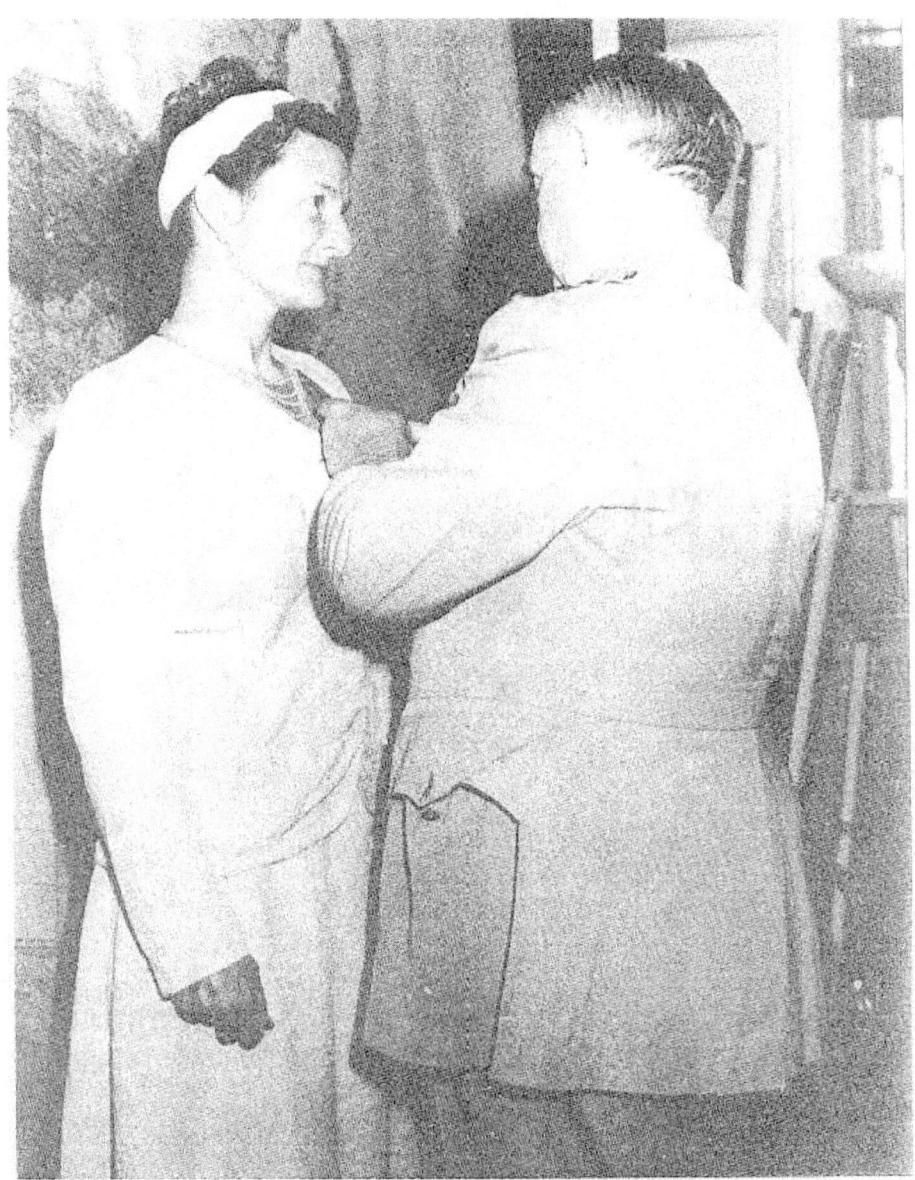

Virginia Hall, OSS Special Operations Branch, one of the most effective OSS spies and resistance leaders, operated behind enemy lines in German-occupied France. The Gestapo called her the limping lady because of her wooden leg and because of her effectiveness offered an enormous reward for her capture as one of the most dangerous Allied agents. This Maryland socialite turned clandestine operative eluded capture and provided valuable intelligence information as well as organizing, arming, and training French Resistance forces that demolished bridges, destroyed supply trains and disrupted enemy communications. Here, Virginia Hall receives the Distinguished Service Cross from Maj. Gen. William J. Donovan in September 1945. She was the only civilian woman to receive the Distinguished Service Cross in World War II. National Archives.

Private Barbara Lauwers, OSS Morale Operations Branch, a member of the U.S. Army's Women's Army Corps (WAC), was stationed with the MO unit in Rome. This Czech-born WAC, working with German Army prisoners of war in Italy, helped develop "Operation Sauerkraut," an innovative psychological program that sent disaffected German POWs back to persuade fellow soldiers to surrender. Many did. In addition, she personally created a program that induced 500 Czech soldiers conscripted into the Germany Army to give up to the Western Allies. Here Lauwers is shown in Rome receiving the Bronze Star Medal. National Archives.

Captain Howard W. Chappell, OSS German Operational Group, a paratrooper from Cleveland, Ohio, who had trained at Areas F and B, had, under direction of the OSS Italian Operational Group headquarters, commanded the "Tacoma" Mission against the German Army in the Italian Alps in 1944-1945. Here, Chappell receives the Silver Star Medal from Maj. Gen. William J. Donovan in front of OSS headquarters in Washington, D.C. in September 1945. He had already received a Purple Heart Medal for being wounded in action. Photo courtesy of Jack R. Chappell.

Sergeant Caesar J. Civitella, OSS Italian Operational Group, a paratrooper from Philadelphia, Pennsylvania, who had trained at Areas F and B, had been sent first to North Africa. From there his OG section was parachuted behind German lines to assist the Allied invasion of southern France in 1944. The "Lafayette" Mission took the surrender of nearly 4,000 German soldiers. Subsequently, his section was part of the "Sewanee" Mission in northern Italy, arming and coordinating partisan forces and overwhelming German outposts and garrisons in the Italian Alps. Civitella, shown here, receives the Bronze Star Medal from General Donovan. Courtesy of Caesar J. Civitella.

Illustrations | 677

In front of OSS headquarters in Washington, D.C., Major General William J. Donovan awards the Legion of Merit to a number of OSS officers who returned from the Far East in 1945. Among them in the front row are three OSS Special Operations officers, from the left, first, **Lieutenant Colonel Arden W. Dow**, OSS Special Operations, an infantry officer from Wenatachee, Washington, who had an instructor at Areas B and A, before being sent to China to head a joint Sino-American training camp for preparing Chinese soldiers as OSS Chinese Commando units to be parachuted behind Japanese lines; second, **Lieutenant Colonel Nicol Smith**, trained at Area B, who commanded OSS-trained Thai agents sent into Thailand from China; and third, **Major Frank A. Gleason** from Wilkes-Barre, Pennsylvania, an Army Engineer and instructor in demolitions at Area B in Maryland and then, deputy commander of the Sino-American training camp headed by Arden W. Dow. Gleason also headed an OSS SO team in China in the fall of 1945 that helped impede and ultimate halt a Japanese Army advance by destroying parts of the transportation network and blowing up warehouses full of arms and munitions to prevent their capture and use by the Japanese. Photo courtesy of Dorothea D. Dow.

As part of the immediately postwar celebration of the O.S.S., Hollywood film studios produced a number of movies about the daring clandestine agents, men and women, who went through OSS training camps and risked their lives behind enemy lines in World War II. One of these was Paramount's feature film, *O.S.S.* starring Alan Ladd and Geraldine Fitzgerald. It was released in 1946 in America and globally, as indicated by this publicity poster from France. The poster shows Ladd using a 1945 Joan-Eleanor voice transmitter.

The Office of Strategic Services (OSS), the predecessor to the Central Intelligence Agency (CIA), was established by direction of President Franklin D. Roosevelt in June of 1942, and existed until September of 1945. Its Director was Major General William J. Donovan. The charter of the OSS was to conduct espionage and covert action against enemy forces during World War II.

This area in the Park, known as Area C, and surrounding buildings, was used by the OSS during that period to conduct communications training for approximately 1500 radio operators and technicians. They were trained in the use of Morse Code and various radio equipments. Following training, these personnel were assigned to overseas theaters of operation to provide communications links in support of clandestine operations conducted by the OSS, U.S. Military and Allied Forces.

This marker furnished by OSS Communications Veterans, 1997

There are several organizations of veterans of the OSS, including the OSS Society, the Detachment 101 Society, and others, but the OSS CommVets, whose members served in the Communications Branch, have maintained a special relationship with Prince William Forest Park, site of the main OSS Communications School at OSS Training Area C. In 1997, the CommVets association dedicated the bronze plaque shown here at Prince William Forest Park. It may be seen in the Visitor Center. Prince William Forest Park.

www.ingramcontent.com/pod-product-compliance
Lightning Source LLC
Chambersburg PA
CBHW071947110526
44592CB00012B/1023